COMPREHENSIVE ANALYTICAL CHEMISTRY

ELSEVIER B.V.	ELSEVIER Inc.	ELSEVIER Ltd	ELSEVIER Ltd
Radarweg 29	525 B Street, Suite 1900	The Boulevard, Langford Lane	84 Theobalds Road
P.O. Box 211, 1000 AE Amsterdam	San Diego, CA 92101-4495	Kidlington, Oxford OX5 1GB	London WC1X 8RR
The Netherlands	USA	UK	UK

© 2005 Elsevier B.V. All rights reserved.

This work is protected under copyright by Elsevier B.V., and the following terms and conditions apply to its use:

Photocopying
Single photocopies of single chapters may be made for personal use as allowed by national copyright laws. Permission of the Publisher and payment of a fee is required for all other photocopying, including multiple or systematic copying, copying for advertising or promotional purposes, resale, and all forms of document delivery. Special rates are available for educational institutions that wish to make photocopies for non-profit educational classroom use.

Permissions may be sought directly from Elsevier's Rights Department in Oxford, UK: phone (+44) 1865 843830, fax (+44) 1865 853333, e-mail: permissions@elsevier.com. Requests may also be completed on-line via the Elsevier homepage (http://www.elsevier.com/locate/permissions).

In the USA, users may clear permissions and make payments through the Copyright Clearance Center, Inc., 222 Rosewood Drive, Danvers, MA 01923, USA; phone: (+1) (978) 7508400, fax: (+1) (978) 7504744, and in the UK through the Copyright Licensing Agency Rapid Clearance Service (CLARCS), 90 Tottenham Court Road, London W1P 0LP, UK; phone: (+44) 20 7631 5555; fax: (+44) 20 7631 5500. Other countries may have a local reprographic rights agency for payments.

Derivative Works
Tables of contents may be reproduced for internal circulation, but permission of the Publisher is required for external resale or distribution of such material. Permission of the Publisher is required for all other derivative works, including compilations and translations.

Electronic Storage or Usage
Permission of the Publisher is required to store or use electronically any material contained in this work, including any chapter or part of a chapter.

Except as outlined above, no part of this work may be reproduced, stored in a retrieval system or transmitted in any form or by any means, electronic, mechanical, photocopying, recording or otherwise, without prior written permission of the Publisher. Address permissions requests to: Elsevier's Rights Department, at the fax and e-mail addresses noted above.

Notice
No responsibility is assumed by the Publisher for any injury and/or damage to persons or property as a matter of products liability, negligence or otherwise, or from any use or operation of any methods, products, instructions or ideas contained in the material herein. Because of rapid advances in the medical sciences, in particular, independent verification of diagnoses and drug dosages should be made.

First edition 2005

Library of Congress Cataloging in Publication Data
A catalog record is available from the Library of Congress.

British Library Cataloguing in Publication Data
A catalogue record is available from the British Library.

ISBN: 0-444-50715-9
ISSN: 0166-526X

∞ The paper used in this publication meets the requirements of ANSI/NISO Z39.48-1992 (Permanence of Paper). Printed in The Netherlands.

Working together to grow
libraries in developing countries

www.elsevier.com | www.bookaid.org | www.sabre.org

ELSEVIER   BOOK AID International   Sabre Foundation

COMPREHENSIVE ANALYTICAL CHEMISTRY

ADVISORY BOARD

Professor A.M. Bond
    Monash University, Clayton, Victoria, Australia

Dr T.W. Collette
    US Environmental Protection Agency, Athens, GA, U.S.A.

Professor M. Grasserbauer
    *Director of the Environment Institute, European Commission'
    Joint Research Centre, Ispra, Italy*

Professor M.-C. Hennion
    Ecole Supérieure de Physique et de Chimie Industrielles, Paris, France

Professor G.M. Hieftje
    Indiana University, Bloomington, IN, U.S.A.

Professor G. Marko-Varga
    AstraZeneca, Lund, Sweden

Professor D.L. Massart
    Vrije Universiteit, Brussels, Belgium

Professor M.E. Meyerhoff
    University of Michigan, Ann Arbor, MI, U.S.A.

Wilson & Wilson's

COMPREHENSIVE ANALYTICAL CHEMISTRY

Edited by

D. BARCELÓ

Research Professor
Department of Environmental Chemistry
IIQAB-CSIC
Jordi Girona 18-26
08034 Barcelona
Spain

Wilson & Wilson's
COMPREHENSIVE ANALYTICAL CHEMISTRY

VOLUME XLIV

BIOSENSORS AND MODERN BIOSPECIFIC ANALYTICAL TECHNIQUES

Edited by

L. GORTON

*Lund University,
Department of Analytical Chemistry,
P.O. Box 124,
SE-221 00 Lund,
Sweden*

2005

ELSEVIER

AMSTERDAM – BOSTON – HEIDELBERG – LONDON – NEW YORK – OXFORD – PARIS
SAN DIEGO – SAN FRANCISCO – SINGAPORE – SYDNEY – TOKYO

CONTRIBUTORS TO VOLUME XLIV

Daniel Andreescu
  Department of Chemistry, State University of New York—Binghamton, P.O. Box 6000, Binghamton, NY 13902, USA
Silvana Andreescu
  Department of Chemistry, State University of New York—Binghamton, P.O. Box 6000, Binghamton, NY 13902, USA
Antje J. Baeumner
  Cornell University, Ithaca, NY 14853-5701, USA
Ursula Bilitewski
  Gesellschaft für Biotechnologische Forschung mbH, Mascheroder Weg 1, 38124 Braunschweig, Germany
Virginia Buchner
  Department of Molecular Microbiology and Biotechnology, Tel-Aviv University, Ramat-Aviv, Israel
Elena Domínguez
  Department of Analytical Chemistry, Faculty of Pharmacy, University of Alcalá 28871, Alcalá de Henares (Madrid), Spain
Jenny Emnéus
  Department of Analytical Chemistry, Lund University, Lund, Sweden
Hiroko I. Karan
  School of Science, Health and Technology, Medgar Evers College, The City University of New York, 1150 Carroll Street, Brooklyn, NY 11225, USA
Laura M. Lechuga
  Microelectronics National Center (CNM), CSIC, Barcelona, Spain
Danila Moscone
  Dipartimento di Scienze e Tecnologie Chimiche, Università di Roma "Tor Vergata", Via della Ricerca Scientifica, 00133 Rome, Italy
Arántzazu Narváez
  Department of Analytical Chemistry, Faculty of Pharmacy, University of Alcalá 28871, Alcalá de Henares (Madrid), Spain
Catalin Nistor
  Department of Analytical Chemistry, Lund University, Lund, Sweden
A.M. Oliveira Brett
  Departamento de Química, Faculdade de Ciências e Tecnologia, Universidade de Coimbra, 3004-535 Coimbra, Portugal
Judith Rishpon
  Department of Molecular Microbiology and Biotechnology, Tel-Aviv University, Ramat-Aviv, Israel

Contributors to Volume XLIV

Omowunmi A. Sadik
Department of Chemistry, State University of New York—Binghamton, P.O. Box 6000, Binghamton, NY 13902, USA

Daniela D. Schlereth
European Patent Office, Landsbergerstrasse 30, D-80335 Munich, Germany

Ulla Wollenberger
Department of Analytical Biochemistry, Institute of Biochemistry and Biologie, University of Potsdam, Karl-Liebknechstrasse 24-25, 14415 Golm, Germany

WILSON AND WILSON'S

COMPREHENSIVE ANALYTICAL CHEMISTRY

VOLUMES IN THE SERIES

Vol. IA	Analytical Processes
	Gas Analysis
	Inorganic Qualitative Analysis
	Organic Qualitative Analysis
	Inorganic Gravimetric Analysis
Vol. IB	Inorganic Titrimetric Analysis
	Organic Quantitative Analysis
Vol. IC	Analytical Chemistry of the Elements
Vol. IIA	Electrochemical Analysis
	Electrodeposition
	Potentiometric Titrations
	Conductometric Titrations
	High-Frequency Titrations
Vol. IIB	Liquid Chromatography in Columns
	Gas Chromatography
	Ion Exchangers
	Distillation
Vol. IIC	Paper and Thin Layer Chromatography
	Radiochemical Methods
	Nuclear Magnetic Resonance and Electron Spin Resonance Methods
	X-Ray Spectrometry
Vol. IID	Coulometric Analysis
Vol. III	Elemental Analysis with Minute Sample
	Standards and Standardization
	Separation by Liquid Amalgams
	Vacuum Fusion Analysis of Gases in Metals
	Electroanalysis in Molten Salts
Vol. IV	Instrumentation for Spectroscopy
	Atomic Absorption and Fluorescence Spectroscopy
	Diffuse Reflectance Spectroscopy
Vol. V	Emission Spectroscopy
	Analytical Microwave Spectroscopy
	Analytical Applications of Electron Microscopy
Vol. VI	Analytical Infrared Spectroscopy
Vol. VII	Thermal Methods in Analytical Chemistry
	Substoichiometric Analytical Methods
Vol. VIII	Enzyme Electrodes in Analytical Chemistry
	Molecular Fluorescence Spectroscopy
	Photometric Titrations
	Analytical Applications of Interferometry
Vol. IX	Ultraviolet Photoelectron and Photoion Spectroscopy
	Auger Electron Spectroscopy
	Plasma Excitation in Spectrochemical Analysis
Vol. X	Organic Spot Tests Analysis
	The History of Analytical Chemistry

Volumes in the series

Vol. XI	The Application of Mathematical Statistics in Analytical Chemistry Mass Spectrometry
	Ion Selective Electrodes
Vol. XII	Thermal Analysis
	Part A. Simultaneous Thermoanalytical Examination by Means of the Derivatograph
	Part B. Biochemical and Clinical Applications of Thermometric and Thermal Analysis
	Part C. Emanation Thermal Analysis and other Radiometric Emanation Methods
	Part D. Thermophysical Properties of Solids
	Part E. Pulse Method of Measuring Thermophysical Parameters
Vol. XIII	Analysis of Complex Hydrocarbons
	Part A. Separation Methods
	Part B. Group Analysis and Detailed Analysis
Vol. XIV	Ion-Exchangers in Analytical Chemistry
Vol. XV	Methods of Organic Analysis
Vol. XVI	Chemical Microscopy
	Thermomicroscopy of Organic Compounds
Vol. XVII	Gas and Liquid Analysers
Vol. XVIII	Kinetic Methods in Chemical Analysis
	Application of Computers in Analytical Chemistry
Vol. XIX	Analytical Visible and Ultraviolet Spectrometry
Vol. XX	Photometric Methods in Inorganic Trace Analysis
Vol. XXI	New Developments in Conductometric and Oscillometric Analysis
Vol. XXII	Titrimetric Analysis in Organic Solvents
Vol. XXIII	Analytical and Biomedical Applications of Ion-Selective Field-Effect Transistors
Vol. XXIV	Energy Dispersive X-Ray Fluorescence Analysis
Vol. XXV	Preconcentration of Trace Elements
Vol. XXVI	Radionuclide X-Ray Fluorescence Analysis
Vol. XXVII	Voltammetry
Vol. XXVIII	Analysis of Substances in the Gaseous Phase
Vol. XXIX	Chemiluminescence Immunoassay
Vol. XXX	Spectrochemical Trace Analysis for Metals and Metalloids
Vol. XXXI	Surfactants in Analytical Chemistry
Vol. XXXII	Environmental Analytical Chemistry
Vol. XXXIII	Elemental Speciation – New Approaches for Trace Element Analysis
Vol. XXXIV	Discrete Sample Introduction Techniques for Inductively Coupled Plasma Mass Spectrometry
Vol. XXXV	Modern Fourier Transform Infrared Spectroscopy
Vol. XXXVI	Chemical Test Methods of Analysis
Vol. XXXVII	Sampling and Sample Preparation for Field and Laboratory
Vol. XXXVIII	Countercurrent Chromatography: The Support-Free Iiquid Stationary Phase
Vol. XXXIX	Integrated Analytical Systems
Vol. XL	Analysis and Fate of Surfactants in the Aquatic Environment
Vol. XLI	Sample Preparation for Trace Element Analysis
Vol. XLII	Non-destructive Microanalysis of Cultural Heritage Materials
Vol. XLIII	Chromatographic-Mass Spectrometric Food Analysis for Trace Determination of Pesticide Residues

Contents

Contributors to Volume XLIV vi
Volumes in the Series viii
Series Editor's Preface xix
Preface xxi

Chapter 1. Biosensors based on self-assembled monolayers 1
    Daniela D. Schlereth
  1.1 Introduction 1
  1.2 Self-assembly of organic monomolecular layers
      on solid supports 2
      1.2.1 Self-assembled monolayers of organosulfur
            compounds on gold surfaces 3
      1.2.2 Protein-resistant surfaces based on
            alkylthiol SAMs 4
      1.2.3 Surface architectures prepared by solid-phase
            derivatization of alkylthiol SAMs 6
  1.3 Affinity biosensors 12
      1.3.1 Site-specific affinity binding of proteins to
            self-assembled molecular architectures 13
      1.3.2 DNA hybridization biosensors 16
      1.3.3 Electrochemical detection of affinity
            interactions 19
  1.4 Amperometric catalytic biosensors 24
      1.4.1 Multipoint attachment of proteins to gold electrode
            surfaces modified with an alkylthiol SAM 25
      1.4.2 Site-specific affinity binding of proteins to gold
            electrode surfaces modified with an
            alkylthiol SAM 32
      1.4.3 Amperometric biosensors based on self-assembled
            multilayered structures 46
  1.5 Switchable self-assembled monolayers 48
  1.6 Outlook 52
  References 53

Contents

Chapter 2. Third generation biosensors—integrating recognition and transduction in electrochemical sensors 65
Ulla Wollenberger
2.1   Introduction 65
2.2   Biosensor generations 66
2.3   Electron transfer processes at electrodes 67
2.4   Direct electron transfer—overview 68
2.5   Heme proteins/heme enzymes 77
    2.5.1   Heme proteins 77
    2.5.2   Heme enzymes 84
    2.5.3   Copper proteins/enzymes 98
    2.5.4   Other enzymes 100
2.6   Selected sensors 103
    2.6.1   Superoxide sensor 103
    2.6.2   Nitric oxide sensor 108
    2.6.3   Peroxide sensors 110
    2.6.4   Immunoassays based on DET indication 117
2.7   Conclusion 118
Acknowledgements 119
References 119

Chapter 3. Enzyme biosensors containing polymeric electron transfer systems 131
Hiroko I. Karan
3.1   Introduction 131
3.2   Glucose biosensors 133
    3.2.1   First approach: oxygen consumption detection . . 133
    3.2.2   Second approach: hydrogen peroxide concentration detection 134
    3.2.3   Third approach: use of non-physiological redox couple 134
    3.2.4   Electron-transfer relay systems: binding electron transfer relays to enzyme proteins 135
    3.2.5   Electrical wiring of redox enzymes with redox polymers 136
    3.2.6   Flexible polymeric electron transfer mediators . . 137
3.3   Development of enzyme biosensors containing polymeric electron relay systems 139
    3.3.1   Membranes 139

	3.3.2	Poly[(vinylpyridine)Os(bipyridine)$_2$Cl$^{2+/3+}$] derivative based redox hydrogels (A) 140

- 3.3.2 Poly[(vinylpyridine)Os(bipyridine)$_2$Cl$^{2+/3+}$] derivative based redox hydrogels (A) 140
- 3.3.3 Amperometric enzyme biosensors using poly[(vinylpyridine)Os(bipyridine)$_2$Cl$^{2+/3+}$] based hydrogels (A) 141
- 3.3.4 Mediator containing flexible polymeric electron transfer systems 145
- 3.3.5 Covalent electropolymerization of glucose oxidase in polypyrrole 157
- 3.3.6 Enzyme biosensors using NAD$^+$/NADH dependent and other dehydrogenases 159
- 3.4 Summary 161
- Acknowledgements 171
- References 171
- Glossary 177

Chapter 4. DNA-based biosensors 179
A.M. Oliveira Brett
- 4.1 Introduction 179
- 4.2 DNA-optical biosensors 182
- 4.3 DNA-acoustic wave biosensors 186
- 4.4 DNA-electrochemical biosensors 189
- 4.5 Conclusions 202
- Acknowledgements 203
- References 203

Chapter 5. Optical biosensors 209
Laura M. Lechuga
- 5.1 Introduction to optical biosensors 209
- 5.2 Optical sensing principles 211
  - 5.2.1 Evanescent wave principle 212
- 5.3 Types of optical biosensors 214
  - 5.3.1 Optical fiber (FO) biosensors 216
  - 5.3.2 Surface plasmon resonance (SPR) sensors ... 219
  - 5.3.3 Integrated optical (IO) sensors 226
  - 5.3.4 Other optical biosensing schemes 235
- 5.4 Biochemical aspects of optical biosensors 237
- 5.5 Applications of optical biosensors 241
  - 5.5.1 Life sciences applications 242
  - 5.5.2 Environmental applications 242

	5.5.3	Chemical and biological warfare	243

	5.5.3	Chemical and biological warfare	243
	5.5.4	Genetic applications	243
5.6	Future trends		243
	5.6.1	Integration	244
	5.6.2	New receptors and new immobilization procedures	245
	5.6.3	Mutianalyte detection	246
	5.6.4	Optical nanobiosensors	246
Acknowledgements			247
References			247

Chapter 6. Bioanalytical microsystems: technology and applications 251

Antje J. Baeumner

6.1	Introduction		251
6.2	Miniaturization techniques		253
	6.2.1	Biosensor fluid flow system (biosensor housing)	253
	6.2.2	Biosensor transducers	263
6.3	Immobilization of biorecognition elements in microbiosensors		266
6.4	Miniaturization and its effect on biosensor performance		267
6.5	Biosensor examples		269
6.6	Conclusions and future outlook		274
References			276

Chapter 7. New materials for biosensors, biochips and molecular bioelectronics 285

Daniel Andreescu, Silvana Andreescu and Omowunmi A. Sadik

7.1	Introduction		285
7.2	Properties of ideal materials for biosensors and bioelectronics		287
7.3	Functionalization of electrode materials		287
	7.3.1	Self-assembled monolayers	288
	7.3.2	Biomaterials developed by covalent binding	290
7.4	Classes of materials for biosensors		292
	7.4.1	Polymers	293
	7.4.2	Biosensing materials containing metal complexes	298
	7.4.3	Sol–gel materials	299

		7.4.4	Nanomaterials	300
		7.4.5	Nanoscaled interdigitated electrodes	308
		7.4.6	Composite materials	308
		7.4.7	Metal oxide semiconductor field effect transistors	312
		7.4.8	Photonic band gap	313
		7.4.9	Zeolites	314
	7.5	Summary and future perspectives		315
	7.6	Conclusions		318
	References			319

Chapter 8. Electrochemical antibody-based sensors 329
Judith Rishpon and Virginia Buchner

	8.1	Electrochemistry merges with immunology		329
	8.2	Principles of conventional immunoassays		330
		8.2.1	Brief historical synopsis	330
		8.2.2	The antigen–antibody reaction	330
		8.2.3	The antibody (immunoglobulin) molecule	332
		8.2.4	Recombinant antibody technology	332
		8.2.5	Immunoassay techniques	333
		8.2.6	Enzyme immunoassays with and without separation	335
	8.3	Principles of electrochemical analysis		336
		8.3.1	Electroanalytical measurements	336
		8.3.2	Screen-printed electrodes	336
		8.3.3	Transducers	336
		8.3.4	The recognition element	343
		8.3.5	Construction of an electroimmunosensor	343
	8.4	Applications		351
		8.4.1	Overview	351
		8.4.2	Diagnostic, biomedical, and veterinary applications	352
		8.4.3	Environmental monitoring and food analysis	359
		8.4.4	Biological warfare agents	365
		8.4.5	Illegal drug screening	366
	8.5	Conclusions		367
	References			368

Chapter 9. Immunoassay: potentials and limitations 375
Catalin Nistor and Jenny Emnéus

	9.1	Introduction	375

9.2	Basic aspects of immunoassay		376
	9.2.1	Antibody–antigen interactions—basic concepts	376
	9.2.2	Immunoassay sensitivity	378
	9.2.3	The principle of antibody occupancy: classification of immunoassay	385
	9.2.4	Immunoassay selectivity	391
9.3	Determinants of sensitivity and selectivity		396
	9.3.1	Immunoreagent quality	397
	9.3.2	Immunoreagent concentration	399
	9.3.3	Incubation time and temperature	401
	9.3.4	Other factors	405
9.4	Particular aspects of immunoassay standardization		415
	9.4.1	Availability of standard materials	415
	9.4.2	Immunoassay validation with emphasis on applications for environmental monitoring	416
9.5	Conclusions		419
References			419
Glossary			425

Chapter 10. Non-affinity sensing technology: the exploitation of biocatalytic events for environmental analysis 429
Elena Domínguez and Arántzazu Narváez

10.1	Introduction		429
10.2	Focus and scope		435
10.3	The upcoming phase in environmental analysis: from research to mandatory monitoring		436
10.4	Catalysis in living systems: from essential natural processes to sensing devices for environmental analysis		437
10.5	Biocatalytic processes for environmental sensing purposes		442
	10.5.1	Complex catalytic pathways or catabolic approaches	442
	10.5.2	Simple catalytic reaction: isolated enzymes	452
10.6	Emerging catalytic molecules for environmental applications		498
	10.6.1	New biocatalysts from natural environments: extremozymes	499
	10.6.2	Catalytic antibodies: abzymes	502
	10.6.3	Redesign of catalytic activities	504

10.7	Concluding remarks		517
Acknowledgements			517
References			518

Chapter 11. Biosensors for bioprocess monitoring 539
Ursula Bilitewski
- 11.1 Introduction 539
- 11.2 Monitoring and control of substrate and product concentrations with enzyme electrodes 543
  - 11.2.1 Enzyme electrodes for alcohols and saccharides 544
  - 11.2.2 Stability of enzyme sensors 545
  - 11.2.3 Specificity of enzyme sensors 546
  - 11.2.4 Sampling devices 548
  - 11.2.5 Control of substrate concentrations 549
- 11.3 Biosensor systems for the evaluation of product quality . . 550
- 11.4 Quantification of plasmid DNA 554
- 11.5 Enzyme-based systems for indicators of cell physiology and metabolic stress 556
  - 11.5.1 Amino acid analysis 557
  - 11.5.2 Nucleotide analysis 558
- 11.6 DNA arrays 560
  - 11.6.1 Fundamentals 561
  - 11.6.2 Gene expression analysis in biochemical engineering 568
- 11.7 Trends 569
- References 571

Chapter 12. Coupling of microdialysis sampling with biosensing detection modes 579
Danila Moscone
- 12.1 Introduction 579
- 12.2 The probe 580
- 12.3 The membrane 583
- 12.4 Sampling considerations 584
- 12.5 On-line coupling to biosensing devices 588
- 12.6 The issue of continuous glucose monitoring 590
  - 12.6.1 In vivo glucose monitoring through microdialysis coupled to electrochemical biosensors 593

Contents

- 12.6.2 In vivo glucose monitoring through microdialysis coupled to other biosensing detection techniques 605
- 12.7 In vivo monitoring of other metabolites 606
  - 12.7.1 Glutamate detection 606
  - 12.7.2 Lactate detection 608
  - 12.7.3 Acetylcholine detection 608
  - 12.7.4 Uric acid and dopamine detection 609
  - 12.7.5 Simultaneous determination of different metabolites 610
- 12.8 Dialysis electrodes 612
- 12.9 Biotechnological applications 615
- 12.10 Conclusion 618
- Acknowledgements 619
- References 619

Index 627

Series Editor's Preface

It is a great pleasure for me to introduce this new book on *Biosensors and Modern Biospecific Analytical Techniques*, edited by Lo Gorton. When I took on the editorship of this series, one of my first ideas was to commission books on sensors and biosensors. The importance of sensors has been already indicated by the recent publication in the Comprehensive Analytical Chemistry series of volume XXXIX, *Integrated Analytical Systems*, edited by Salvador Alegret. I thought immediately to acquire another title for the series and Lo came to mind. My collaboration with his research group began more than 10 years ago. His group is representative of the excellent research in the field of biosensors being done at the moment in Europe. Lo was able to convince some colleagues and prominent experts in the field to contribute to the present work.

This book is an additional step in our coverage of the area of rapid analysis using advanced technological developments. It contains 12 chapters and summarizes the main developments in this field of analytical chemistry over the last 10 years. It offers a comprehensive study on different types of biosensors, including DNA-based, enzymatic, optical, self-assembled monolayers and the third generation of biosensors, which integrate recognition and transduction in electrochemical sensors. Antibody and immunoassay developments also have a prominent place in the book as well as many technological developments on bioanalytical microsystems and new materials for biosensors. The applications covered include environmental analysis, bioprocess monitoring, and biomedicine.

I am convinced that this book will be of great help to analytical chemists in academia, research, industry and government and also for postgraduate students following masters courses in analytical chemistry. To all of them, I would like to announce that this book will certainly not be the last book on sensors published in this series. We are already planning other books covering emerging areas on biosensors and sensor development. I believe that this area will continue to grow due to the societal needs to develop rapid and accurate devices for food, antiterrorism, medical care, and industrial process control.

Finally, I would like to thank the editor and the various authors for their contributions. Several of the chapters are almost "little" books in themselves and I know firsthand how much effort was involved for some of the authors of the book, namely my Spanish colleagues and friends Elena Domínguez, Arántzazu Narváez and Laura Lechuga, to finish their chapters. The appearance of this excellent and unique book on biosensors will certainly be a compensation to all involved in its realization.

D. Barceló

Preface

The biosensor research areas including modern biospecific analytical techniques are developing very rapidly and it is fair to say that analytical chemistry plays a very important role in these developments. New challenges lie ahead with new discoveries and tools opening up novel dimensions in both fundamental research and in possible applications. However, as in many other areas in analytical chemistry for major breakthroughs, we rely on developments in other areas that have contributed and will contribute to new research directions as well as successful improvements in already existing areas.

Many of the pioneers in biosensor research were men (as in most other fields in the natural and engineering sciences) for which they have been rightly recognized, reflected among many other things by special issues of scientific journals when retiring or when reaching the age of 60 or 65. However, behind all these men there was always a group of devoted younger scientists and many of them are women, who greatly contributed to the success of their elder supervisors. As in all other sectors of today's society, it is more difficult for female professionals to become recognized for their contributions and that was one of the major aims of this book (for which I adopted a running title: "Women in Biosensors") to invite female researchers, who by themselves and their own efforts have greatly contributed to the progress in the research on biosensors and related techniques.

"The leaky pipeline" is a phenomenon that describes the gradual loss of women from science throughout the career path, even though women and men go into higher education in equal numbers, the tendency being the same in different scientific disciplines (European Technology Assessment Network report on Women and Science 2000, Science Policies in EU: Promoting excellence through mainstreaming gender equality, http://europa.eu.int/comm/research/science-society/documents_en.html#women). On average 6.4% women compared with 18.8% men reached the top during their academic careers, and this is a common phenomenon to all European member states and Associated countries (She figures, Women and Science, statistics and indicators, 2003, http://europa.eu.int/comm/research/science-society/documents_en.html#women). In order to meet the challenges of excellence and competitiveness in the future, it is time for the scientific community to maximize its total research potential. The stronger presence of women in research will substantially improve the utilization of human resources, while enriching the scientific enterprise in new themes and perspectives.

With this book my intentions were:

- to show the input of women in current biosensor research and related techniques and to stimulate young scientists;
- compile state-of-the-art knowledge of the field and describe emerging technologies;
- provide insight into the frontiers of the field;
- provide examples by relevant applications.

The book is also intended to serve as a general reference for both other researchers and scientists within the bioanalytical field as well as postgraduate students. Each chapter includes references to the corresponding literature to serve as valuable entry points to anyone wanting to move forward in the field, either as a practitioner or for acquiring state-of-the-art knowledge.

Finally, I would like to express my sincere gratitude to the contributing authors for all their time and effort in preparing the chapters. Without their engagement and lively interaction I have had over the last few years to get this project finalized, this book would not have been possible.

Lo Gorton
Lund, Sweden

Chapter 1

Biosensors based on self-assembled monolayers

Daniela D. Schlereth

1.1 INTRODUCTION

The key component of a biosensor is a biological material, such as an antibody, a membrane receptor, a single stranded nucleic acid, an enzyme or even a whole cell, which is usually immobilized on the surface of a suitable transducer. In this system, the biological material is responsible for a specific recognition of the analyte at the transducer surface and the transducer for the conversion of said recognition event into a processable signal. The selective and specific recognition of a given analyte is achieved by a biochemical reaction between said analyte and said biological material, that results either in the formation of a surface-bound affinity complex (affinity biosensors) or in the enzymatic conversion of the analyte into a product (catalytic biosensors) [1–4]. Regardless of the different signal transduction physical principles applied in the construction of biosensors, which may be based on mass sensitive [3,5–7], electrochemical [1–3], or optical evanescent field techniques [8–11], a key issue in the development of any biosensor is to assure a high functional activity of the biological material immobilized on the transducer surface. This is specially relevant for the manufacture of high-throughput microanalytical sensing devices, in which the transducer surfaces are miniaturized to a micro- or nanometer size scale [12–25] and/or the fabrication of bioelectronic molecular devices [26–30] because of the small amount of material that can be deposited thereon. Another aspect which has to be taken into account is the fact that the sensitivity, stability, and reliability of the biosensor response is determined not only by the functional integrity of the immobilized biological material, but also by the accessibility of each individual biological entity within the layer for enabling interaction with the analyte molecules [31].

The capability of many organic molecules to spontaneously self-assemble forming monomolecular layers on several solid substrates provides a useful and convenient synthetic tool for engineering an environment on the surface of the transducer, which has a well-organized molecular architecture that is spatially defined at a molecular level and is resistant to non-specific adsorption, such as to be able to bind selected biological species (i.e., biomolecules or whole cells) merely by site-specific non-covalent interactions without perturbing their native structure. For this reason, the process of self-assembly can be considered as being ideally suited for providing sensor layers, in which the biological material preserves a high functional activity and is easily accessible for analyte molecules [32–37].

1.2 SELF-ASSEMBLY OF ORGANIC MONOMOLECULAR LAYERS ON SOLID SUPPORTS

The process of self-assembly can be described as a spontaneous hierarchical self-association of interlocking components (molecules) to form a well-organized two- or three-dimensional supramolecular structure that represents the thermodynamic equilibrium under a given set of experimental conditions. The self-assembly of organic molecules in solution is driven by an interplay of non-covalent interactions among the self-assembling molecules with themselves and with molecules of solvent, which starts with a recognition-directed spontaneous association of a defined and limited number of molecular components under the intermolecular control of non-covalent interactions that hold them together [38]. Analogously, the spontaneous self-assembly of an organic monomolecular layer on a solid/liquid interface is driven by specific interactions between the head functional groups of the self-assembling molecules and the surface, which is coupled to the displacement of solvent molecules from the surface, and followed by a self-organization of a monomolecular film on the surface which is stabilized by an interplay of non-covalent interactions among the self-assembling molecules [32–34]. This process can be used to modify solid surfaces with an infinite number of stable and well-organized molecular architectures starting from a small number of different molecular building blocks. Moreover, the modification procedure does not necessarily imply a laborious and sophisticated experimental work because the monomolecular film is spontaneously formed just by bringing the surface into contact with a solution of self-assembling molecules [39].

A large amount of work has been devoted during the last 20 years to the study of self-assembled monolayers (SAMs) and their application for building surfaces with a wide variety of physico-chemical properties, such as a controlled hydrophilicity/hydrophobicity (wettability) degree [40–42], electrostatic charge [43,44], electroactivity or electrical conductivity [45–52], catalytic [53–55], and non-linear optical properties [56–59]. In particular, the possibility to tailor at a molecular level the surface properties of a metal and/or a semiconductor, as regards specific molecular recognition and/or chemical activity and its compatibility with microfabrication techniques [16,18–20,23–25,60–66] renders the process of self-assembly a valuable tool for manufacturing highly selective sensor surfaces [35–37,67], ultra-miniaturized nanometer-scale molecular electronic devices [27–30,68–70] and molecular biomaterials [71].

1.2.1 Self-assembled monolayers of organosulfur compounds on gold surfaces

Although there is a large number of organic molecules which are able to form SAMs on a solid surface, the most investigated SAMs are those prepared from surfactant molecules by the Langmuir–Blodgett (LB) technique, and those prepared by chemisorption of organosilanes and organosulfur compounds. LB lipid mono- and bi-molecular layers have been widely used as artificial biomimetic models for biological membranes [72,73]. However, these films are too fragile for practical application in the fabrication of sensor surfaces because they result from a weak physisorption of surfactant head groups on the solid surface. By contrast, self-assembly of organosilanes gives rise to very stable organic monomolecular layers because the driving force of the process is the "in situ" formation of a polysiloxane connected to surface silanol groups. However, the preparation of high-quality organosilane SAMs requires the presence of free hydroxylic groups on the surface and a careful control of the amount of water in solution [32–34], which may be a serious drawback for their practical application in laboratories lacking a suitable infrastructure.

Organosulfur compounds, such as alkylthiols and di-n-alkyl disulfides, self-assemble spontaneously on clean surfaces of several transition metals or semiconductors upon bringing the surface into contact with a solution of the sulfur compound, giving rise to very stable and reproducible monomolecular films [32–34]. While the mechanism of chemisorption of alkylthiols on gold is thought to involve the oxidative addition of the S–H bond to the gold surface followed by a reductive elimination of hydrogen and formation of alkylthiolate ions, the adsorption of di-n-alkyl disulfides is thought to involve a simple oxidative addition of the S–S bond on the gold surface [74–76]. As a consequence of these different reaction mechanisms, the chemisorption of alkylthiols and di-n-alkyl disulfides on gold surfaces gives rise to SAMs which are essentially identical [40], although a higher degree of disorder in di-n-alkyl disulfides SAMs and a higher replacement rate of alkylthiols suggest a stronger interaction of alkylthiols with the gold surface [76]. The self-assembly of alkylthiols on gold surfaces shows a biphasic kinetic behaviour, which starts with a first step that is accomplished within a few minutes, in which the monomolecular layer grows according to a first-order Langmuir adsorption kinetics [77–79]. This fast adsorption step is followed by a slow step that lasts several hours, in which the rate of layer growth depends on several factors, such as the chain length of the alkylthiol, the concentration of alkylthiol in the adsorption solution, and the partial film thickness [80]. Microscopic-level studies of the adsorption mechanism of gaseous alkylthiols on clean Au(111) surfaces carried out with a scanning tunnel microscope show that the formation of a densely packed alkylthiol SAM involves a phase transition from a phase at very low surface coverages in which surface-confined alkylthiol molecules diffuse with their molecular axis aligned with the surface, to a phase in which the thiol molecules realign, after forming stable nucleation islands, with their molecular axis stretching away from the surface [81,82]. The specific molecular packing of the resulting alkylthiol SAM is determined by a balance between interactions between the thiol head groups and the gold lattice, dispersion forces between the alkyl chains and interactions between alkylthiol end groups. Long-alkylthiols having an alkyl chain of more than nine carbon atoms form densely-packed, crystalline-like assemblies on Au(111), in which the alkyl chains are fully extended in an all *trans* conformation, tilted 26–28° from the surface normal and packed in a $\sqrt{3} \times \sqrt{3}$ hexagonal lattice. However, as the length of the alkyl chain decreases, the structure of the monolayer becomes increasingly disordered with a lower packing density and surface coverage [42,45,83–88], which leads to a larger number of defects or pinholes. Indeed, it has been proposed that the higher number of defects in SAMs prepared from short-alkylthiols (length of the alkyl chain shorter than nine carbon atoms) is responsible for their higher electrical permeability [45] as compared with SAMs prepared from long-alkylthiol which behave as electrical insulators [89].

1.2.2 Protein-resistant surfaces based on alkylthiol SAMs

The resistance of a surface to the irreversible non-specific adsorption of proteins is an essential requirement for its application as a substrate in a high-throughput device for carrying out diagnostic assays, in which cell-based assays for screening libraries of drug candidates are performed in a chip format. The protein resistance of the surface allows a cell population patterned thereon to be maintained over a long period of time with a high fidelity [17]. Moreover, the availability of protein-resistant surfaces allows the manufacture of artificial biomimetic environments, which enable the performance of in vitro model experiments for the study of mechanisms involved in the interactions of proteins with cell surfaces [90,91].

It is generally accepted that the main kind of non-covalent interactions which are involved in the irreversible non-specific adsorption of a protein on a substrate are hydrophobic interactions between amino acid side chains on the protein surface and the substrate. For that reason, the manufacture of a protein-resistant surface is usually based on careful control over its degree of hydrophilicity. The hydrophilicity (or hydrophobicity) degree (wettability) of a gold surface modified by chemisorption of a two-dimensional binary mixed alkylthiol SAM (both thiols having the same alkyl chain length) is essentially determined by the hydrophilic (or hydrophobic) nature of the thiol end groups, independent of the length of the alkylthiol underlayer, such that it can be controlled and modulated by using mixtures of alkylthiols having different end groups [40–42]. Analogously, the wettability of a gold surface can also be modulated by chemisorption of alkylthiols having the same end groups but different alkyl chain lengths [92,93], or different alkyl chain lengths and end groups [94–96] because the protruding longest thiols bend over the short covering a part of it with their hydrophobic a considered the simplest approach for prepa compatible) gold surface [96,97], which enal cell growth to be controlled to spatially confin thickness of an alkylthiol SAM, even when pr alkyl chains longer than nine carbon aton hindering non-specific interactions betweer surface. This problem can be overcome by an end-attached long, flexible, hydrophilic spacer, such as an oligo(ethylene oxide)- [98,99] or oligo(propylene sulfoxide)-based [100] tail. These oligomeric end groups are thought to adopt a helical (or amorphous) conformation, which leads to a structural organization of water molecules at the SAM surface, such that the resulting interphase water layer is tightly bound to the monolayer, forming a water cushion that prevents a direct contact of the protein with the surface [101]. More recently, a different approach based on the use of alkylthiol SAMs presenting mannitol end groups [20] has been reported for efficiently preventing protein adsorption and cell adhesion on gold surfaces.

The differences in wettability of mixed alkylthiol SAMs prepared from solutions of a binary mixture of thiols having different molar compositions is a macroscopic property, which does not directly reflect the microscopic-scale spatial distribution of both thiol components in the resulting SAM. Using the scanning tunnel microscope, Allara et al. found that although the average surface composition of thiols in a mixed SAM chemisorbed on Au(111) reflects the composition of the thiol solution, the distribution of thiol components in the SAM is not homogeneous and shows phase segregation of nanometer-scale one-component molecular domains [102]. Phase segregation of separated one-component domains associated with truly mixed alkylthiol SAMs has been observed by other authors in binary SAMs prepared by coadsorption of short- and long-alkylthiols on Au(111) [103–105]. Apparently, some macroscopic properties, which are generally explained as being a consequence of a lower (and presumably homogeneous) surface density of surface-bound functional groups, such as an enhanced reactivity of functional end groups and a higher accessibility of biological ligands for binding to proteins of mixed alkylthiol SAMs [106] as compared to pure one-component alkylthiol SAMs, are not consistent with the presence of phase-segregated well-ordered one-component molecular domains. However, it should be pointed out that most sensor surfaces are made of polycrystalline gold instead of Au(111), wherein a different crystal lattice of the substrate may have a great impact on the degree of order of the alkylthiol SAM. Indeed, the modification of gold surfaces by chemisorption of binary mixed alkylthiol SAMs, in which one of the thiol components has an end-attached biospecific affinity ligand, is a well-established synthetic approach for preparing sensor surfaces on which a reversible site-specific binding of proteins by formation of a surface-bound affinity complex takes place. Some of the examples that have been reported include the use of mixed alkylthiol SAMs having end-attached benzenesulfonamide [107], carbohydrates [108], nitrilo-acetic acid [109,110], biotin [111] and epitopes [112] or haptens [113] for the reversible site-specific binding of carbonic anhydrase, concanavalin A, histidine-tagged proteins, streptavidin, and antibodies (Scheme 1).

1.2.3 Surface architectures prepared by solid-phase derivatization of alkylthiol SAMs

Self-assembly of one- or two-component alkylthiol SAMs provides a convenient tool for tailoring, at a molecular level, the chemical reactivity of a gold surface by choosing and modulating the nature and amount of surface-bound chemically reactive groups (e.g., $-OH$, $-COOH$, or $-NH_2$) in the monolayer. Particularly, mixed alkylthiol SAMs prepared by coadsorption of alkylthiols

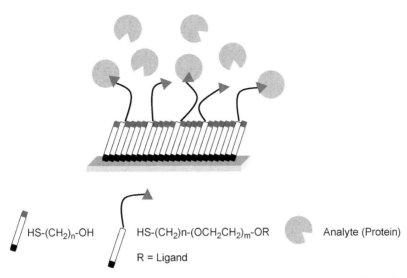

Scheme 1. Idealized sketch of a protein resistant gold surface with specific affinity for an analyte (a protein) prepared by chemisorption of a mixed hydroxylic-terminated alkylthiol and a oligo(ethylene oxide)-terminated alkylthiol having an end-attached affinity ligand [107].

having different chain lengths and different chemically reactive end groups can be used to provide a gold surface with both chemical reactivity and protein resistance. These chemisorbed functionalized alkylthiol SAMs provide a support for further covalent attachment (anchorage) of molecules by using a step-by-step solid-phase synthetic approach similar to that used in peptide chemistry [37]. The most commonly used anchoring reaction is the formation of amide bonds between surface-bound carboxylic end groups of the alkylthiol SAM and free amino groups of the molecule to be anchored. This reaction, which takes place at neutral pH in the presence of 1-ethyl-3-[3-(dimethylamino) propyl]carbodiimide "EDC" and N'-hydroxysuccinimide "NHS" with the formation of an intermediate reactive ester, has been used to anchor catalase [114], benzenesulfonamide [23,115], and membrane receptors [116] to gold surfaces modified with alkylthiol SAMs. More sophisticated anchoring reactions have been reported, which are based on the covalent binding of amino-containing molecules to alkylthiol SAMs having end-attached hydroquinone [117], aldehyde [118] or interchain carboxylic anhydride [119,120] groups; the covalent attachment of maleimide- or sulfhydryl-containing molecules to sulfhydryl-terminated alkylthiol SAMs [121]; and the covalent attachment of aldehyde-containing molecules to amino-terminated alkylthiol SAMs [122] (Scheme 2a–e). Recently, Chidsey et al. have shown the

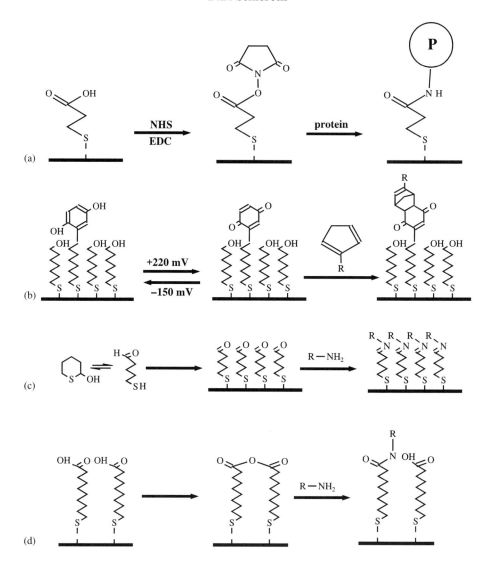

Scheme 2. (a) Covalent binding reaction of amino-containing molecules to carboxylic-terminated alkylthiol SAMs catalysed by 1-ethyl-3-[3-(dimethylamino)-propyl]carbodiimide "EDC" and N'-hydroxysuccinimide "NHS" at neutral pH [114]. (b) Diels–Alder binding reaction of cyclopentadiene-containing molecules to a quinone-terminated alkylthiol SAM. Quinone end groups are formed by electro-chemical oxidation of surface-bound hydroquinone moieties, which allows to modulate electrochemically the concentration of quinone reactive groups on the surface [117]. (c) Covalent binding of alkylamines to aldehyde-terminated alkylthiol SAMs. The monolayer is formed by exploiting the equilibrium between

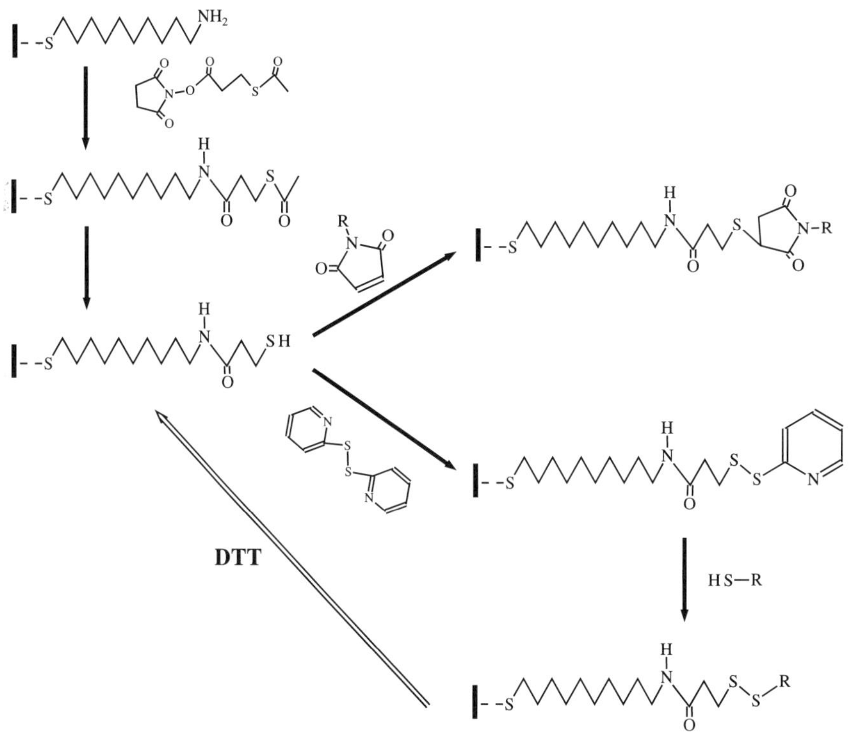

Scheme 2. *(Continued.)*

2-hydroxypentamethylene sulfide and its open-chain aldehyde isomer [118]. (d) Covalent binding of amino-containing molecules to carboxylic-terminated alkylthiols SAMs. Surface-bound reactive interchain anhydride groups are formed by treating the alkylthiol SAM with trifluoroacetic anhydride and triethylamine in anhydrous N,N-dimethylformamide [119]. (e) Covalent binding of maleimide-containing and sulfhydryl-containing molecules to sulfhydryl-terminated alkylthiol SAMs. The reactive sulfhydryl groups are introduced by reacting an amino-terminated alkylthiol SAM with N-succinimidyl S-acetylthiopropionate followed by deprotection of the sulfhydryl group under alkaline conditions. Covalent binding of disulfide-containing molecules gives rise to surface-bound disulfide bond, which can be used in a surface thiol-disulfide exchange reaction for binding sulfhydryl-containing molecules. Furthermore, surface-bound disulfide groups can be cleaved in the presence of dithiothreitol (DTT) to regenerate the reactive sulfhydryl-terminated alkylthiol SAM [121].

applicability of Huisgen 1,3-dipolar cycloadditions ("click" chemistry) as a general methodology for functionalizing surfaces coated with a SAM. The coupling reaction reported by these authors proceeds quantitatively at very mild conditions and is based on the covalent attachment of acetylene moieties to azide-terminated monolayers via the irreversible formation of a triazole ring [123].

These anchoring reactions, however, do not per se (without combining them with micropatterning techniques) allow control of the spatial distribution of the molecules covalently bound to the monolayer at a microscopic scale, such that the attached molecules are non-reproducibly distributed throughout the whole surface. Some attempts to solve this problem have been reported, based on the use of chemically-reactive molecules that have a rigid geometry and act as a template for directing the covalent binding of further molecules to localized locations on the surface. For example, the self-assembly of sulfur-containing helical peptides on a gold surface gives rise to a nanosize-scale array of chemically reactive groups which are separated by a fixed distance defined by the pitch of the helix [124–126]. Analogously, the covalent attachment of a chemically reactive cyclic peptide to an alkylthiol SAM chemisorbed on gold results in an array of functional groups that are bound to the surface with the spatial distribution determined by the molecular shape of the peptide molecule [127]. More recently, Ringler and Schulz have used a biotinylated C_4-symmetric tetrameric aldolase as a rigid four-way connector for producing a quadratic network of streptavidin molecules on a lipid monolayer [128].

Furthermore, gold surfaces modified with an alkylthiol SAM can be used as a chemically reactive support for tethering biomimetic membrane systems [129–132] and ultra-thin composite multilayered architectures by using a solid-phase step-by-step synthetic approach. In general, the construction of a multilayered structure on a gold surface starts with the attachment of a polyfunctional molecule to a chemisorbed alkylthiol SAM, followed by a layer-by-layer assembly of the multilayered structure. The polyfunctional molecule, which can be an amino-containing dendrimer or a positively-charged polyelectrolyte, can be either covalently bound to a carboxylic-terminated alkylthiol SAM [133,134] or electrostatically adsorbed on an oppositely-charged alkylthiol SAM [135,136]. Analogously, the assembly of multiple layers can be organized and stabilized either by electrostatic [137,138] or by affinity interactions [139]. Self-assembled multilayered structures can also be provided with biospecific affinity ligands for binding a desired protein to the multilayered film. When compared with alkylthiol SAMs, these films show a

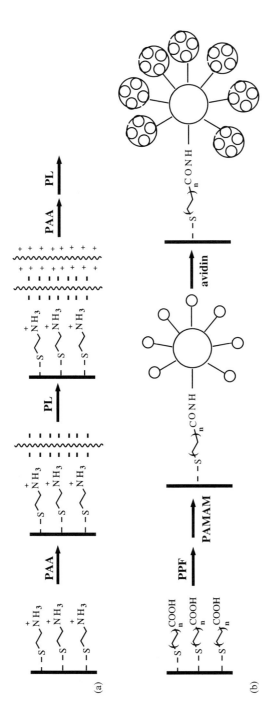

Scheme 3. Idealized sketch of a gold surface modified with a multilayered assembly prepared by (a) alternate deposition of layers of poly(acrylic acid) and poly-L-lysine on a positively charged cystamine SAM (the multilayer assembly is stabilized by electrostatic interactions between adjacent layers, [136]); and by (b) covalent attachment of a layer of G4 poly(amidoamine) dendrimer to a 11-mercapto-1-undecanoic acid SAM after activation of the surface-bound carboxylic groups with pentafluorophenol and EDC. The dendrimer monolayer is afterwards biotinylated by covalent binding of desthiobiotin amidocaproate (not shown) to assemble an overlayer of avidin by affinity interactions [133].

higher loading capacity and a lower extent of non-specific protein adsorption because they provide a thicker organic cushion layer between the surface-bound protein and the metal surface (Scheme 3a,b).

1.3 AFFINITY BIOSENSORS

The most straightforward application of a protein-resistant surface architecture having biospecific recognition ligands (or receptors) is its use as a sensing surface for an affinity biosensor, wherein the selective sensor response arises from the formation of a surface-bound affinity complex between the analyte and the surface-bound ligand. Transducers such as a quartz microbalance (QMB) [5–7] or a surface plasmon resonance (SPR) device [8–11,140], which are able to detect a property of the immobilized layer resulting from the formation of the surface-bound affinity complex, are ideally suited to monitor in real-time affinity binding reactions, with a high sensitivity, and without requiring any additional separation step. Even more interesting is that these techniques enable an affinity-binding surface reaction to be directly detected without the need to attach a reporter label (e.g., fluorescence markers or radioisotope labels) to any one of the affinity partners.[1] SPR- and QMB-based affinity biosensors have been widely used for the study of binding interactions between receptor proteins, antibodies or nucleic acids and their biological affinity partners [107,110,115,120,141], wherein their sensitivity, selectivity, reliability and time-response depend on the capability of the affinity ligand immobilized on the sensor surface to build selectively and reversibly the affinity complex [107,109,142].

---

[1] Although it is beyond the scope of this chapter to give a detailed description of the transduction physical principles of sensor devices based on piezoelectricity or optical evanescent field techniques, it should be mentioned that the QMB and SPR can detect in real time changes in properties (mass and refractive index, respectively) which are characteristic of the film deposited on the sensor surface, independently of the composition of the bulk solution. In the QMB a decrease of resonance frequency of the piezoelectric quartz crystal is correlated with a mass increase of the organic layer covering the crystal according to the Sauerbrey equation [5–7]. SPR is based on the optical excitation of an electromagnetic wave (surface plasmons), that propagates along the interface between a metal and a dielectric to generate an evanescent electromagnetic field that propagates along a short distance (of a few hundred nanometers) into the dielectric medium adjacent to the medium. This is achieved by illuminating a thin metal layer (generally a gold film of ca. 50 nm thickness) with p-polarized light incident in a medium (a prism) under total internal reflection conditions. At a particular angle of incidence (the surface plasmon angle, Θ_p) that is sensitive to the refractive index of the dielectric layer immediately adjacent to the metal surface, the wavevector of the incident light matches the wavevector of the surface plasmon, which is detected by a sharp minimum in reflectance of the metal surface [8–11,140].

1.3.1 Site-specific affinity binding of proteins to self-assembled molecular architectures

It is well known that immobilization techniques which are based on non-specific physical adsorption or on non-specific covalent binding via chemically-reactive amino acid side chains of the protein surface to the solid support result in a multipoint attachment of the protein, which may seriously damage the native structure of the protein. Furthermore, the large fraction of amino acids, which interact non-covalently with the surface, depending on the distribution of chemical moieties on the protein surface and the chemical heterogeneity of the substrate, gives rise to a film in which the orientation of the protein molecules is geometrically random. Protein films in which the adsorbed molecules are anisotropically oriented can be obtained when there is a high-affinity preferential mode of interaction between the surface and a localized contact region of the protein. For example, cytochrome-c adsorbed on clean glass substrates gives rise to monolayer films with a broad distribution of molecular orientations [143], while a narrower distribution of molecular orientations is obtained by modulating the charge distribution of surface active groups [143,144], or the hydrophilicity of the substrate [145]. Moreover, it has been found that for other proteins, such as myoglobin, the molecular orientation distribution in the film is narrower at higher protein surface coverages [145]. More recently Jiang et al. have investigated the effect of the surface charge on the orientation of antibody molecules electrostatically adsorbed on gold surfaces modified with alkanethiol SAMs [146,147].

The general approach for immobilizing a protein on a substrate with a defined geometrical orientation involves the use of a strong interaction between a structurally unique site or region on the protein surface and the substrate, either by a site-directed affinity binding to specific receptors bound to a SAM [31,107,109,148–151], or by specific interaction between a single amino acid of the protein with suitable chemically-active surface groups [152–154]. Both strategies are useful for the manufacture of an affinity biosensor because the surface-bound groups involved in the formation of the affinity complex with the analyte[2] can be either a low molecular weight affinity ligand or a protein affinity partner, wherein in the latter case the surface-bound protein should be immobilized with a geometrical orientation that facilitates the formation of the affinity complex. This is, however, an

---

[2] In affinity sensors, the analyte is usually a high-molecular-weight protein, since its binding to the surface results in a large, easily detectable change in mass and/or in refractive index of the surface-bound layer.

idealized view of the system, because the distribution of molecular orientations in anisotropic protein films also depends on the extent of ordering of the organic underlayer [152,155]. Moreover, the biological performance of the immobilized protein molecules is also influenced by the substrate itself because the binding reaction of immobilized receptors with soluble ligands is modulated not only by ligand–receptor interactions, but also by interactions between the ligand and the supporting matrix [156,157].

A biological affinity system which is particularly useful for the construction of affinity biosensors is the biotin–streptavidin affinity complex. This complex is very specific and has an extremely high affinity constant ($K_a = 10^{15}\, l \times mol^{-1}$). Moreover, streptavidin is a tetramer with one biotin-binding site in each subunit, which can be used as a tetrafunctional anchoring molecule, and biotin is a small ligand that can be used to label other molecules [158]. Due to the irreversibility of the biotin–streptavidin affinity reaction, this system is not particularly interesting as a sensing surface reaction for affinity biosensors. However, it is a very useful tool for the site-specific binding of other proteins to sensor surfaces and for building and stabilizing self-assembled complex multilayered surface architectures. The simplest approach for modifying a gold surface with a protein-resistant organic layer, which provides a molecular-level control of the density and environment of the surface-bound ligands (e.g., biotin), consists in the chemisorption of a binary mixed alkylthiol SAM, in which the biotinyl-attached alkylthiol component is distributed throughout (or "diluted" into) a hydrophilic underlayer comprising a hydroxylic-terminated alkylthiol [158]. In these biotinyl-attached SAMs, the formation of the surface-bound affinity complex is facilitated when the biotinyl ligand is attached to the thiol end group through a long flexible spacer, such that it protrudes from the alkylthiol underlayer.

This system, however, shows some disadvantages, such as a relatively large extent of non-specific protein adsorption, which strongly depends on the extent of order of the alkylthiol underlayer, and a dependency of the kinetic parameters of the affinity-binding reaction on interactions between adsorbed proteins with the surface and with other adsorbed proteins [159]. In this context, it has been proposed that unfavourable steric constraints introduced by tethering (e.g., close packing of ligands induced either by phase segregation of thiols or by a high concentration of ligand on the surface) and restriction of the mobility of the surface-attached biotin molecules may favour desorption of the complex [159].

A more efficient approach for preventing non-specific interactions between tethered proteins and the surface consists in the intercalation of a thick hydrophilic cushion between the end-attached affinity ligands and the alkylthiol underlayer by tethering the ligand through a flexible oligo(ethylene oxide) spacer [107,111,160]. Mrksich et al. found that the specific affinity binding of carbonic anhydrase to binary mixed alkylthiol SAMs having end-attached benzenesulfonamide groups tethered through an oligo(ethylene oxide) spacer proceeds with a ca. 10% extent of non-specific protein adsorption when the molar fraction of ligand-ended thiol in the mixed monolayer is less than 0.1 [107]. The same authors observed an increasingly higher extent of non-specific protein adsorption at higher molar fractions of ligand-terminated thiol, which they attributed to the formation of phase-segregated pure thiol domains in the alkylthiol SAM [107]. Similar results have been reported for the affinity binding of histidine-tagged proteins to binary mixed nitrilotriacetic acid end-attached alkylthiol SAMs [109]. Protein-resistant binary mixed ligand-attached alkylthiol SAMs can also be prepared using a solid-phase synthetic approach, tethering oligo(ethylene oxide) spacer groups to a carboxylic-terminated alkylthiol SAM [120]. Analogously, amino-containing ligands or proteins can be introduced in a binary mixed alkylthiol SAM having carboxylic-terminated oligo(ethylene oxide) end-attached groups by a solid-phase coupling reaction in the presence of EDC and NHS [115].

Protein-resistant surface architectures providing a soft, hydrophilic environment for the immobilized protein and a higher loading capacity, can be prepared by alternate electrostatic adsorption of oppositely-charged polyelectrolyte monolayers tethered by electrostatic interactions to a charged alkylthiol SAM, wherein charged protein molecules are adsorbed onto the multilayered structure by electrostatic interactions with the terminating polyelectrolyte layer [137,138,161]. It has been found, however, that although a larger amount of protein is electrostatically adsorbed on an oppositely-charged polyelectrolyte terminating layer, a certain amount of protein is always adsorbed independent of the charge of the terminating layer. This behaviour can be explained by the diffusion of the protein along the film, which may also explain the large protein surface coverages obtained with multilayered polyelectrolyte films [161]. Since electrostatically-assembled multilayered films are especially sensitive to the pH and ionic strength of the medium, the strong affinity interaction between biotin and streptavidin (or avidin) can be exploited to hold together the whole multilayered assembly. Furthermore, avidin has the additional advantage that it adsorbs strongly on platinum and gold surfaces, which obviates the requirement to tether the multilayer system to the surface via an alkylthiol SAM. Following this approach, several monolayers of biotinyl-labelled poly(ethyleneimine), poly(allylamine) or poly(amidoamine) dendrimer have been deposited on gold surfaces modified with a first adsorbed avidin monolayer and stabilized by intercalating a monolayer of avidin between two biotinylated-polyelectrolyte layers [139,162]. Alternatively, several layers of biotinyl-labelled NH_2-containing polyelectrolytes can be tethered to a gold surface by covalent attachment of a first polyelectrolyte layer to a first carboxylic-terminated alkylthiol SAM, and further intercalation of avidin monolayers between two adjacent polyelectrolyte monolayers [163].

1.3.2 DNA hybridization biosensors

The highly specific affinity-binding reaction between two single-stranded DNA (ssDNA) chains to form a double-stranded DNA (dsDNA) hybrid is widely used as a diagnostic reaction for detecting single point mutations related to genetic diseases, and for high sensitivity detection of the presence of a particular microorganism in a sample [164,165]. The advantage of carrying out a solid-phase hybridization reaction forming a surface-bound dsDNA hybrid is that a very small amount of ssDNA oligonucleotide capture probe is sufficient for selectively detecting a specific oligonucleotide target sequence rapidly, in situ, quantitatively and with a high degree of mismatch discrimination [166]. In particular, SPR- and QMB-based DNA hybridization biosensors are able to detect a hybridization reaction in real time without the need to attach a reporter label to any one of the single-stranded oligonucleotide probes [167]. The simplest approach to prepare a surface for a DNA hybridization sensor is by chemisorption of thiol-containing ssDNA capture probes on a gold surface, such as to form a monolayer of ssDNA on the sensor surface. Alternatively, a thiol-terminated spacer can be covalently attached to the 5′-phosphate end of an oligonucleotide probe using standard phosphoramidite chemistry [167–170], such that the resulting probe can interact with the gold surface preferentially through the thiol group [168]. After the completion of the solid-phase heterogeneous hybridization reaction, the sensor surface may be fully regenerated by thermal or chemical dissociation of the surface-bound dsDNA hybrid. As for all affinity sensors, an optimal performance of a DNA hybridization biosensor is achieved by maximizing the number of functionally-active ssDNA oligonucleotide capture probes which are tethered to the surface, while minimizing non-specific interactions between ssDNA oligonucleotide target probes and the surface. In order to achieve this goal, the capture probes must be accessible for interaction with the target and the underlaying surface must be "inert" towards the capture probes.

Although a thiol-containing ssDNA probe should interact with a gold surface preferentially through the thiol group, an interaction of the nitrogen-containing side nucleotide chains with the surface cannot be excluded.

Indeed, the adsorption of thiolated ssDNA on gold surfaces gives rise to one-component monolayers which are poorly packed and in which the ssDNA probes are oriented with their molecular axis parallel to the surface, thus being unsuitable to form a surface-bound hybrid with a target probe [168]. This scenario can be changed by "diluting" the thiolated ssDNA capture probe into an alkylthiol SAM which is chemically inert to ssDNA, such as an hydroxy-terminated alkylthiol SAM. The dilutor thiol, which is chosen to be of the same chain length as the capture probe and is introduced upon adsorption of the thiolated probe, serves to passivate the non-covered gold surface and to realign the adsorbed thiolated probes with their axis perpendicular to the surface (Scheme 4a), thus making them accessible for hybridization with the

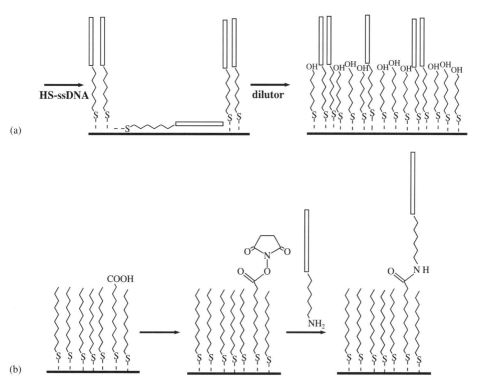

Scheme 4. Idealized sketches of gold surfaces for DNA hybridization sensors prepared by (a) chemisorption of thiolated ssDNA capture probes followed by chemisorption of a dilutor alkylthiol [168]; and by (b) chemisorption of a mixed alkylthiol SAM followed by covalent attachment of an amino-terminated ssDNA capture probe (surface-bound carboxylic groups are activated by reaction with O-(N-succinimidyl)-N,N,N',N'-tetramethyluronium tetrafluoroborate in acetonitrile in the presence of N,N-diisopropylethylamine, [172]).

target ssDNA molecules [168–170]. Although the affinity constant for hybridization and maximal amount of bound target ssDNA do not seem to depend on the length of the spacer separating the probe from the surface, the time to reach the binding equilibrium increases as the spacer gets shorter, which is attributed to a steric hindrance for hybridization near the substrate [171]. A different approach for optimizing the surface density of surface-bound ssDNA accessible for hybridization is based on a step-by-step solid-phase synthesis of a two-dimensional mixed alkylthiol SAM prepared from hydroxylic- or methyl-terminated and carboxylic-terminated thiols, followed by covalent attachment of a 5′-amino-terminated ssDNA capture probe (Scheme 4b). Using this approach, a molar composition of 0.5 mol% of carboxylic-terminated thiol in the monolayer is sufficient to obtain a maximum saturation surface density of ssDNA probe. The efficiency of the hybridization reaction in these monolayers, however, depends on the length of the spacer used to tether the ssDNA probe to the alkylthiol SAM, wherein too long spacers result in lower hybridization yields. This behaviour has been attributed to the linear shape of the ssDNA probe, to steric hindrance among bound probes and/or a wrong orientation of the capture probe with respect to the surface [172]. A different way to prepare an inert surface towards non-specific adsorption of ssDNA target and which presents an homogeneously distributed dense array of easily accessible ssDNA probes is based on the assembly of polyelectrolyte monolayers on a gold surface. For example, 5′-thiol-(or amino-)terminated ssDNA capture probes can be covalently bound to a poly-L-lysine monolayer adsorbed electrostatically onto a carboxylic-terminated alkylthiol SAM [173].

A serious disadvantage of SPR- or QMB-based DNA hybridization affinity sensors arises from the relatively low molecular weight of the ssDNA molecules used as target probes, which results in a very small change in thickness (or mass) of the layer immobilized on the surface sensor associated with the hybridization reaction. The hybridization detection limit of such biosensors can be significantly improved by coupling the surface hybridization reaction with a secondary affinity reaction that involves the binding of a high-molecular-weight molecule, thus leading to an amplification of the sensor signal. For example, the use of biotinylated ssDNA target probes allows the detection of the hybridization surface reaction by sensing the formation of a secondary surface-bound affinity complex between hybridized biotinylated ssDNA target probes and streptavidin (Scheme 5). Further amplification of the signal may be obtained by a successive layer-by-layer deposition of several streptavidin layers, which are held together by intercalated layers of biotinylated dsDNA hybrids [174,175].

A further disadvantage of the sensing methods discussed above is that they require at least the modification of the ssDNA capture probe with an anchoring moiety that allows its attachment to the sensor surface. This involves laborious experimental work in preparing the probes, and moreover the surfaces (and the probes) cannot be regenerated after use. This problem may be overcome by exploiting the strong and specific affinity interactions between non-labelled ssDNA and gold surfaces modified by electrostatic adsorption of a monolayer of a positively-charged bisbenzamidine derivative to a negatively-charged chemisorbed alkylthiol SAM [176, 177]. Since the assembly is held together by electrostatic interactions, the organization of the assembly is pH sensitive, such that a clean alkylthiol-modified gold surface can be regenerated after use simply by adjusting the pH of the solution [176,177]. More recently, a different strategy for preparing ssDNA arrays on a fully-regenerable gold surface has been reported, which is based on the reversible formation of a disulfide bond between a thiolated-terminated ssDNA probes and the sulfhydryl end groups of a SH-terminated alkylthiol SAM (see Scheme 2e) [121].

1.3.3 Electrochemical detection of affinity interactions

In the last decade, a large amount of effort has been devoted to the development of detection methods which are compatible for use in combination with DNA sensor array platforms and are sufficiently sensitive to detect single point mutations using minute amounts of sample [164,166]. Electrochemical detection methods are a particularly interesting alternative to the methods described above due to their simplicity and low cost.

In principle, electrochemical transducers can be used to detect the formation of a surface-bound affinity complex when the affinity-binding reaction is associated with a change in electrical properties (e.g., ion permeability or capacitance) of the layer immobilized onto the electrode surface. For example, the so-called ion-channel sensors detect permeability changes of a film immobilized on an electrode surface to an electroactive molecule, which is used as a redox marker. The formation of a surface-bound affinity complex results in a permeability change, which can be monitored by the change of cyclic voltammetric response of the redox marker.

This methodology has been used for monitoring affinity binding of avidin to gold surfaces modified with a mixed alkylthiol SAM presenting biotinyl- and carboxylic end groups, since the formation of the surface-bound affinity complex screens the negative surface charges of the carboxylic-terminated

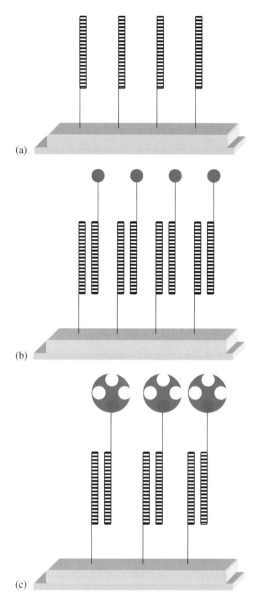

Scheme 5. Idealized sketch of a gold modified surfaces for a SPR-based DNA hybridization sensor. (a) Thiolated ssDNA capture probes are covalently bound through a bifunctional linker to a monolayer of poly(L-lysine) electrostatically adsorbed on a gold surface modified with a 11-mercapto-1-undecanoic SAM. The SPR signal associated with the hybridization surface reaction (b) is amplified by affinity binding of streptavidin to surface-bound biotinylated ssDNA target probes (c) [174].

alkylthiol underlayer. Therefore, the surface net charge is controlled by the charge of bound avidin molecules, which facilitates or hinders the diffusion of charged small redox molecules to the metal surface [178] (Scheme 6a). A similar approach has been reported for detecting immunosensing reactions on gold electrode surfaces modified with a monolayer of thiolated hapten [179]. Ion-channel sensors have been used to detect a one-base single mismatch in ssDNA oligonucleotide targets by monitoring the disappearance of the electrochemical response upon hybridization of small anionic redox markers at gold electrodes modified with a mixed monolayer of hydroxylic-terminated alkylthiol and thiolated peptide nucleic acid, as a consequence of repulsive electrostatic interactions with the phosphate groups of the nucleic acid backbone [180]. Tarlov et al. have reported a DNA hybridization sensor with electrochemical detection, which is based on the capability of small cationic redox markers to interact and screen the negative charges of the phosphate groups of the nucleic acid backbone. Therefore, the surface density of redox marker, which can be calculated by chronocoulometry or by integration of the voltammetric waves at low concentrations of redox marker [181], is directly proportional to the amount of nucleic acid (i.e., number of phosphates) bound to the electrode surface.

Ion-channel sensors have the disadvantage that their electrochemical response is very sensitive to the presence of defects or pinholes in the alkylthiol underlayer, which render the monolayer permanently permeable to some extent to the redox marker. Affinity amperometric sensors based on the electroenzymatic detection of a secondary analyte do not have this drawback, and moreover, they amplify the sensor signal by coupling the affinity reaction with a secondary electroenzymatic reaction [182]. The formation of an immunoaffinity complex between antibody and a surface-bound antigen at gold electrode surfaces modified with a mixed antigen- and ferrocenyl-terminated cystamine SAM can be detected by coupling the affinity reaction to the electroenzymatic oxidation of glucose by the enzyme glucose oxidase. The formation of an immunoaffinity complex hinders the interaction of the enzyme with the surface-bound ferrocenyl molecules, thus hindering the enzymatic oxidation of glucose which results in a loss of electrocatalytic current [182] (Scheme 6b). A similar approach has been used for detecting the affinity binding of avidin to gold electrode surfaces modified with a monolayer of ferrocenyl-tethered biotinylated dendrimer [163].

Affinity sensors with electrochemical detection can also be used for sensing dsDNA, taking advantage of the capability of some organic molecules to intercalate within the double helix forming a complex with dsDNA. Thus, it is

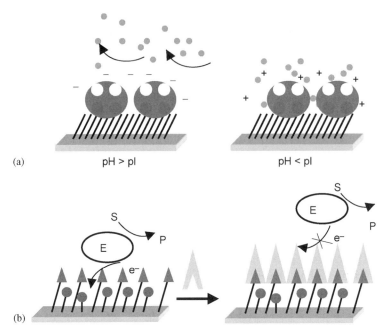

Scheme 6. (a) Ion-channel affinity sensor: the reversibility of the electrochemical response of negatively charged redox probes at a gold surface modified with a mixed biotinyl-terminated and COOH-terminated alkylthiol SAM is modulated by the affinity binding of avidin to the monolayer. At pH > pI of avidin, the avidin monolayer is negatively charged, thus hindering the permeability of the anion probe. The electrochemical behaviour of the anion probe becomes more reversible at pH < pI, in which the avidin monolayer is positively charged [178]. (b) Immunosensor with electroenzymatic detection: the amperometric response associated to the oxidation of glucose by glucose oxidase at gold electrodes modified with a mixed ferrocenyl-terminated and antigen-terminated alkylthiol SAM is blocked by the affinity reaction because the surface-bound antibody molecules are too bulky to enable the interaction of the enzyme with the surface-bound ferrocenyl moieties [182].

possible to build a two-dimensional dsDNA monolayer array on a gold surface by intercalation of dsDNA with a mixed alkylthiol SAM presenting an intercalating molecule (e.g., an acridine derivative) pendant end group [183]. Furthermore, when the intercalating molecule is an electroactive species, such as methylene blue [184–186], doxorubicine [187], 2,6-anthroquinonedisulfonic acid [188,189], or osmium–bipyridyl complexes [190], the electrochemical response of this species can be used for detecting the presence and amount of surface-bound dsDNA. This kind of sensor, however, is not suitable for discriminating single point mutations.

A higher sensitivity can be achieved by using ssDNA label probes, which are tagged with an electroactive species. The use of a "sandwich" assay format is specially advantageous in this context because it obviates the need to tag the oligonucleotide target probe, it does not require a washing step, thus giving rise to a "real time" response and it can be directly coupled to a polymerase chain amplification reaction [166]. This method has, however, a serious drawback because of the extremely small amount of label oligonucleotide probes captured on the electrode surface, which gives rise to an accordingly small electrochemical response. For this reason, electrochemical-based assays for detecting DNA hybridization commonly require a procedural step that is directed to amplify the electrochemical signal response.

The first attempts to improve the sensitivity of electrochemical DNA hybridization sensors were based on coupling of the hybridization reaction with a secondary electroenzymatic reaction that resulted in the formation of an electrochemically detectable compound. For example, Caruana and Heller reported an amperometric DNA sensor based on the use of electrodes modified with a thin layer of redox polymer containing ssDNA capture probes immobilized thereon, and soluble label probes tagged with soybean peroxidase. Only upon hybridization of the label probe to the target is the enzyme sufficiently close to the electrode surface to be able to react with the redox polymer, giving rise to an amperometric response that is proportional to the amount of hybridized ssDNA target [191]. More recently, Kwak et al. have reported an electroenzymatic sandwich assay based on the use of electrodes modified with a layer of partially ferrocenyl-tethered poly(amidoamine) dendrimer containing capture ssDNA probes and attached to a gold electrode via a carboxy-terminated alkanethiol SAM. These authors used biotinylated label probes tagged with an avidin–alkaline phosphatase conjugate. The hybridization reaction is detected by the amount of enzymatically generated p-aminophenol that is electrocatalytically oxidized at the electrode surface [192].

A more elegant strategy for amplifying the electrochemical signal arising from a hybridization reaction is based on labelling the ssDNA label probes with colloidal gold nanoparticles. In such a system, the most simple amplification approach consists of using the gold nanoparticle as a support for tethering (e.g., via an alkanethiol SAM) several electroactive tags to a single probe [193]. Another method for measuring the amount of hybridized target ssDNA captured on the electrode surface involves dissolution of the gold nanoparticles, which are subsequently electrodeposited at the electrode surface [194]. The use of gold nanoparticles enables a higher extent of signal amplification by coupling the hybridization reaction to a "silver enhancement" step [195], which consists of precipitating silver ions onto the gold particles before starting the potentiometric stripping process, thereby increasing the amount of material deposited on the electrode surface [196]. This technique, however, does not give reliable results with gold electrode surfaces due to a serious background silver deposition. In order to solve this problem, Hsing et al. have recently reported a system in which the gold electrode surface is covered with a polyelectrolyte multilayer tethered to the surface via a carboxy-terminated alkanethiol SAM [197].

A considerable simplification of the electrochemical DNA hybridization assay has been reported by Heeger and coworkers, who have developed a "reagentless" DNA sensor based on the use of electroactive molecular beacons as capture (and label) probes. The detection system consists of an electrode surface modified with a layer of self-assembled capture DNA probes having a stem-loop structure and tagged with an electroactive substance, which interacts with the electrode surface. Hybridization to a target ssDNA is coupled with the opening of the stem-loop structure, which results in the electroactive tag moving away from the electrode surface with the subsequent loss of electrochemical signal [198].

1.4 AMPEROMETRIC CATALYTIC BIOSENSORS

Surface architectures which are suitable for the manufacture of an efficient amperometric enzymatic biosensor differ in some aspects from those used for affinity biosensors. While affinity biosensors are generally used to detect high-molecular-weight proteins by a reversible affinity binding of the protein to a surface-bound ligand, catalytic amperometric biosensors are generally used to detect low-molecular-weight molecules by their enzymatic conversion by a surface-bound high-molecular-weight enzyme. Therefore, a stable and reproducible biosensor response can be obtained when the enzyme is irreversibly immobilized on the electrode surface. On the other hand, the enzymatic conversion of analyte can be transduced into a measurable amperometric response only when an efficient electronic pathway has been established between the electrode surface and the enzyme. Therefore, the film layer separating the protein from the electrode surface must be thick enough to minimize non-specific interactions between the protein and the metal, which may lead to protein denaturation, but thin enough not to form an insulating coating that hinders the electron transfer reaction.

1.4.1 Multipoint attachment of proteins to gold electrode surfaces modified with an alkylthiol SAM

Most redox proteins, either in solution or adsorbed at an electrode surface, show a poorly defined electrochemical response at electrode surfaces because their redox site is deeply buried within their polypeptide backbone, which acts as an electrical insulator hindering the electron transfer reaction [199]. However, when the redox site is close to the protein surface and the protein approaches (or is immobilized onto) the electrode surface, such that the redox site faces the surface, the rate of the electron transfer reaction is sufficiently high to give rise to a direct (non-mediated) quasi-reversible electrochemical response [199]. A further requirement, which has to be fulfilled for considering the interaction of a redox protein at an electrode surface as mimicking the electron transfer reaction between two biological redox partners, is that its native conformation is not significantly perturbed upon interaction with the electrode surface. These conditions can be attained with some small redox proteins like cytochrome-c in solution [199] and some high-molecular-weight proteins which have more than one redox site, such as some hydrogenases [200,201] quinoheme [202,203] and flavoheme [204] dehydrogenases when adsorbed at electrode surfaces. However, the reproducibility and stability of the electrochemical response of soluble or physically-adsorbed proteins at an electrode surface depend strongly on the distribution of functional groups present at the metal surface, which depends on the history of the electrode and may be poorly reproducible. Gold electrode surfaces modified with an alkylthiol SAM having end-attached functionalized groups provide a more reproducible surface architecture, in which protein molecules interacting therewith show a narrow distribution of molecular orientations, which is induced by non-covalent interactions between localized patches of the protein surface and the electrode surface [205,206]. Within this context, Gorton et al. have recently reported the use of carboxylic- and hydroxy-terminated alkylthiol SAMs for promoting the direct electron transfer reaction of cellobiose dehydrogenase (a flavoheme-protein) and sulfite oxidase (a heme-protein containing a molybdopterin cofactor), respectively, at gold electrode surfaces [207–209].

Some proteins, such as cytochrome-c [210–214] or cytochrome-c oxidase [214] have a non-homogeneous distribution of charged amino acid side chain groups on their surface and show localized patches with a high density of positively-charged groups. These proteins adsorb by electrostatic interactions on gold surfaces modified with a carboxylic-terminated alkylthiol SAM, giving rise to very stable films in which the molecular orientation of the adsorbed protein molecules favours a direct electron transfer reaction. The amount of adsorbed protein molecules which effectively interacts with the electrode surface can be estimated by using cyclic voltammetry combined with SPR measurements "ex situ". For example, cytochrome-c adsorbed on gold electrode surfaces modified with a monolayer of 3-mercaptopropionic acid (MPA) gives rise to protein films, which surface coverage ($\Gamma_{cyt-c} = 17.0 \times 10^{12}$ mol cm^{-2}) calculated from integration of the voltammetric peaks ($E_{1/2} = +0.045$ V vs. Ag/AgCl at pH 7.0) is slightly lower than that obtained from SPR adsorption experiments ($\Gamma_{cyt-c} = 22.5 \times 10^{12}$ mol cm^{-2}), indicating either a detachment of protein from the electrode surface during voltammetric scanning, or a heterogeneous distribution of orientations of protein molecules at the electrode surface, which results in a surface population of protein molecules that are not electrically wired to the electrode surface [214] (Fig. 1.1A,B). The stability of the adsorbed protein layer under a cyclic voltammetric regimen can be monitored using SPR combined with cyclic voltammetry "in situ".[3] Following the time differential potential-dependent changes in layer thickness under a voltammetric regime in a film of cytochrome-c electrostatically adsorbed on a gold electrode surface modified with a 3-mercaptopropionic acid SAM, a $d\Delta\Theta_{pl}/dt$ vs. E curve is obtained, which resembles its corresponding i vs. E curve obtained by cyclic voltammetry (Fig. 1.2A,B). The protein surface coverage estimated from integration of the $d\Delta\Theta_{pl}/dt$ vs. E curve ($\Gamma_{cyt-c} = 22.0 \times 10^{12}$ mol cm^{-2}) is similar to that calculated from integration of the voltammetric peaks, which suggests a potential-dependent adsorption–desorption of cytochrome-c from the electrode surface that destabilizes the protein film [214].

Cytochrome-c oxidase (COX) is a membrane-bound high-molecular-weight enzyme containing four redox sites (heme a, heme a$_3$, Cu$_A$ and Cu$_B$), which is thought to adsorb on gold surfaces modified with a 3-mercaptopropionic acid SAM forming a monolayer ($\Gamma_{COX} = 1.5 \times 10^{12}$ mol cm^{-2}), in which the

---

[3] Surface plasmon resonance combined with cyclic voltammetry in situ allows the monitoring of cyclic time-differential changes in plasmon angle shift that occur in an electroactive layer deposited on a gold electrode surface as a result of the applied potential ($d\Delta\Theta_p/dt$ vs. E), thus giving information about changes in optical layer thickness of the adsorbed film which are associated with the redox process [214–218]. For electrochemically-induced adsorption–desorption processes the time-differential of the plasmon angle shift ($d\Delta\Theta_p/dt$) correlates with the potential-dependent average layer growth and with the electrochemical current, both in magnitude and sign. The value of $d\Delta\Theta_p/dt$, however, is not influenced by diffusionally controlled processes (e.g., reduction of oxygen traces that may be present in the bulk solution) and reflects solely processes that take place at the electrode surface. Furthermore, changes in optical properties of the adsorbed film can arise from redox-induced conformational changes of the protein or from changes in electronic state of the protein due to a transfer of electrons and/or protons.

Biosensors based on self-assembled monolayers

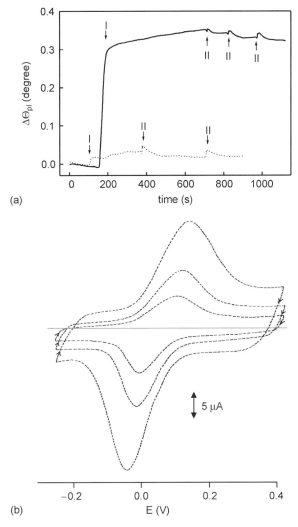

Fig. 1.1. (a) SPR angle shifts ($\Delta\Theta_{pl}$) showing the electrostatical adsorption of cytochrome-c on gold electrode surfaces modified with a 3-mercaptopropionic acid SAM; (I) after addition of cytochrome-c to a final concentration of ca. 2.5×10^{-7} M in 5 mM Na-phosphate pH 7.0 (full line) or in 5 mM Na-phosphate, 100 mM KCl pH 7.0 (dotted line); (II) after rinsing the surface with clean buffer solution. (b) Cyclic voltammograms of a monolayer of cytochrome-c adsorbed on a gold electrode modified with a 3-mercaptopropionic acid SAM recorded in 5 mM Na-phosphate, pH 7.0, at 50, 100 and 200 mV s^{-1}. Reproduced from [214] with permission.

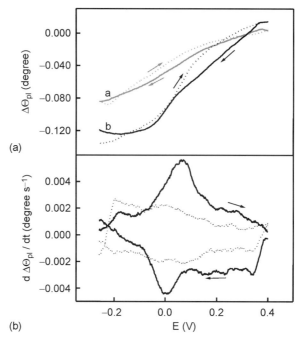

Fig. 1.2. (a) Potential-dependent changes in $\Delta\Theta_{pl}$ in (a) a gold electrode surface modified with a 3-mercaptopropionic acid SAM, and (b) a film of cytochrome-c electrostatically adsorbed on a gold electrode surface modified with a 3-mercaptopropionic acid SAM, recorded at 1 mV s^{-1} in 5 mM Na-phosphate, pH 7.0. Negative (full line) and positive (dotted line) scans. (b) $d\Delta\Theta_{pl}/dt$ vs. E of a gold electrode surface modified with a 3-mercaptopropionic acid SAM (dotted line) and a film of cytochrome-c electrostatically adsorbed on a gold electrode surface modified with a 3-mercaptopropionic acid SAM (full line) under the same experimental conditions. It is noted that $d\Delta\Theta_{pl}/dt$ values have been arbitrarily plotted as negative values for the negative scan, following the same convention as for cathodic currents. Thus, a negative peak in the cathodic scan should be interpreted as an increase of $d\Delta\Theta_{pl}/dt$. Reproduced from [214] with permission.

protein molecules are oriented with their longest axis parallel to the surface (Fig. 1.3A). Cyclic voltammograms recorded with these COX-modified electrodes show a couple of voltammetric waves ($E_{1/2} = +0.130$ V vs. Ag/AgCl at pH 7.0), and a very low catalytic activity of the adsorbed enzyme in the presence of soluble ferrocytochrome-c (Fig. 1.3B). By contrast to cytochrome-c, cytochrome-c oxidase gives rise to very stable protein films when electrostatically adsorbed on gold electrode surfaces modified with a 3-mercaptopropionic acid SAM, showing very small potential-dependent changes in $\Delta\Theta_{pl}$ in clean buffer solutions (Fig. 1.4A).

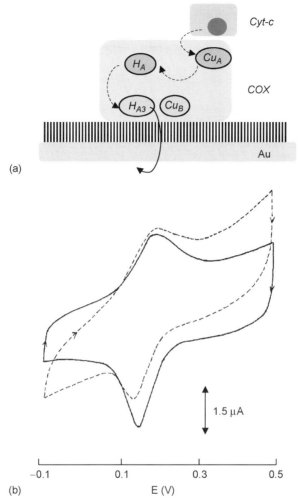

Fig. 1.3. (a) Idealized sketch showing a proposed orientation of cytochrome-c oxidase adsorbed on a gold electrode surface modified with a 3-mercaptopropionic acid SAM and electrocatalytic oxidation of ferrocytochrome-c. (b) Cyclic voltammogram recorded at 50 mV s^{-1} of a monolayer of cytochrome-c oxidase adsorbed on a gold electrode surface modified with a 3-mercaptopropionic acid SAM in 5 mM Na-phosphate pH 7.0 (full line) and after addition of 0.1 mM ferrocytochrome-c (dotted line). Reproduced from [214] with permission.

The presence of ferrocytochrome-c in the buffer solution, however, destabilizes the film leading to significant potential dependent changes of $d\Delta\Theta_{pl}/dt$ associated with two separate oxidoreduction processes at $E_I = +0.200$ ($E_{pc} = +0.105$ V; $E_{pa} = +0.300$ V) and

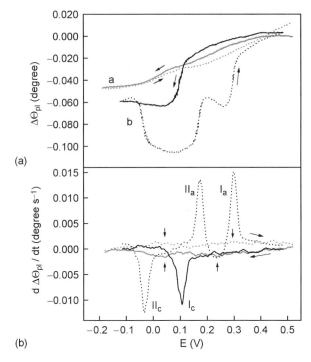

Fig. 1.4. (a) Potential-dependent changes in $\Delta\Theta_{pl}$ in a film of cytochrome-c oxidase electrostatically adsorbed on a gold electrode surface modified with a 3-mercaptopropionic acid SAM, at 1 mV s^{-1} in 5 mM Na-phosphate, pH 7.0 (a) before, and (b) after addition of 0.1 mM ferrocytochrome-c in the bulk solution. Negative (full line) and positive (dotted line) scans. (b) d$\Delta\Theta_{pl}$/dt vs. E of a film of cytochrome-c oxidase electrostatically adsorbed on a gold electrode surface modified with a 3-mercaptopropionic acid SAM at 1 mV s^{-1} in 5 mM Na-phosphate, pH 7.0 (grey), and after addition of 0.1 mM ferrocytochrome-c in the bulk solution (black). Negative (full line) and positive (dotted line) scans. It is noted that d$\Delta\Theta_{pl}$/dt values have been arbitrarily plotted as negative values for the negative scan, following the same convention as for cathodic currents. Thus, a negative peak in the cathodic scan should be interpreted as an increase of d$\Delta\Theta_{pl}$/dt. Reproduced from [214] with permission.

$E_{II} = +0.070$ V ($E_{pc} = -0.030$ V; $E_{pa} = +0.175$ V) vs. Ag/AgCl at pH 7.0, which are attributed to the oxidoreduction of the high (heme a$_3$–Cu$_B$) and the low potential (heme a–Cu$_A$) redox sites of the enzyme (Fig. 1.4B). These potential-dependent changes in optical layer thickness of the COX adsorbed layer have been explained as resulting from a reversible interconversion of COX between the "resting" (stable conformation in the fully oxidized state) and the "pulsed" (stable conformation in the fully reduced state) conformational states during voltammetric scanning under anaerobic conditions in the presence of ferrocytochrome-c [214].

Gold electrode surfaces modified with alkylthiol SAMs can be used to mimic the native environment in which the electron transfer reaction between membrane-bound proteins and cytosolic redox partners takes place, by building a lipophilic surface architecture into which the enzyme is embedded. For example, Kinnear and Monbouquette have reported the direct (non-mediated) electrochemical response of fumarate reductase adsorbed at gold electrode surfaces modified with a mixed octadecyl and dodecyl mercaptane SAM [219], and fructose dehydrogenase at gold electrode surfaces modified with a membrane-like architecture prepared by deposition of a phospholipid onto a chemisorbed mixed alkylthiol SAM composed of octadecyl mercaptane, 3,3′-dithiodipropionic acid and cystamine [220]. In this system, it is thought that the short-alkylthiols provide surface charges for promoting the electron transfer reaction through favourable electrostatic interactions with localized charged patches on the protein surface, while the long-alkylthiol and phospholipid provide the lipophilic environment into which the protein is embedded [220]. More recently, a simplified approach has been reported, which is based on the self-assembly of heterogeneous alkylthiol SAMs prepared by chemisorption of 5-(octyldithio)-2-nitrobenzoic acid on gold surfaces. The system has been used to adsorb D-fructose dehydrogenase, D-gluconate dehydrogenase and L-lactic dehydrogenase (cytochrome b_2) on gold electrode surfaces, and although it does not enable a direct electron transfer reaction between these enzymes and the gold surface, the short-length thioacid provides conductive spots on the monolayer which allow an efficient permeation of small electroactive mediator molecules to the metal surface [221].

As compared with physical adsorption, covalent immobilization of proteins at the metal surface leads to higher irreversible protein surface coverages. It results, however, in a loss of biological functionality, which depends on the protein, the immobilization method and the substrate. Gold surfaces modified with alkylthiol SAMs having chemically-reactive end groups have been widely used as a support for covalent attachment of protein monolayers [222–231]. In this case, surface-bound reactive groups not only provide anchoring sites for protein attachment, but also build a "carpet" of surface charges, which can promote a direct electron transfer reaction between surface-bound proteins, such as cytochrome-c [222] and cytochrome-c oxidase [223] and the electrode surface. In most cases, however, a molecule which is able to diffuse shuttling electrons between the redox site of the protein and the metal surface (a soluble redox mediator) is required to wire the protein to the electrode surface, although their use is not desirable for biosensors used as analytical tools in food technology or for clinical purposes. Gold electrode surfaces modified with a SAM facilitate the integration of a mediator molecule in the surface-bound molecular architecture, either as a pendant end-attached group of the alkylthiol SAM [224,225], or as an electron relay bound to the protein itself [226,227]. In both cases, the effective distance separating different redox sites in the monolayer is significantly decreased, thus resulting in an increase of the rate of the electron transfer reaction, by providing the redox mediator with a flexible longer spacer, which enables a quasi-diffusional movement of the redox mediator over the electrode surface [224–226]. Covalent attachment of mediator molecules, for example to free amino groups of the protein, gives rise to protein species in which the number and location of bound mediator molecules is poorly controlled and may result in a perturbation of the protein structure, loss of functional activity, and a poorly efficient electrical wiring of the protein to the electrode surface because mediator molecules are bound to the more accessible amino acid side chains of the protein surface. Robust proteins, such as glucose oxidase [232] or glutathione reductase [226,227], can be reversibly unfolded in the presence of urea, such that mediator molecules can be bound to less accessible amino acid side chains, which otherwise would be buried into the polypeptide backbone. Using this approach, Willner et al. have electrically wired viologen-modified glutathione reductase to gold electrode surfaces modified with a monolayer of bis(N-hydroxysuccinimidyl) 3,3'-dithiopropionate [226,227] (Scheme 7a). Redox proteins, such as glucose oxidase or D-aminoacid oxidase, the redox site of which (FAD) is not covalently bound to the polypeptide backbone, allow the mediator molecule to be bound directly to the redox site of the protein, thus optimizing the efficiency of inter- and intra-molecular electron shuttling from the redox site of a protein and the electrode surface (Scheme 7b) [232]. This method is especially mild for maintaining a close-to-native conformation of the modified protein, since the apoprotein is reconstituted with the cofactor analogue after covalent binding of the mediator molecule to the released cofactor [232].

1.4.2 Site-specific affinity binding of proteins to gold electrode surfaces modified with an alkylthiol SAM

Multipoint covalent attachment of proteins to electrode surfaces modified with a functionalized alkylthiol SAM gives rise to protein films in which

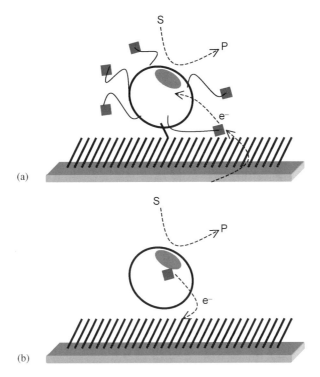

Scheme 7. (a) Idealized sketch showing the electroenzymatic reduction of glutathione at a modified gold electrode surface. The gold surface is modified by chemisorption of a monolayer of bis(N-hydroxysuccinimidyl)3,3′-dithiopropionate, followed by covalent attachment of a mediator-containing glutathione reductase. Prior to binding to the protein to the alkylthiol monolayer, mediator molecules (N-methyl-N′-(carboxyalkyl)-4,4′-bipyridinium salts) are covalently attached to protein amino groups of the protein [226]. (b) Idealized sketch showing the electroenzymatic oxidation of glucose by soluble glucose oxidase at a gold electrode surface modified with a cystamine SAM. Glucose oxidase is modified by release of its FAD cofactor, followed by covalent attachment of a mediator molecule (N-(2-methylferrocene)caproic acid) to the cofactor, and further reconstitution of the apoenzyme [232].

the protein molecules show a broad distribution of molecular orientations, which in the case of a direct non-mediated electron transfer reaction leads to an increase of the effective distance separating the redox site of the protein from the electrode surface. Furthermore, this approach can only be used with some few robust proteins, which after the immobilization procedure still preserve some functional activity. A milder site-directed binding of a protein to an electrode surface may be attained by forming a surface-bound affinity complex between the protein and a ligand-terminated alkylthiol SAM chemisorbed on gold. Suitable affinity ligands for the manufacture of a catalytic amperometric biosensor may be native cofactors, coenzymes or analogues thereof, which give rise to monolayers in which all enzyme molecules are reversibly bound to the electrode surface, having their redox site directly connected thereto through a hydrocarbon spacer.

The preparation of glucose-sensing gold electrode surfaces by reconstitution "in situ" of a monolayer of pyrroloquinoline quinone (PQQ)-dependent glucose dehydrogenase from *Acinetobacter calcoaceticus* is an example of this affinity-binding approach. PQQ is one of the cofactors present in enzymes belonging to the family of quinoproteins, which is easily and reversibly released from the holoenzymes. Although soluble PQQ shows an irreversible electrochemical behaviour at bare gold electrode surfaces, the modification of the surface with an NH_2-terminated short-alkylthiol SAM (e.g., cystamine) promotes a quasi-reversible two-electron transfer reaction at slightly acidic and neutral pHs ($E^0 = -0.125$ V vs. SCE at pH 7.0) [233]. Since the electrochemical response of PQQ at cystamine-modified gold electrode surfaces becomes gradually more irreversible with increasing pH, it is thought that it is influenced by the presence (or absence) of a high density of surface-bound positive charges on the monolayer, which promotes (or hinders) the approach of negatively-charged PQQ molecules to the electrode surface [233]. PQQ can be covalently-bound to a cystamine SAM chemisorbed on a gold electrode surface by formation of an amido bond involving at least one of its three free carboxylic groups, giving rise to PQQ-anchored SAMs ($\Gamma_{PQQ} = 1.1 \times 10^{-10}$ mol cm^{-2}) (Scheme 8a), in which surface-bound PQQ molecules are electrically wired to the metal surface ($E^0 = -0.125$ V vs. SCE at pH 7.0) [233]. Gold electrode surfaces modified with a PQQ-anchored cystamine SAM have been used for electrocatalysing the oxidation of NADH at mild overpotentials (0 V vs. SCE at pH 7.0) [234], as well as a substrate for the site-directed affinity binding of PQQ-dependent glucose dehydrogenase with the aim to construct an amperometric glucose sensor [235]. It was found, however, that the distance separating PQQ from the electrode surface in these monolayers is too small for allowing the formation of the affinity complex with the enzyme, and that PQQ must be bound to the metal surface through a longer hydrocarbon spacer for enabling the reconstitution of glucose dehydrogenase with surface-bound PQQ [235]. This is achieved by solid-phase step-by-step derivatization of the chemisorbed cystamine SAM with glutardialdehyde and a linear alkyldiamine prior to covalent attachment of PQQ, which gives rise to a mixed PQQ-terminated alkylthiol SAM, in which PQQ is bound to

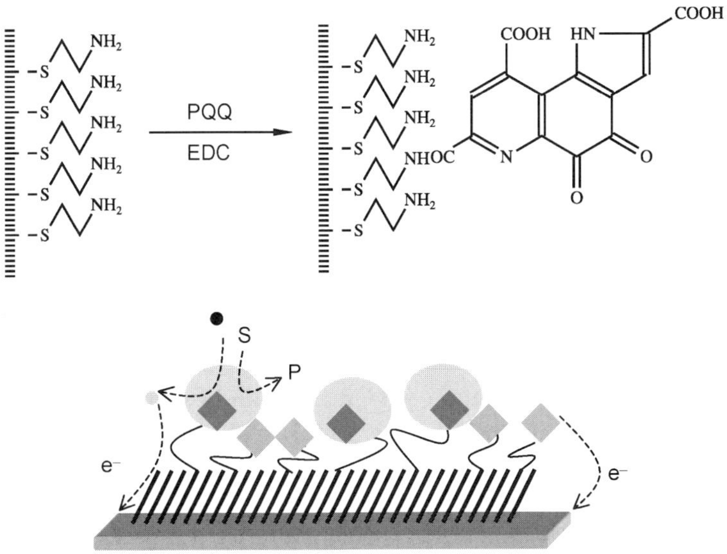

Scheme 8. (a) Covalent attachment of pyrroloquinoline quinone to a gold electrode surface modified with a cystamine SAM [233]. (b) Idealized sketch of a gold electrode surface modified with PQQ-dependent glucose dehydrogenase. The gold electrode surface is modified by chemisorption of a cystamine SAM derivatized with glutardialdehyde and 1,8-diaminooctane, followed by covalent attachment of PQQ, and reconstitution of the apoenzyme. While the oxidoreduction of non-complexed PQQ is observed at -0.125 V (SCE) at pH 7.0; the electroenzymatic oxidation of glucose is observed only in the presence of 2,6-dichlorophenolindophenol used as soluble redox mediator [235].

the metal surface through a long flexible spacer (Scheme 8b) [235]. The resulting glucose dehydrogenase-modified electrodes, however, are not able to electrocatalyse the oxidation of glucose in the absence of a soluble redox mediator, which indicates that the enzyme is not electrically wired to the electrode surface. However, cyclic voltammograms of glucose dehydrogenase-modified electrodes do show the voltammetric waves arising from the oxidoreduction of surface-bound PQQ. This behaviour suggests that the electrochemical response arises from the oxidoreduction of surface-bound PQQ molecules, which are not involved in the formation of the affinity complex with the enzyme and can quasi-diffusionally approach the electrode surface for enabling the electron transfer reaction. By contrast, the quasi-diffusional migration of complexed surface-bound PQQ molecules is hindered by the large polypeptide backbone, which results in a longer separation distance from the electrode surface that hinders the electron transfer reaction (Scheme 8b) [235]. Similar results have been reported recently for horseradish peroxidase-modified electrodes prepared by reconstitution of the apoenzyme on hemin-terminated alkylthiol SAMs chemisorbed on gold electrode surfaces [236].

A similar approach has been used to immobilize pyridine-nucleotide dependent oxidoreductase enzymes on gold electrode surfaces. Although these enzymes do not have a cofactor in their structure, their catalytic reaction involves the formation of an affinity complex between the enzyme and an exogenous soluble pyridine-nucleotide coenzyme. Gold electrode surfaces with biospecific affinity for NAD-dependent dehydrogenases can be prepared by covalent binding of coenzyme-analogue molecules, such as the triazine dyes commercialized under the names of Cibacron and Procion, to a functionalized alkylthiol SAM chemisorbed on the metal surface [237–241]. In particular, Cibacron Blue F3G-A (CB) is known to be a competitive inhibitor of NAD-dependent dehydrogenases due to its ability to bind specifically to the NAD-binding pockets of the enzyme by specific affinity interactions. The structure of this dye shows two chemically reactive moieties, i.e., a chlorine atom attached to the triazin ring and an amino group attached to the anthraquinone ring, which can be used for covalent binding of the dye to amino-, hydroxylic-, or carboxylic-terminated alkylthiol SAMs chemisorbed on gold, giving rise to substrates for the affinity binding of NAD-dependent lactate dehydrogenase (LDH) [237] (Scheme 9).

The surface architecture can be built by covalent binding of Cibacron Blue to a functionalized mixed alkylthiol SAM prepared either by solid-phase step-by-step derivatization of a chemisorbed short-alkylthiol SAM with a bifunctional hydrocarbon spacer (Scheme 10) or by chemisorption of a mixture of short- and long-alkylthiols [237]. When CB is attached to a derivatized short-alkylthiol SAM through the triazine chlorine atom, a minimum chain length of the spacer of six carbon atoms is required for enabling the formation of the surface-bound affinity complex with lactate dehydrogenase [237]. Furthermore, it is found that in CB-anchored monolayers prepared from binary mixed short- and long-alkylthiol SAMs, a molar fraction of the long thiol of 0.1 in the adsorption solution is enough to achieve saturation protein surface coverages, indicating that a very small amount of surface-bound Cibacron Blue is sufficient for binding a densely-packed monolayer of lactate dehydrogenase on the metal surface [237].

LDH-modified electrodes, prepared by affinity binding to CB-anchored alkylthiol SAMs, electrocatalyse the oxidation of L-lactate only in the presence of exogenous NAD^+ and a soluble redox mediator in the reaction mixture,

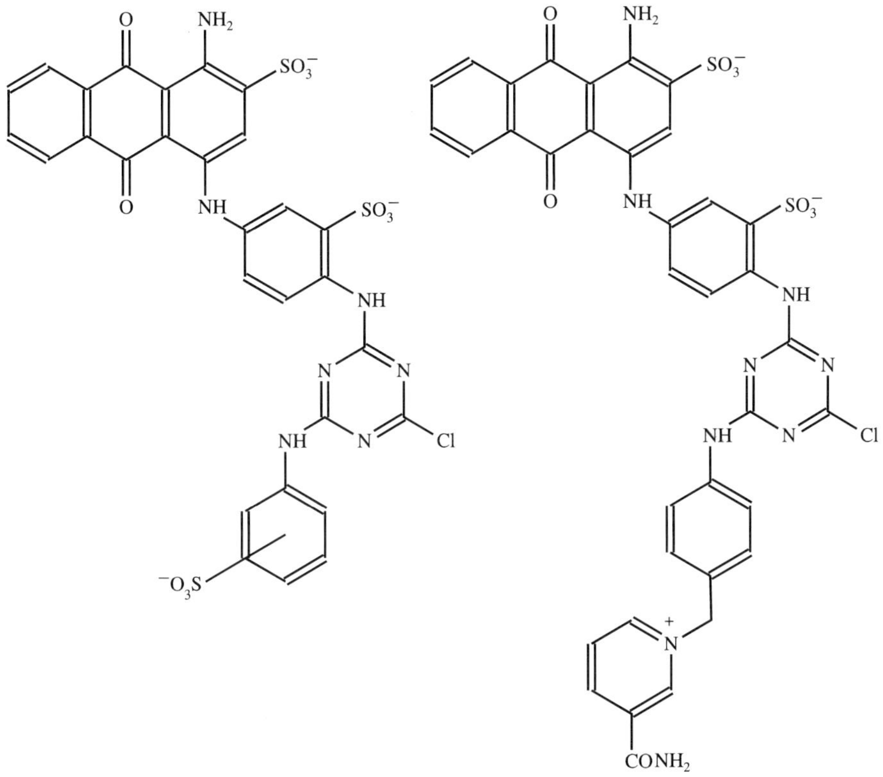

Scheme 9. Chemical structures of the triazine dye Cibacron Blue F3G-A and its analogon L1. Reproduced from [241] with permission.

indicating that the surface-bound enzyme is biologically active, but not efficiently wired to the electrode surface [237]. The mediated amperometric response to oxidation of lactate of LDH-modified gold electrodes depends linearly on the amount of surface-bound protein and the concentration of lactate in the reaction solution ($K_M = 10$ mM) [238]. It should be pointed out that LDH is a tetrameric protein with one NAD-binding pocket in each subunit, wherein the symmetrical distribution of the subunits allows the involvement of two NAD-binding pockets in formation of the surface-bound affinity complex. According to this scheme, two subunits of the tetramer would be blocked (inhibited) for the enzymatic reaction, thus leading to a monolayer of lactate dehydrogenase, in which the enzyme molecules show only one-half of their full enzymatic activity (Scheme 11a). These electrodes show a poor stability, which

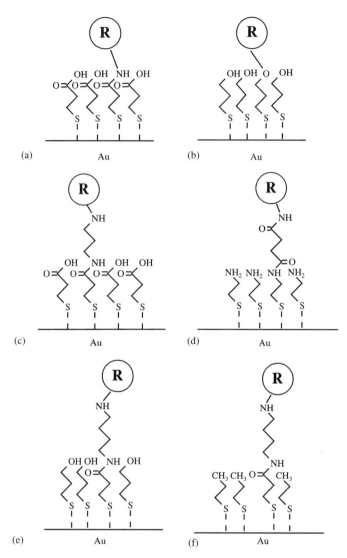

Scheme 10. Idealized sketch of modified gold surfaces prepared by solid-phase covalent binding of a ligand (a triazine dye) to a (a) 3-mercaptopropionic acid SAM; (b) 3-mercaptopropanol SAM; (c) 3-mercaptopropionic acid SAM derivatized with 1,4-diaminobutane; (d) cystamine SAM derivatized with fumaric acid; (e) mixed 3-mercaptopropanol and 3-mercaptopropionic acid SAM derivatized with 1,4-diaminobutane; and (f) mixed propanethiol and 3-mercaptopropionic acid SAM derivatized with 1,4-diaminobutane (approaches (a) and (b) were used to attach triazine dyes to 11-mercapto-1-undecanoic acid and 11-mercapto-1-undecanol SAMs). Reproduced from [240] with permission.

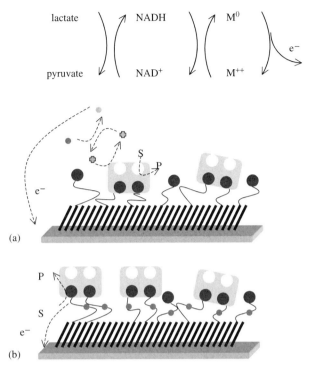

Scheme 11. Idealized sketch showing the electroenzymatic oxidation of L-lactate at gold modified electrode surfaces. (a) Lactate dehydrogenase bound to CB-terminated alkylthiol SAMs prepared by covalent attachment of CB to 3-mercaptopropionic acid SAM derivatized with 1,4-diaminobutane. The electroenzymatic oxidation of lactate is observed only in the presence of soluble coenzyme (NAD^+) and a redox mediator (phenazine methosulfate) [215]. (b) Lactate dehydrogenase bound to NAD-terminated alkylthiol SAMs prepared by covalent attachment of N^6-(2-aminoethyl)-NAD^+ to a cystamine SAM derivatized with pyrroloquinoline quinone. The reconstituted enzyme is electrically wired to the electrode surface via two NAD^+-binding pockets involved in the affinity-binding surface reaction [242].

is thought to arise from the reversible nature of the affinity-binding reaction and from the fact that pyruvate (the product of oxidation of lactate) has a higher affinity for the enzyme than CB. The amperometric response of these electrodes drops sharply after the first shot of lactate and reaches a stable residual level, arising from enzyme molecules which are non-specifically irreversibly bound to the CB-anchored alkylthiol SAM [238]. Spectroscopic analysis of the surface composition of CB-anchored alkylthiol SAMs carried out by Fourier transform infrared reflection absorption spectroscopy combined with amperometric assays and atomic force microscopy imaging of the resulting LDH-modified electrodes shows that the mediated amperometric response of LDH-modified electrodes can be directly correlated to the amount of surface-bound protein, which in turn can be directly correlated to a higher amount of surface-bound ligand in the monolayer. Furthermore, it is found that a higher amount of ligand is covalently bound to mixed, poorly-ordered alkylthiol SAMs in which the ligand is bound to the metal surface through a long and flexible hydrocarbon spacer, which in turn facilitates the formation of the surface-bound affinity complex [239].

However, a direct correlation between amperometric response and protein surface coverage can only be established if the surface architecture used for site-specific affinity binding of the protein is protein resistant. Studying the increase of mediated amperometric response of LDH-modified electrodes prepared by affinity binding to ligand-anchored derivatized short-alkylthiol SAMs (Scheme 10) with the time of protein adsorption, it is found that for adsorption times higher than 10 min, a large amount of enzyme is adsorbed by non-specific interactions (Fig. 1.5A,B). A comparison between the extent of non-specific adsorption of LDH to ligand-modified gold wires prepared by covalent attachment of L1 (a Cibacron Blue derivative, see Scheme 9) to 3-mercaptopropionic acid SAMs derivatized with 1,4-diaminobutane (Fig. 1.5A), and to 2-ethanolamine SAMs derivatized with fumaric acid (Fig. 1.5B), suggests the involvement of strong electrostatic interactions between the thiol underlayer and charged amino acid side chains on the enzyme surface,[4] which compete with site-specific affinity interactions with surface-bound ligand molecules giving rise to higher protein surface coverages at L1-anchored, 1,4-diaminobutane-derivatized 3-mercaptopropionic acid SAMs (MPA-DAB). This is demonstrated by the significant decrease in non-specific adsorbed species when using adsorption buffers with a high ionic strength (50 mM NaCl) for screening the negative surface charges [240]. LDH-modified gold surfaces can also be prepared by covalent binding of ligand (CB or L1) to mixed 11-mercapto-1-undecanol (MUO) and 11-mercapto-1-undecanoic acid SAMs (MUA), wherein non-specific protein adsorption is minimized by binding the ligand to the surface-bound carboxylic groups, which are diluted into a hydrophilic,

---

[4] It is believed that under the pH conditions used for the electroenzymatic reaction (pH 8.0) L1-anchored mercaptopropionic acid SAMs derivatized with 1,4-diaminobutane have an excess of negative surface charges. It is noted that L1 is covalently bound to mercaptopropionic acid SAMs derivatized with 1,4-diaminobutane via the chlorine atom of the triazine ring and to 2-ethanolamine SAMs derivatized with fumaric acid via the anthraquinone amino group [240].

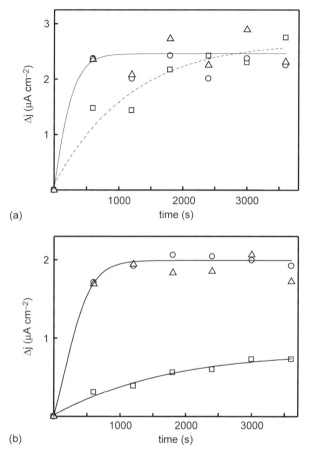

Fig. 1.5. Dependence of the electroenzymatic response (Δj) of lactate dehydrogenase modified gold electrodes on the time of protein layer growth: (a) mixed 3-mercaptopropanol and 3-mercaptopropionic acid SAM derivatized with 1,4-diaminobutane ligand free- (□), CB- (○) and L1-anchored (△); (b) cystamine SAM derivatized with fumaric acid ligand free- (□), CB- (○) and L1-anchored (△). The monolayers were incubated in a 0.36 mg ml^{-1} enzyme solution in 50 mM Na-phosphate buffer, pH 7.0. Reproduced from [240] with permission.

hydroxylic-terminated alkylthiol underlayer [241]. The amperometric response of these LDH-modified electrodes correlates directly with the molar fraction of carboxylic-terminated thiol in the adsorption thiol solution, such that the highest amperometric responses are obtained for electrodes prepared from a pure 11-mercapto-1-undecanoic acid SAM, presumably due to a higher surface concentration of affinity ligand which in turn gives rise

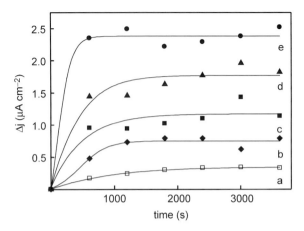

Fig. 1.6. Amperometric response (Δj) vs. time upon addition of 20 mM lactate at 0 mV (SCE) to a 100 mM Tris + HCl, 100 mM KCl, pH 8.0 buffer containing 2.5 mM NAD^+ and 0.25 mM phenazine methosulfate. Lactate dehydrogenase bound to gold wires modified with ligand free $[MUA]_{\chi=1}$ (a) and CB-anchored $[MUA]_{\chi=0.25} - [MUO]_{\chi=0.75}$ (b), $[MUA]_{\chi=0.50} - [MUO]_{\chi=0.50}$ (c), $[MUA]_{\chi=0.75} - [MUO]_{\chi=0.25}$ (d) and $[MUA]_{\chi=1}$ SAMs (e). Reproduced from [241] with permission.

to a higher protein surface coverage (Fig. 1.6) [241].[5] A correlation between amperometric response and protein surface coverage can be obtained by using SPR measurements combined with electrochemistry ex situ [241]. Using this approach, it has been demonstrated that the lower amperometric response observed for LDH-modified gold electrodes prepared from L1-anchored alkylthiol SAMs as compared with their CB-analogue SAMs (Fig. 1.7A) arises from a lower capacity for LDH affinity binding of the former (Fig. 1.7B).[6] This behaviour may be explained either by a lower affinity constant of the complex L1:LDH or by a lower amount of accessible surface-bound ligand molecules on the monolayer. The number of NAD-binding pockets which are involved in the electroenzymatic oxidation of L-lactate by

---

[5] By contrast with CB-anchored SAMs prepared by covalent binding of the dye to the alkylthiol SAM via its chlorine atom, when the dye is bound to the alkylthiol SAM via the antraquinone amino group, it is not necesssary to provide a long flexible hydrocarbon spacer between ligand and alkylthiol underlayer for enabling the formation of the surface-bound affinity complex [241].

[6] Ligand-anchored SAMs prepared from long-alkylthiol SAMs could not be synthesized on SPR gold evaporated substrates, probably due to a better packing of the mixed long-alkylthiol SAM leading to a steric hindrance for covalent binding of the ligand. It is noted that by contrast with the behaviour observed with gold wires, non-specific protein adsorption on SPR gold substrates modified with a CB-anchored 1,4-diaminobutane-derivatized 3-mercaptopropionic acid SAM is negligible, suggesting a strong influence of the nature of the gold substrate itself in the interaction between protein and modified gold surface [241].

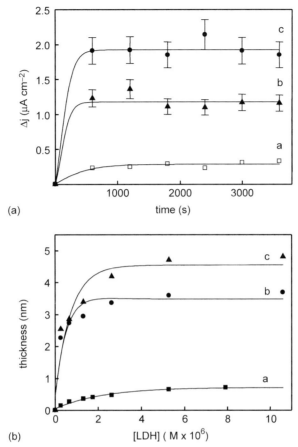

Fig. 1.7. (a) Amperometric response (Δj) vs. time upon addition of 20 mM lactate at 0 mV (SCE) to a 100 mM Tris + HCl, 100 mM KCl, pH 8.0 buffer containing 2.5 mM NAD^+ and 0.25 mM phenazine methosulfate of lactate dehydrogenase bound to gold wires modified with ligand free (a), L1- (b) and CB-anchored (c) $[MUA]_{\chi=1}$ SAMs (error bars indicate the dispersion among data obtained with electrodes prepared in different batches). (b) Adsorption isotherms for LDH dissolved in 50 mM Na-phosphate buffer, pH 7.0 on evaporated gold plates modified with ligand-free (a), L1- (b) and CB-anchored (c) MPA-DAB monolayers (average protein layer thickness was calculated from the observed shifts in the plasmon angle monitored by SPR). Reproduced from [241] with permission.

the surface-bound LDH molecules can be estimated by performing SPR adsorption measurements combined with amperometric assays in situ using the SPR gold evaporated substrates as electrode surfaces. These experiments enable a direct measure of both the increment of anodic current density upon addition of lactate (Δj) and the protein surface coverage (calculated from the average protein layer thickness estimated from $\Delta\Theta_{pl}$) (Fig. 1.8A,B). For the particular case of LDH-modified electrodes, it is found that the ratio $K = \Delta j/\Delta\Theta_{pl}$ varies from 2.4 μA cm^{-2} nm for LDH adsorbed on ligand-free alkylthiol SAMs to 1.2 μA cm^{-2} nm for LDH adsorbed by affinity binding to CB- or L1-terminated alkylthiol SAMs, thus demonstrating

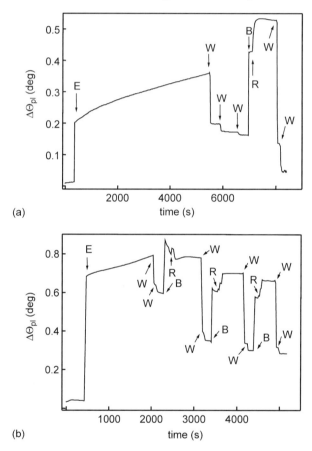

Fig. 1.8. Adsorption events monitored by SPR–electrochemistry "in situ" of LDH on gold evaporated electrodes modified with ligand-free (a) and L1-anchored (B) MPA-DAB monolayers, before and after several runs of electroenzymatic reaction: (E) Addition of a 2.64 × 10^{-6} M LDH solution in 50 mM Na-phosphate buffer, pH 7.0. (W) Washing with 50 mM Na-phosphate buffer, pH 7.0. (b) Changing the buffer to 100 mM Tris–HCl, 100 mM KCl, pH 8.0, switching the potential to +100 mM (Ag/AgCl) and addition of 2.5 mM NAD$^+$ and 0.25 mM phenazine methosulfate. (R) Addition of 20 mM lactate to the reaction mixture at +100 mV (Ag/AgCl). Reproduced from [241] with permission.

that the enzyme is bound to the ligand-anchored alkylthiol SAM via two NAD-binding pockets, which are not involved in the electroenzymatic reaction (Scheme 11a) [241].

Self-contained integrated amperometric lactate sensors based on electrode surfaces modified with a monolayer of NAD-dependent lactate dehydrogenase monolayer have been constructed by affinity binding of the enzyme to gold surfaces modified with a NAD-terminated alkylthiol SAM, wherein the surface-bound pyridine-nucleotide coenzyme is electrically wired to the metal by inserting a redox mediator molecule PQQ within the spacer connecting the coenzyme to the surface [228,229,242,243]. The enzyme is irreversibly attached to the electrode surface by "crosslinking" the surface-bound affinity complex with glutardialdehyde [244]. Similarly to CB-terminated alkylthiol SAMs, the formation of the surface-bound affinity complex on NAD-terminated alkylthiol SAMs involves two NAD-binding pockets, such that the amperometric response to lactate in the absence of any other exogenous reagent arises only from two electrically-wired active sites of the enzyme (Scheme 11b) [242]. The stabilization of surface-bound affinity complexes by covalent "crosslinking" has been used for building monolayer architectures in which a couple of biological protein redox partners are electrically wired to the gold electrode surface [244–248]. The surface architecture is built by chemisorbing an alkylthiol SAM bearing one of the redox partners as a covalently-attached end group on the gold electrode surface, such as to wire electrically the protein to the metal. This is followed by the formation of a surface-bound affinity complex which is stabilized by crosslinking [244]. The ability of microperoxidase-11 (an undecapeptide that mimics the redox site of cytochrome-c) to interact directly with a gold electrode surface after covalent binding to a chemisorbed cystamine SAM ($E_{1/2} = -0.400$ V vs. SCE at pH 7.0) [245] has been exploited to prepare electrode surfaces having specific affinity for nitrate reductase [246] and Co(II)-protoporphyrin IX-reconstituted myoglobin [247], for the manufacture of integrated biosensors for the bioelectrocatalysed reduction of nitrate [246] and the hydrogenation of acetylene [247], respectively. Recently, the same approach has been used to stabilize surface-bound affinity complexes between a surface-bound de novo synthesized hemeprotein and several metalloproteins [248].

Surface chemistry based on self-assembly can be combined with genetic engineering techniques for fabricating surface architectures having "universal" affinity ligands, which can be used for the immobilization of any protein on a gold electrode surface, while maintaining its biological activity virtually intact. An example of this approach has been reported by Fernández et al. for the fabrication of modified gold electrode surfaces prepared by affinity binding of a chimera protein obtained by fusion of β-galactosidase and a polypeptide having the binding domain of choline to a choline-terminated alkylthiol SAM [249].

1.4.3 Amperometric biosensors based on self-assembled multilayered structures

Electrode surfaces modified with a multilayered surface architecture prepared by a layer-by-layer repeated deposition of several enzyme monolayers show a modulated increase of surface-bound protein with a subsequent increase in output current, which is directly correlated with the number of deposited protein layers. The versatility of this approach allows alternate layers of different proteins for the manufacture of electrode surfaces, which are the basis for multianalyte sensing devices with multiple substrate specificities. The surface chemistry used for the manufacture of multilayered electrode surfaces is similar to that previously described for the preparation of affinity sensors, and is based on the stabilization of self-assembled multilayer assemblies by specific affinity interactions, electrostatic attraction, or covalent binding between adjacent monolayers.

Biosensors based on self-assembled multilayered structures for the amperometric detection of glucose (or lactate) can be prepared by a layer-by-layer alternate deposition of monolayers of avidin and biotinylated glucose oxidase (or biotinylated lactate oxidase) on platinum electrode surfaces, in which the oxidation of glucose mediated by ferrocenemethanol is used as the sensing reaction [250]. The layer-by-layer deposition approach allows minimization of interferences due to the presence of ascorbic acid in the sample when the oxidation of water peroxide (at +0.600 V vs. Ag/AgCl) is used as the sensing reaction, by incorporating alternate monolayers of biotinylated ascorbate oxidase in the multilayered assembly [251]. This approach has also been used to fabricate modified platinum electrodes for the amperometric detection of acetylcholine by intercalating a monolayer of avidin between adjacent layers of biotinylated choline esterase and biotinylated choline oxidase [252]. An analogous approach has been reported for the construction of amperometric biosensors for the detection of glucose (or lactate), which is based on the alternate deposition of concanavalin A and mannose-labelled glucose oxidase (or lactate oxidase) on platinum electrode surfaces [253]. Since the affinity binding of a monolayer of avidin intercalated between two layers of biotinylated enzyme requires only two biotin-binding sites of each avidin molecule, it is possible to integrate a biotinylated redox mediator molecule in the multilayered assembly simply by affinity binding to free biotin-binding sites of avidin molecules. This approach has been used to prepare reagentless hydrogenase-based electrodes for the electroenzymatic oxidation of molecular hydrogen by alternate deposition of monolayers of avidin and biotinylated hydrogenase on gold electrode surfaces modified with a biotin-terminated alkylthiol SAM, using biotinylated viologen molecules as integrated redox mediator (Scheme 12a) [254].

A layer-by-layer alternate electrostatic adsorption of monolayers of a positively-charged dendrimer (G4 poly(amidoamine)) and the negatively-charged protein glucose oxidase allows the preparation of multilayered assemblies for the construction of amperometric glucose sensors. This multilayer

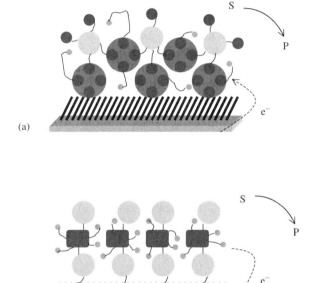

Scheme 12. Idealized sketches showing: (a) The electroenzymatic oxidation of H_2 at gold modified electrode surfaces. A multilayer assembly is built on a gold surface modified with a biotinyl-terminated alkylthiol SAM by alternate deposition of monolayers of avidin and biotinylated hydrogenase. The redox mediator (a biotinylated viologen-terminated poly(ethylene glycol) derivative) is integrated in the assembly by affinity binding to free biotin-binding sites of the avidin layer(s) [254]. (b) The electroenzymatic oxidation of glucose at gold modified electrode surfaces. A multilayer assembly is built on a gold surface modified with a cystamine SAM by alternate deposition of monolayers of glucose oxidase and poly(amidoamine) dendrimer functionalized with ferrocenyl moieties. The multilayered assembly is stabilized by covalent bonds between adjacent layers [256].

assembly can be built on a gold surface modified with a positively-charged cystamine SAM, and can be stabilized by covalent binding of free amino groups of the cystamine and dendrimer monolayers to reactive aldehyde groups of the protein surface obtained by oxidation with periodate. This system gives rise to mediated glucose amperometric biosensors, which are stable over 20 days under day-by-day calibration [255]. Furthermore, it allows the incorporation of ferrocenyl mediator molecules into the multilayered assembly by covalent attachment to free amino groups of the dendrimer for wiring each protein molecule to the electrode surface through a polymer-like multilayered assembly that includes several electron shuttlers (Scheme 12b) [256]. This is, however, an idealized view of the system. It has been found that in amperometric glucose biosensors prepared by alternate deposition of several monolayers of ferrocenyl-tethered poly(allylamine) and glucose oxidase on gold surfaces modified with a SAM of 3-mercapto-1-propanesulfonate, all protein layers contribute equally to the amperometric response. However, only a small fraction of protein molecules is efficiently electrically wired to the electrode surface [257]. A similar behaviour has been reported for gold electrode surfaces modified by alternate deposition of several monolayers of glucose oxidase and poly(allylamine) modified with an electroactive osmium complex [258]. Following a similar approach, Pishko et al. have prepared integrated reagentless glucose (or lactate sensors) by electrostatic adsorption of glucose oxidase (or lactate oxidase) on a layer of poly(allylamine) modified with an osmium complex electrostatically adsorbed on a gold electrode surface modified with a 11-mercapto-1-undecanoic acid SAM and stabilizing the multilayered structure by crosslinking with glutardialdehyde [259]. A simpler approach to integrate the redox mediator in a multilayer assembly while keeping the assembly robust is based on a layer-by-layer covalent binding of monolayers of protein and small redox mediator molecules. The approach has been used to prepare integrated reagentless water peroxide amperometric sensors by intercalating a layer of glutardialdehyde between alternate layers of thionine and horseradish peroxidase on a cysteine-modified gold electrode surface [260].

1.5 SWITCHABLE SELF-ASSEMBLED MONOLAYERS

Surface-bound molecular architectures which display a measurable signal "on-command" can be considered to mimic an electronic switch, which may be integrated in an integrated circuit of a molecular electronics device. Furthermore, if the switchable "signal" is a chemical substance which can be released "on-command" from the surface, the system can be used to start a cascade of chemical processes which mimic the biological cellular mechanisms of signal transduction and amplification.

Willner et al. have developed a system in which an enzymatic electron transfer reaction is switched by illumination, thus providing means for recording and amplifying recorded optical signals. The basic element of this photoswitchable system is a neutral molecule (nitrospiropyrane) which upon illumination with light of $\lambda > 475$ nm photoisomerizes to yield a positively-charged molecule (nitromerocyanine). This reaction is reversible, such that the molecule of nitrospiropyrane is regenerated by further illumination with light of 320 nm $< \lambda <$ 380 nm. Using a similar principle as that applied for ion-channel sensors, a mixed alkylthiol monolayer containing thiopyridine and nitrospiropyrane end-attached moieties can be used to promote the electron transfer of a soluble negatively-charged redox probe, such as cytochrome-c at the gold electrode surface only when the surface-bound photoswitchable molecule is in its positively-charged merocyanine state. Thus, light of a suitable wavelength can be used to enable (or hinder) the electron transfer reaction between cytochrome-c and the modified gold electrode surface, which in turn acts as a redox mediator for a soluble enzyme, such as cytochrome oxidase or L-lactate dehydrogenase (cytochrome b_2), enabling (or hindering) the electroenzymatic conversion of substrate (oxygen or lactate), and amplifying the primary optical signal [261] (Scheme 13). The system can be simplified by using enzymes such as glucose oxidase, which allow the introduction of the spiropyrane compound directly into the redox site of the enzyme [262]. Thus, glucose oxidase can be reconstituted on gold electrode surfaces modified with a spiropyrane-modified FAD-terminated alkylthiol SAM, wherein the photoswitchable electrocatalysed oxidation of glucose in the presence of a soluble redox mediator can be detected amperometrically [263]. The same authors have used a similar approach for constructing photoswitchable amperometric immunosensors, in which the affinity binding of antibody to the antigen-terminated alkanethiol SAM is regulated by illumination light of a suitable wavelength [264].

Mrksich et al. have developed a different kind of switchable surfaces, which are intended to provide "dynamic" substrates that can alter "on command" and in real time the display of ligands, thus altering in turn the interactions of proteins and cells with said switchable substrates. The most simple approach is based on the modification of a gold electrode surface with an alkylthiol SAM and releasing a part of the surface-bound alkylthiol molecules by applying an external electrical potential [265]. A more elegant strategy involves the electrochemically-induced release of pendant groups of the alkylthiol SAM, while keeping the alkanethiol underlayer on the electrode

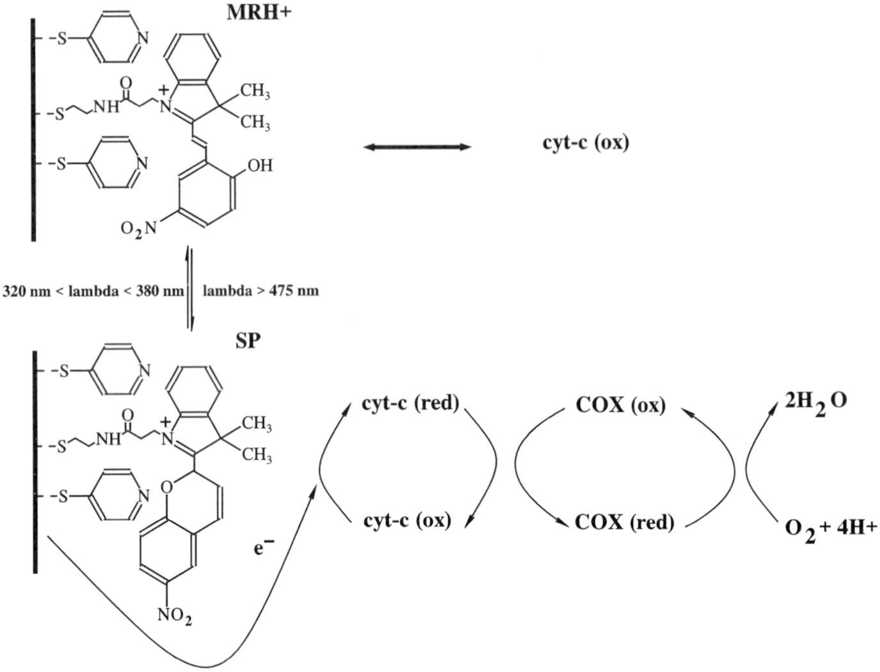

Scheme 13. Photochemical switching of electrical interactions of cytochrome-*c* with a gold electrode surface modified with 4,4′-dipyridyldisulfide and the electroenzymatic reduction of oxygen by cytochrome-*c* oxidase [261].

surface [266–268]. For example, the application of a potential of +0.900 V vs. Ag/AgCl to a gold electrode surface modified with a catechol orthoformate-terminated alkylthiol SAM results in the irreversible oxidation of the catechol end groups to orthoquinone and the release of the orthoformate substituent from the monolayer [266]. Using a similar approach, the same authors have prepared surfaces which release biotin under electrochemical control with the purpose of controlling the affinity binding of streptavidin to these surfaces. The application of a potential of −0.700 V vs. Ag/AgCl to a gold electrode surface modified with a biotinyl-modified paraquinone-terminated alkylthiol SAM leads to the release of biotin from the surface, which results in the suppression of the specific affinity for streptavidin [267] (Scheme 14). Electrochemical switching by applying an external potential can be used to modulate the thickness of a monolayer of dsDNA adsorbed on a gold electrode surface due to a potential-dependent change in orientation of the adsorbed dsDNA helices with respect to the surface. This is the principle for a nanoscale-sized electromechanical switch developed by M.G. Hill et al., who

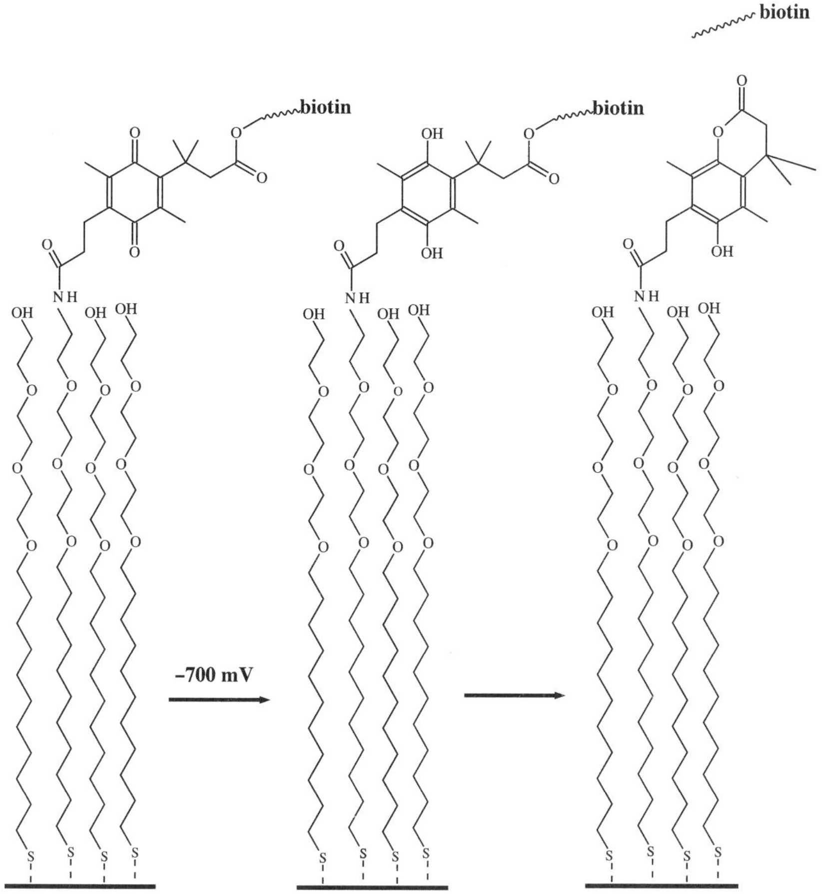

Scheme 14. Electrochemically-induced release of biotin from a gold electrode surface modified with a protein-resistant mixed alkylthiol SAM (biotinyl derivative attached to the alkylthiol SAM via a paraquinone moiety) [267].

by using the atomic force microscope combined with electrochemistry "in situ" succeeded in monitoring the potential-dependent changes in thickness of a monolayer of adsorbed dsDNA. These thickness changes are attributed to a change in orientation of dsDNA helices from an orientation with the molecular axis quasi-perpendicular to the electrode surface (layer thickness of 55 Å at potentials more negative than 0.450 V vs. Ag/AgCl) to an orientation in which the helices align parallel to the electrode surface (lowest layer thickness of 20 Å at potentials more positive than 0.450 V vs. Ag/AgCl) (Scheme 15) [269].

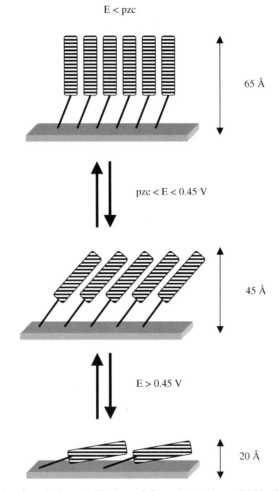

Scheme 15. Electrochemical modulation of the orientation of DNA helices bound to a gold electrode surface [269].

1.6 OUTLOOK

Throughout this chapter the relevance of surface chemistry based on self-assembling techniques has been outlined, especially within the context of the manufacture of analytical biosensing devices. In particular, the combination of self-assembly and microfabrication techniques, such as photopatterning or microcontact printing, allows the manufacture of micrometer-scale arrays of biomolecules distributed throughout a protein-resistant surface (e.g., a surface-modified silicon chip), which are the basis for high-throughput screening and diagnostics devices. It is foreseeable that the role of self-assembly for the manufacture of highly defined surface architectures will further increase with the development of micro- and nanometer scale physical manipulation techniques, such that single molecules would be not only deposited onto a spatially defined location, but also individually addressable.

REFERENCES

1 F. Scheller and F. Schubert, *Biosensors*. Elsevier, Amsterdam, 1992.
2 F. Scheller and R.D. Schmid (Eds.), *Biosensors, Technologies and Applications*, GBF Monographs, Vol. 17. VCH, New York, 1992.
3 A.P.F. Turner, I. Karube and G.S. Wilson (Eds.), *Biosensors. Fundamentals and Applications*. Oxford University Press, Oxford, 1987.
4 H.L. Schmidt, W. Schuhmann, F.W. Scheller and F. Schubert, In: W. Göpel, J. Hesse and J.N. Zemel (Eds.), *Sensors, a Comprehensive Survey*, Vol. 3. VCH, Weinheim, 1992.
5 C. Lu and A.W. Czanderna (Eds.), *Applications of Piezoelectric Quartz Crystal Microbalances*. Elsevier, Amsterdam, 1984.
6 R. Schumacher, *Angew. Chem.*, 102 (1990) 347–361.
7 D.A. Buttry and M.D. Ward, *Chem. Ber.*, 92 (1992) 1355–1379.
8 P. Schuck, *Annu. Rev. Biophys. Biomol. Struct.*, 26 (1997) 541–566.
9 Z. Salamon, H.A. Macleod and G. Tollin, *Biochim. Biophys. Acta*, 1331 (1997) 117–129.
10 Z. Salamon, H.A. Macleod and G. Tollin, *Biochim. Biophys. Acta*, 1331 (1997) 131–152.
11 J. Homola, S.S. Yee and G. Gauglitz, *Sens. Actuators, B*, 54 (1999) 3–15.
12 A.J. Ricco, R.M. Crooks and R.M. Osbourn, *Acc. Chem. Res.*, 31 (1998) 289–296.
13 G.C. Osbourn, J.W. Bartholomew, A.J. Ricco and G.C. Frye, *Acc. Chem. Res.*, 31 (1998) 297–305.
14 S.P.A. Fodor, J.L. Read, M.C. Pirrung, L. Stryer, A.T. Lu and D. Solas, *Science*, 251 (1991) 767–773.
15 R. Kapur, K.A. Giuliano, M. Campana, T. Adams, K. Olson, D. Jung, M. Mrksich, C. Vasudavan and D.L. Taylor, *Biomed. Microdevices*, 2 (1999) 99–109.
16 N. Dontha, W.B. Nowall and W. Kuhr, *Anal. Chem.*, 69 (1997) 2619–2625.
17 M. Mrksich, C.S. Yen, Y. Xia, L.E. Dike, D. Ingber and G.M. Whitesides, *Proc. Natl Acad. Sci. USA*, 93 (1996) 10775–10778.
18 C.S. Chen, M. Mrksich, S. Huang, G.M. Whitesides and D.E. Ingber, *Science*, 276 (1997) 1425–1428.
19 C.D. James, R.C. Davis, L. Kam, H.G. Craighead, M. Isaacson, J.N. Turner and W. Shain, *Langmuir*, 14 (1998) 741–744.
20 Y.Y. Luk, M. Kato and M. Mrksich, *Langmuir*, 16 (2000) 9604–9608.

21 X.H. Liu, H.K. Wang, J.N. Herron and G.D. Prestwich, *Bioconjugate Chem.*, 11 (2000) 755–761.
22 C.D. Tidwell, S.I. Ertel, B. Ratner, B.J. Tarasevich, S. Atre and D.L. Allara, *Langmuir*, 13 (1997) 3404–3413.
23 J. Lahiri, E. Ostuni and G.M. Whitesides, *Langmuir*, 15 (1999) 2055–2060.
24 J.M. Brockman, A.G. Frutos and R.M. Corn, *J. Am. Chem. Soc.*, 121 (1999) 8044–8051.
25 M.L. Amirpour, P. Ghosh, W.M. Lackowski, R.M. Crooks and M.V. Pishko, *Anal. Chem.*, 73 (2001) 1560–1566.
26 C.A. Mirkin and M.A. Ratner, *Annu. Rev. Phys. Chem.*, 43 (1992) 719–754.
27 W. Göpel, *Sens. Actuators, B*, 4 (1991) 7–21.
28 D. Goldhaber-Gordon, M.S. Montemerlo, J.C. Love, G.J. Opiteck and J.C. Ellenbogen, *Proc. IEEE*, 85 (1997) 521–540.
29 G. Joachim, J.K. Gimzewski and A. Aviram, *Nature*, 408 (2000) 541–548.
30 J. Fritz, M.K. Baller, H.P. Lang, I.H. Rothnizen, P. Vettiger, E. Meyer, H.J. Güntherodt, Ch. Berger and J.K. Gimzewski, *Science*, 288 (2000) 316–318.
31 E.L. Schmid, T.A. Keller, Z. Dienes and H. Vogel, *Anal. Chem.*, 69 (1997) 1979–1985.
32 A. Ulman, *An Introduction to Ultrathin Organic Films: from Langmuir–Blodgett to Self-Assembly*. Academic Press, New York, 1991.
33 L.H. Dubois and R.G. Nuzzo, *Annu. Rev. Phys. Chem.*, 43 (1992) 437–463.
34 A. Ulman, *Chem. Rev.*, 96 (1996) 1533–1554.
35 S. Flink, F.C.J.M. van Veggel and D.N. Reinhoudt, *Adv. Mater.*, 12 (2000) 1315–1328.
36 R.M. Crooks and A.J. Ricco, *Acc. Chem. Res.*, 31 (1998) 219–227.
37 V. Chechik, R.M. Crooks and C.J.M. Stirling, *Adv. Mater.*, 12 (2000) 1161–1171.
38 J.M. Lehn, *Supramolecular Chemistry*. VCH, Weinheim, 1995, Chapters 9–10.
39 N.K. Chaki and K. Vijayamohanan, *Biosens. Bioelectron.*, 17 (2002) 1–12.
40 C.D. Bain, E.B. Troughton, Y.T. Tao, J. Evall, G.M. Whitesides and R.G. Nuzzo, *J. Am. Chem. Soc.*, 111 (1989) 321–335.
41 R.G. Nuzzo, L.H. Dubois and D.L. Allara, *J. Am. Chem. Soc.*, 112 (1990) 558–569.
42 C.E.D. Chidsey and D.N. Loiacono, *Langmuir*, 6 (1990) 682–691.
43 K. Takehara and Y. Ide, *Bioelectrochem. Bioenerg.*, 27 (1992) 207–219.
44 K. Takehara and Y. Ide, *Bioelectrochem. Bioenerg.*, 27 (1992) 501–507.
45 M.D. Porter, T.B. Bright, D.L. Allara and C.E.D. Chidsey, *J. Am. Chem. Soc.*, 109 (1987) 3559–3568.
46 C.E.D. Chidsey, C.R. Bertozzi, T.M. Putvinski and A.M. Mujsce, *J. Am. Chem. Soc.*, 112 (1990) 4301–4306.
47 C.E.D. Chidsey, *Science*, 251 (1991) 919–922.
48 D.M. Collard and M.A. Fox, *Langmuir*, 7 (1991) 1192–1197.
49 S.E. Creager and G.K. Rowe, *Anal. Chim. Acta*, 246 (1991) 233–239.
50 Y. Sato, H. Itoigawa and K. Uosaki, *Bull. Chem. Soc. Jpn*, 66 (1993) 1032–1037.
51 L. Zhang, T. Lu, G.W. Gokel and A.E. Kaifer, *Langmuir*, 9 (1993) 786–791.

52 W. Brett, K. Chen, C.A. Mirkin and S.J. Babinek, *Langmuir*, 9 (1993) 1945–1947.
53 D.D. Schlereth, E. Katz and H.L. Schmidt, *Electroanalysis*, 6 (1994) 725–734.
54 D.D. Schlereth and H.L. Schmidt, *J. Electroanal. Chem.*, 380 (1995) 117–125.
55 D.D. Schlereth, E. Katz and H.L. Schmidt, *Electroanalysis*, 7 (1995) 46–54.
56 G. Cuossen, K.E. Drabe, D.A. Wiersma, M.A. Schoondorp, A.J. Schouten, J.B.E. Hulshof and B.L. Feringa, *Langmuir*, 9 (1993) 1974–1977.
57 R. Yam and G. Berkovic, *Langmuir*, 9 (1993) 2109–2111.
58 S.D. Evans, S.R. Johnson, H. Ringsdorf, L.M. Williams and H. Wolf, *Langmuir*, 14 (1998) 6436–6446.
59 H.Z. Yu, S. Ye, H.L. Zhang, K. Uosaki and Z.F. Liu, *Langmuir*, 16 (2000) 6948–6954.
60 A. Kumar and G.M. Whitesides, *Science*, 263 (1994) 60–62.
61 L.F. Rozsnyai and M.S. Wrighton, *J. Am. Chem. Soc.*, 116 (1994) 5993–5994.
62 J.L. Wilbur, A. Kumar, H.A. Biebuyck, E. Kim and G.M. Whitesides, *Nanotechnology*, 7 (1996) 452–457.
63 A. Terfort, N. Bowden and G.M. Whitesides, *Nature*, 386 (1997) 162–164.
64 N. Bowden, A. Terfort, J. Carbeck and G.M. Whitesides, *Science*, 276 (1997) 233–235.
65 J. Tien, A. Terfort and G.M. Whitesides, *Langmuir*, 13 (1997) 5349–5355.
66 T. Kratzmüller, D. Appelhans and H.G. Braun, *Adv. Mater.*, 11 (1999) 555–558.
67 J.J. Gooding, E.A.H. Hall and D.B. Hibbert, *Electroanalysis*, 10 (1998) 1130–1136.
68 W. Göpel, *Biosens. Bioelectron.*, 10 (1995) 35–59.
69 L.A. Bumm, J.A. Arnold, M.T. Cygan, T.D. Dunbar, T.P. Burgin, L. Jones II, D.L. Allara, J.M. Tour and P.S. Weiss, *Science*, 271 (1996) 1705–1707.
70 D.I. Gittins, D. Bethell, D.A. Schiffrin and R.J. Nichols, *Nature*, 408 (2000) 67–69.
71 S. Zhang, *Nat. Biotechnol.*, 21 (2003) 1171–1178.
72 Z. Salamon, D. Huang, W.A. Cramer and G. Tollin, *Biophys. J.*, 75 (1998) 1874–1885.
73 J. Zhang, V. Rosilio, M. Goldmann, M.M. Boissonnade and A. Baszkin, *Langmuir*, 16 (2000) 1226–1232.
74 R.G. Nuzzo, B.R. Zegarski and L.H. Dubois, *J. Am. Chem. Soc.*, 109 (1987) 733–740.
75 H.A. Buebuyck and G.M. Whitesides, *Langmuir*, 9 (1993) 1766–1770.
76 H.A. Buebuyck, C.D. Bain and G.M. Whitesides, *Langmuir*, 10 (1994) 1825–1831.
77 R.C. Thomas, L. Sun, R.M. Crooks and A.J. Ricco, *Langmuir*, 7 (1991) 620.
78 M. Buck, M. Grunze, F. Eisert, J. Fischer and F. Trager, *J. Vac. Sci. Technol., A*, 10 (1992) 926–929.
79 D.S. Karpovich and G.J. Blanchard, *Langmuir*, 10 (1994) 3315–3322.
80 K.A. Peterlinz and R. Georgiadis, *Langmuir*, 12 (1996) 4731–4740.
81 G.E. Poirier, *Chem. Rev.*, 97 (1997) 1117–1127.
82 G.E. Poirier and E.D. Pylant, *Science*, 272 (1996) 1145–1148.

83 C. Schönenberger, J. Jorritsma, J.A.M. Sondag-Huethorst and L.G.J. Fokkink, *J. Phys. Chem.*, 99 (1995) 3259.
84 N. Camillone III, P. Eisenberger, T.Y.B. Leung, P. Schwartz, G. Scoles, G.E. Poirier and M.J. Tarlov, *J. Phys. Chem.*, 101 (1994) 11031–11036.
85 G.E. Poirier and M.J. Tarlov, *Langmuir*, 10 (1994) 2853.
86 E. Delamarche, B. Michel, Ch. Berger, D. Anselmetti, H.J. Güntherodt, H. Wolf and H. Ringsdorf, *Langmuir*, 10 (1994) 2869.
87 C. Schönenberger, J.A.M. Sondag-Huethorst, J. Jorritsma and L.G.J. Fokkink, *Langmuir*, 10 (1994) 611–614.
88 M.A. Walczak, C. Chung, S.M. Stole, C.A. Widrig and M.D. Porter, *J. Am. Chem. Soc.*, 113 (1991) 2370–2378.
89 H. Oevening, M.D. Paddon-Pow, M. Heppener, A.M. Oliver, E. Cotsans, J.W. Verhoeven and N.S. Hush, *J. Am. Chem. Soc.*, 109 (1987) 3258–3269.
90 M. Mrksich, *Cell. Mol. Life Sci.*, 54 (1998) 653–658.
91 M. Mrksich and G.M. Whitesides, *Annu. Rev. Biophys. Biomol. Struct.*, 25 (1996) 55–78.
92 C.D. Bain and G.M. Whitesides, *Science*, 240 (1988) 62–63.
93 P.E. Laibinis, R.G. Nuzzo and G.M. Whitesides, *J. Phys. Chem.*, 96 (1992) 5097–5105.
94 C.D. Bain and G.M. Whitesides, *J. Am. Chem. Soc.*, 110 (1988) 3665–3666.
95 J.P. Folkers, P.E. Laibinis and G.M. Whitesides, *Langmuir*, 8 (1992) 1330–1341.
96 P. Tengvall, M. Lestelius, B. Liedberg and I. Lundström, *Langmuir*, 8 (1992) 1236–1238.
97 M. Lestelius, B. Liedberg, I. Lundström and P. Tengvall, *J. Biomed. Mater. Res.*, 28 (1994) 871–880.
98 K.L. Prime and G.M. Whitesides, *J. Am. Chem. Soc.*, 115 (1993) 10714–10721.
99 K.L. Prime and G.M. Whitesides, *Science*, 252 (1991) 1164–1167.
100 L. Deng, M. Mrksich and G.M. Whitesides, *J. Am. Chem. Soc.*, 118 (1996) 5136–5137.
101 K. Feldman, G. Hähner, N.D. Spencer, P. Harder and M. Grunze, *J. Am. Chem. Soc.*, 121 (1999) 10134–10141.
102 S.J. Stranick, S.V. Atre, A.N. Parikh, M.C. Wood, D.L. Allara, N. Winograd and P.S. Weiss, *Nanotechnology*, 7 (1996) 438–442.
103 H. Munukata, S. Kuwabata, Y. Ohso and H. Yoneyama, *J. Electroanal. Chem.*, 496 (2001) 29–36.
104 T. Sawaguchi, Y. Sato and F. Mizutani, *J. Electroanal. Chem.*, 496 (2001) 50–60.
105 K. Tamada, M. Hara, H. Sasabe and W. Knoll, *Langmuir*, 13 (1997) 1558–1566.
106 B.T. Houseman and M. Mrksich, *Angew. Chem. Int. Ed.*, 38 (1999) 782–785.
107 M. Mrksich, J.R. Grunwell and G.M. Whitesides, *J. Am. Chem. Soc.*, 117 (1995) 12009–12010.
108 D.J. Revell, J.R. Knight, D.J. Blyth, A.H. Haines and D.A. Russell, *Langmuir*, 14 (1998) 4517–4524.

109 G.B. Sigal, C. Bamddad, A. Barberis, J. Strominger and G.M. Whitesides, *Anal. Chem.*, 68 (1996) 490–497.
110 D. Kröger, M. Liley, W. Schiweck, A. Skerra and H. Vogel, *Biosens. Bioelectron.*, 14 (1999) 155–161.
111 K.E. Nelson, L. Gamble, L.S. Jung, M.S. Boeckl, E. Naeemi, S.L. Golledge, T. Sasaki, D.G. Castner, C.T. Campbell and P.S. Stayton, *Langmuir*, (2001) published on the Web http://pub.acs.org.
112 J. Rickert, T. Weiss, W. Kraas, G. Jung and W. Göpel. In: Transducers 95—Eurosensors IX, Stockholm, Sweden, June 25–29, 1995, pp. 528–531
113 M. Liu, Q.X. Li and G.A. Rechnitz, *Electroanalysis*, 12 (2000) 21–26.
114 N. Patel, M.C. Davies, M. Hartshorne, R.J. Heaton, C.J. Roberts, S.J.B. Tendler and P.M. Williams, *Langmuir*, 13 (1997) 6485–6490.
115 J. Lahiri, L. Isaacs, J. Tien and G.M. Whitesides, *Anal. Chem.*, 71 (1999) 777–790.
116 R.M. Nyquist, A.S. Eberhardt, L.A. Silks III, Z. Li, X. Yang and B.I. Swanson, *Langmuir*, 16 (2000) 1793–1800.
117 M.N. Yousaf and M. Mrksich, *J. Am. Chem. Soc.*, 121 (1999) 4286–4287.
118 R.C. Horton Jr., T.M. Herne and D.C. Myles, *J. Am. Chem. Soc.*, 119 (1997) 12980–12981.
119 L. Yan, C. Marzolin, A. Terfort and G.M. Whitesides, *Langmuir*, 13 (1997) 6704–6712.
120 R.G. Chapman, E. Ostuni, L. Yan and G.M. Whitesides, *Langmuir*, 16 (2000) 6927–6936.
121 E.A. Smith, M.J. Wanat, Y. Cheng, S.V.P. Barreira, A.G. Frutos and R.M. Corn, *Langmuir*, 17 (2001) 2502–2507.
122 C.K. Harnett, K.M. Satyalakashmi and H.G. Craighead, *Langmuir*, 17 (2001) 178–182.
123 J.P. Collman, N.K. Devaraj and C.E.D. Chidsey, *Langmuir*, 20 (2004) 1051–1053.
124 A.E. Strong and B.D. Moore, *Chem. Commun.*, (1998) 473–474.
125 A.E. Strong and B.D. Moore, *J. Mater. Chem.*, 9 (1999) 1097–1105.
126 K. Fujita, N. Bunjes, K. Nakajima, M. Hara, H. Sasabe and W. Knoll, *Langmuir*, 14 (1998) 6167–6172.
127 L. Scheibler, P. Dumy, M. Boncheva, K. Leufgen, H.J. Mathieu, M. Mutter and H. Vogel, *Angew. Chem. Int. Ed.*, 38 (1999) 696–699.
128 P. Ringler and G.E. Schulz, *Science*, 302 (2003) 106–109.
129 M. Boncheva, C. Duschl, W. Beck, G. Jung and H. Vogel, *Langmuir*, 12 (1996) 5636–5642.
130 R. Naumann, E.K. Schmidt, A. Jonczyk, K. Fendler, B. Kadenbach, T. Liebermann, A. Offenhäusser and W. Knoll, *Biosens. Bioelectron.*, 14 (1999) 651–662.
131 C. Duschl, M. Liley and H. Vogel, *Angew. Chem. Int. Ed.*, 33 (1994) 1274–1276.

132 S. Lingler, I. Rubinstein, W. Knoll and A. Offenhäuser, *Langmuir*, 13 (1997) 7085–7091.
133 H.C. Yoon, M.Y. Hong and H.S. Kim, *Langmuir*, 17 (2001) 1234–1239.
134 B.L. Frey and R.M. Corn, *Anal. Chem.*, 68 (1996) 3187–3193.
135 C.E. Jordan and R.M. Corn, *Anal. Chem.*, 69 (1997) 1449–1456.
136 V. Pardo-Yissar, E. Katz, O. Lioubashevski and I. Willner, *Langmuir*, 17 (2001) 1110–1118.
137 F. Caruso, K. Niikura, D.N. Furlong and Y. Okahata, *Langmuir*, 13 (1997) 3422–3426.
138 F. Caruso, K. Niikura, D.N. Furlong and Y. Okahata, *Langmuir*, 13 (1997) 3427–3433.
139 J.I. Anzai, Y. Kobayashi, N. Nakamura, M. Nishimura and T. Hoshi, *Langmuir*, 15 (1999) 221–226.
140 H. Raether, In: G. Has, M. Francombe and R. Hoffman (Eds.), *Physics of Thin Films*, Vol. 9. Academic Press, New York, 1977, pp. 145–261.
141 T.B. Dubrovsky, Z. Hou, P. Stroeve and N.L. Abbot, *Anal. Chem.*, 71 (1999) 327–332.
142 M. Mrksich, G.B. Sigal and G.M. Whitesides, *Langmuir*, 11 (1995) 4383–4385.
143 P.L. Edmiston, J.E. Lee, S.S. Cheng and S.S. Saavedra, *J. Am. Chem. Soc.*, 119 (1997) 560–570.
144 A.M. Edwards, J.A. Chupa, R.M. Strongin, A.B. Smith III and J.K. Blasie, *Langmuir*, 13 (1997) 1634–1643.
145 J.E. Lee and S.S. Saavedra, *Langmuir*, 12 (1996) 4025–4032.
146 S. Chen, L. Liu, J. Zhou and S. Jiang, *Langmuir*, 19 (2003) 2859–2864.
147 H. Wang, D.G. Castner, B.D. Ratner and S. Jiang, *Langmuir*, 20 (2004) 1877–1887.
148 S. Terretaz, T. Stora, C. Duschl and H. Vogel, *Langmuir*, 9 (1993) 1361–1369.
149 D. van der Heuvel, R.P.H. Kooyman, J.W. Drijfhout and C.G. Welling, *Anal. Biochem.*, 215 (1993) 223–230.
150 S.M. Amador, J.C. Pachence, R. Fischetti, J.P. McCauley Jr., A.B. Smith III and J.K. Blasie, *Langmuir*, 9 (1993) 812–817.
151 G.E. Wegner, H.J. Lee, G. Marriott and R.M. Corn, *Anal. Chem.*, 75 (2003) 4740–4746.
152 L.L. Wood, S.S. Cheng, P.L. Edmiston and S.S. Saavedra, *J. Am. Chem. Soc.*, 119 (1997) 571–576.
153 H.G. Hong, M. Jiang, S.G. Sligar and P.W. Bohn, *Langmuir*, 10 (1994) 153–158.
154 M. Jiang, B. Nölting, P.S. Stayton and S.G. Sligar, *Langmuir*, 12 (1996) 1278–1283.
155 M.A. Firestone, M.L. Shank, S.G. Sligar and P.W. Bohn, *J. Am. Chem. Soc.*, 118 (1996) 9033–9041.
156 C. Yeung and D. Leckband, *Langmuir*, 13 (1997) 6746–6754.
157 A.A. Kloss, N. Lavrik, C. Yeung and D. Leckband, *Langmuir*, 16 (2000) 3414–3421.

158   L. Häussling, B. Michel, H. Ringsdorf and H. Rohrer, *Angew. Chem. Int. Ed.*, 30 (1991) 569–572.
159   V.H. Perez-Luna, M.J. O'Brien, K.A. Opperman, P.D. Hampton, G.P. Lopez, L.A. Klumb and P.S. Stayton, *J. Am. Chem. Soc.*, 121 (1999) 6469–6478.
160   L.S. Jung, K.E. Nelson, C.T. Campbell, P.S. Stayton, S.S. Yee, V. Pérez-Luna and G.P. López, *Sens. Actuators, B*, 54 (1999) 137–144.
161   G. Ladam, P. Schaaf, F.J.G. Cuisinier, G. Decher and J.C. Voegel, *Langmuir*, 17 (2001) 878–882.
162   J.I. Anzai and M. Nishimura, *J. Chem. Soc., Perkin Trans.*, 2 (1997) 1887–1889.
163   H.C. Yoon, M.Y. Hong and H.S. Kim, *Anal. Biochem.*, 282 (2000) 121–128.
164   J.A. Warrington, S. Dee and M. Trulson, In: M. Schena (Ed.), *Microarray Biochip Technology*. BioTechniques Books, Eaton Publishing, Natick, MA, 2000, pp. 119–148.
165   W. Du, C. Marsac, M. Kruschina, F. Ortigao and C. Florentz, *Anal. Biochem.*, 322 (2003) 14–25.
166   M. Campàs and I. Katakis, *Trends Anal. Chem.*, 23 (2004) 49–62.
167   Y. Okahata, Y. Matsunobu, K. Ijiro, M. Mukae, A. Murakami and K. Makino, *J. Am. Chem. Soc.*, 114 (1992) 8299–8300.
168   T.M. Herne and M.J. Tarlov, *J. Am. Chem. Soc.*, 119 (1997) 8916–8920.
169   K.A. Peterlinz and R.M. Georgiadis, *J. Am. Chem. Soc.*, 119 (1997) 3401–3402.
170   R. Levicky, T.M. Herne, M. Tarlov and S.K. Satija, *J. Am. Chem. Soc.*, 120 (1998) 9787–9792.
171   Y. Okahata, M. Kawase, K. Niikura, F. Ohtake, H. Furusawa and Y. Ebara, *Anal. Chem.*, 70 (1998) 1288–1296.
172   M. Boncheva, L. Scheibler, P. Lincoln, H. Vogel and B. Ackerman, *Langmuir*, 15 (1999) 4317–4320.
173   A.J. Thiel, A.G. Frutos, C.E. Jordan, R.M. Corn and L.M. Smith, *Anal. Chem.*, 69 (1997) 4948–4956.
174   C.E. Jordan, A.G. Frutos, A.J. Thiel and R.M. Corn, *Anal. Chem.*, 69 (1997) 4939–4947.
175   F. Caruso, E. Rodda, D.N. Furlong, K. Niikura and Y. Okahata, *Anal. Chem.*, 69 (1997) 2043–2049.
176   B. Sellergreen, F. Auer and T. Arnebrant, *Chem. Commun.*, (1999) 2001–2002.
177   F. Auer, D.W. Schubert, M. Stamm, T. Arnebrandt, A. Swietlov, M. Zizlsperger and B. Sellergren, *Chem. Eur. J.*, 5 (1999) 1150–1159.
178   H. Kuramitz, K. Sugawara and S. Tonaka, *Electroanalysis*, 12 (2000) 1299–1303.
179   M. Liu, Q.X. Li and G.A. Rechnitz, *Electroanalysis*, 12 (2000) 21–26.
180   H. Aoki, P. Bühlmann and Y. Umezawa, *Electroanalysis*, 12 (2000) 1272–1276.
181   A.B. Steel, T.M. Herne and M.J. Tarlov, *Anal. Chem.*, 70 (1998) 4670–4677.
182   R. Blonder, E. Katz, Y. Cohen, N. Itzhak, A. Riklin and I. Willner, *Anal. Chem.*, 68 (1996) 3151–3157.
183   N. Higashi, M. Takahashi and M. Niwa, *Langmuir*, 15 (1999) 111–115.

184 S.O. Kelley, J.K. Barton, N.M. Jackson and M.G. Hill, *Bioconj. Chem.*, 8 (1997) 31–37.
185 J. Gu, X. Lu and H. Ju, *Electroanalysis*, 14 (2002) 949–954.
186 K. Kerman, D. Ozkan, P. Kara, B. Meric, J.J. Gooding and M. Ozsoz, *Anal. Chim. Acta*, 462 (2002) 39–47.
187 M. Yang, H.C.M. Yau and H.L. Chan, *Langmuir*, 14 (1998) 6121–6129.
188 E. Huang, F. Zhou and L. Deng, *Langmuir*, 16 (2000) 3272–3280.
189 E.L.S. Wong and J.J. Gooding, *Anal. Chem.*, 75 (2003) 3845–3852.
190 H.-X. Ju, Y.-K. Ye, J.-H. Zhao and Y.-L. Zhu, *Anal. Biochem.*, 313 (2003) 255–261.
191 D.J. Caruana and A. Heller, *J. Am. Chem. Soc.*, 121 (1999) 769–774.
192 E. Kim, K. Kim, H. Yang, Y.T. Kim and J. Kwak, *Anal. Chem.*, 75 (2003) 5665–5672.
193 J. Wang, J. Li, A.J. Baca, J. Hu, F. Zhou, W. Yan and D.-W. Pang, *Anal. Chem.*, 75 (2003) 3941–3945.
194 L. Authier, C. Grossiord, P. Brossier and B. Limoges, *Anal. Chem.*, 73 (2001) 4450–4456.
195 T.A. Taton, C.A. Mirkin and R.L. Letsinger, *Science*, 289 (2000) 1757–1760.
196 J. Wang, D. Xu, A.-N. Kawde and R. Polsky, *Anal. Chem.*, 73 (2001) 5576–5581.
197 T.M.-H. Lee, L.-L. Li and I.-M. Hsing, *Langmuir*, 19 (2003) 4338–4343.
198 C. Fan, K.W. Plaxco and A.J. Heeger, *Proc. Natl Acad. Sci. USA*, 100 (2003) 9134–9137.
199 F.A. Armstrong, H.A.O. Hill and J.N. Walton, *Quat. Rev. Biophys.*, 18 (1986) 261–322.
200 D.D. Schlereth, V.M. Fernández, M. Sánchez-Cruz and V.O. Popov, *Bioelectrochem. Bioenerg.*, 28 (1992) 473–482.
201 P. Gros, C. Zaborosch, H.G. Schlegel and A. Bergel, *J. Electroanal. Chem.*, 405 (1996) 189–195.
202 G.F. Khan, H. Shinohara, Y. Ikariyama and M. Aizawa, *J. Electroanal. Chem.*, 315 (1991) 263.
203 T. Ikeda, D. Kobayashi, F. Matsushita, T. Sagara and K. Niki, *J. Electroanal. Chem.*, 361 (1993) 221–228.
204 T. Ikeda, S. Miyaoka and K. Miki, *J. Electroanal. Chem.*, 352 (1993) 267–278.
205 D.D. Schlereth and W. Mäntele, *Biochemistry*, 32 (1993) 1118–1126.
206 D.D. Schlereth, V.M. Fernández and W. Mäntele, *Biochemistry*, 32 (1993) 9199–9208.
207 A. Lindgren, T. Larsson, T. Ruzgas and L. Gorton, *J. Electroanal. Chem.*, 494 (2000) 105–113.
208 A. Lindgren, L. Gorton, T. Ruzgas, U. Baminger, D. Haltrich and M. Schülein, *J. Electroanal. Chem.*, 496 (2001) 76–81.
209 E.E. Ferapontova, T. Ruzgas and L. Gorton, *Anal. Chem.*, 75 (2003) 4841–4850.
210 Z.Q. Feng, S. Imabayashi, T. Kakiuchi and K. Niki, *J. Electroanal. Chem.*, 394 (1995) 149–154.

211 Z.Q. Feng, S. Imabayashi, T. Kakiuchi and K. Niki, *J. Chem. Soc. Faraday Trans.*, 93 (1997) 1367–1371.
212 T. Lu, X. Yu, S. Dong, C. Zhou, S. Ye and T.M. Cotton, *J. Electroanal. Chem.*, 369 (1994) 79.
213 A. El Kasmi, J.M. Wallace, E.F. Bowden, S.M. Binet and R.J. Linderman, *J. Am. Chem. Soc.*, 120 (1998) 225–226.
214 D.D. Schlereth, *J. Electroanal. Chem.*, 464 (1999) 198–207.
215 Y. Iwasaki, T. Horiuchi, M. Morita and O. Niwa, *Sens. Actuators B*, 50 (1998) 145–148.
216 S. Koide, Y. Iwasaki, T. Horiuchi, O. Niwa, E. Tamiya and K. Yokoyama, *Chem. Commun.*, (2000) 741–742.
217 Y. Iwasaki, T. Horiuchi and O. Niwa, *Anal. Chem.*, 73 (2001) 1595–1598.
218 C.B. Brennan, L. Sun and S.G. Weber, *Sens. Actuators B*, 72 (2001) 1–10.
219 K.T. Kinnear and H.G. Monbouquette, *Langmuir*, 9 (1993) 2255–2257.
220 K.T. Kinnear and H.G. Monbouquette, *Anal. Chem.*, 69 (1997) 1771–1775.
221 M. Darder, E. Casero, F. Pariente and E. Lorenzo, *Anal. Chem.*, 72 (2000) 3784–3792.
222 J.M. Cooper, K.R. Greenough and C.J. McNeil, *J. Electroanal. Chem.*, 347 (1993) 267.
223 J. Li, G. Cheng and S. Dong, *J. Electroanal. Chem.*, 416 (1996) 97.
224 S. Rubin, J.T. Chow, J.P. Ferraris and T.A. Zawodzinski Jr., *Langmuir*, 12 (1996) 363–370.
225 T. Hasunuma, S. Kuwabata, E. Fukusaki and A. Kobayashi, *Anal. Chem.*, 76 (2004) 1500–1506.
226 I. Willner, N. Lapidot, A. Riklin, R. Kasher, E. Zahavy and E. Katz, *J. Am. Chem. Soc.*, 116 (1994) 1428–1441.
227 I. Willner, E. Katz, A. Riklin and R. Kasher, *J. Am. Chem. Soc.*, 114 (1992) 10965–10966.
228 E. Katz, V. Heleg-Shabtai, B. Willner, I. Willner and A.F. Bückmann, *Bioelectrochem. Bioenerg.*, 42 (1997) 95.
229 I. Willner, E. Katz, B. Willner, R. Blonder, V. Heleg-Shabtai and A.F. Bückmann, *Biosens. Bioelectron.*, 13 (1997) 337–356.
230 M. Darder, K. Takada, F. Pariente, E. Lorenzo and H.D. Abruna, *Anal. Chem.*, 71 (1999) 5530–5537.
231 J.M. Abad, M. Vélez, C. Santamaría, J.M. Guisán, P.R. Matheus, L. Vázquez, I. Gazaryan, L. Gorton, T. Gibson and V.M. Fernández, *J. Am. Chem. Soc.*, 124 (2002) 12845–12853.
232 A. Riklin, E. Katz, I. Willner, A. Stocker and A.F. Bückmann, *Nature*, 376 (1995) 672–675.
233 E. Katz, D.D. Schlereth and H.L. Schmidt, *J. Electroanal. Chem.*, 367 (1994) 59–70.
234 E. Katz, T. Lötzbeyer, D.D. Schlereth, W. Schuhmann and H.L. Schmidt, *J. Electroanal. Chem.*, 373 (1994) 189–200.

235 E. Katz, D.D. Schlereth, H.L. Schmidt and A.J.J. Olsthoorn, *J. Electroanal. Chem.*, 368 (1994) 165–171.
236 H. Zimmermann, A. Lindgren, W. Schuhmann and L. Gorton, *Chem. Eur. J.*, 6 (2000) 592–599.
237 D.D. Schlereth, *Sens. Actuators B*, 43 (1997) 78–86.
238 D.D. Schlereth, *J. Electroanal. Chem.*, 425 (1997) 77–85.
239 L. Bertilsson, H.J. Butt, G. Nelles and D.D. Schlereth, *Biosens. Bioelectron.*, 12 (1997) 839–852.
240 D.D. Schlereth and R.P.H. Kooyman, *J. Electroanal. Chem.*, 431 (1997) 285–295.
241 D.D. Schlereth and R.P.H. Kooyman, *J. Electroanal. Chem.*, 444 (1998) 231–240.
242 A. Bardea, E. Katz, A. Bückmann and I. Willner, *J. Am. Chem. Soc.*, 119 (1997) 9114–9119.
243 I. Willner, V. Heleg-Shabtai, R. Blonder, E. Katz and G. Tao, *J. Am. Chem. Soc.*, 118 (1996) 10321–10322.
244 I. Willner, *Acta Polym.*, 49 (1998) 652–662.
245 I. Willner, E. Katz, F. Patolsky and A.F. Bückmann, *J. Chem. Soc. Perkin Trans. 2*, (1998) 1817–1822.
246 F. Patolsky, E. Katz, V. Heleg-Shabtai and I. Willner, *Chem. Eur. J.*, 4 (1998) 1068–1073.
247 V. Heleg-Shabtai, E. Katz, S. Levi and I. Willner, *J. Chem. Soc. Perkin Trans. 2*, (1997) 2645–2651.
248 I. Willner, V. Heleg-Shabtai, E. Katz, H.K. Rau and W. Haehnel, *J. Am. Chem. Soc.*, 121 (1999) 6455–6468.
249 J. Madoz, B.A. Kuznetzov, F.J. Medrano, J.L. García and V.M. Fernández, *J. Am. Chem. Soc.*, 119 (1997) 1043–1051.
250 J.I. Anzai, Y. Kobayashi, Y. Suzuki, H. Takeshita, Q. Chen, T. Osa, T. Oshi and X.Y. Du, *Sens. Actuators B*, 52 (1998) 3–9.
251 J.I. Anzai, H. Takeshita, Y. Kobayashi, T. Osa and T. Hoshi, *Anal. Chem.*, 70 (1998) 811–817.
252 Q. Chen, Y. Kobayashi, H. Takeshita, T. Hoshi and J.I. Anzai, *Electroanalysis*, 10 (1998) 94–97.
253 J.I. Anzai, Y. Kobayashi and N. Nakamura, *J. Chem. Soc. Perkin Trans. 2*, (1998) 461–462.
254 H.C. Yoon and H.S. Kim, *Anal. Chem.*, 72 (2000) 922–926.
255 H.C. Yoon, M.Y. Hong and H.S. Kim, *Anal. Chem.*, 72 (2000) 4420–4427.
256 A.L. de Lacey, M. Detcheverry, J. Moiroux and C. Bourdillon, *Biotechnol. Bioeng.*, 68 (2000) 1–10.
257 J. Hodak, R. Etchenique, E.J. Calvo, K. Singhal and P.N. Bartlett, *Langmuir*, 13 (1997) 2708–2716.
258 E.J. Calvo, R. Etchenique, L. Pietrasanta, A. Wolosiuk and C. Danilowicz, *Anal. Chem.*, 73 (2001) 1161–1168.
259 K. Sirkar, A. Revzin and M.V. Pishko, *Anal. Chem.*, 72 (2000) 2930–2936.
260 C. Ruan, F. Yang, C. Lai and J. Deng, *Anal. Chem.*, 70 (1998) 1721–1725.

261 I. Willner, M. Lion-Dagan, S. Marx-Tibbon and E. Katz, *J. Am. Chem. Soc.*, 117 (1995) 6581–6592.
262 I. Willner, R. Blonder, E. Katz, A. Stocker and A.F. Bückmann, *J. Am. Chem. Soc.*, 118 (1996) 5310–5311.
263 R. Blonder, E. Katz, I. Willner, V. Wray and A.F. Bückmann, *J. Am. Chem. Soc.*, 119 (1997) 11747–11757.
264 E. Kaganer, R. Pogreb, D. Davidov and I. Willner, *Langmuir*, 15 (1999) 3920–3923.
265 X. Jiang, R. Ferrigno, M. Mrksch and G.M. Whitesides, *J. Am. Chem. Soc.*, 125 (2003) 2366–2367.
266 C.D. Hodneland and M. Mrksich, *Langmuir*, 13 (1997) 6001–6003.
267 C.D. Hodneland and M. Mrksich, *J. Am. Chem. Soc.*, 122 (2000) 4235–4236.
268 W.-S. Yeo, M.N. Yousaf and M. Mrksich, *J. Am. Chem. Soc.*, 125 (2003) 14994–14995.
269 S.O. Kelley, J.K. Barton, N.M. Jackson, L.D. McPherson, A.B. Potter, E.M. Spain, M.J. Allen and M.G. Hill, *Langmuir*, 14 (1998) 6781–6784.

Chapter 2

Third generation biosensors—integrating recognition and transduction in electrochemical sensors

Ulla Wollenberger

2.1 INTRODUCTION

An enzyme electrode is basically a dense package of dialyzer, enzyme reactor, and electrode (detector). Enzymes introduce analytical selectivity due to the specificity of the signal-producing interaction of the enzyme with the analyte. They enhance the equilibrium formation of chemical reactions. For example, splitting of H_2O_2 is accelerated by a factor of 3×10^{11} in the presence of catalase. Turnover numbers can be as fast as $6 \times 10^5 \, s^{-1}$ (carbonic anhydrase) where k_{cat}/K_m approaches the diffusion limited value of $10^9 \, M^{-1} \, s^{-1}$.

Amperometric devices exploit the electron exchange between these biocatalytic systems and electrodes. A typical example is a glucose or lactate electrode comprising the appropriate oxidase entrapped in or bound to a membrane that is fixed at an oxygen or hydrogen peroxide detecting electrode. But the selective recognition process followed by catalytic substrate conversion and electrode reaction can easily be interfered with by other redox active species or influenced by physico-chemical parameters. The exchange of the natural co-reagents by mobile or bound small redox mediators results in a favourable working potential and, in some cases, an extended measuring range. The most successful blood glucose sensors are based on this principle [1]. However, these redox mediators react not only with the biomolecule of interest but also with accessible solution species of respective redox potential. The fast direct transfer of electrons between enzyme and electrode opens the avenue to highly selective sensors virtually free of interference. Sensors based on this principle are regarded as third generation biosensors (Fig. 2.1).

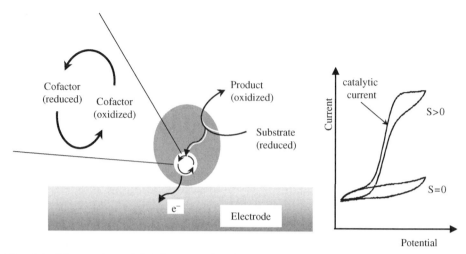

Fig. 2.1. Illustration of third generation biosensors illustrating the direct electron transfer contact between an enzyme and an electrode and catalytic current generated in the presence of substrate S.

2.2 BIOSENSOR GENERATIONS

The term *generation* was originally created to describe the stages of integration in biosensors. Biosensors with mediated response mainly generated by membrane-bound or membrane entrapped biocomponents (*first generation*), were followed by virtually reagentless biosensors where biocomponents are fixed directly to the sensor surface and either the reaction of coimmobilized cosubstrates (bound to the electrode, polymer or enzyme itself) or the direct heterogeneous electron transfer between the prosthetic group and the electrode is exploited (*second generation*). While the immobilization of the receptors directly on an electronic circuit leads to systems with integrated signal generation, transduction and processing (*third generation*) having the potential of considerable miniaturization [2]. Urease on a MOSFET is a third generation biosensor in the above sense. The recent developments of biochips and microarrays, however, lead to an altered view on the generation picture, paying attention to miniaturization and parallelization of the recognition [3].

In the last few years the term *generation* has mainly been used to differentiate between the modes of signal transfer between a redox enzyme and an electrode, i.e., via the natural secondary substrates and products of the enzyme catalyzed reaction (*first generation*), artificial electron mediators instead of the natural cosubstrates (*second generation*) or in direct electronic contact (*third generation*) [4–8]. In this sense a typical first generation biosensor is a glucose sensor with gel entrapped glucose oxidase on a Clark-type electrode [9,10] and also many variants of coupled enzymes on oxygen-sensitive electrodes [11]. The use of glucose oxidase together with ferrocene derivatives on screen-printed electrodes [12], immobilization of the flavin group to an electrode [13] or the wiring of glucose oxidase in a redox polymer [14,15] are examples of the second generation, while the transfer of electrons directly from the electrode to a peroxidase [8,16–18], for example, belongs to the third generation. The most successful devices for self-control of blood glucose are based on the combination of the enzymes glucose oxidase and PQQ-dependent or NAD^+-dependent glucose dehydrogenase with different redox mediators.

This review is a survey of the research on the direct electron transfer (DET) between biomolecules and electrodes for the development of reagentless biosensors. Both the catalytic reaction of a protein or an enzyme and the coupling with further reaction have been used analytically. For better understanding and a better overview, this chapter begins with a description of electron transfer processes of redox proteins at electrodes. Then the behaviour of the relevant proteins and enzymes at electrodes is briefly characterized and the respective biosensors are described. In the last section sensors for superoxide, nitric oxide and peroxide are presented. These have been developed with several proteins and enzymes. The review is far from complete, for example, the large class of iron–sulfur proteins has hardly been touched. Here the interested reader may consult recent reviews and work cited therein [1,19].

2.3 ELECTRON TRANSFER PROCESSES AT ELECTRODES

Electron transfer reactions between and within proteins play a central role in biological energy conversion, metabolism and regulation ranging from cell respiration, photosynthesis, redox driven conformation change to regulation of gene activity. A vectorial electron transfer with a relatively small driving force is guaranteed by spatial arrangements and structural complementarity of the interacting protein partners. The charge is transferred either by redox mediators, or by direct intermolecular interaction of enzymes. Small redox proteins are often the carriers of the redox equivalents in electron transfer chains. Similar principles are used in the electronic communication between biomaterials and redox electrodes. This has been an attractive research area because of its potential application for the creation of virtually interference-free reagentless biosensors, effective biofuel cells and selective biosynthesis routes. Furthermore, these systems may also increase our basic knowledge of biological redox reactions. A number of excellent comprehensive reviews have been published [1,8,20–24].

The electron transfer in biology usually involves initial protein–protein complex formation based on the complementarity of the docking sites. Efficient protein–electrode reactions appear to have some similarities to the way in which proteins act with their natural redox partner [22]. Therefore, methods for chemically modifying electrode surfaces as to mimic the biological situation were developed. The heterogeneous electron transfer between proteins and electrodes may be coupled with other reactions where the proteins act as vectorial mediators [25,26].

The prerequisites for a DET can be derived from Marcus Theory [27,28]. The highly specific and directional protein-mediated electron transfer in biological systems is governed by factors such as the distance and the bonds between the redox centres, the redox-potential difference between donor and acceptor, an appropriate association of the redox couple and protein-structure dynamics coupled with electron transfer [24,27,29].

The Marcus Theory can also be applied for heterogeneous electron transfer reaction at electrode surfaces [24 and references therein]. The electronic coupling between the protein and the electrode can be varied using different self-assembled monolayers controlling the orientation of the redox active protein on the surface and the distance between the redox active site of the protein and the electrode. The driving force is related to the applied potential and the redox potential of the protein. In many cases the rate of electron transfer across the protein–electrode interface is limited by conformational reorganization. This has focussed the efforts of many groups on tailored interaction between proteins and enzymes and electrode surfaces.

2.4 DIRECT ELECTRON TRANSFER—OVERVIEW

The first reports on a reversible DET between redox proteins and electrodes were published in 1977 showing that cytochrome c is reversibly oxidized and reduced at tin-doped indium oxide [30] and gold in the presence of 4,4'-bipyridyl [31]. Only shortly after these publications appeared, papers were published describing the DET between electrode and enzyme for laccase and peroxidase [32,33]. It was observed that the overpotential for oxygen reduction at a carbon electrode was reduced by several hundred millivolts compared to the "uncatalyzed" reduction when laccase was adsorbed. This reaction could be inhibited by azide. The term bioelectrocatalysis was introduced for such an acceleration of the electrode process by an enzyme [34,35]. The direct communication with an electrode generally termed as DET between the electrode material and redox active biomolecule is a mediatorless heterogeneous electron transfer process.

For biosensors of the third generation DET to small redox proteins is of particular interest as they show interaction with reactive (oxygen) species, while enzymes in direct electric contact are suitable for reagentless metabolite measurement. Peroxidase, catalase and superoxide dismutase are also relevant to the determination of reactive oxygen species and their scavengers.

Self-assembling monolayers (SAM) of functionalized thiols on metals, polyelectrolyte and surfactant films on carbon, electron-transfer promoters and metal oxides were shown to be suitable tools [1,15,25,36–40]. Particularly SAMs spontaneously formed on clean gold surfaces from solutions of thiols with functional head groups are suitable to facilitate the aniosotropic immobilization of proteins and enzymes [15,39,40]. The distance between the functional head group and the electrode surface and thus the distance over which the electron has to travel can be controlled by the chain length of the alkyl spacer between thiol and head group [41,42]. The mobility of the protein is an important factor for a further coupling to a protein/enzyme partner [26,43]. Recent works also show the potential of carbon nanotubes because of their small size and remarkable electrocatalytic activity [44]. Progress has been made not only in surface chemistry but also in engineering of the protein/ enzyme itself [45–50] to achieve rapid and efficient direct electron-transfer reactions. Monolayers of proteins with determined orientation can also be produced with direct binding of surface exposed cysteine(s) on the protein shell. Such attachment sites can specifically be engineered.

Redox proteins are relatively small molecules. In biological systems they are membrane associated, mobile (soluble) or associated with other proteins. Their molecular structure ensures specific interactions with other proteins or enzymes. In a simplified way this situation is mimicked when electrodes are chemically modified to substitute one of the reaction partners of biological redox pairs. The major classes of soluble redox active proteins are heme proteins, ferredoxins, flavoproteins and copper proteins (Table 2.1). In most cases they do not catalyze specific chemical reactions themselves, but function as biological (natural) electron carriers to or between enzymes catalyzing specific transformations. Also some proteins which are naturally not involved in redox processes but carry redox active sites (e.g., hemoglobin and myoglobin) show reversible electron exchange at proper functionalized electrodes.

Even though they do not comprise enzymatic functions a few of these proteins show specific interaction with small signal molecules, an interaction

TABLE 2.1

Classes of redox proteins, examples, and some reaction partners

Redox active group	Protein	Enzyme/protein partner
Heme	Cytochrome c	Cytochrome c oxidase, cytochrome c peroxidase, cellobiose dehydrogenase, nitrate reductase, sulphite oxidase, theophylline oxidase, cytochrome b2
	Cytochrome c550	QH amineDH
	Cytochrome b562	QH glucoseDH
	Cytochrome c3	Fumarate DH
	Cytochrome b5	Myoglobin
Iron–sulfur complex (ferredoxins)	Putidaredoxin	P450cam, putidaredoxin reductase
	Adrenodoxin	P450scc, 15-β-hydroxylase, adrenodoxin reductase
Copper	Azurin	Cytochrome c3
	Plastocyanin	Cytochrome bf-complex
	Amicyanin	Methylamine DH
Flavoproteins	Flavodoxin	Reductases

that can be used for biosensing (Table 2.2). Biosensing capabilities were exploited in particular for the analysis of reactive oxygen species such as superoxide [51–55], NO [56–59], and NO_2^- [60] and scavengers of these [54]. The reaction is a particular feature of the central iron atom. Therefore, heme proteins are studied most intensively.

Among the roughly 3000 (wild type) *enzymes* currently known, of which about 1060 are oxidoreductases, only a small number of enzymes is capable of communicating directly with an electrode [7,8,21,25,61,62]. The reason is found in the structure of the enzyme. To ensure high selectivity towards its substrate the catalytic centre is deeply buried in the polypeptide and thus it is almost impossible to transfer electrons directly to or from an electrode without any conformational change. The enzymes capable of being directly (without additional redox relays) contacted to electrodes fall into the class of extrinsic enzymes [21] having proteins such as azurin (cytochrome c3), cytochrome c (fructose dehydrogenase, cytochrome c peroxidase, cytochrome c oxidase), and putidaredoxin (P450cam) as natural redox partners. If the electrode is designed to resemble the surface characteristics of the protein (in the simple way the net charge or hydrophobicity) the enzyme may bind to such a surface without dramatic structural changes and if the distance between the redox

TABLE 2.2

Direct electron transfer with proteins and analytical application

Protein	Analyte
Cytochrome c	H_2O_2
	O_2^-
	Superoxide scavenger
	Lactate, inhibitors of Cyt. c oxidase, theophylline (in connection with the respective substrate oxidase)
Cytochrome c'	NO
Cytochrome c3	Fumarate
Cytochrome b562	Glucose (in conjunction with QH GDH)
Haemoglobin	NO
	NO^{2-}
	CN (binding to MetHb)
	CO
	H_2O_2
Myoglobin	Styrene
	Haloorganics
	NO
	H_2O_2
Azurin	Amines
Putidaredoxin	Camphor (with P450cam)

centre and the electrode is small enough direct electrochemistry may be observed [22].

A collection of redox enzymes for which efficient DET with electrodes has been observed is given in Table 2.3. Most of them are metalloenzymes containing iron or copper. Many of these enzymes are part of electron transfer chains, i.e., have macromolecular redox partners, or react on large substrates. The evidence for DET has not always been presented by direct electrochemical measurements. In many cases the DET has been proved indirectly by measurement of a substrate dependent catalytic current. Various metabolites ranging from sugars such as fructose, cellobiose and gluconate [6], amines like methylamine and histamine [123], lactate [91], p-cresol [93] and drugs such as benzphetamine [74] can be measured with enzymes in direct contact to an electrode. The bioelectrocatalytic reaction of peroxide is one of the most important reactions not only for the determination of peroxide(s) in various media but also substrates of coupled oxidase [8] and enzyme inhibitors [130, 252]. Furthermore, enzyme immunoassays have been developed based on DET of peroxidase and laccase and electrodes [7,131,132].

TABLE 2.3
Redox enzymes for which direct mediator-free reactions with electrodes have been shown

Enzymes, EC-number	Prosthetic groups	Substrates	Comments	References
Ascorbate oxidase, 1.10.3.3	Cu	O_2	Au/bipy, Au/ion exchanger, $E'_0 = 310$ mV (NHE, pH 5.5)	[63,64]
Laccase, 1.10.3.2, *Rhus vernicifera*, *Coriolus hirsitus*, *Polyporus versicolor*	Cu	O_2	C, PG, G, Au/MPA, Au/TMPC, T1 Cu accessible, $E'_0 = 790$–410 mV (NHE, pH5.5), application for fuel cells, immuno assays or phenol sensors (not DET)	[32,64,67,105, 130,132]
Superoxide dismutase, 1.15.1.1	Cu–Zn, Fe, Mn	O_2^-	Au/cystamine, Au/MPA, $E'_0(Cu) = -60$ to $+47$ mV $E'_0(Fe) = -158$ mV (SCE)	[68–70]
Diaphorase, 1.6.99.–	FMN	NADH		[66]
Glucose oxidase, 1.1.3.4	FAD	Glucose	Carbon nano tubes $E'_0 = -659$ mV, Au/DTSSP $E'_0 = -395$ mV (Ag/AgCL), response to mM glucose	[80,81]
Methylamine dehydrogenase, 1.4.99.3	TTQ	Methylamine	PGE/ promoter: polyamines, aminoglycosides, Au/dithioglycolate, $E'_0 = -148$ mV	[117]
Phospholipidhydroperoxide glutathione peroxidase, 1.11.1.12	Selenocysteine	Glutathione, H_2O_2	Au/MUA/MU	[118]

Enzyme	Cofactor	Substrate	Electrode/notes	Ref
Catalase, 1.11.1.6	4 Heme	H_2O_2/O_2	GC	[71]
Cytochrome P450, 1.14.x.x	Heme b	O_2, Aminopyrin, Benzphetamin, Cholesterol	Hg, HOPG, GC/various modifiers: surfactants, clay, lipids, polyelectrolytes, reversible 1 e transfer, oxygen sensitive various types: P450cam, 2B4, 3A4, 1A2, scc	[20,72–74, 76,129]
Methane monooxygenase, 1.14.13.25	Binuclear heme	Acetonitrile, methane	Au/hexapeptide	[79]
Peroxidases 1.11.1.7				
Chloroperoxidase	Heme	H_2O_2	Carbon	[83]
Cytochrome c peroxidase, (Baker's yeast) 1.11.1.5	Heme b	H_2O_2	PGE, comp I at +750 mV	[84–86]
Arthromyces ramosus peroxidase	Heme	RHO_2/H_2O_2	G coimmobil. oxidases, carbon fiber/CDI	[87,88]
Horseradish peroxidase	Heme	RHO_2/H_2O_2	C, PG, SG, CP, SPE, Au/modifier, also recombinant HRP, application for peroxide detection in various media, coupled to oxidases and peroxide consuming catalysts, immunoassays	[17,18,33, 96,130,131, 228,229]

continued

TABLE 2.3 (Continuation)

Enzymes, EC-number	Prosthetic groups	Substrates	Comments	References
Lignin peroxidase	Heme	H_2O_2	G, adsorbed, $H_2O_2 < 20$ μM	[98]
Manganese peroxidase	Heme	H_2O_2	G, adsorbed	[98]
(Microperoxidase)	Heme	H_2O_2	C, Au	[89,90]
Peanut peroxidase	Heme	H_2O_2	G, adsorbed	[100]
Soybean peroxidase	Heme	H_2O_2	G, adsorbed	[102]
Sweet potato peroxidase	Heme	H_2O_2	G, adsorbed, most efficient DET of plant POD on G	[102]
Tobacco peroxidase	Heme	H_2O_2	G, adsorbed, Au	[102,103]
Multi centre enzymes				
Amine dehydrogenase, 1.4.99.3	CTQ, 2 heme c	Amines	From *Paracoccus denitrificans* Au-bipy, reacts with cyt. c_{550}	[124]
Amine oxidase, 1.4.3.6	Cu, topa quinone	Amines	SG, adsorbed, histamine and various biogene amines	[123]
Cytochrome c oxidase, 1.9.3.1	Cu_A, Cu_B, heme $a3$,	cyt. c, O_2	Au/MPA, Au/Lipid	[127,128]
Nitrite reductase, 1.6.6.4	Cu, multi heme	Nitrite (NO_2^-)	Au/bipy, T1-site visible $E_{1/2} = 260$ mV (NHE)	[65]
Cellobiose dehydrogenase, From various rot, 1.1.99.18	FAD, heme	Cellobiose, lactose	G, adsorbed, Au/cystamine, Au/MPA, Au/MUA, DET to single domains, flavin domain has also catalytic activity	[26,109,110, 225,226]

Enzyme	Cofactors	Substrate	Notes	Refs
p-Cresolmethylhydrolase, 1.17.99.1	FAD, heme c	p-cresol	PGE, polyamines, aminoglycosides, Au/negative charged modifier,	[93]
L-Lactate dehydrogenase (Flavocytochrome b_2), 1.1.2.3	FMN, heme b_2	Lactate	Reacts also via Cyt. c550, Cyt. c with electrode for lactate measurement	[91]
Flavocytochrome c_{522}	FAD, heme	Sulphide		[92]
Flavocytochrome $c3$, (Succinate dehydrogenase *Shewanella* sp., 1.3.99.1)	FAD, 4 heme c	Fumarate Succinate	PGE, adsorbed monolayer, All sites accessible E'_0(heme) = −102, −146, −196, −238 mV, E'_0(FAD) = −152 mV (SHE), mechanism	[108]
Fumarate reductase	FAD, Fe−S	Fumarate	PGE, polymyxin	[94,95,97]
D-Gluconate dehydrogenase, 1.1.99.3	FAD, heme c, Fe−S	D-gluconate	CPE, PG, Au, Pt	[125,126]
Alcohol dehydrogenase, 1.1.99.8	PQQ, 4 heme c	Ethanol	C, CPE, SPE, G, Au/SAM, LDL 10 µM ethanol	[114−116,119]
D-Fructose dehydrogenase, 1.1.99.11	PQQ, heme c	Fructose	CPE, Pt, Au, applied to real samples: juice, AscorbateOD coimmobilized to reduce interferents	[99,111,113]
Hydrogenase, 1.98.1.1	Ni−Fe, (Se), Fe−S	H_2, H^+, NAD	PG	[82,107]

continued

TABLE 2.3 (Continuation)

Enzymes, EC-number	Prosthetic groups	Substrates	Comments	References
Aconitase, EC 4.2.1.3	[4Fe-4S]	Citrate	PG	[104]
DMSO-reductase	Mo-pterin, Fe–S	DMSO	G	[101]
Nitrate reductase, 1.9.6.1	Mo-pterin, Fe–S	Nitrate	PGE, neomycin sulphate	[252]
Sulphite oxidase, 1.8.3.1	Mo-pterin Heme b5	Sulphide	Au/HS-x,	[120,121]
Sulphite dehydrogenase, 1.8.2.1	Mo-pterin, Heme c	Sulphide	PG/DDAB	[122]

PG: pyrolytic graphite, PGE: pyrolytic graphite edge, G: graphite, GC: glassy carbon, SPG: spectrographic graphite, PLL: Poly L-lysine, CDI: carbodiimide, Au: gold, bipy-bispyridine, MPA: mercaptopropionic acid, MUA: mercaptoundecanoic acid, MU: mercaptoundecanol.

The following sections will concentrate on the analytical application of redox proteins and redox enzymes for biosensing. The "biomolecule" will be briefly introduced, the major route for its direct electric contact to electrodes outlined and the analytical application discussed. The bioelectrochemical studies on structure–function relationship and their role in biological redox processes will not be covered in detail in this review.

2.5 HEME PROTEINS/HEME ENZYMES

Heme proteins are widely distributed in nature. Some of these proteins have oxygen carrier (haemoglobin) and storage (myoglobin) function, while many others are part of redox reaction chains, either as electron transfer mediators (e.g., cytochrome $b5$, cytochrome c) or as catalyst (e.g., peroxidase, catalase, cytochrome P450). Several enzymes contain heme(s) along with other cofactors such as flavin, copper, iron–sulfur cluster(s) or PQQ. Here the heme mediates or collects electrons (storage). Examples of these multi-cofactor enzymes, which have been coupled with electrodes, are the flavohemoproteins flavocytochrome $b2$, flavocytochrome $c3$, p-cresol-methylhydroxylase, and cellobiose dehydrogenase and the quinohemoproteins fructose dehydrogenase, gluconate dehydrogenase and alcohol dehydrogenase (see below and Table 2.3). A prominent example of the copper-containing heme enzymes is cytochrome c oxidase.

Common to all hemeproteins is the iron porphyrin prosthetic group (Fig. 2.2). Four of the six coordination positions of the heme iron are occupied by nitrogens from the porphyrin ring. The other two coordination positions on either side of the heme plane are available for further ligands which strongly influence the redox potential and reactivity of the heme protein [20].

2.5.1 Heme proteins

2.5.1.1 Cytochrome c
Mitochondrial cytochrome c is the most widely investigated heme protein with respect to its electrochemical properties. It is active in electron transfer pathways such as the respiratory chain in the mitochondria where it transfers electrons between membrane bound cytochrome reductase complex III and cytochrome c oxidase. The active site is an iron porphyrin (heme) covalently linked to the protein at Cys14 and Cys17 through thioether bonds (heme c). The iron itself lies in the plane of the porphyrin ring, the two axial positions

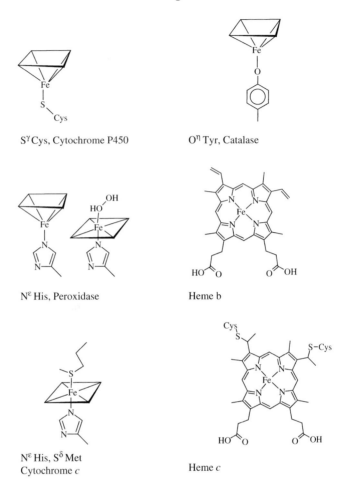

Fig. 2.2. Illustration of heme c and b, heme coordination in cytochrome c, P450, catalase and peroxidase.

of the iron are occupied by sulphur of Met80 and nitrogen of His18 [133] (Fig. 2.2).

Cytochrome c from horse heart is a small globular protein of 12.4 kDa with 108 amino acids in one polypetide chain. At pH 7 cytochrome c bears an overall positive charge of +7, due to an excess of basic lysine residues. There is also a large dipole moment because of the charge distribution on the protein. In the vicinity of the solvent exposed heme edge is a cluster of lysine residues whose charge is largely uncompensated by acidic side chains. This domain is thought to be the interaction site for physiological redox partners.

The rest of the heme is packed in a hydrophobic environment which is responsible for the considerable positive redox potential (+260 vs. NHE) of cytochrome c compared to isolated heme in aqueous milieu.

When an unmodified electrode is brought into contact to cytochrome c strong adsorption to the surface is considered to block the DET to cytochrome c and the electrode often exhibits an irreversible voltammetric response.

Various electrode modifiers allow a reversible or quasi reversible conversion of cytochrome c in solution [134,135]. The ideal promoter has a functional group for surface attachment (e.g., thiols or disulfide for binding to Au, Pt or Ag), groups which minimize structural changes of the protein (hydrophilic groups) and terminal group(s) for the attraction of cytochrome c in appropriate orientation (e.g., negatively charged groups). Among the many modifiers of gold electrodes which facilitate a reversible fast electrode reaction of cytochrome c are 4,4'-dithiodipyridine, cysteine, mercaptopurine, and mercaptopropionic acid, oligonucleotides, mercaptoundecanoic acid (MUA) and many others [41,42,136,137]. The ability of cytochrome c to communicate with these electrode surfaces is mainly based on electrostatic orientation of the heme site towards the electrode. The interaction with the modified surface does not dramatically influence the redox potential of cytochrome c which indicates that the protein keeps its native configuration and the modifier does not bind to the heme. However, the heterogeneous electron transfer rates for cytochrome c in solution vary over a large region depending on the monolayer, i.e., $k_{het} = 4 \times 10^{-6}$ cm s^{-1} (MUA) and $k_{het} = 4 \times 10^{-3}$ cm s^{-1} (cysteine).

Comparable stable electrodes have been developed with ω-terminated carboxyalkylthiol promoter layers that enable a direct and reversible electron transfer to cytochrome c [41,42]. At low ionic strengths cytochrome c is strongly adsorbed to the carboxylic head groups. For example, the formal potential of cytochrome c at MUA modified gold electrode is 30 mV (vs. Ag/AgCl/1 M KCl) in 10 mM phosphate buffer at pH 7.0 [43]. Compared with the $E^{0\prime}$-value in solution, a negative potential shift of about 40 mV was observed as was also indicated by other authors [41]. The same chemistry can be used with thick-film array electrodes (Fig. 2.3) and microfabricated thin-film interdigitated electrodes [138]. The latter offer an easy characterization of the dynamic properties of the protein on the modified surface.

The interactions with the surface and reaction kinetics have been studied in detail using various techniques, such as voltammetry, electroreflectance measurements and surface enhanced Raman spectroscopy [139,140]. For monolayers on gold assembled from long chain thiols of the structure HS–$(CH_2)_n$–COOH (with $n > 9$) the interfacial electron transfer rate exponentially decreases with chain length and the tunnelling parameter β is in the

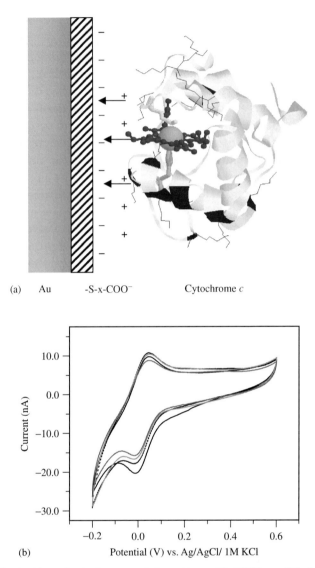

Fig. 2.3. (a) Illustration of cytochrome c interaction with MUA-modified electrode and (b) cyclic voltammograms of a thick-film array gold electrode with immobilized cytochrome c. Electrode modifier: mercaptoundecanoic acid (MUA).

order of $\beta = 1.09$ per CH_2-unit [41,42]. This shows that here the rate of electron tunnelling through the alkanethiol layer is limiting the overall process of electron transfer between cytochrome c and the gold electrode. For modifiers of shorter chains the electron transfer is approaching saturation.

Experimentally determined rates are up to $3000\ s^{-1}$. The strength of interaction is limiting the rate of rotation of the protein at the monolayer, and therefore, influences the reaction kinetics. Dilution of the carboxylic acid terminated thiols with methyl or hydroxyl terminated alkanethiols (similar chain length) enhanced electron transfer rates by factors of 5–6 [141,142]. Also for the mixed SAM of mercaptopropionic acid/mercaptopropanol the electron transfer is much faster [55]. At 1:3 mixed HS–$(CH_2)_{10}$–COOH/ HS–$(CH_2)_{11}$–OH (MUA/MU) Ge and Lisdat [143] found a $k_s = 70\ s^{-1}$ compared to $k_s = 23\ s^{-1}$ for only MUA. Covalent coupling of cytochrome c to the surface reduced the rate by a factor of 2–6 and shifted the formal potential slightly toward negative potentials. The covalently coupled cytochrome c on a tight layer has, however, advantages for the biosensing application as will be discussed below.

Alternatively cytochrome c was entrapped in organic and inorganic polyelectrolytes. It was reported that cytochrome c spontaneously incorporates into the cation exchanger Eastman AQ [144] and sodium montmorillonite [145]. In the latter case the electron transfer was quasi-reversible with a formal potential of 15 mV (vs. Ag/AgCl/1 M KCl). Interestingly, the protein was still mobile as indicated by the observation that the peak current grows linearly with the square root of the scan rate. Much higher loading of electroactive protein is possible when multilayers of cytochrome c and the weak polyelectrolyte polyaniline sulfonate are formed on gold electrodes using the LBL-approach [146]. In this case the electron transfer is proposed to involve protein–protein self-exchange. From a practical point of view the highest loading of cytochrome c with fast electrode reaction will be the ultimate goal for a sensitive sensor.

An interesting aspect of cytochrome c electrochemistry is the possibility to couple it with other redox reactions such as enzymes [25,26,43] or short-lived species [25,147]. Analytical application of this bioelectrochemical process has been reported for the determination of superoxide anion, hydrogen peroxide and antioxidants (see below).

2.5.1.2 Cytochrome c'
Cytochrome c' is a dimeric heme protein which has been found in phototropic, sulphur oxidizing and denitrifying bacteria [148]. In some species cytochrome c' dimers show the tendency to dissociate upon ligand binding. Cytochrome c' from *Chromatium vinosum* and *Rhodocyclus gelatinosus* are typical examples. The heme moiety is bound to two cysteines, similar to the binding found in mitochondrial cytochrome c. However, the iron atom of cytochrome c' is mainly high-spin pentacoordinated. The sixth ligand site of ferrous

cytochrome c' is deeply buried within the protein and closely surrounded by aromatic and hydrophobic amino acid residues restricting access to exogenous compounds except to small molecules such as carbon monoxide and nitric oxide [150].

The quasi-reversible redox responses of cytochrome c' from *Rhodospirillum rubrum* in solution was observed on a surface-modified gold electrode [149]. Progress in NO-sensor development was achieved after adsorption of cytochrome c' from *Chromatium pinosum* and *Rh. gelatinosus* on a mercaptosuccinic acid modified gold electrode [56]. The adsorption was found to be dominated by non-ionic interactions since adsorption could be observed at higher ionic strength (100 mM) and was facilitated at low pH (5-6) where both the promoter and the protein become less charged. This electrode showed a quasi-reversible, diffusionless electrochemical redox behaviour of the surface adsorbed protein with a formal potential of about -132 mV vs. Ag/AgCl. The heterogeneous electron transfer rate constant of adsorbed cytochrome c was determined in the range of 30–50 s^{-1} [151]. The charge transferred was calculated from the cyclic voltammogram to be 4.5 μC/cm^2 for the total amount of bound protein of 16 pmol/cm^2. These values support that both parts of the cytochrome c' molecule dimer are electroactive.

When the cytochrome c' functionalized electrode was exposed to a *nitric oxide* an increase in the reduction current was observed due to coordination and subsequent oxidation followed by an electrocatalytic reduction of the biomolecule. This can also be used in an amperometric mode for NO-biosensing [56,151].

2.5.1.3 Myoglobin
Myoglobin is an oxygen-carrying protein located in the muscles of vertebrates. The 3D structures of a large number of vertebrate Mbs in various states are known. It is a monomeric heme protein of about 16 kDa [152,153]. The heme group is located in a crevice of the molecule and polar propionate side chains of the heme are on the surface. Inside the molecule the heme group is surrounded almost entirely by nonpolar residues except two histidines. The native state of myoglobin features a proximal histidine bound to the fifth heme iron coordination site and the other axial heme iron position remains essentially free for O_2-coordination. A distal histidine stabilizes a water-ligand to ferric iron and in the ferrous state this conformation stabilizes bound oxygen as the sixth ligand and suppresses autoxidation of the heme [154]. The outside of the protein has both polar and apolar residues with an isoelectric point of 7.0.

Direct electrochemistry of myoglobin has been of interest for a long time and numerous papers have been published and reviewed [1]. Among the electrode modifications are various films on carbon electrodes, particularly surfactants, clay and polyelectrolyte films in which Mb exhibits direct electrochemistry for the redox transition of the heme group [155–159]. The role of surfactants in this process is suggested to be partly orientation of the protein and inhibition of oligomeric adsorbates. By alternating adsorption of protein and surfactant or polyelectrolyte multilayers with higher protein loading can be created. The formal potential of myoglobin is sensitive to the nature of the film material used and also to the presence of heme ligands. For example, for Mb-films on carbon electrodes values were determined between -0.1 and -0.21 V vs. SCE for the anionic surfactants didodecyldimethyl-ammoniumbromide (DDAB), sodiumdodecyl sulphate (SDS) or cetyltri-methylammonium bromide (CTAB) [155,159,160], and -0.380 vs. SCE for montmorillonite [59,156] and -335 mV for polyacrylamide [158] (values in pH 7.0 buffers). Intermediate values are found for combinations of surfactant and clay. For myoglobin in solution a formal potential of -0.297 V vs. Ag/AgCl was reported [20]. A negative shift has also been found for myoglobin in bentonite and surfactant films [161]. The rates of electron transfer of myoglobin adsorbed to clay are fast at high ionic strength of the background solution. In 200 mM Tris buffer k_s values of 450 s^{-1} were estimated [59]. The fast electron transfer rate which was observed although the protein is naturally not an electron transfer protein and the variation of the formal potential compared to the value in solution strongly suggested differences in the conformation due to the interaction with the modifier. From studies on His64 mutants of sperm whale myoglobin Van Dyke et al. [154] concluded that a partly hydrophobic environment of a surfactant can disrupt the hydrogen bonding network of His64 and thus lower the reorganization energy for the electron transfer.

Myoglobin on electrodes possesses pseudo-enzymatic activity towards O_2, H_2O_2, nitrite and organohalides. Under aerobic conditions oxymyoglobin (MbO$_2$) is also active in oxygenation reactions. Thus peroxide reduction, styrene oxygenation and dehalogenation of pollutants can be followed electrocatalytically on myoglobin-modified electrodes. Furthermore, NO can bind to ferric Mb in the absence of oxygen and MbO$_2$ reacts rapidly and stoichiometrically with nitric oxides [59,158,159].

2.5.1.4 Haemoglobin
Haemoglobin (Hb), functions physiologically in the storage and transport of molecular oxygen in mammalian blood. The molecule is nearly spherical with

a diameter of 55 Å with the four chains packed close together in a tetrahedral array. Human Hb is a 64,456 kDa protein and comprises four subunits (two alpha and two beta chains), each of which has a heme within molecularly accessible crevices [162]. The heme is five-coordinated (four nitrogen ligands and one histidine), while the sixth site is available for binding oxygen.

The large three-dimensional structure of Hb, the resulting inaccessibility of the heme centre, and the subsequent electrode passivation due to protein adsorption make it difficult to obtain DET between Hb and electrodes. Numerous efforts have been made to improve the electron transfer characteristics by using mediators or promoters [163]. The most efficient way is to modify preferentially carboneous electrodes with polymeric and membrane forming films, for example, surfactants [38], clay [57], and composite films of surfactant and bentonite. The redox potential of immobilized haemoglobin was determined between -100 and -380 mV vs. SCE.

We could demonstrate that haemoglobin adsorbs to colloidal clay nanoparticles prepared from sodium montmorillonite and a Pt-polyvinyl alcohol colloid onto glassy carbon electrodes [57] Hb displayed a quasi-reversible one-electron transfer process with an $E^{0\prime} = -370$ mV (vs. Ag/AgCl), $\Delta E_P = 130$ mV and $k_s = 70$ s^{-1}. The linear dependence of the peak currents from the scan rate is indicative of a surface process (Fig. 2.4a).

The addition of carbon monoxide to the high-spin ferrous Hb yields the heme ferrous-CO complex, which has a slightly higher redox potential. The affinity to CO is about 200 times as high as for oxygen. Therefore, the observation of a positive potential shift after the introduction of CO was used as indication that the DET of Hb was indeed from its heme iron [57]. Other authors exploited the same reaction as an indicator for CO. Although Hb does not play a role as an electron carrier in biological systems, it has been shown to possess enzyme-like catalytic activity [164–166]. Common to all Hb-modified electrodes is their ability for electrocatalytic oxygen and peroxide reduction (Table 2.2). Furthermore, ferrous oxygenated Hb (HbO$_2$) reacts with NO that yields nitrate and MetHb, which can be electrochemically reduced and then either bind again oxygen or NO (Fig. 2.4b). These reactions were explored for peroxide and NO-sensors (see Sections 6.2 and 6.3).

2.5.2 Heme enzymes

2.5.2.1 Cytochromes P450

Cytochromes P450 (P450) form a large family of heme enzymes that catalyze a diversity of transformations including epoxidation, hydroxylation and heteroatom oxidation. The enzymes are involved in the metabolism of many

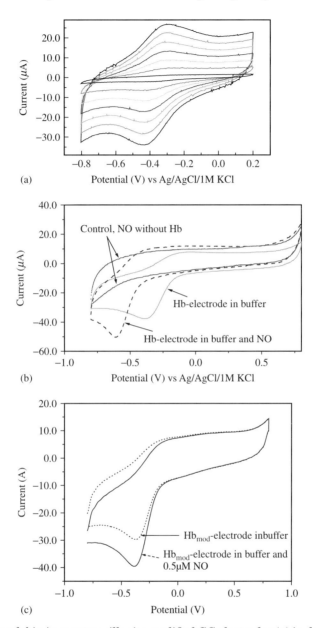

Fig. 2.4. Haemoglobin in montmorillonite modified GC electrodes (a) in degassed 0.1 M K-phosphate buffer, pH 7.0 and (b) effect of NO on the redox behaviour of heme iron. and (c) effect of NO after blocking cysteines with iodoacetate at a scan rate of $2\,\mathrm{V\,s^{-1}}$. ((b) Reproduced from Ref. [57] with permission of Springer-Verlag, Heidelberg.)

drugs and xenobiotics and are responsible for bioactivation. Furthermore, their appearance indicates the presence of a respective class of compounds as P450 enzymes are induced on their exposure [167–169]. Since the first three-dimensional structure of the bacterial P450cam was elucidated by Poulos [170] several other structures have been resolved, including P450 2B4 [171,172].

The active centre is the iron-protoporphyrin IX with an axial thiolate of a cysteine residue as fifth ligand iron (Fig. 2.2). Resting P450 is in the hexa-coordinate low-spin ferric form.

The overall reaction of substrate hydroxylation of a monooxygenase is insertion of one atom of an oxygen molecule into a substrate RH, the second atom of oxygen being reduced to water and consumption of two reducing equivalents under the formation of ROH:

$$RH + O_2 + 2e^- + 2H^+ \rightarrow ROH + H_2O \tag{2.1}$$

The electrons are delivered by flavoproteins or ferredoxin-like proteins and NAD(P)H in a complex electron transfer chain. The most generally accepted mechanism for substrate hydoxylation by P450 includes the following steps although some details remain still unsolved [173]: substrate binding to the hexa-coordinate low-spin ferric enzyme excludes water from the active site, which causes a change to the 5-coordinate high-spin state. The decrease of polarity is accompanied by a positive shift of the redox potential of about 130 mV that makes the first electron transfer step thermodynamically favourable. The transfer of one electron from a redox partner reduces the ferric iron to the ferrous enzyme. This can now bind molecular oxygen forming a ferrous-dioxy ($Fe^{II}-O_2$) complex. The second electron is transferred along with a proton gaining an iron-hydroperoxo ($Fe^{III}-OOH$) intermediate. The O–O bond is cleaved to release a water molecule and a highly active iron-oxo ferryl intermediate. This intermediate abstracts one hydrogen atom from the substrate to yield a one-electron reduced ferryl species ($Fe^{IV}-OH$) and a substrate radical. Then follows immediately enzyme-product complex formation and release of product ROH to regenerate the initial low-spin state.

P450s are highly relevant to the bioanalytical area. The main problem, however, is the complexity of the monooxygenase systems which require flavoproteins or ferredoxin-like proteins and NAD(P)H as electron supplying components. Furthermore, the low redox potential along with the need for oxygen makes an unfavourable working potential necessary at which oxygen reduction also takes place.

For a potential application of cytochromes P450 in bioreactors or biosensors an interesting alternative is to substitute the biological electron donors by artificial ones like electrochemical [73,74,77,174] or photochemical systems [175,176]. Mediated spectroelectrochemistry based on the use of antimony-doped tin oxide electrodes has been used to determine the influence of mutations on the redox potentials [177]. Although some electrochemical aspects of P450 were reported more than 20 years ago [20, 178] the direct, non-mediated electrochemistry of P450 is rather difficult to obtain. On unmodified electrodes enzymes tend to denature and to passivate the electrode. However, P450s are naturally involved in electron transport pathways of protein redox partners, which require specific docking sites. Therefore, electrical contact to P450-enzymes should be possible at suitable surface modifications of electrodes.

In recent years reversible one-electron transfer could be achieved with P450s assembled in or at biomembrane-like films, inorganic and organic polyion layers and surfactant modifiers.

For example, the enzyme has been immobilized on electrodes modified with sodium montmorillonite [72] and additional detergent [74], synthetic membrane film (didodecyldimethylammonium bromide) [75,179] and dimyristoyl-L-α-phosphatidylcholine film [129] and in thin protein-polyanion film on mercaptopropane sulfonate coated Au electrodes [180]. In most cases carboneous electrode material has been used, but in a few cases gold with self-assembled modifier layers and tin oxide electrodes were reported. However, on gold-electrodes modified with monolayers of short thiols, which were most successful for other heme proteins such as cytochrome c and peroxidase, the reduction of P450 is very slow [181], and therefore, not yet applicable for biosensor preparation. DET has also been observed using multilayer modified Au and pyrolytic graphite electrodes based on the use of alternating P450cam/polycation layers [222].

Reversible direct one-electron transfer has been reported for P450cam [73] and various mutants [254] at edge-plane graphite by using P450cam in solution. DET has been shown with different P450-types such as P450cam [72,75,129,179–181], P450 1A1 [182], P450 1A2 [77,183], P450 3A4 [76], P450 4A1 [184], P450 2B4 [74], P450 BM3 [185], P450cin [186] and P450scc [77].

The potential for the non-mediated first electron-redox transition in a number of P450s and at different electrodes is in the range of -450 to -100 mV vs. Ag/AgCl. At montmorillonite modified glassy carbon electrodes [72] reversible and very fast heterogeneous redox reaction of substrate-free cytochrome P450cam with a formal potential of -361 mV (vs. Ag/AgCl) was obtained. The heterogeneous electron transfer rate constants reached values as high as $152\ s^{-1}$ for P450cam comparing to rates between 27 and $84\ s^{-1}$ reported for the transfer of the first electron from putidaredoxin to

P450cam [187,188]. This similarity suggests that the negatively charged clay [189] obviously mimics the electrostatics of the natural redox partner putidaredoxin and may hold the P450 in a productive orientation. In this orientation the active site of the adsorbed P450cam is still accessible for small iron ligands like CO and dioxygen and also the larger metyrapone P450 indicated by a positive shift of the formal potential. The apparent surface coverage of the electroactive P450cam was calculated to be about 3.54×10^{-12} mol cm^{-2}, which was about 35% of the total amount of immobilized P450. The observed linearity of the peak current with the scan rate is characteristic for a surface process.

The formal potential of substrate-free P450cam is approximately 160 mV more positive than the solution value, but close to the value of the camphor-bound species in solution ($E^{0\prime} = -407$ mV vs. Ag/AgCl 1 M KCl) [174]. Thus the interaction with the matrix may force displacement of solvent in the local environment of the heme or conformational changes. Changes of the secondary structures, however, were not identified with IR-spectroscopy [72].

In a spectroelectrochemical experiment at modified gold for substrate-bound P450cam an $E^{0\prime}$ of -373 mV was found. The surface interaction and direct electrochemical transformation does not affect the enzyme structure as was confirmed spectroscopically. Both, upon direct electrochemical reduction and upon ligand binding the spectral changes clearly indicated the native state of P450cam during reversible reduction and oxidation [181].

The liver microsomal phenobarbital induced P450 2B4 has also been incorporated in montmorillonite on glassy carbon electrodes [74]. In contrast to P450cam this enzyme has a flavoenzyme as redox partner and does not need an iron–sulfur helping protein for delivery of electrons. Using cyclic voltammetry at low scan rates a reduction peak is observed at around -430 mV (vs. Ag/AgCl), which disappears at higher rates. The electron transfer reaction is obviously very slow. However, this process is enhanced in the presence of a non-ionic detergent such as Tween 80 (Fig. 2.5). P450 2B4 is a membrane bound enzyme and detergent is needed to monomerize P450 2B4 [190] as was confirmed also by AFM-studies. From the cyclic voltammograms the amount of electroactive protein of 40.5 pmol/cm^2 was calculated. Cyclic voltammetry also demonstrates a reversible one-electron surface redox reaction with a formal potential of about -295 vs. Ag/AgCl 1 M KCl and a heterogeneous electron transfer rate constant of 80 s^{-1}. As in many of the published cases the formal potential of P450 determined by the heterogeneous redox reaction is more positive than the redox potential in solution. Furthermore, the values $E^{0\prime}$ in films containing P450cam show a distribution ranging from -231 to -350 mV vs. Ag/AgCl 1 M KCl [129]. A positive shift of

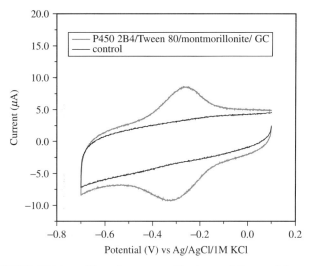

Fig. 2.5. CV of P450 2B4 in a film of montmorillonite and Tween 80 on GC-electrode.

the redox potential may be indicative of low to high spin-state conversion ascribed to strong interaction of P450 with surfaces [139]. The positive shifts of the redox potential are generally observed when water is excluded from the heme pocket as in the case of camphor binding [191,192], and therefore, we suggested that the adsorption process leads to a dehydration of the P450 structure. A similar effect of detergent can be discussed.

The studies of direct heterogeneous electron transfer have been carried out in most cases using cyclic and square wave voltammetry. In these studies the first of the two electrons required for the catalytic reaction has been transferred although the authors do not see the shift of the reduction potential upon substrate addition as has been reported in Ref. [73] and is known for the reaction in solution. In all cases catalytic oxygen reduction is observed but only rarely could catalytic substrate conversion be achieved.

P450cam predominantly catalyzes the regio- and stereo-specific hydroxylation of (1R)-camphor to exclusively 5-exo-hydroxycamphor. Other compounds than camphor, such as compounds of environmental and industrial interest have also been identified as substrates for P450cam. During catalysis electrons are transferred from NADH to P450cam through putidaredoxin reductase (PdR) and putidaredoxin (Pdx). The negatively charged group of Asp38 in Pdx forms a salt bridge with Arg112 in the positively charged patch (Arg112 and Arg109, Arg79) in P450cam to shuttle electrons between PdR and P450cam [193]. Thus a ferredoxin in DET contact to an electrode may deliver the reducing equivalents to P450. On indium–tin oxide electrodes P450cam conducts camphor hydroxylation mediated by putidaredoxin [194] and dehalogenation of haloalkanes with spinach ferredoxin in the presence of polylysine as promoter proceeds [195].

Using only the P450-component, electrochemically driven epoxidation of styrene was observed at polyion multilayers containing P450cam [180] or P4501A2 [183]. The mechanism involves a peroxide activated reaction step. Epoxidation is initiated by a single electron reduction of the heme. In an electrocatalytic oxygen reduction peroxide is formed, which activates the enzyme for styrene epoxidation. The role of peroxide has been proved by lack of styrene oxide formation in the presence of catalase [78,183]. Acceleration of styrene epoxidation and dehalogenation of hexachloroethane, carbon tetra-chloride and other polyhalomethanes was successful with mutated P450cam [177,196,197].

Human P4503A4 (quinine 3-monooxygenase) is electrocatalytically active at a polycationic film-loaded gold electrode prepared with polydimethyldiallyl-ammonium chloride adsorbed to mercaptopropane sulfonate activated gold [76]. This immobilization causes a drastic anionic potential shift of enzyme to about $+98$ mV (vs. NHE) indicating conformational changes. For drug sensing dealkylation of verapamil, midazolam, progesterone, and quinidin was followed. Addition of millimolar amounts of verapamil or midazolam to oxygenated solution increased the reduction current. In the case of the human P4503A4 peroxide had only a minor effect. Furthermore, product analysis after electrolysis at -500 mV under aerobic conditions confirmed the demethylation and dealkylation of verapamil at a rate of about $4-5$ min^{-1}.

Attempts have been made to contact P450-enzymes to electrodes by introducing electroactive bridges covalently coupled to the protein [198,254]. Such redox relay has been introduced at specifically selected sites generated by protein engineering or randomly. In the case of the P450 2B4 the native redox partner is a flavoenzyme. Therefore, riboflavin was proposed to be a possible electron donor for P450 2B4. Riboflavin was covalently attached to the enzyme and immobilized on rhodium–graphite screen-printed electrodes [198]. At a constant potential of -500 mV (vs. Ag/AgCl) substrates such as aminopyrine, aniline, and 7-pentoxyresorufin were converted.

We succeeded in developing biosensors based on mediator-free P450 2B4 catalysis by immobilizing monomerized P450 2B4 in montmorillonite [74]. When aminopyrine was added to air saturated buffer solution, there was an increase in the reduction current. In chronoamperometry the detection limit is 1 mM aminopyrine (Fig. 2.6) and 1.2 mM benzphetamine. The reaction was inhibited by metyrapone. This indicates that P450 2B4 possesses catalytic

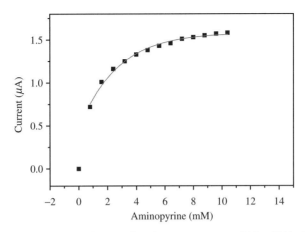

Fig. 2.6. Dependence of the catalytic reduction current at −500 mV (Ag/AgCl/1 M KCl) on aminopyrine in 50 mM phosphate buffer, 50 mM KCl, pH 7. The sensor contains P450 2B4, tween 80 and montmorillonite.

activity in the presence of substrate. Further evidence was delivered by product analysis. After 1 h of controlled potential electrolysis at −500 mV vs. Ag/AgCl formaldehyde was measured. The apparent catalytic rate related to the amount of electroactive protein is $k'_{cat} = 1.54$ min^{-1} which is comparable to the value $k_{cat} = 3.5$ min^{-1} of the microsomal system [77].

2.5.2.2 Catalase

Catalase (EC 1.11.1.6) is a heme containing oxidoreductase that acts on peroxides liberating oxygen and water. The enzyme is a very fast biocatalyst, i.e., *Pichia pastoris* catalase has a turnover number of 3×10^9 H_2O_2 s^{-1} [199]. The enzyme consists of multiple subunits containing each heme as active site.

The enzyme catalyzes the disproportionation of H_2O_2. In deoxygenated solutions a slow DET occurs with $k_s = 3$ s^{-1} between catalase and carbon electrodes. In oxygenated solutions both oxygen and peroxide reductions proceed [71]. Another paper [200] reported the assembly of polyelectrolyte encapsulated catalase microcrystals onto gold electrodes to form enzyme thin films for biosensing. The polyelectrolyte coating around the microcrystals provided a regular surface charge, thus facilitating the stepwise film growth, and effectively prevented catalase leakage from the assembled films. DET between catalase molecules and the gold electrode was achieved without the aid of any electron mediator. In pH 5.0 phosphate buffer solution, the apparent formal potential of catalase was -0.131 V (vs. Ag/AgCl). For peroxide measurement it is more convenient to use it in combination with an oxygen electrode.

2.5.2.3 Peroxidase

The majority of peroxidases are glycoproteins of 20–70 kDa molecular weight, which contain ferric protoporphyrin IX as a prosthetic group [201]. They are ubiquitous in nature and named after their sources, e.g., horseradish, tobacco, and soybean peroxidase. In some rare cases they were also named after their substrates such as cytochrome c, lignin and chloroperoxidase. Some of the peroxidases are well investigated, e.g., the intracellular (class I) cytochrome c peroxidase from baker's yeast is the first peroxidase of which a high-resolution crystal structure has been obtained [202]. More recently, high resolution structures of some extracellular plant peroxidases (peroxidase class III), e.g., peanut peroxidase [203], horseradish peroxidase [204], and class II peroxidases, namely, lignin peroxidase [205] and *Arthromyces ramosus* peroxidase [206] have been resolved.

Peroxidases (EC 1.11.1.7) catalyze the reduction of hydrogen peroxide or alkyl hydroperoxides while a wide range of substrates act as electron donors. The mechanism of peroxidase catalyzed reactions has been intensively studied (see reviews [201,207–210]). The kinetics of catalysis reveals a ping–pong mechanism. In the first step, the peroxide binds to a free coordination site of iron (Fe^{III}) and is reduced to water (or an alcohol ROH) in a rapid two-electron process, whereby compound I is formed as the stable primary intermediate:

$$POD(Fe^{III})P + ROOH \xrightarrow{k_1} POD(Fe^{IV}=O)P^{\cdot+} + ROH \tag{2.2}$$

$$POD(Fe^{IV}=O)P^{\cdot+} + AH_2 \xrightarrow{k_2} POD(Fe^{IV}=O)P + AH^{\cdot} \tag{2.3}$$

$$POD(Fe^{IV}=O)P + AH_2 \xrightarrow{k_3} POD(Fe^{III})P + AH^{\cdot} \tag{2.4}$$

Compound I is the oxy-ferryl species (($Fe^{IV}=O)P^{\cdot+}$) formed with one oxygen atom from the peroxide, one electron from iron and one electron withdrawn from the heme group to form a porphyrin π cation radical. In the next step the porphyrin π cation radical is reduced to compound II (($Fe^{IV} = O)P$) which is subsequently reduced to the resting ferric enzyme (Eq. (2.4)). This reaction needs electron donor substrates AH_2.

Formation of compound I is fast with an apparent second order rate constant in the order of 10^6–10^7 M^{-1} s^{-1} [207]. Furthermore, the oxidation of reducing substrates by compound I (Eq. (2.3)) is 10–100 times faster than by compound II (Eq. (2.4)). Therefore, in the presence of excess of peroxide and AH$_2$-limitation, the overall reaction is controlled by k_3. Depending on the nature of the peroxidase, a number of compounds reduce the higher oxidation state intermediates of the enzyme back to its native form. With the most commonly used HRP, virtually any organic and inorganic reducing agent may react in this way. These substrates bind at approximately the same site in the vicinity of the heme, in an orientation perpendicular to the heme plane and interact mainly with the exposed part of the heme [211]. Among the best electron donors are hydroquinone and o-phenylenediamine with reaction rates of 3×10^6 and 5×10^7 M^{-1}s^{-1} [212]. HRP is active in a number of organic solvents.

For the specific case of an appropriately immobilized peroxidase on an electrode electrons may be delivered by the electrode. The reduction of compound I (Eqs. (2.3) and (2.4)):

$$POD(Fe^{IV}=O)P^{\cdot+} + 2e^- + 2H^+ \xrightarrow{k_s} POD(Fe^{III})P + H_2O \qquad (2.5)$$

Thus the electrode can be considered as substrate of the enzyme and peroxide is electrocatalytically reduced. For peroxidases of different sources with the majority being HRP reaction (2.5) has been demonstrated (see Tables 2.3 and 2.4) (for review see Refs. [8,255]). In most publications carboneous material [18,33,96,217,256] and gold [46,213,216,218] was used. The rate constant for the heterogeneous electron transfer of native HRP on graphite and modified gold electrodes is 1–2 s^{-1} and can be enhanced when the non-glycosylated HRP is applied [46]. The glycosylation obviously restricts the DET. This is also supported by the observation that native HRP bound via the sugar moiety to boronic acid activated gold surface does only catalyze peroxide reduction in the presence of soluble mediators [230]. Non-glycosylated (recombinant wild-type) HRP adsorbed on graphite reached values of 7.6 s^{-1}. The rate of the recombinant wild-type HRP $k_s = 18$ s^{-1} [213] is higher at freshly cleaned gold electrodes and can be further increased when terminal His-tags are introduced to the protein. Also the catalytic rate (k_1) is higher for the recombinant enzymes adsorbed on gold compared to graphite electrodes and native HRP. Systematic studies on how the DET properties of HRP can be influenced involved electrode material and pH [214], single point mutations and glycosylation [215], cysteine mutants and introduction of a (His)6-tag [46,47,213]. Further investigations include also other plant and fungal peroxidases adsorbed onto graphite and modified gold electrodes [98,100,102,220]. The transfer of electrons from a SAM-modified gold electrode to tobacco peroxidase is much faster than to HRP. The most effective DET of the native enzyme was reported for sweet potato peroxidase where 91% efficiency of DET was reached [46]. Higher efficiency was demonstrated for the HRP active site mutants Asn70Val and Asn70Asp, which showed an almost 100% effective DET electron transfer. This behaviour could lead to more selective *peroxide biosensors*. A method and a model of kinetics of the bioelectrochemical process can be found in Ref. [219].

2.5.2.4 Quinoproteins, quinohemoproteins and flavohemoproteins
Due to their analytical relevance attempts are being made to exploit oxygen and NAD(P)H independent dehydrogenases. Many of them contain more than one subunit, and at least one subunit consists of a heme and another flavin or a quinone cofactor (Table 2.3). Usually the substrate reacts at the flavin or quinone cofactor and the electrons are then transferred by an intramolecular electron transfer to the heme site of the enzyme. In vivo the heme group donates the electrons to an external (protein) redox partners [221]. The heme is accessible for a large redox partner, and therefore, this class of enzymes can be coupled to an appropriately treated electrode. This has been discussed in several reviews [6,8,15,25]. A limited number of quinoenzymes show DET mainly at carbon and modified gold electrodes. The direct electrochemistry at unmodified carbon material requires in most cases the presence of soluble promotors, such as polyamines, Mg^{2+}, or aminoglycosides depending on the nature of the enzyme. The approach is applicable to other flavoheme and quinoheme enzymes.

The majority of the electrochemical investigations of quinoproteins focus on the understanding of the mechanism and biological function of these enzymes. The enzymes find application in mediated biosensors for the main substrates of these enzymes and the measurement of redox dyes and phenolic compounds. The few cases of DET of these enzymes for biosensor applications are presented here.

Mediatorless amperometric measurement of gluconate, alcohol and fructose was demonstrated in the simple configuration of a carbon paste electrode when gluconate dehydrogenase, alcohol dehydrogenase and fructose dehydrogenase, respectively, were immobilized behind a membrane [6].

Bioelectrocatalytic D-gluconate oxidation was obtained with gluconate dehydrogenase modified glassy carbon electrodes [125,126]. *Gluconate dehydrogenase* (EC 1.1.99.3) from *Pseudomonas fluorescens* is a trifunctional membrane-bound enzyme of MW 130 kDa that catalyzes the oxidation of D-gluconate to 2-keto-D-gluconate. This process is measurable when gluconate is in the concentration range of $\mu M–mM$.

It has been proposed that the substrate conversion is at the flavin site and the electrons are then transferred via Fe–S– and the heme to the electrode (ubiquinone in the native system). The electron transfer at the heme moiety has been confirmed with another quinohemoprotein using electroreflectance [114]. Since then a number of examples supporting this model have appeared (Fig. 2.7).

Detailed mechanistic studies were also published by Gorton's group for the flavohemoprotein cellobiose dehydrogenase [8,26,109,110,225,226].

Cellobiose dehydrogenase (EC 1.1.99.18, CDH) is an extracellular flavo-heme-glycoprotein from white rot fungi [223,224]. It catalyses the oxidation of cellobiose and related oligosaccharides by a number of acceptors including cytochrome c. Typically, CDH is a monomeric protein with a molecular mass of about 100 kDa and an acidic pI of around 4. The three-dimensional structure of CDH has one 55 kDa domain carrying FAD and a second 35 kDa domain carrying the cytochrome b type heme. The distance between the two prosthetic groups is less than 15 Å, which is within acceptable limits for intramolecular electron transfer.

DET was obtained for CDH from different origins [255].

Initial and most detailed studies were made with CDH from *Phanerochaete chrysosporium*. The enzyme was immobilized by simple adsorption onto graphite electrodes and in the presence of cellobiose or lactose but without external mediator, a response current to cellobiose was registered providing indirect proof of DET [109,225].

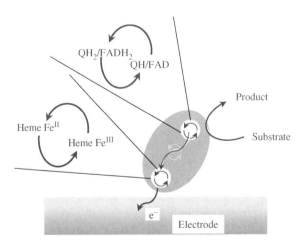

Fig. 2.7. Scheme of the electron transfer in multi-centre enzymes containing heme according to Refs. [7,8].

Some direct indications of electrochemical activity on graphite were deduced from square wave voltammetry experiments, which were not possible when using cyclic voltammetry [225]. However, on cystamine modified gold electrodes clear independent evidence of DET was shown [110,227]. To increase the local concentration of CDH the modified Au electrode was mounted with a dialysis membrane.

The behaviour of intact CDH was compared with the single separate domains. Interestingly, the voltammograms of the intact CDH and the heme domain were close to similar, whereas the flavo domain did not reveal any electroactivity within the investigated potential range. However, in the presence of cellobiose a typical catalytic cyclic voltammogram could only be seen for the intact enzyme. Thus it was concluded that the electrode communicates with the heme domain and electrons flow from the catalytic flavo domain to the heme and further to the electrode. The catalytic current is very much pH-dependent because the intramolecular electron transfer between FAD and heme is blocked above pH 6 [110]. Further support for the electroactivity of the heme domain of intact CDH comes from spectroelectrochemical investigations of CDH in an aldrithiol modified gold capillary electrode [226].

D-*Fructose dehydrogenase* (EC 1.1.99.11) from *Gluconobacter industrius* is a 140 kDa membrane-bound quinohemoprotein with a PQQ and a heme c-containing subunit.

Bioelectrocatalytic properties were obtained for FDH at carbon and gold and platinum electrodes [111,113]. The catalytic oxidation current of FDH-modified carbon paste electrodes approached a maximum value at +0.5 V vs. Ag/AgCl. At this potential and under optimum conditions, i.e., pH 4.5, fructose can be measured between 0.2 and 20 mM in fruit juices. Most importantly the fructose sensor was insensitive to ambient oxygen.

Unfortunately at the working potential of +0.5 V reducing compounds such as ascorbic acid may interfere with the fructose signal in citrus juice, and therefore, the authors proposed for real sample measurements the introduction of an ascorbate oxidase anti-interference layer [111]. Another fructose sensor contains an ascorbate blocking phosphilipid coenzyme Q-film [99]. The electron transfer between enzyme and electrode is rate limiting, and therefore, higher sensitivity is achieved when additionally mediators are introduced as was demonstrated for a mediator containing polypyrrole matrix or a carbon paste with a $Os(bpy)_2Cl_2$ redox polymer [99].

Ethanol sensors were fabricated using the membrane-bound quinohemoprotein alcohol dehydrogenase (EC 1.1.99.8) from different sources and a variety of electrode materials such as C, CPE, PG, GC, Au, Ag, Pt, [6,15,114–116,119]. This *alcohol dehydrogenase* adsorbs strongly to surfaces with preferentially the heme domain faced to the electrode [114]. Although in a cyclic voltammogram there is hardly any redox wave visible catalytic oxidation current can be recognized in the presence of ethanol without the addition of artificial mediators. The K'_m value was between 1 and 1.5 mM with the highest current density at hydrophobic gold surface. This is obviously because alcohol dehydrogenase is a membrane-bound enzyme. Furthermore, this is also the reason why detergents can accelerate desorption from the surface. Ethanol as low as 10 μM can be measured [116] with QH ADH on modified gold electrodes.

Bacterial *methylamine dehydrogenases* (EC 1.4.99.3, MADH) are quinoheme or quinoproteins with covalently bound tryptophan tryptophylquinone (TTQ) or cysteine tryptophylquinone (CTQ). Some of them have CTQ and heme c-domains [233]. In the presence of cationic promotors such as spermidine or hexaamonium salts MADH/pyrolytic or modified gold electrodes respond to 10–200 μM methylamine [117]. Other amine oxidases (EC 1.4.3.6) contain copper and a quinonoic cofactor (topaquinone). This amine oxidase was adsorbed onto spectroscopic graphite and studied [123]. Catalytic currents appear on histamine addition. The copper depleted enzyme does not respond. Various biogene amines were indicated up to a concentration of about 200 μM [123].

Flavocytochrome c3 (EC 1.3.99.1) isolated from *Shewanella frigidimarina* is a unique fumarate reductase of 63.8 kDa MW in a single subunit composed of two domains. The active site is located in the flavin domain. The heme domain contains four c-type hemes, each with a bis–His axial ligation. It has been proposed that this domain is similar to cytochrome $c3$ from *Desulfovibrio desulfuricans*. On pyrolytic graphite (edge) electrodes in the presence of polymyxin the single redox centres were examined. Fumarate addition is followed by a catalytic current [108].

p-Cresolmethylhydroxylase (EC 1.17.99.1) from *Pseudomonas putida* is a tetrameric oxygenase of 115 kDa with every two units containing FAD and heme c have been shown to display (bio)electrocatalytic properties at edge-plane oriented pyrolytic graphite in the presence of aminoglycoside promoters [186]. In the presence of *p*-cresol and promoter the catalytic current showed a linear relationship to enzyme concentration up to 1.2 μM. The oxygenation of *p*-cresol to *p*-hydroxybenzaldehyde can be followed for concentrations up to 0.5 mM. The reaction does also proceed at gold electrodes with negative modifiers when a redox-inactive cation act as bridging molecule between the negatively charged surface and the negatively charged binding sites of the heme subunit.

Recently the proof of principle for a third generation sulphite biosensor could be delivered using the two-redox cofactor enzyme sulphite oxidase [121, 122]. *Sulphite oxidase* from chicken liver (EC 1.8.3.1) is a soluble mitochondrial enzyme. It is composed of a molybdopterin cofactor (MoCo)-containing subunit connected by a flexible polypeptide chain region to a second subunit with a b-type heme [231]. The 2e$^-$ oxidation of sulphite to sulphate occurs at the MoCo-site with concomitant reduction of Mo^{VI} to Mo^{IV}. The reducing equivalents are then passed on from the MoCo domain to the heme domain and from there to cytochrome c. The last electron acceptor can be replaced by an electrode. If this reaction is fast enough catalytic currents will flow in the presence of sulphite. Catalytic sulphite oxidation was demonstrated at pyrolytic graphite, mercapto-6-hexanol modified polycrystalline gold [121] and 11-mercaptoundecanol (MU) or MU/mercaptoundecanamine [120].

$$SO_3^{2-} + H_2O \rightarrow SO_4^{2-} + 2H^+ + 2e^- \quad (2.6)$$

In the absence of substrate at MU-modified Au electrode oxidation and reduction was measured which corresponded to the Fe^{II}/Fe^{III} redox couple of the heme site with $E^{0\prime} = -120$ mV [120]. The heterogeneous electron transfer rate constant is in the order of $k_s = 15$ s^{-1}. The oxidation peak current increases proportionally to sulfite concentrations between 10 and 100 μM. The current increase is approaching saturation at concentrations higher than 3 mM.

2.5.3 Copper proteins/enzymes

Copper proteins can be classified according to their copper centre into type I blue copper proteins including small blue proteins which act as electron carriers (azurin, plastocyanin) and blue oxidases containing type I and trinuclear copper (laccase, ascorbate oxidase, ceruloplasmin) and reduce oxygen to water, type II enzymes (CuZn–SOD, nitrite reductase), type III copper containing oxygenases (tyrosinase) and non-blue oxidases with two copper atoms also catalyzing oxygen reduction to water (cytochrome c oxidase).

For a number of copper proteins DET has been reported [63,232,281] and some of them were used to mediate enzyme reactions. Copper enzymes were investigated in DET contact (see below) and their ability to oxidize a large variety of substrates caused researchers to explore these enzymes for biosensing [10,239]. The analytically most interesting application of the DET to type I copper oxidases is the reagentless signal generation in binding assays [7,132].

Laccases (EC 1.10.3.2) are copper containing "blue" oxidases, which couple four one-electron oxidation processes of *p*-diphenols and catechols to the four-electron reduction of dioxygen to water. The structure, properties, catalytic mechanism, and application of laccases have been described in a number of recent articles [234–239,255]. Laccases are widely distributed, though not ubiquitously, in plants [240], fungi [241–242], and insects [243] but are apparently absent from higher organisms. Laccases are glycoproteins; the carbohydrate content is about 10–45% of the total molecular weight, which is 55–90 kDa for fungal laccases and 110–140 kDa for enzymes from plants. Some structures of laccases have been solved to date, e.g., from *Coprinus cinereus* [234] *Trametes versicolor* [245] and *Pycnoporus cinnabarinus* [246]. Typically, laccases are negatively charged with isoelectric points around 3.5. The substrate binding site is a small negatively charged cavity near the T1 copper site [245].

Most laccases are single chain "blue" copper proteins which contain four copper ions classified as T1, T2 and T3 [246]. The type-1 copper (T1) is EPR positive and shows an intensive absorption at ca. 600 nm, thus being responsible for the blue colour of the enzyme. The Type-2 copper (T2) binding site also contains a single copper ion, while Type-3 copper (EPR-inactive T3) consists of a tightly coupled Cu^{II} ion pair that is also coupled with T2 copper in a trinuclear arrangement. A reaction mechanism for laccase has been proposed based on studies of the reduction of laccase from *Rhus vernicifera* by hydroquinone and ascorbic acid [247]. The electrons are entered at the T1-site and are rapidly transferred to the trinuclear site. The T3 reduces oxygen to water via a tightly bound peroxide intermediate while T2 facilitates the breakage of the oxygen–oxygen bond in the latter.

The electron donor can be substituted by an electrode (Eq. (2.7)), thereby enabling mechanistic studies, estimations of ET rates and bioelectrocatalysis. The latter has currently gained particular interest for the development of biofuel cells [248,249] and virtually reagentless immunosensors [7,250].

$$O_2 + 4e^- + 4H^+ \rightarrow 2H_2O \tag{2.7}$$

The redox potential of the T1 Cu-site has been determined using potentiometric titrations with redox mediators for a large number of different laccases and varies between 410 mV vs. NHE for *Rhus vernicifera* [67] and 790 mV for laccases from *Polyporus versicolor* and *Coriolus hirsutus* [244,251]. The T2 and T3 sites have higher potentials [251].

Laccases from different sources have been coupled to carbon material and recently also to modified gold electrodes [64,67,105,257,258]. For example, laccase was coupled with carbodiimide to a SAM of mercaptopropionic acid gold. A midpoint potential of the immobilized laccase of 410 mV vs. NHE was observed consistent with the potential for the T1 site [67]. DET is most obvious from the electrocatalytic oxygen reduction (Eq. (2.7)). The inhibition by halides or azide is a clear evidence of the participation of laccase.

Furthermore, the DET to another blue oxidase *ascorbate oxidase* has been studied [64]. The $E^{0\prime}$ of the T1-site is close to the value of Rhus vernificera laccase and oxygen was catalytically reduced. Analytical application can be found in the electrochemical immunoassays described below.

2.5.4 Other enzymes

2.5.4.1 Superoxide dismutase

Superoxide dismutase (SOD, EC 1.15.1.1) is a scavenger of the superoxide anion, and therefore, provides protection against oxidative stress in biological systems [259]. Most SODs are homodimeric metalloenzymes and contain redox active Fe, Ni, Mn or Cu. The superoxide dismutation by SOD is among the fastest enzyme reactions known. The rate constant for CuZnSOD is $k = 2 \times 10^9 \, \text{M}^{-1} \text{s}^{-1}$ [260], FeSOD is about one order of magnitude slower. SOD has an overall negative charge at pH 7 but positive charges are found around the active site. FeSOD and CuZnSOD showed quasi-reversible oxidation and reduction at gold electrodes modified with mercaptopropionic acid [68] with formal potentials in the range of −154 and 47 mV (Ag/AgCl) and k_s-values around 35 and 65 s^{-1}, respectively. CuZnSOD gold electrodes adsorbed to cysteine-modified Au-electrodes had a formal potential of 60 mV [69,70]. The electron transfer was found to be associated with the redox reaction of the copper active site of SOD. Such electrodes were used as superoxide detectors.

2.5.4.2 Glutathione peroxidases

Phospholipid hydroperoxide glutathione peroxidase (PHGPx, EC 1.11.1.12) is a monomeric selenoprotein of 18 kDa. In contrast to all other glutathione peroxidases, PHGPx also reduces hydroperoxides from complex lipids like phospholipids and cholesterol even when the peroxides are in membranes. The catalytic activity is based on the redox reaction of the selenol of selenocysteine, which forms the catalytic centre together with tryptophan and glutamine [261]. The isoelectric point of pig heart PHGPx is around 8.2, which results in a slightly positive overall charge in neutral buffer solution [262].

The overall reaction is oxidation of glutathione at the expense of hydrogen peroxide, organic hydroperoxides, and fatty acid hydroperoxides, respectively, according to the following reaction:

$$2GSH + ROOH \rightarrow GSSG + H_2O + ROH \tag{2.8}$$

PHGPx is known to act on lipid peroxides in membranes and should, therefore, be particularly suited for surface reactions at electrodes. In a detailed study the surface properties of gold electrodes were varied to determine conditions for the interaction with the enzyme [112]. For this purpose glutathione and ω-alkylthiols with either terminal carboxylic (N-acetylcysteine or MUA) or methyl (dodecanethiol) group were used to obtain either a hydrophobic or a negatively charged layer. A positive surface charge was generated by adsorption of polyallylamine onto the MUA layer.

These modified electrodes did not create voltammetric signals in the presence of PHGPx alone or after incubation of PHGPx with hydrogen peroxide as measured between +500 and −600 mV. However, a distinct reduction current was induced by the addition of GSH to PHGPx preincubated with hydrogen peroxide. This was manifested by the appearance of a wave in the cyclic voltammogram at a potential of −280 mV (Fig. 2.8). This is in contrast to the behaviour of heme peroxidases.

The effect of GSH on the reduction current was studied chronoamperometrically at −350 mV.

On plain gold electrodes or on electrodes modified with GSH or N-acetylcysteine no reduction current was obtained in the presence of PHGPx. The current response was highest (at least 5-fold that of dodecanthiol or polyallylamine onto MUA) when the electrodes were modified with MUA. The adsorption of PHGPx on the electrode was verified by measuring

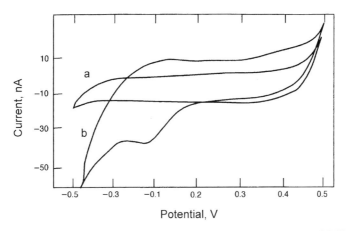

Fig. 2.8. Cyclic voltammograms of a modified gold electrode in 5 mM Tris–HCl + 0.1 mM EDTA, pH 7.6 in the presence of PHGPx + 100 μM H_2O_2 and: (a) 1 mM GSH; (b) 3 mM GSH at a scan rate of 50 mV s^{-1}. Reproduced from Ref. [112] with permission from Wiley-VCH.

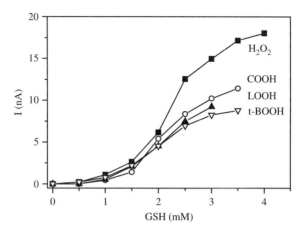

Fig. 2.9. Effect of GSH on reduction of hydroperoxides by PHGPx-modified gold electrode. (Adapted from [118]).

spectrophotometrically the adsorbed enzymatic activity, which was 27 mU/cm^2.

In the presence of PHGPx, the current showed sigmoidal dependence on the GSH concentration starting from 1.2 mM and reaching a plateau at about 4 mM (Fig. 2.9). Increasing amounts of hydrogen peroxide also increased the current. Incubation of PHGPx alone, hydrogen peroxide plus GSH in the absence of PHGPx or GSSG alone did not result in any signal indicating that the system works only in the presence of all three components (PHGPx, hydrogen peroxide, and GSH). Other hydroperoxides are also substrates of PHGPx. From inhibition studies with iodoacetate we know that the observed reduction current is related to the catalytic activity.

The reduction current observed with PHGPx must be derived from the formation of one of the intermediates of the catalytic cycle, because a measurable reduction current was only observed with the complete system, e.g., PHGPx plus hydroperoxide and GSH and the iodoacetate-inactivated enzyme did not yield any reduction current. The strict requirement for GSH to obtain a reduction current demonstrates that a glutathione-containing intermediate was reduced. The only feasible candidate for the reduction is the selenadisulfide (E–Se–SG) produced in reaction where the selenium is in the −1 oxidation state. Selenadisulfides have been shown to have a less negative reduction potential than disulfides like GSSG and could more easily be reduced at the electrode. It has thus been proposed that the reduction current results from the reduction of the intermediate E–Se–SG at the electrode, whereby the ground state enzyme is regenerated [118].

Third generation biosensors—integrating recognition

Apart from the analytical application of a PHGPx modified electrode for measuring hydroperoxides the findings might lead to the discovery of new biological roles for PHGPx. An active site, which easily transfers oxidation equivalents to an electrode, might also do so in a biological context. The data support the view that the main role of PHGPx is to catalyze redox reactions at surfaces not easily accessible by their tetrameric congeners.

2.6 SELECTED SENSORS

Sensors for metabolites were already introduced where the respective enzymes are presented, in particular glucose, ethanol, fructose, and gluconate sensors.

2.6.1 Superoxide sensor

Free radical biosensors may help to investigate the role of O_2^- and NO as cellular messengers by the direct, real-time measuring of free radical production directly as cell signal and in relation to stimuli to which the cell is exposed. Two types of biosensors have been developed. The first exploits the highly specific reaction of superoxide dismutase [69], while the other type is based on the cytochrome c reduction by O_2^- [53–55,143,147]. A further alternative uses protoporphyrin IX adsorbed to carbon material [263]. Here, however, peroxide is a strong interferent.

In most cases superoxide anion sensors contain cytochrome c in direct contact to gold electrodes, where cytochrome c is covalently coupled to succinimide activated surface modifiers (DSP and DTSSP, respectively), MPA/MP or MUA/MU (Table 2.5).

In most studies the reaction of xanthine oxidase (XOD) is exploited as a model superoxide generator:

$$\text{Hypoxanthine } 2O_2 + H_2O \xrightarrow{\text{XOD}} \text{uric acid} + O_2^{-\cdot} + H_2O_2 \xrightarrow{k_2} \text{uric acid} + O_2 + H_2O_2 \qquad (2.9)$$

$$O_2^{-\cdot} + \text{Cyt } c(\text{Fe}^{\text{III}}) \xrightarrow{k_3} \text{Cyt } c(\text{Fe}^{\text{II}}) + O_2 \qquad (2.10)$$

Two competitive processes are taking place relevant to the measurement. These are the generation of the superoxide and the dismutation (Fig. 2.10). As long as hypoxanthine and oxygen are not limiting the equilibrium between generation and dismutation will be established resulting in a stable concentration of superoxide in solution. Ge and Lisdat [143] found a

TABLE 2.4
Superoxide sensors

Electrode/Modifier	Cytochrome c coupling	Sensor parameter, detection range	Application/comment	Reference
Cytochrome c				
Au, N acetylcyteine	EDC	0–0.48 μM XOD 2.2 μA/min cm^{-2} μM	O_2^- produced by neutrophils	[51]
Pt PACE, ads. SDS	Adsorbed	58 M mm^2/A min	In brain tissue during hypoxemia	[278]
Au, DTSP	Covalent	0.4–2.5 μM		[277]
Au, DTSSP	Covalent	nmolar range 0.5 pA/pM	Real-time measurements in various cell cultures	[147]
Au, MUA	Adsorbed		Efficiency lower than with free cyt. c Peroxide may interfere	[43]
Au, MUA	EDC		Ischaemia/reperfusion, real-time in vivo	[53]
Au, MUA/MU	EDC	9 nA/cm^2 [XOD]$^{1/2}$	SOD-activity, 10–200 mU/ml	[54]
Au, MUA/MU	EDC	300–1200 nM	Calibration	[143,276]
Au, MUA/MU	EDC		Application to organic solvents Determination of antioxidants	[264]
Au, MPA/MP	EDC	tres 15 s		[55]
Au, MPA/MP	EDC, coimmob. XOD	10–800 mU/ml	SOD activity	[55]

HEMIN				
PG	Adsorbed	tres 10–15 s, 23 nA/cm^2 [XOD]$^{1/2}$	Calibration via XOD, responds also to peroxide 5 h stability	[263]
SOD				
Au, MPA	Adsorbed or EDC		CuZnSOD Amperometric spike observed when O_2^- was generated from 50 μM HX, 50 mU XOD	[68]
Au, cysteine	Adsorbed	LDL 5 nM	CuZnSOD	[69, 70]
Au, MPA	Adsorbed or EDC		FeSOD	[68]

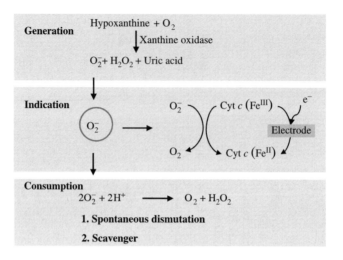

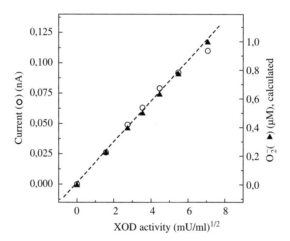

Fig. 2.10. (a) Illustration of the reactions while measuring superoxide anion with a cytochrome c electrode and (b) Superoxide anion measurement. Cytochrome c was immobilized on MUA/MU-modified gold electrode. The working potential is $+130$ mV. Superoxide was generated by XOD in buffer with 100 μM hypoxanthine. ((b) Reproduced from Ref. [143] with permission from Elsevier.)

dependence between the stationary current and the square root of XOD-activity and developed a model for the calibration of the sensor (Fig. 2.10B). In most reports, however, the current is related to superoxide generation rate but not to the actual concentration in the solution. Therefore, the sensors cannot be compared directly.

Mixed monolayers of MUA are most attractive for sensor construction, because they form not only a negatively charged surface for orientation of cytochrome c (Fig. 2.3A) but also a rather dense SAM and thus provide an efficient barrier for potentially interfering substances. The protein can be adsorbed at low ionic strength and further covalently linked to the carboxylic groups by activation with N-(3-dimethyl-aminopropyl)-N-ethylcarbodiimide (EDC) [43,151]. Superoxide radicals react quickly with immobilized cytochrome c. It has been reported that this reaction is faster when cytochrome c is immobilized to mixed monolayers of MUA/MU compared to a monolayer of only MUA with values of $k_{MUA/MU} = 3.1 \times 10^4$ and $k_{MUA} = 1.4 \times 10^4$ M^{-1} s^{-1}, respectively [143]. Superoxide reduces the immobilized cytochrome c, which subsequently oxidized at an appropriate electrode potential. At a constant potential of +150 mV the cytochrome c electrode is sensitive to superoxide in the nanomolar concentration range. Changes are indicated within seconds. The specificity of the sensor signal can easily be tested by additions of superoxide dismutase, which is the most effective natural scavenger of superoxide radicals. Well-prepared sensors showed an instantaneous decline of the sensor current to the baseline. Superoxide sensors were applied, for example, in various cell cultures and brain [147] and to follow in vivo the superoxide level during ischemia and reperfusion [53].

SOD-modified sensors also were demonstrated to respond to superoxide addition. Using either 3-mercaptopropionic acid [68] or cysteine [70] as a promoter on Au-electrodes superoxide sensors could be constructed where FeSOD and CuZnSOD in direct contact to the electrode acts as catalyst for the highly specific dismutation of O_2^- to O_2 and H_2O_2. Either Fe or Cu of SOD are oxidized and reduced (Fe^{II}/Fe^{III}; Cu^{II}/Cu^{I}) at the modified gold electrodes. Both, anodic and cathodic peak currents increase in the presence of O_2^-. At a potential of 300 or −200 mV O_2^--generation could be recognized with detection limits of 5 and 6 nM, respectively [70].

The mechanism is as follows:

Reduction:

$$Cu^{I}ZnSOD + O_2^- \rightarrow Cu^{II}ZnSOD + H_2O_2 \tag{2.11}$$

$$Cu^{II}ZnSOD + e^- \rightarrow Cu^{I}ZnSOD \tag{2.12}$$

Oxidation:

$$Cu^{II}ZnSOD + O_2^- \rightarrow Cu^{I}ZnSOD + O_2 \tag{2.13}$$

$$Cu^{II}ZnSOD \rightarrow Cu^{II}ZnSOD + e^- \tag{2.14}$$

Another important field of application of superoxide-sensors is the determination of antioxidative efficiency due to the increasing use of antioxidants in food products, drugs and cosmetics. This has been done simply by measuring the depletion rate of superoxide with the cytochrome c-sensor in the presence of antioxidants such as additives to cosmetic formulars [264], herbs and SOD [54]. The sensitivity of the sensor is critically limited by the amount of electroactive protein. One way to avoid this limitation is multilayer arrangement of cytochrome c on gold electrodes using polyaniline sulfonic acid as electrostatic glue between the cytochrome c layers [146].

2.6.2 Nitric oxide sensor

The elucidation of the potential biological role of NO has led to intensive work on techniques which can provide a fast, sensitive and selective detection of NO. The binding of NO to heme centres of *cytochrome c'* [43,56], *myoglobin* [59], and *haemoglobin* [57] is the basis for NO biosensors.

An amperometric NO detector was constructed with cytochrome c' from *Chromatium vinosum* immobilized to mercaptosuccinic acid modified gold electrodes [56].

If the *cytochrome c'* electrode is polarized at -220 mV, the amperometric detection of NO is feasible. The addition of the NO-liberating Glyco-SNAP-2 resulted in a current which increased proportionally up to 1000 mM Glyco-SNAP-2, which corresponds to a NO concentration of 500 nM (Fig. 2.11). The electrode reaction was found to be rather fast with response times of a few seconds. The interaction of NO with *cytochrome c'* in solution was also confirmed spectroscopically.

In another NO sensor the reaction of Hb is exploited. NO reacts rapidly with Hb in the presence and in the absence of O_2. The direct reaction between NO and ferrous oxygenated Hb (HbO_2) with a rate of $k = 3.7 \times 10^7$ M^{-1} s^{-1} [265] yields nitrate and MetHb, which can be electrochemically reduced and then again bind either oxygen or NO.

With almost the same rate ($k = 2.5 \times 10^7$ M^{-1} s^{-1}) [265] NO binds to ferrous deoxyhaemoglobin, which yields nitrosylhaemoglobin (HbNO).

$$Hb(Fe(II))O_2 + NO \rightarrow Hb(Fe(III)) + NO_3^- \qquad (2.15)$$

$$Hb(Fe(II)) + NO \rightarrow Hb(Fe(II))NO \qquad (2.16)$$

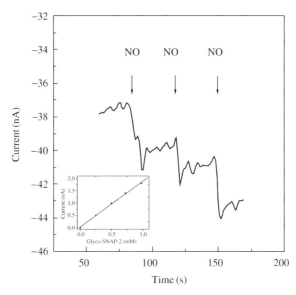

Fig. 2.11. NO-sensor. Amperometric response of a cytochrome c CVCP gold wire electrode on injection of NO. 5 mM Tris pH 7.45, purged with N2, E vs. AgrAgCl/1 M KCl, E = 220 mV, NO donor: 240 mM Glyco-SNAP-2. The inset shows the dependence of the current signal on the concentration of Glyco-SNAP-2. (Reproduced from Ref. [56] with permission from Elsevier.)

This is the basis of a spectroscopic technique for NO detection [266]. The electrochemical method could be an elegant alternative for the measurement of NO concentration. The voltammetric picture is changed after the introduction of NO (Fig. 2.4), whereas the control using an electrode modified with only clay, but without Hb, shows no effect. The original MetHb-reduction peak is clearly cathodically shifted, which is most likely the nitrosylhaemoglobin. This effect is only observed when the NO concentration exceeds 50 nM. At lower concentrations a slight effect on the peak current is obtained. The shift of the cathodic peak potential due to formation of nitrosylhaemoglobin (Eq. 2.16) was also observed and confirmed at a Hb/SnO$_2$-electrode [58].

The NO heme reactions predominate in the deoxygenated blood to form either Hb(III) or Hb(II)NO [267]. However, the presence of O$_2$ causes formation of S-nitrosohaemoglobin [268] leading to a shift of the Hb reduction potential. Although this reaction (rate constant: $6 \pm 2 \times 10^6$ M^{-2} s^{-1}) is not as fast as the oxidation of HbO$_2$ to MetHb it may interfere in the determination of NO by causing a shift in reduction potential [267]. To avoid this problem we used iodoacetate to block the reactive SH groups on the Hb. The modification does not change the formal potential but results in a slight increase in k_s to a value of 185 s^{-1}. A catalytic response on NO is observed with a detection limit of 0.5 µM NO [Bistolas and Wollenberger, unpublished].

Myoglobin-modified electrodes respond in a similar way as Hb electrodes. The modifier providing productive immobilization of Hb can also be used for Mb [72].

In the presence of small amounts of NO under nitrogen atmosphere around -870 mV a reversible redox couple appears that is most likely the Fe^{II}/Fe^{I}-couple. At higher NO concentrations a catalytic reduction current is generated around -700 mV. It is assumed that this is the result of N$_2$O formation.

$$Mb(Fe^{II}) + NO \rightarrow Mb(Fe^{II})NO \tag{2.17}$$

$$Mb(Fe^{II})NO + e^- \rightarrow Mb(Fe^{II})NO^- \tag{2.18}$$

$$Mb(Fe^{II})NO^- + e^- \rightarrow Mb(Fe^{I}) + NO^- \tag{2.19}$$

$$NO^- + 2NO \rightarrow N_2O + NO_2^- \tag{2.20}$$

$$Mb(Fe^{II})NO^- + O_2 \rightarrow Mb(Fe^{II}) + NO_3^- \tag{2.21}$$

In oxygenated solutions nitrate is formed (according to Eq. (2.21)) and Mb(Fe^{II}) liberated can either bind NO (Eq. (2.17)) or oxygen which have reaction orders of the same order of magnitude. Thus in oxygenated solutions sensitivity can be provided for micromolar NO concentrations [59]. In oxygen-free solutions Mb/clay-modified electrode can be applied for the NO detection in the nanomolar concentration range by evaluating of the shift of the formal potential of the Mb(Fe^{II})/Mb(Fe^{III}) redox couple. Upon exposure of the Mb electrode to higher NO concentrations (>2.5 µM) the reduction peak at -750 mV increased. The current at -0.75 V was found to be linearly dependent on NO concentrations in the micromolar range.

Nitrite is also catalytically reduced with Mb or Hb modified electrodes. A catalytic reduction peak current was found at -750 mV (pH 5.5). It increased with the concentration of NO^{2-} in the linear range of 0.17–3.2 mM [158].

2.6.3 Peroxide sensors

The measurement of peroxides, for example, has attracted the interest of clinical chemists and environmentalists. Hydrogen peroxide is among the reactive oxygen species generated during pathological processes, e.g., pulmonary diseases. Peroxide concentration in the upper nanomolar range is indicative of pathological situations. Thus very sensitive detectors are required. In other fields of applications such as photobleaching and disinfection higher concentrations of peroxides are present.

The reaction of peroxide with ferrous heme iron is the basis of electrocatalytic peroxide sensors. A selection that gives a representative overview of the biomolecules and transducers is included in Table 2.5. Peroxidase, catalase, haemoglobin, cytochrome c, microperoxidase and hemin can all be explored for peroxide measurement. Most papers on DET-based biosensor are related to peroxide detection in a variety of environments with peroxidases.

Peroxidases from different sources were used and recently also genetically engineered enzymes. Not only hydrogen peroxide but also organic hydroperoxides [96] are indicated with third-generation peroxidase biosensors, although the highest reaction rate is obtained for hydrogen peroxide.

In the last 15 years numerous publications have described peroxide sensors based on HRP in DET contact. In most cases carboneous material was used for HRP and the oxygen functionalities of the carbon material were suggested to be the reason for the efficiency of the DET process on such surfaces [217,269]. The material ranges from carbon black, graphite powder, spectroscopic and pyrolytic graphite, epoxygraphite to carbon fibres and fullerenes to which peroxidases are physically adsorbed or covalently coupled. When incorporated into a carbon paste it is important to adsorb the enzyme before binder material, usually oil or grease, is added [96,256]. Various additives have proved to enhance the sensitivity of the composite electrodes. Among the most effective are polyelectrolytes and polyalcohols such as polyethyleneimine and lactitol. Gold has been applied as gold colloid [216] or as disk and wire electrodes [213,218,220]. If the electrode reaction and the enzyme reaction are very fast mass transport limits the sensitivity of the peroxide electrode. Assuming a diffusion coefficient of 1.6×10^5 cm^2 s^{-1} the upper limit of sensitivity in a stationary mode was estimated to be 1 A M^{-1} cm^{-2} [219].

Many of the developed sensors based on peroxidases have lower detection limits in the range 10–100 nM (Table 2.5). For cytochrome c which has a considerably lower catalytic rate with peroxide the detection limits are around 10 μM. Only for a multilayer approach with a conducting underlayer peroxide as low as 3 nM could be indicated [270].

For the measurement a moderate reduction potential between -100 and $+100$ mV vs. Ag/AgCl is applied (Fig. 2.12). In this region the potential for electrochemical interferences is very low. However, the biggest problems arise from the high reactivity of compounds I and II with reducing substrates (electron donors), which compete with the electrode for the reduction of peroxidase. Ascorbic acid, naturally occurring phenolics and aromatic amines are among those compounds. The competitive reaction of reductants should be

TABLE 2.5
A selection of third generation peroxide sensors and their parameters

Enzymes	Electrodes	k_s (s^{-1})	Sensor parameter, detection range for H$_2$O$_2$	Application/Comments	References
Peroxidases[a]					
HRP	SG adsorbed				[17]
HRP	CPE/PEI		0.1–500 μM	Heat treated graphite combined with AAOD	[229]
HRP	Pt/polypyrrole		μM-range	Coupled to GOD Glucose—1.4 mM	[18]
HRP	PG/polypyrrole		50–1700 μM	Coupled to GOD Glucose—1.4 mM	[18]
HRP	CPE/graphite epoxy/HSA		~400 μM life time > 40 d	Also organic hydroperoxides: different sensitivity	[96]
HRP	Carbon fiber μ-El.		20–2500 μM	Coimmobilized GOD for glucose measurement	[88]
HRP	Screen printed activated carbon		0.2–150 μM	Very stable (235 d) immunoassay	[147,217]
HRP			CumeneOOH	Biomimetic sensor for octane	[106]
HRP	PG, SPAN/PSS		LDL 3 nM	Multilayers by LBL SPAN may have wiring capability	[270]
HRP	Graphite	1.9	LDL 50 nM		[102]
HRP	Au-DTSP		LDL 100 nM		[90]
HRP	Au-cystamine-colloidal gold		1.4 μM–2.8 mM	48% in DET	[216]

Third generation biosensors—integrating recognition

Enzyme	Surface	Value	LDL/Range	Notes	Ref
HRP	Au-DTSP/dithiopropionic acid-hemin, Apo-HRP		μM-range	Reconstitution of Holoenzyme on surface, PQQ-wire improves ET	[218]
ARP	PG-carbon black glutardialdehyde			Potentiometric, AchE/COD for organophosphates	[253]
ARP	Graphite, carbodiimide		LDL 20 nM		[87]
Cytochrome c CCOP	PG				[85]
Lignolytic LiPOD	graphite	1.6		9.4% DET	[98]
MnPOD	graphite	1.6		27.4% in DET	[98]
Peanut PNP	graphite	1.3	LDL 60 nM	44% in DET	[102]
Sweet potato SPP	graphite	4.8	LDL 40 nM	91%-Highest percentage of DET of native plant POD,	[102]
Tobacco TOP	graphite	2.6	LDL 110 nM	68% in DET	[102]
TOP	Au-aminoalkylthiols adsorbed		μM-range	17.9 μA/M 0.033 cm^2	[220]
Recombinant peroxidases[a]					
HRP	Au, adsorbed	1	0.1–40 μM		[46]
Rec, wt	Au, adsorbed	18		58% in DET	[213]
C(his)$_6$ –	Au, adsorbed	500–33.2[b]		75% in DET	[214]
N(his)$_6$ –	Au, adsorbed	32.7		63% in DET	[46]

continued

TABLE 2.5 (Continuation)

Enzymes	Electrodes	k_s (s^{-1})	Sensor parameter, detection range for H_2O_2	Application/Comments	References
Asn70Asp	Au, adsorbed/	2.2	0.01–0.5 μM	No mediator effect-100% in DET	[215]
Asn70Val	G adsorbed	2.4		Low response to potential interferants	[271]
C(Strep)-HRP	Au	88//115.1–18.7^b	0.01–1 μM		[47]
C(Strep)-HRP -cys309	Au	451–28.8^b		64% in DET Reaction used also for immunosensing	[273]
Other molecules					
Hemin	Adsorbed	$k_s = 15$ s^{-1}			[263]
Hemin	Au-DIDS	$k_s = 15$ s^{-1}			[90]
Cytochrome c	CPE, Colloidal Au	$k = 1.2$ s^{-1}	10 μM H_2O_2,		[281]
Cytochrome c	GC, montmorillonite		LDL 20 μM H_2O_2		[156]
Cytochrome c	Au, MU/MUA-EDC		10–1000 μM		[276]
Cytochrome c	Au, DTSP				[90]
Myoglobin	PG, SPAN/PSS		LDL 3–500 nM	Multilayers by LBL SPAN may have wiring capability	[270]
MP-11	Au-DTSP	$k = 4$ s^{-1}			[90]

DIDS: diaminodiethyl disulfide; PEI-polyethylene imine; AAOD: Amino acid oxidase; ARP: Arthromyces ramosus; LDL: Lower limit of detection; LBL: Layer by layer; SPAN: Sulfonated polyaniline.
^{a}Catalytic reduction typically between −50 and +50 mV vs SCE.
^{b}pH-dependent, first value at pH 7.4, second value at pH 6.0.

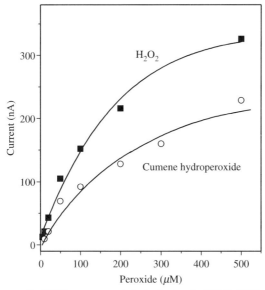

Fig. 2.12. Response of a HRP-screen printed electrode (CLS, UK) on hydrogen peroxide and cumene hydroperoxide. The working potential is −50 mV. (Reproduced from Ref. [106] with permission from Elsevier.)

minimized when all enzyme molecules are in DET contact to the electrode. From the comparison of peroxidases from different sources sweet potato peroxidase showed the highest percentage of DET on graphite electrodes. Still less interference is found for Asn70Asp–mutated HRP on graphite [215,271]. Unfortunately, the stability of the sensors with recombinant enzymes was only a few hours. Longer shelf life is obtained when peroxidase is in a composite or covalently attached. The most stable peroxide detector has been described in Ref. [217], where HRP is embedded with activated carbon in a screen-printed paste (Fig. 2.12). The material is extremely rich in oxygen. Real application was reported for peroxide measurement in various samples including on-line monitoring of cerebral peroxide [272].

The use of peroxide sensors has been extended to the sensing of glucose, lactate, alcohol, oxalate, glutamate and other amino acids by coimmobilization of the respective oxidase on top of the peroxidase [7,8,228,229], and furthermore, to affinity-based assays [130,131,217,273]. An octane sensor was created by layering a porphyrin type P450-mimics on a screen-printed HRP-modified carbon electrode (Fig. 2.13) [106]. The biomimetic catalyst (iron(III)-meso-tetrakis-(pentafluorophenyl)-β-tetrasulfonatoporphyrin chloride) was linked to the electrode with polyallylamine on the basis

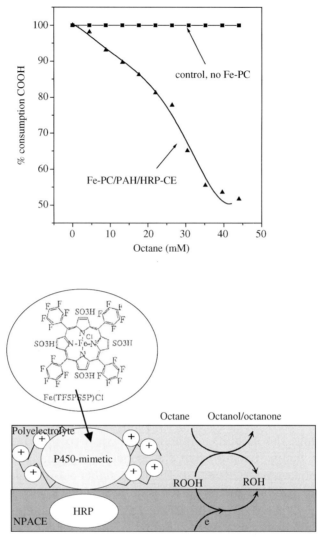

Fig. 2.13. Measurement of octane with a Fe-porphyrin biomimetic (Fe-PC) immobilized on HRP-screen-printed electrode by polyallylamine (PAH). The response represents consumption of cumene hydroperoxide. Working potential was −50 mV. Adapted from Ref. [106].

of the strong electrostatic interaction between the anionic polyelectrolyte and the polyhalogenated iron porphyrin. (Fig. 2.13). The basis of sensing is a competition between peroxidase and the biomimetic for cumene hydroperoxide. This sensor was investigated for its applicability to octane determination.

A decreasing reduction current due to peroxide consumption could be observed when increasing amounts of octane were injected into the peroxide containing solution. The graph shows a sigmoidal behaviour with a levelling off at about 40 mM octane. A decrease of about 5 μM peroxide per mM octane is obtained at up to 20 mM octane (Fig. 2.13) [106].

An H_2O_2 biosensor consisting of a layer of the polyelectrolyte encapsulated *catalase* [200] displayed higher and more stable electrocatalytic responses than did corresponding electrodes made of nonencapsulated catalase or solubilized catalase. The mechanism of the electron transfer, however, remains unclear. The current response was proportional to the H_2O_2 concentration in the range 3.0 μM–10 mM. For this sensor an increase in either the number of "precursor" layers between the gold electrodes and the catalase microcrystal layers in the film or the number of layers encapsulating the catalase microcrystals was found to decrease the electrocatalytic activity of the electrode. This was attributed to the increasing difficulty of electron transfer and substrate diffusion limitations.

2.6.4 Immunoassays based on DET indication

An approach to the development of amperometric electrochemical immunosensors capable of separation-free measurement has been to exploit the direct electrochemistry of enzymes [7,130–132,273] at electrode surfaces using enzyme channelling or proximity effects. The basic principle is to combine the interfacial DET and affinity interaction of antigen and antibody [131,147,217]. Antibodies are loaded on an electrode surface to which peroxidase labelled antigen binds indirectly proportionally to (non-labelled) antigen in the bulk. Upon addition of peroxide catalytic reduction is obtained for the portion of peroxidase in close proximity to the surface while the bulk components are silent. A more effective approach is to coimmobilize peroxidase and antibody and measure H_2O_2 formed by enzyme labelled species bound to electrode surfaces via specific recognition events. In the presence of catalase in solution only the surface bound label is measured.

These generic systems have been applied to a number of analytes including a separation-free non-competitive assay for TSH. The TSH assay relies on H_2O_2 formation by an alkaline phosphatase label in the presence of the substrate 5-bromo-4-chloro-3-indolyl phosphate. Peroxide generated by the bound enzyme label is detected at the electrode surface by reduction with immobilized HRP followed by DET from the activated carbon electrode to the HRP. The reduction current is generated only by the enzyme close to the surface. Enzyme channelling of H_2O_2 does not occur in the bulk solution.

Alternatively, glucose oxidase label was used and glucose addition is triggering the signal generation [273,274]. A channelling sensor for simazine [273] was composed of engineered HRP optimized for binding to gold electrodes [46] and already 1 ng l^{-1} simazine gave a clearly distinguishable signal. The sensor needs, however, further optimization.

Another approach uses the property of laccase to catalyze the electroreduction of oxygen for a mediator-less electrochemical enzyme immunoassay [7,132,250]. Here laccase-labelled antigens are attracted to the antibody-loaded electrode surface. Laccase reaching the vicinity of the electrode surface can be detected by electrocatalytic oxygen reduction measurement. If the bulk contains a sufficient amount of antigen the signal decreases. For the model analyte immunoglobulin G a detection limit of 0.3 nM was achieved. The assay can also be performed as competitive assay with bound antigen and labelled antibody as was demonstrated for the determination of carcinoma antigen 125 [275].

One of the great advantages of the bioelectrocatalytic approach is that the enzyme needs no addition of a specific substrate (other than molecular oxygen).

2.7 CONCLUSION

With most enzymes in DET, with the main exception of peroxidases, substrate measurements can be made only in the range of 10^{-4}–10^{-3} M. This restriction is due to a kinetic limitation of the heterogeneous electron transfer which can be overcome by engineering specific attachment regions and electron transfer bridges.

In recent years we have profited from molecular biology, nanotechnology and surface chemistry because they provide tools for the molecular design of proteins and nanostructuring of surfaces. Important impact is also coming from novel methods of spectroscopy and in situ imaging techniques for mapping and control of single-molecule activity or visualization of molecules during their redox processes [279]. The discovery of microorganisms loaded with cytochromes in their outer membranes [282] will broaden the scope of DET sensors to whole cell-base assays.

The understanding of the molecular mechanism leads to tailor-made catalytic systems. These do not necessarily have to be of pure native origin. Recent work indicates the potential of engineering sites for surface binding and redox active dyes, de novo designed redox proteins and genetic chimeras [45,48,50]. Thus tailor-made systems are foreseen.

Acknowledgements

The help of present and former colleagues and partners of the Department of Analytical Biochemistry at the University of Potsdam is gratefully acknowledged. My special thanks go to Frieder Scheller for his encouragement and many stimulating ideas and Andrea Kühn for her technical assistance over many years.

REFERENCES

1 F.A. Armstrong and G.S. Wilson, *Electrochim. Acta*, 45 (2000) 2623.
2 F.W. Scheller, F. Schubert, R. Renneberg, H.G. Müller, M. Jänchen and H. Weise, *Biosensors*, 1 (1985) 135.
3 F. Scheller, U. Wollenberger and F. Lisdat, Electrical connected enzymes in biosensors. In: I. Willner, E. Katz (Eds.), *Bioelectronics: From Theory to Applications*, Wiley, in press.
4 C.R. Lowe, B.F.Y. Yon Hin, D.C. Cullen, S.E. Evans, L.D.G. Stephens and P.J. Maynard, *J. Chromatogr.*, 510 (1990) 347.
5 C.G.J. Koopal, *Third Generation Amperometric Biosensors*, Thesis, Kath. Univ., Nijmegen, 1992.
6 T. Ikeda, In: F.W. Scheller, F. Schubert and J. Fedrowitz (Eds.), *Frontiers in Biosensorics I, Fundamental Aspects*. Birkhäuser Verlag, Basel, 1997, pp. 243–266.
7 A.L. Ghindilis, P. Atanasov and E. Wilkins, *Electroanalysis*, 9 (1997) 661–674.
8 L. Gorton, A. Lindgren, T. Larsson, F.D. Munteanu, T. Ruzgas and I. Gazaryan, *Anal. Chim. Acta*, 400 (1999) 91.
9 S.J. Updike and G.P. Hicks, *Nature*, 214 (1967) 986.
10 F.W. Scheller and F. Schubert, *Biosensors*. Elsevier, Amsterdam, 1992.
11 U. Wollenberger, F. Schubert, D. Pfeiffer and F. Scheller, *Trends Biotechnol.*, 11 (1993) 255.
12 A. Cass, G. Davis, G. Francis, H.A.O. Hill, W. Aston, I.J. Higgins, E. Plotkin, L. Scott and A.P.F. Turner, *Anal. Chem.*, 56 (1984) 667.
13 I. Willner and E. Katz, *Angew. Chem. Int. Ed.*, 39 (2000) 1180.
14 Y. Degani and A. Heller, *J. Am. Chem. Soc.*, 111 (1989) 2357.
15 W. Schuhmann, *Rev. Mol. Biotechnol.*, 82 (2002) 425–442.
16 F.A. Armstrong and A.M. Lannon, *J. Am. Chem. Soc.*, 109 (1987) 7211.
17 G. Jönsson and L. Gorton, *Electroanalysis*, 1 (1989) 465.
18 U. Wollenberger, V.A. Bogdanovskaya, S. Bobrin, F. Scheller and M.R. Tarasevich, *Anal. Lett.*, 23 (1990) 1795.

19 F.A. Armstrong, In: G. Wilson, A.J. Bard and M. Stratmann (Eds.), *Bioelectrochemistry*, Encyclopedia of Electrochemistry, Vol. 9. Wiley-VCH, Weinheim, 2002.
20 G. Dryhurst, K.M. Kadish, F.W. Scheller, R. Renneberg, *Biological Electrochemistry*, Vol. 1. Academic Press, New York, 1982.
21 F.A. Armstrong, *Struct. Bonding*, 72 (1990) 137.
22 L.H. Guo and H.A.O. Hill, *Adv. Inorg. Chem.*, 36 (1991) 341.
23 G. Wilson, A.J. Bard and M. Stratmann (Eds.), *Bioelectrochemistry in Encyclopedia of Electrochemistry Vol. 9*. Wiley-VCH, Weinheim, 2002.
24 L.J.C. Jeukens, *Biochim. Biophys. Acta*, 1604 (2003) 67.
25 F.W. Scheller, U. Wollenberger, C. Lei, W. Jin, B. Ge, C. Lehmann, F. Lisdat and V. Fridman, *Rev. Mol. Biotechnol.*, 82 (2002) 425.
26 V. Fridman, U. Wollenberger, V. Bogdanovskaya, F. Lisdat, T. Ruzgas, A. Lindgren, L. Gorton and F.W. Scheller, *Biochem. Soc. Trans.*, 28 (2000) 63.
27 R. Marcus and N. Sutin, *Biochim. Biophys. Acta*, 811 (1985) 265.
28 C.C. Moser, J.M. Keske, K. Warncke, R.S. Farid and P.L. Dutton, *Nature*, 355 (1992) 796.
29 A.M. Bond, *Inorg. Chim. Acta*, 226 (1994) 293.
30 P. Yeh and T. Kuwana, *Chem. Lett.*, (1977) 1145.
31 M.J. Eddowes and H.A.O. Hill, *J. Chem. Soc., Chem. Commun.*, (1977) 771.
32 S.D. Varfolomeev and I.V. Berezin, *J. Mol. Catal.*, 4 (1978) 387.
33 A.I. Yaropolov, V. Malovik, S.D. Varfolomeev and I.V. Berezin, *Dokl. Akad. Nauk. SSSR*, 249 (1979) 1399.
34 M.R. Tarasevich, *Bioelectrochem. Bioenerg.*, 6 (1979) 587.
35 J.J. Kulys and A.S. Samalius, *Bioelectrochem. Bioenerg.*, 13 (1984) 163.
36 C.D. Bain and G.M. Whitesides, *Angew. Chem. Int. Ed.*, 28 (1989) 506.
37 A. Ulman, *Chem. Rev.*, 96 (1996) 1533.
38 J.F. Rusling, *Acc. Chem. Res.*, 31 (1998) 363.
39 J.J. Gooding and D.B. Hibbert, *Trends Anal. Chem.*, 18 (1999) 525.
40 H.O. Finklea, In: A.J. Bard (Ed.), *Electroanalytical Chemistry*, Vol. 19. Marcel Dekker, New York, 1996, p. 109.
41 S. Song, R.A. Clark, E.F. Bowden and M.J. Tarlov, *J. Phys. Chem.*, 97 (1993) 6564.
42 Z.Q. Feng, S. Imabayashi, T. Kakiuchi and K. Niki, *J. Chem. Soc. Faraday Trans.*, 93 (1997) 1367.
43 W. Jin, U. Wollenberger, E. Kärgel, W.H. Schunck and F.W. Scheller, *J. Electroanal. Chem.*, 433 (1997) 135.
44 J.J. Gooding, R. Wibowo, J. Liu, W. Yang, D. Losic, S. Orbons, F.J. Mearns, J.G. Shaper and D.B. Hibbert, *J. Am. Chem. Soc.*, 125 (2003) 9006.
45 G. Gilardi, A. Fantuzzi and S.J. Sadeghi, *Curr. Opin. Struct. Biol.*, 11 (2001) 491.
46 G. Presnova, A. Egorov, T. Ruzgas, A. Lindgren, L. Gorton and T. Börchers, *Faraday Discuss.*, 116 (2000) 281.

47 E. Ferapontova, K. Schmengler, T. Börchers, T. Ruzgas and L. Gorton, *Biosens. Bioelectron.*, 17 (2002) 953.
48 G. Gilardi, Y.T. Meharenna, G.E. Tsotsou, S.J. Sadeghi, M. Fairhead and S. Giannini, *Biosens. Bioelectron.*, 17 (2002) 133.
49 L. Andolfi, D. Bruce, S. Cannistraro, G.W. Canters, J.J. Davis, H.A.O. Hill, J. Crozier, M.P. Verbeet, C.L. Wrathmell and Y. Astier, *J. Electroanal. Chem.*, 565 (2004) 21.
50 T.S. Wong and U. Schwaneberg, *Curr. Opin. Biotechnol.*, 14 (2003) 590.
51 J.M. Cooper, K.R. Greenough and C.J. McNeil, *J. Electroanal. Chem.*, 347 (1993) 267.
52 C.J. McNeil and K.A. Smith, *Free Rad. Res. Commun.*, 7 (1989) 89.
53 W. Scheller, W. Jin, E. Ehrentreich-Förster, B. Ge, F. Lisdat, R. Büttemeier, U. Wollenberger and F.W. Scheller, *Electroanalysis*, 11 (1999) 703.
54 F. Lisdat, B. Ge, E. Ehrentreich-Förster, R. Reszka and F.W. Scheller, *Anal. Chem.*, 71 (1999) 1359.
55 V. Gobi and F.J. Mizutani, *Electroanal. Chem.*, 484 (2000) 172.
56 B. Ge, T. Meyer, M.J. Schöning, U. Wollenberger and F. Lisdat, *Electrochem. Commun.*, 2 (2000) 557.
57 C. Lei, U. Wollenberger, N. Bistolas, A. Guiseppi-Elie and F.W. Scheller, *Anal. Bioanal. Chem.*, 372 (2002) 235.
58 E. Topoglidis, Y. Astuti, F. Duriaux, M. Grätzel and J.R. Durrant, *Langmuir*, 19 (2003) 6894.
59 S. Kröning, F.W. Scheller, U. Wollenberger and F. Lisdat, *Electroanalysis*, 16 (2004) 253.
60 Z. Dai, S. Liu, H. Ju and H. Chen, *Biosens. Bioelectron.*, (2003) 18.
61 H.A.O. Hill and N.I. Hunt, *Methods Enzymol.*, 227 (1993) 501.
62 T. Ikeda, *Bull. Electrochem.*, 8 (1992) 145.
63 T. Sakurai, *Chem. Lett.*, (1996) 481.
64 R. Santucci, T. Ferri, L. Morpurgo, I. Savini and L. Avigliano, *Biochem. J.*, 332 (1998) 611.
65 T. Kohzuma, S. Shidara, K. Yamaguchi, N. Nakamura and S. Suzuki, *Chem. Lett.*, (1993) 2029.
66 D. Kobayashi, S. Qzawa, T. Mihara and T. Ikeda, *Denki Kagaku*, 60 (1992) 1056.
67 D.L. Johnson, J.L. Thompson, S.M. Brinkmann, K.A. Schuller and L.L. Martin, *Biochemistry*, 42 (2003) 10229.
68 B. Ge, F.W. Scheller and L. Lisdat, *Biosens. Bioelectron.*, 18 (2003) 295.
69 Y. Tian, M. Shioda, S. Kasahara, T. Okajima, L. Mao, T. Hisaboli and T. Ohsaka, *Biochim. Biophys. Acta*, 1569 (2002) 151.
70 Y. Tian, L. Mao, T. Okajima and T. Ohsaka, *Anal. Chem.*, 74 (2002) 2428.
71 M.E. Lai and A. Bergel, *Bioelectrochemistry*, 55 (2002) 157.
72 C. Lei, U. Wollenberger, C. Jung and F.W. Scheller, *Biochem. Biophys. Res. Commun.*, 268 (2000) 740.

73   J. Kazlauskaite, A.C.G. Westlake, L.L. Wong and H.A.O. Hill, *Chem. Commun.*, (1996) 2189.
74   V.V. Shumyantseva, Y.D. Ivanov, N. Bistolas, F.W. Scheller, A.I. Archakov and U. Wollenberger, *Anal. Chem.* 76 (2004) 6046.
75   E.I. Iwuoha, S. Joseph, Z. Zhang, M.R. Smyth, U. Fuhr and P.R. Ortiz de Montellano, *J. Pharm. Biomed. Anal.*, 17 (1998) 1101.
76   S. Joseph, J.F. Rusling, Y.M. Lvov, T. Friedberg and U. Fuhr, *Biochem. Pharmacol.*, 65 (2003) 1817.
77   V.V. Shumyantseva, T.V. Bulko, S.A. Usanov, R.D. Schmid, C. Nicolini and A.I. Archakov, *J. Inorg. Biochem.*, 87 (2001) 185.
78   B. Munge, C. Estavillo, J.B. Schenkman and J.F. Rusling, *Chem. Biochem.*, 4 (2003) 101.
79   Y. Aster, S. Balendra, H.A.O. Hill, T.J. Smith and H. Dalton, *Eur. J. Biochem.*, 270 (2003) 539.
80   L. Jiang, C.J. McNeil and J.M.J. Cooper, *Chem. Soc., Chem. Commun.*, (1995) 1293.
81   Y.D. Zhao, W.D. Zhang, H. Chen and Q.M. Luo, *Anal. Sci.*, 18 (2002) 939.
82   C. Leger, A.K. Jones, S.P.J. Albracht and F.A. Armstrong, *J. Phys. Chem.*, 106 (2003) 13058.
83   T. Ruzgas, L. Gorton, J. Emnéus, E. Csöregi and G. Marko-Varga, *Anal. Proc.*, 6 (1995) 207.
84   M.S. Mondal, H.A. Fuller and F.A. Armstrong, *J. Am. Chem. Soc.*, 118 (1996) 293.
85   R.M. Paddock and E.F. Bowden, *J. Electroanal. Chem.*, 260 (1989) 487.
86   D.L. Scott, R.M. Paddock and E.F. Bowden, *J. Electroanal. Chem.*, 341 (1992) 307.
87   J. Kulys and R.D. Schmid, *Bioelectrochem. Bioenerg.*, 24 (1990) 305.
88   E. Csöregi, L. Gorton and G. Marko-Varga, *Electroanalysis*, 6 (1994) 925.
89   V. Razumas, J. Kazlauskaite, T. Ruzgas and J. Kulys, *J. Bioelectrochem. Bioenerg.*, 28 (1992) 159.
90   T. Lötzbeyer, W. Schuhmann and H.L. Schmidt, *Sens. Actuators B*, 33 (1996) 50.
91   J.J. Kulys and G.Y. Svirmitskas, *Dok Phys. Chem.*, 245 (1979) 208.
92   L.H. Guo, H.A.O. Hill, D.J. Hopper, G.A. Lawrance and G.S. Sanghera, *J. Biol. Chem.*, 265 (1990) 1958.
93   L.H. Guo, H.A.O. Hill, G.A. Lawrance, G.S. Sanghera and D. Hopper, *J. Electroanal. Chem.*, 266 (1989) 379.
94   A. Sucheta, B.A.C. Ackrell, B. Cochran and F.A. Armstrong, *Nature*, 356 (1992) 361.
95   A. Sucheta, R. Cammack, J. Weiner and F.A. Armstrong, *Biochemistry*, 32 (1993) 5455.
96   U. Wollenberger, J. Wang, M. Ozsoz, E. Gonzalez Romero and F. Scheller, *Bioelectrochem. Bioenerg.*, 26 (1991) 287.

97   J. Hirst, A. Sucheta, B.A.C. Ackrell and F.A. Armstrong, *J. Am. Chem. Soc.*, 118 (1996) 5031.
98   E.E. Ferapontova, N.S. Reading, S.D. Aust, T. Ruzgas and L. Gorton, *Electroanalysis*, 14 (2002) 1411.
99   K.T. Kinnear and H.G. Monbouquette, *Anal. Chem.*, 69 (1997) 1771.
100 F.D. Munteanu, A. Lindgren, J. Emnéus, L. Gorton, T. Ruzgas, E. Csöregi, A. Ciucu, R.B. van Huystee, I.G. Gazaryan and L.M. Lagrimini, *Anal. Chem.*, 70 (1998) 2596.
101 K. Heffron, C. Leger, R.A. Rothery, J.H. Weiner and F.A. Armstrong, *Biochemistry*, 40 (2001) 3117.
102 A. Lindgren, T. Ruzgas, L. Gorton, E. Csöregi, G.B. Ardila, I.Y. Sakharov and I.G. Gazaryan, *Biosens. Bioelectron.*, 15 (2000) 491.
103 A. Lindgren, J. Emnéus, T. Ruzgas, L. Gorton and G. Marko-Varga, *Anal. Chim. Acta*, 347 (1997) 51.
104 J. Tong and B.A. Feingerg, *J. Biol. Chem.*, 269 (1994) 24920.
105 A.I. Yaropolov, A.N. Kharybin, J. Emnéus, G. Marko-Varga and L. Gorton, *Bioelectrochem. Bioenerg.*, 40 (1996) 49.
106 U. Wollenberger, B. Neumann and F.W. Scheller, *Electrochim. Acta*, 43 (1998) 3581.
107 P. Bianco and J. Haladian, *J. Electrochem. Soc.*, 139 (1992) 2428.
108 K.L. Turner, M.K. Doherty, H.A. Heering, F.A. Armstrong, G.A. Reid and S.K. Chapman, *Biochemistry*, 38 (1999) 3302.
109 T. Larsson, S.-E. Lindquist, M. Elmgren, M. Tessema, L. Gorton and G. Henriksson, *Anal. Chim. Acta*, 331 (1996) 207.
110 A. Lindgren, T. Larsson, T. Ruzgas and L. Gorton, *J. Electroanal. Chem.*, 494 (2000) 105.
111 T. Ikeda, F. Matsushita and M. Senda, *Biosens. Bioelectron.*, 6 (1991) 299.
112 C. Lehmann, U. Wollenberger, R. Brigelius-Flohé and F.W. Scheller, *Electroanalysis*, 13 (2001) 364.
113 G.F. Khan, H. Shinohara, Y. Ikariyama and M. Aizawa, *J. Electroanal. Chem.*, 315 (1991) 263.
114 T. Ikeda, D. Kobayashi, F. Matsushita and K. Niki, *J. Electroanal. Chem.*, 361 (1993) 221.
115 A. Ramanavicius, K. Habermüller, V. Laurinavicius, E. Csöregi and W. Schuhmann, *Anal. Chem.*, 71 (1999) 3581.
116 J. Razumiene, M. Niculescu, A. Ramanavicius, V. Laurinavicius and E. Csöregi, *Electroanalysis*, 14 (2002) 1.
117 A.L. Burrows, H.A.O. Hill, T.A. Leese, W.S. Mcintire, H. Nakayama and G.S. Sanghera, *Eur. J. Biochem.*, 199 (1991) 73.
118 C. Lehmann, U. Wollenberger, R. Brigelius-Flohé and F.W. Scheller, *J. Electroanal. Chem.*, 455 (1998) 259.
119 W. Schuhmann, H. Zimmermann, K. Habermüller and V. Laurinavicius, *Faraday Discuss.*, 116 (2000) 245.

120 E.E. Ferapontova, T. Ruzgas and L. Gorton, *Anal. Chem.*, 75 (2003) 4841.
121 S.J. Elliot, A.E. McElhaney, C. Feng, J.H. Enemark and F.A. Armstrong, *J. Am. Chem. Soc.*, 124 (2002) 11612.
122 K.F. Aguey-Zinsou, P.V. Bernhardt, U. Kappler and A.G. McEwan, *J. Am. Chem. Soc.*, 125 (2003) 530.
123 M. Niculescu, T. Ruzgas, C. Nistor, I. Frebort, M. Sebela, P. Pec and E. Csöregi, *Anal. Chem.*, 72 (2000) 5988.
124 N. Fujieda, M. Mori, K. Kano and T. Ikeda, *Biochemistry*, 41 (2002) 13736.
125 T. Ikeda, F. Fusimi, K. Miki and M. Senda, *Agric. Biol. Chem.*, 52 (1988) 2655.
126 T. Ikeda, S. Miyaoka and K. Miki, *J. Electroanal. Chem.*, 352 (1993) 267.
127 J.D. Burgess, M.C. Rhoten and F.M. Hawkridge, *J. Am. Chem. Soc.*, 120 (1998) 4488.
128 J. Li, G. Cheng and S.J. Dong, *Electrochemistry*, 416 (1996) 97.
129 Z. Zhang, A.E.F. Nassar, Z. Lu, J.B. Schenkman and J.F. Rusling, *J. Chem. Soc. Faraday Trans.*, 93 (1997) 1769.
130 A.L. Ghindilis, T.G. Morzunova, A.V. Barman and I.N. Kurochkin, *Biosens. Bioelectron.*, 11 (1996) 873.
131 C.J. McNeil, D. Athey and R. Renneberg, In: F.W. Scheller, F. Schubert and J. Fedrowitz (Eds.), *Frontiers in Biosensorics II, Fundamental Aspects*. Birkhäuser Verlag, Basel, 1997, pp. 17.
132 B.A. Kuznetsov, G.P. Shumakovich, O.V. Koroleva and A.I. Yaropolov, *Biosens. Bioelectron.*, 16 (2001) 73.
133 R.A. Scott and A.G. Mauk (Eds.), *Cytochrome c: A Multidisciplinary Approach*. University Science Books, Sausalito, 1996.
134 M. Fedurco, *Coord. Chem. Rev.*, 209 (2000) 263.
135 P.M. Allen, H.A.O. Hill and N.J. Walton, *J. Electroanal. Chem.*, 178 (1984) 69.
136 K. Taniguchi, H. Toyosawa, K. Yamaguchi and K. Yasukouchi, *J. Electroanal. Chem.*, 140 (1982) 187.
137 J.E. Frew and H.A.O. Hill, *Eur. J. Biochem.*, 172 (1988) 261.
138 M. Paeschke, R. Hintsche, U. Wollenberger, W. Jin and F. Scheller, *J. Electroanal. Chem.*, 393 (1995) 131.
139 K. Niki, Bioelectrochemistry. In: G. Wilson, A.J. Bard and M. Stratmann (Eds.), *Encyclopedia of Electrochemistry Vol. 9*. Wiley-VCH, Weinheim, 2002, pp. 341.
140 D.H. Murgida and P. Hildebrandt, *Angew. Chem. Int. Ed.*, 40 (2001) 728.
141 A.E. Kasmi, J.M. Wallace, E.F. Bowden, S.M. Binet and R.J. Linderman, *J. Am. Chem. Soc.*, 120 (1998) 225.
142 S. Arnold, Z.Q. Feng, K. Kakiuchi, W. Knoll and K. Niki, *J. Electroanal. Chem.*, 438 (1997) 91.
143 B. Ge and F. Lisdat, *Anal. Chim. Acta*, 454 (2002) 53.
144 P. Bianco, A. Taye, J. Haladjian, *J. Electroanal. Chem.*, 377 (1994) 299.
145 C. Lei, F. Lisdat, U. Wollenberger and F.W. Scheller, *Electroanalysis*, 11 (1999) 274.

146 M. Beissenhirtz, F.W. Scheller, W. Stöcklein and F. Lisdat, *Angew. Chem. Int. Ed.*, 43 (2004) 4357.
147 C.J. McNeil and P. Manning, *Rev. Mol. Biotechnol.*, 82 (2002) 443.
148 S.L.R. Barker, R. Kopelman, T.E. Meyer and M.A. Cusanovich, *Anal. Chem.*, 70 (1998) 971.
149 T. Erabi, S. Ozawa, S. Hayase and M. Wada, *Chem. Lett.*, 11 (1992) 2115.
150 Z. Ren, T. Meyer and D.E. McRee, *J. Mol. Biol.*, 234 (1993) 433.
151 F. Lisdat, B. Ge, M.E. Meyerhoff and F.W. Scheller, *Probe Microscopy*, 2 (2001) 113.
152 L. Stryer, *Biochemistry*, 4th ed., Freeman, New York, 1995.
153 Protein Database 1mbn.pdb
154 B.R. Van Dyke, P. Saltman and F.A. Armstrong, *J. Am. Chem. Soc.*, 118 (1996) 3490.
155 A.E.F. Nassar, W.S. Willis and J.F. Rusling, *Anal. Chem.*, 67 (1995) 2386.
156 C. Lei, U. Wollenberger and F. Scheller, *Quim. Anal.*, 19 (2000) 28.
157 M. Bayachou, R. Lin, W. Cho and P.J. Farmer, *J. Am. Chem. Soc.*, 120 (1998) 9888.
158 L. Shen, R. Huang and N. Hu, *Talanta*, 56 (2002) 1131.
159 R. Lin, M. Bayachou, J. Greaves and P.J. Farmer, *J. Am. Chem. Soc.*, 119 (1997) 12689.
160 G.N. Kamau, M.P. Guto, B. Munge, V. Panchagnula and J.F. Rusling, *Langmuir*, 19 (2003) 6976.
161 Y. Zhou, N. Hu, Y. Zeng and J.F. Rusling, *Langmuir*, 18 (2002) 211.
162 M. Weissbluth, *Molecular Biology: Biochemistry and Biophysics*, Vol. 15. Springer, New York, 1974.
163 S. Dong and T. Chen, In: F.W. Scheller, F. Schubert and J. Fedrowitz (Eds.), *Frontiers in Biosensorics I*. Birkhäuser, Basel, 1997, p. 209.
164 P.R. Ortiz de Montellano and C.E. Catalano, *J. Biol. Chem.*, 260 (1985) 9265.
165 R.S. Wade and C.E. Castro, *J. Am. Chem. Soc.*, 95 (1973) 231.
166 C. Giulivi and K.J.A. Davies, *J. Biol. Chem.*, 265 (1990) 19453.
167 T.L. Poulos, *Curr. Opin. Struct. Biol.*, 5 (1995) 767.
168 P.R. Ortiz de Montellano (Ed.), *Cytochrome P450: Structure, Mechanism, and Biochemistry*, Plenum Press, New York, 1995.
169 D.F.V. Lewis, *Cytochromes P450—Structure, Function and Mechanism*, Taylor & Francis, London, 1996.
170 T.L. Poulos, B.C. Finzel, I.C. Gunsalus, G.C. Wagner and J. Kraut, *J. Biol. Chem.*, 260 (1985) 16122.
171 E.E. Scott, Y.A. He, M.R. Wester, M.A. White, C.C. Chin, J.R. Halpert, E.F. Johnson and C.D. Stout, *Proc. Natl Acad. Sci. USA*, 100 (2003) 13196.
172 D. Werck-Reichhart and R. Feyereisen, *Genome Biol.*, 1 (2000) 3003.1.
173 K. Auclair, Z. Hu, D.M. Little, P.R. Ortiz de Montellano and J.T. Groves, *J. Am. Chem. Soc.*, 124 (2002) 6020.
174 R.W. Estabrook, K.M. Faulkner, M.S. Shet and C.W. Fisher, *Adv. Enzymol.*, 272 (1996) 44.

175 M.J. Hintz and J.A. Peterson, *J. Biol. Chem.*, 255 (1980) 7317.
176 J. Contzen and C. Jung, *Biochemistry*, 38 (1999) 16253.
177 V. Reipa, M.P. Mayhew, M.J. Holden and V.L. Vilker, *Chem. Commun.*, (2002) 318.
178 F.W. Scheller, R. Renneberg, G. Strnad, K. Pommerening and P. Mohr, *Bioelectrochem. Bioenerg.*, 4 (1977) 500.
179 E.I. Iwuoha and M.R. Smyth, *Biosens. Bioelectron.*, 18 (2003) 237.
180 X. Zu, Z. Lu, Z. Zhang, J.B. Schenkman and J.F. Rusling, *Langmuir*, 15 (1999) 7372.
181 N. Bistolas, A. Christenson, T. Ruzgas, C. Jung, F.W. Scheller and U. Wollenberger, *Biochem. Biophys. Res. Commun.*, 314 (2004) 810.
182 M. Hara, Y. Yasuda, H. Toyotama, H. Ohkawa, T. Nozawa and J. Miyake, *Biosens. Bioelectron.*, 17 (2002) 173.
183 C. Estavillo, Z. Lu, I. Jansson, J.B. Schenkman and J.F. Rusling, *Biophys. Chem.*, 104 (2003) 291.
184 K.M. Faulkner, M.S. Shet, C.W. Fisher and R.W. Estabrook, *Proc. Natl Acad. Sci. USA*, 92 (1995) 7705.
185 B.D. Flemming, Y. Tian, S.G. Bell, L.L. Wong, V. Urlacher and H.A.O. Hill, *Eur. J. Biochem.*, 270 (2003) 4082.
186 K.F. Aguey-Zinsou, P.V. Bernhardt, J.J. De Voss and K.E. Slessor, *Chem. Commun.*, 3 (2003) 418.
187 M.T. Fisher and S.G. Sligar, *J. Am. Chem. Soc.*, 107 (1985) 5018.
188 C. Mouro, A. Bondon, C. Jung, G. Hui Bon Hoa, J.D. De Certaines, R.G.S. Spencer and G. Simonneaux, *FEBS Lett.*, 455 (1999) 302.
189 D. Ege, P.K. Ghosh, J.R. White, J.F. Equey and A.J. Bard, *J. Am. Chem. Soc.*, 107 (1985) 5644.
190 O.I. Kiselyova, I.V. Yaminsky, Y.D. Ivanov, I.P. Kanaeva, V.Y. Kuznetsov and A.I. Archakov, *Arch. Biochem. Biophys.*, (1999) 371.
191 T.L. Poulos, B.C. Finzel and A.J. Howard, *Biochemistry*, 25 (1986) 5314.
192 C. Jung, S.A. Kozin, B. Canny, J.C. Chervin and G. Hui Bon Hoa, *Biochem. Biophys. Res. Commun.*, 312 (2003) 197.
193 A.E. Roitberg, M.J. Holden, M.P. Mayhew, I.V. Kurnikov, D.N. Beratan and V.L. Vilker, *J. Am. Chem. Soc.*, 120 (1998) 8927.
194 V. Reipa, M.P. Mayhew and V.L. Vilker, *Proc. Natl Acad. Sci. USA*, 94 (1997) 13554.
195 M. Wirtz, J. Klucik and M. Rivera, *J. Am. Chem. Soc.*, 122 (2000) 1047.
196 M.P. Mayhew, V. Reipa, M.J. Holden and V.L. Vilker, *Biotechnol. Prog.*, 16 (2000) 610.
197 M.E. Walsh, P. Kyritsis, N.A.J. Eady, H.A.O. Hill and L.L. Wong, *Eur. J. Biochem.*, 267 (2000) 5815.
198 V.V. Shumyantseva, T.V. Bulko, T.T. Bachmann, U. Bilitewski, R.D. Schmid and A.I. Archakov, *Arch. Biochem. Biophys.*, 376 (2000) 43.

199 M.V. Potapovich, A.N. Eryomin, I.M. Artzukevich, I.P. Chernikevich and D.I. Metelitza, *Biochemistry (Moscow)*, 66 (2001) 646.
200 A. Yu and F. Caruso, *Anal. Chem.*, 75 (2003) 3031.
201 H.B. Dunford, *Heme Peroxidases*. Wiley-VCH, New York, 1999.
202 T.L. Poulos, S.T. Freer, R.A. Alden, S.L. Edwards, U. Skoglund, K. Takio, B. Eriksson, N.H. Xuong, T. Yonetani and J. Kraut, *J. Biol. Chem.*, 255 (1980) 575.
203 D.J. Schnuller, N. Ban, R.B. van Huystee, A. McPherson and T.L. Poulos, *Structure*, 4 (1996) 311.
204 M. Gajede, D.J. Schnuller, A. Henriksen, A.T. Smith and T.L. Poulos, *Nature Struct. Biol.*, 4 (1997) 1032.
205 T. Choinowski, W. Blodig, K.H. Winterhalter and K. Piontek, *J. Mol. Biol.*, 286 (1999) 809.
206 N. Kunishima, F. Amada, K. Fukuyama, M. Kawamoto, T. Matsunaga and H. Matsubara, *FEBS Lett.*, 378 (1996) 291.
207 H. Anni and T. Yonetani, In: H. Siegel and A. Siegel (Eds.), *Metal Ions in Biological Systems*. Marcel Dekker, New York, 1992, pp. 219.
208 J. Everse, K.E. Everse and M.B. Grisham (Eds.), *Peroxidases in Chemistry and Biology*, Vol. 1/2. CRC Press, Boca Raton, 1992.
209 T.L. Poulos, *Curr. Opin. Biotechnol.*, 4 (1993) 484.
210 O. Ryan, M.R. Smyth and C. O'Fagain, *Essays Biochem.*, 28 (1994) 129.
211 M.A. Ator and P.R. Ortiz de Montellano, *J. Biol. Chem.*, 262 (1987) 1542.
212 T.E. Barman (Ed.), *Enzyme Handbook*, Vol. 1. Springer, Berlin, 1992.
213 E.E. Ferapontova, V.G. Grigorenko, A.M. Egorov, T. Börchers and L. Gorton, *J. Electroanal. Chem.*, 509 (2001) 19.
214 E.E. Ferapontova and L. Gorton, *Bioelectrochemistry*, 55 (2002) 83.
215 A. Lindgren, M. Tanaka, T. Ruzgas, L. Gorton, I. Gazaryan, K. Ishimori and I. Morishima, *Electrochem. Commun.*, 1 (1999) 171.
216 A.L. Crumbliss, S.C. Perine, J. Stonehuerner, K.R. Tubergen, J.G. Zhao and R.W. Henkens, *Biotechnol. Bioeng.*, 40 (1992) 483.
217 W.O. Ho, D. Athey, C.J. McNeil, H.J. Hager, G.P. Evans and W.H. Mullen, *J. Electroanal. Chem.*, 351 (1993) 185.
218 H. Zimmermann, A. Lindgren, W. Schuhmann and L. Gorton, *Chem. Eur. J.*, 6 (2000) 592.
219 T. Ruzgas, L. Gorton, J. Emnéus and G. Marko-Varga, *J. Electroanal. Chem.*, 391 (1995) 41.
220 S. Gaspar, H. Zimmermann, I. Gazaryan, E. Csöregi and W. Schuhmann, *Electroanalysis* (2001) 285.
221 J.A. Duine, *Biosens. Bioelectron.*, 10 (1995) 17.
222 Y.M. Lvov, Z. Lu, J.B. Schenkman, X. Zu and J.F. Rusling, *J. Am. Chem. Soc.*, 120 (1998) 4073.
223 D. Lehner, P. Zipper, G. Henriksson and G. Petterson, *Biochim. Biophys. Acta*, 1293 (1996) 161.

224 B.M. Hallberg, T. Bergfors, K. Backbro, G. Pettersson, G. Henriksson and C. Divne, *Structure*, 8 (2000) 79.
225 T. Larsson, A. Lindgren, T. Ruzgas, S.-E. Lindquist and L. Gorton, *J. Electroanal. Chem.*, 482 (2000) 1.
226 T. Larsson, A. Lindgren and T. Ruzgas, *Bioelectrochemistry*, 53 (2001) 243.
227 A. Lindgren, L. Gorton, T. Ruzgas, U. Baminger, D. Haltrich and M. Schülein, *J. Electroanal. Chem.*, 496 (2001) 76.
228 V. Kacaniklic, K. Johansson, G. Marko-Varga, L. Gorton, G. Jönsson-Petterson and E. Csöregi, *Electroanalysis*, 6 (1994) 381.
229 J. Kulys, U. Bilitewski and R.D. Schmid, *Sens. Actuators*, 3 (1991) 227.
230 P.A. Paredes, J. Parella, V.M. Fernandez, I. Katakis and E. Dominguez, *Biosens. Bioelectron.*, 12 (1997) 1233.
231 C. Kisker, H. Schindeline, A. Pacheco, W.A. Wehbi, R.M. Garrett, K.V. Rajagopalan, J.H. Enemark and D.C. Rees, *Cell*, 91 (1997) 973.
232 L.J.C. Jeukens, J.P. McEvoy and F.A. Armstrong, *J. Phys. Chem. B*, 106 (2002) 2304.
233 Y. Datta, K. Mori, K. Takagi, K. Kawaguchi, Z. Chen, T. Okajima, S. Kuroda, T. Ikeda, K. Kano, K. Tanizawa and F.S. Methews, *Proc. Natl Acad. Sci. USA*, 98 (2001) 14268.
234 V. Ducros, A.M. Brzozowski, K.S. Wilson, S.H. Brown, P. Ostergaards, P. Schneider, D.S. Yaver, A.H. Pedersen and G. Davies, *J. Nat. Struct. Biol.*, 5 (1998) 310.
235 A. Messerschmidt and L. Huber, *Eur. J. Biochem.*, 187 (1990) 341.
236 E.I. Solomon, M.J. Baldwin and M.D. Lowery, *Chem. Rev.*, 92 (1992) 521.
237 C.F. Thurston, *Microbiology*, 140 (1994) 19.
238 A.I. Yaropolov, O.V. Skorobogatko, S.S. Vartanov and S.D. Varfolomeyev, *Appl. Biochem. Biotechnol.*, 49 (1994) 257.
239 M. Peter and U. Wollenberger, In: F.W. Scheller, F. Schubert and J. Fedrowitz (Eds.), *Frontiers in Biosensorics II*. Birkhäuser Verlag, Basel, 1997, pp. 63–82.
240 A.M. Mayer, *Phytochemistry*, 26 (1987) 11.
241 F. Pelaez, M.J. Martinez and A.T. Martinez, *Mycol. Res.*, 99 (1995) 37.
242 C. Raghukumar, S. Raghukumar, A. Chinnaraj, D. Chandramohan, T.M. Dsouza and C.A. Reddy, *Bot. Marina*, 37 (1994) 515.
243 S.O. Andersen, In: G.P. Kerkut and L.I. Gilbert (Eds.), *Comparative Insect Physiology, Biochemistry, and Pharmacology*, Vol. 3. Pergamon Press, Oxford, 1985, p. 59.
244 F. Xu, W. Shin, S.H. Brown, J.A. Wahleithner, U. Sundaram and E.I. Solomon, *Biochem. Biophys. Acta*, 1292 (1996) 303.
245 K. Piontek, M. Antorini and T. Choinowski, *J. Biol. Chem.*, 277 (2002) 37663.
246 M. Antorini, I. Herpoel-Gimbert, T. Choinowski, J.G. Sigoillot, M. Asther, K. Winterhalter and K. Piontek, *Biochim. Biophys, Acta*, 1594 (2002) 109.
247 L.E. Andreasson and B. Reinhammar, *Biochim. Biophys. Acta*, 558 (1979) 145.
248 N. Mano and A. Heller, *J. Electrochem. Soc.*, 150 (2003) 1136.

249 I. Willner and E. Katz, *Angew. Chem. Int. Ed.*, 42 (2003) 4576.
250 S.D. Varfolomeev, I.N. Kurochkin and A.I. Yaropolov, *Biosens. Bioelectron.*, 11 (1996) 863.
251 B.R.M. Reinhammar, *Biochim. Biophys. Acta*, 275 (1972) 245.
252 L.J. Anderson, D.J. Richardson and J.N. Butt, *Faraday Disc.*, 116 (2000) 155.
253 J. Diehl-Faxon, A.L. Ghindilis, P. Atanasov and E. Wilkins, *Sens. Actuators*, 35 (1996) 448.
254 K.K. Lo, L.L. Wong and H.A.O. Hill, *FEBS Lett.*, 451 (1999) 342.
255 A. Christensson, N. Dimcheva, E.E. Ferapontova, L. Gorton, T. Ruzgas, L. Stoicha, S. Shleev, A.I. Yaropolov, D. Haltrich, R.N.F. Thorneley and S.D. Aust, *Electroanalysis* (2004) in press.
256 T. Ruzgas, E. Csöregi, J. Emnéus, L. Gorton and G. Marko-Varga, *Anal. Chim. Acta*, 330 (1996) 123.
257 M.H. Thuesen, O. Farver, B. Reinhammar and J. Ulstrup, *Acta Chem. Scand.*, 52 (1998) 555.
258 M.R. Tarasevich, V.A. Bogdanovskaya, L.N. Kuznetsova and Russian, *J. Electrochem.*, 37 (2001) 833.
259 I. Fridovich, *Science*, 201 (1978) 875.
260 G. Rotilio, R.C. Bray and E.M. Fielden, *Biochim. Biophys. Acta*, 268 (1972) 605.
261 L. Flohé, In: D. Dolphin, R. Poulson and O. Avramovic (Eds.), *Glutathione—Chemical, Biochemical, and Medical Aspects*. Wiley, New York, 1989, p. 643.
262 M. Maiorino, K.D. Aumann, R. Brigelius-Flohé, D. Doria, J. van der Heuvel, J. McCarthy, A. Roveri, F. Ursini and L. Flohé, *Biol. Chem. Hoppe-Seyler*, 376 (1995) 651.
263 J. Chen, U. Wollenberger, F. Lisdat, B. Ge and F.W. Scheller, *Sens. Actuators B*, 70 (2000) 115.
264 M. Beissenhirtz, F.W. Scheller and F. Lisdat, *Electroanalysis*, 15 (2003) 1425.
265 A.J. Gow, B.P. Luchsinger, J.R. Pawlowski, D.J. Singel and J.S. Stamler, *Proc. Natl Acad. Sci. USA*, 96 (1999) 9027.
266 M.E. Murphy and E. Noack, *Methods Enzymol.*, 233 (1994) 240.
267 D.A. Wink, R.W. Nims, J.F. Darbyshire, D. Christodoulou, I. Hanbauer, G.W. Cox, F. Laval, J. Laval, J.A. Cook, M.C. Krishna, W.G. DeGraff and J.B. Mitchell, *Chem. Res. Toxicol.*, 7 (1994) 519.
268 A.J. Gow and J.S. Stamler, *Nature*, 391 (1998) 169.
269 E. Csöregi, G. Jönsson-Pettersson and L. Gorton, *J. Biotechnol.*, 30 (1993) 315.
270 X. Yu, G.A. Sotzing, F. Papadimitrakopoulos and J.F. Rusling, *Anal. Chem.*, 75 (2003) 4565.
271 E. Dock, A. Lindgren, T. Ruzgas and L. Gorton, *Analyst*, 126 (2001) 1929.
272 L.Q. Mao, P.G. Osborne, K. Yamamoto and T. Kato, *Anal. Chem.*, 74 (2002) 3684.
273 J. Zeravik, T. Ruzgas and M. Franek, *Biosens. Bioelectron.*, 18 (2003) 1321.
274 D. Ivnitzky and J. Rishpon, *Biosens. Bioelectron.*, 11 (1996) 409.

275 Z. Dai, F. Yan, J. Chen and H. Ju, *Anal. Chem.*, 75 (2003) 5429.
276 A. Krylov, M. Beissenhirtz, H. Adamzig, F.W. Scheller and F. Lisdat, *J. Anal. Bioanal. Chem.*, 378 (2004) 1327.
277 K. Tammeveski, T.T. Tenno, A.A. Mashirin, W.E. Hillhouse, P. Manning and C.J. McNeil, *Free Rad. Biol. Med.*, 25 (1998) 973.
278 R.H. Fabian, D.S. deWitt and T.A. Kent, *J. Cereb. Blood flow Metab.*, 15 (1995) 242.
279 J. Zhang, Q. Chi, A.M. Kuznetsov, A.G. Hansen, H. Wackerbarth, H.E.M. Christensen, J.E.T. Andersen and J. Ulstrup, *J. Phys. Chem. B*, 106 (2002) 1131.
280 H. Ju, S. Liu, B. Ge, F. Lisdat and F.W. Scheller, *Electroanalysis*, 14 (2002) 141.
281 J.J. Davis, D. Bruce, G.W. Canters, J. Crozier and H.A.O. Hill, *Chem. Commun.*, (2003) 576.
282 G. Tayhas and R. Palmore, *Trends Biotechnol.*, 22 (2004) 99.

Chapter 3

Enzyme biosensors containing polymeric electron transfer systems

Hiroko I. Karan

3.1 INTRODUCTION

An enzyme biosensor consists of an enzyme as a biological sensing element and a transducer, which may be amperometric, potentiometric, conductimetric, optical, calorimetric, etc. Enzyme biosensors have been applied to detecting various substrates (Table 3.1), which are selectively oxidized or reduced in enzyme-catalyzed processes depending on the nature of substrates and enzymes used (oxidases or reductases) to construct sensors.

Enzyme biosensors containing polymeric electron transfer systems have been studied for more than a decade. One of the earlier systems was first reported by Degani and Heller [1,2] using electron transfer relays to improve electrochemical assay of substrates. Soon after Okamoto, Skotheim, Hale and co-workers reported various flexible polymeric electron transfer systems applied to amperometric enzyme biosensors [3–16]. Heller and co-workers further developed a concept of "wired" amperometric enzyme electrodes [17–27] to increase sensor accuracy and stability.

To date various polymeric electron transfer systems have been incorporated to enzyme sensors for glucose, lactate, aldose, pyruvate, ascorbate, choline, acetylcholine, cholesterol, formaldehyde, phenolics, nitrate, sulfite and histamine.

The most important and most studied applications of enzyme biosensors are to detect and monitor blood glucose, followed by lactate, because of the medical applications of such sensors. Thus, by initially detailing the development of glucose biosensors we can better understand and trace the general development of enzyme biosensors containing polymeric electron transfer systems.

TABLE 3.1

Applications of amperometric enzyme biosensors

Examples of substrates that can be detected by amperometric enzyme sensors		
Carbohydrates	Glucose	Galactose
	Lactose	Sucrose
	Maltose	Cellobiose
	Fructose	Mannose
	Xylose	Arabinose
Amino acids	Aspartate	Glutamate
	Sarcosine	L-Phenylalanine
	N-Benzoyl-L-tyrosine	
Carboxylic acids	Ascorbate	Oxalate
	Pyruvate	Glycolate
	Lactate	Tartarate
	Malate	Fumarate
Alcohols and phenols	Ethanol	Phenol
	Methanol	p-Cresol
	Cholesterol esters	Catechol
	Cholesterol	o-Aminophenol
	Bilirubin	o-Cresol
		Dopamine
Amines and heterocycles	Acetylcholine	Xanthine
	Uric acid	Histamine
	Hypoxanthine	Putrescine
	Choline	
Aldehyde	Formaldehyde	
Inorganic ions, oxide and peroxide	Nitrate	Potassium ion
	Sulfite	Nitric oxide
	Oxygen	Hydrogen peroxide
Quinone	Vitamin K	
Steroids	Testosterone	Bile acid
	Estradiol	
Azide	Sodium azide	
Hemoprotein	Cytochrome c	

3.2 GLUCOSE BIOSENSORS

Diabetes is a worldwide problem afflicting approximately 5% of the adult population of industrialized nations. In the USA 18.2 million people (13 million diagnosed, 5.2 million undiagnosed), approximately 6% of the US population is afflicted with diabetes. Over 40% of the population with diabetes is age 60 or older and 90–95% of all diagnosed cases of diabetes are type II diabetes (non-insulin dependent diabetes mellitus or adult-onset diabetes) [28]. There is a great need to develop the best possible glucose biosensors to continuously, accurately, painlessly and safely monitor blood glucose level to improve the lives of all diabetics.

3.2.1 First approach: oxygen consumption detection

Almost 40 years ago Clark and Lyons [29] first reported an in vitro glucose-monitoring electrode that incorporated the enzyme glucose oxidase. In their approach, glucose oxidase was first reduced by the substrate glucose then reoxidized by oxygen (illustrated in Fig. 3.1). An amperometric oxygen electrode was used as a second electrode to measure the change in oxygen concentration to deduce the concentration of glucose in the blood. Unfortunately, the second non-enzymatic oxygen electrode is very sensitive to variations in oxygen partial pressure within the fluid to be tested, which is in contact with the electrode. This leads to inherent misreading of oxygen consumption, particularly in vivo, where physiological and/or pathological fluctuations of oxygen partial pressures are expected [30,31].

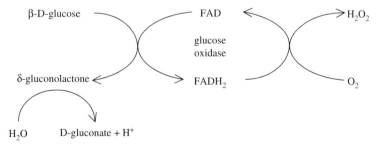

Fig. 3.1. Reaction scheme of glucose oxidation by glucose oxidase and oxygen (the natural enzymatic reaction).

3.2.2 Second approach: hydrogen peroxide concentration detection

A second approach is based on hydrogen peroxide detection (Fig. 3.1). This requires an applied potential of 650–700 mV, which is sufficient to oxidize other endogenous, interfering species such as ascorbate, urate and acetaminophene. These sensors also depend on a minimum oxygen concentration. Insufficient oxygen concentration in the media surrounding the sensor cannot oxidize all the glucose oxidase that is reduced by glucose, such that the concentration of hydrogen peroxide (produced by oxidation of the reduced glucose oxidase) measured will not accurately represent the concentration of glucose in the physiological samples.

3.2.3 Third approach: use of non-physiological redox couple

Because the redox centers in glucose oxidase are well insulated within the enzyme molecule, direct electron transfer to the surface of a conventional electrode does not occur to any measurable degree. The most common method of indirectly measuring the amount of glucose oxidase, a measure that correlates to the amount of glucose, relies on the natural enzymatic reaction of oxygen as the electron acceptor for reduced glucose oxidase as shown in Fig. 3.1. However, detecting oxygen concentration change or the amount of hydrogen peroxide produced has the drawbacks stated in the previous sections. To avoid these problems, a third approach incorporating non-physiological electron mediators [32–43] to enzyme sensors in order to shuttle electrons between flavin adenine dinucleotide (FAD) centers within glucose oxidase and the electrode surface was developed (shown in Fig. 3.2). An optimal mediator lowers required applied potential below that for hydrogen peroxide concentration detection causing less misreading of interfering species oxidized at higher applied potentials. Unfortunately, freely diffusing mediators invariably leach from the chemically

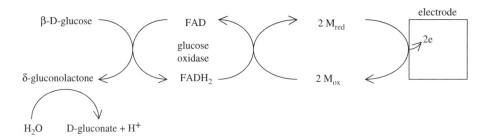

Fig. 3.2. Reaction scheme for a glucose biosensor with electron transfer mediators. The mediating species M_{red}/M_{ox} is assumed to be a one-electron couple.

modified electrode into bulk solution, precluding use of these electrodes as implantable probes for clinical applications as these mediators are usually toxic to biologic tissue and restricting their use in long-term in situ measurements in the food industry. Examples of redox mediators that have been used to construct in vitro enzyme sensors are ferrocene and its derivatives, quinone derivatives, tetrathiafulvalene (TTF), tetracyanoquinodimethanide (TCNQ), organic dye molecules, osmium complexes, potassium ferricyanide, octacyanotungstens and ruthenium complexes.

3.2.4 Electron-transfer relay systems: binding electron transfer relays to enzyme proteins

To prevent redox mediators diffusing into the bulk media, Degani and Heller attempted to chemically modify the enzyme by covalently bonding redox mediators such as ferrocene derivatives to the periphery of the enzyme. In their first attempt, ferrocenecarboxylic acid reacted with [1-[3-(dimethylamino)propyl]-3-ethylcarbodiimide hydrochloride] and the product was mixed with glucose oxidase and urea adjusting solution pH to 7.2–7.3 causing attachment of electron mediators, ferrocenecarboxylic acids to the glucose oxidase protein forming amide links with free enzyme amines such as those of lysine. When an average of 12 molecules of ferrocenecarboxylic acid was covalently bonded to a glucose oxidase molecule, measurable electrical communications between the FAD centers of glucose oxidase and gold, platinum or carbon electrodes were detected at 0.5 V (vs. SCE) [1]. Similar attempts were made to covalently attach ferrocylacetic acid or ruthenium pentaamine complexes of isonicotinic acid to lysine, tyrosine or histidine sites of glucose oxidase (GOx) or D-amino acid oxidase [2]. Amperometric glucose and D-amino acid sensors were constructed with the chemically modified enzymes and conventional metal or carbon electrodes.

These chemically modified enzyme relay systems successfully transferred electrons from the FAD/FADH$_2$ centers of enzyme to the electrode surface producing measurable current. Glucose concentration dependence of the current at 0.5 V (vs. SHE) for the ferrocenylacetamide-modified glucose oxidase solution showed linearity up to 5 mM of glucose concentration. GOx used in this study was modified by forming amides between approximately 13 of its free amines and ferrocylacetic acid. Degani and Heller also reported [2] that the residual enzymatic activity in glucose oxidase after attachment of relays exceeded 60% of the native, unmodified enzyme. Sensors constructed with redox mediator attached enzymes lost 10% of initial catalytic oxidation current within 2 h for ferrocene derivative and 10 h for ruthenium complexes. Enzyme deactivation was inevitable when covalently attaching a large number of relay molecules to the inner protein part of the enzyme which were unable to reproducibly yield quantitative attachment of electron transfer relays to the enzyme's protein. Another problem is that effective electron transfer occurs only when the enzyme aligns a specific zone of its surface to the electrode. If a relay-modified enzyme is covalently bound to an electrode surface in random fashion, only a small fraction of the enzymes can properly orient and the currents produced are too small to measure. Also oxygen interference cannot be prevented [18].

3.2.5 Electrical wiring of redox enzymes with redox polymers

To get around these problems, Degani and Heller expanded electrical communication between $FAD/FADH_2$ centers of enzymes and conventional metal or carbon electrodes by chemically modifying enzymes by wiring the enzyme with a high molecular weight (MW ~ 60,000) cationic redox polymer, a segment of which was bound to the electrodes. The polymer then forms an electrostatic complex between the polyanionic enzyme, proteins of glucose oxidase or other enzymes with a negative surface charge (i.e., due to excess glutamate and aspartate over lysine and arginine) at physiological pH resulting in electron transfer distance being reduced [17,18]. This allowed preparation of a stable polycationic electron-relay system by complexing commercial 50,000 Da poly(vinylpyridine) with $[Os(2,2'-bipyridine)_2Cl]^{+1/+2}$ and the N-methylating part of the uncomplexed pyridine residue. The polycationic electron relay was then covalently bonded to lysine amines of the enzyme in such manner as to keep the enzyme properly wired even at high ionic strength and at a sufficiently high redox polymer to enzyme ratio. Wired enzyme films of ~1 μm thickness sufficiently conducted electrons from enzyme to the electrode to allow measurement [18]. Degani and Heller selected osmium-based redox polycations over ferrocene derivatives as redox polymers because they react faster, are more durable and have redox potential in the range of 0.2–0.5 V (vs. SCE) [44]. They observed that electrical communication through these films was consistent with the electrical properties of the pure redox polymer. These results indicated that redox polymer wired enzymes could be useful to build amperometric enzyme sensors.

3.2.6 Flexible polymeric electron transfer mediators

Co-extensive with the development and application of redox enzyme wired with redox polymer to amperometric glucose biosensors by Degani and Heller, Okamoto, Skotheim, Hale and their co-workers reported a similar approach but using different redox polymer to prevent free mediator diffusion into the bulk media [3].

Okamoto et al. prepared electron transfer systems where the electron mediating species is chemically bound to a flexible polymer backbone that allows close contact between the $FAD/FADH_2$ centers of the enzyme and the mediator yet prevents the enzyme indirectly from diffusing away from the electrode surface. Methyl(ferrocenylethyl)siloxane homopolymer and methyl(ferrocenylethyl)dimethyl-siloxane co-polymers with molecular weight approximately ≥ 4000 were synthesized. Results of utility tests for carbon paste electrodes containing glucose oxidase and either of these flexible and water insoluble redox polymers showed they can act as an efficient electron-transfer relay system between $FAD/FADH_2$ centers of glucose oxidase and the carbon paste electrode. Steady-state current density measurement at 0.4 V (vs. SCE) indicated a linear response to glucose concentration over a clinically useful range (5–6 mM). The unique flexibility of the polysiloxane backbone which has a very small barrier to rotation [45] allowed these relay moieties to interact intimately with the enzyme molecule and achieve close contact with the $FAD/FADH_2$ centers. They also noted that ferrocene-modified siloxane co-polymers resulted in more efficient electron transfer than that of homopolymer. Okamoto et al. attributed this increased electron transfer efficiency to polymer flexibility. A correlation which appeared to be supported by the demonstration that redox polymer such as poly(vinylferrocene), a rather rigid polymer could not achieve close contact to enzyme redox centers and consequently could not serve as effective electron-transfer relay systems [3,46].

The ratio of ferrocene-modified siloxane subunits to unsubstituted siloxane subunits (m:n ratio) was varied as was the length (x) of the alkyl side chain onto which the ferrocene moiety was attached as shown in Fig. 3.3. The electrode containing co-polymer with m:n ratio of 1:1 or 1:2 was the more efficient electron relay systems. The ferrocene-modified homopolymer on the other hand loses flexibility due to steric hindrance caused by the side chain substitution by ferrocene, preventing efficient electron transfer from the enzyme to the electrode. The length of the alkyl side chain onto which the ferrocene moiety is attached was also found to influence the electron transfer efficiency of the electron relay system. Maximal current density was measured

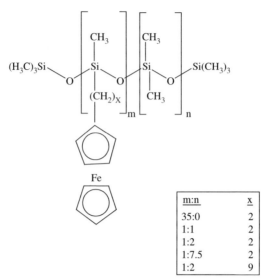

Fig. 3.3. Structure of the ferrocene-containing siloxane random block co-polymer. Reprinted with permission from [9]. © American Chemical Society.

with sensors containing ferrocene modified polysiloxane co-polymer ($m:n = 1:2$) and a longer alkyl chain ($x = 9$) [9]. The authors postulated that the longer alkyl chain might facilitate a more intimate interaction between the ferrocene moieties and the FAD/FADH$_2$ centers of glucose oxidase and/or it could create more efficient electrical communication between the relays themselves.

There is some debate whether these flexible polymeric electron transfer mediators may diffuse away from the electrode surface, as these polymers have relatively low molecular weight. Enzyme/polymer-modified carbon paste electrodes showed less than a 20% decrease in the maximum current densities after storage under dry conditions at 4°C for over 7 months and 50–60% of initial glucose response when it was stored in phosphate buffer (pH 7.0) at 4°C over the same period of time. This observation suggests the loss is most likely due to loss of water-soluble enzyme, not polymeric electron transfer mediators into the buffer solution [9]. When the electrode surfaces were coated by membranes, stability of sensors increased greatly as described in Section 3.3.1.

Unfortunately, the response of sensors containing glucose oxidase and ferrocene modified electron-transfer relay system began to deviate from linearity below 10 mM of glucose concentration. Additionally, sensor sensitivity toward oxygen and oxidation of interferent species at operating potential has yet to be solved.

3.3 DEVELOPMENT OF ENZYME BIOSENSORS CONTAINING POLYMERIC ELECTRON RELAY SYSTEMS

Although much fine tuning must be done before constructing an optimal glucose biosensor, development of polymeric electron relay systems such as the redox polymer wired enzymes of Degani and Heller, and the flexible and water insoluble polymeric electron transfer mediators and with glucose oxidase entrapped in its matrix in carbon paste by Okamoto, Skotheim, Hale and co-workers has contributed toward constructing more efficient amperometric glucose biosensors by showing that constructing electrodes with these polymeric electron transfer systems are quite general and applicable to other enzyme biosensor preparations, clearly marking trails for future enzyme biosensors development to move along.

Relevant issues still to be addressed in constructing amperometric enzyme sensors either using the electrical "wiring" of enzymes with redox polymers or with flexible polymeric electron mediators are sensor efficiency, accuracy, reproducibility, selectivity, insensitivity to partial pressure of oxygen, detectivity (signal-to-noise ratio) as well as sensor lifetime and biocompatibility [47]. Then we can address manufacturability and the cost of use of either in vitro or in vivo sensors.

3.3.1 Membranes

One approach to extend sensor stability and lifetime is to use a membrane to coat the electrode surface to prevent enzyme and/or polymer-mediator loss over time. Emr and Yacynych [48] have already published an excellent review on use of polymer films in amperometric biosensors and it is not the intent of the present article to detail work they have previously explored. Therefore, in this section a few unique membrane systems of improving performance of amperometric glucose sensors will be noted. Gorton et al. [49] reported use of commercially available Eastman AQ-29D, a water-soluble poly(ester-sulfonic acid) cation-exchanger as coating material. Carbon paste glucose electrodes consisting of a ferrocene-containing siloxane polymer and glucose oxidase were coated with AQ-29D and sensor efficiency was tested. The authors observed improvement of sensor performance through exclusion of anionic interferents and reduced electrode fouling by biological materials such as bovine serum albumin. The apparent Michaelis–Menten constant, K_m^{app} of the electrodes was reported to be ~ 200 mM. The coating improved the useful concentration range of the glucose sensor with no decrease in response current and substantially increased sensor lifetime by glucose oxidase being no longer free to diffuse away from the electrode.

Binyamin, Chen and Heller reported that "wired" enzyme electrodes constituted of glassy carbon electrodes coated with poly(4-vinylpyridine) complexed with [Os(bpy)$_2$Cl]$^{+/2+}$ and quaternized with 2-bromoethylamine or poly[(N-vinylimidazole) complexed with [Os(4,4′-dimethyl-2,2′-bypyridine)$_2$Cl]$^{+/2+}$ or poly(vinylpyridine) complexed with [Os(4,4′-dimethoxy-2,2′-bypyridine)$_2$Cl]$^{+/2+}$ quaternized with methyl groups lost their electrocatalytic activity more rapidly in serum or saline phosphate buffer (pH 7.2) in the presence of urate and transitional metal ions such as Zn^{2+} and Fe^{2+} than in plain saline phosphate buffer (pH 7.2). It was reported that as much as two-thirds of the current is lost in 2 h in some anodes. However, when a composite membrane of cellulose acetate, Nafion, and the polyaziridine-cross-linked co-polymer of poly(4-vinyl pyridine) quaternized with bromoacetic acid was applied, the glucose sensor stability in serum was improved and maintained for at least 3 days [27,50].

3.3.2 Poly[(vinylpyridine)Os(bipyridine)$_2$Cl$^{2+/3+}$] derivative based redox hydrogels (A)

Gregg and Heller further developed their concept of "wiring" by using a novel method for making cross-linked redox polymers that resemble hydrophilic epoxy cements, which they called hydrogels [51,52]. These hydrogels promote enzyme stability in the polymer matrix by providing a hydrophilic environment for enzyme to covalently bind to the polymer film. In addition, the authors reported fast diffusion of substrate and product through the polymer film, rapid electron exchange in the redox polymer and fast electrical communication with the active site of the oxidoreductases. Immobilization of enzymes in redox polymer film no longer required the use of membrane to keep either enzyme or redox mediators in the domain of the electrode surface.

Applying this concept, glucose microelectrodes were prepared [19] with glucose oxidase immobilized in a redox epoxy formed from poly(vinylpyridine)-containing Os(bipyridine)$_2$Cl, partially quaternized with bromoethylamine and polyethylene glycol diglycidyl ether (Fig. 3.4) on beveled carbon-fiber microdisk (7 μm diameter). These sensors exhibited high steady-state current densities (0.3 mA/cm$_2$) and sensitivity (20 mA/cm^2/M) at 5 mM glucose and an approximately linear glucose concentration dependence up to 6 mM glucose concentration. Pishko et al. reported that compared to macroelectrodes, these microelectrodes showed a 10-fold increase in current density, improved signal-to-noise ratios, lower glucose detection limits, significantly reduced sensitivity

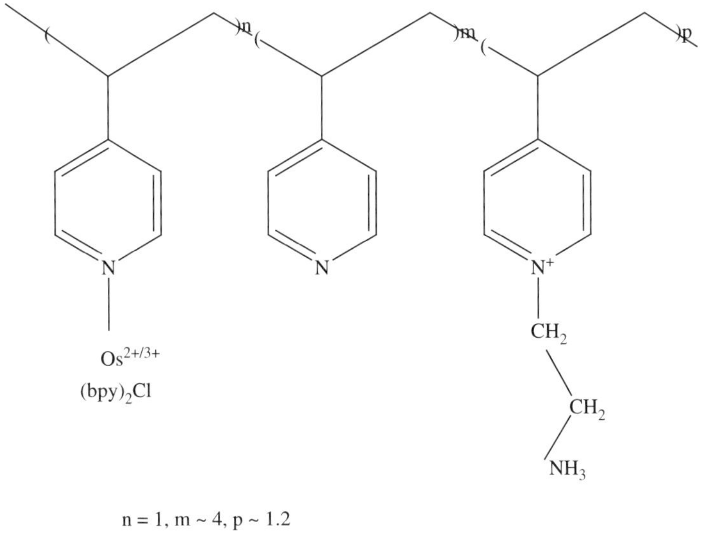

n = 1, m ~ 4, p ~ 1.2

Fig. 3.4. Structure of the poly([vinylpyridine)Os(bipyridine)$_2$Cl$^{2+/3+}$]/polyamine. Reprinted with permission from [19]. © American Chemical Society.

to oxygen and rapid response to changes in glucose concentration. This poly[(vinylpyridine)Os(bipyridine)$_2$Cl$^{2+/3+}$] derivative based redox hydrogels plays a major role in their subsequent development of amperometric enzyme biosensors.

3.3.3 Amperometric enzyme biosensors using poly[(vinylpyridine)Os(bipyridine)$_2$Cl$^{2+/3+}$] based hydrogels (A)

Poly[(vinylpyridine)Os(bipyridine)$_2$Cl$^{2+/3+}$] will be abbreviated as poly[(VP)Os(bpy)$_2$Cl$^{2+/3+}$] or hydrogel (A) hereafter.

3.3.3.1 Glucose and lactate biosensors
Since Heller and co-workers reported the use of [Poly(VP)Os(bpy)$_2$Cl$^{2+/3+}$] based hydrogels to construct amperometric glucose sensors in 1991, various researchers reported use of redox hydrogels to construct various enzyme biosensors. Researchers also took interest in developing efficient lactate sensors as lactate detection and monitoring is of great interest and importance to clinical chemistry, food processing, biotechnology and sports medicine. An efficient and accurate lactate sensor offers a critical tool to diagnose myocardial infarction, congestive heart failure, pulmonary edema, septicemia and hemorrhage. Hence, development of optimal glucose and lactate sensors

has been of particular interest to many researchers because of their medical and commercial importance.

Heller and co-workers reported subcutaneously inserted needle-type glucose biosensors using redox hydrogels by incorporating added layers of lactate oxidase and horseradish peroxidase (HRP) film to the electrode surface to eliminate common oxidizable interferents, i.e., ascorbate, urate and acetaminophene [20]. In this system blood lactate and oxygen react to produce pyruvate and hydrogen peroxide; hydrogen peroxide oxidizes peroxidase, and oxidized peroxidase oxidizes to eliminate interferents. They indicated that the reported redox hydrogel where all components are bound in a single giant three-dimensional structure would not have components leach into the biological fluid and they succeeded in miniaturizing sensors (i.e., an epoxy-embedded graphite-fiber, 7 μm diameter, in a glass capillary) by achieving high enough current densities (as high as 10^{-3} A/cm^2) and sensitivities reaching 1 A/cm^2/M with a macroelectrode. However, these sensors still face obstacles, particularly for in vivo use. The stability of wired enzymes at body temperature over extended periods of time and a method for recalibrating sensors in vivo have yet to be accomplished, and sensor fouling by blood components must be eliminated as well as sensitivity reduction associated with motion [50].

Katakis and Heller [21] also applied the redox hydrogel to construct L-lactate and L-α-glycerophosphate sensors, where they observed electro-catalytic oxidation of both lactate and glycerol-3-phosphate at electrodes coated with three-dimensional redox polymer epoxy cross-linked networks which incorporated the respective oxidases.

Ohara et al. [23,53] reported glucose and lactate sensors based on the enzyme "wiring" hydrogel formed by cross-linking water soluble poly(1-vinylimidazole) complexed with Os(bpy (a) or 4,4'-dimethylbpy (b))$_2$Cl with poly(ethylene glycol) diglycidyl ether (PEGDGE) using gold or vitreous carbon electrodes. They reported that the polymer was simpler and easier to make than previously reported for poly[(VP)Os(bpy)$_2$Cl$^{2+/3+}$] hydrogels and also lowered the operating potential at which glucose and lactate electrodes were selective for their substrates in the presence of physiologic concentrations of electrooxidizable interferents from 300 to 200 mV (vs. SCE). Coating the electrodes with Nafion did not improve sensor selectivity but reducing substrate mass transport, increased the linear sensing range from 6 to 30 mM for glucose and 4 to 7 mM for lactate.

Recently, Kenausis et al. reported novel glucose and lactate sensors by constructing four polymer coated layers on vitreous carbon, based on "wired" thermostable soybean peroxidase and redox hydrogels (A) [54]. The first layer was cross-linked thermostable soybean peroxidase and redox polymer, the second layer was insulating and H_2O_2 transport controlling cellulose acetate, the third layer was immobilized glucose oxidase or lactate oxidase and the fourth layer was a substrate transport controlling cellulose acetate layer. The glucose sensors were operated at 0.00 V (vs. SCE) and were relatively insensitive to motion and interferents. The current was independent of oxygen partial pressure above 15 Torr and the sensors maintained stable output under continuous operation at 37°C for 12 days. Lactate electrodes showed similar results to those of glucose sensors except they were more influenced by low oxygen concentrations. These sensors considerably improved overall sensor performance compared to the reported bilayer glucose sensor consisting of HRP in an electron conducting redox hydrogel (first layer) and the avidin cross-linked and immobilized biotin labeled glucose oxidase (second layer) [55]. The bilayer sensors were affected by common interferents and instability even at ambient temperature.

However, actual commercial application of biosensor technology has been limited to home glucose test meters and blood-gas instruments designed to detect glucose and lactate. Therefore, development of miniature sensors that can detect multiple analytes in vivo is of great commercial interest. These sensors must contain a large concentration of enzymes to provide high signal level, must be easily and reproducibly fabricated and spatially distinct. Hence, nanoscale assembly for this type of biosensor has become an important issue. To achieve this, generation of biotin/avidin/enzyme nanostructure with maskless photolithography on micrometer-sized domains of a carbon surface to measure neurotransmitter dynamics [56,57], and a GOx monolayer electrode via reconstitution of apo-protein with pyrroloquinoline quinone (PQQ)/FAD monolayer with potential application for a miniaturized enzyme electrode for continuous glucose monitoring in vivo have been reported [58, 59]. More recently Pishko et al. reported glucose, lactate and pyruvate enzyme biosensors based on poly[(VP)Os(bpy)$_2$Cl]-co-allylamine nanocomposite thin films [60,61] using techniques of self-assembly and polyion adsorption on gold electrode on SiO_2/Si or flexible Mylar substrate with 50 μm in diameter and 10 μm in width. Sensor performance still needs to be improved, but this clearly shows the future direction for enzyme biosensor fabrication (miniaturizing) and mass production in both invasive and non-invasive multianalytes sensing. The same trend is noted for recently reported applications of enzyme biosensors using electron-conducting polymers to construct miniature biofuel cells [62–64].

Recent advances and progress in electrically contacted, layered enzyme electrodes in the potential applications for amperometric biosensors, sensor arrays, logic gates and optical memories can be found in the review by Willner and Katz [65].

3.3.3.2 Applications of hydrogel family to other enzyme biosensors
Other examples of sensors constructed with hydrogels (A) are amperometric peroxide, choline and acetylcholine sensors based on electron transfer between HRP and a redox polymer [66]. More recently, Niculescu et al. reported histamine (a biomarker for food freshness) detection by amine oxidase based amperometric biosensors constructed with [Os(4,4'-dimethylbpy)$_2$Cl]$^{+/2+}$ complexed poly(vinylimidazole) hydrogels [67] which exhibited better electrochemical selectivity and sensitivity compared to those based only on the enzyme amine oxidase. Application of the redox hydrogels family for amperometric graphite sugar sensors with oligosaccharide dehydrogenase in a flow injection system [68], carbon ink screen-printed formaldehyde detecting sensors [69] and respiratory toxins such as azide [70] has been reported. More recently, Barton et al. reported an application of a redox hydrogel of poly-N-vinylimidazole with one-fifth of its ring coordinated to [Os(tpy)(dme-bpy)]$^{2+/3+}$ (where tpy is 2,2';6',2''-terpyridine and dme-bpy is 4,4'dimethyl-2,2'-bipyridine) to wire *Pleurotus ostreatus* laccase to construct oxygen sensors [71].

3.3.3.3 Enzyme immobilization methods other than using the hydrogels (A)
Since applications of the redox hydrogels (A) to glucose sensors was first reported by Pishko et al. in 1991, various other methods have been reported to immobilize redox polymers to construct biosensors as indicated (see Section 3.3.3.2). Other recent examples are as follows.

Amperometric glucose biosensors based on co-immobilization of GOx with [Os(bpy)$_2$(4-VP)$_{10}$Cl]Cl in electrochemically generated polyphenol film [72]. A bilayer electrode in which Os-polymer is first adsorbed on a Pt electrode and then electrochemically deposited on polyphenol-GOx film showed sensitivity of 1.63–1.79 μA/cm^2/mM to 20 mM glucose at 0.4 V and K_m^{app} 6–7 mM. Low background current and less interference by common electroactive compounds were noted.

Jezkova et al. reported the addition of polyethyleneimine (PEI) as stabilizer to a carbon paste glucose biosensor based on [Os(bpy)$_2$(4-VP)$_{10}$Cl]Cl polymer [73] increased sensor sensitivity and stability compared with sensors without PEI. Addition of PEI apparently enhances the partition and diffusion coefficients of the substrate within the reaction region.

Another enzyme immobilization technique uses sol–gel film. Recently, amperometric glucose [74], lactate [75], phenolics [76] and hydrogen peroxide

[74,77] sensors using [Os(bpy)$_2$(VP)$_{10}$Cl]Cl as polymeric electron transfer system coated by sol–gel film derived from methyltriethoxysilane have been reported. Generally, sol–gel immobilization of enzyme appears to increase sensitivity, K_m^{app} and sensor stability.

Scheller et al. reported amperometric pyruvate sensors by potentiostatic co-polymerization of Os(bipy)$_2$pyridineCl-modified pyrrole monomer and thiophene on platinized glassy carbon electrodes on which pyruvate oxidase was adsorbed [78]. This polythiophene based redox polymer was reported to have excellent electron transfer properties with significantly improved stability compared with polypyrrole as they are not affected by oxygen [79]. However, notable interference by ascorbate needs to be eliminated.

Recently, Schuhmann et al. reported ethanol biosensors by entrapping quinohemoprotein alcohol dehydrogenase and Os-complex-modified poly(vinyl imidazole) during the electrochemically induced deposition of the poly(acrylate)-based resin [80]. The sensor exhibited its efficiency and also sufficient stability for practical applications. Author claims that the reported sensor preparation process is simple, easy to control, oxygen insensitive and can be applicable to other enzyme sensors.

3.3.4 Mediator containing flexible polymeric electron transfer systems

When ferrocene-containing polysiloxane proved to be an efficient electron-transfer relay system, further modification of this type redox polymer was investigated to develop optimal enzyme biosensors. Attempts were made to synthesize redox polymers with different mediators and/or different polymer backbones and/or different side chains through which mediators are attached to the polymer backbone. Resulting redox polymers were tested to construct different types of enzyme sensors.

3.3.4.1 Ferrocene modified flexible polymeric electron transfer systems
Ferrocene and its derivatives are readily available and commonly used organometallic redox mediators, so it is quite natural that they were selected first to synthesize mediator modified polymeric electron transfer systems. Siloxane polymers are flexible but aqueous insoluble polymers. As previously indicated, a flexible polymer backbone allows close contact between the redox center(s) of the enzyme and the mediator, and the water insoluble property of the polymer prevents not only redox polymer from leaching into bulk media but also prevents enzyme diffusion away from the electrode surface by entrapping it in the polymer/carbon paste matrix. Therefore, ferrocene and

polysiloxane combination appeared to form a desirable polymeric electron transfer systems that could efficiently interact with enzyme at the electrode. Various ferrocene modified polymeric electron transfer systems have been prepared with varying polymer flexibility, polymer backbone, spacing between ferrocene molecules on the redox polymer, nature of pendant chains through which mediators are attached to polymer backbone.

As indicated previously, Hale et al. initially reported a new class of amperometric glucose biosensors in 1989 incorporating methyl(ferrocenylethyl)siloxane homopolymer (approximately 35 subunits each containing a bound ferrocene) and methyl(ferrocenylethyl)dimethylsiloxane co-polymer (random co-polymer ratio $m:n = 1:2$, Fig. 3.3) [3] and showed that these polymeric electron transfer mediators efficiently shuttled electrons from FAD/FADH$_2$ centers of glucose oxidase to the electrode surface. Systematic tests of glucose sensors using carbon paste or graphite rod or graphite rod disk electrodes with glucose oxidase and various ferrocene-modified polysiloxanes were carried out ($E = 300$ mV vs. SCE) [4,5,7,9]. Varying the $m:n$ ratios of co-polymers showed $m:n = 1:1$ or 1:2 to be more efficient electron transfer systems than homopolymer. Apparently, adequate spacing between ferrocene attached to polymers maintains polymer flexibility with adequate ferrocene site density in the polymer matrix. Longer alkyl side chain through which ferrocene was attached to the siloxane backbone increased sensor sensitivity, perhaps because increased flexibility of the longer side chains provided better interaction between FAD/FADH$_2$ centers in GOx and ferrocene moieties. Sensors with polymers ($m:n = 1:2$, $x = 2$ in Fig. 3.3) stored under dry conditions at 4°C lost less than 20% sensitivity, when stored in phosphate buffer (pH 7.0) at 4°C, they lost approximately 50–60% of initial response over the same period of time. An increased loss is most likely due to loss of GOx into buffer. However, the fact that sensors stored in buffer retained over 20% of initial response after more than 2 years indicates that some enzyme is irreversibly adsorbed to the polysiloxane-modified carbon paste.

These ferrocene modified polysiloxane polymers were also used to construct glycolate [6,7], lactate [7], acetylcholine [12,81], glutamate [12] and cholesterol [81] sensors. All these electrodes showed that ferrocene containing siloxane polymers efficiently shuttled electrons between redox center(s) of enzyme and the electrode surface.

One way to increase interaction between redox polymers and enzyme (i.e., glucose oxidase) known to have surface charges is to increase hydrophilicity of the polymer. To test the hypothesis, ferrocene containing polysiloxane with ethylene oxide pendant chains was prepared and their efficiency tested as glucose and acetylcholine sensors [12,81]. Glucose and acetylcholine sensors constructed with ferrocene-ethylene oxide-siloxane responded better to set amounts of substrate than those constructed with ferrocene-ethyl-siloxane at $E = 300$ and 250 mV (vs. SCE), respectively. The next logical step was to tailor redox polymers to increase hydrophilicity of polymer backbone and test the effect on relaying electrons from the enzyme redox center(s) to the electrode surface. Okamoto and co-workers chose this path to prepare ferrocene-containing poly(ethylene oxide) (Fig. 3.5, $m:n = $ 1:10 and 1:15) to test efficiency compared with ferrocene containing polysiloxane. These redox polymers effectively facilitated electron flow from FAD/FADH$_2$ centers of glucose oxidase to a conventional carbon paste electrode even at $E = 100$ mV (vs. SCE). They further found that polymers with a random co-polymer ratio $m:n = 1{:}10$ were more efficient electron relay systems than those with an $m:n = 1{:}15$ [8]. Although increase in polymer backbone hydrophilicity had a positive effect on lowering the sensor operating potential and minimizing interferent detection, the expected increase in sensor response was unfortunately not observed because the density of ferrocene was not sufficient due to rather large $m:n$ ratios. Swelling of these carbon paste electrodes was also observed after prolonged use in phosphate buffer at pH 7.0.

Other ferrocene containing polymeric electron relay systems were investigated by Casado et al. and Watanabe et al. for application to glucose

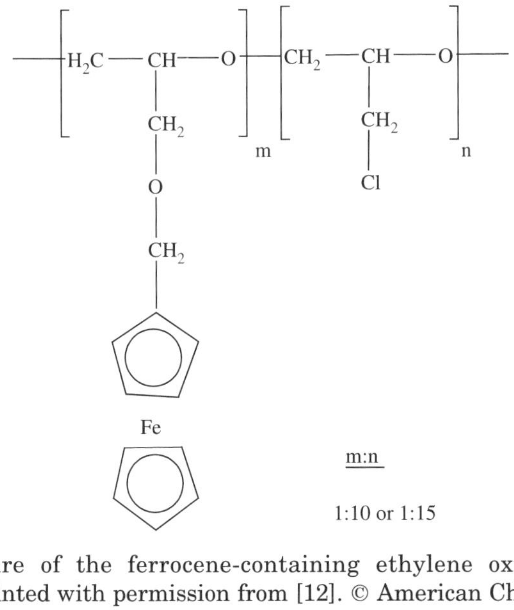

Fig. 3.5. Structure of the ferrocene-containing ethylene oxide random block co-polymers. Reprinted with permission from [12]. © American Chemical Society.

biosensors. Casado et al. synthesized both siloxane polymers in which ferrocene was covalently attached pendant like and ferrocene was incorporated as part of polymer chains [82]. The former [[{η^5-C$_5$H$_5$}Fe{η^5-C$_5$H$_4$CH$_2$-CH$_2$NHC(O)CH$_2$CH$_2$CH$_2$}]CH$_3$SiO−] was found to be an effective electron relay system comparable to other reported ferrocene modified poly(siloxane) glucose sensors. The latter became very insoluble in all organic solvents due to possible cross-linking and was not practical for sensor construction. Watanabe et al. [83] synthesized a series of flexible novel co-polymers, poly[ferrocenylmethyl methacrylate-co-methoxy-oligo(ethylene oxide)methacrylate (M_w = 30,000–78,000). Carbon paste glucose electrodes prepared with GOx and these ferrocene containing polymers efficiently transferred electrons from the GOx redox centers to the electrode surface with linear response to glucose concentration up to 50 mM at $E = 0.5$ V (vs. SCE). Calvo et al. also demonstrated direct electrical communication of glucose oxidase with glassy carbon electrodes coated with acrylamide–acrylic acid ferrocene co-polymer [84] or ferrocene-allylamine polymer (ferrocene based hydrogels) [85]. However, additional studies on sensor stability, sensitivity to oxygen and elimination of interferent detection are necessary to determine the utility of these polymeric electron relay systems for optimal glucose and other biosensor development.

A ferrocene modified siloxane redox polymeric electron transfer system in carbon paste electrodes for aldose biosensors using PQQ-dependent aldose dehydrogenase was reported by Smolander et al. [86]. Polymethyl(11-ferrocenyl-4,7,10-trioxa-undecanyl)methyl(12-amino-4,7,10-trioxa-dodecyl)-siloxane (1:1 random co-polymer) (Fig. 3.6) was found to be an efficient electron transfer system yielding better electrode operational stability than those constructed with dimethylferrocene free mediator. The hydrophilic nature of the pendant chain and side chain on dimethyl siloxane units favorably interact with enzyme causing efficient electron transfer from coenzyme PQQ of aldose dehydrogenase to the electrode surface.

Casado et al. reported preparation of ferrocene-containing dendritic macromolecules in which the ferrocenyl units are at the end of long flexible silicon-containing branches. When these macromolecules were tested as mediators in amperometric biosensors [87–89], sensors constructed with the redox dendritic macromolecules and glucose oxidase in carbon paste were reported to be stable and exhibited a higher sensitivity than ferrocene-modified polymer mediated electrodes. More recently this group reported platinum electrodes coated with a new family of siloxane-based poly(ferrocenyl)dendronized molecules [90] for glucose and hydrogen peroxide

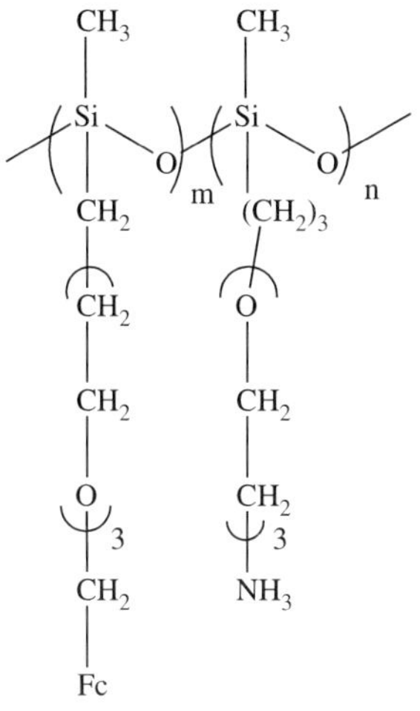

Fig. 3.6. Structure of polymethyl(11-ferrocenyl-4,7,10-trioxaundecanyl)methyl(12-amino-4,7,10-trioxa-dodecyl)siloxane. Reprinted with permission from [86].

detection [91]. These electrodes showed comparable efficiency to other glucose biosensors based on ferrocene-modified polymeric electron transfer systems.

Researchers turned their attention to applications of silica gel as a new electrode material. Silica gel, which has a three-dimensional structure with high specific surface area and is electroinactive in an aqueous medium can be used as a support for electroactive species during their formation and/or enzymes by adsorption or entrapment [92,93]. Patel et al. recently reported application of polyvinyl ferrocene immobilized on silica gel particles to construct glucose sensors. Efficiency of carbon paste electrodes prepared with these polymeric electron mediators and GOx was comparable to electrodes constructed with other ferrocene based polymeric electron transfer systems. The fact that 70% of initial anodic current was retained after a month when electrodes were kept in the buffer at room temperature shows that polymerization of monomer vinylferrocene in the pores of silica gel and entrapping GOx in the matrix of polyvinyl ferrocene appears to have added stability to the sensors [94].

As was indicated in Section 3.3, an issue to be addressed before glucose or other biosensors is a commercially practical sensor fabrication. An easier and simpler sensor fabrication method was recently investigated using ferrocene modified redox polymer hydrogels. Sirkar and Pishko reported amperometric biosensors based on oxidoreductase immobilization in UV-photopolymerized redox polymer hydrogels (precursors consisted of vinylferrocene, 2,2′-dimethoxy-2-phenylacetophenone and poly(ethylene glycol) diacrylate) [95] on 1.6 mm diameter gold electrodes. Amperometric glucose biosensors based on trapping glucose oxidase in this hydrogel were reported to be linear in the range of 2–20 mM of glucose with a sensitivity of 0.5 μA/cm^2/mM at $E = 0.3$ V (vs. Ag/AgCl). These studies demonstrated that enzyme-containing redox polymer hydrogels could be rapidly fabricated using UV-initiated free radical photopolymerization. Further expansion of sensor fabrication methods for glucose, lactate and pyruvate was recently reported by these investigators (see Section 3.3.3.1).

3.3.4.1.1 Hydrogen peroxide detection using horseradish peroxidase
Ferrocene modified redox polymers were tailored to increase sensor efficiency, reproducibility and to lower operating potential to minimize oxidation of interferents such as ascorbate, urate and acetaminophene in biological fluid and minimize sensitivity to ambient oxygen. Application of membranes such as AQ-29D was tested to improve sensor accuracy and sensitivity. However, no single or combination of methods addressed all the issues related to achieving desired biosensors. One of the major obstacles encountered was difficulty synthesizing ideal redox polymers. Simple and easy redox polymer synthetic methods have yet to be developed.

Instead of using direct coupling of redox mediators to oxidoreductases, researchers reconsidered detection of hydrogen peroxide, which is produced by substrate–oxidoreductases reaction with oxygen. To lower the high operating potential necessary to detect hydrogen peroxide in previous hydrogen peroxide detection methods, an attempt was made to develop sensors by coupling oxidoreductases with H_2O_2-converting enzymes like peroxidases [96–98] that could enable H_2O_2 detection at considerably lower potentials, thereby greatly reducing interference by common oxidizable species (Fig. 3.7).

Amperometric biosensors with bienzyme systems were developed by co-immobilizing HRP and oxidases in carbon paste (graphite powder was

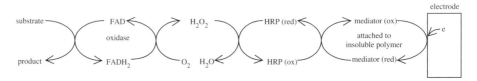

Fig. 3.7. Redox cycles for bilayer-bienzyme electrodes incorporating a flavoenzyme, horseradish peroxidase and a polymeric electron relay system.

activated by a carbodiimide derivative) and studied for detection of glucose, alcohol, amino acid and lactate (under a flow injection system) [99,100] at lower operating potential (-50 mV ~ 0 mV vs. Ag/AgCl) than reported for monoenzyme electrodes. These bienzyme sensor responses were reported as very fast with interference by the direct electrochemical oxidation of easily oxidizable species such as urate, paracetamol and NaN_3 minimized. However, these sensors suffered from low linear response range of substrate, sensor instability and substrate inhibition of HRP at substrate concentration higher than physiologic range. These shortcomings were lessened by covering the electrode surface with electropolymerized layers of o-phenylenediamine or a mixture of p-phenylenediamine and resorcinol.

Alternative ways to improve sensor sensitivity and stability of the bienzyme system reported use of redox polymers for bienzyme systems, specifically to interact with HRP instead of with oxidoreductases.

Vreeke et al. reported amperometric hydrogen peroxide and β-nicotinamide adenine dinucleotide sensing electrodes based on the electrical connection of HRP to the electrode through a three-dimensional electron relaying polymer network [101]. In this system, a quinoid mediator, 5-methylphenazonium methyl sulfate is reduced by NADH and the reduced mediator is reoxidized by dissolved oxygen, which is, in turn, reduced to hydrogen peroxide. The hydrogen peroxide is then reduced by HRP, which is electrically wired through poly[(VP)Os(bpy)$_2$Cl$^{2+/3+}$] based polymeric electron transfer systems to the electrode surface. The sensitivity of the resulting hydrogen peroxide sensor is 1 A/cm^2/M at 0.00 V (vs. SCE) with detection of NADH in a range of 1×10^{-7}–2×10^{-4} M.

Use of Os-complexed redox polymer based hydrogel for bienzyme systems to detect glucose, peroxide, choline and acetylcholine was reported by Garguilo et al. [66]. In this case redox mediators interact with HRP instead of with oxidase. Their glucose sensors operating at 0.00 V (vs. SCE) resulted in highly efficient glucose detection that was comparable to monoenzyme glucose detection at $E = 0.45$ V (vs. SCE) under 5 mM glucose concentration. Choline sensors operating at -0.1 V (vs. SCE) showed linear correlation between choline concentration and current output below 100 μM choline, well within biological relevant concentrations.

Another example of this type of sensor (multilayer thin film bienzyme amperometric biosensors using graphite rod electrodes, based on ferrocene containing siloxane polymers with ferrocene covalently attached through seven units of the ethylene oxide side chain [102]) was reported by Boguslavsky et al. The sensors operated at 0.00 mV (vs. Ag/AgCl) and were used to test for glucose, ethanol, choline and glycerol-3-phosphate. Calibration curves for these sensors indicated linearity up to the 16, 15, 5 and 2 mM range for respective substrates (well within the relevant physiologic concentration range). Better insulation from one enzyme layer to another in these electrodes seems to prevent short-circuiting, i.e., substrate inhibition of HRP, and direct interaction of oxidase and redox polymer. These sensors effectively reduced interference by common oxidizable species.

More recently, Danilowicz et al. reported bienzyme single layer modified electrodes with HRP and GOx or lactate oxidase and $Os(bpy)_2ClPyCH_2NH$-poly(allylamine), depositing and cross-linking on glassy carbon disk [103]. These glucose and lactate bienzyme sensors were reported to have good operational stability and could be used in flow injection analysis. Also, Gorton et al. reported amperometric bienzyme alcohol sensors by co-immobilizing commercially available alcohol oxidase and HRP which was first "wired" by osmium containing three-dimensional redox hydrogel, in a carbon paste matrix [104]. The authors reported that the oxidase-wired peroxidase electrodes exhibited higher sensitivity for ethanol detection and better storage and operational stability than those based on adsorption onto graphite. Similar bienzyme glutamate sensors with chip integrated enzyme microreactors reported by Csöregi et al. have shown a potential off-line in vivo glutamate measurement without interference from ascorbate [105]. Castillo et al. recently reported sensors based on redox hydrogels constructed with relevant oxidases using sweet potato peroxidase instead of HRP to detect glucose, alcohol and putrescine. Sensors constructed with amine oxidase and sweet potato peroxidase were reported to exhibit higher sensitivity and lower detection limit for putrescine detection than those with HRP [106].

3.3.4.2 Quinone modified polymeric electron transfer systems
The redox mechanism of quinone (two electron–proton acceptor/donor) is pH dependent and somewhat more complicated than for ferrocene or osmium (one electron accepter/donor). However, quinones are naturally occurring redox mediators and therefore, many researchers have studied their application to biosensors [107–109].

Karan et al. [10] reported glucose sensors using quinone modified polysiloxane (Fig. 3.8a-A) and acrylonitrile–ethylene (Fig. 3.8a-B) co-polymers and glucose oxidase. Sensors constructed with glucose oxidase and quinone modified polysiloxane were considerably more efficient than those using acrylonitrile–ethylene system to transfer electrons from reduced glucose oxidase to a conventional carbon paste electrode. Their results coincide with those described previously for the ferrocene-modified polysiloxane system. The excellent flexibility of poly(siloxane) allows it to function as an efficient

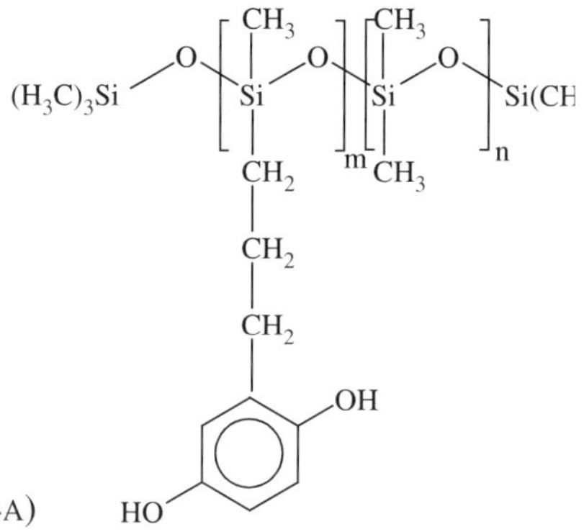

Fig. 3.8. (a) A: Structure of the hydroquinone-containing poly(siloxane). B: Structure of the hydroquinone-containing poly(acrylonitrile-ethylene). The $m:n$ ratio is approximately 1:2 for both systems. Reprinted with permission from [10]. (b) Structure of the poly(ether amine quinone) polymers. Reprinted with permission from [15]. © 1994 American Chemical Society.

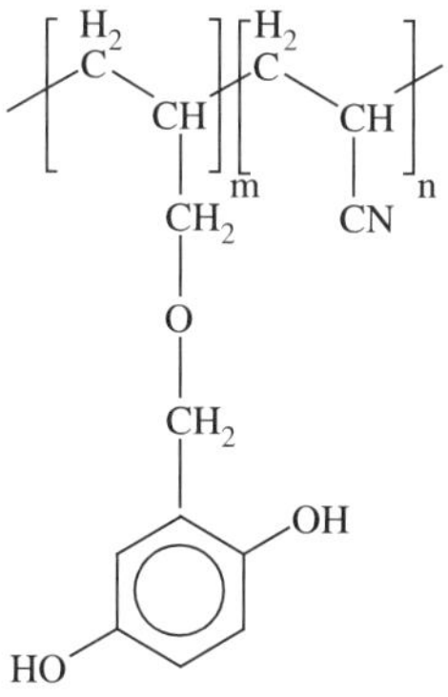

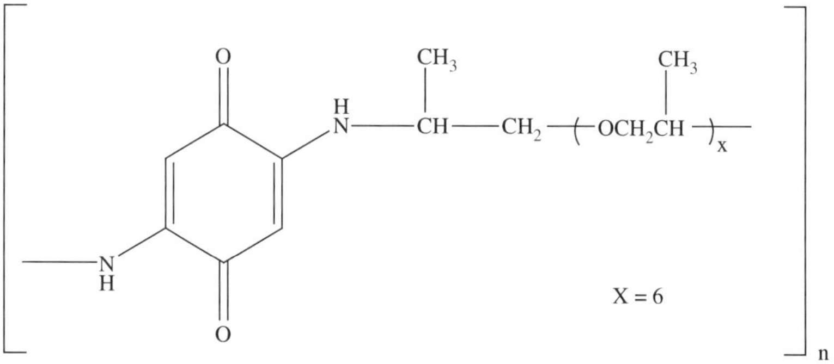

electron transfer system. However, decrease in sensor response due to oxygen interference could not be completely eliminated.

Kaku et al. subsequently reported using poly(ether amine quinone)s [13, 15,16] as the polymeric electron transfer system for glucose sensors. This polymeric electron transfer system contains redox mediator as part of the main polymer chain instead of as side chains. This system efficiently transferred electron from FAD/FADH$_2$ centers of glucose oxidase to the electrode surface (Fig. 3.8b). When poly(ether amine quinone) polymers were cross-linked, the stability of carbon paste electrode consisting of cross-linked polymer and glucose oxidase was markedly improved, though sensor sensitivity was somewhat compromised. The authors reported that sensors maintained their original efficiency over 50 days, which they attributed to entrapping glucose oxidase in the cross-linked polymer matrix that prevented enzyme leaching out to bulk media. Tessema et al. also reported construction of oligosaccharide sensors by doping poly(ether amine quinone) and oligosaccharide dehydrogenase into the bulk of carbon paste [110], which showed higher sensitivity than those constructed with benzoquinone.

Recently, Dordick and co-workers reported amperometric glucose sensors using enzymatically prepared flexible poly(hydroquinone) (PHQ) as a polymeric electron transfer mediator [111]. A pellet (cross-sectional area of 0.3 cm^2 with 1 mm thickness) formed from a mixture of graphite, PHQ and GOx was placed on the surface of a glassy carbon electrode to form a sensor. The resulting glucose sensors demonstrated rapid response time to glucose and could be operated at lower operating potential (<200 mV), thereby minimizing interference from common endogenous and oxidizable interferents. However, to achieve long-term sensor stability, GOx had to be immobilized because of the hydrophilic nature of PHQ, and the linearity of the calibration curve at lower operating potentials must be improved before any practical application could be attempted.

More recently, Ng et al. reported a sulfite biosensor based on sulfite oxidase covalently immobilized onto a matrix of co-polymer chitosan-poly(hydroxyethyl methacrylate) (chitosan-HEMA) via UV curing process using a platinum disk electrode. *p*-Benzoquinone was coupled to the polymer network [112] and hydrogen peroxide generated by the action of sulfite oxidase was reduced by hydroquinone. Sensors using this system successfully transferred electrons to the electrode surface but sensor sensitivity decreased more than 20% after overnight storage in a 0.1 M phosphate buffer at pH 8. Since chitosan is inherently biodegradable, biocompatible and non-toxic, further improvement and development of this polymeric redox system is expected.

3.3.4.3 Viologen containing polymeric electron transfer systems

As previously indicated, ferrocene, quinone and other organic and inorganic derivatives are known to be efficient electron transfer mediators and have been applied by many researchers as mediators to construct efficient biosensors. Unfortunately, due to the relatively high oxidation potential of these mediators, sensors using these mediators must be operated at potentials where several common interferents in biological fluids are directly oxidized. Water-soluble viologen derivates that have higher oxidation potentials than FAD can mediate electrons efficiently from the FAD centers of glucose oxidase to the electrode surface. Another advantage of using viologen derivatives for biosensors is that the oxidation of viologen derivatives occurs at lower potential than for derivatives of ferrocene or quinone resulting in sensors that can operate at potentials where the oxidation of common interferents such as ascorbate, urate and acetaminophene in biological fluid is minimized. Hale and co-workers used water-soluble poly(o- and p-xylylviologen) to construct glucose biosensors and reported that these sensors efficiently mediated electrons from reduced glucose oxidase to a conventional carbon paste electrode [4,113]. Okamoto et al. synthesized viologen-containing polysiloxane, where viologens are covalently attached to siloxane co-polymer yielding a redox polymer, which is water insoluble (Fig. 3.9) [14]. Glucose sensors constructed with viologen modified siloxane polymers efficiently mediated electron transfer from FAD centers in glucose oxidase to the conventional carbon paste electrode at an applied potential of -100 mV (vs. Ag/AgCl) with minimal oxidation of common biological interferents. However, it must be noted that purification of the redox polymer is a painstaking task.

Application of viologen containing redox polymers to construction of amperometric nitrate biosensors by electropolymerization of nitrate reductase and amphilic pyrrole viologen mixtures [114–116], and hydrogen sensors consisting of glassy carbon electrodes coated with viologen substituted polyvinylpyridine redox polymer to which enzyme (hydrogenase) was cross-linked by PEGDGE have also been reported [117].

3.3.4.4 Tetrathiafulvalene containing polymeric electron transfer systems

Tetrathiafulvalene (TTF) is another mediator many researchers have utilized as an electron mediator in constructing amperometric biosensors. Electrodes constructed of TTF efficiently oxidize glucose oxidase [118,119], lactate oxidase [120] and choline oxidase [121]. These sensors like those using viologen containing polymeric electron transfer systems have the

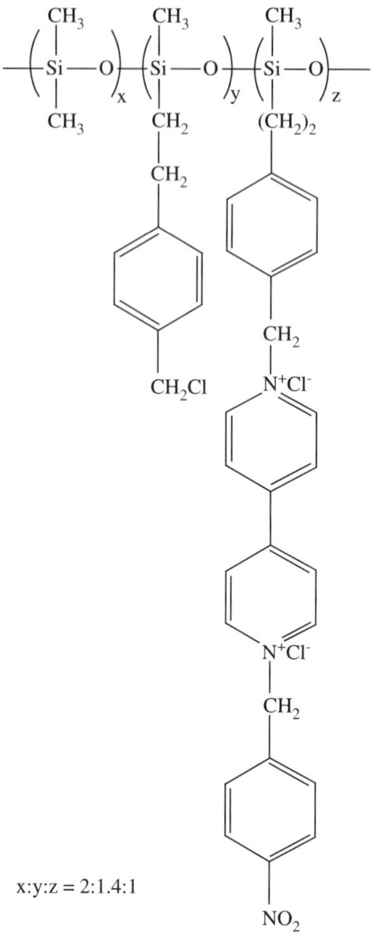

Fig. 3.9. Structure of the viologen-containing poly(siloxane). Reprinted with permission from [14]. © 1994 American Chemical Society.

advantage of operating at much lower potential (150 mV vs. SCE) than those based on H_2O_2 detection (700 mV vs. SCE) or sensors using mediators such as ferrocene or quinones (300–400 mV vs. SCE). Okamoto and co-workers synthesized TTF-containing siloxane polymer by covalently attaching TTF moiety to a highly flexible siloxane polymer (Fig. 3.10). Amperometric glucose sensors constructed with the insoluble polymeric electron-transfer relay system facilitated electron flow from the enzyme to the electrode and functioned efficiently over a clinically significant range of glucose concentration at $E = 200$ mV vs. SCE [122].

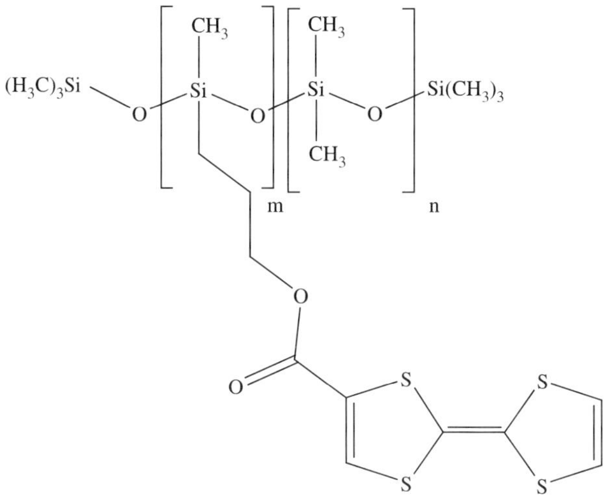

m:n = 1:2

Fig. 3.10. Structure of the TTF-containing poly(siloxane). Reprinted with permission from [122].

3.3.5 Covalent electropolymerization of glucose oxidase in polypyrrole

When constructing biosensors, which are to be used continuously in vivo or in situ, maintaining sensor efficiency while increasing sensor lifetime are major issues to be addressed. Researchers have attempted various methods to prevent enzyme inactivation and maintain a high density of redox mediators at the sensor surface. Use of hydrogels, sol–gel systems, PEI and carbon paste matrices to stabilize enzymes and redox polymers was mentioned in previous sections. Another alternative is to use conductive polymers such as polypyrrole [123–127], polythiophene [78,79] or polyaniline [128] to immobilize enzymes and mediators through either covalent bonding or entrapment in the polymer matrix. Application to various enzyme biosensors has been tested.

Lowe and co-workers investigated covalent immobilization of enzyme in polypyrrole film [123,124]. First, pyrrole-modified enzyme was prepared by reacting glucose oxidase with either N-(3-aminopropyl)pyrrole or N-(2-carboxyethyl)pyrrole, then electropolymerized enzyme films, which covalently immobilize enzyme were deposited at platinum disk electrodes from solutions of free pyrrole and native or pyrrole modified enzyme of equivalent activity. They observed that sensors constructed of covalently electropolymerized GOx films demonstrated higher enzyme activity than those using entrapped native enzymes in the film, indicating pyrrole covalently bound to GOx was more accessible to electron transfer. However, direct electron transfers from FAD/FADH$_2$ centers of GOx through conductive polymer to the electrode surface are not generally efficient, causing sensor sensitivity to glucose to be considerably lower than for those having the flexible polymeric electron transfer systems that were discussed in the previous sections.

Yoneyama and co-workers reported simultaneous immobilization of glucose oxidase (GOx) and hydroquinonesulfonate (HQS) ions in polypyrrole (PPy) films [125] and investigated glucose sensitivity of the resulting polymer films. They observed efficient amperometric responses of the GOx/HQS/PPy electrodes to glucose, however glucose sensitivity depended on the deposition charge of polypyrrole films and hence optimum deposition charge needed to be determined as a function of the diffusion of glucose in the polypyrrole films and the rate of the electron transfer between GOx and HQS.

Subsequently, Schuhmann reported GOx entrapment by ferrocene derivatives attached to the enzyme outer surface via long flexible polyethylene spacer chains within polypyrrole films [126]. However, this system still did not significantly improve sensor sensitivity and the author pointed out that modifications are needed to increase the concentration of redox relays to enhance the sensor efficiency.

Recent development in multilayer sensor architecture using sequential electrochemical polymerization of pyrrole and pyrrole derivatives to entrap enzymes was tested on a tyrosinase-based phenol sensor [127]. A phenothiazine dye, thionine served as redox mediator and was covalently attached to the thin, functionalized first polypyrrole layer on Platinum disk electrodes. Then, a second layer of polypyrrole with entrapped tyrosinase was electrochemically deposited. The phenol sensor constructed in this manner effectively transferred electron from enzyme to the electrode surface. As all steps in preparation, including deposition of the enzyme-containing layer are carried out electrochemically, this technique may prove to be applicable for mass production of miniature sensors.

Sharma et al. reported application of poly(2-fluoroaniline) films which was electrochemically deposited on ITO coated glass plates to produce glucose sensors. GOx was immobilized on the polymer films by physical adsorption methods. Sensors constructed by this method showed efficient electron transfer between the adsorbed GOx and the electrode surface and were found to be stable up to 32 days [128].

Palmisano et al. recently reported a comprehensive review for amperometric biosensors based on electrosynthesized polymeric films [129]. Potential development of so-called "third generation biosensors", namely sensors without mediators have been reported by electrosynthesizing conducting polymers to provide an electrically conducting matrix for simultaneous immobilization and electrical wiring of the enzyme [130–135]. However, direct electron transfer from enzyme active site through conductive polymer chains, such as polypyrrole, can be easily interfered with by the presence of oxygen, residual mediators and oligomeric conducting polymers in the film matrix [130,133,136,137], which make biosensors based on conductive polymers impractical. A major breakthrough has been noted in biosensors based on non-conducting polymers with built-in permselectivity. An overoxidized-polypyrrole (PPYox) modified Pt glucose sensor showed that major common interferents and fouling species, i.e., ascorbate, urate, paracetamol and cystine, were largely blocked by a controlled electrochemical overoxidation process [138,139]. The polymer chains of PPYox were broken during the overoxidation process causing loss of conjugation and formation of a negatively charged barrier that is mainly responsible for blocking negatively charged interferents. Biosensors with optimum built-in permselectivity can be produced by controlling the degree of overoxidation. This is normally performed through a short potential pulse at 1 V vs. Ag/AgCl until reaching the optimum composition between interferents rejection layer (the outermost oxidized layer) and conductivity of the bulk polymer. Since sensor sensitivity and linearity can be modulated by controlling film thickness, tailoring membrane to construct optimum sensors is obtained by fine tuning built-in permselectivity and film thickness when polymer is fully characterized. The review also described application of poly(1,2-diaminobenzene), 1,3-diaminobenzene/resorcinol co-polymers or mutilayered structures to construct sensors with polymers with built-in permselectivity [129].

More recently, Olivia et al. reported glucose biosensors using Pt nanoparticles deposited boron-doped diamond microfiber (BDDMF) electrode on which GOx was immobilized by overoxidized polypyrrole [140]. Authors reported that amperometric measurement of glucose with the GOx/overoxidized polypyrrole/Pt-modified BDDMF electrode showed a linear response in the range of 1–70 mM and was free from ascorbic acid interference.

3.3.6 Enzyme biosensors using NAD^+/NADH dependent and other dehydrogenases

Enzyme biosensors are usually constructed with enzymes that are oxidoreductases. The largest known group of oxidoreductases is dehydrogenases, of which more than 250 depend on the soluble coenzyme nicotinamide adenine dinucleotide, NAD^+/NADH couple. Excellent review for electrocatalytic oxidation of NAD(P)H at mediator modified electrodes based on polymers can be found in a paper by Gorton and Domínguez [141].

The electrochemistry of both NAD^+ and NADH at clean electrodes occurs at high overvoltages (of the order of 1 V) and hence causes unwanted side reactions, which tend to foul the electrode [142–144]. One way around this problem is to use mediators (two electron–proton acceptor/donors) such as o-quinone [145–147] or p-phenylenediamine derivatives [148], arylnitro derivative such as 2-nitro-9-fluorenone [149], PQQ with Ca^{2+} [150] and polyaromatic dye molecules, i.e., phenazine, phenoxazine and phenothiazine derivatives [11] that substantially lower the voltage for needed NADH oxidation.

However, use of even one of the more promising groups of mediators such as phenoxazine, phenothiazine or phenazine has some drawbacks as they are pH sensitive, have restricted adsorption stability on graphite and suffer from chemical instability in light [141]. With this in mind, Gorton and co-workers [11,151] reported feasibility studies of electron transfer between aqueous insoluble dye-modified polymeric electron transfer mediators, e.g., siloxane polymer containing covalently attached Meldola Blue and NAD^+/NADH by using carbon paste electrodes to co-immobilize them. Studies showed that dye-modified polymeric electron transfer systems efficiently shuttle electrons from NAD^+/NADH to electrode surface and hence can be useful transducers in dehydrogenase-based biosensors.

Gorton et al. reported carbon paste electrodes based on Toluidine Blue O (TBO)-methacrylate co-polymers or ethylenediamine polymer derivative and NAD^+ with yeast alcohol dehydrogenase for the analysis of ethanol [152,153] and with D-lactate dehydrogenase for the analysis of D-lactic acid [154]. Use of electrodes prepared with dye-modified polymeric electron transfer systems and NAD^+/NADH to detect vitamin K and pyruvic acid has also been reported by Okamoto et al. [153]. Although these sensors showed acceptable performances, insensitivity to ambient oxygen concentration, sensor stability and lifetime still need to be improved to obtain optimal dehydrogenase based enzyme biosensors.

Herschkovitz et al. ([69] Section 3.3.3.2) reported formaldehyde, a major air pollutant, detection in solution phase based on the coupling of a biosensor measuring device and a flow injection system. Amperometric formaldehyde sensor was constructed using screen-printed carbon ink electrodes modified with osmium-based hydrogel and then placing formaldehyde dehydrogenase immobilized nylon membrane directly onto the electrode. The sensor is selective, inexpensive, stable over several days, easy to construct and disposable. The authors pointed out that this system could be easily adapted to other substrates using their corresponding dehydrogenases.

Recently, researchers showed great interest in application of PQQ-dependent dehydrogenases that have a quinone cofactor instead of NAD^+, to biosensors. Sensors constructed with PQQ–GDH are reported to be oxygen independent, do not require addition of soluble cofactor and do not involve high overpotentials, which are associated with the electrochemical detection of the reduced form of NADH [155]. Haemmerle et al. reported sensors where PQQ–GDH was wired on a glassy carbon with Os-complex containing redox-conducting epoxy network. Sensors constructed in this manner exhibited high limiting current density, 1.8 mA/cm^2 that suggested fast electron transfer from quinone to the Os complex center of the polymer [155]. Niculescu et al. reported ethanol, glucose and glycerol sensors with PQQ-dependent dehydrogenases using Os complex modified redox hydrogels and PEGDGE as cross-linking agent [156]. Sensors were tested for simultaneous detection of ethanol, glucose and glycerol in wines using a flow injection system. These sensors exhibited fast reactions and detection of substrates by these sensors showed good agreement with values obtained by corresponding spectrophotometric methods. The authors noted that all three biosensors could be operated in similar conditions regarding carrier buffer and applied potential (~200 mV), important factors for simultaneous, on-line measurements of three analytes in a sample.

3.4 SUMMARY

This chapter reviewed the development of polymeric electron transfer systems and discussed their applications to enzyme biosensors. Two major developments of polymeric electron transfer systems, $poly[(VP)Os(bpy)_2Cl^{2+/3+}]$ and its derivatives based hydrogels and mediator containing flexible polymeric electron transfer systems in a carbon paste and their applications to enzyme sensors were reviewed in terms of

- immobilizing free electron mediators as well as enzymes to increase sensor stability and efficiency
- minimizing oxidizable interferent detection in the substrate sample fluid by lowering sensor operating potential
- minimizing or eliminating oxygen interference
- hydrogen peroxide detection using bienzyme systems.

The systems have been developed and improved over the last decade and the knowledge obtained from the studies has been useful to construct practical sensors. Recent reports of sensor fabrication techniques such as nanocomposite thin films will be useful to commercially mass-produce practical reliable sensors. Table 3.2 summarizes amperometric enzyme biosensors that have been reported to utilize various polymeric electron transfer systems.

When we review recent developments in polymeric electron transfer systems, we can also review recent glucose sensor development. Use of electron transfer mediators to construct amperometric glucose sensors leads to commercial availability of smaller (palm or pencil size) sensor units that require much smaller blood samples to monitor blood glucose levels than previously. However, these sensors are invasive and monitor blood glucose levels infrequently. To monitor glucose levels accurately and continuously, development of in vivo sensors became the focus of studies. Many studies of polymeric electron transfer systems focused on in vivo and in situ sensors. To increase sensor lifetime, entrapping enzymes in hydrogels, carbon paste, conductive polymer films or cross-linked polymer matrix have been extensively studied during the past decade. However, there are many other problems to overcome before invasive, implantable (either in vivo or subcutaneous) sensors are practical. Sensor calibration, preventing sensor fouling by biological materials in blood, sensor lifetime and biocompatibility of sensor materials are critical issues that must be addressed. Use of implantable sensors also require physician care, therefore, many researchers have shifted their attention from implantable sensors to non-invasive glucose sensors that are small enough for patients to carry and which can monitor glucose level non-invasively and continuously. To pursue this idea, transdermal and/or near infrared spectrophotometric detection of glucose concentration has been studied. Most recently, the first non-invasive home glucose monitor, GlucoWatch by Cygnus, Inc. has been developed based on hydrogen peroxide detection. It uses reverse iontophoresis to collect interstitial fluid samples through intact skin onto gel collection discs that are part of a single-use sensor cartridge. It can measure glucose concentrations every 20 min for 12 h after a single-point calibration obtained from a traditional finger stick meter [157–159]. This monitor is currently under review by the US FDA and many diabetic patients are waiting for this type of sensor to be commercially available. The knowledge obtained from studies and application of polymeric electron transfer systems have contributed to the development and improvement of amperometric glucose sensors leading to more practical and patient friendly products. It is also noted that application of polymeric electron transfer systems can be used to improve other enzyme sensors, such as cholesterol, lactate, acetylcholine and pyruvate sensors for medical use, mono-, di- and oligosaccharide sensors for food and fermentation industry and nitrate sensor for water quality control. Very recently, Vostiar et al. reported

TABLE 3.2
Amperometric enzyme biosensors using polymeric electron transfer systems

Enzyme biosensors containing polymeric electron transfer systems

Redox polymer	Electrode material	Enzyme used	Substrate	Applied potential	Reference
1) Poly[(4-Vp)Os(bpy)$_2$Cl$^{2+/3+}$] quaternized with BEA	GC	GOx	Glucose	+400 mV vs. Ag/AgCl	[50]
2) Poly[(N-VI)Os(dme-bpy)$_2$Cl$^{2+/3+}$]	GC	GOx	Glucose	+250 mV vs. Ag/AgCl	[50]
3) Poly[(4-VP)Os(dmo-bpy)$_2$Cl$^{2+/3+}$] quaternized with methyl groups	GC	GOx	Glucose	+150 mV vs. Ag/AgCl	[50]
4) Poly[(1-VI)Os(dme-bpy)$_2$Cl$^{2+/3+}$]	Graphite	Amine oxidase	Histamine	+200 mV vs. Ag/AgCl	[67]
Poly[(N-VI)Os(tpy)(dme-bpy)]$^{2+/3+}$	GC	Laccase	Oxygen	+400 mV vs. Ag/AgCl	[71]
Hydrogel A partially quaternized with BEA cross-linked with PEGDGE	Carbon fiber	GOx	Glucose	+400 mV vs. SCE	[19]
Hydrogel A partially quarternized with BEA cross-linked with PEGDGE	VC	GOx/LOx/ peroxidase	Glucose	+400 mV vs. SCE	[20]
Hydrogel A N-derivatized with BEA	GC disk	LOx	L-lactate	+450 mV vs. SCE	[21]
	GC disk	GPO	L-α-glycero-phosphate	+450 mV vs. SCE	[21]

continued

TABLE 3.2 (Continuation)

Redox polymer	Electrode material	Enzyme used	Substrate	Applied potential	Reference
Hydrogel A + soybean peroxidase four polymer layer system	VC	GOx	Glucose	0.00 mV vs. SCE	[54]
Hydrogel A − co-allylamine nano-composite thin film	Gold	GOx	Glucose	+300 mV vs. Ag/AgCl	[60,61]
	Gold	LOx	Lactate	+300 mV vs. Ag/AgCl	[60,61]
Poly[(VI$_{15}$)Os(dme-bpy)$_2$Cl$^{2+/3+}$] cross-linked with PEGDGE	GC	GOx	Glucose	+200 mV vs. SCE	[23]
	GC	LOx	Lactate	+200 mV vs. SCE	[23]
Poly[(VI)$_{13}$Os(dme-bpy)$_2$Cl$^{2+/3+}$] PEGDGE	Graphite	PQQ–GDH	Glucose	+200 mV vs. Ag/AgCl	[156]
		PQQ–ADH	Ethanol	+200 mV vs. Ag/AgCl	[156]
		PQQ–GlyDH	Glycerol	+200 mV vs. Ag/AgCl	[156]
Poly[(VI$_{10}$)Os(dme-bpy)$_2$Cl$^{2+/3+}$] cross-linked with PEGDGE	Solid graphite rod	ODH	Sugars	+200 mV vs. Ag/AgCl	[68]
Poly[(VP)Os(bpy)$_2$Cl$^{2+/3+}$] partially quaternized with BEA	Carbon ink screen printed	FDH	Formaldehyde	+350 mV vs. Ag/AgCl	[69]
	Carbon ink screen printed	ADH	Ethanol	+350 mV vs. Ag/AgCl	[69]
	Carbon ink screen printed	SDH	Sorbitol	+350 mV vs. Ag/AgCl	[69]

Enzyme biosensors containing polymeric electron transfer systems

System	Electrode	Enzyme	Analyte	Potential	Ref.
Os-poly(VI) redox hydrogel	Graphite	GH–ADH	Ethanol	+300 mV vs. AG/AgCl	[80]
Poly[(VI)$_{10}$Os(bpy)$_2$Cl]Cl (B) with PEG	GC	Tyrosinase	Azide	~ +300 mV vs. Ag/AgCl	[70]
(B) in polyphenol film	Pt	GOx	Glucose	+400 mV vs. Ag/AgCl	[72]
(B) with PEI	CP	GOx	Glucose	+350 mV vs. Ag/AgCl	[73]
(B) coated with sol–gel film + derived from methyltriethoxysilane	GC	LOx	Lactate	+400 mV vs. Ag/AgCl	[75]
	GC	HRP	Phenolics	0.00 mV vs. Ag/AgCl	[76]
	GC	HRP	H_2O_2	−50 mV vs. Ag/AgCl	[77]
5) [Os(bpy)$_2$(4-[pyrrole-1-ylhexyl]pyridine)Cl]$^+$ PF$_6$ and thiophene co-polymer (C)	Platinized GC	Pyruvate oxidase	Pyruvate	+400 mV vs. Ag/AgCl	[78]
[Methyl(2-ferrocenylethyl)] siloxane (D) homopolymer	CP	GOx	Glucose	+400 mV vs. SCE	[3,4,7,9]
	CP	Glycolate oxidase	Glycolate	+300 mV vs. SCE	[6,7]
(D) dme-siloxane co-polymer (E)	CP	GOx	Glucose	+400 mV vs. SCE	[3,5,7,12,81]
	CP	LOx	Lactate	+200 mV vs. SCE	[7]
	CP	Glycolate Ox	Glycolate	+300 mV vs. SCE	[6,7]
	CP	ChOx + AChE	Acetylcholine	+250 mV vs. SCE	[12,81]

continued

TABLE 3.2 (Continuation)

Redox polymer	Electrode material	Enzyme used	Substrate	Applied potential	Reference
[Methyl(2,2'-dimethylferrocenyl)] siloxane-dme-siloxane co-polymer (F)	CP	GOx	Glucose	+300 mV vs. SCE	[9]
Ferrocene-ethylene oxide-siloxane polymer (G)	Graphite rod	Glutamate Ox	Glutamate	+350 mV vs. SCE	[12]
	CP	GOx	Glucose	+300 mV vs. SCE	[12,81]
	CP	ChOx + AChE	Acetylcholine	+250 mV vs. SCE	[12,81]
Ferrocene-poly(ethylene oxide) (H)	CP	GOx	Glucose	+100 mV vs. SCE	[8,12,81]
	CP	ChOx + AChE	Acetylcholine	+250 mV vs. SCE	[12]
Ferrocene-poly(methylsiloxane)	CP	GOx	Glucose	+350 mV vs. SCE	[82]
Poly[ferrocenylmethyl MA-co-methoxy-oligo(EO)MA]	CP	GOx	Glucose	+500 mV vs. SCE	[83]
6)Ferrocene-acrylamide-acrylic acid co-polymer	GC	GOx	Glucose	+600 mV vs. SCE	[84]
	GC	HRP	H_2O_2	+150 mV vs. SCE	[84]
Ferrocene-allylamine polymer hydrogel cross-linked with epichlorohydrin	GC	GOx	Glucose	+350 mV vs. SCE	[85]

Polymer	Electrode	Enzyme	Analyte	Potential	Ref.
Poly[methyl(11-ferrocenyl-4,7,10-trioxa-undecanyl)methyl(12-amino-4,7,10-trioxa-dodecyl)]siloxane (1:1 co-polymer)	CP	ALDH (PQQ based)	Glucose, xylose	+200 mV vs. Ag/AgCl	[86]
Polyvinylferrocene-silica gel	CP	GOx	Glucose	+350, +200 mV vs. Ag/AgCl	[94]
Ferrocenyl dendrimers	CP	GOx	Glucose	+350 mV vs. SCE	[89]
Poly(ferrocenyl) functionalized siloxane	Pt	GOx	Glucose	+300 mV vs. SCE	[91]
Hydroquinone–polysiloxane	CP	GOx	Glucose	+400 mV vs. SCE	[10]
Poly(etheramine methylquinone)	CP	GOx	Glucose	+200 mV vs. Ag/AgCl	[13]
Poly(etheramine quinone) (I)	CP	GOx	Glucose	+200 mV vs. Ag/AgCl	[13]
	CP	GOx	Glucose	+300 mV vs. Ag/AgCl	[15]
	CP	ODH	Sugars	+400 mV vs. SCE	[110]
(I) cross-linked by heating	CP	GOx	Glucose	+300 mV vs. Ag/AgCl	[15]
Poly(hydroquinone)	Graphite pellet on GC	GOx	Glucose	+200 mV vs. SCE	[111]

continued

TABLE 3.2 (Continuation)

Redox polymer	Electrode material	Enzyme used	Substrate	Applied potential	Reference
p-Benzoquinone-chitosan-poly(hydroxyethylMA)	Platinum disk	Sulfite oxidase	Sulfite	+650 mV vs. Ag/AgCl	[112]
Poly(o- and p-xylylviologen dibromide)	CP	GOx	Glucose	−100 mV vs. SCE	[4,113]
Viologen-polysiloxane	CP	GOx	Glucose	−100 mV vs. Ag/AgCl	[14]
Poly(pyrrole-viologen)film	GC disk	Nitrate reductases	Nitrate	−700 mV vs. SCE	[114,115]
Viologen-polyvinylpyridine	GC	Hydrogenase	Hydrogen	0.0 mV vs. SCE	[117]
TTF-siloxane polymer	CP	GOx	Glucose	+200 mV vs. SCE	[122]
Polypyrrole carboxy- or lysyl-modified GOx	Platinum disk	GOx	Glucose	+700 mV vs. SCE (H_2O_2 detection)	[124]
7) Thionine–polypyrrole layer topped with enzyme entrapped polypyrrole layer	Platinum disk	Tyrosinase	Phenol	250 mV vs. SCE	[127]
Poly(2-fluoroaniline)	ITO coated glass plates	GOx	Glucose	+200 mV vs. Ag/AgCl	[128]
Hydrogel A cross-linked with PEGDGE	VC	HRP	H_2O_2	0.00 mV vs. SCE	[101]

Polymer	Electrode	Enzyme	Analyte	Potential	Ref.
Hydrogel A cross-linked with PEG	GC	GOx	Glucose	0.0 mV vs. SCE	[66]
	GC	HRP	H_2O_2	−100 mV vs. SCE	[66]
	GC	ChOx + HRP	Choline	−100 mV vs. SCE	[66]
	GC	AChE + ChOx + HRP	Acetylcholine	−100 mV vs. SCE	[66]
Co-polymer (E)	Carbon rod disk	ChOx + HRP	Cholesterol	−100 mV vs. SCE	[81]
Os(bpy)$_2$ClPyCH$_2$NHPoly(allylamine) cross-linked with PEG-400	GC	GOx + HRP	Glucose	+50 mV vs. SCE	[103]
8) Poly[(1-VI)$_7$Os(dme-bpy)$_2$Cl]$^{+/2+}$ with PEGDGE	CP	Alcohol Ox + HRP	Ethanol	−50 mV vs. Ag/AgCl	[104]
9) Poly[(1-VI)$_7$Os(dme-bpy)$_2$Cl]$^{2+/3+}$ with PEGDGE	Graphite rod	Ascorbate Ox + HRP	Glutamate	−50 mV vs. Ag/AgCl	[105]
10) Poly[(VI)$_7$Os(dme-bpy)$_2$Cl]$^{2+/3+}$ with PEGDGE	Graphite	GOx + SWPP	Glucose	−50 mV vs. Ag/AgCl	[106]
		AlcOx + SWPP	Ethanol	−50 mV vs. Ag/AgCl	[106]
		AO + SWPP	Putrescine	−50 mV vs. Ag/AgCl	[106]
Poly(ethylene glycol)diacrylate-vinylferrocene co-polymer	Au	GOx	Glucose	+300 mV vs. Ag/AgCl	[95]

continued

TABLE 3.2 (Continuation)

Redox polymer	Electrode material	Enzyme used	Substrate	Applied potential	Reference
Meldola Blue-Polysiloxane	Graphite	Dehydrogenase	NADH	0.00 mV vs. SCE	[151]
Toluidine Blue O (TBO)-PMMA	CP	ADH	Ethanol	+100 mV vs. Ag/AgCl	[152,153]
	CP	NADH	Vitamin K_3	−400 mV vs. Ag/AgCl	[153]
TBO-polyethylenimine	CP	D-LDH	D-Lactate	−50 mV vs. Ag/AgCl	[154]
Hydrogel A; bilayer (1) HRP in hydrogel layer (2) adivin cross-linked and immobilized biotin labeled GOx layer	VC	GOx + HRP	Glucose	+100 mV vs. Ag/AgCl	[55]

electrical wiring of whole *Gluconobacter oxydans* cells with a flexible polyvinylimidazole osmium functionalized polymer [160]. It is noteworthy as the use of entire microorganisms as microreactors for detection of glucose, ethanol and glycerol is a new concept for enzyme sensors using polymeric electron mediators.

Finally, it should be noted that the recent development of so-called "third generation" biosensors to achieve direct electron transfer from redox enzyme, oxidoreductase to the electrode without mediators, but through a series of enzyme cofactors or conductive polymers to transfer electrons from the enzyme redox center to the electrode surface [161–164]. This concept with the current technology for preparing miniature sensors with nanotechnology is of great interest to many researchers trying to develop practical sensors in clinical, environmental and industrial analysis. Whether with mediators or without, research for optimum sensor development for various purposes will be intensive in the future.

Acknowledgements

The author thanks Professor Lo Gorton for inviting her to contribute to this chapter. She also thanks Professors Yoshiyuki Okamoto and Gorton for their useful comments in preparing this manuscript and Dr. Harini Patel for assisting in library research.

REFERENCES

1   Y. Degani and A. Heller, *J. Phys. Chem.*, 91 (1987) 1285.
2   Y. Degani and A. Heller, *J. Am. Chem. Soc.*, 110 (1988) 2615.
3   P.D. Hale, T. Inagaki, H.I. Karan, Y. Okamoto and T.A. Skotheim, *J. Am. Chem. Soc.*, 111 (1989) 3482.
4   P.D. Hale, L.I. Boguslavsky, T. Skotheim, H.I. Karan, H.L. Lan and Y. Okamoto, *Mol. Cryst. Liq. Cryst.*, 190 (1990) 259.
5   P.D. Hale, L.I. Boguslavsky, T. Inagaki, H.S. Lee, T. Skotheim, H.I. Karan and Y. Okamoto, *Mol. Cryst. Liq. Cryst.*, 190 (1990) 251.
6   P.D. Hale, T. Inagaki, H.L. Lee, H.I. Karan, Y. Okamoto and T. Skotheim, *Anal. Chim. Acta*, 228 (1990) 31.
7   P.D. Hale, H.I. Karan, T. Inagaki, H.S. Lee, Y. Okamoto and T. Skotheim, In: E.F. Bowden, R.P. Buck, W.E. Hatfield and M. Umana (Eds.), *Biosensor Technology: Fundamentals and Application*. Marcel Dekker, New York, 1990, pp. 195.
8   P.D. Hale, H.L. Lan, L.I. Boguslavsky, H.I. Karan, Y. Okamoto and T. Skotheim, *Anal. Chim. Acta*, 251 (1991) 121.

9   P.D. Hale, L. Boguslavsky, T. Inagaki, H.I. Karan, H.S. Lee, T. Skotheim and Y. Okamoto, *Anal. Chem.*, 63 (1991) 677.
10 H.I. Karan, P.D. Hale, H.L. Lan, H.S. Lee, L.F. Liu, T. Skotheim and Y. Okamoto, *Polym. Adv. Technol.*, 2 (1992) 229.
11 L. Gorton, B. Persson, P.D. Hale, L.I. Boguslavsky, H.I. Karan, H.S. Lee, T.A. Skotheim, H.L. Lan and Y. Okamoto, In: P.G. Edelman and J. Wang (Eds.), *Biosensors and Chemical Sensors*, ACS Symposium Series, Vol. 487, 1992, pp. 56.
12 P.D. Hale, L.I. Bougslavsky, T.A. Skotheim, L.F. Liu, H.S. Lee, H.I. Karan, H.L. Lan and Y. Okamoto, In: P.G. Edelman and J. Wang (Eds.), *Biosensors and Chemical Sensors*, ACS Symposium Series, Vol. 487, 1992, pp. 111.
13 H.L. Lan, T. Kaku, H.I. Karan and Y. Okamoto, In: A. Usmani and N. Akmal (Eds.), *Diagnostic Biosensor Polymers*, ACS Symposium Series, Vol. 556, 1994, pp. 124.
14 H.I. Karan, H.S. Lan and Y. Okamoto, In: A. Usmani and N. Akmal (Eds.), *Diagnostic Biosensor Polymers*, ACS Symposium Series, Vol. 556, 1994, pp. 169.
15 T. Kaku, H.I. Karan and Y. Okamoto, *Anal. Chem.*, 66 (1994) 1231.
16 T. Kaku, L. Charles, W. Holness, H.I. Karan and Y. Okamoto, *Polymer*, 36 (1995) 2813.
17 Y. Degani and A. Heller, *J. Am. Chem. Soc.*, 111 (1989) 2357.
18 A. Heller, *Acc. Chem. Res.*, 23 (1990) 128.
19 M.V. Pishko, A.C. Michael and A. Heller, *Anal. Chem.*, 63 (1991) 2268.
20 A. Heller, *J. Phys. Chem.*, 96 (1992) 3579.
21 I. Katakis and A. Heller, *Anal. Chem.*, 64 (1992) 1008.
22 R. Maidan and A. Heller, *Anal. Chem.*, 64 (1992) 2889.
23 T.J. Ohara, R. Rajagopalan and A. Heller, *Anal. Chem.*, 66 (1994) 2451.
24 E. Csöregi, C.P. Quinn, D.W. Schmidtke, S. Lindquist, M.V. Pishko, L. Ye, I. Katakis, J. Hubbell and A. Heller, *Anal. Chem.*, 66 (1994) 3131.
25 L. Yang, E. Janle, T. Huang, J. Gitzen, P.T. Kissinger, M. Vreeke and A. Heller, *Anal. Chem.*, 67 (1995) 1326.
26 T. de Lumley-Woodyear, P. Rocca, J. Lindsay, Y. Dror, A. Freeman and A. Heller, *Anal. Chem.*, 67 (1995) 1332.
27 E. Csöregi, D. Schmidtke and A. Heller, *Anal. Chem.*, 67 (1995) 1240.
28 National Diabetes Fact Sheet, National Center for Chronic Disease Prevention and Health Promotion: Diabetes Public Resource, Center for Disease Control, USA, 2003.
29 L.C. Clark Jr. and C. Lyons, *Ann. N.Y. Acad. Sci.*, 102 (1962) 29.
30 D. Velho, G. Reach and D.R. Thevenot, In: A.P.F. Turner, I. Karube and G. Wilson (Eds.), *Biosensors: Fundamentals and Applications*, Vol. 22. Oxford Scientific Publications, Oxford, 1987, pp. 398.
31 G. Reach and G. Wilson, *Anal. Chem.*, 64 (1992) 381A.
32 A.E.G. Cass, G. Davis, G.D. Francis, H.A.O. Hill, W.J. Aston, I.J. Higgins, E.V. Plotkin, L.D.L. Scott and A.P.F. Turner, *Anal. Chem.*, 56 (1984) 667.

33   A.L. Crumbliss, H.A.O. Hill and D.J. Page, *J. Electroanal. Chem. Interfacial Electrochem.*, 206 (1986) 327.
34   M.A. Lange and J.Q. Chambers, *Anal. Chim. Acta*, 175 (1985) 89.
35   C. Iwakura, Y. Kajiya and H. Yoneyama, *J. Chem. Soc., Chem. Commum.*, (1988) 1019.
36   D.A. Gough, J.Y. Lucisano and P.H.S. Tse, *Anal. Chem.*, 57 (1985) 2351.
37   A.P.F. Turner, *World Biotech. Rep.*, 1 (1985) 181.
38   L. Gorton, F. Scheller and G. Johansson, *Stud. Biophys.*, 109 (1986) 199.
39   J. Kulys and N.K. Cenas, *Biochim. Biophys. Acta*, 744 (1983) 57.
40   J. Albery, P.N. Bartlett and D.H. Craston, *J. Electroanal. Chem.*, 194 (1985) 223.
41   K. Mckenna and A. Brajter-Toth, *Anal. Chem.*, 59 (1987) 954.
42   P.D. Hale and R.M. Wightman, *Mol. Cryst. Liq. Cryst.*, 160 (1988) 269.
43   P.D. Hale and T.A. Skotheim, *Synth. Met.*, 28 (1989) 853.
44   E.M. Kober, J.V. Casper, J.V. Sullivan and T.J. Meyer, *Inorg. Chem.*, 27 (1988) 4587.
45   R. Anderson, B. Arkles and G.L. Larson, *Silicon Compounds Review and Register*. Petrarch Systems, Bristol, PA, 1987, pp. 259.
46   J.A. Chambers and N.J. Walton, *J. Electroanal. Chem.*, 250 (1988) 417.
47   M. Gerritsen, J.A. Jansen and J.A. Lutterman, *Neth. J. Med.*, 54 (1999) 167.
48   S.A. Emr and A.M. Yacynych, *Electroanalysis*, 7 (1995) 913.
49   L. Gorton, H.I. Karan, P.D. Hale, T. Inagaki, Y. Okamoto and T. Skotheim, *Anal. Chim. Acta*, 228 (1990) 23.
50   G. Binyamin, T. Chen and A. Heller, *J. Electroanal. Chem.*, 500 (2001) 604.
51   B.A. Gregg and A. Heller, *J. Phys. Chem.*, 95 (1991) 5970.
52   B.A. Gregg and A. Heller, *J. Phys. Chem.*, 95 (1991) 5976.
53   T.J. Ohara, R. Rajagopalan and A. Heller, *Anal. Chem.*, 65 (1993) 3512.
54   G. Kenausis, Q. Chen and A. Heller, *Anal. Chem.*, 69 (1997) 1054.
55   M.S. Vreeke and P. Rocca, *Electroanalysis*, 8 (1995) 55.
56   N. Dontha, W. Nowall and W. Kuhr, *Anal. Chem.*, 69 (1997) 2619.
57   J. Anzai, Y. Kobayashi, N. Nakamura, M. Nishimura and T. Hoshi, *Langmuir*, 15 (1999) 221.
58   I. Willner, V. Heleg-Shabtai, R. Blonder, E. Katz and G. Tao, *J. Am. Chem. Soc.*, 118 (1996) 10321.
59   I. Willner, E. Katz and B. Willner, *Electroanalysis*, 9 (1997) 965.
60   K. Sirkar, A. Revzin and M. Pishko, *Anal. Chem.*, 72 (2000) 2930.
61   A.F. Revzin, S. Kaushik, A. Simonian and M. Pishko, *Sens. Actuators, B*, 81 (2002) 359.
62   T. Chen, S.C. Barton, G. Binyamin, Z. Gao, Y. Zhang, H.-H. Kim and A. Heller, *J. Am. Chem. Soc.*, 123 (2001) 8630.
63   S.C. Kelley, G.A. Deluga and W.H. Smyrl, *Electrochem. Solid-State Lett.*, 3 (2000) 407.
64   E. Katz, I. Willner and A.B. Kotlyar, *J. Electroanal. Chem.*, 479 (1999) 64.
65   I. Willner and E. Katz, *Angew. Chem. Int. Ed.*, 39 (2000) 1180.

66 M.G. Garguilo, N. Huynh, A. Proctor and A.C. Michael, *Anal. Chem.*, 65 (1993) 523.
67 M. Niculescu, I. Frebort, P. Pec, P. Galuszka, B. Mattiasson and E. Csöregi, *Electroanalysis*, 12 (2000) 369.
68 M. Tessema, E. Csöregi, T. Ruzgas, G. Kenausis, T. Solomon and L. Gorton, *Anal. Chem.*, 69 (1997) 4039.
69 Y. Herschkovitz, I. Eshkenazi, C.E. Campbell and J. Rishpon, *J. Electroanal. Chem.*, 491 (2000) 182.
70 F. Daigle, F. Trudeau, G. Robinson, M.R. Smyth and D. Leech, *Biosens. Bioelectron.*, 13 (1997) 417.
71 S.C. Barton, M. Pickard, R. Vazquez-Duhalt and A. Heller, *Biosens. Bioelectron.*, 17 (2002) 1071.
72 M. Pravda, C.M. Jungar, E.I. Iwuoha, M.R. Smyth, K. Vytras and A. Ivaska, *Anal. Chim. Acta*, 304 (1995) 127.
73 J. Jezkova, E.I. Iwuoha, M.R. Smyth and K. Vytras, *Electroanalysis*, 9 (1997) 978.
74 S. Dong and X. Chen, *Rev. Mol. Biotechnol.*, 82 (2002) 303.
75 T.M. Park, E.I. Iwuoha, M. Smyth, R. Freaney and A.J. McShane, *Talanta*, 44 (1997) 973.
76 S.A. Kane, E.I. Iwuoha and M.R. Smyth, *Analyst*, 123 (1998) 2001.
77 T.M. Park, *Anal. Lett.*, 32 (1999) 287.
78 N. Gajovic, K. Habermüller, A. Warsinke, W. Schuhmann and F.W. Scheller, *Electroanalysis*, 11 (1999) 1377.
79 S.-K. Cha, *J. Polym. Sci.*, 35 (1997) 165.
80 A. Vilkanauskyte, T. Erichsen, L. Marcinkeviciene, V. Laurinavicius and W. Schuhmann, *Biosens. Bioelectron.*, 17 (2002) 1025.
81 L. Boguslavsky, P.D. Hale, L. Geng, T.A. Skotheim and H.S. Lee, *Solid State Ionics*, 60 (1993) 189.
82 C.M. Casado, M. Moran, J. Losada and I. Cuadrado, *Inorg. Chem.*, 34 (1995) 1668.
83 H. Nagasaka, T. Saito, H. Hatakeyama and M. Watanabe, *Denki Kagaku*, 63 (1995) 1088.
84 E.J. Calvo, C. Danilowicz and L. Diaz, *J. Chem. Soc., Faraday Trans.*, 89 (1993) 377.
85 J. Calvo, R. Etchenique, C. Danilowicz and L. Diaz, *Anal. Chem.*, 68 (1996) 4186.
86 M. Smolander, L. Gorton, H.S. Lee, T. Skotheim and H.L. Lan, *Electroanalysis*, 7 (1995) 941.
87 L. Losada, I. Cuadrado, M. Moran, C.M. Casado, B. Alonso and M. Barranco, *Anal. Chim. Acta*, 251 (1996) 5.
88 J. Losada, I. Cuadrado, M. Moran, C.M. Casado, B. Alonso and N. Barranco, *Anal. Chim. Acta*, 338 (1997) 191.
89 M. Casado, I. Cuadrado, M. Moran, B. Alonso, B. Garcia, B. Gonzalez and J. Losada, *Coord. Chem. Rev.*, 185–186 (1999) 53.
90 B. Alonso, B. Gonzalez, B. Garcia, E. Ramirez-Oliva, M. Zamora, C.M. Casado and I. Cuadrado, *J. Organomet. Chem.*, 637–639 (2001) 642.

91 M.P.G. Armada, J. Losada, I. Cuadrado, B. Alonso, B. Gonzalez, E. Ramirez-Oliva and C.M. Casado, *Sens. Actuators, B*, 88 (2003) 190.
92 A. de S. Santos, L. Gorton and L. Kubota, *Electrochim. Acta*, 47 (2002) 3351.
93 F.-D. Munteanu, Y. Okamoto and L. Gorton, *Anal. Chim. Acta*, 476 (2003) 43.
94 H. Patel, X. Li and H.I. Karan, *Biosens. Bioelectron.*, 18 (2003) 1073.
95 K. Sirkar and M. Pishko, *Anal. Chem.*, 70 (1998) 2888.
96 T. Ruzgas, E. Csöregi, J. Emnéus, L. Gorton and G. Marko-Varga, *Anal. Chim. Acta*, 330 (1996) 123.
97 T. Ruzgas, L. Gorton, J. Emnéus and G. Marko-Varga, *J. Electroanal. Chem.*, 391 (1995) 41.
98 S. Gasper, K. Habermüller, E. Csöregi and W. Schuhmann, *Sens. Actuators, B*, 72 (2001) 63.
99 L. Gorton, G. Jönsson-Pettersson, E. Csöregi, K. Johansson, E. Domínguez and G. Marko-Varga, *Analyst*, 117 (1992) 1235.
100 U. Spohn, D. Narasaiah, L. Gorton and D. Pfeiffer, *Anal. Chim. Acta*, 319 (1996) 79.
101 M. Vreeke, R. Maidan and A. Heller, *Anal. Chem.*, 64 (1992) 3084.
102 L. Boguslavsky, H. Kalash, Z. Xu, D. Beckles, L. Geng, T. Skotheim, V. Laurinavicius and H.S. Lee, *Anal. Chim. Acta*, 311 (1995) 15.
103 C. Danilowicz, E. Corton, F. Battaglini and J. Calvo, *Electrochim. Acta*, 43 (1998) 3525.
104 A.R. Vijayakumar, E. Csöregi, A. Heller and L. Gorton, *Anal. Chim. Acta*, 327 (1996) 223.
105 A. Collins, E. Mikeladze, M. Bengtsson, M. Kokaia, T. Lairell and E. Csöregi, *Electroanalysis*, 13 (2001) 425.
106 J. Castillo, S. Gasper, I. Sakharov and E. Csöregi, *Biosens. Bioelectron.*, 18 (2003) 705.
107 T. Ikeda, H. Hamada and S. Senda, *Agric. Biol. Chem.*, 50 (1986) 883.
108 T. Ikeda, T. Shibata and S. Senda, *J. Electroanal. Chem.*, 261 (1989) 351.
109 J.J. Kulys and N.K. Cenas, *Biochim. Biophys. Acta*, 744 (1983) 57.
110 M. Tessema, T. Ruzgas, L. Gorton and T. Ikeda, *Anal. Chim. Acta*, 310 (1995) 161.
111 P. Wang, S. Amarasingle, J. Leddy, M. Arnold and J.S. Dordick, *Polymer*, 39 (1998) 123.
112 L.-T. Ng, Y.J. Yuan and H. Zhao, *Electroanalysis*, 10 (1998) 1119.
113 P.D. Hale, L.I. Boguslavsky, H.I. Karan, H.L. Lan, H.S. Lee, Y. Okamoto and T.A. Skotheim, *Anal. Chim. Acta*, 248 (1991) 155.
114 S. Cosnier, C. Innocent and Y. Jouanneau, *Anal. Chem.*, 66 (1994) 3198.
115 S. Cosnier, B. Galland and C. Innocent, *J. Electroanal. Chem.*, 433 (1997) 113.
116 G. Ramsay and S.M. Wolpert, *Anal. Chem.*, 71 (1999) 504.
117 L.H. Eng, M. Elmgren, P. Komlos, M. Nordling, S.-E. Lindquist and H.Y. Neujahr, *J. Phys. Chem.*, 98 (1994) 7068.
118 H. Gunasingham and C.-H. Tan, *Analyst*, 115 (1990) 35.

119   H. Gunasingham, C.-H. Tan and T.-C. Aw, *Anal. Chim. Acta*, 234 (1990) 321.
120   G. Palleschi and A.P.F. Turner, *Anal. Chim. Acta*, 234 (1990) 459.
121   P.D. Hale, L.-F. Liu and T.A. Skotheim, *Electroanalysis*, 3 (1991) 751.
122   H.S. Lee, L.-F. Liu, P.D. Hale and Y. Okamoto, *Heteroat. Chem.*, 3 (1992) 303.
123   S.E. Wolowacz, B.F.Y. Yon Hin and C.R. Lowe, *Anal. Chem.*, 64 (1992) 1541.
124   B.F.Y. Yon-Hin, M. Smolander, T. Crompton and C.R. Lowe, *Anal. Chem.*, 65 (1993) 2067.
125   Y. Kajiya, H. Sugai, C. Iwakura and H. Yoneyama, *Anal. Chem.*, 63 (1991) 49.
126   W. Schuhmann, *Biosens. Bioelectron.*, 10 (1995) 181.
127   Y. Krantz, H. Wohlschlager, H.-L. Schmidt and W. Schuhmann, *Electroanalysis*, 10 (1998) 546.
128   A.L. Sharma, S. Annapoorni and B.D. Malhotra, *Curr. Appl. Phys.*, 3 (2003) 239.
129   F. Palmisano, P.G. Zambonin and D. Centonze, *Fresenius J. Anal. Chem.*, 366 (2000) 586.
130   M. Trojanowicz and T. Krawczynski vel Krawczyk, *Mikrochim. Acta*, 121 (1995) 167.
131   W. Schuhmann, *Mikrochim. Acta*, 121 (1995) 1.
132   S. Cosnier, *Electroanalysis*, 9 (1997) 894.
133   P.N. Bartlett and J.M. Cooper, *J. Electroanal. Chem.*, 362 (1993) 1.
134   A.L. Ghindilis, P. Atanasov and E. Wilkins, *Electroanalysis*, 9 (1997) 661.
135   S.I. Yabuki, H. Shinohara and M. Aizawa, *J. Chem. Soc., Chem. Commun.*, (1989) 945.
136   P. de Taxis du Poët, S. Miyamoto, T. Murakami, J. Kimura and I. Karube, *Anal. Chim. Acta*, 235 (1990) 255.
137   P.J.H.J. van Os, A. Bult, C.G.J. Koopal and W.P. van Bennekom, *Anal. Chim. Acta*, 335 (1996) 209.
138   F. Palmisano, D. Centonze, A. Guerrieri and P.G. Zambonin, *Biosens. Bioelectron.*, 8 (1993) 393.
139   F. Palmisano, D. Centonze, C. Malitesta and P.G. Zambonin, *Anal. Chem.*, 67 (1995) 2207.
140   H. Olivia, B.V. Sarada, K. Honda and A. Fujishima, *Electrochim. Acta*, 49 (2004) 2069.
141   L. Gorton and E. Domínguez, *Rev. Mol. Biotechnol.*, 82 (2002) 371.
142   P.N. Bartlett, P. Tebbutt and R.G. Whitaker, *Prog. React. Kinet.*, 16 (1991) 55.
143   H.K. Chenault and G.M. Whitesides, *Appl. Biochem. Biotechnol.*, 14 (1987) 147.
144   T. Matsue, M. Suda, I. Uchida, I. Kato, U. Akiba and T. Osa, *J. Electroanal. Chem.*, 234 (1987) 163.
145   B.W. Carlson and L.L. Miller, *J. Am. Chem. Soc.*, 107 (1985) 479.
146   C. Degrand and L. Miller, *J. Am. Chem. Soc.*, 102 (1980) 5728.
147   A.N.K. Lau and L. Miller, *J. Am. Chem. Soc.*, 105 (1983) 5271.
148   A. Kitani, Y.-H. So and L.L. Miller, *J. Am. Chem. Soc.*, 103 (1981) 7636.
149   N. Mano and A. Kuhn, *J. Electroanal. Chem.*, 498 (2001) 58.

150 E. Katz, T. Lötzbeyer, D.D. Schlereth, W. Schuhmann and H.-L. Schmidt, *J. Electroanal. Chem.*, 373 (1994) 189.
151 P.D. Hale, H.S. Lee and Y. Okamoto, *Anal. Lett.*, 26 (1993) 1073.
152 E. Dominguez, H.L. Lan, Y. Okamoto, P.D. Hale, T.A. Skotheim, L. Gorton and B. Hahn-Hägerdal, *Biosens. Bioelectron.*, 8 (1993) 229.
153 Y. Okamoto, T. Kaku and R. Shundo, *Pure Appl. Chem.*, 68 (1996) 1417.
154 H.-C. Shu, B. Mattiasson, B. Persson, G. Nagy, L. Gorton, S. Sahni, L. Geng, L. Boguslavsky and T. Skotheim, *Biotechnol. Bioeng.*, 46 (1995) 270.
155 L. Ye, M. Haemmerle, A.J.J. Olsthoorn, W. Schuhmann, W. Schmidt, H. Ludwig, J.A. Duine and A. Heller, *Anal. Chem.*, 65 (1993) 238.
156 M. Niculescu, R. Mieliauskiene, V. Laurinavicius and E. Csöregi, *Food Chem.*, 82 (2003) 481.
157 Clinical Laboratory News, March 2000, pp. 10
158 M. Tierney, H. Kim, J. Tamada and R. Potts, *Electroanalysis*, 12 (2000) 666.
159 J. Wang, *Electroanalysis*, 13 (2001) 983.
160 I. Vostiar, E.E. Ferapontova and L. Gorton, *ElectroChem. Commun.*, 6 (2004) 621.
161 K. Habermüller, M. Mosbach and W. Schuhmann, *Fresenius J. Anal. Chem.*, 366 (2000) 560.
162 W. Schuhmann, *Rev. Mol. Biotechnol.*, 82 (2002) 425.
163 F.W. Scheller, U. Wollenberger, C. Lei, W. Jin, B. Ge, C. Lehmann, F. Lisdat and V. Fridman, *Rev. Mol. Biotechnol.*, 82 (2002) 411.
164 S. Cosnier, D. Fologea, S. Szunerits and R. Marks, *Electrochem. Commun.*, 2 (2000) 827.

GLOSSARY

AchE: Acetylcholinesterase
ADH: Alcohol dehydrogenase
AlcOx: Alcohol oxidase
ALDH: Aldose dehydrogenase
AO: Amine oxidase
ChOx: Choline oxidase
FDH: Formaldehyde dehydrogenase
GDH: Glucose dehydrogenase
GOx: Glucose oxidase
GlyDH: Glycerol dehydrogenase
HRP: Horseradish peroxidase
GPO: L-α-Glycerol phosphate oxidase
D-LDH: D-Lactate dehydrogenase

LOx: Lactate oxidase
ODH: Oligosaccharide dehydrogenase
Ox: Oxidase
SDH: Sorbitol dehydrogenase
SWPP: Sweet potato peroxidase
bpy: 2,2′-bipyridine
BEA: 2-bromoethylamine
dme-: dimethyl-
dme-bpy: 4,4′-dimethyl-2,2′-bipyridine
dmo-bpy: 4,4′-dimethoxy-2,2′-bipyridine
tpy: 2,2′:6′,2″-terpyridine
EO: Ethylene oxide
MA: Methacrylate
Hydrogel A: Poly[(VP)Os(bpy)$_2$Cl$^{2+/3+}$] based hydrogel
PEG: Polyethylene glycol
PEGDGE: Poly(ethylene glycol) diglycidyl ether
PEI: Polyethylenimine
PQQ: Pyrroloquinoline quinone
PMMA: Poly(methylmethacrylate)
TTF: Tetrathiafulvalene
TBO: Toluidine blue O
VI: Vinylimidazole
VP: Vinylpyridine
CP: Carbon paste
GC: Glassy carbon
SCE: Standard calomel electrode
VC: Vitreous carbon
ITO: Indium tin oxide

Chapter 4

DNA-based biosensors

A.M. Oliveira Brett

4.1 INTRODUCTION

There are hundreds of compounds which bind and interact with DNA. These reactions cause changes in the structure of DNA and the base sequence leading to perturbations in DNA replication. It is very important to explain the factors that determine affinity and selectivity in binding molecules to DNA, because, as described by Larsen [1], a quantitative understanding of the reasons that determine the selection of DNA reaction sites is useful in designing sequence-specific DNA binding molecules for application in chemotherapy and in explaining the mechanism of action of neoplastic drugs. Research on metal ion nucleic acid complexes advanced when antitumour activities of platinum(II) compounds were discovered.

During the transfer of genetic information, the interactions between DNA and the divalent ions play an essential role in promoting and maintaining the nucleic acid functionalities. Some are recognized for their carcinogenicity as they damage DNA molecules and alter the fidelity of DNA synthesis. Nickel, as well as chromium and cadmium, seems to be recognized as the most effective carcinogens. Many inorganic nickel compounds have been tested and their effect on a cell or tissue has been established as an interaction with the base donor systems, especially within unwound parts of nucleic acids.

Chemical modification of DNA bases is called mutagenesis. Substitutions, deletions and insertions cause mutations in the base sequence of DNA, according to Saenger [2]. Exposure to toxic chemicals is the cause of many human cancers; these carcinogens act by chemically damaging the DNA. Thus, it is important to identify these chemicals and ascertain their potency so that human exposure to them can be minimized.

In a health preventing perspective the need for the analysis of gene sequences, oxidative damage to DNA and the understanding of DNA interactions with molecules or ions led to the development of DNA-based biosensors [3–8]. The DNA-based biosensor is a device that incorporates immobilized DNA, as molecular recognition element in the biological active layer on the surface, and measures specific binding processes with DNA mainly using electrochemical, optical and piezoelectric transducers. The fact that the DNA sequences are unique to each organism means that any self-replicating biological organism can be discriminated. The DNA-based biosensor is also a complementary tool for the study of biomolecular interaction mechanisms of compounds with double-stranded DNA (dsDNA) enabling the screening and evaluation of the effect caused to dsDNA by health hazardous compounds and oxidizing substances.

However, the structure–function relationship of DNA–ligand interaction on interfaces has to be established and the adsorption of monolayers and multilayers of DNA, ligands and DNA–ligand complexes on the transducer investigated to characterize and optimize biosensor operation. The chemical selectivity of biosensors is derived from biological materials interfaced to the surface of transducing devices. Molecular recognition events led to macroscopic response suitable for analytical measurements. The techniques of ellipsometry, fluorescence microscopy, electron microscopy and scanning tunnelling microscopy (STM) mentioned by Krull et al. [9] are important to investigate aspects of the interfacial physical structure of monolayers and multilayers of biological materials at biosensor interfaces.

Many DNA-based biosensors (genosensors) are based on the ability of complementary nucleic acid strands to selectively form hybrid complexes. The complementary strands anneal to one another in a Watson–Crick manner of base pairing. Hybridization methods used today, such as microlitre plates or gel-based methods, are usually quite slow, requiring hours to days to produce reliable results, as described by Keller and Manak [10]. Biosensors offer a promising alternative for much faster hybridization assays.

The basis of operation of a DNA hybridization biosensor is the complementary coupling between the specific single-stranded DNA (ssDNA) sequences (*the target*) within the analyte, which also contains non-complementary ssDNA strands, and the specific ssDNA sequences (*the probe*) immobilized onto the solid support (*the transducer*). The specific and selective detection of DNA sequences, with a single-base mismatch detection ability, is a major challenge in DNA biosensing. Real-time in situ hybridization analysis offers the opportunity to obtain biological information, such as the specificity and kinetics of binding of biomolecules. These will give an insight into the relationship between the molecule structures and functions.

Biosensors presently being developed for the detection of DNA hybridization are mostly based on optical, surface acoustic wave and electrochemical transducers. The detection of specific DNA sequences, using

DNA hybridization, is very important since many inherited diseases are already known and is a procedure useful for the detection of microorganisms in medical, environmental and food control. The detection of genetically modified organisms (GMOs) is of great relevance for food analysis. Results using three DNA biosensors based on electrochemical, piezoelectric and optical strategies were compared by Mascini and co-workers [11] and showed the great advantages of biosensor technology, much simpler than ethidium bromide electrophoresis, the reference method in GMOs analysis.

The immobilization of dsDNA and ssDNA to surfaces can be attained very easily by adsorption. No reagents or DNA modifications occur since the immobilization does not involve formation of covalent bonds with the surface. Surface immobilization of ssDNA by covalent binding, described by Thompson and co-workers [12], is convenient in DNA-hybridization sensing because it enables probe structure flexibility with respect to changes in its conformation to occur, such that hybridization can take place without the probe being removed from the sensor surface. However, non-selective adsorption of non-complementary ssDNA, added to bulk solution, also occurs at multiple sites in the interstitial regions on the sensor surface between immobilized ssDNA strands. The effects of selective and non-selective binding influence the detection of hybridization of the immobilized strands, as found by Krull and co-workers [13].

Two large proteins avidin and streptavidin (70 kDa), each containing four binding sites with very high affinity ($K_d = 10^{-15}$ M) for biotin, forming an avidin/streptavidin–biotin complex, have been extensively used to attach an ssDNA sequence to the surface of various transducers. The protein is first adsorbed onto the surface of the transducer and is then exposed to an aqueous solution of biotinylated nucleic acid. The great stability of the avidin–biotin complex makes the system easy to employ, but the presence of the underlying protein layer yields possible multiple sites for non-specific adsorption and according to Thompson and co-workers [14] not all biotin bonding sites are available.

The use of peptide nucleic acid (PNA), originally developed by Nielsen et al. [15] as a nucleic acid sequence-specific reagent, as recognition layer in DNA biosensors, holds wide biological potential and diagnostic applications. PNA is a nucleic acid analogue in which the sugar–phosphate backbone of natural nucleic acid has been replaced by a synthetic peptide backbone formed from N-(2-amino-ethyl)-glycine units, resulting in a chiral and uncharged mimic. PNA is chemically stable, resistant to hydrolytic (enzymatic) cleavage and thus not expected to be degraded inside a living cell, but the cellular uptake is very slow according to Ray and Nordén [16]. PNA can hybridize with complementary DNA or RNA sequences following the Watson–Crick hydrogen bonding scheme. These hybrid complexes exhibit good thermal stability and have been used as a sequence-specific hybridization probe by Wang [17] with electrochemical, optical and piezoelectric transducers.

Other less common DNA sensors have been described, such as the use of magnetic microbeads on a solid substrate, to detect and characterize many individual biomolecular interaction events simultaneously using the concept of measuring changes in the intermolecular forces by arrays of microfabricated magnetoresistive DNA sensors, as developed by Baselt et al. [18].

The development of miniaturized sensors, based on hybridization and base pairing, is the basis of the technology of producing DNA microarrays described by Schena [19]. This oligonucleotide DNA arrays consist of an orderly arrangement of oligonucleotides in a single chip, enabling automated and quick monitoring of known and unknown DNA samples. Chip-based microsystems for genomic analysis are enabling a large number of reactions to be studied within a very small area and in a very short time as shown by Sanders and Manz [8]. As a consequence, the concept of a miniaturized total analysis systems (μ-TAS) for biological analysis, combining microelectronics and molecular biology, has been developed and applied to a variety of chemical and biological problems as described by Mello and co-workers [20].

Irrespective of how DNA-based biosensors are fabricated, according to Steel et al. [21], a greater understanding of the factors influencing the structure of immobilized DNA layers is needed to design surfaces exhibiting greater biological activity and selectivity. Shorter ssDNA tends to organize in a high surface density whereas longer ssDNA leads to a decrease in surface coverage with probe length.

The DNA-based sensors should have a rapid response time and should be quantitative, sensitive, suitable for automation, cost effective, disposable and solve analytical problems in a wide range of industrial contexts in order to be commercially viable. The developments attained so far by different transducing systems are going to be reviewed.

4.2 DNA-OPTICAL BIOSENSORS

Several optical methods have been used for DNA sensing, such as luminescence, fluorescence Raman or optical waveguide structure spectroscopy, and surface plasmon resonance (SPR). Various immobilization strategies to attach ssDNA to surfaces with the aim of attaining maximum selectivity and sensitivity have been described.

Two methods for the covalent immobilization of single-stranded DNA onto fused silica optical fibres using various linkers for the development of biosensors were reported by Krull and co-workers [22]. One method involved a hydrophobic and the other a hydrophilic spacer arm and it was determined which linker would provide the best immobilization efficiency, hybridization kinetics, and minimal non-specific adsorption to the surface.

Fluorescent intercalating and groove binding dyes that associate with dsDNA were used for the detection of hybridization. However, a specific example of a situation where dye–dye interactions occur at concentrations of dye that are relevant to biosensor use and can lead to unexpected and undesired emission wavelength shifts and fluorescence quenching interactions was described by Krull and co-workers [23], and suggests cautions and considerations for the development of quantitative biosensors and biochips. Thermodynamics of two DNA-binding domains with and without conjugated cyanine dyes were studied by Thompson and Woodbury [24] using fluorescence techniques to determine the contribution to specific and non-specific binding in terms of polyelectrolyte and hydrophobic effects, indicating that non-specific binding is more sensitive to changes in salt concentration. The characteristic thermodynamic parameters of specific and non-specific DNA binding by each of the DNA-binding domains and their respective conjugates were presented. An imaging system called the scanning near field optical/atomic force microscope (SNOM/AFM), using a bent optical fibre as cantilever, to obtain images for the interaction of a dye with ssDNA molecules was used by Kim et al. [25].

A fibre-optic biosensor for the fluorimetric detection of T/AT triple-helical DNA formation was used by Krull and co-workers [26]. Hybridization between immobilized decaadenylic acid oligonucleotides on functionalized fused optical fibres and complementary oligonucleotides from the solution phase was detected by fluorescence from ethidium bromide. A correlation between the triplex melting temperature (T_m below 25°C) and the temperature at which the temperature coefficient of the fluorescence intensity changes from negative to positive was investigated and enabled the detection of reverse-Hoogsteen T/AT triplex helix formation.

Molecular beacons (MBs) are hairpin-shaped oligonucleotides that report the presence of specific nucleic acids. The MBs have been immobilized by Tan and co-workers [27] onto ultrasmall optical fibre probes through avidin–biotin binding. The MB-DNA biosensor detected its target DNA molecules, in real time, with selectivity for a single base-pair mismatch. This MB-DNA biosensor was used by Perlette and Tan [28] for real-time monitoring of mRNA–DNA hybridization inside a living cell.

A microfluidic sensor to detect fluorescently labelled DNA was developed by Khur and co-workers [29]. Photopatterning was obtained with carbene-generating photobiotin and allowed fabrication of homogeneous regions of immobilized biotin and the control of the spatial distribution of DNAs in a microchannel-flow-based sensor.

The suitability of three-dimensional flow-through microchannel glass substrates for multiplexed, heterogeneous nucleic acid fluorescence hybridization assays was demonstrated by Cooper and co-workers [30]. Optical tweezers on a microbead–single DNA molecule–cover slip constructed to study the kinetics of the reaction between dsDNA and formamide were described by Bhattacharyya and Feingold [31] showing that partially denatured DNA was an intermediate product. The single molecule kinetics was obtained from the change in the contour length of the DNA.

Real-time hybridization of 5′-fluorescein-labelled target oligonucleotides was described using a fibre-optic DNA sensor by Ehrat [32], Lu [33] and co-workers, and a fibre-optic DNA sensor array capable of positively identifying a point mutation of a biotin-primer-labelled PCR product was reported by Walt [34].

A DNA optical sensor system was proposed by Cass and co-workers [35] based on the combination of sandwich solution hybridization, magnetic bead capture, flow injection and chemiluminescence for the rapid detection of DNA hybridization. Sandwich solution hybridization uses two sets of DNA probes, one labelled with biotin, the other with an enzyme marker and hybridization is performed in solution where the mobility is greater and the hybridization process is faster, rather than on a surface. The hybrids were bound to the streptavidin-coated magnetic beads through biotin–streptavidin binding reaction. A chemiluminescence fibre-optic biosensor for the detection of hybridization of horseradish peroxidase-labelled complementary DNA to covalent immobilized DNA probes was developed by Zhou and co-workers [36].

A DNA biosensor based on various porous silicon layers was fabricated using an oxidized microcavity resonator design developed by Chan et al. [37], the porous silicon containing silicon nanocrystals that can luminesce efficiently in the visible.

Ellipsometry quantification and rapid visual detection of DNA hybridization, using biotinylated target sequences, on amorphous silicon and diamond-like carbon deposited on a porous polycarbonate membrane, creating a gold-coloured reflective test surface coated with immobilized ssDNA capture probe were described by Ostroff et al. [38].

The construction of biomolecular arrays by two photolithographic methods was carried out by Chrisey and co-workers [39] for the formation of patterned single or multiple DNA species on SiO_2 substrates. The biotinylated DNA patterns were visualized using a streptavidin–horseradish peroxidase conjugate.

Micro-reaction chambers for the thermal cycling of DNA amplification by the polymerase chain reaction (PCR) in silicon by bulk machining using anisotropic wet etching have been fabricated by Moore and co-workers [40]. The rapid temperature cycling and the small size of the chambers may be combined with an appropriate transducer in a miniaturized analytical system.

SPR measures the change in refractive index of the analyte molecules near a surface that occurs during complex formation or dissociation based on the anomalous diffraction due to the excitation of the surface plasmon waves and has been reviewed by Gauglitz [41]. SPR is a charge-density oscillation that may exist at the interface of two media with dielectric constants of opposite signs, for instance, a metal and a dielectric. The instrumentation available is capable of characterizing binding reactions in real time without labelling requirements as shown by Rich and Myszka [42].

SPR-based DNA biosensing using the high-affinity streptavidin–biotin system to immobilize DNA fragments has been applied by Nygren and co-workers [43] to real-time monitoring of hybridization and DNA strand separation as well as techniques including enzymatic action, such as ligation, endonuclease cleavage and DNA synthesis. Another application, by Mascini and co-workers [44], was to the detection of GMOs using immobilized probes specific for promoter and terminator sequences characteristic of GMOs. Prior to SPR detection the PCR amplified samples, which were dsDNA, were dissociated at high pH and physical separation of the two strands was obtained by the use of magnetic particles coated with immobilized streptavidin.

SPR imaging measurements of DNA microarrays fabricated on gold surfaces to monitor DNA–DNA, RNA–DNA and protein–DNA interactions down to nanomolar concentrations were used by Corn and co-workers [45]. Microfluidic channels in SPR imaging experiments for the detection of DNA and RNA adsorption onto chemically modified gold surfaces were also employed by Corn and co-workers [46]. The DNA probes were thiol-modified oligonucleotides. The microfluidic channels were used for the fabrication of 1D DNA line arrays for hybridization. Fabrication of 2D hybridization arrays of dsDNA probe spots was possible after attaching a second set of microchannels to the surface perpendicular to the 1D DNA line array to deliver target molecules to the surface, minimizing the total volume of the sample used to 1 μl.

SPR detection of hybridization and denaturation kinetics for tethered ssDNA thiol on gold was achieved by monitoring the gain or loss of DNA at the interface in the presence of an applied electrostatic field. Redox reactions were avoided and the current measured was limited to the capacitive, non-faradaic charging current, at selected potentials applied to the gold electrode interface, as described by Georgiadis and co-workers [47]. The specific DNA thiol monolayer films were robust and could be reused.

The resonant mirror is an evanescent wave sensor, which has been designed to combine the construction of SPR devices with the enhanced sensitivity of waveguide devices, and was applied by Watts et al. [48] for real-time detection and quantification of DNA hybridization using several biotinylated oligonucleotide immobilized probes. A comparison of probes indicated that the relative position of complementary sequence and the length of probe affected the hybridization response obtained. Regeneration of the surface-immobilized probe was possible, allowing reuse without a significant loss of hybridization activity.

Restriction enzymes are known to bind and cleave dsDNA and are highly specific at their recognition site. Binding and catalytic activity of the type II restriction endonuclease *Eco* RI on immobilized DNA has been observed in real time by Scheller and co-workers [49], using three different evanescent wave biosensors, grating coupler, SPR and surface-generated fluorescence and two different immobilization techniques, the streptavidin–biotin coupling reaction or the covalent attachment. The combination of different evanescent wave techniques gave access to the catalytic mechanism and allowed the determination of the rate-determining step, the cleavage of the second DNA strand.

4.3 DNA-ACOUSTIC WAVE BIOSENSORS

DNA-acoustic wave biosensors have been employed to study the duplex formation at the sensor surface and for monitoring a wide variety of processes involving nucleic acid chemistry at the solid–liquid interface without the need of labels such as radiochemical or fluorescent agents. The theory and applications of acoustic wave technology were reviewed by Thompson [50,51], Ziegler [52] and co-workers.

Piezoelectric phenomena are related to the reversible electric polarization generated by mechanical strain in crystals that do not display a centre of symmetry. The signal produced by acoustic wave devices is generated by bulk or surface acoustic waves launched by metal transducers at ultrasonic frequencies. Such waves are propagated through piezoelectric materials, usually quartz, where properties such as the orientation and thickness of the crystal, as well as the geometry of the transducer, determine the characteristics of the wave motion. A thickness-shear-mode (TSM) acoustic wave sensor consists of an AT-cut quartz wafer and metal electrodes on its both sides for generating an oscillating electric field of operational frequency range 5–20 MHz [51,52]. The measured change in frequency is determined by mass loading and by the particular boundary conditions that exist at the sensor–liquid interface, such as interfacial free energy, liquid structure and coating film properties. The formation of an adherent biomolecular layer at the interface produces changes in frequency through viscous losses together with alteration of energy and dissipation processes. The TSM acoustic wave sensor has been used extensively as a mass response sensor, the quartz crystal microbalance (QCM) acoustic wave sensor, using an oscillating electric field of 5–10 MHz operational frequency.

The kinetics of interfacial hybridization was studied by Thompson and co-workers [53] with a TSM sensor using a PdO surface to immobilize RNA. It was not clear whether the RNA strands were fixed on a few points or whether significant portions of each molecule are held in fixed positions. The entanglement of long RNA molecules, forming loops and coils, could be responsible for restricted free diffusion of strands and a low rate for the hybridization reaction. This rate is also influenced by the coverage of the surface by RNA and other interfacial effects. The interpretation of the influence of the blocking agents and ionic strength on the kinetics of nucleic acid hybridization and the mechanism of strand association with respect to solid-support hybridization were considered.

Attention has also been paid to the TSM detection of the interaction of surface-bound nucleic acids with small molecules such as specific-binding *cis*- and *trans*platin anticancer drugs. The results showed two distinct kinetic processes that were interpreted in terms of nucleic acid binding of the hydrolysis products of the two drugs by Thompson and co-workers [54].

Immobilization of DNA on silver surfaces of TSM through modification was reported by Yao and co-workers with thioglycollic acid [55] and didodecyl ditiono-oxamide and bovine serum albumin [56]. The Langmuir–Blodgett (LB) technique was used to deposit ssDNA-containing films on QCM crystals by Nicolini et al. [57] and the nanogravimetric hybridization assay was confirmed by fluorescent measurements.

The use of atomic force microscopy (AFM) and flow injection QCM in tandem provided important information about the surface coverage and orientation on gold of a thiolated DNA probe, as reported by Zhou and co-workers [58]. The effect of using a different alkanethiol to reorient the preformed film for a higher extent and efficiency of the hybridization was examined. The AFM images clearly indicate that DNA hybrid formation at the heterogeneous sensor preferentially occurred at the gold grain boundaries, suggesting that the hindrance by the adjacent DNA probe molecules plays an important role in governing the amount of hybridization. Whereas AFM allowed direct visualization of the orientation of both probe and target DNA molecules, QCM provided a means to semiquantitatively measure the amount of immobilized probe molecule and that of the hybridized duplexes, both studies indicating low hybridization efficiency.

Poly(pyrrole/pyrrole-ODN) films electrosynthesized by Vieil and co-workers [59] were also used to modify QCM electrodes to monitor DNA hybridization.

The characteristics of QCM sensors containing mono- or multilayered DNA probe constructed by direct chemical bonding, avidin–biotin interaction or electrostatic adsorption on polyelectrolyte films were compared by Zhou et al. [60]. The use of the polyethyleneimine adhesion, glutaraldehyde cross-linking (PEI–Glu) method to immobilize hepatitis B virus DNA onto gold QCM quartz crystals, enabling the sensor to be regenerated five times, was reported by Hu and co-workers [61].

The avidin–biotin procedure has been extensively used in hybridization studies. Since the concentration of all analysed DNAs used is identical, the total concentration of the analysed samples is high. The amplification of the base-mismatch recognition event is necessary to improve sensitivity. The use of oligonucleotide-functionalized liposomes or biotin-labelled liposomes as probes for the dendritic amplification of DNA-sensing processes was characterized by Willner and co-workers [62] and showed better performance using QCM than impedance spectroscopy measurements.

Three different indirect methods, Fig. 4.1, to detect and amplify a single-base mismatch were described by Willner et al. [63]. All methods were based on a surface treatment, involving immobilization of the thiolated DNA, hybridization and biotinylation in the presence of polymerase I of the sensing interface, resulting in functionalized Au–quartz crystals, performed outside the QCM cell.

The three amplification procedures used with this functionalized Au–quartz crystal interface consisted of detection first using avidin and biotin-labelled liposomes, secondly using avidin–Au-nanoparticle conjugate and the catalysed electroless deposition of gold, and thirdly avidin–alkaline phosphatase interaction with 5-bromo-4-chloro-3-indolyl phosphate causing the biocatalysed precipitation of the insoluble product on the piezoelectric crystal. The separation of surface treatment outside the QCM cell coupled with

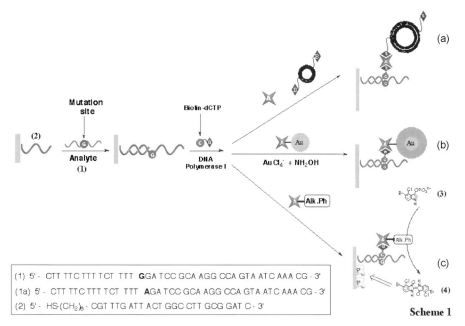

Fig. 4.1. Amplified detection paths of a single-base mismatch in nucleic acids: (a) using avidin and biotin-labelled liposomes; (b) using an avidin–Au-nanoparticle conjugate and the catalysed deposition of gold; (c) using an avidin–alkaline phosphatase bioconjugate and the biocatalysed precipitation of the insoluble product 4. (Reproduced from Ref. [63] with permission from Elsevier.)

amplified detection led to minimized non-specific adsorption of avidin or the liposomes and higher sensitivity on the functionalized sensing interface.

An overview of studies applying TSM mode devices for monitoring interfacial nucleic acid interactions by Cavic and Thompson [51] shows the advantages of this biosensing configuration and its suitability for being integrated into the design of high-density arrays of DNA probes.

4.4 DNA-ELECTROCHEMICAL BIOSENSORS

The aim in developing DNA-modified electrodes was to study the interaction of DNA immobilized on the electrode surface with analytes in solution and to use the DNA biosensor to evaluate and to predict DNA interactions and damage by health hazardous compounds based on their binding to nucleic acids. DNA would act as a promoter between the electrode and the biological molecule under study. Electrochemical techniques have the advantage in DNA-biosensor design of being rapid, sensitive, cost effective and enable in situ generation of reactive intermediates and detection of DNA damage. Comprehensive descriptions of research on DNA and DNA sensing [64–73] show the great possibilities of using electrochemical transduction in DNA diagnostics.

Electrochemical research on DNA is of great relevance to explain many biological mechanisms and the DNA biosensor is a very good model for simulating nucleic acid interaction with cell membranes, potential environmental carcinogenic compounds and to clarify the mechanisms of action of drugs used as chemotherapeutic agents.

When compared with optical, piezoelectric or other transducers, the electrochemical transduction is dynamic in that the electrode is itself a tuneable charged reagent as well as a detector of all surface phenomena, which greatly enlarges the electrochemical DNA biosensing capabilities. However, it is necessary that the analyte is electroactive, i.e., capable of undergoing electron transfer reactions, in order to use an electrochemical transducer. To design heterogeneous DNA-based biosensors, it is essential to understand the surface structures of the modified surfaces and so it is important to know which DNA groups are electroactive.

The double helical structure of DNA deduced by Watson and Crick consisted of two polynucleotide chains running in opposite directions and made up of a large number of deoxyribonucleotides, each composed of a base, a sugar and a phosphate group. The two chains of the double helix are held together by hydrogen bonds between purine, adenine (A) and guanine (G), and pyrimidine bases, cytosine (C) and thymine (T), and it is possible to identify a major and a minor groove, the latter with a higher negative density charge. The bases are always paired: adenine with thymine and guanine with cytosine and are on the inside of the helix, whereas the phosphate and deoxyribose units are on the outside.

The electrochemical behaviour of DNA and adsorption at different types of electrodes have been investigated for a number of years first using a dropping mercury electrode and more recently solid electrodes [64–80]. Electrochemical reduction of natural and biosynthetic nucleic acids at a dropping mercury electrode [64] showed that adenine and cytosine residues as well as guanine residues in a polynucleotide chain are reducible. Electrochemical oxidation of natural and synthetic nucleic acids on carbon electrodes [74–80] showed that all bases—guanine (G), adenine (A), cytosine (C) and thymine (T)—can be oxidized, Fig. 4.2, following a pH-dependent mechanism.

This figure, by Oliveira Brett and Matysik [80], showed for the first time that the pyrimidine bases thymine and cytosine can undergo oxidation on glassy carbon electrodes albeit for very positive potentials and that the

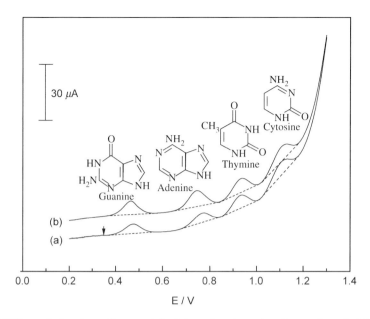

Fig. 4.2. Differential pulse voltammetric determination of purine and pyrimidine bases guanine (2×10^{-5} M), adenine (3×10^{-5} M), thymine (3×10^{-4} M) and cytosine (3×10^{-4} M) in borate buffer (pH 10.02), (a) with ultrasonic pretreatment (power intensity, 72 W cm^{-2}; horn tip-electrode separation, 5 mm), (b) successive scan without ultrasonic pretreatment. Scan rate, 5 mV s^{-1}; amplitude, 50 mV. (Reproduced from Ref. [80] with permission from Elsevier.)

concentration of the pyrimidine bases necessary was 10 times higher than the concentration of the purine bases. This means that, since the ratios of adenine to thymine and of guanine to cytosine are very close to 1.0 in all species, it is difficult to detect cytosine and thymine.

The electrocatalytic oxidation of DNA in the sugar and amine moiety, at a copper electrode surface, was studied by Singhal and Kuhr [81] using a very different detection procedure.

As mentioned, the electroactive bases in dsDNA are on the inside of the double helix and their distance and accessibility to the electrode surface are determinant for nucleic acids' electrochemical behaviour. In Figs. 4.3 and 4.4 some differential pulse voltammograms obtained with dsDNA and ssDNA at a glassy carbon electrode are presented.

The electrochemical signal due to ssDNA is always higher than that of dsDNA, see Refs. [64,82]. This illustrates the greater difficulty for the transition of electrons from the inside of the double-stranded rigid form of dsDNA to the electrode surface than from the flexible single-stranded form

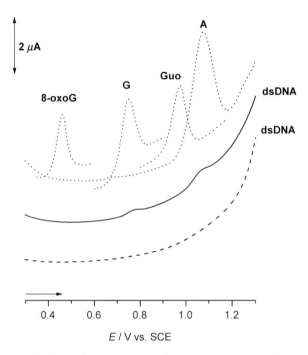

Fig. 4.3. Differential pulse voltammograms obtained with a bare GCE in pH 4.5 0.1 M acetate buffer of: (···) 15 μM 8-oxoguanine (8-oxoG); (···) 15 μM guanine (G); (···) 15 μM guanosine (Guo); (···) 100 μM adenine (A); and 60 μg ml^{-1} dsDNA (- - -) 1st and (—) 40th voltammogram. Pulse amplitude, 50 mV; pulse width, 70 ms; scan rate, 5 mV s^{-1}. (Reproduced from Ref. [123] with permission from Elsevier.)

where the guanine and adenine residues are in close proximity to the electrode surface.

This has been justified by the stereochemistry of the DNA double helix in comparison with the ribbon-like feature of the ssDNA that can follow the electrode contours more easily. The roughness of a solid electrode surface also means that double helix DNA has some difficulty in following the surface contours, whereas single-stranded unwound DNA molecules fit more easily into the grooves on the surface of the electrode because of their greater flexibility.

The adsorption of dsDNA onto gold electrode surfaces shown by STM and AFM images obtained by Lindsay et al. [83–86] demonstrates that DNA molecules adsorb on the surface reversibly and can undergo conformation transitions. Additionally, the electrochemical deposition and in situ potential control during imaging can be used to obtain reproducible STM images of the internal structure of adsorbed ssDNA and dsDNA. Spectroscopic techniques

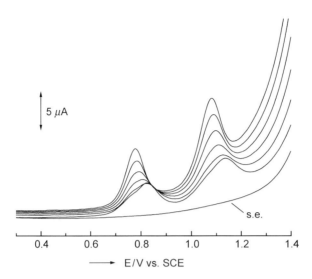

Fig. 4.4. Successive differential pulse voltammograms of the clean glassy carbon electrode, immersed in the ssDNA solution during modification; s.e.—supporting electrolyte: 0.1 M acetate buffer pH 4.5. Pulse amplitude, 50 mV; pulse width, 70 ms; scan rate, 5 mV s^{-1}. (Reproduced from Ref. [70] with permission from Elsevier.)

such as infrared spectroelectrochemistry used by Dong and co-workers [87] and surface-enhanced Raman spectroscopy (SERS) by Otto et al. [88] have also been used to investigate the adsorption of DNA onto electrode surfaces.

Electrode surface characteristics represent an important aspect in the construction of sensitive DNA electrochemical biosensors for rapid detection of DNA interaction and damage. A critical issue in the development of a DNA-electrochemical biosensor is the sensor material and the degree of surface coverage that influences directly the sensor response.

Using ex situ MAC mode AFM, Oliveira Brett et al. [89–91] have been able to visualize directly the surface characteristics of the dsDNA films prepared on a highly oriented pyrolytic graphite (HOPG) electrode. It was found that different immobilization methodologies lead to structural changes on the DNA-biosensor surface and consequently different sensor response. AFM images of an HOPG substrate modified by a thick and a thin layer of dsDNA are shown in Fig. 4.5. The HOPG surface is extremely smooth, which enables the identification of the topography changes when the surface is modified with dsDNA. Two different immobilization procedures of double-stranded DNA at the surface of an HOPG electrode were evaluated, a thin dsDNA adsorbed film forming a network structure, with holes not covered by the molecular film exposing the electrode surface, and a thick dsDNA film completely covering

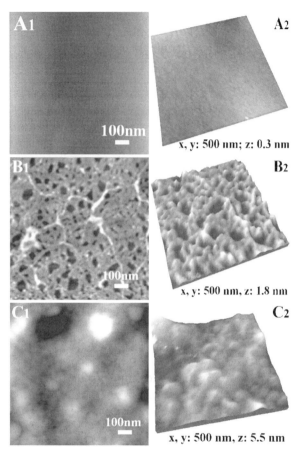

Fig. 4.5. MAC mode AFM topographical images in air of: (A1 and A2) clean HOPG electrode surface; (B1 and B2) thin-film dsDNA-biosensor surface, prepared onto HOPG by 3 min free adsorption from a solution of 60 μg ml^{-1} dsDNA in pH 4.5 0.1 M acetate buffer electrolyte; (C1 and C2) thick-film dsDNA-biosensor surface, prepared onto HOPG by evaporation from solution of 37.5 μg ml^{-1} dsDNA in pH 4.5 0.1 M acetate buffer electrolyte; (A1, B1 and C1) two-dimensional view 1 μm × 1 μm scan size and (A2, B2 and C2) three-dimensional view 500 nm × 500 nm scan size. (Reproduced from Ref. [91] with permission from Elsevier.)

the electrode surface with a uniform multilayer film, presenting a much rougher structure.

The dsDNA networks formed at the HOPG electrode during the formation of thin dsDNA layers define different active surface areas on the DNA electrochemical biosensor. The uncovered regions may act as a system of microelectrodes with nanometre or micrometre dimensions. The two-dimensional dsDNA networks form a biomaterial matrix to attach and study other molecules. The major problem encountered with the HOPG electrodes modified by a thin film of dsDNA is the fact that the electrode is not completely covered allowing the diffusion of molecules from the bulk solution to the surface and their non-specific adsorption. This leads to two contributions to the electrochemical signal, one from the simple adsorbed compound and other from the damage to immobilized dsDNA, and it is difficult to distinguish between the two signals [123].

The big advantage of the thick film of dsDNA is that the HOPG surface is completely covered by dsDNA so that the undesired binding of molecules to the electrode surface is impossible. The DNA-biosensor response is thus only determined by the interaction of the compound with the dsDNA in the film, without any contribution from the electrochemical reaction of the compound at the HOPG substrate surface.

Evidence that thiol-derivatized dsDNA molecules on gold surfaces, adsorbed under potential control, suffer a morphology change in which the helices either stand up straight or lie flat down on the metal surface, depending on the applied potential relative to the potential of zero charge (pzc), has been described by Barton and co-workers [92]. At positive applied potentials the helical axis becomes parallel to the electrode surface, the base pairs being vertically oriented against the electrode surface, leading to the conclusion that the thickness of a monolayer of adsorbed dsDNA at the electrode surface is less than 2 nm. The possibility of controlling the coverage of the electrode surface and selecting the DNA orientation by using small changes in potential opens new perspectives for the development and applications of DNA-electrochemical biosensors. A silicon chip with an array of platinum electrodes was used by Heller and co-workers [93] for biotin–streptavidin mediated immobilization of oligonucleotides and hybridization was controlled by adjustment of the electric field strength.

The electrode surface can be modified by immobilizing ssDNA or dsDNA. The electrochemical ssDNA biosensor uses short oligonucleotide sequences for the surface recognition layer capable of identifying a complementary nucleotide sequence of a target DNA through hybridization. The electrochemical dsDNA biosensor can predict the mechanism and detect the damage caused to DNA by health hazard compounds that cause strand breaks and makes possible electrochemical detection of the exposed bases. Immobilization procedures and chemical treatment of an electrode by chemical reactions or by chemisorption alter its surface that should be characterized by surface techniques. The tuning of the electrode potential enables the in situ generation of radicals from electroactive compounds that interact with DNA followed by the electrochemical detection of the damage.

Electrochemical biosensing of DNA sequences using direct electrochemical detection of DNA hybridization, adsorptive striping analysis, metal complex hybridization indicators, organic compound electroactive hybridization indicators and renewable DNA probes have been considered [65,67,72,73]. With metal complexes and organic compound electroactive hybridization indicators, non-specific adsorption can influence the results [68,94]. Chronopotentiometric detection was used to monitor the hybridization onto screen-printed carbon electrodes by following the oxidation of the guanine peak, which decreases in the presence of the complementary strand [64,68,73].

DNA has four different potential coordination sites for binding with metal ions: the negatively charged phosphate oxygen atoms, the ribose hydroxyls, the base ring nitrogens and the exocyclic base keto groups. Metal ions such as Na^+, K^+, Mg^{2+} and Ca^{2+} exist in the body in high concentrations and the nucleic acids and nucleotides occur as complexes coordinated with these ions. Because of the great potential of electrochemical methods for studying the adsorption and reactions of biological molecules at electrified interfaces, they have been used by researchers who have been studying metal ion–DNA interactions.

Studies of redox-active metallointercalation agents in the presence of dsDNA have been done with solutions containing the redox complexes of cobalt, iron and osmium [64,68,72,95]. Osmium tetroxide complexes with tertiary amines (Os, L) have been used as a chemical probe of DNA structure. The simultaneous determination, based on a sufficient peak separation on the potential scale of (Os, L)–DNA adducts and free (Os, L), was obtained by Fojta et al. [96] using a pyrolytic graphite electrode.

The binding of ligand and ligand substituents in complexes of ruthenium(II) has been investigated by Barton and co-workers [97,98] in a systematic fashion and the binding parameters for the series compared in order to determine the different ligand functionalities and sizes in binding with DNA, i.e., intercalation and surface binding. It was found that if one compares the various factors that contribute to stabilizing the metal complexes of ruthenium on the DNA helix, it appears that the most significant factor is that of molecular shape.

The complexes that fit most closely to the DNA helical structure, those in which the van der Waals interactions between complex and DNA are maximized, display the highest binding affinity. The binding was interpreted by Barton and co-workers [99] in terms of the interplay of electrostatic interactions of the metal coordination complexes with the charged sugar–phosphate backbone and the intercalative, hydrophobic interactions within the DNA helix, i.e., the π-stacked base pairs. These metallointercalation agents have often been used as electroactive hybridization indicators based on their different interaction with dsDNA and ssDNA.

Gold electrodes modified with DNA probes have been used to detect hybridization using electrochemically active indicators [100,101]. A sandwich-type ternary complex with a target DNA has been described by Maeda and co-workers [102] for electrochemical detection of hybridization based on a ferrocene-oligonucleotide conjugate. Colloidal Au was used by Fang and co-workers [103] to enhance the ssDNA immobilization on a gold electrode and the hybridization was carried out by exposure of the ssDNA containing gold electrode to ferrocenecarboxaldehyde-labelled complementary ssDNA. A drawback is that the oxidation of DNA bases occurs at higher positive potentials and can never be observed with gold electrodes.

Electrochemical detection of DNA hybridization on colloidal Au nanoparticles has been combined by Wang et al. [104] with an advanced biomagnetic processing technology that couples efficient magnetic removal of non-hybridized DNA with low-volume magnetic mixing developed by Limoges and co-workers [105]. The hybridization of a target oligonucleotide to magnetic bead-linked oligonucleotides was followed by binding of the streptavidin-coated metal nanoparticles to the captured DNA, dissolution of the nanometre-sized gold tag, and potentiometric stripping measurements of the dissolved metal tag at single-use thick-film carbon electrodes. This electrochemical genomagnetic hybridization assay was also applied to streptavidin–alkaline phosphatase (AP) using an enzyme-linked sandwich solution hybridization by Wang et al. [106].

Enzyme DNA hybridization assays with electrochemical detection can offer enhanced sensitivity and reduced instrumentation costs in comparison with their optical counterparts. Efforts to prevent non-specific binding of the codissolved enzyme and to avoid fouling problems by selecting conditions suitable to amplify the electrode response have been reported by Heller and co-workers [107]. A disposable electrochemical sensor based on an ion-exchange film-coated screen-printed electrode was described by Limoges and co-workers for an enzyme nucleic acid hybridization assay using alkaline phosphatase [108] or horseradish peroxidase [109]. In another methodology to improve sensitivity, a carbon paste electrode with an immobilized nucleotide on the electrode surface and methylene blue as hybridization indicator was coupled, by Mascini and co-workers [110], with PCR amplification of DNA extracted from human blood for the electrochemical detection of virus.

Originally developed as a gene-targeting drug [15–17], PNA, as mentioned previously, demonstrates hybridization properties towards complementary DNA sequences and the adsorption behaviour was studied by means of AC impedance measurements at a hanging mercury drop electrode (HMDE) by Palecek and co-workers [111]. The differential capacity of the electrode double layer was measured as a function of potential. The adsorption of PNA, with a neutral backbone, was considerably different from that of DNA, with no PNA desorption occurring even after the application to the electrode of a high negative potential for a long period of time.

Electrochemical impedance measurements were also used to detect the hybridization of DNA on Si/SiO$_2$ chips and great emphasis has been put by Cloarec et al. [112] on the immobilization of single strands on the substrates in order to obtain reproducible sensors. The adsorption of dsDNA and nucleotides on a glassy carbon surface has also been evaluated by electrochemical impedance spectroscopy by Oliveira Brett et al. [113,114].

The utilization of a dsDNA matrix to aid enzyme immobilization is also a promising alternative for the development of DNA biosensors and its application in flow injection analysis (FIA) systems. A DNA-tyrosinase carbon paste electrode described by Serrano and co-workers [115] showed excellent performance for the detection of catechol in an FIA system and suggests that other enzymatic biosensors can benefit from the presence of DNA.

The cytostatic activity of various platinum drugs has shown that platinum coordination complexes cause irreversible inhibition of DNA synthesis due to covalent binding with DNA. This causes the treatment to be often accompanied by adverse reactions. Differential pulse voltammetry was used to investigate the interactions of platinum coordination complexes with strong anticancer agents in solution with DNA by Oliveira Brett et al. [116,117]. The DNA interacting drugs prevent cell growth, but not only cancer cell growth; the cytotoxic effect also blocks the growth of normal cells. The lack of selectivity of cancer drugs is one of the main problems in cancer chemotherapy and the DNA biosensors are an important tool for the investigation of the chemical and biological mechanism of drugs active against cancer cells.

Nitroimidazoles are among the most important nitroheterocyclic drugs of interest in cancer chemotherapy. It was observed that adenine and guanine interact with intermediates generated during nitroimidazole reduction, causing irreversible damage to DNA and suggesting mutagenic properties of these compounds. The mechanism of reduction of a group of nitroimidazoles was investigated by a new approach by Oliveira Brett et al. [118–122] using the DNA biosensor. The analyte was pre-concentrated on the electrode surface containing DNA and either the reduction or the oxidation of the reduction products retained on the electrode surface was studied. It was possible to follow their reduction, the reversible oxidation of the hydroxylamine derivative (RNHOH) to the corresponding nitroso derivative (RNO), the condensation reaction between the hydroxylamine and nitroso derivatives to form the azoxycompound (RNO:NR) and the interaction with DNA.

Electrochemical voltammetric in situ detection of dsDNA oxidative damage caused by reduced adriamycin, an antibiotic of the family of anthracyclines, intercalated into DNA, was carried out using a DNA biosensor. Oxidation and reduction of adriamycin molecules intercalated in dsDNA were investigated by Oliveira Brett et al. [123] in order to understand the in vivo mechanism of action of DNA with this anti-neoplasic drug. However, it is not possible to detect the adriamycin–DNA damage by monitoring only changes in the adriamycin oxidation peak. The damage to immobilized dsDNA causes the appearance of oxidation peaks from DNA bases and this should always be measured and taken into account, Fig. 4.5. The results showed that the interaction of adriamycin with dsDNA is potential dependent causing contact between DNA guanine and adenine bases and the electrode surface such that their oxidation is easily detected, Fig. 4.6. A mechanism for adriamycin reduction and oxidation in situ when intercalated in dsDNA immobilized onto the glassy carbon electrode surface was proposed. This mechanism leads to the formation of the mutagenic 8-oxoguanine, whose redox behaviour was studied by Oliveira Brett et al. [124,125].

This information is very relevant because the mechanism of interaction of DNA–drug at charged interfaces mimics better the in vivo DNA–drug complex situation, where it is expected that DNA will be in close contact with charged phospholipid membranes and proteins, rather than when the interaction is in solution. Oxidative damage to DNA was demonstrated by Barton and co-workers [126] to depend upon oxidation potential. Oxidative damage can be promoted from a remote site as a result of electron hole migration through the DNA π-stack: the hole migrates down the double helix to damage guanine, a site sensitive to oxidative nucleic acid damage within the cell.

Similarly, the detection of DNA damage involving strand breaks was observed by Palecek and co-workers [127] using an HMDE. Extensive cleavage of electrode-confined DNA by reactive oxygen species (ROS) was obtained in the absence of chemical reductants when redox cycling of the metal (iron/DNA complex) was controlled. Not only the cleaving agents were detected but also the DNA cleavage was modulated, by generating the DNA-damaging species electrochemically.

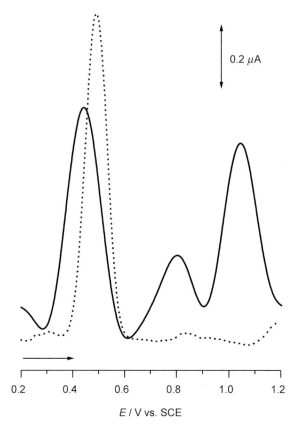

Fig. 4.6. Background-subtracted differential pulse voltammograms in pH 4.5 0.1 M acetate buffer obtained with a thick layer dsDNA-modified GCE after being immersed during 10 min in a 1 μM adriamycin solution and rinsed with water before the experiment in buffer: (···) without applied potential; (—) subsequent scan after applying a potential of −0.6 V during 120 s. Pulse amplitude, 50 mV; pulse width, 70 ms; scan rate, 5 mV s^{-1}. (Reproduced from Ref. [123] with permission from Elsevier.)

DNA adsorbed on a glassy carbon electrode was also used as an effective electron promoter enabling electron transfer via hopping conduction through electrode/base pair/cytochrome c by Ikeda et al. [128]. Gold electrodes modified with short oligonucleotides immobilized via thiol chemisorption were described by Lisdat et al. [129] to study the promotion of electron transfer to cytochrome c.

The detection of chemicals that cause irreversible damage to DNA is very important, as it may lead to hereditary or carcinogenic diseases. DNA biosensors have been used for trace measurements of toxic amine compounds, phenothiazine compounds with neuroleptic and antidepressive action as well as detection of radiation-induced DNA damage by Wang and co-workers [130]. Screening tests for carcinogens based on voltammetric measurements were developed to study in vitro damage to DNA caused by the action of pollutants by Oliveira Brett and Silva [131], pathogens by Palecek and co-workers [67] and detection of DNA-adduct formation that start the carcinogenic process, such as benzo[a]pyrene–DNA adducts by Ozsoz and co-workers [132]. The application of DNA-electrochemical biosensors to detection in food of bacterial and viral pathogens responsible for disease, due to their unique nucleic acid sequences, is attractive but still has some drawbacks when compared to immunobiosensing techniques by Limoges [133], Wilsins [134] and co-workers.

Microfabricated capillary electrophoresis (CE) chips with integrated injection and indirect electrochemical detection were used for high-sensitivity DNA restriction fragment and PCR product sizing by Mathies and co-workers [135]. The coupling of this and other electrochemical DNA detection schemes with microfluidic devices holds great promise for genetic testing as described by Wang [136].

The development of electrochemical DNA biosensors opened a wide perspective using a particularly sensitive and selective method for the detection of specific interactions. The possibility of foreseeing the damage that these compounds cause to DNA integrity arises from the possibility of pre-concentration of either the starting materials or the redox reaction products on the DNA-biosensor surface, thus permitting the electrochemical probing of the presence of short-lived intermediates and of their damage to DNA.

Effectively, the DNA-electrochemical biosensor enables pre-concentration of the drug investigated onto the electrode sensor surface and in situ electrochemical generation of radicals, which cause damage to the DNA immobilized on the glassy carbon electrode surface and can be detected electrochemically. However, non-uniform coverage of the electrode surface by DNA and adsorption of drug on the bare glassy carbon, Fig. 4.5, may lead to contributions from both simple adsorbed analyte and from products of damage to immobilized DNA, which need to be carefully distinguished. The damage to immobilized DNA always leads to the appearance of oxidation peaks from DNA bases or even 8-oxoguanine as shown by Oliveira Brett et al. [123–125], which should be monitored and taken into account.

The use of DNA-electrochemical biosensors for the understanding of DNA interactions with molecules or ions exploits the use of voltammetric techniques for in situ generation of reactive intermediates and is a complementary tool for the study of biomolecular interaction mechanisms.

Voltammetric methods are an inexpensive and fast detection procedure. Additionally, the interpretation of electrochemical data can contribute to elucidation of the mechanism by which DNA is oxidatively damaged by such substances, in an approach to the real action scenario that occurs in the living cell.

4.5 CONCLUSIONS

The DNA biosensors described clearly show the relevance and importance of different types of transducer for the study of the biological mechanism of DNA interactions and of the structure–activity relationships. DNA sensing can be carried out by several different detection strategies and, as shown, each DNA biosensor has some clear advantages and all the approaches are important to achieve a correct response.

New DNA biosensors will combine information from more than one transducer for characterization and analysis of oligonucleotides and single-stranded DNA at interfaces, using new immobilization procedures and better knowledge of the surface coverage in order to avoid non-specific adsorption affecting hybridization efficiency. A hybridization assay for the direct detection of a specific genomic sequence of an infectious disease would need, according to Diamandis [137], a sensitivity allowing detection of 10^4–10^5 copies of a specific target DNA which is well below the present sensitivities obtainable with the aid of amplification systems prior to detection such as PCR. Further routes to enhance detection of DNA will be necessary but in the future it is almost certain that DNA biosensors will be part of diagnostic kits.

The understanding of the mechanism of action of drugs that interact with DNA will explain the differences in reactivity between similar compounds. This knowledge can be used as an important parameter for quantitative structure–activity relationships (QSAR) and/or molecular modelling studies, as a contribution to the design of new structure-specific DNA-binding drugs, and for the possibility of pre-screening the damage they may cause to DNA integrity.

DNA biosensors have a great potential for numerous applications which will include new improvements for the detection of single nucleotide polymorphisms (SNPs) analysis in personalized medicine, pathogenic organisms in the field of food control, and toxic pollutants in environmental monitoring.

DNA biosensors will continue to exploit the remarkable specificity of biomolecular recognition to provide analytical tools that can measure the presence of a single molecular species in a complex mixture, pre-screen hazard compounds that cause damage to DNA and help to explain DNA−protein interactions.

Acknowledgements

Financial support from Fundação para a Ciência e Tecnologia (FCT), projects POCTI (co-financed by the European Community fund FEDER), ICEMS (Research Unit 103), and European Projects ERBICT15-CT98-0915, QLK3-2000-01311, and HPRN-CT-2002-00186 are gratefully acknowledged.

REFERENCES

1. I.K. Larsen, In: P. Krosgaard-Larsen and H. Bundgaard (Eds.), *A Textbook of Drug Design and Development*. Harwood Academic Publishers, New York, 1991, pp. 192.
2. W. Saenger, In: C.R. Cantor (Ed.), *Principles of Nucleic Acid Structure*, Springer Advanced Texts in Chemistry, Springer, New York, 1984.
3. M.E.A. Downs, S. Kobayashi and I. Karube, *Anal. Lett.*, 20 (1987) 1897.
4. M. Thompson and U.J. Krull, *Anal. Chem.*, 63 (1991) 393A.
5. M. Yang, M.E. McGovern and M. Thompson, *Anal. Chim. Acta*, 346 (1997) 259.
6. Z. Junhui, C. Hong and Y. Ruifu, *Biotechnol. Adv.*, 15 (1997) 43.
7. J. Wang, *Nucleic Acids Res.*, 28 (2000) 3011.
8. G.H.W. Sanders and A. Manz, *Trends Anal. Chem.*, 19 (2000) 364.
9. U.J. Krull, R.S. Brown and E.T. Vandenberg, *J. Electron Microsc. Tech.*, 18 (1991) 212.
10. G.H. Keller and M.M. Manak (Eds.), *DNA Probes*. Stockton Press, New York, 1993.
11. M. Minunni, S. Tombelli, E. Mariotti and M. Mascini, *Fresenius J. Anal. Chem.*, 369 (2001) 589.
12. M. Yang, M.E. McGovern and M. Thompson, *Anal. Chim. Acta*, 346 (1997) 259.
13. J.H. Watterson, P.A.E. Piunno, C.C. Wust, S. Raha and U.J. Krull, *Fresenius J. Anal. Chem.*, 369 (2001) 601.
14. L.M. Furtado, H. Su, M. Thompson, D.P. Mack and G.L. Hayward, *Anal. Chem.*, 71 (1999) 1167.
15. P.E. Nielsen, M. Egholm, R. Berg and O. Buchardt, *Science*, 254 (1991) 1497.
16. A. Ray and B. Nordén, *FASEB J.*, 14 (2000) 1041.
17. J. Wang, *Biosens. Bioelectron.*, 13 (1998) 757.
18. D.R. Baselt, G.U. Lee, M. Natesan, S.W. Metzger, P.E. Sheehan and R.J. Colton, *Biosens. Bioelectron.*, 13 (1998) 731.

19   M. Schena, *DNA-Microarrays—A Practical Approach*. Oxford University Press, Oxford, 1999.
20   S.C. Jakeway, A.J. Mello and E.L. Russel, *Fresenius J. Anal. Chem.*, 366 (2000) 525.
21   A.B. Steel, R.L. Levicky, T.M. Herne and M.J. Tarlov, *Biophys. J.*, 79 (2000) 975.
22   L. Henke, P.A.E. Piunno, A.C. McClure and U.J. Krull, *Anal. Chim. Acta*, 344 (1997) 201.
23   D. Hanfi-Bagby, P.A.E. Piunno, C.C. Wust and U.J. Krull, *Anal. Chim. Acta*, 411 (2000) 19.
24   M. Thompson and N.W. Woodbury, *Biophys. J.*, 81 (2001) 1793.
25   J.H. Kim, T. Ohtani, S. Sugiyama, T. Hirose and H. Muramatsu, *Anal. Chem.*, 73 (2001) 5984.
26   A.H. Uddin, P.A.E. Piunno, R.H.E. Hudson, M.J. Damha and U.J. Krull, *Nucleic Acids Res.*, 25 (1997) 4139.
27   X. Liu, W. Farmerie, S. Schuster and W. Tan, *Anal. Biochem.*, 283 (2000) 56.
28   J. Perlette and W. Tan, *Anal. Chem.*, 73 (2001) 5544.
29   L.M. Shamansky, C.B. Davis, J.K. Stuart and W.G. Khur, *Talanta*, 55 (2001) 909.
30   V. Benoit, A. Steel, M. Torres, Y. Yu, H. Yang and J. Cooper, *Anal. Chem.*, 73 (2001) 2412.
31   A.J. Bhattacharyya and M. Feingold, *Talanta*, 55 (2001) 943.
32   A.P. Abel, M.G. Weller, G.L. Duveneck, M. Ehrat and H.M. Widmer, *Anal. Chem.*, 68 (1996) 2905.
33   H. Lu, Y. Zhao, J. Ma, W. Li and Z. Lu, *Colloids Surf. A*, 175 (2000) 147.
34   B.G. Healey, R.S. Matson and D.R. Walt, *Anal. Biochem.*, 251 (1997) 270.
35   X. Chen, X.-E. Zhang, Y.-Q. Chai, W.-P. Hu, Z.-P. Zhang, X.-M. Zhang and A.E.G. Cass, *Biosens. Bioelectron.*, 13 (1998) 451.
36   G. Zhang, Y. Zhou, J. Yuan and S. Ren, *Anal. Lett.*, 32 (1999) 2725.
37   S. Chan, Y. Li, L.J. Rothberg, B.L. Miller and F.M. Fauchet, *Mater. Sci. Eng. C*, 15 (2001) 277.
38   R.M. Ostroff, D. Hopkins, A.B. Haeberli, W. Baouchi and B. Polisky, *Clin. Chem.*, 45 (1999) 1659.
39   L.A. Chrisey, C.E. O'Ferrall, B.J. Spargo, C.S. Dulcey and J.M. Calvert, *Nucleic Acids Res.*, 24 (1996) 3040.
40   J.H. Daniel, S. Iqbal, R.B. Millington, D.F. Moore, C.R. Lowe, D.L. Leslie, M.A. Lee and M.J. Pearce, *Sens. Actuators A*, 71 (1998) 81.
41   J. Homola, S.S. Yee and G. Gauglitz, *Sens. Actuators B*, 54 (1999) 3.
42   R.L. Rich and D.G. Myszka, *Curr. Opin. Biotechnol.*, 11 (2000) 54.
43   P. Nilsson, B. Persson, M. Uhlén and P. Nygren, *Anal. Biochem.*, 224 (1995) 400.
44   E. Mariotti, M. Minunni and M. Mascini, *Anal. Chim. Acta*, 453 (2002) 165.
45   B.P. Nelson, T.E. Grimsrud, M.R. Liles, R.M. Goodman and R.M. Corn, *Anal. Chem.*, 73 (2001) 1.
46   H.J. Lee, T.T. Goodrich and R.M. Corn, *Anal. Chem.*, 73 (2001) 5525.

47 R.J. Heaton, A.W. Peterson and R.M. Georgiadis, *Proc. Natl Acad. Sci. USA*, 98 (2001) 3701.
48 H.J. Watts, B. Yeung and H. Parkes, *Anal. Chem.*, 67 (1995) 4283.
49 F.B. Bier, F. Kleinjung, P.M. Schmidt and F.W. Scheller, *Anal. Bioanal. Chem.*, 372 (2002) 308.
50 B.A. Cavic, G.L. Hayward and M. Thompson, *Analyst*, 124 (1999) 1405.
51 B.A. Cavic and M. Thompson, *Anal. Chim. Acta*, 469 (2002) 101.
52 M. Kaspar, H. Stadler, T. Weiss and Ch. Ziegler, *Fresenius J. Anal. Chem.*, 366 (2000) 602.
53 H. Su, S. Chong and M. Thompson, *Biosens. Bioelectron.*, 12 (1997) 161.
54 H. Su, P. Williams and M. Thompson, *Anal. Chem.*, 67 (1995) 1010.
55 H. Zhang, R. Wang, H. Tan, L. Nie and S. Yao, *Talanta*, 46 (1998) 171.
56 H. Zhang, H. Tan, R. Wang, W. Wei and S. Yao, *Anal. Chim. Acta*, 374 (1998) 31.
57 C. Nicolini, V. Erokhin, P. Facci, S. Guerzoni, A. Ross and P. Paschkevitsch, *Biosens. Bioelectron.*, 12 (1997) 613.
58 E. Huang, M. Satjapipat, S. Han and F. Zhou, *Langmuir*, 17 (2001) 1215.
59 N. Lassalle, A. Roget, T. Livache, P. Mailley and E. Vieil, *Talanta*, 55 (2001) 993.
60 X.C. Zhou, L.Q. Huang and S.F.Y. Li, *Biosens. Bioelectron.*, 16 (2001) 85.
61 X. Zhou, L. Liu, M. Hu, L. Wang and J. Hu, *J. Pharm. Biomed. Anal.*, 27 (2002) 341.
62 F. Patolsky, A. Lichtenstein and I. Willner, *J. Am. Chem. Soc.*, 123 (2001) 5194.
63 I. Willner, F. Patolsky, Y. Weizmann and B. Willner, *Talanta*, 56 (2002) 847.
64 E. Palecek, *Talanta*, 56 (2002) 809.
65 S.R. Mikkelsen, *Electroanalysis*, 8 (1996) 15.
66 A.M. Oliveira Brett and S.H.P. Serrano, In: P. Frangopol, D.P. Nikolelis and U.J. Krull (Eds.), *Current Topics in Biophysics, Biosensors*. A.I. Cuza University Press, Iasi, Romania, 1997, pp. 223, Chapter 10.
67 J. Wang, G. Rivas, X. Cai, E. Palecek, P. Nielsen, H. Shiraishi, N. Dontha, D. Luo, C. Parrado, M. Chicharro, P.A.M. Farias, F.S. Valera, D.H. Grant, M. Ozsoz and M.N. Flair, *Anal. Chim. Acta*, 347 (1997) 1.
68 J. Wang, X. Cai, G. Rivas, H. Shiraishi and N. Dontha, *Biosens. Bioelectron.*, 12 (1997) 587.
69 E. Palecek, M. Fojta, F. Jelen and V. Vetterl, Electrochemical analysis of nucleic acids. In: A.J. Bard and M. Stratmann (Eds.), *The Encyclopedia of Electrochemistry, Bioelectrochemistry*, Vol. 9. Wiley-VCH, Weinheim, 2002, pp. 365–429, and references therein.
70 A.M. Oliveira Brett, S.H.P. Serrano and J.A.P. Piedade, In: R.G. Compton (Ed.), *Applications of Kinetic Modelling*, Comprehensive Chemical Kinetics, Vol. 37. Elsevier, Amsterdam, 1999.
71 M.I. Pividori, A. Merkoci and S. Alegret, *Biosens. Bioelectron.*, 15 (2000) 291.
72 M. Mascini, I. Palchetti and G. Marrazza, *Fresenius J. Anal. Chem.*, 369 (2001) 15.
73 E. Palecek and M. Fojta, *Anal. Chem.*, 73 (2001) 75A.

74 G. Dryhurst and P.J. Elving, *Talanta*, 16 (1969) 855.
75 T. Tao, T. Wasa and S. Mursha, *Bull. Chem. Soc. Jpn*, 51 (1978) 1235.
76 J.M. Hall, J. Moore-Smith, J.V. Bannister and I.J. Higgins, *Biochem. Mol. Biol. Int.*, 32 (1994) 21.
77 C.M.A. Brett, A.M. Oliveira Brett and S.H.P. Serrano, *J. Electroanal. Chem.*, 366 (1994) 225.
78 A.M. Oliveira Brett and F.-M. Matysik, *Electrochim. Acta*, 42 (1997) 945.
79 A.M. Oliveira Brett and F.-M. Matysik, *J. Electroanal. Chem.*, 429 (1997) 95.
80 A.M. Oliveira Brett and F.-M. Matysik, *Bioelectrochem. Bioenerg.*, 42 (1997) 111.
81 P. Singhal and W.G. Kuhr, *Anal. Chem.*, 69 (1997) 4828.
82 A.M. Oliveira Brett and S.H.P. Serrano, *J. Braz. Chem. Soc.*, 6 (1995) 1–6.
83 S.M. Lindsay, T. Thundat, L. Nagahara, U. Knipping and R.L. Rill, *Science*, 244 (1989) 1063.
84 S.M. Lindsay, N.J. Tao, J.A. DeRose, P.I. Oden, Yu.L. Lyubchenko, R.E. Harrington and L. Shyakhtenko, *Biophys. J.*, 61 (1992) 1570.
85 N.J. Tao, J.A. DeRose and S.M. Lindsay, *J. Phys. Chem.*, 97 (1993) 910.
86 S.M. Lindsay and N.J. Tao, In: M. Amrein and O. Marti (Eds.), *STM and SFM in Biology*. Academic Press, London, 1993, pp. 229–257, Chapter 5.
87 Z. Wang, D. Liu and S. Dong, *Bioelectrochemistry*, 53 (2002) 175.
88 C. Otto, F.P. Hoeben and J. Greve, *J. Raman Spectrosc.*, 22 (1991) 791.
89 A.M. Oliveira Brett and A.-M. Chiorcea, *Electrochem. Commun.*, 5 (2003) 178.
90 A.M. Oliveira Brett and A.-M. Chiorcea, *Langmuir*, 19(9) (2003) 3830.
91 A.M. Chiorcea and A.M. Oliveira Brett, *Bioelectrochemistry*, 63 (2004) 229.
92 S.O. Kelly, J.K. Barton, N.M. Jackson, L.D. McPherson, A.B. Potter, E.M. Spain, M.J. Allen and M.G. Hill, *Langmuir*, 14 (1998) 6781.
93 R.G. Sosnowski, E. Tu, W.F. Butler, J.P. O'Connel and M. Heller, *Proc. Natl Acad. Sci. USA*, 94 (1997) 1117.
94 K. Hashimoto, K. Ito and Y. Ishimori, *Anal. Chim. Acta*, 286 (1994) 219.
95 H.H. Thorp, *Trends Biotechnol.*, 16 (1998) 117.
96 M. Fojta, L. Havran, R. Kizek and S. Billová, *Talanta*, 56 (2002) 867.
97 A.M. Pyle, J.P. Rehmann, R. Meshoyrer, C.V. Kumar, N.J. Turro and J.K. Barton, *J. Am. Chem. Soc.*, 111 (1989) 3051.
98 C.J. Murphy, M.R. Arkin, Y. Jenkins, N.D. Ghatlia, S.H. Bossmann, N.J. Turro and J.K. Barton, *Science*, 262 (1993) 1025.
99 S.O. Kelly, J.K. Barton, N.M. Jackson and M.G. Hill, *Bioconjugate Chem.*, 8 (1997) 31.
100 X. Sun, P. He, S. Liu, J. Ye and Y. Fang, *Talanta*, 47 (1998) 487.
101 S. Takenaka, K. Yamashita, M. Takagi, Y. Uto and H. Kondo, *Anal. Chem.*, 72 (2000) 1334.
102 M. Nakayama, T. Ihara, K. Nakano and M. Maeda, *Talanta*, 56 (2002) 857.
103 H. Cai, C. Xu, P. He and Y. Fang, *J. Electroanal. Chem.*, 510 (2001) 78.
104 J. Wang, D. Xu, A. Kawde and R. Polsky, *Anal. Chem.*, 73 (2001) 5576.
105 M. Dequaire, C. Degrand and B. Limoges, *Anal. Chem.*, 72 (2000) 5521.

106 J. Wang, D. Xu, A. Erdem, R. Polsky and M.A. Salazar, *Talanta*, 56 (2002) 931.

107 T. Lumley-Woodyear, C.N. Campbell and A. Heller, *J. Am. Chem. Soc.*, 118 (1996) 5504.

108 O. Bagel, C. Degrand, B. Limoges, M. Joannes, F. Azek and P. Brossier, *Electroanalysis*, 12 (2000) 1447.

109 F. Azek, C. Grossiord, M. Joannes, B. Limoges and P. Brossier, *Anal. Biochem.*, 284 (2000) 107.

110 B. Meric, K. Kerman, D. Ozkan, P. Kara, S. Erensoy, U.S. Akarca, M. Mascini and M. Ozsoz, *Talanta*, 56 (2002) 837.

111 M. Fotja, V. Vetterl, M. Tomschik, F. Jelen, P. Nielsen and E. Palecek, *Biophys. J.*, 72 (1997) 2285.

112 J.P. Cloarec, J.R. Martin, C. Polychronakos, I. Lawrence, M.F. Lawrence and E. Souteyrand, *Sens. Actuators B*, 58 (1999) 394.

113 C.M.A. Brett, A.M. Oliveira Brett and S.H.P. Serrano, *Electrochim. Acta*, 44 (1999) 4233.

114 A.M. Oliveira Brett, L.A. Silva and C.M.A. Brett, *Langmuir*, 18 (2002) 2326.

115 P. Dantoni, S.H.P. Serrano, A.M. Oliveira Brett and I.G.R. Gutz, *Anal. Chim. Acta*, 366 (1998) 137.

116 A.M. Oliveira Brett, S.H.P. Serrano, T.R.A. Macedo, D. Raimundo, M.H. Marques and M.A. La-Scalea, *Electroanalysis*, 8 (1996) 992.

117 V. Brabec, *Electrochim. Acta*, 45 (2000) 2929.

118 A.M. Oliveira Brett, S.H.P. Serrano, I. Gutz and M.A. La-Scalea, *Bioelectrochem. Bioenerg.*, 42 (1997) 175.

119 A.M. Oliveira Brett, S.H.P. Serrano, I. Gutz and M.A. La-Scalea, *Electroanalysis*, 9 (1997) 110.

120 A.M. Oliveira Brett, S.H.P. Serrano, I. Gutz, M.A. La-Scalea and M.L. Cruz, *Electroanalysis*, 9 (1997) 1132.

121 A.M. Oliveira Brett, S.H.P. Serrano, I. Gutz, M.A. La-Scalea and M.L. Cruz, In: L. Packer (Ed.), *Methods Enzymol.*, 300-B (1999) 314.

122 M.A. La-Scalea, S.H.P. Serrano, E.I. Ferreira and A.M. Oliveira Brett, *J. Pharm. Biomed. Anal.*, 29 (2002) 561.

123 A.M. Oliveira Brett, M. Vivan, I.R. Fernandes and J.A.P. Piedade, *Talanta*, 56 (2002) 959.

124 A.M. Oliveira Brett, J.A.P. Piedade and S.H.P. Serrano, *Electroanalysis*, 12 (2000) 969.

125 I.A. Rebelo, J.A.P. Piedade and A.M. Oliveira Brett, *Talanta*, 63 (2004) 323.

126 D.B. Hall, R.E. Holmlin and J.K. Barton, *Nature*, 382 (1996) 731.

127 M. Fojta, T. Kubičárová and E. Palecek, *Biosens. Bioelectron.*, 15 (2000) 107.

128 O. Ikeda, Y. Shirota and T. Sakurai, *J. Electroanal. Chem.*, 287 (1990) 179.

129 L. Lisdat, B. Ge, B. Krause, H. Bienert and F.W. Scheller, *Electroanalysis*, 13 (2001) 1225.

130 J. Wang, G. Rivas, M. Ozsoz, D.H. Grant, X. Cai and C. Parrado, *Anal. Chem.*, 69 (1997) 1457.
131 A.M. Oliveira Brett and L.A. Silva, *Anal. Bioanal. Chem.*, 373 (2002) 717–723.
132 K. Kerman, B. Meric, D. Ozkan, P. Kara, A. Erdem and M. Ozsoz, *Anal. Chim. Acta*, 450 (2001) 45.
133 O. Bagel, C. Degrand, B. Limoges, M. Joannes, F. Azek and P. Brossier, *Electroanalysis*, 12 (2000) 1447.
134 D. Ivnitski, I. Abdel-Hamid, P. Atanasov, E. Wilskins and S. Stricker, *Electroanalysis*, 12 (2000) 317.
135 A.T. Woolley, K. Lao, A.N. Glazer and R. Mathies, *Anal. Chem.*, 70 (1998) 684.
136 J. Wang, *Talanta*, 56 (2002) 223.
137 E.P. Diamandis, *Clin. Chim. Acta*, 194 (1990) 19.

Chapter 5

Optical biosensors

Laura M. Lechuga

5.1 INTRODUCTION TO OPTICAL BIOSENSORS

Optical biosensors can be defined as sensor devices which make use of optical principles for the transduction of a biochemical interaction into a suitable output signal. The biomolecular interaction on the sensor surface modulates the light characteristics of the transducer (i.e., intensity, phase, polarization, etc.), and the biosensing event can be detected by the change in diverse optical properties such as absorption, fluorescence, luminescence or refractive index, among others.

Optical biosensors have had, and still are having, an increasing impact on analytical technology for the detection of biological and chemical species. Optical biosensing technology can be an alternative and/or a complement to conventional analytical techniques as it avoids expensive, complex and time-consuming detection procedures. For this reason it has been the subject of active research for many years [1–6].

The optical sensing approach offers many advantages over its electrical counterpart, such as the absence of risk of electrical shocks or explosions, its immunity to electromagnetic interferences, in general a higher sensitivity and a wider bandwidth. Moreover, by using optical fibers to guide light towards and out from the device, remote sensing is also possible. In addition, optical transducers have a potential for parallel detection, making array or imaging detection possible.

The heart of any optical biosensor is an optical waveguide. Although optical waveguides were originally developed for applications in the telecommunications field, their mechanical stability, flexible geometry, noise immunity and efficient light-conducting over long distances make them well suited for implementation in sensor applications. Up to now, optical waveguide sensors have mainly been based on optical fibers due to their low cost, small size and flexible geometry [1]. A disadvantage is that optical fibers cannot be designed to fit a specific application, and one has to adapt to what is available on the market. For this reason, we now see that the use of planar or channel optical waveguides for sensor development [3] has increased dramatically and is even surpassing the use of optical fiber sensors.

The field of applications for optical biosensors is wide, covering clinical, industrial control processes, veterinary, food, environmental monitoring, among others [1]. For all these applications, it is desirable to have a compact sensor of high sensitivity, fast response time and which is able to perform real-time measurements. These requirements can be achieved mainly with optical sensors, due to the intrinsic nature of optical measurements that accommodate a great number of different techniques based on emission, absorption, fluorescence, refractometry or polarimetry.

The advantages of optical sensing are significantly improved when this approach is used in an integrated scheme [7]. The technology of integrated optics allows the integration of many passive and active optical components (including fibers, emitters, detectors, waveguides and related devices, etc.) onto the same substrate, allowing the flexible development of miniaturized compact sensing devices, with the additional possibility of fabrication of multiple sensors on one chip. Other advantages are miniaturization, robustness, reliability, potential for mass production with consequent reduction of production costs, low energy consumption, and simplicity in the alignment of the individual optical elements. The latest developments in the field of integrated optics have resulted in an innovative class of microoptical sensors exhibiting biosensing sensing capabilities comparable to those of sophisticated analytical laboratory instrumentation.

The detection principle of most optical biosensors is an application of evanescent field detection [3]. The evanescent wave principle allows the direct monitoring of small changes in the optical properties that are particularly useful in the direct affinity detection of biomolecular interaction. The direct detection method is not as sensitive as indirect ones (i.e., fluorescence, radiolabeling or enzyme amplification) but it generally requires no prior sample preparation, and can be used in real-time evaluations, allowing the determination of biomolecule concentration, kinetic constants and binding specificity. Moreover, disturbances from conjugated labels or handling of radioactive materials is avoided. Among the hundreds of examples investigated using this technique are antigen–antibody interactions, antibody engineering, receptor–ligand, protein–DNA, DNA–intercalator, and DNA–DNA interactions.

Since the first label-free optical biosensor was commercialized (Biacore®) in 1990, a rising number of publications have demonstrated the benefits of direct biomolecular interaction analysis for biology and biochemistry. In 1998, Nellen et al. published their work on grating couplers as biochemical sensors, also resulting in a commercial device in 1991 called BIOS-1. Several integrated optical interferometric techniques have also been applied to biosensing, including the Young interferometer, the Mach-Zehnder Interferometer, and the Difference interferometer. Another waveguide sensor, the Resonant Mirror, was published in 1993 by Cush et al. and commercialized in the same year by Affinity Sensors. In recent years, other biosensors schemes have been introduced in the literature and in the market.

In order to familiarize the reader with the different optical biosensors that will be reviewed in the following sections, Table 5.1 provides a list of the main optical biosensor technologies developed until now.

In this chapter, an overview is presented of the main optical biosensors, the operating principle of the different devices, the design of the sensors, the technology of fabrication, the resolution, the dynamic range and detection limit of each device, the most important applications and the commercial devices on the market. Finally, an outlook of future prospects for this technology is given.

Because of the extent of the subject matter, the reader is referred to more specific books and reviews on the theme [1–10] for more specific details.

5.2 OPTICAL SENSING PRINCIPLES

It is common practice to classify the optical sensors according to the type of effect produced by the receptor–analyte interaction in the incoming light beam.

TABLE 5.1

Optical biosensors

Optical fibers
Surface plasmon resonance sensors (SPR)
Waveguide based SPR
Integrated interferometers (Mach-Zehnder and Young interferometers)
Differential mode interferometry
Resonant mirror
Grating coupler
Bidifractive couplers
Optical waveguide lightmode spectroscopy system (OWLS)
Reflactometric interference spectroscopy (RIfS)

Two main approaches have been used: the estimation of the change in the real part of the refractive index or the estimation of the change in the imaginary part of the refractive index. Most developments on optical biosensors exploit the change in the real part of the refractive index of the optical medium in contact with the analyte. Any change of this magnitude induces a corresponding change in the propagation constant of a given waveguide mode, thus producing a measurable change in the guided light characteristics, mainly due to a phase shift. Transducers relying on changes in the propagation of guided light are more interesting for the sensitivity that can be reached. The effects involved include phase change, coupling effects between waveguide structures, surface plasmon resonance, grating couplers, dual beam and mode beat interferometry. Besides, refractometric techniques like either ellipsometry or multiple reflection interference (white light interference) at thin films can be applied.

Some optical sensors based on the change of the imaginary part of the refractive index (as absorbance sensors) have also been reported [7,11] but these sensors have, in general, less sensitivity and are poorly selective if no reference can be used. An important group of intrinsic optical sensors is based upon the luminescent radiation produced in the sensing region. In these sensors, the sensing layer absorbs light of a concrete wavelength from the waveguide and emits light of a different wavelength, which is then collected by the same waveguide. The luminescent properties of the material forming the sensing layer vary as a result of its interaction with the analyte. Therefore, the characteristics of the emitted radiation are related to the nature and concentration of the analyte, and consequently the analysis of this radiation yields the desired analytical information.

5.2.1 Evanescent wave principle

The most common principle used in optical biosensors for detection in the real part of the refractive index is evanescent wave detection, where the transducer optics are modified by changes in optical parameters of the medium in contact with the sensor surface via the interaction with the evanescent light wave penetrating into the ambient medium. As the evanescent wave decays exponentially from the surface, the most sensitive detection is just at the transducer surface.

An optical waveguide is a high refractive index layer situated between two materials of lower refractive index, as it is shown in Fig. 5.1.

When a light beam has an incident angle exceeding the critical angle, a total internal reflection (TIR) occurs at the interfaces and, as a consequence, the

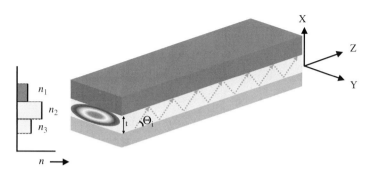

Fig. 5.1. Schematic representation of a planar optical waveguide of refractive index n_2, surrounded by layers of lower refractive index (n_1 and n_3). The light, confined within the structure by TIR, travels through guided modes.

light travels inside the waveguide, confined within the structure [3]. A detailed study of how the light travels inside the waveguide shows that the light is transmitted through a model of the electromagnetic field called the "guided mode" (as it is shown in Fig. 5.2). In a waveguide, both transverse electric (TE) and transverse magnetic (TM) modes can propagate [9]. Although light is confined inside the waveguide, there is a part of it (evanescent field) that travels through a region that extends outward, around a hundred nanometers, into the medium surrounding the waveguide (see Fig. 5.2). This fact can be used for sensing purposes.

When a receptor layer is immobilized onto the waveguide, as shown in Fig. 5.2, exposure of such a surface to the partner analyte molecules produces a (bio)chemical reaction that takes place on the surface of the waveguide and induces a change in its optical properties that is detected by the evanescent wave. The extent of the optical change will depend on the concentration of the analyte and on the affinity constant of the interaction, so obtaining a quantitative sensor of that interaction. The evanescent wave decays exponentially as it penetrates the outer medium and, therefore, only detects those changes that take place on the surface of the waveguide, because the intensity of the evanescent field is much higher in this region [3]. It is therefore not necessary to carry out a prior separation of non-specific components (which is necessary in conventional analysis) because any change in the bulk solution will hardly affect the sensor response [1]. In this way, evanescent wave sensors are selective and sensitive devices for the detection of very low levels of chemicals and biological substances and for the measurement of molecular interactions in situ and in real time.

The extent of the evanescent field can be tailored by the design of the waveguide (thickness and refractive index), the wavelength, its cladding and

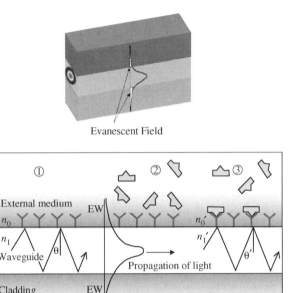

Fig. 5.2. (a) Evanescent field of the fundamental propagation mode in an optical waveguide. (b) Interaction of the evanescent wave with a biomolecular reaction for sensing purposes. The adsorption of the receptor layer and the recognition process produces a change of the effective refractive index of the waveguide inside the evanescent field and this change is quantitatively related with the concentration of the analyte to be measured.

by any layer in proximity with the waveguide. This means that the structure of a waveguide can be tailored to a specific measuring situation by selecting appropriate materials, wavelength, geometry and manufacturing technologies. The materials commonly used for the fabrication of the waveguides are glass, silicon and related materials, polymers, lithium niobate, and III–V compounds. The techniques usually used for waveguide fabrication are ion-exchange, spin or dip-coating, chemical vapour deposition, sol–gel and plasma polymerization, among others. For the processing of the waveguides special installations are usually required (mainly Clean Room facilities) which lowers the accessibility to work with this type of sensor [3,7].

5.3 TYPES OF OPTICAL BIOSENSORS

Several evanescent wave transduction schemes have been proposed, depending on the physical effect used to measure the optical variation. We will focus on some of the following sensors that have been reported in the literature and are widely used and even commercialized: surface plasmon resonance sensors [1,12,13], grating couplers [1,9], resonant mirror [1,9], Mach-Zehnder interferometer [14,15], difference interferometers [14,15], directional couplers [1,9], reflectometric interference spectroscopy (RIfS) [1], etc. Special mention is made of the optical fiber biosensors [16–18], the first ones to be implemented, although no direct measurement is possible with them. All these sensor principles have advantages and disadvantages but one common characteristic is that they all require careful sensor design, fabrication and testing. In Table 5.2 a comparison among all these different technologies is shown.

Despite the differences in generation of the evanescent field, the basic binding experiment is basically the same for all the optical biosensors (see Fig. 5.3). One of the interacting partners, the receptor, is attached to the sensor surface while the analyte binds to the receptor from free solution. As the sensor monitors refractive index changes occurring in real time, the amount of receptor, analyte and the rate of binding can be determined. Indeed, the estimation of the interaction kinetics is one of the key advantage of this technique.

The sensitivity of an optical biosensor depends on two main factors: the capacity of the sensing layer to bind the analyte and the optical detection limit of the device (minimum amount of analyte able to trigger a signal). The first depends on the affinity of the interaction and the number of binding sites accessible to the analyte. The second depends on the molecular weight of the analyte, the signal-to-noise ratio and drift.

Although an evanescent wave biosensor can be explained in a simple way, a range of complex physical phenomena underlies this apparent simplicity.

TABLE 5.2

Comparison of sensitivities for different optical biosensors [15]

Sensing principle	Limit of detection (pg/mm^2)
SPR	1–5
Waveguide-SPR	2
Resonant mirror	5
Grating coupler	1–10
Mach-Zehnder interferometer	0.1
Differential mode interferometer	1
Young interferometer	0.7
Reflectometric interference spectroscopy (RifS)	1–5

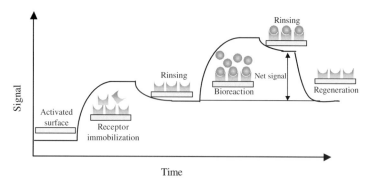

Fig. 5.3. The basic response curve in real time for any evanescent wave biosensor.

On the other hand, in the development of a complete biosensor instrument it is necessary to develop associated technologies, as the whole system usually consists of four integrated parts: (i) a sensitive transducer for detection of the interaction, (ii) a fluid delivery system for sample handling, (iii) surface immobilization chemistries for receptor attachment and for regeneration, and (iv) control electronics, acquisition software and data evaluation.

Because of the broad range of optical biosensors, the classification in a few types is difficult and, in some way, arbitrary. The classification we adopt in this chapter is made according to the main groups of optical biosensors that exist: (i) optical fiber biosensors, (ii) surface plasmon resonance sensors, and (iii) integrated sensors (or sensors based on planar/channel waveguides). Some degree of overlap between the different categories chosen in this chapter exists. In the following, a detailed description of each type of optical biosensor will be provided.

5.3.1 Optical fiber (FO) biosensors

Since the early 1980s, considerable research effort has been devoted to the development of fiber optic (FO) biosensors because their potential sensitivity, detection speed, and adaptability to a wide variety of assay conditions. The area of optical fibers biosensors is quite wide and numerous applications have been described in the literature, mainly via the evanescent wave detection [1,16–19]. A number of possibilities for FO sensing have been proposed and some of them have reached commercial development [1,16].

Fiber optical biosensors are based on the transmission of the light along a fiber strand to the place of analysis. FO biosensors can be classified in two

Optical biosensors

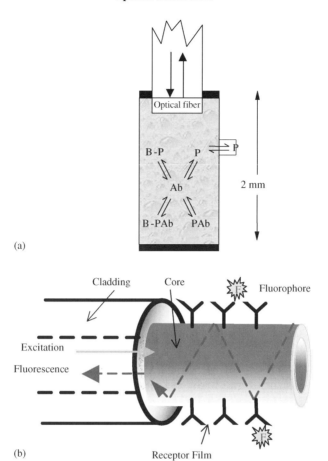

Fig. 5.4. Optical Fiber biosensor. (a) Extrinsic: optical fiber is used for the guiding the light to and from the sensor area. (b) Intrinsic: the receptor molecules are immobilized on the fiber core after decladding of the fiber. The detection is based on fluorescence labels.

main groups: extrinsic and intrinsic, as shown in Fig. 5.4. Extrinsic sensors make use of a single fiber only for guiding incoming and outcoming light and intrinsic sensors are based on evanescent wave detection.

In the extrinsic technique, light from a suitable source travels along the FO to the distal end where an immobilized sensing layer is located. Reflected, scattered or emitted light returns from the sample by a second fiber or by bifurcation of the original fiber. The emitted light is interpreted at the detector and is a measure of the concentration of the analyte of interest. The simplest FO biosensor uses absorbance measurements to determine any changes in the concentration of the analyte that absorbs a given wavelength. It is also possible to do a fluorescent measurement using a label (see Fig. 5.4a).

In the intrinsic sensors, generally a tapered, fused-silica fiber is used, on which and after decladding of the core, specific biological receptors are immobilized via a well-establish chemical procedure. Changes in the absorbance, luminescence, polarization or refractive index are detected. But direct measurement is not usually possible and competitive configurations are employed using fluorescent labels (see Fig. 5.4b).

For example, several strategies have been used for immunoassay techniques with fiber-optic biosensors. In the sandwich format, the receptor is immobilized on the surface of the fiber waveguide and a secondary or "tracer" antibody (which is labelled with a fluorescent dye) is added to the solution. In the absence of the analyte, the tracer remains in solution and little fluorescence is observed. However, after addition of the analyte, a molecular sandwich is formed on the sensor surface within the evanescent excitation volume. The sandwich assay is usually more sensitive than a competitive-binding assay because the fluorescence intensity increases with analyte concentration.

The development of fiber communication technology has provided extremely low-loss and stable fiber components and fiber with standard interfaces. With the advent of optical transducers, better electronics, and improved immobilization methods, FO biosensors are now being increasingly applied to industrial process and environmental monitoring, food processing, and clinical applications. Fiber optics biosensors can be easily miniaturized and integrated for the determination of different target compounds. This has made possible the development of commercial FO biosensors. The FO biosensor has evolved from a single-channel instrument [17] into a compact, portable, multichannel instrument such as the Analyte 2000 developed by Research International (WA, USA). The Analyte 2000 is an automatic prototype which successfully performs four simultaneous fluoroimmunoassays at the surface of tapered, fused silica optical fibers to which specific antibodies were previously attached. Actually, there are other two FO biosensors on the market. The first one is called Endotect™ (ThreeFold sensors, Ann Arbor, MI) [1], designed for single-use in clinical applications and measures binding rates in competitive assays. The second one is a portable, fully automatic sensor called RAPTOR-Plus (Research International, Woodinville, WA) [1] able to perform discrete measurements repeatedly. This sensor is mainly applied for military purposes.

For FO biosensors, the detection limit can be femptomolar. Devices containing an array of individual fibers with different sensitivities can provide multisensing detection and a portable sensor system can be developed. As fiber sensors are made of glass they are environmentally rugged and can tolerate high temperatures, vibrations, shock, and other harsh conditions. They are also relatively safe and biocompatible for use within the human body. The development of multianalyte FO biosensors is relatively recent [1].

With the advent of the nanotechnology, submicron fiber-optic antibody based biosensors have been developed by Vo-Dinh and co-workers [10] for the measurement of biochemical compounds inside a single cell. Nanometer-scale FO biosensors were also used for monitoring biomarkers related to human health effects that are associated with exposure to polycyclic aromatic hydrocarbon, using a specific antibody for detection of benzopyrene tetrol (BPT) (a metabolite of the carcinogen benzopyrene). Excitation light is launched into the fiber and the resulting evanescent field at the tip of the fiber is used to excite any of the ligand molecules bound to the antibodies. Using this nanosensor a detection limit for BPT of ca. 300 zeptomol (10^{-21} moles) have been reported. These nanosensors allow the probing of cellular and subcellular environments.

5.3.2 Surface plasmon resonance (SPR) sensors

Since this technique was applied to biosensing of an antibody–antigen interaction in the landmark paper of Liedberg et al. in 1983 [20,21], SPR has become a very-well known and established method of detecting biomolecular interactions, showing a great potential for affinity biosensors. Since then, SPR sensors have received much attention, with hundreds of publications in the literature every year [12]. Several commercial devices are on the market and new prototypes are appearing continuously. More extended reviews on SPR sensors can be found in Refs. [1,4,12,13,22–24].

5.3.2.1 How it works
The SPR is an optical phenomenon due to a charge density oscillation at the interface of a metal and a dielectric, which have dielectric constants of opposite signs [25]. Optical excitation of a surface plasmon can be achieved when a light beam (p or TM polarized) incidents at the interface between a thin metal layer and a dielectric media at a defined angle, called the angle of resonance (see Fig. 5.5). When resonance occurs, a minimum in the intensity of the reflected light for the resonance angle is observed, a plot of incident angle versus reflectivity shows the dip at that angle (see Fig. 5.5)

The surface plasmon wave is excited when the resonant condition is fulfilled, i.e., the propagating vectors of both the surface plasmon (κ_{sp}) and the

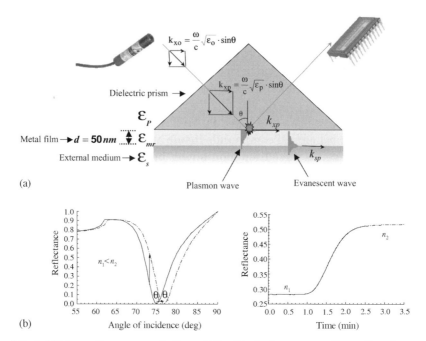

Fig. 5.5. (a) Schematic representation of the Kretschmann configuration for a SPR sensor showing the excitation through the evanescent field. (κ = wave vector of the incident light, θ = incident angle, ε = dielectric constant). (b) Refractive index change at the outer media of a SPR sensor can be detected by measuring the change in intensity of the reflected beam (*right*) as a function of the angle of incidence (*left*) as a function of time at a fixed angle of incidence.

incident electromagnetic waves ($\kappa_{x,d}$) are equal. When wavevector matching occurs, its component parallel to the surface ($\kappa_{x,d}$) must verify the resonant condition:

$$\kappa_{x,d} = \frac{\omega}{c}\sqrt{\varepsilon_d}\cdot\sin\theta = \frac{2\pi}{\lambda}\sqrt{\frac{\varepsilon_m\cdot\varepsilon_d}{\varepsilon_m + \varepsilon_d}} \tag{5.1}$$

where θ is the incident angle, ε_m is the dielectric constant of the metal and ε_d is the dielectric constant of the prism. From Eq. (5.1) it is clear that the SPR propagation can be supported only if $\varepsilon_{mr} < -\varepsilon_d$. This means that the surface plasmon can only exist if the dielectric permeability of the metal and dielectric medium are of opposite sign. This condition is only achieved at frequencies in the infrared to visible part of the spectrum by several metals of which gold and silver are the most commonly employed.

The thickness of the metal film is critical for the minimum reflectance value and the optimal thickness depends on the optical constants of the boundary media and on the wavelength of light [25]. For gold, the optimal thickness is 45 nm at $\lambda = 790$ nm.

The resonant angle is very sensitive to variations of the refractive index of the medium adjacent to the metal surface, which is within sensing distance of the plasmon field and then any change of the refractive index such as a homogeneous change of material (e.g., gas) or a chemical interaction can be detected through the shift in the angular position of the plasmon resonance angle. In both cases, the SPR curve shifts towards higher angles. This fact can be used for sensing applications (see Fig. 5.5b). The general SPR sensor can be converted in a highly specific biosensor to detect biospecific interactions by using a functionalized sensor surface specific for a particular analyte.

Usually, there are two ways of optical excitation to achieve the resonant condition: total reflection in prism-coupler structures [12] and diffraction at diffraction gratings [13]. The most commonly used is the first one due to its simplicity, and it is called the Kretschmann configuration, already shown in Fig. 5.5a.

When monochromatic light is used to excite the SPR response in the Kretschmann configuration, there are two means of measurement: to follow the variation of the coupling resonance angle or to follow the intensity of the reflected light at a fixed angle, as is schematically shown in Fig. 5.5b. In the first (angular SPR), the sample and the detector are fixed upon a rotating table in such a way that the detector moves at twice the angular speed of the sample. The resonant condition is observed as a very sharp minimum of the light reflectance when the angle of incidence is varied. When a (bio)chemical reaction takes place, a shift in the resonance curve is observed. This shift can be related quantitatively to the analyte of interest. Angular scans cannot offer real-time measurements, as a single scan takes several minutes. In the second one, by choosing an angle of incidence at the half width of the resonant dip and measuring the intensity of the reflected light at that constant angle, close to the plasmon resonance, real-time changes in the refractive index due to the process of adsorption of molecules onto the metal surface can be measured with high sensitivity. This kind of measurement can be applied only when small changes in the refractive index are produced as the linear region is rather small, e.g., in a biomolecular interaction. Continuous monitoring at the same angle provides a real-time analysis of the binding events involved in the reaction. A great deal of information (e.g., specificity, concentration, kinetics) can be obtained from this plot.

It is also possible to measure the reflectance curve using a convergent light beam (range of incident angles) and detecting the signal by photodiode array.

With additional electronics and software, the resonance angle can be determined from the photodetector signal. If we use polychromatic light for the measurement, one more parameter is added (the wavelength of the intensity dip) making wavelength modulated detection possible [12,22–24], thereby increasing the operational range of the sensor (spectral SPR).

An example of an experimental set-up used for the SPR measurements is shown in Fig. 5.6. Usually, lasers or LEDs are used as light sources and photodiodes are used for detection. Some modern SPR devices use linear arrays of charge-coupled devices (CCDs) to detect reflected light from the surface. In this way detection is possible through a wide range of angles and avoids the need for mechanically controlling the angular position of the detector. An example of an SPR response can be see in Fig. 5.7 for the detection of chemical pesticides using specific monoclonal antibodies as receptors.

5.3.2.2 Types of SPR sensors
Several types of SPR sensors have been developed. The most common one (and the basis of most of the commercial devices) is based on bulk optics using a

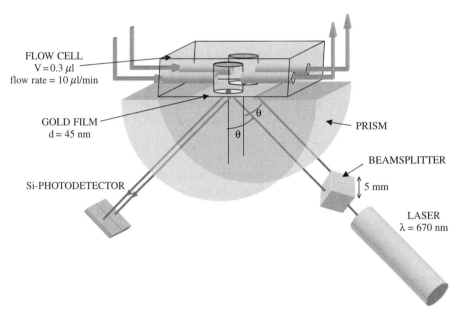

Fig. 5.6. An SPR experimental set-up based on the Kretschmann configuration. A glass hemicylinder is covered with a glass slide coated with gold (45 nm), using a matching oil and then exposed to the sample solutions using a flow cell with two channels. The reflected intensities of both channels are measured in a photodiode. (More details in the figure itself).

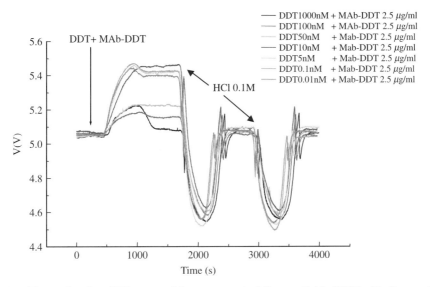

Fig. 5.7. Example of an SPR curve. Measurement of the pesticide DDT with its specific monoclonal antibody. Two cycles of regeneration of the receptor sensor surface are also shown.

prism coupler (the already-mentioned Kretschmann configuration) because it is more suitable for sensing. Other SPR sensors configurations are based on a grating coupler [12], optical fibers [23,24] or integrated optical waveguides [26,27].

In the grating coupler configuration, if a metal–dielectric interface is periodically distorted, the incident light is diffracted forming a series of beams directed away from the surface at a variety of angles that can be coupled to a plasmon wave [22]. This sensor has been used for monitoring biomolecular interactions in an aqueous environment [23]. The optical interrogation system is almost the same as that used for the prism-coupler SPR, but in the grating-based SPR sensor it is not necessary to accurately control the thickness of the metal as in the prism-coupler, but then an accurate control of the grating depth is required. Another drawback of this sensor is that the light must be incident through the sample, and thus the flow cell and samples must be transparent at the wavelength used. Laser or LEDs are used as light sources.

The optical fiber SPR is the smaller device of this technology, allowing the use of this technique in distant locations. Sensors based on monomode and multimode optical fibers have been reported [23,24]. The cladding of the fiber is partially removed and a gold layer is deposited symmetrically around the exposed fiber core. This type of fabrication limits the interaction area to a few millimeters. Another drawback of using fiber SPR sensors is the difficulty of obtaining homogeneous coatings deposition and a good chemical functionalization of the sensor surface because the modal light distribution is affected by mechanical and surface changes.

SPR sensors based on integrated optical waveguides have also been developed [22,23]. These sensors combine the resonant coupling of guided light modes inside the waveguide with SPR at a gold-coated surface. Homola et al. [24] used a prism to couple monochromatic light into and out of a waveguide with a narrow strip of gold along the optical path. Similar to monomode optical-fiber SPR devices, these sensors have a quite limited operating range [26]. A multichannel SPR sensor with improved performance has been recently presented [27,28].

The sensitivity and resolution obtained by the SPR sensors have been extensively discussed theoretically and experimentally [22,29,30]. The sensitivity of SPR using a prism coupler is higher than the devices using grating coupler [22]. Refractive-index resolution of the SPR sensors based on prism coupler ranges, generally, from 2×10^{-5} to 5×10^{-5} RIU (refractive index units) [24] although a refractive index resolution better than 3×10^{-7} has been reached. In general SPR allows the detection of adsorbed thin films of subnanometer thickness. A great variety of applications have been shown using the SPR technology [31–36]. The detection limit of actual SPR devices is below 1 pg/mm^2 of analysed biomaterial [33], which is still not enough for direct detection of low concentrations of low weight molecules.

The main limitation of this technique is that the sensitivity depends on the molecular weight of the analyte, implying that low concentrations of small molecules cannot be detected in a direct way. In these cases a sandwich or competitive assay can be employed. Some improvements of surface chemistry and sensitivity of SPR are now allowing the direct detection of low molecular weight analytes in some cases.

SPR imaging or SPR microscopy uses a fast array of photodetectors and can allow for the simultaneous measurement of a complete array of immobilized molecules. In this technique, a collimated light beam from a polychromatic source passes through a prism and after incidents on the SPR surface. The reflected light is detected by a CCD camera after passing a band interference filter. This technique provides excellent spatial resolution at the same sensitivity as the standard SPR sensors. These features make SPR imaging a promising alternative for the development of biochips. No commercial SPR imaging instrument is currently available.

5.3.2.3 Commercial SPR devices

The first commercial SPR was launched by Pharmacia Biosensor AB (presently Swedish BIAcore AB) in 1990. Since then, the device has been refined and now BIAcore [37] offers several models (BIACORE® 3000, BIACORE® 2000, BIACORE® 1000, BIACORE® X, J, Q, S51, and C models). The biosensors of BIACORE 1000 to 3000 are fully automated instruments, with a disposable sensor chip, an optical detection unit, an integrated microfluidic cartridge, an autosampler, method programming and control software. Less expensive manually controlled alternatives are the BIACORE® x and BiacoreQuant™.

A wide number of companies across the world are offering SPR devices, as it is shown in Table 5.3. Some of these companies are Jandratek [38], GWC [39], Ibis [40], Leica [41], Autolab, NLE (SPR670) [42], HTS Biosystems [43], Texas Instruments (Spreeta) [44], DKTOA, Analytical Microsystems (Biosuplar), Sensia S.L., Vir Biosensor. Texas Instruments (Dallas, USA) was the first in the development of a miniaturized integrated SPR sensor called TI-SPR-1 Spreeta™ [44]. The device of Texas Instruments is a novel miniature

TABLE 5.3

Commercial optical biosensors

Optical biosensor	Commercial devices
Surface plasmon resonance (SPR) (different versions)	BIAcore (BIAcore AB), www.biacore.com Jandratek, Leica, Ibis, Autolab, Texas Instruments (Spreeta), DKTOA, Analytical Microsystems (Biosuplar), NLE (SPR670), HTS Biosystems, Applied Biosystems, GWC, Sensia S.L., Vir Biosensor
Grating couplers (GC)	ASI (Artificial Sensing Instrument) Reflection Coupler (IPM Freiburg) (http://www.ipm.fhg.de/gfelder/bioanalytik)
OWLS	MicroVacuum Ltd. (http://www.owls-sensors.com)
Resonant mirror (RM)	IAsys (Afinity Sensors) (http://www.affinitysensors.com)
Interferometers	Interferometric Biosensor IBS 101 (http://www.ipm.fhg.de/gfelder/bioanalytik/ AnaLight Bio200, Farfield Sensors (http://www.farfield-sensors.co.uk)
Total internal reflection fluorescence (TIRF)	ZEPTOSENS (www.zeptosens.com)
Reflectrometric interference spectroscopy (RIfS)	Modified SPECOL 1100PM (Zeis)

and compact SPR (the total volume is about 7 cm^3 and the weight about 7 g). Spreeta has all the components die mounted, wire bonded onto a miniature moulded transducer, with multiwavelength source and angle deflection measured by a photodetector array. The complete apparatus is provided in a very small package, produced at low cost and in high volume. The sensor performance is restricted by the S/N ratio due to the electronics components, limiting the system sensitivity. Although the sensitivity of the state-of-the-art SPR sensors cannot be reached by Spreeta, this sensor can be configured for industrial, environmental and biological applications and is cost-effective and small.

5.3.3 Integrated optical (IO) sensors

We define IO biosensors as those sensors based on planar/channel waveguides which could include, on the same substrate and together with the light guiding structure, some other optical components such as grating, dividers, combiners, etc. Contrary to FO biosensors (which uses FO from the telecom area), the integrated optical sensors must be designed and fabricated fit for purpose, mainly using more sophisticated techniques (as Clean Room silicon processing) than those used, e.g., in the fabrication of SPR sensors. There are three main types of integrated optical biosensors (interferometric devices, grating coupler devices, and resonant mirror devices) which will be treated in detail in the following.

5.3.3.1 Interferometric devices
The interferometric arrangement for biosensing is highly sensitive and is the only one that provides an internal reference for compensation of refractive-index fluctuations and unspecific adsorption [14,15]. Several interferometric devices have been described, e.g., the Mach-Zehnder [45–47], the difference interferometer [48] or the Young interferometer [49]. Interferometric sensors have a broader dynamic range than most other types of sensors and show higher sensitivity as compared to other integrated schemes as shown in Table 5.2, where a comparison of the different sensor technologies as a function of the limits of detection (in pg/mm^2) is presented.

Optical interferometer techniques can achieve resolution far beyond 1 nm. Because of this high sensitivity of the interferometer sensor, the direct detection of small molecules at low concentrations should be possible [15]. Detection is generally limited by electronic and mechanical noise, thermal drift, light source instabilities, and chemical noise. But interferometric devices have an intrinsic reference channel which offers the possibility of reducing common mode effects like temperature drifts and non-specific adsorptions. Detection limits of 10^{-7} in refractive index (or better) can be achieved with these devices, which opens the possibility of development of highly sensitive devices for in situ detection [14].

The type of interferometer that is most commonly employed for biosensing is the Mach-Zehnder device [50]. In a Mach-Zehnder interferometer (MZI) device the light from a laser beam is split by a Y-junction into two identical beams that travel the MZI arms (sensor and reference areas) and are recombined again into a monomode channel waveguide, giving a signal which is dependent on the phase difference between the two beams. Any change in the sensor area (in the region of the evanescent field) produces a phase difference (and therein a change of the effective refractive index of the waveguide) between the reference and the sensor beam and then in the intensity of the outcoupled light. A schematic of this sensor is shown in Fig. 5.8.

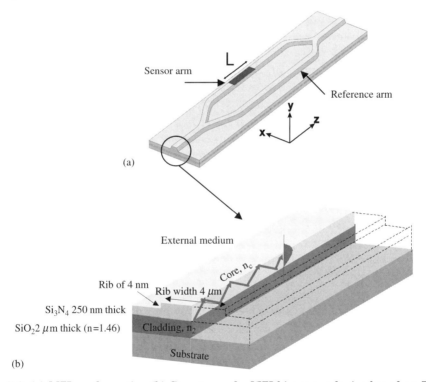

Fig. 5.8. (a) MZI configuration (b) Structure of a MZI biosensor device based on TIR waveguides. Note the dimensions of the rib channel (4 nm) for monomode and high sensitivity waveguides.

When a chemical or biochemical reaction takes place in the sensor area, only the light that travels through this arm will experience a change in its effective refractive index. At the sensor output, the intensity (I) of the light coming from both arms will interfere, showing a sinusoidal variation that depends on the difference of the effective refractive indexes of the sensor ($N_{\text{eff,S}}$) and reference arms ($N_{\text{eff,R}}$) and on the interaction length (L):

$$I = \frac{1}{2} I_0 \left[1 + \cos\left(\frac{2\pi}{\lambda} (N_{\text{eff,S}} - N_{\text{eff,R}}) L \right) \right] \qquad (5.2)$$

where λ is the wavelength. This sinusoidal variation can be directly related to the concentration of the analyte to be measured. The visibility factor gives the contrast of the interference signal (difference between the maximum and minimum intensity) and depends on the coupling factor of the divisor and on the propagation losses of the guided mode in the interferometer arms. To obtain a maximum visibility factor it is important to design a divisor or Y-junction with a coupling factor of 3 dB which allows input light to be equally divided in each branch of the interferometer. Moreover, propagation losses in the sensor and reference arm should be identical.

An attractive aspect of this device is the possibility of using long interaction lengths, in this way increasing the sensitivity of the device. An MZI offering a phase resolution of 0.002π, which corresponds to a refractive-index resolution of 10^{-6}, has been described [51]. A theoretical study shows that the Mach-Zehnder interferometer sensor seems to be one of the more promising concepts [15] for detection of low concentrations of small molecules without labels (10^{-12} M or even lower).

Owing to the evanescent sensing approach employed, the optical waveguides must be monomode. If several modes were propagated through the structure, each of them would detect the variations in the characteristics of the outer medium and the information carried by all the modes would interfere between them. Using a conventional planar waveguide structure we have developed [46,52,53] an MZI sensor for immunological purposes with a sensitivity of 10^{-3} nm in the thickness of the adsorbed layer (which corresponds to an adsorbed molecular layer of 1 pg/mm^2). Using integrated channel waveguides and for monomode behavior the height of the waveguide (rib) has to be less than 4 nm (for a more detailed discussion about the design, see Ref. [51]). These devices have shown a surface sensitivity of 2×10^{-4} nm^{-1}, close to the maximum reported up to now. In Fig. 5.8b, the cross-section of the MZI TIR waveguide is shown and in Fig. 5.9 a photograph of some of the devices can be seen.

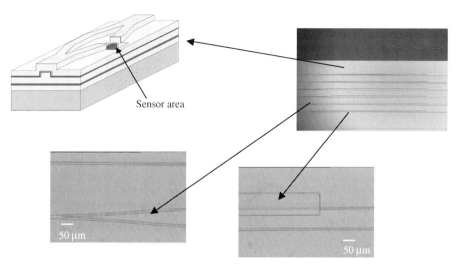

Fig. 5.9. Photograph of an integrated MZI: details of the MZI Y-divider and sensor area.

An antiresonant reflecting optical waveguide (ARROW) has also been used for the fabrication of an MZI sensor [54]. The attractive characteristics of the ARROW technology include low optical losses, compatibility with standard silicon technology, an effective single-mode propagation, a selective behavior in polarization and wavelength and a high tolerance for the refractive index and thickness of the layers used in the construction of the structure. This waveguiding concept solved the two main limitations of conventional waveguides: the reduced dimensions for monomode behavior (an important subject for further technological development and mass production of the sensors) and the high insertion-losses in the optical-interconnects fiber-waveguide. Both designs, TIR and ARROW, have advantages and disadvantages, and depending on the specific application one or the other can be employed.

The MZI devices can be evaluated using an experimental set-up as the one shown in Fig. 5.10. For the TIR device, the lower detection limit measured was $\Delta n_{o,\min} = 7 \times 10^{-6}$, equivalent to an effective refractive index of $\Delta N = 4 \times 10^{-7}$, which means a lowest phase shift measurable around $0.01 \times 2\pi$. A typical curve from an MZI device is shown in Fig. 5.11. The detection limit corresponds to a surface sensitivity around 2×10^{-4} nm^{-1}. The MZI biosensor has been extensively used for different applications such as environmental pollutant detection [55,56], protein–protein interaction or detection of proteins in blood samples [57].

Fig. 5.10. Optical bench for the optical and biochemical characterization of the MZI devices.

The main drawback in the development and commercialization of the integrated MZI device is the complexity of the design, fabrication and optical adjustments: the overall procedure for MZI fabrication is rather laborious and monomode waveguides are required, thereby further increasing the complexity of the technology. But, recently, with the utilization of micro/nanotechnology for their integration on silicon, these devices could offer some advantages such as better control of the light path by the use of optical waveguides, mechanical stability, higher sensitivity, miniaturization and the possibility of mass-production.

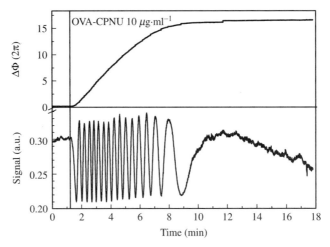

Fig. 5.11. MZI response in real time during the covalent attachment of a biological receptor to a previous functionalized sensor surface.

Another interesting interferometer device is the Young interferometer, formed by an integrated optical Y-junction acting as a beam splitter, as can be seen in Fig. 5.12 [58–60]. It is a variation of the integrated MZI as the sensor and the reference arms are not recombining again inside the structure but the light coming out from the two arms are made to interfere at the output. Light is end-coupled in the device. At the output of the sensor, the light emitted forms two cones, which superimpose and produce an interference pattern. The intensity distribution is detected by a detector array.

The light coupled out of the two branches generates an interference pattern on a screen or CCD detector with a cosine intensity distribution function. The phase difference of the two interfering rays is given by:

$$\Phi = \frac{2\pi}{\lambda}\left(\frac{d \cdot x}{f} - \left(N_{\text{eff},S} - N_{\text{eff},R}\right)L\right) \quad (5.3)$$

where d is the distance between the two branches, f is the distance between the output sensor and the screen and x denotes the position on the screen. In one of the arms (sensor arm) and during a certain interaction length L, a change in the optical characteristics of the outer medium is induced, which produces a variation in the effective refractive index in one arm respective to the other ($N_{\text{eff},S} - N_{\text{eff},R}$). Under the influence of the adsorption of biomolecules on the sensor branch, the fringe pattern moves laterally. One disadvantage of the Young device is the distance needed from the output to the detector in order to get a maximum resolution. The advantages of this type of interferometer include the simplicity of the arrangement, the detection of the complete intensity distribution and the identical length of the arms, which allow to side effects arising from temperature and wavelength drift to be avoided.

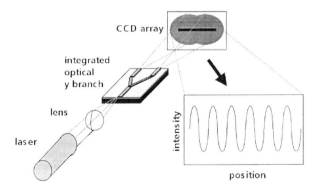

Fig. 5.12. Working principle of a Young interferometer biosensor.

With this device a theoretical detection limit of 9×10^{-8} in the effective refractive index can be achieved and an experimental limit of 50 ng/ml for the detection of proteins has been measured.

Recently, a commercial biosensor device based on interferometry (Ana-Light Bio200) produced using planar waveguides was introduced by the company Farfield sensors [61].

5.3.3.2 Resonant mirror sensor

The resonant mirror device (IAsys) [62–64] is a leaky planar waveguide sensor that uses frustrated TIR to coupled light into and out of the waveguide layer. It is similar to the SPR device but in this sensor the metal film is replaced by a dielectric resonant layer of high refractive index (e.g., titania or hafnia) with a thickness of approximately 100 nm, separated from a glass prism by a dielectric layer of low refractive index (SiO_2) and thickness of 0.5–1 μm (sandwich configuration of high–low–high n). This layer is thin enough to allow light to couple into the resonant layer via the evanescent field when incident light is focused onto the prism–silica interface. Efficient coupling occurs for certain angles where phase matching with the resonant guided modes in the waveguide is achieved. A schematic of the sensor is shown in Fig. 5.13. As the waveguide layer acts as a resonant cavity (for which the device has been termed *resonant mirror*), the light reflected from the RM device undergoes a full 2π phase change across the resonance in either angle (for a fixed input wavelength) or wavelength (for a fixed input angle).

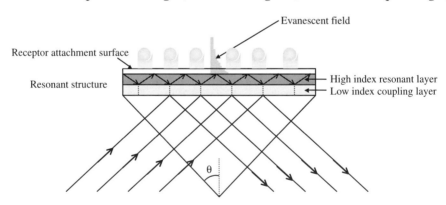

Fig. 5.13. The resonant mirror device. The light from the source is coupled through a prism and is totally reflected at the interface with the low refractive index layer, generating an evanescent field which couples light into the high refractive index waveguide layer: The light transmitted through the waveguide also generates a evanescent field which interacts with the receptor layer.

The phase change due to a biosensing reaction can be visualized using crossed polarizer to produce a peak in intensity at the resonance angle or wavelength.

The IAsys commercial biosensor [65] uses this optical platform for real-time measurement of biomolecular interactions. The first commercial instrument, launched in 1993, was a single channel one with manual sample injection. With the fabrication technology achieved in this commercial device the sensor chips can be made in large quantities with reasonable manufacturing tolerances and cost. This device has been used for studies of DNA hybridization, kinetics of Ag–Ab interaction and kinetics of protein–protein interaction [66,67], for diagnostics, protein folding studies, immunoresponse evaluation or vaccine development, among other applications. The IAsys utilizes a single channel but the model IAsys Auto + has dual-channel flow cells for independent measurements and a most advanced version, called AUTO + *Advantage* was introduced in 1998. Kinetics software is also provided.

5.3.3.3 Grating coupler systems
This sensor, proposed first by Tiefenthaler and Lukosz [68,69] in 1989, is based on an optical grating prepared on a thin waveguide deposited over a glass substrate, as depicted in Fig. 5.14. The grating facilitates the direct input of the laser light onto the waveguide at an angle which excites a guide mode when the incoupling condition is fulfilled. The same can happen for the output beam. As the in(out)coupling angle is very sensitive to any variation in the refractive index above the grating surface, the change produced in the

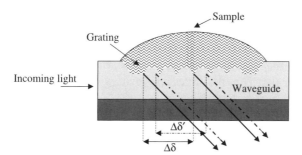

Fig. 5.14. Schematic illustration of a grating coupler device (the outcoupler one is shown, see text for details). The light is incoupling by end-fire. A diffraction grating outcoupled the light onto a photodiode array (detector). The displacement in the position of the outcoupled beam is a quantitative measurement of the interaction that takes place on the grating sensor region.

coupled angle as a consequence of any reaction onto the grating can be used for sensing.

Varying the angle of incidence, a mode spectrum can be obtained and the effective refractive indexes calculated for both TE and TM polarizations. Adequate software allows real-time monitoring of any surface process, providing information on refractive index, thickness of the adsorbed molecular layer and adsorbed mass per unit surface. Measuring time is about 2 min per spectrum with an accuracy of 0.005° and a surface density detection limit around 1 pg/mm^2, which is enough for measuring small molecules.

The grating coupler is made on a waveguide of high refractive index (SiO_2–TiO_2 or Ta_2O_2 made by sol–gel technology, $n = 1.75$–1.82) and a thickness of 170–220 nm, situated over a substrate of low refractive index (usually glass). This device is commercialized by Artificial Sensing Instruments, Zurich ASI AG. The commercial device (BIOS-1) employs an integrated optical scanner to determine the angle at which a guided mode is excited. The incident angle of the light is continuously varied and the angle of coupling is detected by the photodiodes placed at the end of the waveguide. The GKR 102 model has a resolution in effective refractive index of 1×10^{-6} and temperature control.

The grating coupler has been used for monitoring drugs, pesticides and metal ions, for studies of affinity reactions and for measurement of adhesion and spreading kinetics of living cells.

Fattinger et al. reported in 1995 [70,71] the development of a *bidiffractive grating coupler system*. In this biosensor, the transducer consist of a plastic substrate covered by a very thin waveguide film of amorphous TiO_2 which supports only zero-order modes and it is structured with a submicron grating relief. This microrelief is composed of two superimposed uniform gratings that have different periodicities, forming the so-called "bidiffractive grating coupler" that serves as both an input port and an output port for coupling and decoupling light beams to and from the planar waveguide, respectively. The bidiffractive grating biosensor uses a differential measurement approach where two modes with orthogonal, linear polarization (TE_0 and TM_0) are coupled into the waveguide simultaneously using the dual grating structure. The response of the device is obtained by measuring the small differences between the decoupled beam angles of the two modes. Some advantages in comparison with the grating coupler are obtained [71]. Typical values obtained for interfacial mass loading range from several ng/mm^2 to few pg/mm^2.

More refinement of this sensor concept has been the subject of further developments [72–75] mainly related to the fabrication of more compact and miniaturized integrated optical sensors as, e.g., the *replicated integrated optical sensor* [73]. This further miniaturization and integration is quite important for real applications, as it allows reduction of the cost of the sensor chips and the whole system as well as an increase in the density of the sensing pad on a single chip for multianalyte detection and for achieving enhanced specificity and accuracy.

The biosensing technique using this sensor is also called *optical waveguide lightmode spectroscopy* (OWLS) [76–79]. The detection is based on the amount and polarizability of the adsorbed molecules. Depending on that, the light is reflected with a certain phase shift. Owing to the phase shift, the light intensity of both the TE and the TM modes have a maximum at certain angles (coupling angles) between the chip and the laser beam. The instrument OWLS 100 is commercialized by Micro Vacuum Ltd. (Hungary) [80]. OWLS has been successfully applied in the study of protein–DNA interactions, lipid bilayers, biomembranes, biomaterial, monitoring of environmental pollutants, the interaction of surfaces with blood plasma and serum and interaction with cells, among others.

5.3.4 Other optical biosensing schemes

5.3.4.1 Total internal reflection fluorescence (TIRF)
In TIRF, the incident light excites molecules near the sensor surface, which, in turn, creates a fluorescent evanescent wave [81]. This couples back (re-enters) into the waveguide and the emerging fluorescence is detected. When molecules with an absorption spectrum including the excitation wavelength are located in the evanescent filed, they absorb energy leading to an attenuation in the reflected light of the waveguide. This phenomenon is known as attenuated total reflection. The sensitivity reached usually is not enough and thus it is necessary to make use of labeled molecules that are able to re-emit the absorbed evanescent photons at a longer wavelength as fluorescence. Part of this emission is coupled back to the waveguide and in this way is transmitted to the detector. This phenomenon is known as TIRF and is the basic principle of several optical biosensors. One example of this device is the *fluorescence capillary fill device* (FCFD). In TIRF sensors using a fluorescence-label analyte, the detection limit has been reduced to a few fM (10^{-15} M) by employing an optical fiber tapered loop and a channel etched thin film waveguide.

TIRF is currently the most developed technique for the study of 2D arrays of biomolecules immobilized on the surface of planar waveguides. A portable biosensor array based on this technique has been developed at the Naval Research Labs [1].

5.3.4.2 Reflectometric interference spectroscopy (RifS)
The basic principle of this optical biosensing technique is based on the wavelength-dependent modulations that occur at thin transparent films. A light beam passing the interface between two media of different refractive index will be partially reflected. Therefore, a thin transparent film will produce an array of reflected beams when the reflectance of the interfaces is small (<0.05). As this pattern is due to the superposition of reflected beams the method is called *reflectometric interference spectroscopy* (RifS) [1,8]. A schematic of this technique can be seen in Fig. 5.15. These beams will have a phase difference, which is directly related to the thickness of the layer. If the product of the wavelength and phase difference is below the coherence length of the light source, the two beams will interfere, leading to a modulation of the reflected light intensity as a consequence of constructive and destructive interference. Changes in the film thickness can thus be determined by changes in the interference spectrum. This method allows a very sensitive detection of biomolecular interaction at the transducer surface, resolving an average change of the physical thickness of the layer as low as 1 pm. An advantage of RIfS compared to evanescent wave methods is its smaller dependency on temperature.

An overview of all the optical biosensor described in the above paragraphs can be found in Table 5.3.

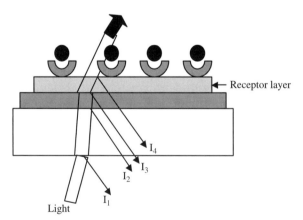

Fig. 5.15. Illustration of the basic principle of the reflectometric interference spectroscopy (RifS) biosensor. See text for details of the working principle.

5.4 BIOCHEMICAL ASPECTS OF OPTICAL BIOSENSORS

Although different detection techniques are applied in optical biosensors all of them must measure effects at a liquid–solid interface. Then, the immobilization of the receptor molecule on the sensor surface is a key point for the performance of the sensor. The chosen immobilization method must retain the stability and activity of the bound biological receptor. Generally, direct adsorption is not adequate, giving significant losses in biological activity and random orientation of the receptors. Despite these difficulties, direct adsorption is widely employed since it is simple, fast and does not required special reagents.

The immobilization methods available are divided in two general categories: (i) covalent coupling and (ii) affinity non-covalent interactions. Covalent coupling gives a stable immobilization as the receptors do not dissociate from the surface or exchange with other proteins in solution.

Various surface chemistries have been developed based on affinity immobilization and/or covalent bonding. In affinity bonding, a high affinity-capture ligand is non-reversibly immobilized on the sensor surface: for example, streptavidin monolayers using biotinylated biomolecules for recognition. Another approach is to form a self-assembled monolayer (SAM) of alkylsilanes and then the receptor can be coupled using the end of the SAM via a functional group ($-NH_2$, $-COOH$, etc.). The affinity binding can be performed with biotin-conjugated ligands to avidin, streptavidin or neutravidin modified surfaces. The advantage of this procedure is an oriented binding ensuring equal binding sites. The functionalisation of the transducer also depends on the surface. Transducers with gold are usually modified with thiols, forming SAMs. Oxidized surfaces can easily be modified with silane chemistry.

An overview of the different strategies used for immobilization is shown in Fig. 5.16 and one example of protein immobilization for an oxidized sensor surface is shown in Fig. 5.17.

For evanescent wave sensors, it is possible to increase the sensitivity of the sensor by immobilizing more receptors in a three-dimensional matrix, using in a more efficient way the whole volume of the evanescent field. In addition, in a three-dimensional network the ligand has more freedom to bind the receptor at a right orientation. The use of a polymer matrix maximizes the interaction volume probed by the evanescent field, increasing greatly the surface capacity and therefore the sensitivity of the device. The method of the carboximethyldextran hydrogel presented by Löfas et al. [82] has been the most widely employed: the idea is to obtain a surface of general application

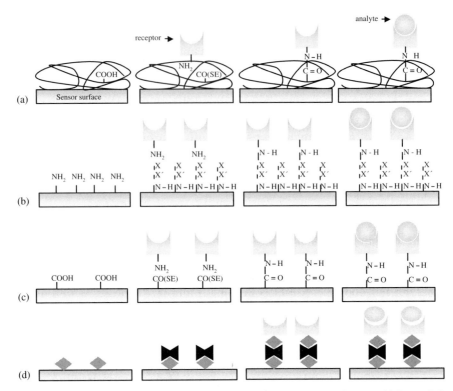

Fig. 5.16. Different immobilization strategies for receptor attachment (Adapted from Ref. [4] with permission from John Wiley). Binding is achieved using: (a) dextran (b) amino (c) carboxylate, and (d) biotin surfaces. Each cartoon represents from the initial activated sensor surface to the covalent bonding with the receptor.

for a wide variety of receptors. The surface is suitable for a range of covalent bonding chemistries that yield a fast and simple methodology and avoiding manipulation of the biomolecules. In this dextran approach, the receptor molecules are attached to flexible dextran chains and are freely accessible in a three-dimensional space thus minimizing steric hindrance and increasing the sensitivity. A variety of surface activation chemistries can be used to couple the receptor to the hydrogel via amine, thiol, disulphide or aldehyde groups [37], that yield a fast and simple methodology for producing active surface for different ligands. Typically, surface concentrations of 1–5 ng/mm^2 of receptor are coupled, depending on the application. A review of methods for controlled coupling to carboximethyldextran surfaces can be found in Ref. [37].

The immobilization of nucleic acid probes is also important. Various schemes can be used for attaching the DNA probes to the surface. These include the use of thiolated DNA for self-assembly onto gold surfaces, covalent

Optical biosensors

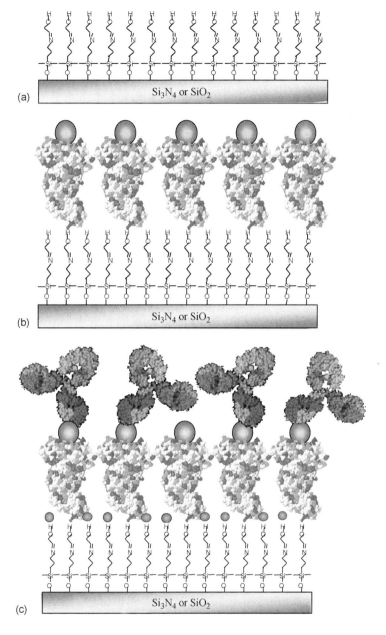

Fig. 5.17. Scheme of one immobilization procedure on the nanometer scale employed for the covalent attachment of the receptor at a sensor surface of SiO_2 or Si_3N_4.

linkage to gold via functional alkanethiol-based monolayers, the use of biotylated DNA from complex formation with a surface-confined avidin or streptavidin, covalent (carboiimide) coupling to functional groups on the sensor surface or simple physical adsorption.

The commercial sensors BIACORE [37] and IAsys [65] offer a variety of advanced surface chemistries for attachment of the different receptor layers as can be seen in Table 5.4.

Another important issue is the possibility of regeneration of the receptor sensor surface after the interaction. It is desirable to use reusable sensors for performing continuous measurements, thereby reducing costs and handling as well as enhancing the reproducibility. The covalent or affinity based binding of the ligands allows the regeneration of the biosensor surface and thus repeated measurements. For example, for regeneration of the SPR sensor surface after an Ab–Ag interaction, we can apply denaturing conditions to break the antigen–antibody bond. Usually it can be achieved using 10–100 mM HCl or 10 mM glycine pH 1.7–2.2. With this treatment more than 100 analyses can be performed using the same receptor.

Some identified problems in the performance of the optical biosensors, related to the biochemical aspects, are discussed in the following.

TABLE 5.4

Commercial chips for optical biosensors

	Application
Biacore surfaces (SPR)	
CM5—carboxymethyl dextran	Routine analysis
SA—streptavidin	Biotin conjugation
NTA—nickel quelation	His-tagged conjugation
B1—low charge	Reduces non-specific binding
C1—flat carboxymethylated	No dextran
F1—short dextran	Large analytes
J1—bare gold surface	Custom design
L1—lipophilic dextran	Capture liposomes
HPA—hydrophobic monolayer	Create hybrid lipid bilayers
IAsys surfaces (RM)	
CM5—carboxymethyl dextran	Routine analysis
Hydrophobic planar	Create lipid monolayers
Amino planar	Covalent coupling
Carboxylate planar	No dextran
Biotinylated planar	Streptavidin conjugation

Optical biosensors

The susceptibility to interference from adsorption of non-specific components is a difficult and complex task, including signal amplification, sample dilution, surface blocking treatments and signal referencing. Systems are often drift rather that noise limited: in particular evanescent field methods are highly sensitive to temperature changes and minor alterations in the buffer composition. Another source of drift is the instability of the light source or detector caused by aging or fluctuations of the power supply. Temperature changes can be minimized by efficient thermostatting, whereas the introduction of a reference channel allows one not only to minimize instabilities of all units but also changes in the buffer composition. A reference channel also allows one to eliminate effects of non-specific binding, a problem that is common to all label-free detection systems.

5.5 APPLICATIONS OF OPTICAL BIOSENSORS

Despite the differences in the optical biosensors covered in this chapter, the general design of the experiments is essentially the same for each instrument (see Fig. 5.3). One of the interacting partners, the receptor, is attached to the sensor surface (by the chosen immobilization procedure). The analyte binds to the receptor from free solution. For most of the optical biosensors, the device monitors the refractive index changes in real time and then, the amount of analyte and the binding rate can be determined. With current instruments the detection of mass changes of the sensor surface in the picogram range is possible. The experiments can be performed within minutes and need only very small amounts of sample. The possibility of observing interactions directly without the use of labels is one of the key advantage of this technology, facilitating kinetic analysis. Characterization of the kinetics and thermodynamics of macromolecular interactions with biosensor techniques thus contributes to the understanding of the molecular basis of biological recognition. Hence, understanding the forces that determine molecular recognition helps to elucidate the mechanisms of important biological processes and facilitates the discovery of innovative biotechnological methods and materials for therapeutics, diagnostics and separation science.

Optical biosensors have been applied extensively in many fields, such as the life sciences, for biotechnology quality control, in clinical analysis, environmental control, fermentation monitoring, product control in the food and beverage industry, just to name a few [83–96]. They can be used to study a wide variety of biological systems interactions from proteins, oligonucleotides, oligosaccharides, and lipids to small molecules, phage, viral particles and cells [83], determination of bacterial and viral concentrations interactions, between proteins and nucleic acids, carbohydrates, lipid and eukaryotic cell binding partners, combinatorial library screening, micropurification, and peptides tandem analysis.

Evanescent wave sensors are particularly suitable for immunoassays and most of the applications of these sensors have been as immunosensors, coupling an antibody at the sensor surface and measuring the binding of the specific antigen contained in the samples. Owing to the possibility of obtaining specific antibodies not only for any antigen but also for low molecular-weight organic molecules, e.g., a pollutant or any other hapten, the immunodetection has been extended from clinical to other areas such as environmental monitoring or food industry control. Now, these sensors can be applied to continuous monitoring of critical-care analytes in serious illness, accepted levels of toxic or explosive species, as well as control of environmental pollutants. Several pharmaceutical companies are using immunosensors routinely to screen and characterize monoclonal antibodies.

5.5.1 Life sciences applications

Optical biosensors are highly sensitive. For example, the high signal-to-noise ratio achieved by SPR permits the detection of binding of molecules as small as 200 Da. In addition, kinetic models have been applied successfully to molecules as small as 1500 Da (Morton, 98). With the Biacore SPR system it is possible to evaluate equilibrium constants in the range 10^{-4}–10^{-12} M, association constants in the range 10^3–10^7 M^{-1}s^{-1} and dissociation constants from 10^{-6}–10^{-1} s^{-1}. These results are impressive considering that no labelling of the biomolecules is required. In life sciences, this technique has been applied in biomolecular engineering, drug design, monoclonal antibody characterization, epitope mapping, phage display libraries, and virus–protein interaction among others interesting problems.

5.5.2 Environmental applications

The application of optical immunosensors for environmental monitoring started some years ago. The use of sensors for the measurement of pollutants is a viable alternative in environmental control where it is important to develop sensors of small size, that are reliable, sensitive and selective, for operation in situ and produced by a low cost technology. Several applications have been reported mainly applied to pesticide detection (herbicides, biocides, etc.). The SPR technology has been applied to many environmental problems, mainly for the detection of pollutants in the aquatic environment, such as the detection of phenols, pesticides (atrazine, herbicides) in the 0.05–5.0 µg/ml range [92,93].

5.5.3 Chemical and biological warfare

This is a new area of application for the optical biosensing techniques and is due to the current world situation. The best defence against these agents is the early detection and/or identification. A critical need exists for a field deployable biosensor to detect biological and chemical warfare agents in air and water samples, both rapidly and with a high sensitivity and sensitivity approaching standard laboratory procedures. For example, one biosensor based on the bidiffractive grating device has been developed for the simultaneous measurement of four biological threat agents (*Staphlococcus aureus* enterotoxin B, *Francisella turarensis*, Ricin and *Clostridium botulinum* toxin). Detection in the 1 ng/ml range has been achieved. Six biohazardous agents using a planar waveguide array biosensor have been also reported [94].

5.5.4 Genetic applications

DNA biosensors and gene chips are of considerable interest due to their potential for obtaining sequence-specific information in a faster, simpler and cheaper manner compared to traditional hybridization assays. DNA optical biosensors, based on nucleic acid recognition processes, are rapidly being developed towards the goal of simple, rapid and inexpensive testing of genetic and infectious diseases and for detection of DNA damage and interactions.

SPR sensors have been used to monitor in real time the binding of low molecular weight ligands to DNA fragment that were irreversibly bound to the sensor surface. The sensor was able to detect binding effects between 10 and 400 pg/mm^2. Binding rates and equilibrium coverage were determined for various ligands by changing the ligand concentration. In addition, affinity constants, association rates and dissociation rates were also determined for these various ligands.

5.6 FUTURE TRENDS

The success of the optical biosensor technology is seen in the increasing number of commercially available instruments. This also reflects the growing acceptance of the methodology. But although various technological developments have impacted the optical biosensor markets, commercialization is not simple, due to the high costs, problems of stability and sensitivity, quality assurance and the dominance of competitive technologies. Optical biosensing technology must overcome a number of market-related and technical obstacles to ensure commercial viability in the highly competitive area of field analytical methods. For example, there is still a long way to go to replace completely the conventional immunoassays by optical biosensors in clinical laboratories. On the other hand, the DNA biosensor technology is rapidly advancing and applications ranging from genetic testing to gene expression and drug discovery have been demonstrated. Further scaling down, particularly of the support instrumentation, should lead to hand-held DNA analyzers.

Actually, there is a need for biosensors capable of detecting very low levels of a great number of chemical and biochemical substances in the areas of environmental monitoring, industrial and food process, health care, biomedical technology, clinical analysis, etc. In the environmental field the demand for new sensors is increasing continuously due to strict legislation and control, and for improving living standards through waste management and remediation programs, for example. Biological and chemical warfare is an increasingly important area which also needs biosensors. Ideally, we must achieve the fabrication of optical biosensors with the following characteristics: very high sensitivity and selectivity, broad dynamic range, immunity to matrix effects, capable of simultaneous multianalyte determination, fast, reversible, stable, simple to operate, robust, cheap and of small size (for making a portable system for spatial mapping over large or remote areas).

For reaching all the objectives described above, the future research and development in optical biosensors must focus on the following key points.

5.6.1 Integration

Most of the commercial sensors are rather bulky and the prices for the instrument and/or the sensor chips are very high. We need to develop compact integrated optical sensors. Some attempts have already been made or are ongoing. The high degree of development of optical fibers and integrated optic technology, including fibers, emitters, detectors, waveguides and related devices, could allow the flexible development of miniaturized compact devices. These can be included in a remote control optical fiber network, for chemical analysis in standard analytical chemical and medical laboratories, as well as for environmental monitoring or control in chemical industry manufacturing

Optical biosensors

processes. While size and weight are less important considerations for instruments intended for laboratory use, devices for on-site analysis or point-of-care operations must be geared for portability, ease of use and low cost. Integrated optical devices have a compact structure and will allow for fabricating optical sensor arrays on a single substrate for simultaneous detection of multiple analytes [97]. Mass-production of sensors will also be possible with the fabrication of miniaturized devices that integrate the electronics and optics (lab-on-a-chip in which the light source, photodiodes and sensor waveguides are combined on a single semiconductor package) (see Figs. 5.18 and 5.19), the flow system and the reagent deposition (by ink-jet, screen-printed or other technology). A complete system fabricated with integrated optics will offer low complexity, robustness, standardized device and portability.

5.6.2 New receptors and new immobilization procedures

Biological receptors can include chemically and genetically modified enzymes, new types of antibodies with high affinity and selectivity for small molecules, natural or artificial receptors or complex biological recognition elements.

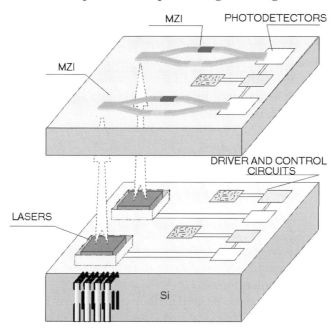

Fig. 5.18. A scheme for lab-on-a-chip integration including interferometric optical biosensors, lasers, photodetectors and CMOS electronics.

245

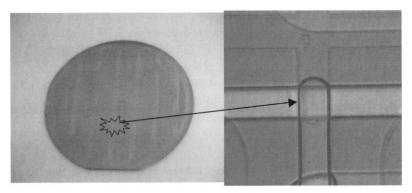

Fig. 5.19. One step for the integration of an optical biosensor into a-lab-on-a-chip: a Si wafer containing the devices is bonding with a Pyrex wafer, which has been previously processed with SU-8 polymer to define the microfluidics.

More reproducible and stable methods for immobilization of the receptor layers are also needed. Self-assembled methods at the nanometer scale, allowing stable and ordered receptor binding, must be further developed.

5.6.3 Mutianalyte detection

Multianalyte detection is becoming the crucial issue for biosensing development. Many areas demand a multianalyte operation: environmental screening, with thousands of samples per year to be analysed, genomics and proteomics, pharmaceutical screening, etc. Direct optical detection with evanescent wave sensors could be a possibility but a parallel detection of as many sites as possible is necessary [98].

5.6.4 Optical nanobiosensors

With the advent of nanotechnology new and exciting transducers of very high sensitivity can be developed. Improved detection limits must be achieved for most of the applications, as for example in the environmental field [99]. In addition, using micro- and nanotechnologies, optical biosensors could be integrated in "lab-on-a-chip" microsystems which could be used in real applications in many different scenarios (home, patient office, work, etc.) for real-time and on-line monitoring.

The latest developments on optical biosensors have been presented at specific conferences [100]. There is still a lot of ground to cover in the optical biosensing field, but this is a very active area of research and new and exciting developments will be achieved in the near future.

Acknowledgements

The author would like to express her gratitude for the continued support on biosensors research by the Spanish Ministry of Education and Science and the European Union. The author thanks specially all the team members belonging to the Biosensors Group from CNM in Madrid and Barcelona (Spain) for their work.

REFERENCES

1   F.S. Ligler and C. Rowe Taitt (Eds.), *Optical Biosensors: Present and future*. Elsevier, Amsterdam, 2002.
2   R.A. Potyrailo, S.E. Hobbs and G.M. Hieftje, *Fresenius J. Anal. Chem.*, 362 (1998) 349–373.
3   R.P.H. Kooyman and L.M. Lechuga, In: E. Kress-Rogers (Ed.), *Handbook of Biosensors: Medicine, Food and the Environment*. CRC Press, FL, USA, 1997, pp. 169–196.
4   G. Ramsay (Ed.), *Commercial Biosensors*. John Wiley, New York, 1998.
5   G. Canziani, W. Zhang, D. Cines, A. Rux, S. Willis, G. Cohen, R. Eisenberg and I. Chaiken, *Methods*, 19 (1999) 253–269.
6   P.D. Hansen and A. Usedom, In: F.W. Scheller, F. Schubert and J. Fedrowitz (Eds.), *Frontiers in Biosensorics II*. Birkhäuser Verlag, Switzerland, 1997, pp. 109-120.
7   C. Domínguez, L.M. Lechuga and J.A. Rodríguez, In: S. Alegret (Ed.), *Integrated Analytical Systems*. Elsevier, Amsterdam, 2003, pp. 541–584, ISBN: 0-444-51037-0.
8   G. Gauglitz, In: W. Göpel and J. Hesse (Eds.), *Sensors Update*. Wiley-VCH, Weinheim, 1999.
9   L.M. Lechuga, *Quím. Anal.*, 19 (2000) 61–67.
10 T. Vo-Dinh and B. Cullen, *Fresenius J. Anal. Chem.*, 366 (2000) 540–551.
11 I. Garcés, F. Villuendas, I. Salinas, J. Alonso, M. Puyol, C. Domínguez and A. Llobera, *Sens. Actuators B*, 60 (1999) 191–199.
12 J. Homola, *Anal. Bioanal. Chem.*, 377 (2003) 528–539.
13 L.M. Lechuga, A. Calle and F. Prieto, *Quím. Anal.*, 19 (2000) 54–60.
14 D.P. Campbell and J. McCloskey, In: F. Ligler and C. Rowe Taitt (Eds.), *Optical Biosensors: Present and Future*. Elsevier, Amsterdam, 2002, pp. 277–304.
15 L.M. Lechuga, F. Prieto and B. Sepúlveda, In: R. Narayanaswamy and O.S. Wolfbeis (Eds.), *Optical Sensors. Industrial, Environmental and Diagnostic Applications*. Springer, Heidelberg, 2003, pp. 227–248, ISBN: 3-540-40886-X.
16 G. Boisde and A. Harmer, *Chemical and Biochemical Sensing with Optical Fibers and Waveguides*. Artech House, Boston, 1996.
17 O.S. Wolfbeis, *Anal. Chem.*, 72 (2000) 81R–89R.
18 M. Mehrvar, C. Bis, J.M. Scharer, M. Moo-Young and H. Luong, *Anal. Sci.*, 6 (2000) 677–692.
19 G.P. Anderson and N.L. Nerurkar, *J. Immunol. Methods*, 271 (2002) 17–24.

20 C. Nylander, B. Liedberg and T. Lind, *Sens. Actuators*, 3 (1982) 79–88.
21 B. Liedberg, C. Nylander and I. Lundström, *Sens. Actuators*, 4 (1983) 299–304.
22 J. Homola, S.S. Yee and G. Gauglitz, *Sens. Actuators B*, 54 (1999) 3–15.
23 C.R. Lawrence and N.J. Geddes, In: E. Kress-Rogers (Ed.), *Handbook of Biosensors: Medicine, Food and the Environment*. CRC Press, FL, USA, 1997, pp. 149–167.
24 J. Homola, S. Yee and D. Myszka, In: F.S. Ligler and C. Rowe Taitt (Eds.), *Optical Biosensors: Present and Future*. Elsevier, Amsterdam, 2002, pp. 207–251.
25 H. Raether, Surface plasmons oscillations and their applications. *Physics of Thin Films*, Vol. 9. Academic Press, FL, USA, 1977, pp. 145–262.
26 J. Homola, J. Ctyroky, M. Skalsky, J. Hradilova and P. Kolarova, *Sens. Actuators B*, 38/39 (1997) 286–290.
27 J. Dostálek, J. Ctyroky, J. Homola, E. Brynda, M. Skalsky, P. Nekvindovà, J. Skvor and J. Schröfel, *Sens. Actuators B*, 3722 (2001) 1–5.
28 J. Homola, H.B. Lu, G.G. Nenninger, J. Dostálek and S.S. Yee, *Sens. Actuators B*, 3786 (2001) 1–8.
29 A.A. Kolomenskii, P.D. Gershon and A. Schuessler, *Appl. Opt.*, 36 (1997) 6539–6547.
30 T. Chinowsky, L. Jung and S. Yee, *Sens. Actuators B*, 54 (1999) 89–97.
31 W.M. Mullett, E.P.C. Lai and J.M. Yeung, *Methods*, 22 (2000) 77–91.
32 T. Akimoto, S. Sasaki, K. Ikebukuro and I. Karube, *Biosens. Bioelectron.*, 15 (2000) 355–362.
33 R.L. Rich and D.G. Myszka, *Curr. Op. Biotechnol.*, 11 (2000) 54–61.
34 R.L. Rich and D.G. Myszka, *J. Mol. Recognit.*, 14 (2001) 273–294.
35 K. Johnasen, R. Stalberg, I. Lünstrom and B. Liedberg, *Meas. Sci. Technol.*, 11 (2000) 1630–1638.
36 G.C. Nenninger, M. Piliarik and J. Homola, *Meas. Sci. Technol.*, 13 (2002) 2038–2046.
37 http://www.biacore.com
38 www.jandratek.de/produkt/produkt_plasmonic.html
39 www.gwcinstruments.com/SPRproducts.html
40 www.ibis-spr.nl/products/spr/index.htm
41 www.leica-microsystems.com
42 www.nle-lab.co.jp/biosensor/doc/products/index.htm
43 www.htsbiosystems.com/products/index.htm
44 http://www.ti.com/spr
45 B.J. Luff, J.S. Wilkinson, J. Piehler, U. Hollenbach, J. Ingenhoff and N. Fabricius, *J. Lightwave Technol.*, 16(4) (1998) 583–592.
46 L.M. Lechuga, A.T.M. Lenferink, R.P.H. Kooyman and J. Greve, *Sens. Actuators B*, 24 (1995) 762–765.
47 E.F. Schipper, A.M. Brugman, C. Domínguez, L.M. Lechuga, R.P.H. Kooyman and J. Greve, *Sens. Actuators B*, 40 (1997) 147–153.
48 Ch. Fattinger, H. Koller, D. Schlatter and P. Wehrli, *Biosens. Bioelectron.* 8, (1993) 99–107.
49 A. Brandenburg, R. Krauter, C. Künzel, M. Stefan and H. Schulte, *Appl. Optics*, 39 (2000) 6396–6405.
50 R.G. Heideman and P.V. Lambeck, *Sens. Actuators B*, 61 (1999) 100–127.

51 F. Prieto, B. Sepúlveda, A. Calle, A. Llobera, C. Domínguez, A. Abad, A. Montoya and L.M. Lechuga, *Nanotechnology*, 14 (2003) 907–912.
52 E.F. Schipper, A.J.H. Bergevoet, R.P.H. Kooyman and J. Greve, *Anal. Chim. Acta*, 341 (1997) 171–176.
53 A.T.M. Lenferink, E.F. Schipper and R.P.H. Kooyman, *Rev. Sci. Instrum.*, 68 (1997) 1582–1586.
54 F. Prieto, B. Sepúlveda, A. Calle, A. Llobera, C. Domínguez and L.M. Lechuga, *Sens. Actuators B*, 92 (2003) 151–158.
55 M. Weisser, G. Tovar, S. Mittler-Neher, W. Knoll, F. Brosinger, H. Freimuth, M. Lacher and W. Ehrfeld, *Biosens. Bioelectron.*, 14 (1999) 405–411.
56 L.M. Lechuga, B. Sepúlveda, A. Llobera, A. Calle and C. Domínguez, *Microtechnologies for the New Millennium: Bioengineered and Bioinspired Systems*, Proceedings of SPIE (The International Society for Optical Engineering), 5119 (2003) 140–148.
57 B.H. Scheneider, E.L. Dickinson, M.D. Vach, J.V. Hoijer and L.V. Howard, *Biosens. Bioelectron.*, 15 (2000) 597–604.
58 E. Brynda, M. Houska, A. Brandenburg, A. Wikerstal and J. Skvor, *Biosens. Bioelectron.*, 14 (1999) 363–368.
59 S. Busse, V. Scheumann, B. Menges and S. Mittler, *Biosens. Bioelectron.*, 17 (2002) 704–710.
60 S. Busse, M. DePaoli, G. Wenz and S. Mittler, *Sens. Actuators B*, 80 (2001) 116–124.
61 www.farfield-sensors.com
62 R. Cush, J.M. Cronin, W. Stewart, C.H. Maule, J. Molloy and N.J. Goddard, *Biosens. Bioelectron.*, 8 (1993) 347–353.
63 P.E. Buckle, R.J. Davies, T. Kinning, D. Yeung, P.R. Edwards and D. Pollard-Knight, *Biosens. Bioelectron.*, 8 (1993) 355–363.
64 P.R. Edwards, A. Gill, D.V. Pollard-Knight, M. Hoare, P.E. Buckle, P.A. Lowe and J. Leatherbarrow, *Anal. Biochem.*, 231 (1995) 210–217.
65 http://www.affinity-sensors.com
66 S. Nicholson, J.L. Gallop, P. Law, H. Thomas and A.J.T. George, *Lancet*, 353 (1999) 808.
67 T. Kinning and P. Edwards, In: F.S. Ligler and C. Rowe Taitt (Eds.), *Optical Biosensors: Present and Future*. Elsevier, Amsterdam, 2002, 253–276.
68 K. Tiefenthaler and W. Lukosz, *Opt. Soc. Am.*, B6 (1989) 209–220.
69 W. Lukosz, D. Cerc, M. Nellen, Ch. Stamm and P. Weiss, *Biosens. Bioelectron.*, 6 (1991) 227–232.
70 D. Schlatter, R. Barner, Ch. Fattinger, W. Huber, J. Höbscher, J. Hurst, H. Koller, C. Mangold and F. Müller, *Biosens. Bioelectron.*, 8 (1993) 109–116.
71 Ch. Fattinger, C. Mangold, M.T. Gale and H. Schütz, *Opt. Eng.*, 34 (1995) 2744–2753.
72 R.F. Kunz, *Sens. Actuators B*, 38-39 (1997) 13–28.
73 R.E. Kunz, In: E.J. Murphy (Ed.), *Integrated Optical Circuits and Components*. M. Dekker, New York, USA, 1999, pp. 335–380.
74 T. O'Brien, et al., *Biosens. Bioelectron.*, 14 (2000) 815–828.
75 M. Wiki, H. Gao, M. Juvet and E. Kunz, *Biosens. Bioelectron.*, 16 (2001) 37–45.
76 J.J. Ramsden, *Chimia*, 53 (1999) 67–74.

77   R. Horváth, G. Fricsovsszky and E. Papp, *Biosens. Bioelectron.*, 18 (2003) 415–428.
78   J. Vörös, J.J. Ramsden, G. Csúcs, I. Szendro, S.M. de Paul, M. Textor and N.D. Spencer, *Biomaterials*, 23 (2002) 3699–3710.
79   T.S. Hug, J.E. Prenosil and M. Morbidelli, *Biosens. Bioelectron.*, 16 (2001) 865–874.
80   www.owls-sensors.com
81   K.E. Sapsford, C. Rowe Taitt and F.S. Ligler, In: F.S. Ligler and C. Rowe Taitt (Eds.), *Optical Biosensors: Present and Future*. Elsevier, Amsterdam, 2002, pp. 95–121.
82   S. Löfås, B. Johnsson, Å. Edström, A. Hansson, G. Lindquist, R.-M. Müller Hillgren and L. Stigh, *Biosens. Bioelectron.*, 10 (1995) 813–822.
83   T. Weimar, *Angew. Chem. Int. Ed.*, 39 (2000) 1219–1221.
84   T.A. Morton and D.G. Myszka, *Methods Enzymol.*, 295 (1998) 268–294.
85   S. Marose, C. Lindemann, R. Ulber and T. Scheper, *Trends Biotechnol.*, 17 (1999) 30–34.
86   R.L. Rich and D.G. Myszka, *Trends Microbiol.*, 11 (2003) 124–133.
87   E. Ehrentreich-Förster, F.W. Scheller and F.F. Bier, *Biosens. Bioelectron.*, 18 (2003) 375–380.
88   E. Brynda, M. Houska, A. Brandenburg and A. Wikerstal, *Biosens. Bioelectron.*, 17 (2002) 665–675.
89   J. Wang, *Nucleic Acids Res.*, 28 (2000) 3011–3016.
90   A.N. Naimushin, S.D. Soelberg, D.K. Nguyen, L. Dunlap, D. Bartholomew, J. Elkind, J. Melendez and C.E. Furlong, *Biosens. Bioelectron.*, 17 (2002) 573–584.
91   V. Silin and A. Plant, *Trends Biotechnol.*, 15 (1997) 353–359.
92   C. Barzen, A. Brecht and G. Gauglitz, *Biosens. Bioelectron.*, 17 (2002) 289–295.
93   S.K. Van Bergen, I.B. Bakaltcheva, J.S. Lundgren and L.C. Shriver-Lake, *Environ. Sci. Technol.*, 34 (2000) 704.
94   C.A. Rowe-Taitt, J.W. Hazzard, K.E. Hoffman, J.L. Cras, J.P. Golden and F.S. Ligler, *Biosens. Bioelectron.*, 15 (2000) 579–589.
95   R.J. Green, R.A. Frazier, K.M. Shakesheff, M.C. Davies, C.J. Roberts and S.J.B. Tendler, *Biomaterials*, 21 (2000) 1823–1835.
96   P. Guedon, T. Livache, F. Martin, F. Lesbre, A. Roget, G. Bidan and Y. Levy, *Anal. Chem.*, 72 (2000) 6003–6009.
97   L.M. Lechuga, B. Sepúlveda, J. Sánchez del Río, F. Blanco, A. Calle and C. Domínguez, In: D.J. Robbins and G.E. Jabbour (Eds.), *Optoelectronic Integration on Silicon*, Proceedings of Photonics West (SPIE), 5357 (2004) 96–110.
98   P.I. Nikitin, M.V. Valeiko and B.G. Gorshkov, *Sens. Actuators B*, 90 (2003) 46–51.
99   A.N. Naimushin, S.D. Soelberg, D.U. Bartholomew, J.L. Elkind and C.E. Furlong, *Sens. Actuators B*, 96 (2003) 253–260.
100 Proceedings of EUROT(R)ODE VII Conference, April 4–7 2004, Madrid, Spain.

Chapter 6

Bioanalytical microsystems: technology and applications

Antje J. Baeumner

6.1 INTRODUCTION

The quest for rapid, reliable, sensitive and inexpensive devices in applications such as medical diagnostics, food safety, analysis of chemical and biological warfare agents, environmental monitoring, drug discovery, genomics and proteomics is the driving force behind efforts to miniaturize the bioanalytical laboratory. Miniaturization in most of these applications is the answer to many problems. The analysis of large numbers of assays required in screening procedures is only cost efficient, if small amounts of reagents are used. For instance, the analysis of 25,000 gene sequences (or more) is now possible simultaneously with one DNA microarray chip, thus allowing for more analyzed sequence per cost and time involved. Similarly, rapid analysis of thousands of environmental samples per day to monitor the presence of biological warfare agents becomes feasible with miniaturized biosensors incorporated micro-Total Analysis Systems (μTAS).

In addition to the impact on high throughput screening, the miniaturization of biosensors opens up new alleys for multi-analyte assays. Parallel processing of samples, combination of several biorecognition elements with one or a cassette of transducers is feasible, while keeping the size of the biosensor in the range of 2 cm × 2 cm. For example, one food sample could be analyzed for all important food pathogens in one assay, one air sample could be analyzed for all known biological warfare agents in one assay, and one blood sample could be analyzed for all clinical markers of interest in one step.

While being single or multi-analyte and/or high throughput devices, miniaturized biosensors give a new dimension to portability, i.e., pocket size and implantable whole biosensor systems are possible. Thus, the availability for on-site and in-field testing can be achieved by almost any biosensor design, as long as it is a miniaturized version. This is important not only for medical diagnostic, but also for food safety and environmental monitoring with new users being patients and small doctor's offices, consumers, food processors, restaurants, first responders, farmers, etc.

Biosensors were originally defined as a self-contained integrated device that is capable of providing specific quantitative or semi-quantitative analytical information using a biological recognition element (biochemical receptor), which is in direct spatial contact with a transduction element [1]. They were also thought to detect analytes directly in field, i.e., without any sample pretreatment steps, and to be continuous and/or reversible. Only very few biosensors described in the literature fulfill all of these requirements, however, in the age of miniaturization and the development of μTAS all of this can become feasible for a vast number of complicated analytes and sample matrices.

A distinction has to be made between a truly miniaturized biosensor (i.e., self-contained microbiosensors) and a biosensor in which the bioassay is miniaturized but a complex and typically large transduction system is required for signal detection and analysis. Many of the currently described miniaturized biosensors fall under this latter category, such as DNA microarrays using confocal microscopes and CCD cameras for data acquisition, surface plasmon resonance (SPR) biosensors, cantilever based biosensors using an atomic force microscope (AFM), etc. Thus, these biosensing systems take advantage of the enhanced molecular performance of micro/nanobioassays, however, they can typically not be considered portable, inexpensive and easy to use.

This chapter intends to give an insight into miniaturization techniques currently applied in the fabrication of novel biosensing systems. The techniques described are by no means inclusive, but instead are meant as a representation of possibilities available for the design of biosensors with dimensions in the micro and nanometer size range. Novel immobilization procedures of biorecognition elements and sensing layers will be described briefly. The effect miniaturization has on the biosensor performance will be discussed. Several examples of successfully miniaturized biosensors (and miniaturized bioassays in combination with an external signal detection unit) are chosen from current literature. Finally, the conclusions will give an outlook into the future of miniaturized biosensor. It shall be pointed out here, that no sample preparation procedures for miniaturized biosensors are described in this chapter. However, especially if talking about μTAS, these are important and critical for the success of the new class of biosensors. The reader is advised to refer to current and pertinent literature in this field to obtain an insight and overview. Typical sample preparation steps spanning from filtration to extraction are as important in miniaturized analytical devices as in their macro counter parts—if not even more.

6.2 MINIATURIZATION TECHNIQUES

An excellent overview and in-depth description of miniaturization techniques is given in Ref. [2]. Here, only a few examples of how miniaturized biosensors can be fabricated are described. A clear distinction can be made between the miniaturization of the biosensor fluid flow system and the use of a miniaturized transducer. Both result in different advantages for the biosensor assay, as it will be discussed below. In the case of biosensor fluid flow system, microfabrication tools used are by far fewer than those applied to the fabrication of micro/nanotransducers, which can span all of the micro/nanofabrication techniques developed.

6.2.1 Biosensor fluid flow system (biosensor housing)

Three substrates are frequently used for the development of a biosensor fluid flow system, or simply called biosensor housing: silicon, glass, and polymers. In the case of glass, normal borosilicate and quartz glass are applied, in the case of polymers, acrylic, polycarbonate, poly(methyl methacrylate) (PMMA) and poly(dimethylsiloxane) (PDMS) are very often used. Patterns are created in the substrates using a variety of different techniques, among which lithography is most widely used for biosensor housing microfabrication, which will be described in more detail. Other processes such as laser ablation will be discussed only briefly.

6.2.1.1 Design and preparation of a fluid flow system using lithography
Independent of the final substrate used as biosensor housing, most processes begin with the CAD assisted design of the fluid flow system pattern. Subsequently, masks are prepared using a number of different techniques, such as UV- or X-ray photolithography, e-beam, or ion-beam lithography. Photomasks are typically made of quartz coated with chromium. The chromium layer is coated with a resist that can be patterned by one of the lithographic techniques in order to create a pattern in the mask. The preparation of masks bearing the desired design is often a lengthy procedure since the pattern has to be created in a serial exposure process. Depending on the lithographic principle, feature sizes in the submicrometer range can be made. For example, e-beam and ion-beam lithography enable the fabrication of very small features (i.e., below 0.1 μm). Also, X-ray, e-beam and ion-beam

lithography offer a deep focus depth (i.e., 50 μm to 1 mm). UV-photolithography has been used traditionally in the IC industry, and it is comparably easier to use (i.e., e-beam and ion-beam lithography need to work in a vacuum) but focal depth (1 μm) and feature sizes are more limited. However, it is by far the most often used lithographic technique in order to fabricate biosensor housing and fluid control systems [2,3]. The two main drawbacks of UV-photolithography, i.e., two-dimensional microfabrication and physical size limitations, are not of the greatest concern in this application field. In addition, new developments of photoresists allow the design of topographically more three-dimensional structures [2,4,5]. Also, X-ray, e-beam and ion-beam lithography can be used for a mask design with feature sizes in submicrometer range, while the subsequent pattern generation can then be done using UV-photolithography.

Some of the more important components in the UV-photolithography microfabrication are exposure wavelength, exposure time, photoresist and thickness of the photoresist. A detailed description of all of these aspects is given elsewhere [2]. The focus in this chapter is more on the understanding of the underlying principle and subsequent application of biosensor design. Photoresists come typically in two forms, positive and negative. Both of them contain a photosensitive chemical, a polymer and a casting solvent. In general, positive photoresists render more soluble upon light exposure, while negative photoresists become more solid. One frequently applied positive photoresist is PMMA, in which the polymer chains scission under exposure of deep UV light (Fig. 6.1). In Fig. 6.2, the principle of photoresist patterning is shown. Photoresist is spun onto a flat surface (such as a silicon or glass wafer). The volume used and spinning parameters such as speed and time determine the thickness of the photoresist layer. Subsequently, light of a specific wavelength (adjusted to the sensitizer in the photoresist used) is shined through a mask, which bears the desired pattern. Thus, only certain areas of the photoresist are exposed to the light. Upon exposure, a chemical reaction changes the photoresist and renders it either more soluble (positive photoresist) or less

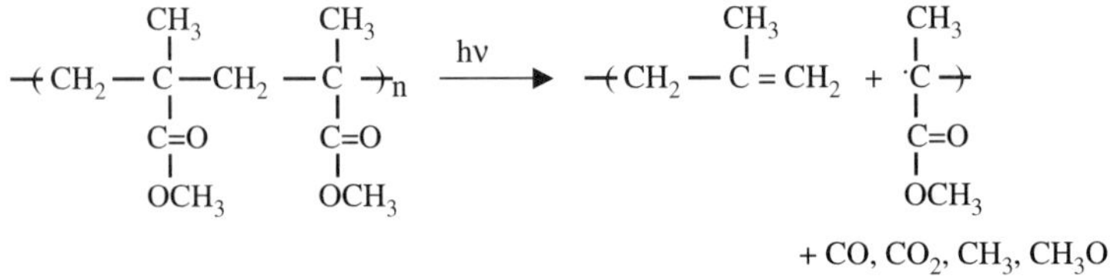

$$-(CH_2-\underset{\underset{OCH_3}{\overset{\overset{CH_3}{|}}{C=O}}}{\overset{|}{C}}-CH_2-\underset{\underset{OCH_3}{\overset{\overset{CH_3}{|}}{C=O}}}{\overset{|}{C}})_n \xrightarrow{h\nu} -(CH_2-\underset{}{\overset{\overset{CH_3}{|}}{C}}=CH_2 + \underset{\underset{OCH_3}{\overset{\overset{CH_3}{|}}{C=O}}}{\overset{|}{\cdot C}})-$$

$$+ CO, CO_2, CH_3, CH_3O$$

Fig. 6.1. Poly(methyl methacrylate) (PMMA), a positive photoresist. The polymer chains scission under exposure of deep UV light.

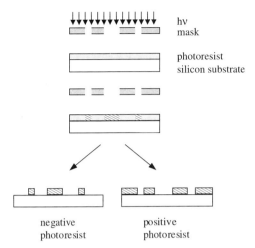

Fig. 6.2. Patterning of positive (a) and negative (b) photoresists using UV-photolithography.

soluble (negative photoresist) when immersed into a specific organic solvent (developer). Thus, a pattern is developed in the photoresist (Fig. 6.2).

Depending on the exposure time, the pattern can appear as "overcut", straight, or "undercut" (Fig. 6.3). These three profiles can be advantageous for a variety of further uses of the pattern. For example, the "overcut" is frequently used for lift-off procedures (see Section 6.2.2). In general, ideal exposure times have to be determined for each new protocol. However, guideline values are provided by the manufacturers of photoresists and developers. More information on exposure effects on photoresists such as light scattering effects, over exposure, etc., can be found in specialized microfabrication books. After a pattern has been designed and developed in the photoresist, it can be used for patterning of the substrate on which the photoresist was spun. For example, a silicon wafer can be dry or wet-etched,

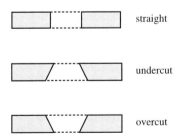

Fig. 6.3. Exposure profiles of positive photoresists.

metals can be deposited, further photoresists can be spun on the design, etc. Some examples will be discussed below.

6.2.1.2 Silicon, glass and polymer substrates

After a resist pattern has been created on a wafer, dry and wet etching can be used for the fabrication of patterns in silicon or glass. Wet etching provides a higher degree of selectivity than dry etching techniques. Solutions containing mixtures of nitric acid (HNO_3) and hydrofluoric acid (HF) can be used for isotropic etching (often with acidic additives), which typically shows diffusion limitations. Thus, stirring in an isotropic etching bath can influence the shape and rate of the process. The overall reaction of such a solution with silicon is as follows:

$$Si + 6HF + HNO_3 \rightarrow H_2SiF_6 + HNO_2 + H_2O + H_2. \tag{6.1}$$

Anisotropic wet etching is carried out using solutions containing KOH, NaOH, LiOH, CsOH, NH_4OH sometimes with the addition of alcohols. Anisotropic etching is typically reaction limited, and can result in structures defined by the crystallographic planes. Dry etching processes are typically much slower, and result in less defined structures than wet etching. It is typically applied for resist removal (resist stripping) and isotropic features. A variety of different techniques can be used, ranging from reactive gas plasma to inert gas plasma, inert gas ion beam and reactive gas ion beam etching processes. A good overview of all of these techniques is given in Ref. [2].

Applying these etching processes to a silicon wafer, the resist designed pattern can then be etched into the silicon resulting, for example, in channels of the biosensor microfluidic system. This is shown in Fig. 6.4a. Similarly, channel structures can be made in glass substrates. Alternatively, the negative image of channels can be etched into silicon (Fig. 6.4b). This "mold" can then be used for embossing or imprinting into hard plastics (such as acrylic, PMMA, polycarbonate) or for soft lithography, in which case an elastomer would be cast on the silicon pattern which is described in more detail below. Alternatively, molding can be applied as well. All three examples will be discussed in the following paragraphs since they are of great importance in the fabrication of microfluidic channel systems for microbiosensors.

Imprinting into plastic materials can help to overcome two main disadvantages of silicon-based microfluidic systems: expense of fabrication and brittleness of the material. Imprinting can be carried out at elevated temperatures [6,7] or at room temperature [8]. While heating of the plastic material can result in better feature aspect ratios, it is limited by the breaking of silicon templates during the cooling process due to the different thermal

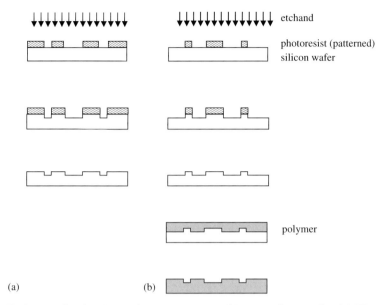

Fig. 6.4. Etching a fluidic channel pattern into a silicon or glass wafer. (a) The channel is created in the substrate. (b) The negative of the channel is created in silicon and used subsequently as "mold" for imprinting of soft lithography.

expansion properties of silicon and the plastic material. Heating conditions are, for example, 155°C with a pressure of 13.8 MPa applied for 1.5 h. Thus, Xu et al. found that in hot-imprinting processes, one silicon template could only be used for about 10 times, while a room temperature process avoided this problem. In addition, they reported a reduction in overall fabrication time to only 5 min. Feature sizes reported in room temperature embossing were with 30 μm depth and 90 μm width comparable to those for hot embossing, however typically high aspect ratio structures can be more reliably fabricated using hot embossing. Chou and co-workers have described a compression based imprinting method. Metal patterns with a feature size of 25 nm and a period of 70 nm were designed and fabricated using resist templates (created by imprint lithography) in combination with lift-off processes [9].

Similar to these hard plastic applications, the use of elastomers has a number of advantages over silicon as biosensor housing material. Besides fabrication cost and brittleness as described above (PDMS is about 50 times less expensive than silicon on a per volume basis), moving parts are extremely difficult to make in the stiff silicon material without increasing the overall size significantly. Also, valves, which will be discussed in more detail below, need a soft material as valve seat to close completely (thus, elastomers have to be incorporated in silicon processing, which is not always straightforward). Finally, the immobilization of biorecognition elements in biosensors often needs a specific surface chemistry, which is much more difficult to achieve with the high temperatures required to bond silicon and glass [10]. A variety of different techniques and applications of elastomers have been introduced by George Whitesides and co-workers in recent years. These are mainly micromolding in capillaries, microtransfer molding, replica molding, and microcontact printing, which are used for the creation of self-assembled monolayers (SAM) [11–15]. Replica molding is the most similar to previously described imprinting techniques. Feature sizes as low as 30 nm can be created when casting an UV curable polymer against an elastomeric mold. Simultaneously, the overall size of a molded piece can be in the centimeter range with a few millimeter thickness.

PDMS is a popular and frequently used elastomer not only because of its cost-effectiveness, but also because of the ease of use and its specific molding characteristics. For instance, PDMS can form conformal contact with a substrate and it can be released again easily from complex or fragile substrates. In addition, due to its low interfacial free energy (ca. 21.6 dyn cm^{-1}) and its low reactivity, it does not adhere irreversibly to other substrates such as polymers, silicon, etc. [14,15]. It does, however, bind to photoresist residues on the silicon template. Thus, careful photoresist removal is advised before casting the elastomer on its template.

The LIGA process [German acronym for X-ray lithography (X-ray Lithographie), electrodeposition (Galvanoformung), and molding (Abformtechnik)] can be used to fabricate microstructures via a molding procedure with high aspect ratios and relatively large feature sizes around 10 μm. It involves the use of thick X-ray resist layers (micrometer to centimeter), and results in the fabrication of three-dimensional resist structures. Electrodeposition is used to fill the resist mold, the resist is removed and a freestanding metal mold is created. This can either be used directly, or it can function as a mold for accurate microinjection molding [16,17]. The process offers the possibility to produce structures with very high aspect ratios and is therefore interesting to the microbiosensor field. Similar procedures are described to be either better than the LIGA process in respect to sensitivity [18], or adjusted for cost concerns [7].

Laser ablation is a process in which an intense burst of energy is used to remove a small amount of material from the surface [19]. It is the basis for excimer laser micromachining. It can be used with a variety of different materials, ranging from silicon to polymers and ceramics. Feature sizes in the low micrometer range can be realized. The advantage of laser ablation is the creation of truly three-dimensional structures without the need of a mask. Software programs are commercially available to assist in the design and fabrication process. Disadvantageous in comparison to the soft lithography and molding processes described above, is the serial fabrication, i.e., each device needs the same processing time, while in molding a much faster throughput with small feature sizes can be achieved. In addition, the chemical surface composition of the substrate is often being changed during the ablation process. More recently, Locascio and co-workers described the use of an UV excimer laser (KrF, 248 nm) to create microchannels and subsequently, characterized the surface of laser-ablated polymers used for microfluidics. It was found that ablated surfaces of PMMA were very similar to native PMMA surfaces. No change was found independent of the ablation atmosphere, which was assumed to be due to the low absorption at 248 nm light by this polymer. In contrast, poly(ethylene terephthalate glycol), poly(vinyl chloride), and poly(carbonate) showed significant changes in surface chemistry when the ablation atmosphere was varied [20]. This has a major impact on the use of laser-ablated microfluidic channels when using electroosmotic flow as the fluid control system (described further below). Girault and co-workers have studied laser-ablated microfluidic channels for the use in electrophoresis with electroosmotic flow and optimized the fabrication process to obtain reliable and well-working channel surfaces and dimensions [21–23]. Finally, Locascio and co-workers described the use of laser ablation for the postmodification of polycarbonate microchannels. A series of slanted wells was created at the junction of a T-microchannel. These wells lead to a high degree of lateral transport in the channels and thus rapid mixing of two confluent streams [24].

6.2.1.3 Fabrication of active fluidic elements
In order to understand the *fluid mechanics* in microchannel systems one has to calculate the Reynolds numbers in a given system. The dimensionless Reynolds number (Re) is the ratio of the inertial forces to the viscous forces acting on a small element of fluid, and can be seen as the ratio of shear stress due to turbulence to shear stress due to viscosity.

$$Re = \rho v L / \eta \tag{6.2}$$

with the viscosity η (Pa s), the average velocity v (m s^{-1}), the density ρ (g l^{-1}), and the characteristic length L (m). The length L can be calculated from the ratio of fluid volume to surface area of the walls that bound it. For example, a microchannel that is 100 mm long, 100 μm wide and deep has a volume of 10^{-9} m^3 (1 μl), a surface area of approximately 2×10^{-5} m^2, and thus a characteristic length of $L = 50$ μm. Calculating the Reynolds number for water flowing with a velocity of 10 mm s^{-1} in said microchannel ($\rho = 1$ g l^{-1} and $\eta = 0.001$ Pa s or $\eta = 1$ g m^{-1} s^{-1}) it is below $Re < 1$ (i.e., $Re = 0.5$). Since Reynolds numbers below 2000 are typically found in laminar flow, Reynolds numbers near or below 1 can safely be considered laminar. Thus, a velocity of more than 200 m s^{-1} in the described microchannel would only lead to potentially turbulent events in the channels.

The small microchannel sizes used in microbiosensor systems (around 50 µm) result in the fact that diffusion can often be sufficient as a *mixing* element. However, in many cases, active mixing is required in order to increase reaction speed, enhance contact time between two immiscible liquids, or create a homogeneous solution within a certain length of the microchannel biosensor. Thus, a need of mixers is eminent. Two main mixing principles are applied: mechanical and static mixing. In the case of mechanical mixing, moving parts are required such as rotating rods [25] plus electronics and power sources to control and empower the mixer. Barbic et al. have demonstrated that using an electromagnetic micromotor that a motor can spin with 250 rpm in water in a microchannel system. In contrast, static mixers are based on the design of channel patterns that promote the mixing of two fluids. Excellent examples are given by Johnson et al. who created a number of slanted wells in a microchannel near a T-junction to improve mixing. Stroock et al. developed microfluidic channels with ridges in one wall of the channel to create transverse flows that stir the fluid as it passes through the channel. He et al. based their mixer on a packed-bed column flow principle and simulated the packed bed with a series of channels [24,26]. Ajdari, Stroock et al., and Erickson and Li have demonstrated that modifying the surface to bear heterogeneity in a channel with electroosmotic fluid control can cause localized circulation in the bulk laminar flow [27–29].

In addition, *pumps and valves* can be required as active fluid element. The integration of these elements into the microchannel design is important, if portability and small size of an overall biosensor are desired. Often, off-chip pumps, vacuums, and voltages are applied, however, their use renders the entire microbiosensor non-portable. An excellent overview of controlling fluids in microfluidic systems is given by Polson and Hayes [30]. In addition, interesting fluid control systems are described by Prins et al. [31], Zhao et al. [32], Leventis and Gao [33], Vahey et al. [34], and McKnight et al. [35], ranging from electroosmotically induced hydraulic pumping, to magnetohydrodynamic pumps, and to electrocapillary pressure fluid control. Here only a couple of examples are described: electroosmotic flow and valve-based micropumps.

Electroosmotic flow is the most popular of the electrokinetic pumping techniques. One of the main reasons for its popularity with microfluidic devices is the comparable ease of pumping in comparison to the high pressure needed for a volume-forced flow by pressure. In addition, it has a flat flow profile, dispersion is limited and its mechanism works in very small channels, limited in size only by the relative thickness of the electrical double layer (10s of nanometer) [30]. In contrast, volumetric-forced flow depends on the channel diameter. In fact, the average fluid velocity reduces with the square of the channel radius, which can be derived from the Hagen–Poiseuille equation that describes the slow flow of a fluid in a capillary

$$Q = \pi r^4 \Delta p/(8\eta L) \tag{6.3}$$

with volumetric flow rate Q (m^3 s^{-1}), radius r (m), pressure drop Δp (N m^{-2}) over the length L (m), and the fluid viscosity η (N s m^{-2}). The average velocity v (m s^{-1}) relates to the volumetric flow rate Q with

$$v = Q/(\pi r^2). \tag{6.4}$$

Thus,

$$\Delta p = 8\eta L v/r^2 \quad \text{or} \quad v = \Delta p r^2/(8\eta L). \tag{6.5}$$

The two equations above indicate that the pressure drop increases and the average velocity decreases dramatically with decreasing channel radius. Thus, it is obvious, why electrokinetic pumping procedures find widespread application in fluid systems with micrometer dimensions.

Electroosmotic flow is based on the movement of charged particles in a fluid due to an electric field. In macroscopic systems, this leads to electrophoresis, in narrow capillaries to electroosmosis. Applied in a fused capillary column, it was first reported by Jorgenson and Lukacs in 1981 [36]. In glass channels, for example, an electric double layer is generated by negative glass surface charge and positively charged ions in the solution. In an applied electric field, these cations are moving toward the cathode. Due to the viscosity of the solution, the entire fluid in the capillary is dragged along and an electroosmotic pumping is established. The solution moves with the electroosmotic flow velocity u_{eo} (m s^{-1}), which is proportional to the voltage per channel length [or electrical field E (V m^{-1})] and the electroosmotic mobility μ_{eo} (m^2 s^{-1} V^{-1}). Thus,

$$u_{eo} = \mu_{eo} V/L = \mu_{eo} E \text{ with } \mu_{eo} = \varepsilon_0 \varepsilon \zeta/\eta \text{ it is } u_{eo} = E\varepsilon_0 \varepsilon \zeta/\eta$$

where ε_0 is permittivity of the free space (C V^{-1} m^{-1}), ε is the dielectric constant of the solution, η (Pa s) is the solution viscosity, and ζ (V) is the zeta potential [37]. Thus, the electroosmotic flow velocity is directly proportional to the zeta potential and the electrical field strength.

The generated flow profile is flat, in contrast to the parabolic profile generated in pressure-driven flow, thus, minimizing dispersion. In addition, one channel or a section of a channel can pump for the rest of the microfluidic device, so that the flow control is only a matter of controlling the electric field strength. This, however, can prove to be more complicated since it depends on the channel surface composition, buffer characteristics, and external voltage fields [30]. High flow rates such as 1 cm s^{-1} have been achieved in 20 μm glass channels [38] or 3.6 μl min^{-1} for a 2 kV applied potential [39]. In contrast to fused glass capillaries, uniform and reliable electroosmotic flow in polymer microchannel is difficult to achieve due to the non-uniform chemical surface after the imprinting or embossing fabrication process. Polyelectrolyte multilayers have been suggested to overcome this problem and allow a well-controlled electroosmotic flow in polymer-based microchannels [40,41].

The use of *micropumps* as fluid control system also has been demonstrated successfully. In most cases, micropumps are the miniaturized version of their macro counter parts, which are characterized by mechanical valves connected to a chamber with oscillation volume. The principle of a valve type micropump is shown in Fig. 6.5. Membrane deflection in a micropump displaces fluid through one of two integrated check valves. The driving force of the membrane deflection, i.e., its actuation, can be thermopneumatic [42], electrostatic [43, 44], piezoelectric [45–50], etc. The main advantage of micropumps over their electrokinetical counterparts is the independence from solution characteristics (such as pH, ionic strength, and type) and channel surface properties. Maximum flow rates described range around 300–560 μl min^{-1}. Examples of valveless micropumps are given by Böhm et al. demonstrating an integrated micromachine pump based on an electrochemical reaction [51] and Gong et al. and Leventis et al. with a microelectromagnetic pump [33,52]. Many pumps

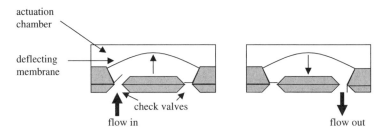

Fig. 6.5. Principle of a valve-based micropump. Membrane deflection in a micropump displaces fluid through one of two integrated check valves.

have been suggested, patented, but their competitiveness and use in real world biosensor sample detection and long-term use still needs to be proven. Nonetheless, valve type or valveless micropumps that can be integrated into the microchip and are independent of the solution characteristics will be ideal for portable miniaturized biosensors.

Finally, *microfilters* are described as part of the microchannel system in order to assist sample preparation. Channels with diameters less than 30 μm can easily be blocked by particulates in the sample, or crystals formed in solutions held in a micro-reservoir. While prefiltration outside of the microchannel system can separate most particles out of the sample solution prior to addition to the biosensor system, volumes required for "macro-filtration" are much larger than the volume finally applied to the biosensor. In addition, crystal or aggregate formation inside the channel cannot be avoided. Thus, the use of microfilters inside the microchannel system will be an important element in miniaturized biosensors. Filters have been described for trapping cells from blood [53], percolation filters for filtering solvents containing particulate materials ranging from dust to cells [54], and nanofilters that can separate particles as small as 44 nm [55].

Two different approaches can be followed, the fabrication of porous membranes by constructing pillars or channels within the microchannel or by creating a shallow microchannel covered with a slide. Comb-shape, array of posts and weir-type filters are described by Wilding and co-workers [56–58].

6.2.2 Biosensor transducers

As it was mentioned earlier, a much larger range of micro/nanofabrication tools are utilized in the design and fabrication of micro/nanotransducers than applied in the fabrication of microfluidic systems. Standard photolithography as described above is only one of the many techniques used. Three examples will be described here: the fabrication of freestanding transducers, UV-photolithography with lift-off technique for preparation of interdigitated ultramicroelectrode arrays (IDUAs), and more traditional techniques for microtransducer preparations.

6.2.2.1 Free-standing transducers
Freestanding or suspended nanostructures can be fabricated using a combination of electron-beam lithography and etching. With the help of e-beam lithography, objects with feature sizes below 1 μm can easily be designed, etching will subsequently remove material beneath the object and result in freestanding or suspended objects. The removal process can be done either by

isotropic etching or by another defined lithographic step. This process is often referred to as surface-micromachining [2,59]. For example, nanosized cantilevers can be made with this process. They have been used first in the mid-1980s in an AFM. In the case of an AFM, the tip of a cantilever is run across an object under investigation. Intermolecular forces between the cantilever and the surface deflect the tip up and down over. A reflected laser beam monitors this motion and the signal is converted into an image of the surface. Examples on how these cantilevers can be used as transducer surface will be described below.

6.2.2.2 Photolithography with image reversal and lift-off techniques
Standard photolithography can be used for the fabrication of micro-sized electrochemical transducer. When designing metal structures in the ultra-micrometer size range (i.e., feature sizes of a few micrometer), a process called image reversal and lift-off is added. The process described here is used in our laboratory for the fabrication of IDUAs [60]. A schematic of an IDUA is shown in Fig. 6.6a and an optical micrograph in Fig. 6.6b. Starting with normal mask design and fabrication (as described above), a silicon or glass wafer as substrate is coated with a positive photoresist. The photoresist is patterned using photolithography (patterns are generated either as direct image of the mask or are reduced 5 or 10 times in size using a 5 × or 10 × steppers). Subsequently, the wafers are baked in an NH_3 oven for 90 min. During this process, the exposed positive photoresist hardens and renders insoluble in solvents that can dissolve the unexposed photoresist. Thus, during the development step, photoresist previously not exposed will be washed away and the

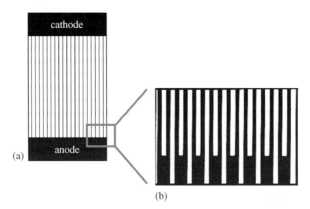

Fig. 6.6. Interdigitated ultramicroelectrode arrays (IDUAs). (a) schematic, (b) optical micrographs with 1000 × magnification. IDUAs were fabricated using standard photolithography and lift-off techniques on silicon and were made out of gold.

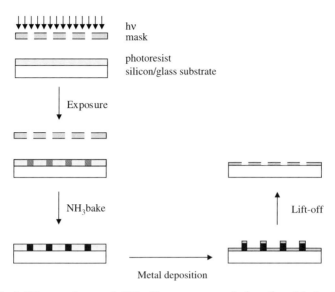

Fig. 6.7. Photolithography and lift-off process used for the fabrication of gold interdigitated ultramicroelectrode arrays on silicon or glass substrates.

exact image of the mask remains on the wafer (Fig. 6.7). Then, a thin layer of a metal is deposited via evaporation on the patterned wafer. Finally, all remaining photoresist is dissolved and lifted off the wafer. Thus, the reversed image of the mask remains as metal on the wafer surface (Fig. 6.7). Image reversal and lift-off techniques are used in order to exploit exposure patterns of a positive photoresist (Fig. 6.3). The undercut is very favorable in creating well-defined metal structures in the ultramicrometer size range on a substrate, since no continuous metal layer can be formed during the evaporation process.

6.2.2.3 The use of traditional fabrication techniques

Nanosensors for electrochemical detection have been made for years using more traditional fabrication methods, e.g., pulled platinum strings and carbon fibers. Carbon fibers can be purchased with diameters in the low μm range. These can subsequently be etched in an Ar^+ beam until conically shaped tips are produced with tip diameters between 100 and 500 nm [61]. Similarly, a platinum wire can be heated and pulled in order to create tips of similar diameters. Thick film electrodes made by screen printing [62] have also been shown to find application as transducer in microchannel systems [63].

Regarding optical transducers, a small prism can, for example, be coated with a thin gold layer in order to construct a SPR transducer, which subsequently is connected to a microchannel system. Also, optical fibers are available as simple miniaturized transducers as shown by Walt and co-workers [64].

6.3 IMMOBILIZATION OF BIORECOGNITION ELEMENTS IN MICROBIOSENSORS

According to the IUPAC biosensor definition, a direct immobilization of the biorecognition element onto the transducer is needed. However, this necessity becomes obsolete in miniaturized biosensors in which the signal can either be generated directly above the transducer, or at some distance and then transported in the fluid flow to the transducer for the signal generation. Immobilization techniques used are very similar to those applied in macrobiosensors, i.e., based on covalent binding, adsorption, matrix entrapment, or via a biological function such as antibody–antigen binding, streptavidin–biotin binding, etc. Challenges derive from small volumes and spaces used, and micropattern requirements, i.e., site-specific immobilization. A variety of interesting immobilization techniques have been used that provide site selectivity that will be described here briefly.

One of the fastest growing new techniques exploits photo-inducible reaction mechanisms. Since photolithography processes are available for microbiosensor fabrication, these can also be used for immobilization of the biorecognition element. Foder and co-workers demonstrated this technique in the early 1990s. They used photolithography for in situ chemical synthesis of biochemicals directly on a silicon substrate. Nucleotides modified with photosensitive linkers were used as the building blocks [65,66]. Initially, 256 probes were fabricated on a 1.28 cm^2 chip [67]. The same procedure is now used frequently in commercial DNA microarray fabrication with 20,000–45,000 probe sites in a 1.28 cm^2 chip area [68]. An example synthesis of a 3 × 3 array is shown in Fig. 6.8. In a similar approach, surfaces were photoactivated for site-specific immobilization of proteins. Nivens and co-workers have recently developed a photoactive poly(ethylene glycol) organosilane film for site-specific protein immobilization [69]. Liu and co-workers have developed a thiol-reactive, photoactivatable linker [70]. Somewhat differently, photoactivatable polymers have also been used for site-selective immobilization of antibodies, antigens [71], enzymes [72], and DNA molecules [73].

A different approach of site-specific polymerization and thus immobilization of a biorecognition element was shown by several research groups by utilizing electropolymerization [74–76]. In combination with microelectrodes in microchannel systems, a site directed simple immobilization of the biorecognition element could be achieved. Patterns can also be created by

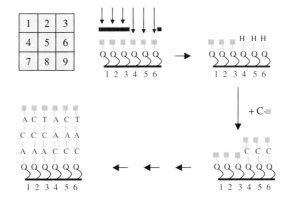

Fig. 6.8. In situ synthesis of DNA probes on a silicon substrate. An example of a 3 × 3 array is shown in which 3-mers are synthesized in three different locations of the array. Defined locations on the surface of a silicon wafer are activated via a photolithography process using a mask, and subsequently exposed to an activated nucleic acid nucleotide which bears at the same time a photoactivatable protection group for further coupling reactions. With each exposure step a new nucleotide can be synthesized into a defined location.

ink-jet based deposition processes as found in industrial applications already, or via other types of picoliter dispensing mechanisms [77]. De Rooij and co-workers have demonstrated the immobilization of antibodies in picoliter wells using electroosmotic flow [78].

As a final example, it should be pointed out that it is also possible to circumvent the need of immobilization of the biorecognition element in the microchannel system. Instead, it can be immobilized on superparamagnetic beads, silica beads, latex particles, etc. [79–81]. These beads are applied together with the sample into the microchannel system and can be collected on or near the transducer via magnetic or membrane separation.

6.4 MINIATURIZATION AND ITS EFFECT ON BIOSENSOR PERFORMANCE

Clearly, miniaturized biosensors have a number of advantages over their macro counterparts, but they also have some hurdles to overcome and in fact, might not always be the best analytical system for every problem. Sample and reagent volumes are one of the first items to consider. In general, analyzing a small sample volume is advantageous in clinical settings, for single nucleotide polymorphism (SNP) analyses, genomics and proteomics. Either only tiny sample volumes are available, or many analyses have to be performed from one sample. In addition, the consumption of costly reagents such as antibodies, cofactors, receptors, hormones, nucleotides, and oligonucleotides can be reduced. However, in the case of environmental and food samples, it might not be necessary to analyze only 100 nl of sample, because 1 g of food or 10 ml of water is easily available. In fact, in those cases only the analysis of a larger sample volume would make sense, since a representative quantity has to be investigated to provide a statistically significant result.

Related to the volume processed, the detection limit of biosensors has to be considered. Take, for example, a hypothetical biosensor for HIV detection that has a detection limit of one virus per analysis. If only 100 nl of sample are analyzed, then the detection limit of biosensor cannot be below 10,000 virus particles per milliliter of sample, which is neither a stunning nor an acceptable limit of detection for rapid detection early in the disease. Thus, in this case, a sample preparation step that reduces a 1 ml blood sample to 100 nl is of imminent importance (i.e., now, reduced reagent costs make the miniaturized biosensor advantageous over a macroscopic counter part). The same problem is encountered in most environmental and food sample analyses. Thus, the incorporation of sample pretreatment steps into the bioanalysis is very important, and for miniaturized biosensors even more than for their macro counterparts.

Due to the small volumes and feature sizes, reaction rates are found to be quite different in the microbiosensor system in comparison to their macro counter part. Most of this is due to the fact that diffusion is not the limiting factor in a reaction any longer. For example, the diffusion time of a particle with a diffusion coefficient $D_e = 10^{-9}\,\mathrm{m}^2\,\mathrm{s}^{-1}$ is 15 min to travel a distance of 1 mm, but only 10 s to travel 100 μm and only 0.1 s to travel 10 μm [37]. Therefore, DNA hybridization reactions, antibody–antigen binding events, and enzyme–substrate catalytic reactions take place in a fraction of the time required earlier. DNA hybridization can be accomplished in a matter of seconds in a microchannel system, while it takes in the order of an hour when employing standard Northern or Southern Blotting techniques with a piece of nylon membrane soaking in several milliliters of hybridization solution.

Small dimensions are also advantageous with respect to heat transfer in reactions such as the polymerase chain reaction (PCR) that are highly dependent on rapid temperature changes and that are frequently used as sample preparation step prior to biosensor analysis. Reactions taking typically in the order of hours in a standard thermocycler have been reduced to a few minutes in microchannels [82].

6.5 BIOSENSOR EXAMPLES

There are hundreds if not thousands of miniaturized biosensors published in literature today. Thus, a selection of only a few of them for a brief description is a difficult task. While the biosensors described here are exceptional examples of miniaturized systems, there are many others that would have deserved a description as well, if the space had been available. A selection has been made to give an overview of interesting biosensors such as DNA microarrays, biosensors coupled with capillary electrophoresis (CE), cantilever-based biosensors, electrochemical systems, optical biosensors, and visions of a μTAS. The examples are described only briefly, for a complete understanding of the work published, the reader is advised to refer to the original publication. Hopefully, this overview gives a grasp of the interesting biosensors developed in the new miniature world.

The fabrication of DNA microarray chips had been described earlier. While they meant a revolution in genomic sequence analysis just a decade ago, they are now routinely used for genome sequencing [83], mRNA expression monitoring, mutation analysis [84–86], detection of disease markers [87], drug discovery [88], etc. Costs of fabrication are still high, nonetheless, some companies provide their customers with custom-made arrays for independent experiments. DNA microarrays do not fall into the category of simple to use, inexpensive and portable biosensor, however, they allow a high throughput analysis of entire genomes in one experiment and utilize microfabrication techniques for the site-selective immobilization of their biorecognition elements. DNA probes are immobilized on the microarrays in one of two ways: they are positioned via dropping a picoliter volume into each location, or they are synthesized directly on the array surface as described further above. Each single experiment will produce a large number of data that have to be interpreted using software programs; often, mathematical models have to be applied in order to understand the results. While it is advantageous to obtain the many data within one experiment, each experiment is limited to one specific condition (which might not be ideal for each spot location in the microarray). Thus, data have to be interpreted carefully. A good overview of the microarray industry is given by Schena et al. [89], and a book providing excellent details on the microarray technology, as well as a book on the proteomic applications was published more recently by the same author [90,91].

Biosensors based on cantilever action were briefly mentioned above. Nanosized cantilevers have been first used in AFM. Here, intermolecular forces between the cantilever and a surface are detected by monitoring the motion of the cantilever tip. In the case of biosensor applications, they are used similarly to the quartz crystal microbalance, in which the change in frequency of the cantilever motion is measured upon a change of mass. Craighead and co-workers, for example, measured masses in the order of 1 pg [92,93]. They coated a cantilever with antibodies specific for *Escherichia coli* O157:H7, immersed it into an *E. coli* sample and subsequently measured the resonant frequency shift of the cantilever vibrating in air. One cell was sufficient to cause a measurable signal. A different approach is described by Gimzewski and co-workers [94]. The bending of the cantilever is monitored (no mass is measured). The cantilever is coated with short DNA oligonucleotides. Upon binding of a complementary sequence, intermolecular forces are exerted and expand the coating on the cantilever surface. Thus, the cantilever bends downward similar to a bimetallic strip. Gimzewkski and co-workers report the detection of single mismatches in 12-mer DNA oligonucleotides and believe that cantilever setups could be potential rivals to DNA microarrays. A third approach for cantilever biosensors is described by Rudnitsky et al. [95]. Here, ferromagnetic tips are used in magnetic force microscope cantilever arrays. They describe the fundamental physics of the approach in their publication, which should allow the detection of single molecules. In a fourth approach, 0.05 nM concentrations of DNA and a single base-pair mismatch can be detected, by labeling DNA with gold particles and subsequent silver nucleation. A frequency shift upon this reaction is measured [96]. All four approaches for cantilever biosensors give an idea about the impressive sensitivity that can be achieved. However, besides the disadvantage of highly complex signal reporting mechanisms, these biosensors will suffer significantly from non-specific binding of interfering components in a sample and effort is required to purify a "real world sample" and ensure that the cantilever surface repels non-analyte molecules.

Biosensors based on optical fibers as transduction element have been recently reviewed by Wolfbeis [97]. Optical biosensors based on miniaturized SPR or on evanescent field monitoring are not as often found in miniaturized biosensors, especially in comparison to miniaturized electrochemical transducers, yet. Two examples will be given here: a miniaturized SPR biosensor by Cullen and co-workers [98] and an evanescent based microchip biosensor by Börchers and co-workers [99]. The best-known SPR biosensor is the BIAcore device from Pharmacia Company, Sweden. It has been on the market for over a decade and is routinely used for hybridization kinetic analyses, specificity analyses, etc. Cullen and co-workers have incorporated a commercially available miniaturized SPR transducer into a field analyzer and developed a competition and inhibition assay for an estrogenic compound in water samples that function as endocrine disrupting compounds (EDCs).

Low detection limits of 10 μg l^{-1} were established. Similar approaches for the miniaturization of SPR can be found in literature, however, similar to the miniaturized cantilever biosensor, any surface-active interfering compound in samples will cause significant analytical challenges. Börchers and co-workers used a microchip evanescent waveguide for the detection of real-time DNA hybridization events. A lower detection limit of 0.21 nmol l^{-1} was demonstrated. The authors also showed multi-analyte detection capabilities of their system and suggested that this strategy can be utilized in real-time DNA array format with analysis times as short as 2 min.

A word shall be spent on quantum dots, fluorescent semiconductor nanocrystals, which have found their way into optical biosensor research in recent years. A number of groups currently investigate, how these inorganic materials can be used in bioanalysis. The inorganic nanocrystals have a characteristic spectral emission, which is tunable to a desired wavelength by controlling the size of the crystal. In addition, the nanocrystals have narrow emission bandwidths with broad absorption bands so that a number of different particles can be excited using a common wavelength. This makes them to ideal probes for multiplexing. While their fluorescence characteristics are immensely interesting, they suffer from the inability to be coupled to biorecognition elements and special coating layers are required to make them water soluble, avoid aggregation and provide surfaces that allow bonding to biological molecules. Solving many of these problems, the inorganic nanoparticles find their way into biosensing. A few examples have shown that they can be used in immunoassays, ligand binding assays, DNA hybridization assays, etc. [100–102]. Goldman and colleagues have also shown, how universal "bio" quantum dots can be generated for biosensor use by coating the surface of CdSe–ZnS core shell quantum dots with either antibody or streptavidin molecules via an electrostatic/hydrophobic self assembly [103].

Electrochemical transducers have a clear advantage over optical transducers in respect to simple microfabrication and simple miniaturization techniques. While many "miniaturized" biosensors have optical detection mechanisms, these are seldomly incorporated into the microchip as mentioned previously. In contrast, a variety of suggested miniaturized biosensors based on electrochemical detection are found in current literature. One detailed fabrication example had been given above, for the preparation of IDUAs in our lab. It is used for the amperometric detection of reversible redox couples such as potassium ferro/ferrihexacyanide in a two-electrode setup. However, instead of describing a couple of examples in more detail, a large number of biosensors are listed below and it is suggested to the reader to refer to the publications for more detail. Bridan and co-workers demonstrated how DNA microarrays can be fabricated using 48 addressable microelectrodes [104], and Wilson and co-worker studied the limitation of independently addressable microbiosensor arrays [105]. Ertl and Mikkelsen use electrochemical biosensor arrays for the identification of microorganisms [106]. Thin-film biosensor arrays and the effect of thin films on the electrode response are investigated by a number of groups [107–111]. Not only ultramicroelectrodes but also nanoelectrodes are being developed [112,113]. Integration of traditional ELISA techniques with microchip-based detection is described by Girault and co-workers [114], as well as a traditional acetylcholine esterase assay with micro-flow injection biosensors by Rishpon and co-workers [115]. Fritsch and co-workers have shown, how a combination of enzyme amplification, electrodes, and microcavities can be used to detect nucleic acid sequences at extremely low levels, i.e., 146 nM of a synthetic target sequence were detected [116]. Finally, Wang gives an overview of the utilization of amperometric biosensors for clinical and therapeutic drug monitoring [117].

A chapter about miniaturized biosensors would not be complete, if not one paragraph would be spent on novel miniaturized glucose biosensors. As the oldest and commercially most successful biosensor, the glucose sensor has exceptional examples in current literature. Again, only a very few examples are picked. Research is carried out on new permselective films that can be used in miniaturized sensors [74]. Carbon-fiber nanosensors are developed for glucose detection [61]. Also, amperometric miniaturized glucose biosensors [117,118], and fiber optic biosensors are described [119]. More complex systems have multi-enzyme microsystems for real-time analysis of more than only glucose [120], are in a biosensor array format [121], or are in the format of combined microdialysis and detection system for continuous biomedical monitoring [122].

μTAS or "lab-on-a-chip" devices are the biosensor systems of the future as will be described further below. Some commercial μTAS can be found in today's market. These devices typically have a number of miniaturized analytical steps, but are operated using computers, pumps and sometimes even more complex workstations. These are very interesting devices, however, they are missing the true portability, ease of use, etc., that should be seen in a miniaturized biosensor system, and also in a μTAS. For a good overview of these commercial systems, please refer to Ref. [123]. Different from most other commercial systems is the i-STAT blood analyzer device. It is a single use, disposable cartridge that contains microfabricated thin film electrodes. Depending on the analyte, signals are measured with amperometric (glucose, pO_2), potentiometric (Na^+, K^+, Cl^-, ionized calcium, pH, pCO_2), or conductometric (hematocrit) circuits [124,125]. The very first microanalytical devices used CE in combination with mainly optical but also electrochemical detection; and CE is the technique most often found in current μTAS. Strictly seen, this is not a biosensor system, since most of the time no biorecognition element is involved. However, it was used as one of the first separation methods in miniaturized analytical systems since the separation efficiency in CE does not depend on capillary dimensions but mainly on the applied potential. Thus, a high number of theoretical plates can be achieved in a miniaturized CE device. Manz and Harrison pioneered CE in the early 1990s and published interesting work applying CE to μTAS [38,126–128]. Armstrong and co-workers have shown just recently a CE approach with up to 1,600,000 theoretical plates/m for the separation of bacteria [129]. A very good overview of miniaturized CE is given by Bossi et al. [130]. Combining CE with biological reactions results in a biosensor based μTAS. For example, the integration of miniaturized CE with a PCR in one microchip device is described by Lagally et al. [131]. They demonstrate the amplification and analysis of a single DNA molecule in an integrated microfluidic device. No sample preparation steps are included, however a PCR reaction, CE for DNA separation, thin film heaters, valves and hydrophobic vents were all included in the microdevice. Detection was done off-chip using a laser-excited confocal fluorescence detection system. Ramsey's group had shown earlier integration of PCR and CE with higher detection limits [132]. Lunte and co-workers have recently published a work on microchip CE with in-channel electrochemical detection [133]. These in-channel types of detection systems will be the direction into which research for miniaturized CE-biosensor systems will go in the future. Other researchers describe elements of μTAS such as catheter tip sensor chips for clinical applications [134], the segregation of micrometer-biosensors elements [135], and three-way microvalves for medical μTAS [136]. Our research group works on the development of μTAS modules with an electrochemical and optical microbiosensor as core module that is fabricated using soft lithography [81]. A review of most recent developments in the field of μTAS has been published by Manz and colleagues [137]. The review provides an overview not only of the different modules of a μTAS, but also applications including immunoassays and nucleic acid sequence-based detection.

As a final example molecular imprinted polymers (MIPs) are mentioned here. Mosbach and co-workers were among the pioneers of developing these polymers and applying them to biosensor analysis in the mid-1990s [138].

However, only in the last few years the field of MIPs grew significantly, spreading into areas of chemistry, biochemistry, material science and engineering and being applied to miniaturized biosensing systems. The polymers are either used as abiotic recognition element mimicking a biorecognition element or are immobilized in capillaries moving from CE to capillary chromatography. Preparation, stability and design toward new analytes are their clear advantages over biological recognition elements, and it will be seen in future years, if they can actually compete with their biological counterparts in bioanalysis. Examples are given in the following publications, ranging from MIPs for protein recognition [139] to MIPs applied in piezoelectric biosensors for aminopyrine detection [140], or applied in biomimetic bulk acoustic wave sensors for paracetamol in clinical samples [141], and to MIPs used in electrochemical biosensors [142,143].

6.6 CONCLUSIONS AND FUTURE OUTLOOK

It is still a long way for truly miniaturized biosensors and bioanalytical microsystems that fulfill all of the requirements of portability, reliability, accuracy, ruggedness, long-term stability, and ease of use. However, micro/nanofabrication techniques allowed in recent years a significant improvement of biosensor systems regarding these characteristics. As mentioned earlier, a distinction in miniaturized biosensors has to be made between a miniaturized flow system and the use of a miniaturized transducer. Only the integration of both can result in a microbiosensor with true advantages over its macro counter part. Besides decrease in reaction time and significant decrease of sample and solution volume, microbiosensors have the chance for in vivo studies by direct implanting in the body. While this has been demonstrated for glucose sensors, it is clear that many medically relevant biosensors will be developed as implantable sensor system in the near future. Similarly, material research will have a significant impact on future miniaturized biosensors. Hydrogels and sol–gel matrices are well known in the biosensor world, molecular imprinting recently had a new boost of research and the interest in "smart materials" for medical and security applications will become more imminent in the near future. Thus, miniaturization combined with novel biocompatible materials will lead this field into a new era.

The simple and inexpensive electrochemical detection systems are easier to miniaturize and included within the biosensor chip than their optical counterparts. They will thus lead to truly miniaturized systems in the near future. However, improvements in the optoelectronic technology will allow optical transducers to play an important role as well. Currently, though, they are typically utilized in their macroscopic versions such as the CCD camera microscope combination used for signal detection of DNA and protein microarrays. Miniaturized SPR or evanescent field transducers are suggested and fabricated, but still need computer controlled hardware to monitor signals generated [98,99]. Similarly, mass spectrometry and gas chromatography are targeted for miniaturization, which can then be used in conjunction with a bioanalytical separation or detection mechanism.

Miniaturized biosensors combined with miniaturized sample pretreatment steps will offer one of the true advantages of biosensors over other analytical and bioanalytical methods: portability. These future bioanalytical microsystems will have the size of a calculator, and will be much smaller and lighter than a laptop computer. Thus, research and development are needed not only for the miniaturized biosensor part of these bioanalytical microsystems but also for miniaturized sample pretreatment steps. The concentration of sample volumes (especially for environmental and food samples), the purification and extraction of the analyte from a complex sample are as important as the detection of the analyte itself. Finally, integrated pump systems that are independent from the sample components (such as pH and ionic strength), that require little energy and are stable and rugged are still needed.

Regarding the design of the entire bioanalytical microsystem, a modular system is easier to develop and fabricate than a complete system and thus more likely to be successful. Thus, instead of designing sample pretreatment, detection and pumping elements on one chip, each element will be designed separately as a module and subsequently integrated together to form the bioanalytical microsystem. A generic housing platform will then contain the respective "dropped in" modules. Thus, versatile bioanalytical microsystems can use the same principles, but can easily be adapted to their respective set of analytical problems. For example, a rather simplistic example, in the case of an immunoassay for environmental sample analysis, the sample is first purified from particulates via a microfilter. Subsequently, antibodies immobilized on magnetic beads are mixed with the filtered samples and are captured on a magnet. Bound analyte can then be detected. Using the same principle for RNA analysis, DNA probes are immobilized on magnetic beads, which are again captured on a magnet to quantify bound RNA. However, the filter is substituted by a cell lysis and RNA extraction module. Therefore, a variety of detection modules can be coupled with a variety of sample pretreatment modules, all in one miniaturized bioanalytical microsystem, hence, being the ultimate biosensor.

REFERENCES

1 D.R. Thévenot, K. Toth, R.A. Durst and G.S. Wilson, Electrochemical biosensors: recommended definitions and classification, *Pure Appl. Chem.*, 71 (1999) 2333–2348.
2 M. Madou, *Fundamentals of Microfabrication*. CRC Press, Boca Raton, FL, 1997.
3 G. Urban, Microstructuring of organic layers for microsystems, *Sens. Actuators, A*, 74 (1999) 219–224.
4 N.C. LaBianca, J.D. Gelorme, E. Cooper, E. O'Sullivan and J. Shaw, *High aspect ratio optical resist chemistry for MEMS applications*, JECS 188th Meeting, Chicago, IL, USA, 1995, pp. 500–501.
5 G. Engelmann, O. Ehrmann, J. Simon and H. Reichl, *Fabrication of high depth-to-width aspect ratio microstructures*, Proceedings, IEEE Microelectro Mechanical Systems, (MEMS'92), Travenmünde, Germany, 1992, pp. 3–98.
6 L. Martynova, L.E. Locascio, M. Gaitan, G. Kramer, R.G. Christensen and W.A. MacGrehan, Fabrication of plastic microfluid channels by imprinting methods, *Anal. Chem.*, 69 (1997) 4783–4789.
7 H. Becker and U. Heim, Hot embossing as a method for the fabrication of polymer high aspect ratio structures, *Sens. Actuators, A*, 83 (2000) 130–135.
8 J. Xu, L. Locascio, M. Gaitan and C.S. Lee, Room-temperature imprinting method for plastic microchannel fabrication, *Anal. Chem.*, 72 (2000) 1930–1933.
9 S.Y. Chou, P.R. Krauss and P.J. Renstrom, Imprint lithography with 25-nanometer resolution, *Science*, 272 (1996) 85–87.
10 S.R. Quake and A. Scherer, From micro- to nanofabrication with soft materials, *Science*, 290 (2000) 1536–1540.
11 A. Kumar, N.A. Abbott, E. Kim, H.A. Biebuyck and G.M. Whitesides, Patterned self-assembled monolayers and mesoscale phenomena, *Acc. Chem. Res.*, 28 (1995) 219–226.
12 E. Kim, Y.N. Xia and G.M. Whitesides, Polymer microstructures formed by molding in capillaries, *Nature*, 376 (1995) 581–584.
13 X.-M. Zhao, Y. Xia and G.M. Whitesides, Fabrication of three-dimensional microstructures: Microtransfer molding, *Adv. Mater.*, 8 (1996) 837.
14 Y. Xia, E. Kim, X.-M. Zhao, J.A. Rogers, M. Prentiss and G.M. Whitesides, Complex optical surfaces formed by replica molding against elastomeric substrates, *Science*, 273 (1996) 347–349.
15 Y. Xia, E. Kim and G.M. Whitesides, Micromolding of polymers in capillaries: applications in microfabrication, *Chem. Mater.*, 8 (1996) 1558–1567.
16 B. Löchel, A. Maciossek, H.J. Quenzer and B. Wagner, Ultraviolet depth lithography and galvanoforming for micromachining, *J. Electrochem. Soc.*, 143 (1996) 237–244.
17 R.M. McCormick, R.J. Nelson, M.G. Alonso-Amigo, D.J. Benvegnu and H.H. Hooper, Microchannel electrophoretic separations of DNA in injection-molded plastic substrates, *Anal. Chem.*, 69 (1997) 2626–2630.

18 V. White, R. Ghodssi, C. Herdey, D.D. Denton and L. McCaughan, Use of photosensitive polyimide for deep X-ray lithography, *Appl. Phys. Lett.*, 66 (1995) 2072–2073.
19 T. Lizotte and O. Ohar, Laser drilling speeds BGA packaging, *Solid State Technol.*, 39 (1996) 120–125.
20 D.L. Pugmire, E.A. Waddell, R. Haasch, J.J. Tarlov and L.E. Locascio, Surface characterization of laser-ablated polymers used for microfluidics, *Anal. Chem.*, 74 (2002) 871–878.
21 J.S. Rossier, A. Schwarz, F. Reymond, R. Ferrigno, F. Bianchi and H.H. Girault, Microchannel networks for electrophoretic separations, *Electrophoresis*, 20 (1999) 727–731.
22 F. Bianchi, Y. Chevolot, H.J. Mathieu and H.H. Girault, Photomodification of polymer microchannels induced by static and dynamic excimer ablation: effect on the electroosmotic flow, *Anal. Chem.*, 73 (2001) 3845–3853.
23 F. Bianchi, F. Wagner, P. Hoffmann and H.H. Girault, Electroosmotic flow in composite microchannels and implications in microcapillary electrophoresis systems, *Anal. Chem.*, 73 (2001) 829–836.
24 T.J. Johnson, D. Ross and L.E. Locascio, Rapid microfluidic mixing, *Anal. Chem.*, 74 (2002) 45–51.
25 M. Barbic, J.J. Mock, A.P. Gray and S. Schultz, Electromagnetic micromotor for microfluidic applications, *Appl. Phys. Lett.*, 79 (2001) 1399–1401.
26 B. He, B.J. Burke, X. Zhang, R. Zhang and F.E. Reginer, A picoliter-volume mixer for microfluidic analytical systems, *Anal. Chem.*, 73 (2001) 1942–1947.
27 A. Ajdari, Electro-osmosis on inhomogeneously charged surfaces, *Phys. Rev. Lett.*, 75 (1995) 755–758.
28 A.D. Stroock, M. Weck, D.T. Chiu, W.T.S. Huck, P.J.A. Kenis, R.F. Ismagilov and G.W. Whitesides, Patterning electro-osmotic flow with patterned surface charge, *Phys. Rev. Lett.*, 84 (2000) 3314–3317.
29 D. Erickson and D. Li, Influence of surface heterogeneity on electrokinetically driven microfluidic mixing, *Langmuir*, 18 (2002) 1883–1892.
30 N.A. Polson and M.A. Hayes, Controlling fluids in small places, *Anal. Chem.*, 73 (2001) 313A–319A.
31 M.W.J. Prins, W.J.J. Welters and J.W. Weekamp, Fluid control in multi-channel structures by electrocapillary pressure, *Science*, 291 (2001) 277–280.
32 B. Zhao, J.S. Moore and D.J. Beebe, Surface-directed liquid flow inside microchannels, *Science*, 291 (2001) 1023–1026.
33 N. Leventis and X. Gao, Magnetohydrodynamic electrochemistry in the field of Nd–Fe–B magnets. Theory, experiment, and application in self-powered flow delivery systems, *Anal. Chem.*, 73 (2001) 3981–3992.
34 P.G. Vahey, S.A. Smith, C.D. Costin, Y. Xia, A. Brodsky, L.W. Burgess and R.E. Synovec, Toward a fully integrated positive-pressure driven microfabricated liquid analyzer, *Anal. Chem.*, 74 (2002) 177–184.

35 T.E. McKnight, C.T. Culbertson, S.C. Jacobson and J.M. Ramsey, Electroosmotically induced hydraulic pumping with integrated electrodes on microfluidic devices, *Anal. Chem.*, 73 (2001) 4045–4049.
36 J.W. Jorgenson and K.D. Lukacs, Zone electrophoresis in open-tubular glass capillaries, *Anal. Chem.*, 53 (1981) 1298–1302.
37 A. Manz and J.C.T. Eijkel, Miniaturization and chip technology. What can we expect?, *Pure Appl. Chem.*, 73 (2001) 1555–1561.
38 D.J. Harrison, A. Manz, Z. Fan, H. Lüdi and H.M. Widmer, Capillary electrophoresis and sample injection systems integrated on a planar glass chip, *Anal. Chem.*, 64 (1992) 1926–1932.
39 S. Zeng, C.-H. Chen, J.C. Mikkelsen Jr. and J.G. Santiago, Fabrication and characterization of electroosmotic micropumps, *Sens. Actuators, B*, 79 (2001) 107–114.
40 L.E. Locascio, S.L.R. Barker, D. Ross, J. Xu, S. Robertson, M. Tarlov and M. Gaitan, *Fabrication and characterization of plastic microfluidic devices modified with polyelectrolyte multilayers*, Proceedings—electrochemical society, 2000-19 (Microfabricated Systems and MEMS V), 2000, pp. 72–79.
41 S.L.R. Barker, M.J. Tarlov, D. Ross, T. Johnson, E. Waddell and L.E. Locascio, *Fabrication, derivatization, and application of plastic microfluidic devices*, Proceedings of SPIE—The International Society for Optical Engineering, 4205 (Advanced Environmental and Chemical Sensing Technology), 2001, pp. 112–118.
42 A. Wego and L. Pagel, A self-filling micropump based on PCB technology, *Sens. Actuators, A*, 88 (2001) 220–226.
43 R. Zengerle, A. Richter and H. Sandmaier, *Proc. MEMS'92*, 1990, pp. 19–24.
44 O. Français and I. Dufour, Enhancement of elementary displaced volume with electrostatically actuated diaphragms: application to electrostatic micropumps, *J. Micromech. Microeng.*, 10 (2000) 282–286.
45 S. Shoji, M. Esashi, B.H. van der Schoot and N.J. de Rooij, A study of a high-pressure micropump for integrated chemical analyzing systems, *Sens. Actuators, A*, 32 (1992) 335–339.
46 S. Shoji and M. Esashi, Microfabrication and microsensors, *Appl. Biochem. Biotechnol.*, 41 (1993) 21–34.
47 R. Zengerle, J. Uhlrich, S. Kluge, M. Richter and A. Richter, A bidirectional silicon micropump, *Sens. Actuators, A*, 50 (1995) 81–86.
48 C.J. Morris and F.K. Forster, Optimization of a circular piezoelectric bimorph for a micropump driver, *J. Micromech. Microeng.*, 10 (2000) 459–465.
49 H. Andersson, W. van der Wijngaart, P. Nilsson, P. Enoksson and G. Stemme, A valve-less diffuser micropump for microfluidic analytical systems, *Sens. Actuators, B*, 72 (2001) 259–265.
50 Y. Suzuki, K. Tani and T. Sakuhara, Development of a new type piezoelectric micromotor, *Sens. Actuators, A*, 83 (2000) 244–248.
51 S. Böhm, W. Olthuis and P. Bergveld, An integrated micromachined electrochemical pump and dosing system, *J. Biomed. Microdevices*, 1 (1999) 121–130.

52 Q. Gong, Z. Zhou, Y. Yang and X. Wang, Design, optimization and simulation on microelectromagnetic pump, *Sens. Actuators, A*, 83 (2000) 200–207.
53 P. Wilding and L.J. Kricka, Mesoscale sperm handling devices. US Patent # 05427946, 1995.
54 B. He, L. Tan and F. Regnier, Microfabricated filters for microfluidic analytical systems, *Anal. Chem.*, 71 (1999) 1464–1468.
55 J.K. Tu, T. Huen, R. Szema and M. Ferrari, Filtration of sub-100 nm particles using a bulk-micromachined, direct-bonded silicon filter, *J. Biomed. Microdevices*, 1 (1999) 113–119.
56 P. Wilding, J. Pfahler, H.H. Bau, J.N. Zemel and L.J. Kricka, Manipulation and flow of biological fluids in straight channels micromachined in silicon, *Clin. Chem.*, 40 (1994) 43–47.
57 J. Cheng, R. Fortina, S. Sorrey, L.J. Kricka and P. Wilding, Microchip-based devices for molecular diagnosis of genetic diseases, *Mol. Diagn.*, 1 (1996) 183–200.
58 L.J. Kricka, P. Wilding and J. Cheng, HPCE 97, January 26–30th, Anaheim, CA, USA, 1997.
59 H.G. Craighead, Nanoelectromechanical systems, *Science*, 290 (2000) 1532–1535.
60 J. Min and A. Baeumner, Characterization and optimization of interdigitated ultramicroelectrode arrays as electrochemical biosensor transducers, *Electroanalysis*, 16 (2004) 724–729.
61 X. Zhang, J. Wang, B. Ogorevc and U.E. Spichiger, Glucose nanosensor based on Prussian-blue modified carbon-fiber cone nanoelectrode and an integrated reference electrode, *Electroanalysis*, 11 (1999) 945–949.
62 A.J. Bäumner and R.D. Schmid, Development of a new immunosensor for pesticide detection: a disposable system with liposome-enhancement and amperometric detection, *Biosens. Bioelectron.*, 13 (1998) 519–529.
63 J. Wang, M. Pumera, M.P. Chatrathi, A. Escarpa, R. Konrad, A. Griebel, W. Dorner and H. Lowe, Towards disposable lab-on-a-chip: poly(methylmethacrylate) microchip electrophoresis device with electrochemical detection, *Electrophoresis*, 23 (2002) 596–601.
64 K.J. Albert and D.R. Walt, Optical multibead arrays for simple and complex odor discrimination, *Anal. Chem.*, 73 (2001) 2501–2508.
65 S.P. Fodor, J.L. Read, M.C. Pirrung, L. Stryer, A.T. Lu and D. Solas, Light-directed, spatially addressable parallel chemical synthesis, *Science*, 251 (1991) 767–773.
66 S.P. Fodor, R.P. Rava, X.C. Huang, A.C. Pease, C.P. Homes and C.L. Adams, Multiplexed biochemical assays with biological chips, *Nature*, 364 (1993) 555–556.
67 A.C. Pease, D. Solas, E.J. Sullivan, M. Cronin, C.P. Holmes and S.P.A. Fodor, Light-generated oligonucleotide arrays for rapid DNA sequence analysis, *Proc. Natl Acad. Sci. USA*, 91 (1994) 5022–5026.

68 R. McGlennen, Miniaturization technologies for molecular diagnostics, *Clin. Chem.*, 47 (2001) 393–402.
69 D.A. Nivens and D.W. Conrad, Photoactive poly(ethylene glycol) organosilane films for site-specific protein immobilization, *Langmuir*, 18 (2002) 499–504.
70 X.-H. Liu, H.-K. Wang, J.N. Herron and G.D. Prestwich, Photopatterning of antibodies on biosensors, *Bioconjugate Chem.*, 11 (2000) 755–761.
71 M.S. Sanford, P.T. Charles, S.M. Commisso, J.C. Roberts and D.W. Conrad, Photoactivatable cross-linked polyacrylamide for the site-selective immobilization of antigens and antibodies, *Chem. Mater.*, 10 (1998) 1510–1520.
72 F.M. Andreopoulos, M.J. Roberts, M.D. Bentley, J.M. Harris, E.J. Beckman and A.J. Russell, Photoimmobilization of organophosphorus hydrolase within a PEG-based hydrogel, *Biotechnol. Bioeng.*, 65 (1999) 579–588.
73 K.G. Olsen, D.J. Ross and M.J. Tarlov, Immobilization of DNA hydrogel plugs in microfluidic channels, *Anal. Chem.*, 74 (2002) 1436–1441.
74 M. Quinto, I. Losito, F. Palmisano and C.G. Zambonin, Needle-type glucose microbiosensor based on glucose oxidase immobilised in an overoxidised polypyrrole film (an *in-vitro* study), *Fresenius J. Anal. Chem.*, 367 (2000) 692–696.
75 X. Chen, N. Matsumoto, Y. Hu and G.S. Wilson, Electrochemically mediated electrodeposition/electropolymerization to yield a glucose microbiosensor with improved characteristics, *Anal. Chem.*, 74 (2002) 368–372.
76 C. Kurzawa, A. Hengstenberg and W. Schuhmann, Immobilization method for the preparation of biosensors based on pH shift-induced deposition of biomolecule-containing polymer films, *Anal. Chem.*, 74 (2002) 355–361.
77 M. Mosbach, H. Zimmermann, T. Laurell, J. Nilsson, E. Csöregi and W. Schuhmann, Picodroplet-deposition of enzymes on functionalized self-assembled monolayers as a basis for miniaturized multi-sensor structures, *Biosens. Bioelectron.*, 16 (2001) 827–837.
78 A. Dodge, K. Fluri, E. Verpoorte and N.F. deRooij, Electrokinetically driven microfluidic chips with surface-modified chambers for heterogenous immunoassays, *Anal. Chem.*, 73 (2001) 3400–3409.
79 M.A. Hayes, N.A. Polson, A.N. Phayre and A.A. Garcia, Flow-based microimmunoassay, *Anal. Chem.*, 73 (2001) 5896–5902.
80 O.D. Velev and E.W. Kaler, In situ assembly of colloidal particles into miniaturized biosensors, *Langmuir*, 15 (1999) 3693–3698.
81 S. Kwakye and A. Baeumner, A microfluidic biosensor based on nucleic acid sequence recognition, *Anal. Bioanal. Chem.*, 376 (2003) 1062–1068.
82 M.U. Kopp, A.J. de Mello and A. Manz, *Science*, 280 (1998) 1046–1048.
83 M. Chee, R. Yang, E. Hubbell, A. Berno, X.C. Huang, D. Stern, J. Winkler, D.J. Lockhart, M.S. Morris and S.P.A. Fodor, Accessing genetic information with high-density DNA arrays, *Science*, 274 (1996) 610–614.
84 M.J. Kozal, N. Shah, N. Shen, R. Yang, R. Fucini, R. Merigan, D. Richman, D. Morris, E. Hubbell, M.S. Chee and T.G. Gingeras, Extensive polymorphisms

observed in HIV-1 clade B protease gene using high-density oligonucleotide arrays, *Nat. Med.*, 2 (1996) 753–759.
85 R.G. Sosnowski, E. Tu, W.F. Butler, J.P. O'Connell and M.J. Heller, Rapid determination of single base mismatch mutations in DNA hybrids by direct electric field control, *Proc. Natl Acad. Sci. USA*, (1997) 1119–1123.
86 N.P. Gerry, N.E. Witowski, J. Day, R.P. Hammer, G. Barany and F. Barany, Universal DNA microarray method for multiplex detection of low abundance point mutations, *J. Mol. Biol.*, 292 (1999) 251–262.
87 N.E. Witowski, C. Leiendecker-Foster, N.P. Gerry, R.C. McGlennen and G. Barany, Microarray-based detection of select cardiovascular disease markers, *Biotechniques*, 29 (2000) 936–939.
88 C. Debouck and P.N. Goodfellow, DNA microarrays in drug discovery and development, *Nat. Genet.*, 21 (1999) 48–50.
89 M. Schena, R.A. Heller, T.P. Theriaul, K. Konrad, E. Lachenmeier and R.W. Davis, Microarrays: biotechnology's discovery platform for functional genomics, *Trends Biotechnol.*, 16 (1998) 301–306.
90 M. Schena, *Microarray Analysis*. Wiley, New York, 2002, ISBN 0-471-41443-3.
91 M. Schena, *Protein Microarrays*. Jones and Bartlett Publishers, Sudbury, MA, 2004, ISBN 0-763-73127-7.
92 B. Ilic, D. Czaplewski, H.G. Craighead, P. Neuzil, C. Campagnolo and C. Batt, Mechanical resonant immunospecific biological detector, *Appl. Phys. Lett.*, 77 (2000) 450–452.
93 Z.J. Davis, G. Abadal, O. Kuhn, O. Hansen, F. Grey and A. Boisen, Fabrication and characterization of nanoresonating devices for mass detection, *J. Vac. Sci. Technol. B*, 18 (2000) 612–616.
94 J. Fritz, M.K. Baller, H.P. Lang, H. Rothuizen, P. Vettiger, E. Meyer, H.J. Guntherodt, C. Gerber and J.K. Gimzewski, Translating biomolecular recognition into nanomechanics, *Science*, 288 (2000) 316–318.
95 R.G. Rudnitsky, E.M. Chow and T.W. Kenny, Rapid biochemical detection and differentiation with magnetic force microscope cantilever arrays, *Sensors and Actuators A. Physical*, 83 (2000) 256–262.
96 M. Su, S. Li and V.P. Dravid, Microcantilever resonance-based DNA detection with nanoparticle probes, *Appl. Phys. Lett.*, 82 (2003) 3562–3564.
97 O.S. Wolfbeis, Fiber-optic chemical sensors and biosensors, *Anal. Chem.*, 76 (2004) 3269–3284.
98 A.M. Sesay and D.C. Cullen, Detection of hormone mimics in water using a miniaturized SPR sensor, *Environ. Monit. Assess.*, 70 (2001) 83–92.
99 C. Peter, M. Meusel, F. Grawe, A. Katerkamp, K. Cammann and T. Börchers, Optical DNA-sensor chip for real-time detection of hybridization events, *Fresenius J. Anal. Chem.*, 271 (2001) 120–127.
100 M.G. Bawendi, F.V. Mikulec and V.C. Sundar, Biological applications of quantum dots. US Patent # 160,454, 2001, pp. 32.

101 E.F. Duijs, F. Findeis, R.A. Deutschmann, M. Bichler, A. Zrenner, G. Abstreiter, K. Adlkofer, M. Tanaka and E. Sackmann, Influence of thiol coupling on photoluminescence of near surface InAs quantum dots, *Phys. Status Solidi B: Bas. Res.*, 224 (2001) 871–875.

102 S. Nie, M. Han and X. Gao, *Lab-on-a-bead: optically encoded microspheres for massively parallel analysis of genes and proteins*, Abstract Paper—American Chemical Society, 221st IEC-019, 2001.

103 E.R. Goldman, E.D. Balighian, M.K. Kuno, S. LaBrenz, P.T. Tran, G.P. Anderson, J.M. Mauro and H. Mattoussi, Luminescent quantum dot-adaptor protein–antibody conjugates for use in fluoroimmunoassays, *Phys. Status Solidi B*, 229 (2002) 407–414.

104 G. Bidan, M. Billon, K. Galasso, T. Livache, G. Mathis, A. Roget, L.M. Torres-Rodriguez and E. Vieil, Electropolymerization as a versatile route for immobilizing biological species onto surfaces: application to DNA biochips, *Appl. Biochem. Biotechnol.*, 89 (2000) 183–193.

105 P. Yu and G.S. Wilson, An independently addressable microbiosensor array: what are the limits of sensing element density?, *Faraday Discuss.*, 116 (2000) 305–317.

106 P. Ertl and S.R. Mikkelsen, Electrochemical biosensor array for the identification of microorganisms based on lectin-lipopolysaccharide recognition, *Anal. Chem.*, 73 (2001) 4241–4248.

107 I. Moser, G. Jobst, P. Scasek, M. Varahram and G. Urban, Rapid liver enzyme assay with miniaturized liquid handling system comprising thin film biosensor array, *Sens. Actuators, B*, 44 (1997) 277–380.

108 A. Steinschaden, D. Adamovic, G. Jobst, R. Glatz and G. Urban, Miniaturized thin film conductometric biosensors with high dynamic range and high sensitivity, *Sens. Actuators, B*, 44 (1997) 365–369.

109 V. Tvarozek, T. Hianik, I. Novotny, V. Rehacek, W. Ziegler, R. Ivanic and M. Andel, Thin films in biosensors, *Vacuum*, 50 (1998) 251–262.

110 V. Tvarozek, I. Novotny, V. Rehacek, R. Ivanic and F. Mika, Thin-film electrode chips for microelectrochemical sensors, *Nexus Res. News*, 1 (1998) 15–17.

111 V. Yegnaraman and A.M. Yacynych, Enzyme immobilization on ultramicroelectrodes through electropolymerization: effect of polymeric film thickness on the amperometric response of the electrode, *Bull. Electrochem.*, 16 (2000) 21–24.

112 O.D. Velev, D. Orlin and E.W. Kaler, *Miniaturized biosensors by in situ assembly of colloidal particles onto micropatterned electrodes*, Book of Abstracts, 219th ACS National Meeting, San Francisco, CA, March 26–30, 2000, 2000.

113 P. Van Gerwen, W. Laureyn, A. Campitelli, P. Jacobs, P. Detemple, K. Baert, W. Sansen and R. Mertens, Cost effective realization of nanoscaled interdigitated electrodes, *J. Micromech. Microeng.*, 10 (2000) N1–N5.

114 J.S. Rossier and H.H. Girault, Enzyme linked immunosorbent assay on a microchip with electrochemical detection, *Lab Chip*, 1 (2001) 153–157.

115 T. Neufeld, I. Eshkenazi, E. Cohen and J. Rishpon, A micro flow injection electrochemical biosensor for organophosphorus pesticides, *Biosens. Bioelectron.*, 15 (2000) 323–329.

116 Z.P. Aguilar and I. Fritsch, Immobilized enzyme-linked DNA hybridization assay with electrochemical detection for *Cryptosporidium parvum* hsp70 mRNA, *Anal. Chem.*, 75 (2003) 3890–3897.

117 J. Wang, Amperometric biosensors for clinical and therapeutic drug monitoring: a review, *J. Pharm. Biomed. Anal.*, 19 (1999) 47–53.

118 M.A. McRipley and R.A. Linsenmeier, Fabrication of a mediated glucose oxidase recessed microelectrode for the amperometric determination of glucose, *J. Electroanal. Chem.*, 414 (1996) 235–246.

119 Z. Rosenzweig and R. Kopelman, Analytical properties of miniaturized oxygen and glucose fiber optic sensors, *Sens. Actuators, B*, 35–36 (1996) 475–483.

120 J. Perdomo, H. Hinkers, C. Sundermeier, W. Seifert, O. Maertinez Morell and M. Knoll, Miniaturized real-time monitoring system for L-lactate and glucose using microfabricated multi-enzyme sensors, *Biosens. Bioelectron.*, 15 (2000) 515–522.

121 G. Jobst, I. Moser, P. Svasek, V. Varahram, Z. Trajanoski, P. Wach, P. Kotanko, F. Skrabal and G. Urban, Mass producible miniaturized flow through a device with a biosensor array, *Sens. Actuators, B*, 43 (1997) 121–125.

122 M.M. Rhemrev-Boom, J. Korf, K. Venema, G. Urban and P. Vadgama, A versatile biosensor device for continuous biomedical monitoring, *Biosens. Bioelectron.*, 16 (2001) 839–847.

123 P. Mitchell, Microfluidics—downsizing large-scale biology, *Nat. Biotechnol.*, 19 (2001) 717–721.

124 T. Mock, D. Morrison and R. Yatscoff, Evaluation of the i-STAT system: a portable chemistry analyzer for the measurement of sodium, potassium, chloride, urea, glucose, and hematocrit, *Clin. Biochem.*, 28 (1995) 187–192.

125 i-STAT http://www.istat.com/products/docs/biosenso.pdf, 2002.

126 A. Manz, D.J. Harrison, E. Verpoorte, J.C. Fettinger, A. Paulus, H. Lüdi and H.D. Widmer, Planar chips technology for miniaturization and integration of separation techniques into monitoring systems—capillary electrophoresis on a chip, *J. Chromatogr.*, 593 (1992) 253–258.

127 A. Arora, A.J. de Mello and A. Manz, Sub-microliter electrochemiluminescence detector—a model for small volume analysis systems, *Anal. Commun.*, 34 (1997) 393–395.

128 A. Arora, J.C.T. Eijkel, W.E. Morf and A. Manz, A wireless electrochemiluminescence detector applied to direct and indirect detection for electrophoresis on a microfabricated glass device, *Anal. Chem.*, 73 (2001) 3282–3288.

129 D.W. Armstrong, G. Schulte, J.M. Schneiderheinze and D.J. Westenberg, Separating microbes in the manner of molecules. 1. Capillary electrokinetic approaches, *Anal. Chem.*, 71 (1999) 5465–5469.

130 A. Bossi, S.A. Piletsky, E.V. Piletski, P.G. Righetti and A.P.F. Turner, Capillary electrophoresis coupled to biosensor detection, *J. Chromatogr. A*, 892 (2000) 143–153.

131 E.T. Lagally, I. Medintz and R.A. Mathies, Single molecule DNA amplification and analysis in an integrated microfluidic device, *Anal. Chem.*, 73 (2001) 565–570.

132 L.C. Waters, S.C. Jacobson, N. Kroutchinina, J. Khandurina, R.S. Foote and J.M. Ramsey, Multiple sample PCR amplification and electrophoretic analysis on a microchip, *Anal. Chem.*, 70 (1998) 5172–5176.

133 R.S. Martin, K.L. Ratzlaff, B.H. Huyn and S.M. Lunte, In-channel electrochemical detection of microchip capillary electrophoresis using an electrically isolated potentiostat, *Anal. Chem.*, 74 (2002) 1136–1143.

134 P. Bergveld, Bedside clinical chemistry: from catheter tip sensor chips towards micro total analysis systems, *J. Biomed. Microdevices*, 2 (2000) 185–195.

135 S.A. Brooks, N. Dontha, C.B. Davis, J.K. Stuart, G. O'Neill and W. Kuhr, Segregation of micrometer-dimension biosensor elements on a variety of substrate surfaces, *Anal. Chem.*, 72 (2000) 3253–3259.

136 T. Ohori, S. Shoji, K. Miura and A. Yotsumoto, Partly disposable three-way microvalve for a medical micro total analysis system (μTAS), *Sens. Actuators, A*, 64 (1998) 57–62.

137 T. Vilkner, D. Janasek and A. Manz, Micro total analysis systems: recent developments, *Anal. Chem.*, 76 (2004) 3373–3386.

138 O. Brüggemann, K. Haupt, L. Ye, E. Yilmaz and K. Mosbach, New configurations and applications for molecularly imprinted polymers, *J. Chromatogr. A*, 889 (2000) 15–24.

139 A. Bossi, S.A. Piletsky, P.G. Righetti and A.P.F. Turner, Surface-grafted molecularly imprinted polymers for protein recognition, *Anal. Chem.*, 73 (2001) 5281–5286.

140 Y. Tan, L. Nie and S. Yao, A piezoelectric biomimetic sensor for aminopyrine with a molecularly imprinted polymer coating, *Analyst*, 126 (2001) 664–668.

141 Y. Tan, Z. Zhou, P. Wang, L. Nie and S. Yao, A study of a biomimetic recognition material for the BAW sensor by molecular imprinting and its application for the determination of paracetamol in the human serum and urine, *Talanta*, 55 (2001) 337–347.

142 C. Malitesta, I. Losito and P.G. Zambonin, Molecularly imprinted electrosynthesized polymers: new materials for biomimetic sensors, *Anal. Chem.*, 71 (1999) 1366–1370.

143 S. Zhang, G. Wright and Y. Yang, Materials and techniques for electrochemical biosensor design and construction, *Biosens. Bioelectron.*, 15 (2000) 273–282.

Chapter 7

New materials for biosensors, biochips and molecular bioelectronics

Daniel Andreescu, Silvana Andreescu and Omowunmi A. Sadik

7.1 INTRODUCTION

During the last two decades, advances in materials chemistry have remarkably influenced the design of analytical sensor and biosensor devices. Recent progress in biosensor technology involves interdisciplinary studies including physics, chemistry and surface chemistry, biology, biochemistry, material science, nanotechnology and computer science. Due to the need for rapid, cheap and easy-to-use methods, biochemical sensors have emerged as a dynamic technique for qualitative and quantitative determination of different analytes, important in many areas of environmental, clinical, agricultural, food or military investigations. Biosensor devices consist of *a biological recognition system* (e.g., enzyme, antibody, nucleic acid, cells, tissue) in conjunction with a *physicochemical transducer*, which converts biological recognition event into an electronic signal (Fig. 7.1). Biosensors are expected to possess several characteristics that make them comparable or even better than traditional analytical systems. They must be simple to handle, small, cheap and able to provide reliable information in real time [1–7]. The exceptional combination of a biological element in direct contact with a physical transducer makes it possible to fulfill all these requirements.

When designing biosensors, it is essential to study and understand each component that constitutes this complex system as well as all the factors that influence its unique performance and limitations [5,6]. Despite an important number of publications in the biosensor field, several aspects require further optimization and improvements; many of these could be related to inadequate materials and insufficient understanding of the underlining mechanism. Since the properties and the type of the material used are largely connected with the transducer and the transducer/detector interface, a considerable attention must be paid to the nature of electrode material. The rapid

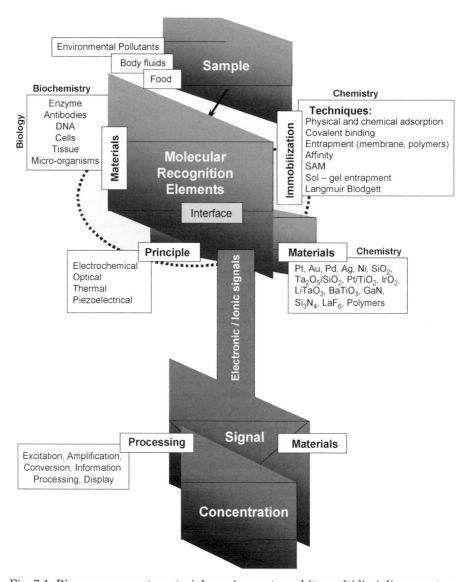

Fig. 7.1. Biosensor concept, material requirements and its multidisciplinary nature.

technological development of new electrode materials has expanded the range and classes of detectable compounds. Biosensor devices are fabricated from a large variety of materials, essential not only for simplifying assay, but also for ensuring the requirements of reproducibility, biocompatibility, cost and suitability for mass production.

Novel materials are thus needed to improve the mechanical and chemical stability of the sensor for practical applications in various conditions and, on the other hand, to improve the immobilization scheme in order to ensure sensor stability and the spatial control of biomolecules. The most important materials for chemical and biochemical sensors include organic polymers, sol–gel systems, semiconductors and other various conducting composites. This chapter reviews the state-of-the-art biosensing materials and addresses the limitations of existing ones.

7.2 PROPERTIES OF IDEAL MATERIALS FOR BIOSENSORS AND BIOELECTRONICS

Recently, an emerging area of research is the implementation of new materials for the development of a novel class of biosensors, biochips, nanosensors and other transducers. Many efforts are focused on finding the ideal material to ensure optimal characteristics of the sensor system. Basically, in the construction of biosensors there are several requirements for selecting an electrode material: (i) biocompatibility with the biological element, (ii) absence of diffusion barriers, (iii) stability with changes in temperature, pH, ionic strength or macroenvironment, (iv) sufficient sensitivity and selectivity for the analyte of interest as well as (v) low cost and ease of mass production. Moreover, these materials should either possess the necessary functional groups needed for the attachment of biomolecules or they should be able to be easily functionalized. Some sensors require operation in harsh conditions and, in this case, materials with special mechanical and chemical resistance, almost inert, are needed. Additionally, an ideal electrode material must be characterized by a good conductivity to ensure a rapid electron transfer.

7.3 FUNCTIONALIZATION OF ELECTRODE MATERIALS

The basic requirement in biosensor development is ascribed to the successful attachment of the recognition material, a process governed by various interactions between the biological component and the sensor interface. Advanced immobilization technologies capable of depositing biologically active material onto or in close proximity of the transducer surface have been reported. In this context, the choice of a biocompatible electrode material is essential. The material surfaces (support) include almost all material types: metals, ceramics, polymers, composites and carbon materials [8]. In most cases, when a native material does not meet all the requirements for bio-immobilization, further functionalization steps are necessary. Thus, an appropriate selection of the electrode material and/or functionalization method is dictated by several parameters such as the nature of biomolecules and/or support, detection technique and/or the final application of the biosensing device.

Native material surfaces can serve as support matrices for biomolecule attachment. This is achieved almost exclusively through *physical adsorption* via the weak bonds such van der Waals forces, ionic binding or hydrophobic interactions [4,9]. However, such biosensors are characterized by poor analytical characteristics especially with regard to operational and storage stability. In addition, adsorbed biomaterials are subject to variation when exposed to changes in pH, temperature and ionic strength. Despite these major disadvantages, this process has been successfully used to couple proteins to various solid supports including glass, plastics and rubber. The binding of biological components could be monitored directly using QCM and SAW devices, by correlating the changes in the frequency to the amount of substances adsorbed onto the surface [10]. SPR and ellipsometry techniques are also used to analyze physical adsorption process when the refractive index at the thin metal film surface is changed as a result of biomolecule adsorption [11–13]. Due to its simplicity and low cost, the direct adsorption of proteins onto polystyrene wells for use in enzyme-linked immunosorbent assays (ELISAs) constitutes the most common immobilization method [4].

Functionalization of surfaces with photopolymerized monomers or sol–gel matrices has also been used. In this case, the biomolecules can be *physically entrapped* within the supporting matrix. The method is applicable for almost all types of surfaces, it is easy to handle, and the large spectrum of monomer precursors commercially available permits the immobilization of basically all biological elements. In this case, the absence of chemical bond formation helps to preserve the activity of the bioreagent during the immobilization process. However, several drawbacks such as leaking of biocomponent and possible diffusion barriers restrict the performance of biosensor devices fabricated using this procedure.

7.3.1 Self-assembled monolayers

Noble metal surfaces, particularly gold, have been usually functionalized using *self-assembled monolayer* (SAM) techniques [14]. The formation of stable and aligned packed monolayers is carried out directly onto the transducer surface via bifunctional molecules. Two methods are used to deposit molecular layers onto solid substrates such as glass or metal: Langmuir–Blodgett and self-assembly [15]. The most studied self-assembled modified materials have been developed by monolayer formation involving *strong adsorption* of alkenesilanes or disulfides (R–S–S–R), sulfides (R–S–R) and thiols (R–SH) on the metal surfaces (notably gold). Sulfur donor atoms coordinate strongly on the gold support while the van der Waals forces between methylene groups orient and stabilize the monolayer. The method is generally developed with a long-chain (C_{12} or even higher) *n*-alkylthiols or silanes with different functional groups that are easily linked to the gold substrate, thus utilizing the strong affinity of thiol groups for gold [15–17]. For alkenethiols ($n > 5$) simple adsorption stops at a monolayer level. The formation of a multilayer has been observed after 6 days of immersion [18]. The use of a shorter chain length leads to less stable and ordered structures [19]. SAMs can also be obtained using an *electrodeposition* procedure. In this case, the monolayer formed onto the surface can be reversibly adsorbed and desorbed by applying an electrical potential [20].

In practical applications, the protocol for attaching short nucleic acid probe, proteins and oligonucleotides ranges from the self-assembly of thiols or disulfide moieties on gold, to the use of adjunct layers of polymers such as polyacrylamide [21,22] or polypyrrole [23]. For instance, the use of SAM with further covalent binding of proteins [24] and oligonucleotides [25] onto gold piezoelectric quartz crystal has been investigated. Then, the analytical characteristics of the resulting immunosensors have been compared with another procedure such as simple adsorption and avidin/biotin coupling. SAMs are also useful as materials in nanotechnology, membrane biochemistry, biophysics and cell biology [14].

Modifying a sensor surface with SAMs generates various functionalized materials with specific properties or functions that are advantageous in electrochemical, SPR and (E)QCM sensors. This method ensures the orientation and spatial control of the biomolecules in the immobilization process, as well as the absence of diffusion barriers. These are important parameters that make SAM a convenient and promising approach for metal surface functionalization. Despite these advantages, the reproducible layering of biomolecules remains a major problem in the development of self-assembled biomaterials. This has been attributed to the difficulties in achieving homogenous surface coverage, reliable orientation of the molecules and accessibility of the major functional groups (e.g., base pairs in nucleic acids) for optimal sensing surface [26].

7.3.2 Biomaterials developed by covalent binding

The most employed method for developing stable materials for biosensor application is via *covalent attachment*. There are two routes to carry out this process: (i) the first involves direct immobilization of biomolecules onto the transducer surface that contains residual groups (carboxyl, aldehyde, hydroxyl). Despite its simplicity, this procedure is characterized by poor reproducibility. (ii) The second and mostly used route involves the derivatization of the transducer surface with a cross-linker such as glutaraldehyde, carbodiimide/succinimide or avidin/biotin pathways [27]. Commercial pre-activated membranes are available and can be utilized as support to immobilize biomolecules.

Biosensors developed by covalent bio-functionalization are generally characterized by good stability and a short response time. However, the covalent procedure is more complex, requiring special experimental conditions (e.g., pH, ionic strength), is difficult to reproduce and, in most cases, the amount of the biocomponent is very high. This technique is only suitable for functionalized membranes or onto materials that allow the attachment of functional groups. Another disadvantage can be the toxicity of cross-linkers, which can induce a partial denaturation of the biomolecule. The most utilized cross-linkers are carbodiimides (e.g. *N*-ethyl-[*N*-dimethylaminopropyl]carbodiimide (EDC)) and glutaraldehyde (GA) [1,28–30]. The carbodiimide reacts with activated carboxyl groups on the support and then with the amine group of biomolecules. In many cases, the EDC is utilized together with *N*-hydroxysuccinimide (NHS) that permits the formation of a more stable complex. This procedure is mostly used in commercial SPR biosensing device for developing biofunctionalized materials [11]. GA is coupled via amino groups of the support, generating an anchoring that contains an aldehyde active group (Schiff's base) capable of binding to the amino groups of biomolecules. Other coupling agents are available including cyanobromide, epibromohydrin and epichlorohydrin. These can be used to modify the electrode material surface via other groups such hydroxyl. In biosensor development this approach has been utilized to immobilize various enzymes such as: tyrosinase via EDC and GA [31,32], glutathion-peroxidase [33] and acetylcholinesterase with GA [34,35] or EDC [36], lactate oxidase [37] with GA. Amino-modified oligonucleotide probe via covalent attachment has also been used for manufacturing a DNA microarray onto glass and silicon surfaces using 1,4 phenylene-diisothiocyanate (PDITC) as cross-linker agent [38].

Covalent functionalization of glass surfaces can be achieved via mercaptosilane that allowed subsequent attachment of a bifunctional succinimidyl cross-linker (Fig. 7.2). For instance, a glass capillary tube modified in this way was then used to covalently attach a dsDNA molecule for a fluorescence biosensor [39]. Approaches for monitoring the interactions of DNA with other ligands depend largely on hybridization reactions [40–42]. Similar procedure could be used to immobilize different types of antibodies onto the surface of identical capillary [43]. In this case, avidin was covalently attached to the internal surface of Tygon capillary and then treated with the appropriate biotin conjugate capture antibodies. The major advantage of the waveguide properties of the capillary is to integrate the signal over an increased surface area without simultaneously increasing the background noise from the detector [40].

A gold quartz crystal surface can be easily functionalized with an avidin monolayer (Fig. 7.3) that will then allow immobilization of any biotinylated biomolecule through strong avidin–biotin interaction. The only requirement for the development of a successful biosensor device is the availability of biotin-tagged molecules. We have used this strategy to achieve DNA electrode materials without the need for hybridization [44–46]. The dsDNA serial assembly was achieved by first linearizing circular plasmid

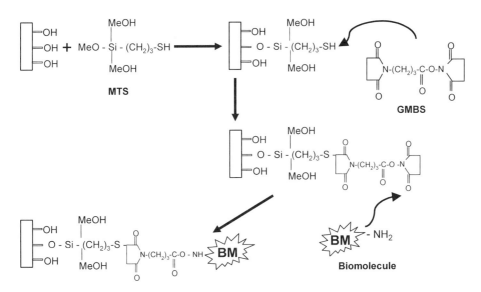

Fig. 7.2. Functionalization of a silica surface using 3-mercaptopropyl trimethoxysilane (MTS) and the bifunctional agent N-(g-maleimidobutyryloxy)-succinimide ester (GMBS) and immobilization of biomolecules (BM).

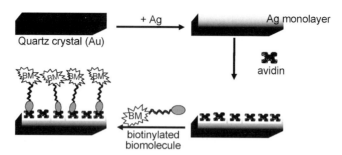

Fig. 7.3. Schematic representation of immobilization of biotinylated reagents via an avidin monolayer deposited onto the Au surface of a quartz crystal.

dsDNA using *Bam* HI endonuclease enzyme followed by bisulfite-catalyzed transamination reaction in order to biotinylate the dsDNA. This DNA material was used as a ligand at gold electrodes containing avidin. The immobilized dsDNA material was used for the determination of small molecular weight organics such as cisplatin, daunomycin, bisphenol-A and chlorinated phenols [45]. Also, covalent attachment onto an Au or Pt electrode surface through the amino bond using 3-aminopropyltriethoxysilane (APTS) was used to construct an immunosensor for anti-cyanazine detection (Fig. 7.4) [47]. Despite poor reproducibility, biomaterials developed by covalent binding are currently the most used functionalization procedure for biosensors.

7.4 CLASSES OF MATERIALS FOR BIOSENSORS

In order to develop a reliable biosensor device, a straightforward engineering materials approach must be considered. Commonly used coating materials include polymers, metals and alloys, ceramics, inorganic semiconductors,

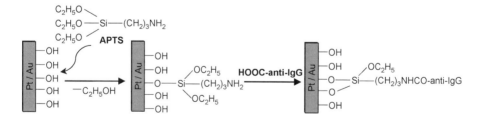

Fig. 7.4. Functionalization of Pt/Au surface via 3-aminopropyltriethoxysilane (APTS) for the development of an immunosensor for the detection of anti-cyanazine.

molecular crystals, membranes, composites and are usually thermodynamically stable [48].

7.4.1 Polymers

7.4.1.1 Conductive polymers
Polymers were the first materials in sensor development for the purpose of increased sensitivity and selectivity. Conducting electroactive polymers (CEPs) are the most used in advanced biosensors and are recognized for their biocompatibility, allowing the microfabrication of low cost and disposable devices [49]. CEPs have been described and used for their remarkable sensing characteristics, by being able to be reversibly oxidized or reduced after applying an electrical potential [47,50–54]. Generally, polymeric films are developed at fixed potentials or by cyclic voltammetry. In this case, the growth of the resulting conducting layer corresponds to an increase in the current intensity, suggesting the electrochemical deposition of an electroactive polymer. Therefore, their morphology depends upon the deposition condition, which may affect the final biosensor characteristics. In this direction, a variety of different biological components and transduction mechanisms have been developed depending on particular application [55–59].

Several techniques are available for applying polymer films: solvent casting, spin coating, chemical polymerization and electropolymerization [60]. Of these, electrochemical polymerization is the most common. This involves entrapment of biomolecules in organic polymers during electrochemical oxidation of an aqueous monomer solution. The immobilization of biomolecules in electropolymerized films allows the easy electrochemical control of various parameters such as the thickness of polymeric layer, the biocomponent loading or location [56]. Also, it is possible to generate a polymer coating over a small electrode surface with complex geometry. The principal advantage of electropolymerized materials is that the resulting materials are easily prepared in a single step procedure (galvanostatic, potentiodynamic). However, the high concentration of monomer and biomolecule required during the electropolymerization process is a major limitation of this procedure. In addition, the small quantity of molecular recognition elements that remains active can be affected by the shift of the pH value, which can occur near the electrode surface during the electrodeposition process.

Due to the low oxidation potential required for its synthesis, pyrrole (PPy) has been mostly used for biosensing applications compared to other conducting polymers. The ease of adding a biomolecule into the monomer solution for subsequent entrapment simplifies the biosensor fabrication process. Basically, almost all pyrrole-based materials have been obtained following an electrochemical procedure. A typical graph registered during electropolymerization of PPy using a galvanostatic method is presented in Fig. 7.5. However, in view of mass production and practical bioanalytical techniques, pyrrole-based bioimmobilization techniques require chromatographic purification/separation of the monomer and deoxygenation of the polymerization solution. Most importantly, the nucleophilic groups present on the biomolecule to be immobilized sometimes quench the polymerization process.

Another approach to deposit conducting polymers can be achieved by photochemical polymerization of the monomer precursors. This procedure provides a means by which different composites (metals and/or various alloy materials with or without biomolecules) can be deposited from an electrolyte onto a non-conducting surface. Such a procedure was optimized and applied for polymerization of pyrrole in the presence of metal nanoparticles [61]. Photopolymerized films containing metals analyzed by environmental scanning electron microscopy (SEM) appeared to be typical of amorphous polypyrrole in which bright Ag particles were found on the surface (Fig. 7.6).

Conducting polymer films are commonly characterized using cyclic voltammetric techniques, while their morphology and the presence of specific functional groups are achieved using SEM, NMR and FTIR spectroscopy.

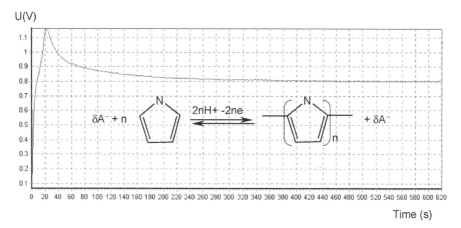

Fig. 7.5. Galvanostatic electrodeposition of PPy containing anti-IgG onto gold electrode (0.5 M Py, 0.1 mg ml^{-1} anti-IgG in PBS, pH 7.2, current density = 1 mA cm^{-2}).

New materials for biosensors, biochips and molecular bioelectronics

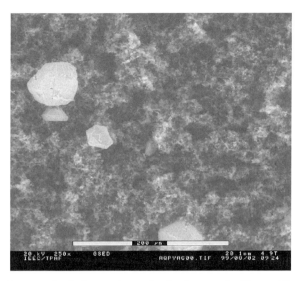

Fig. 7.6. SEM micrograph of photopolymerized polypyrrole film on fiberglass. Electron dispersive elemental analysis confirms that bright spots are silver. Scale bar is 200 μm.

Such structural information could provide strong arguments concerning the chemical structures of the polymers and can be extremely important for bioimmobilization purposes. In addition, it offers valuable information about the distribution and geometry of the resulting layer. On the other hand, the stability of this film and the active surface area strongly affect the performance (especially reproducibility and stability) of the final biosensor device, particularly when electrochemical transduction schemes are employed. Such parameters can be determined using electrochemical procedures.

Applications. Polyaniline (PAni) [62,63], polypyrrole (PPy) [64,65], polythiophene (PTf) [66] and their derivatives have been of much interest for scientific applications. Immobilization of glucose oxidase (GOX) in a PPy matrix represents the first biosensor containing biomolecules entrapped in an electropolymerized film [67]. The influence of polymerization conditions, the effect of film thickness on activity and stability of the immobilized enzyme have been studied in detail [68,69]. In another application, anti-IgG was successfully entrapped onto a PPy-modified Pt support obtained by electrodeposition of Py following a galvanostatic procedure (Fig. 7.4) [47]. The mechanism involves the interaction between the negatively charged Ab at neutral pH, plus a delocalized positive charge along CEPs chain, and the adsorption of the antigen [53]. By coupling conducting polymers and pulsed electrochemical detection (PED) systems to a flow injection analysis (FIA), we

295

generate a unique approach for obtaining practically useful Ab–Ag signals [70,71]. In this case, regeneration step is achieved by using oscillating potential without denaturation of the immobilized antibodies, while the assays performed were more than in the cases of other regeneration protocols involving change in pH [72].

7.4.1.2 Functionalized polymers
Functionalized polymer films represent another class of materials used in biosensor design, suitable for improving their analytical characteristics. In this case, manufacturing of biosensors may involve a two-step procedure: (i) direct (electro)polymerization of monomers containing functional groups resulting in a complex of functionalized polymers, followed by (ii) (electro)chemical post-functionalization of deposited conductive polymer films [73,74]. The main advantage of this sequential procedure, namely, *electropolymerization and binding* lies in the possibility of using optimal condition for each step. More advanced approach involves the electrodeposition of polymers that are functionalized by amino groups [74–76] followed by subsequent coupling of biomolecules via the activation of the carboxylic groups of the protein via carbodiimide cross-linker.

Based on the high affinity of avidin–biotin interaction (association constant $K_a = 10^{15}$ M^{-1}), a new class of biotinylated polymer materials is being developed. Cosnier discussed in detail the attachment of bimolecular groups to biotinylated polymers through avidin–biotin affinity interactions [56]. Basically, this procedure involves the formation of the biotinylated PPy, subsequently used as anchoring points for the immobilization of avidin units. These grafting units can be regenerated by breaking the biotin–avidin bridge, thus allowing the reuse of the matrix [77]. The method allows the control of polymer/biomolecule layer, but could also lead to the denaturation of the active element during the immobilization process while the immobilized biomolecules are restricted to a monolayer at the interface polymer–solution [56]. The method is suitable for the fabrication of biosensors with small surface area, opening new ways for the development of microbiosensors. Also, the commercial availability of a variety of biotinylated biomolecules and avidin-conjugated biomolecules offers a great future for this method.

Applications. The most common application of electrofunctional polymers is in the development of new materials in the (bio)sensors field [78]. For instance, a pyrrole-modified biotin allows successive attachment of avidin and biotin-labeled glucose oxidase, resulting in an efficient glucose biosensor [77,79]. Oureghi et al. [80] developed an impedimetric immunosensor for

IgG detection based on the electrostatic binding between the terminal cyano group of a N-substituted PPy bearing CN and the hydroxyl groups on the heavy chains of the Ab.

7.4.1.3 Polymeric sorbents

Polymeric sorbents with selectivity for a particular analyte or groups of similar analytes have been developed as biosensing materials using molecular imprinting technique [81]. This is a copolymerization process between monomers of desired functional groups and the target analyte, which acts like a molecular template. Initially, the functional monomer forms a complex with the imprinted molecule and, following the polymerization process, its functional groups are held by the highly cross-linked polymeric structure. The presence of a molecule of interest during the polymerization process allows its entrapment within the resulting rigid polymeric network, with high specificity. After polymerization, the imprinted molecules can be removed from the polymer by simple solvent extraction or chemical cleavage. The resulting polymeric bridge shows the sites that are complementary in size and shape to the analyte. In this way, a molecular memory is introduced into the polymer capable of selective rebinding to the analyte.

The majority of molecularly imprinted polymers used in biosensors are synthesized by radical polymerization from monomers containing acryl or vinyl groups that interact with a templating ligand to create low energy interactions, with concomitant polymerization. Among these, methyl methacrylate, 4-vinylpyridine and N-vinyl-α-pyrrolidone led to polymers from which the imprinted molecule can be rapidly and quantitatively extracted. Polymers generated from methacrylic acid and trifluoromethylacrylic acid retained the template more strongly [82]. For the design of materials based on the molecular-imprinted polymers, the most used analytical techniques are the mass sensitive acoustic transducers SAW [83], wave oscillator [84] and QCM [85,86]. Selective and reversible binding of the template to the imprinted polymer could be observed on-line by infrared evanescent-wave spectroscopy (IR-EWS). Biosensors based on polymer-imprinted materials provide more reliable substrate recognition and screening properties compared to conventional systems.

Applications. An example of practical approach based on polymer-imprinted materials is the development of biosensors for the detection of the herbicides atrazine [87] and 2,4-dichlorophenoxyacetic acid (2,4-D) [88]. The first one is based on conductimetric detection and it uses an atrazine selective polymeric matrix generated from triethylene glycol dimethacrylate and oligourethane acrylate. The second one was prepared on a ZnSe surface and reported to be able to detect 2,4-D in the ppb concentration range.

7.4.2 Biosensing materials containing metal complexes

Another important group of materials with application in biosensors is the polymers that contain different metal (Os, Ru, Ni, etc.) centers coordinated within the polymeric network. The electroactivity of these polymers is associated with the reversible redox behavior of the metals. The electrodeposition of redox polymers results in the formation of coordinative bonds that crosslink the chains. Gao et al. [89] showed that when the initially reversible adsorbed water-soluble redox polymers are electrochemically cycled, they are irreversibly electrodeposited and crosslinked. As the crosslinking caused by ligand exchange proceeds, the dissolved biomolecules with amine or heterocyclic nitrogen functions coordinating transition metals are conveniently co-electrodeposited. Ru^{2+} complexes containing labile ligands (ex. water, etc.) are capable of binding proteins by exchanging their ligand with a protein-bound histidine.

Applications. A biotinylated GOX-based biosensor was developed based on a new electropolymerized material consisting of a polypyridyl complex of ruthenium(II) functionalized with a pyrrole group [90]. Because histidine, lysine and arginine functions also coordinate Os^{2+}/Os^{3+}, biosensors based on co-electrodeposited GOX, HRP, soybean peroxidase (SBP) and laccase with redox Os^{2+}/Os^{3+} polymer have been developed [89]. A metal chelate formed by nickel and nitrilotriacetic acid was used to modify a screen-printed electrode surface. The functionalized support allowed stable attachment of acetylcholinesterase and the resulting biosensor was used for sensitive detection of organophosphorus insecticides [91]. This method is attractive because it ensures a controlled and oriented enzyme immobilization, considerably improving the sensitivity and the detection limit.

Nieuwenhuizen and Harteveld [92] have realized a nerve agent dosimeter gas sensor based on the strong interaction between certain metal ions and organophosphorus compounds. In this case, the sensor material contains La(III) 2-bis(carboxymethyl)amino hexadecanoic acid and different factors such as humidity, concentration and layer thickness have been studied and optimized. Using a combination of a metallic complex with a molecular-imprinted polymer, a very sensitive sensor was developed for the detection of soman, a chemical warfare agent (the detection limit was 7 ppt) [93]. The biosensing material is based on a polymer coated onto a fiber-optic probe modified with a luminescent europium detection complex. This complex was ligated by divinylmethyl benzoate and the analyte pinacoyl methylphosphonate was copolymerized with styrene. The analyte molecule was removed by washing and then the rebinding process was evaluated by laser-excited luminescent spectra.

7.4.3 Sol–gel materials

Sol–gel processes have long been used for the powderless processing of glasses and ceramics. Intermediate between glasses and polymers, sol–gel materials have opened new possibilities as biosensor material proving a versatile way for bioimmobilization. Thus, antibodies, DNA, RNA, antigens, plant and animal cells and bacteria have been attached on various supports using this method. The sol–gel technique is based on the ability to form a solid metal or semi-metal oxide via the aqueous processing of hydrolytically labile precursors, rather than by high temperature chemistry [94]. With rapid advances in sol–gel precursors, nanoengineering polymers, encapsulation protocols and fabrication methods, novel sol–gel nanocomposites have been developed during the past few years.

In biosensors, this technique could result in development of novel strategies to generate advanced materials for the immobilization of biological receptors within silica, metal oxide, organosiloxane and hybrid sol–gel polymers [95]. It has led to the development of functionalized coatings, optical devices, chemical sensors and biosensors. Basically, the technique involves two steps: (i) the hydrolysis of a precursor (metallic salts of Al(III), Ti(IV), V(V), Sn(IV) or esters of silicic (IV) or polysilicic (IV)) in aqueous medium to generate a *sol* soluble hydroxylated oligomers, followed by (ii) polymerization and phase extraction in order to form a hydrate oxide *hydrogel*. Among sol–gel sensor materials, the mostly used for practical applications are those derived from glasses, characterized by chemical inertness, thermal, photochemical and structural stability and excellent optical transparency [96].

A crucial feature of sol–gel polymerization is that the metal salts, complexes and organic dyes can be incorporated before hydrogel formation, to give the sol–gel composite. On the other hand, the sol–gel process has the advantage that it can be carried out at room temperature and the procedure is not severe enough to denature the biomolecules, thus retaining activity to a large degree [97]. These have proved to be invaluable for use in sensors, waveguides and biosensors [98]. However, problems such as diffusional limitations through the porous network, reproducibility and control of pore sizes require further optimization for biosensing purpose.

Applications. Dong et al. synthesized a self gelatinizable grafting copolymer of poly(vinyl alcohol) and 4-vinylpyridine (PVA-*g*-PVP) and hybridized this polymer into a silica sol [99] or a titanium oxide [100] generating inorganic–organic hybrid thin film composites. Based on this hybrid material, amperometric glucose [100,101] and tyrosinase biosensors [100] have been developed. Recently, low temperature sol–gels have been used for enzyme immobilization on different supports [102], including screen-printed electrodes [103–105]. The coupling of the sol–gel technique with screen printing allows a very simple one-step fabrication of the electrodes, since enzymes, mediators, graphite powder and additives were incorporated in the same sol–gel ink [106]. Wang [110] recently summarized the trends in the development of sol–gels as sensing materials in biosensors, including doped sol–gel films [107,108], and multilayer "sandwich" configuration [109].

7.4.4 Nanomaterials

Since the discovery of carbon nanotubes (CNTs) in 1991 [111], these materials have been the target of numerous investigation as nanoscale building blocks for analytical devices [112]. The ability to control the particle size and morphology is of crucial importance in nanostructured materials for chemical sensors and biosensors for both fundamental and industrial applications. Because of the high electrochemically accessible surface area of porous nanotube arrays, combined with their high electronic conductivity and useful mechanical properties, these materials are especially attractive for electrochemical biosensor devices. Current synthetic techniques and patterning methods used to develop nanomaterials are self-assembly, mechanochemistry, chemistry by microwave, lithography or template- and membrane-based synthesis [113]. To afford the production, cheap, clean, reliable and durable materials with controlled properties are needed for realistic and practical applications of nanotechnology. Moreover, the request for mass production of thin film devices is one of the most important issues in their development.

7.4.4.1 Carbon nanotubes
The first important group of nanomaterials with application in sensors and biosensors field is represented by the CNTs [114]. The nanodimensions, surface chemistry and electronic properties of CNT make this material an ideal candidate for chemical and biochemical sensing. There are two main types of CNTs: (i) *single-walled nanotubes* (SWNTs) consisting of single graphite rolled seamlessly wrapped into a cylindrical tube (1–2 nm diameter) represents undoubtedly one of the more remarkable discoveries in the

material chemistry in recent years. (ii) *Multi-walled nanotubes* (MWNTs) comprise an array of concentric and closed nanotubes with multiple layers of graphite sheet defining a hole, typically from 2 to 25 nm separated by a distance of approximately 0.34 nm [115,116]. The behavior of CNTs can be associated with those of metals or semiconductors, function of structure–diameter and helicity [117]. Their excellent electrocatalytic properties in the redox activity of different biomolecules constitute the most important characteristic for biosensors.

Similar to conventional carbon-based electrode, nanotubes can be used as electrodes for detecting molecules in solution. In order to allow stable attachment of bioreceptors, the CNTs need several external modification steps. These include interaction with polymers, ionic/covalent functionalization and adsorption of gases [117]. To develop and optimize new types of CNT miniature devices for biological applications, it is important to study the interface between biological molecules and nanomaterials. CNT electrodes exhibit useful voltammetric properties with direct electrical communication between a redox active biomolecule and the delocalized π-system of its CNT support [117]. Usually, before the functionalization of CNTs, a purification process is developed in order to remove any impurity, such as amorphous carbon and catalyst particle generated during the synthesis process. In this case, treatment with oxidizing acids, followed sometimes by a gas-phase oxidation process, application of microfiltration techniques and chromatographic procedures are the most widely used purification methods [118].

Non-covalent functionalization. Non-covalent functionalization has the advantage of a non-destructive process, by conserving the physical properties of CNTs, improving the solubility and also allowing their purification. Different routes for non-covalent modification of SWNTs have been reported, including polymer wrapping and adsorption [119], adsorption of amines [120] and molecules with large π-systems [121], complexation with organometallic compounds [122] and radio frequency glow-discharge plasma modification [123]. CNT can be non-destructively oxidized along their sidewalls or ends and then covalently functionalized with colloidal particles or polyamine dendrimers via carboxylate chemistry [117]. Duesberg et al. and Krstic et al. [124,125] showed that CNTs interact with certain molecules such as sodium dodecylsulfate (SDS) resulting in micelles in which the nanotube materials are surrounded by hydrophobic moieties. These hydrophobic parts react with amphiphilic compounds and the process becomes stronger when the latter contain an aromatic group. This is a consequence of favorable π–π interaction with the walls of CNTs [126].

The formation of supramolecular systems between SWNTs and polymers has been utilized in non-covalent functionalization processes. The properties of these supramolecular complexes are different from those of the individual components [127]. For instance, suspended CNTs in the presence of poly(p-phenylenevinylene) type polymer molecules wrap around the nanotube, the resulting complex showed an increased electrical conductivity. Functionalization of CNTs by surfactant and polymer species significantly reduces non-specific protein adsorption, while co-immobilization of ligands or antibodies can impart specificity in protein binding on the nanotube surface. Immobilization of proteins and small biomolecules onto non-covalently functionalized SWNTs represents an easy method to generate biomaterials for biosensor's construction. Proteins can be adsorbed individually, strongly and non-covalently along the lengths of the nanotubes or can be covalently attached to the CNTs that are initially derivatized with specific functional groups [128–131]. The versatility of this simple approach can be extended beyond biological molecules and could be useful for the construction of many other systems based on small molecules with desired properties, polymerizable compounds or inorganic nanoparticles.

Applications. The ability of CNTs to promote the electron transfer reactions of NADH and hydrogen peroxide has been used for the development of dehydrogenase- and oxidase-based biosensors [132]. SWNT modified with phthalocyanine is another example of CNTs with applicability in biosensors [118]. Functionalization of SWNTs by co-adsorption of a surfactant (Triton-X) and poly(ethylene glycol) was found to be effective in resisting non-specific adsorption of streptavidin. Specific binding of streptavidin onto SWNTs has been achieved by co-functionalization of nanotubes with biotin and protein-resistant polymers [133]. The streptavidin/biotin system was also used to investigate the adsorption behavior of proteins on the sides of SWNTs. Biomolecules such as DNA can be linked to M(S)WNTs via non-covalent interactions [134–137]. Functionalization of CNTs with 1-pyrenebutanoic acid, succinimidyl ester has been used for immobilizations of proteins [138,139] and anti-fullerene IgG monoclonal antibody has been bound to SWNTs [140].

Functionalization of CNTs by covalent chemistry. Covalent functionalization of CNTs has attracted a great interest for biosensors development. This type of functionalization is expected to play a crucial role in tailoring the properties of materials and the engineering of CNT biosensing devices. The first step in a covalent functionalization process involves a chemical treatment of CNTs under oxidizing conditions, such as sonication in a mixture of sulfuric and nitric acids or treatment with piranha (sulfuric acid/hydrogen peroxide) [141]. These procedures leave the CNTs open ended, the ends containing somewhat indeterminate number of oxygen functionalities such as carboxylic acids, anhydrides, quinones and esters [142,143]. Such functional groups can also be introduced onto CNT's surface by treatment with ozone [144]. Several other methodologies include fluorination [145], electrophilic addition of chloroform [146], esterification [147], protein [148], carbenes [149], azomethine [150] and nucleic acid functionalization [151] via diimide-activated amidation and electrochemical reduction of aryl diazonium salts [152]. More recently, electrochemical modification of SWNTs under oxidative conditions has been reported [153]. In this case, substrates deposited on Si/SiO_2 are reacted with aminoaryl compounds and, under longer reaction times, polymerization of the radicals takes place, leading to the formation of a multilayer coating.

It may be possible to generate reliable biosensing devices by combining the surface assembly and molecular recognition properties of proteins and enzymes with the molecular electronic properties of CNTs. CNTs represent a promising material for chemical sensors for detecting molecules in the gas phase and biosensors for probing biological processes in solutions. Nevertheless, significant effort is required in order to understand interactions between nanotubes and biomolecules and how to impart specificity and selectivity to nanotube-based bioelectronic devices. Although dissolution and manipulation remain problematic, a number of physical and chemical modifications of SWNTs and MWNTs can be developed in order to improve their characteristics to ensure a good stability of biological materials. The key for the development of CNT-based biosensor is the ability to understand and control interactions at the relevant surfaces and to ensure the biocompatibility of the support.

Applications. Combining physical surface modification and covalent attachment of quantum dots to SWNTs, Azamian et al. [154] deposited silver colloids from a solution phase of organometallic complex onto oxidized SWNT. Davis et al. [117] controlled the oxidation of SWNTs along their lengths by utilizing carboxylates introduced to bind colloidal gold particles (~3 nm) via 2-aminoethanethiol linkages. Using a native CNT electrode and bromoform as binder, Britto et al. [155] reported an enhanced electrochemical behavior of dopamine.

Usually, biomodification of S(M)WNTs electrodes occurs by carbodiimide coupling chemistry when the amine groups of the biomolecules (DNA, proteins, enzymes) are coupled to the carboxylate moieties of nanotubes in aqueous solution. Based on this principle, cytochrome *C*, ferritin, catalase, GOX, azurin were immobilized on SWNTs. In this case, biomolecular coverage was dependent on the protein concentration, quality of nanotube suspension and incubation time. The adsorption process for these proteins is pH and ionic strength independent (to at least 1 M KCl) and cannot be significantly improved by the addition of anionic or neutral surface-active agents (SDS or Triton X). The magnitude of the catalytic response of a GOX nanotube-based biosensor was more than one order of magnitude greater than obtained with an activated carbon macroelectrode, under identical experimental conditions. Recent studies with SWNTs glassy carbon-modified electrodes showed an enhancement of the electrochemical behavior of catecholamine neurotransmitters, cytochrome C, ascorbic acid, NADH and hydrazine compounds [156–159]. Wohlstadter et al. [160] reported the use of CNT as electrode and immobilization phase for the construction of an electrochemiluminescent (ECL) immunoassay for α-fetoprotein (AFP). In the first step, the CNTs were extruded with poly(ethylene vinylacetate) (EVA) to produce CNT–EVA composites followed by a chemical treatment with a solution of chromic or sulfuric acid in order to create a dense network of exposed nanotube and to introduce carboxylic groups onto the surface. Then, the NHS/EDC coupling was used for the attachment of streptavidin (8.1 pmol cm^{-2}) [160]. The immobilization of anti-AFP onto the streptavidin-coated nanotube composite was carried out using biotinylated antibodies. The resulting sensor allowed detection of 0.1 nM AFP.

A major barrier for developing CNT-based biosensing device is the insolubility of CNT in most solvents. Wang et al. developed CNT/Teflon composite electrodes based on the dispersion of CNT within a Teflon binder. Then the CNT/Teflon serves as a matrix for the incorporation of GOX and alcohol dehydrogenase/NAD$^+$. The resulting amperometric sensor showed an enhanced electrocatalytic activity toward hydrogen peroxide and NADH, allowing effective detection of glucose and ethanol at low operating potential. In this case, the accelerated electron transfer was coupled with minimization of surface fouling and renewability [132].

By using a multistep procedure, DNA molecules have been covalently attached to SWNTs. First, the purified SWNTs were oxidized to form carboxylic acid groups at the ends and sidewalls, followed by reaction with thionyl chloride and ethylenediamine to produce amine-terminated sites. The amines were then reacted with the heterobifunctional cross-linker succinimidyl 4-(N-maleimidomethyl)cyclohexane-1-carboxylate (SMCC), leaving the surface terminated with maleimide groups. Finally, thiol-terminated DNA reacted with these groups to produce DNA-modified SWNTs [161]. When DNA is covalently attached to SWNTs, a better stability, accessibility and selectivity are expected during competitive hybridization.

7.4.4.2 Nanocrystalline films

Another class of nanomaterials currently investigated comprises metal oxides such as TiO_2, ZnO, Fe_2O_3, RuO_2, IrO_2 or different colloidal particles on various substrates, including conducting tin oxide glass, Si wafers, polymers and Teflon [162]. One of the problems related to these materials is that the agglomeration and aggregation have been very difficult to control. These difficulties are generated by the presence of strong particle–particle interaction involving various types of chemical bonds, such as ionic, hydrogen or strong dipole interactions. To overcome these problems, these interactions have to be reduced to van der Waals forces or bonds that are easily controllable in an appropriate media. For this reason, a surface modification of the nanoparticles would be needed in order to successfully apply these nanomaterials for biosensing device development. Following this modification, the reactivity of nanomaterials can change in many ways and, in consequence, the process needs to be carefully selected. For example, they can be developed by polymerization to form bonds to organic molecules such as monomers, biochemical molecules (enzymes, antibodies, proteins, etc.) or via organic spacers. The surface modification of nanoparticles by polymerizing groups is the most used method to incorporate nanoparticles into polymers matrix with high nanomaterial distribution, which also ensures sufficient dispersion [163]. On the other hand, incorporating metal nanoparticles into polymeric membranes is known to enhance their conductivity [164]. These nanostructured composites have been used as novel materials in biosensors [165]. The construction of such a device is usually based on a combination of an electrochemical biosensor array with novel synthetic organic polymers. In this case, the polymer serves as a matrix for the incorporation of metal particles followed by subsequent protein immobilization via the metal ions. However, the number of applications is still relatively small and the potential of this method has not been fully exploited. A new trend in biosensor manufacturing is to find novel nanocomposite materials that contain free functional groups precisely located onto the surface and available for specific biomolecule binding in a controlled and oriented manner.

Applications. As an example, the incorporation of metal nanoparticles was carried out in organic conducting polymers [61]. For the first time, electrically conducting metallic-nanoparticles modified polymers were prepared by a systematic photopolymerization of polypyrrole simultaneous with the incorporation of Group 1B metal particles. Numerous substrates were used including polystyrene, acrylics, ceramics, glass, fiberglass, printed wire board and Nalgene tubing. A mechanistic insight associated with structural and morphological properties of the incorporated nanoparticles was proposed with respect to the role of the metal ions, nature of the substrate, exposure time and monomer counterion ratios.

Recently, our research effort is focused on the development of novel composites of poly(amic acid (PAA))–metallic silver nanoparticles prepared using pyromellitic dianhydride (PMDA) and 4,4′-oxydianiline (ODA) solutions containing the salts of the metal ions [166]. ^{1}H-NMR measurements have been performed to clarify the chemical structure of PAA–metal composite films (Fig. 7.7). The composites were prepared onto carbon vitreous electrodes following an electrochemical procedure combined with thermal curing. SEM and electron dispersive elemental analysis have been performed for PAA and PAA metal composite films to determine the morphology of films and to identify the nature and distribution of metallic particles in PAA films (Fig. 7.8). The electrode potential increases with the increase in the applied current. Simultaneously, the increase in the current density produces an enhancement in the polymer weight gain. This behavior suggests that the process obeys Faraday's law of electrolysis. The novelty of this approach lies in the incorporation of silver nanoparticles within electropolymerized PAA at low temperature, by preventing the imidization process. Thus, by avoiding the

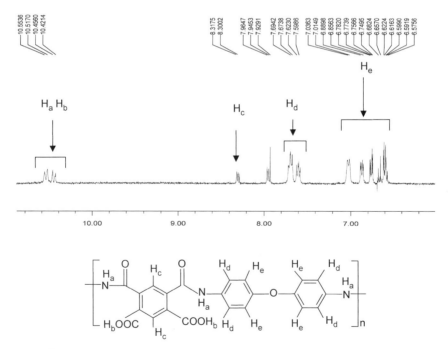

Fig. 7.7. ^{1}H-NMR spectra of Ag–PAA/MeCN film in d$_6$-DMSO [R = H, N$^+$(C$_2$H$_5$)$_3$].

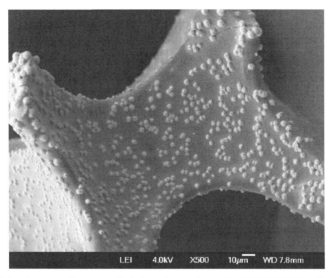

Fig. 7.8. SEM micrographs recorded for electrodeposited PAA/THF silver composite film developed in PBS. Experimental conditions: $Dc = 1.0$ mA cm^{-2} and 5 min electrodeposition time.

imidization process, the electrogenerated silver–PAA films can constitute an attractive material for biomolecules immobilization via the free carboxylic groups.

Velev and Kaler [167] developed an array of immunosensors by in situ assembly of gold colloidal particles onto micropatterned electrode. In this case, latex microspheres from suspension have been collected between planar electrodes and the assembled particulate was fixed by changing the colloidal interaction to induce the coagulation. The biospecific agent (protein A) attached to the latex particles will bind the target immunomolecule (IgG). The target analyte was tagged with colloidal gold and conjugated with its specific antibody (anti-IgG). This yields a layer of strongly and specifically attached colloidal gold particles around the latex microspheres. The gold particles were supported by a silver layer via nucleation process ("silver enhancement") and the detection was carried out by simply measuring the resistance between the electrodes.

Using a nanoporous TiO_2 film, an optical biosensor for the determination of dissolved CO from aqueous solutions was developed [168]. The redox states of immobilized proteins like Cyt-c and Hb were modulated by the application of an electrical bias potential to the TiO_2 film, without the addition of electron transfer mediator. The protein immobilization occurred from aqueous solution at 4°C with high binding stability, while no detectable denaturation process was observed [169]. The successful biofunctionalization of organoplatinum(II) sites by covalently linking the N-terminus of amino acids represents a very promising way for various applications in environmental biosensors and diagnostic biomarkers [163]. This organoplatinum marker combines an effective labeling system with a selective molecular recognition site and was used for the detection of SO_2 [170] and NO [171].

7.4.5 Nanoscaled interdigitated electrodes

The principal goal for the development of interdigitated electrodes (IDEs) is for application in miniature and sensitive affinity biosensors. IDEs comprise electrodes with widths and spacing ranging from 3 to 300 nm, in different configurations, and are composed of Pd, Au or Ti electrodes onto a SiO_2 substrate [172]. The major advantages of IDEs over conventional electrodes for biosensing are (i) the improved sensitivity compared to macro-electrodes and (ii) the fabrication procedure by lithography allowing reproducible and low cost device [173]. The most used detection method is electrochemical impedance spectroscopy (EIS). Recently, impedimetric Au, Pd and Ti IDE immunosensors have been developed [172]. The immobilization of biological molecules has been carried out using the silanization procedure and the EDC/NHS coupling kit. When studying GOX immobilization, Laureyn demonstrated that the sensitivity of different IDEs was in the following order: Pd IDEs > non-oxidized Ti IDEs > oxidized Ti IDEs. It was also shown that by doping TiO_2 with $RuCl_3$, the electron transfer through material could be facilitated. When TiO_2 on SiO_2 was used, the covalent linkage (via Ti–O–Si bonds) is present, together with crosslinking between silanes (via Si–O–Si bonds). This covalent bond between TiO_2 and aminosilanes enables a full covalent immobilization of biomolecules and inherently stable biomaterials.

7.4.6 Composite materials

Biosensors containing composite materials have proved to be extremely useful for bulk modification with various modifiers such as enzyme and inorganic catalysts. These electrodes are formed by the combination of the powdered electronic conductor and a binding agent. The most used composite is made from carbon (generating the well-known carbon paste electrode—CPE) and its derivative compositions that provide good electrical conductivity and homogeneity. Metal powders are also available and can be used [174,175]. Carbon composites generated probably the greatest number of biocomposite materials for biosensing applications. The biocomposite not only acts as a reservoir of biological material but may also contain catalysts, mediators and cofactors that could generate an enhanced response from the resulting biosensor system. From an electrochemical point of view, there is a preference for CPEs because of the low noise and background currents. The level of sensitivity obtainable has been linked to viscosity and binder composition used [176]. Basically, a CPE consists of a homogenous mixture of a carbon powder and a pasting liquid. The carbon powder must have uniform particle size and high chemical purity. The pasting liquid has to be inert, pure and electro-inactive and also characterized by a low volatility and solubility in the analyte solution. Paste materials include aliphatic hydrocarbons, aromatic compounds and silicone oils. The CPE was first described by Adams [177] with the goal of producing a dropping, renewable carbon electrode. Gorton [178] presented a comprehensive overview on CPEs including bulk modification with biological species.

Applications. Recently, a simple amperometric tyrosinase–carbon paste biosensor was used for the measurement of endocrine disrupting chemicals. Tyrosinase was immobilized onto the working electrode through a simple inclusion of the enzyme powder into a carbon paste made of carbon powder and mineral oil [179]. The mixture was thoroughly mixed in a mortar and then packed into the end of a glass capillary tube. A copper wire was inserted into the capillary to make the electrical contact. The electrode was preserved at 4°C in dry state and was carefully polished on fine paper before each use in order to obtain a reproducible working surface. The analytical characteristics of the sensor have been studied for the detection of a polyphenol quercetin, by optimizing enzyme loading and geometrical area of the electrode at three different configurations (Table 7.1). The biosensor resulted in rapid, simple

TABLE 7.1

Optimization of a tyrosinase carbon paste biosensor for the detection of the polyphenol quercetin

Configuration	Tyr (U mg^{-1} graphite)	Graphite (mg)/mineral oil (mg)	Geometrical area (cm^2)	Sensitivity (mA M^{-1})[a]	Linear range (μM)	R^2
A	189.23	120/90	0.15	1.87 ($\pm$0.2)	1–10	0.9882
B	521.81	120/90	0.15	3.66 ($\pm$0.43)	1–15	0.9689
C	521.81	120/90	0.05	6.95 ($\pm$2.16)	1–25	0.9892

[a] Measurements with $n = 5$ sensors under the same experimental conditions (-150 mV vs. Ag/AgCl reference electrode; phosphate buffer solution 0.1 M at pH 6.5).

and accurate measurement of various phenolic estrogens with varying degrees of sensitivity, selectivity and response times. A detection limit of 0.15 μM was obtained for bisphenol A [179].

Alegret et al. [180] developed carbon biocomposite materials by mixing graphite powder with a non-conducting polymer resin and lyophilized enzymes (acetylcholinesterase, GOX, peroxidase). In this case, epoxy, silicone, methacrylate and polyester polymers have been used to prepare a rigid conducting composite material for the development of thick film biosensors. Thick film deposition, namely, *screen-printed* technique seems to be one of the most promising technology allowing biosensors to be placed in large scale on the market in the near future. The major advantages of this technology are its versatility, low cost and suitability for mass production. Typically, these thick film configurations consist of depositing of successive layers of special inks (or paste) onto an insulating support or substrate. It can be developed with conducting carbon ink or other metal pastes with or without mediator and/or catalyst and then applied onto a suitable substrate. In addition, biomolecules can be readily incorporated in the screen-printed composite material or they can be attached by different immobilization techniques after the deposition of the first carbon-conducting layer [91,106]. Recently, Albareda-Sirvent et al. [103] revised the principal configurations used for the fabrication of screen-printed biosensors. The first one is a multiple layer process, involving manual deposition of biomolecule through a covalent procedure or by electropolymerization. The second one is a combined configuration (two-step procedure) based on composite inks or pastes containing the enzyme, printed over the electrode surface. Finally, the third one is a one-step deposition of a total biocomposite ink. Using a disposable copper screen-printed carbon electrode, Zen et al. [181] achieved the selective determination of *o*-diphenols (e.g., catechol, dopamine and pyrogallol) in the presence of other isomers and phenols.

Wang et al. [182] reported recently a very interesting approach for biocomposite material used for the immobilization of dsDNA onto a working modified screen-printed surface. The attachment was realized through the 5'-phosphate groups of dsDNA by the formation of a phosphoramidate bond with the amino groups of a self-assembled cystamine monolayer. A significant improvement for the immobilization of dsDNA onto solid substrates was made by facilitating the attachment of long pieces of unmodified dsDNA via a simple biotinylation step.

Composite biomaterials formed between small molecules (ligands, metals and proteins) and DNA capable of generating analytical signals have also been used to design selective biosensors [183–185]. In this case, a monolayer of silver can be deposited onto the gold electrode or other conducting substrates (platinum, glassy carbon) and the electrochemical oxidation of silver, accompanied by a reversible redox signal can be monitored using cyclic voltammetry technique (Fig. 7.9) [185]. If dsDNA is present at the surface, the silver ions are dispersed and are held electrostatically. Upon reduction, the silver ions return to the surface and a reduction current is measured. If a DNA binding low molecular weight organic molecules is introduced into the solution, structural changes in the DNA molecule occurs, which is observed by a simultaneous change in the redox current from silver (a decrease in current). This decrease is proportional with the concentration of DNA binding molecule. Based on this principle, we have developed a new biocomposite material using DNA-monodispersed silver ions for the detection of cisplatin−DNA interactions [185].

Recent trend in biosensor manufacturing is to use synthetic procedures to incorporate a label indicative of a change in a physicochemical property upon binding to biomolecule. An example is the preparation of a novel, fluorescently active, carboxylic acid derivative of 2-pyridylazo compounds (PAR) [186].

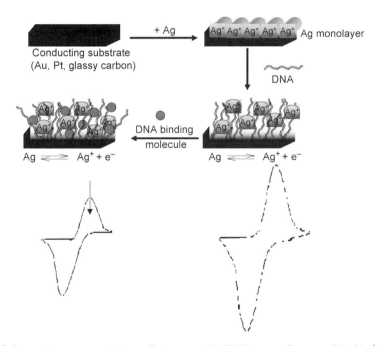

Fig. 7.9. Schematic representation of biocomposite DNA-monodispersed Ag/Ag$^+$ couple and electrochemical detection concept.

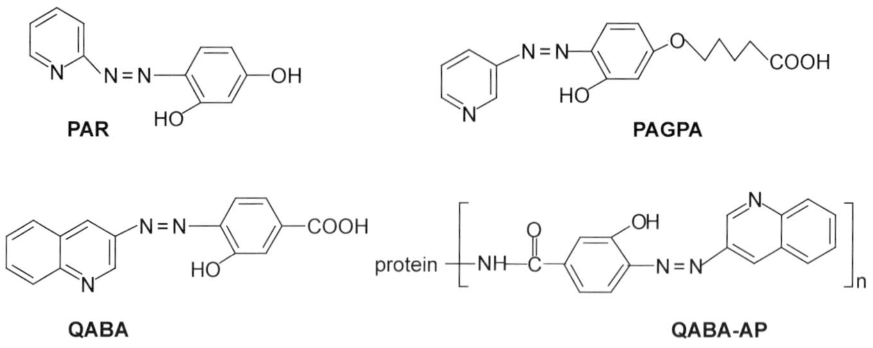

Fig. 7.10. Structures of PAR (2-pyridylazo) and PAR derivative (2-pyridylazo) (3-hydroxy-5-phenoxy)ethylpentanoic acid—PAGPA), QABA: 4-(3-quinolinoazo)-3-hydroxybenzoic acid) and QABA protein labeled chelator (QABA-AP).

The synthesis was carried out using a water-soluble carbodiimide and the NHS coupling technique [187]. The chemical structures of PAR and PAR derivatives developed in our group are shown in Fig. 7.10. These conjugates were tested for the determination of Ga(III) using a sandwich chelate assay format and showed very high selectivity relative to several other metals ions such as Al^{3+}, In^{3+}, Zn^{2+}, Cu^{2+}, Co^{2+}, Fe^{2+}, Pb^{2+}, Ni^{2+}. Based on these compounds, biomaterials containing ovalbumin, bovine serum albumin and alkaline phosphatase have been developed by conjugation. Thus, the use of novel conjugate derivatives such as the PAR ligand as a recognition molecule could be a new approach for designing a new class of biomaterials.

In other works, the electrochemical oxidation of a hydrosulfite cluster $[Fe_4S_4(SH)_4]^{2-}$ on gold, platinum or vitreous carbon in a methyl cyanide electrolyte has led to the growth of a conducting film, thus generating a new polyferredoxin-based electrode material [188]. Even though the electronic behavior of polyferredoxin materials is not very well understood, a conduction along the polymer chains involving $\{4Fe4S\}^{2+}$ and $\{4Fe4S\}^{3+}$ was suggested.

7.4.7 Metal oxide semiconductor field effect transistors

Metal oxide semiconductor field effect transistors (MOSFETs) constitute other materials with applicability in the development of biosensors. Usually, a MOSFET structure consists of a metal gate on top of an oxide layer, typically SiO_2 [189]. The catalytic properties of these sensors depend upon the type of the gate metal as well as the temperature at which the MOSFET is operated. The most used catalytic metals used as gate materials are: Pd (is a good hydrogen sensor) and Pt, Ir (have sensitivities for ammonia and ethanol). The first MOSFET sensor was developed for hydrogen by Lundström et al. [190] and contains palladium as metal gate. During the time, several other MOSFETs variants were developed: GASFETs (the same structure as a MOSFET but the classical aluminum gate is replaced with a catalytic active metal), open gate field effect transistor (OGFET), ion sensitive field effect transistors (ISFETs which contain an electrolyte solution as the conducting layer) and CHEMFETs (ISFETs coated with an organic membrane) [191].

The remarkable progress made in the last decade in the field of wide-gap semiconductors such as Si, GaAs, SiC, diamond, GaN or the II–IVs has opened new possibilities for novel (bio)sensors and complex integrated systems development. Stutzmann [192] discussed the potential applications of AlGaN/GaN heterostructures in biosensors, characterized by excellent thermal, chemical and mechanical stability. In this case, two-dimensional electron gas is formed at the AlGaN/GaN interface due to the difference in the spontaneous polarization of the two adjacent III-nitride layers, which has been shown to respond very sensitively to the change in the electrostatic boundary conditions caused by the adsorption of ions, wetting by polar liquids, exposure to gases. These sensors function on the principle of an integrated electronic or optoelectronic device—high frequency/high electron mobility transistors (HEMTs), micromechanical systems (MEMS) as well as passive surface acoustic wave (SAW) devices. These characteristics opened a possible way for the integration of both sensor functions with analog/digital data processing and transmission.

7.4.8 Photonic band gap

Photonic band gap (PBG) materials represent a class of composites that are designed to monitor the properties of photons in much the same manner as semiconductors manipulate the electrons' properties. These composites have been developed from ordered arrays of self-assembled nanometer-sized polystyrene spheres that have been repeatedly encapsulated with various polymeric systems. However, it has been recognized that PBG materials also support surface electromagnetic waves–optical modes that propagate at the interface of a PBG crystal [193].

The PBG materials were used to produce high-sensitivity, versatile and robust biosensors. Some examples investigated include PBG-biosensors constructed from directional couplers [194], Mach–Zehnder interferometers [195] and various optical configurations that use SPR technique [196,197]. Among these, the SPR is the most used. In this case, the metal that generates the surface plasmons optical waves (typically gold, aluminum or silver) was replaced with a PBG material. Usually, the SPR biosensors operate by measuring small shifts in resonance position or by measuring enhanced fluorescence. Using surface plasmon in metals, narrower surface-wave resonances generated by the loss in dielectric property can be obtained. For PBG materials without any metal content, the dielectric loss can be very low. The sharp resonance in PBG arrays led to the enhancement of the analytical response with one order of magnitude. Another advantage in PBG-based SPR analysis is the ability to eliminate the effect of corrosive environments and high temperature against the active metals. Because the band gap of a PBG array is set by the periodicity of the sample, suitable surface-active PBG materials can be designed to operate at any target wavelength. The use of non-metallic surfaces for the sensing layers may lead to new opportunities to generate chemically and biologically activated sensing surfaces.

A composite biomaterial formed by Pd metal, carbon–ceramic mixture and oxidoreductase enzymes constitutes a new type of renewable surface biosensor with a controllable size reaction layer [198]. The carbon provides the electrical conductivity, the enzymes are used for biocatalyst process, metallic palladium is used for electrocatalysis of biochemical reaction product and the porous silica provides a rigid skeleton. The hydrophobicity of this composite material allows only a limited section of the electrode to be wetted by the aqueous analyte, thus providing a controlled thickness reactive layer. Another biocomposite material containing enzyme-modified boron-doped diamond was used in the development of biosensors for the determination of phenol derivatives [199], alcohol [200] and glucose [201].

7.4.9 Zeolites

Zeolites represent another composite materials class involved in chemically modified electrodes (CMEs) with analytical application. Zeolite-modified electrodes (ZMEs) form a subcategory of the so-called "CMEs" and was promoted by Murray [202]. Walcarius classified the ZMEs in five categories, function of the detection process: thin layer characterization, permselective membranes, charge storage devices, electrocatalysis and electroanalysis [203]. Zeolites are hydrated aluminosilicate minerals with general formulation—$(C^{n+})_x[(AlO_2^-)_{n \times x}(SiO_2)_y]m(H_2O)$, where C^{n+} is charge compensating cations. They are crystalline in nature and made of solid particles, which are not electronically conductive. For this reason, the application of zeolite in electrochemistry, particularly in biosensors development is not technically easy. In order to obtain conductive ZME composite materials numerous preparation/modification methods have been used. The preparation of ZMEs can be classified in four classes: (1) the dispersion of zeolite particles within a conductive composite matrix, (2) the compression of zeolites onto a conductive substrate, (3) the formation of zeolite films on solid electrodes and (4) the covalent binding of zeolite particles to an electrode surface [204]. Kotte et al. developed the first biosensor based on a zeolite and it was utilized for the detection of phenolic compounds. An example of ZME was achieved by screen-printed technique by coating a mediator-modified carbon surface containing zeolite particles with a biocomposite material containing polyurethane hydrogel and an enzyme (tyrosinase) [205]. Wang and Walcarius used zeolites to improve the performance of first generation oxidase-based carbon paste biosensors. Thus, the use of this sensor for the amperometric detection of glucose resulted in an extended linear range and increased sensitivity [206,207]. The combination between the ion exchange capacity and their unique size selectivity at the molecular level, together with the doping process with an electron transfer cofactor make ZMEs good candidate to construct efficient amperometric sensors for non-electroactive species.

7.5 SUMMARY AND FUTURE PERSPECTIVES

Novel electrode materials have emerged in the last decades as a consequence of new requirements for reliable, sensitive and selective biosensor devices. This paper reviewed the most important materials used for the development of biosensors, biochip and supporting matrices. A summary of the principal electrode materials and methods is presented in Table 7.2. The work focused primarily on the characterization of these materials, the most important techniques used for electrode modification and their applicability for the construction of sensors and biosensors. Current and future trends in material science for biosensors have been also discussed.

The real challenges for the future concern the development of good electrode material, biocompatible and suitable for integration with a biomolecule, with possibilities of miniaturization, stability and real-time measurements. Currently, the major limitations in the development of biosensors are due to the compatibility between the material of the transducer surface and the biological element. Initially, sensor materials served as support matrix for the incorporation of biological recognition elements. In the last two decades, new methods for immobilizing enzymes, chromophores, antibodies and electron transfer catalysts have been reported due to the advances in materials science. Therefore, the main directions in biosensor research could be focused to the rational design of new electrode (bio)materials

TABLE 7.2
Principal methods used for functionalization of biomolecules

Electrode material/procedure	Advantages	Limitations	Refs.
Native material/adsorption	Simple; short response time	Poor operational and storage stability; sensitive to changes in pH, temperature, ionic strength	[4,9–13]
Au/self-assembled monolayers (SAMs)	Stability; controlled and oriented mono layers; absence of diffusion barriers	Complex and difficult to reproduce; possible electrode fouling	[14–26]
Pre-activated or functionalizable surfaces/covalent binding	Stable; absence of diffusion barriers; short response time	Poor reproducibility; high amount of bioreceptor; possible denaturation of the biomolecules	[27–38,47]
Polymers/electro-polymerization	Easy electrochemical control of the surface; single step; suitable for the fabrication of micro sensors; control and orientation of biomolecules	High monomer and bioreceptor concentration; denaturation of the procedure for bioreceptor	[47,49–69,73]
Polymers/physical entrapment	Simple; many biocompatible polymers available; suitable for a large variety of bioreceptors	The presence of diffusion barriers	[106]
Metal complexes/electrodeposition, affinity	Reusable surface	Need the presence of specific groups in the bioreceptor molecule (e.g., histidine, lysine, arginine)	[89–93]

New materials for biosensors, biochips and molecular bioelectronics

Sol–gel/entrapment	Chemical, thermal, photochemical and structural stability; one-step procedure at ambient or low temperature	Diffusion barriers; problems of reproducibility and control of pore size	[94–98,106]
Nanomaterials, carbon nanotubes, metal nanoparticles/adsorption, electro- and photopolymerization	Control of pore size and morphology; high electro chemical active surface area; high conductivity; suitable for development of miniaturized biosensor devices	Generally require further functionalization steps; agglomeration and aggregation of nanoparticles are difficult to control	[114–118]
Nanoscale interdigitated/covalent binding	Suitable for miniaturization; enhanced sensitivity; low cost production (usually lithography)		[172,173]
Composite/simple incorporation	Extremely simple, single-step procedure	Possible linkage; high amount of bioreceptor	[176–179]

that would allow the systematic tailoring of the material properties for a wide range of analytes. Such rational approach could lead to more powerful biosensing devices. One solution for the future is to use a so-called *"hybrid device"* as an intermediate approach in which molecular functionality is used to establish the characteristics of electronic device.

Currently, many laboratories have focused their efforts on the development of *innovative materials*: new polymers such as molecular imprinted [208–210], specific bioconjugates and catalytic materials that further permit biomolecule immobilization in a controlled manner [91,211–214], complementary metal oxide semiconductor, array microchip [215] or SAMs in alternate mixture (thiols with long and short chain) [186,216]. In addition, there is a continuous need for new *functionalization strategies*: magnetic permeability measurements [217], SAM—three-dimensional sol–gel network [218], layered double hydroxides [219]. Such material would most likely result in enhanced analytical characteristics of the biosensors, particularly with respect to response time, stability and lifetime. Furthermore, the successful use of biosensors in practical applications is strongly dependent on how efficient the biological molecule is attached onto the transducer surface.

For instance, Cassageneau and Caruso designed a new type of polymerization strategy named conjugated polymer inverse opals for potentiometric biosensing. In this case, the enzyme creatinine deiminase (CD) was distributed within ordered arrays of hollow PPy capsules of wall thickness 70–75 nm on ITO electrode. Initially, colloidal crystals were prepared and used as a template to electrochemically grow CD-doped PPy within the interstice prior to the dissolution of polystyrene template in THF solution [220]. In this case, an enhancement of the sensing performance was found independent of the film thickness. Inverse opal technique is expected to be suitable for a broad spectrum of sensing components, yielding highly sensitive biosensors [221].

7.6 CONCLUSIONS

Currently, new advances in material science with application in the fabrication of biologically functional surface demand very sophisticated procedures for both preparation and characterization, including dry and wet chemical layers deposition and nano- and microfabrication [222]. While most biosensors are designed to work in mild conditions, the development of new devices for harsh environment is a necessity. However, whatever the final application of the biosensor device is, the electrode material used for their construction plays a very important role for achieving the desired performance.

Multiple sensor arrays containing single or multiple selective bioagents are expected to play a very important role in environmental analysis [223–225], clinical [226], drug discovery [227], genetically modified organisms [228] and bio-warfare agents [229,230]. There is no doubt that biosensors represent the solution for many important problems encountered in conventional analytical chemistry. However, these devices require further optimization with respect to the biomaterial stability and/or minimization of the signal dependence upon the physico-chemical parameters of transducers. Recent developments in the area of new materials can considerably improve the analytical performance of the existing biosensors.

REFERENCES

1   D.R. Thévenot, K. Toth, R.A. Durst and G.S. Wilson, *Biosens. Bioelectron.*, 16 (2001) 121–131.
2   A.P.F. Turner, I. Karube and G.S. Wilson, In: A.P.F. Turner (Ed.), *Biosensors: Fundamentals and Applications*. Oxford University Press, Oxford, 1987.
3   P.D. Patel, *Trends Anal. Chem.*, 21(2) (2002) 96–115.
4   A.F. Collings and F. Caruso, *Rep. Prog. Phys.*, 60 (1997) 1397–1445.
5   K.R. Rogers and J.N. Lin, *Biosens. Bioelectron.*, 7 (1992) 317–321.
6   K.R. Rogers, *Biosens. Bioelectron.*, 10 (1995) 533–541.
7   E.C. Alocilja and S.M. Radke, *Biosens. Bioelectron.*, 18 (2003) 841–846.
8   B. Kasemo, *Surf. Sci.*, 500 (2002) 656–677.
9   M. Mehrvar, C. Bis, J.M. Scharer, M. Moo-Young and J.H. Luong, *Anal. Sci.*, 16 (2000) 677–692.
10   A.A. Suleiman and G.G. Guilbault, *Analyst*, 119 (1994) 2279–2284.
11   C.L. Baird and D.G. Myszka, *J. Mol. Recognit.*, 14 (2001) 261–268.
12   J.E. Pearson, A. Gill and P. Vadgama, *Ann. Clin. Biochem.*, 37 (2000) 119–145.
13   E. Brynda, M. Houska, A. Brandenburg and A. Wikerstal, *Biosens. Bioelectron.*, 17 (2002) 665–675.
14   H. Ti Tien, *Mater. Sci. Eng.*, C3 (1995) 7–12.
15   (a) J.C. Love, D.B. Wolfe, R. Haasch, M.L. Chabinyc, K.E. Paul, G.M. Whitesides and R.G. Nuzzo, *J. Am. Chem. Soc.*, 125(9) (2003) 2597–2609; (b) G.B. Sigal, C. Bamdad, A. Barberis, J. Strominger and G.M. Whitesides, *Anal. Chem.*, 68(3) (1996) 490–497; (c) Th. Wink, S.J. van Zuilen, A. Bult and W.P. van Benneckom, *Analyst*, 122 (1997) 43R–50R.
16   V.M. Mirsky, M. Riepl and O.S. Wolfbeis, *Biosens. Bioelectron.*, 12(9–10) (1997) 977–989.
17   J.J. Gooding and D.B. Hibbert, *Trends Anal. Chem.*, 18(8) (1999) 525–532.

18 Y.T. Kim, R.L. McCarley and A.J. Bard, *Langmuir*, 9 (1993) 1941–1944.
19 M. Riepl, V.M. Mirsky, I. Novotny, V. Tvarozek, V. Rehacek and O.S. Wolfbeis, *Anal. Chim. Acta*, 392 (1999) 77–84.
20 D.E. Weisshaar, B.D. Lamp and M.D. Porter, *J. Am. Chem. Soc.*, 114 (1992) 5860–5862.
21 M.C. Pirrung, J.D. Davis and A.L. Odenbaugh, *Langmuir*, 16 (2000) 2185–2191.
22 L.M. Shamansky, C.B. Davis, J.K. Stuart and W.G. Kuhr, *Talanta*, 55 (2001) 909–918.
23 R.J. Willicut and R.L. McCarley, *Adv. Mater.*, 7 (1995) 759–763.
24 S. Storri, T. Santoni, M. Minunni and M. Mascini, *Biosens. Bioelectron.*, 13(3–4) (1998) 347–357.
25 S. Tombelli, M. Mascini and A.F.P. Turner, *Biosens. Bioelectron.*, 17(11–12) (2002) 929–936.
26 X.C. Lee, J.Z. Xing, J. Lee, S.A. Leadon and M. Weinfled, *Science*, 280 (1998) 1060–1069.
27 L.D. Mello and L.T. Kubota, *Food Chem.*, 77 (2002) 237–256.
28 Y.G. Li, Y.X. Zhou, J.L. Feng, Z.H. Jiang and L.R. Ma, *Anal. Chim. Acta*, 382 (1999) 277–282.
29 X.D. Dong, J. Lu and C. Cha, *Bioelectrochem. Bioenerg.*, 42 (1997) 63–69.
30 J.E.T. Anderson, K.G. Olesen, A.I. Danilov, C.E. Foverskov, P. Moller and J. Ulstrup, *Bioelectrochem. Bioenerg.*, 44 (1997) 57–63.
31 J. Parellada, A. Narvaez, M.A. Lopez, E. Domínguez, J.J. Fernandez, V. Pavlov and I. Katakis, *Anal. Chim. Acta*, 362 (1998) 47–57.
32 I.C. Nistor, J. Emnéus, L. Gorton and A. Ciucu, *Anal. Chim. Acta*, 387 (1999) 309–326.
33 L. Rover, L. Tatsuo Kubota and N. Fenalti Hoehr, *Clin. Chim. Acta*, 308 (2001) 55–67.
34 P. Skladal, M. Fiala and J. Krejci, *Int. J. Environ. Anal. Chem.*, 65 (1996) 139–148.
35 H.-S. Lee, Y.A. Kim, Y.A. Chao and Y.T. Lee, *Chemosphere*, 46 (2002) 571–576.
36 J.J. Rippeth, T.D. Gibson, J.P. Hart, I.C. Hartley and G. Nelson, *Analyst*, 122 (1997) 1425–1429.
37 F. Palmisano, G.E. De Benedetto and C.G. Zambonin, *Analyst*, 122 (1997) 365–369.
38 M. Manning, S. Harvey, P. Galvin and G. Redmond, *Mater. Sci. Eng.*, C23 (2003) 347–351.
39 M.A. Breimer, Y. Gelfand and O.A. Sadik, *Biosens. Bioelectron.*, 18 (2003) 1135–1147.
40 J. Wang, *Anal. Chim. Acta*, 469 (2002) 63–71.
41 M. Mascini, *Pure Appl. Chem.*, 73(1) (2001) 23–30.
42 J. Wang, *Anal. Chem.*, 71 (1999) 328–332.
43 F.S. Ligler, M.A. Breimer, J.P. Golden, D.A. Nivens, J.P. Dodson, T.M. Green, D.P. Haders and O.A. Sadik, *Anal. Chem.*, 74 (2002) 713–719.

44 F. Yan and O.A. Sadik, *J. Am. Chem. Soc.*, 123 (2001) 11335–11340.
45 F. Yan and O.A. Sadik, *Anal. Chem.*, 73 (2001) 5272–5280.
46 F. Yan, A. Erdem, B. Meric, K. Kerman, M. Ozsoz and O.A. Sadik, *Electrochem. Commun.*, 3 (2001) 224–228.
47 O.A. Sadik, H. Xu, E. Gheorghiu, D. Andreescu, C. Balut, M. Gheorghiu and D. Bratu, *Anal. Chem.*, 74 (2002) 3142–3150.
48 W. Göpel, *Biosens. Bioelectron.*, 13 (1998) 723–728.
49 V. Tvarozek, T. Hianik, I. Novotny, V. Rehacek, W. Ziegler, R. Ivanic and M. Andel, *Vacuum*, 50(3–4) (1998) 251–262.
50 M.E.G. Lyons, In: M.E.G. Lyons (Ed.), *Electroactive Polymer Chemistry Part 1 Fundamentals*. Plenum Press, New York, 1994, Chapter 1.
51 S. Thanakhasai, S. Yoshida and T. Watanabe, *Anal. Sci.*, 19 (2003) 665–669.
52 O.A. Sadik, *Electroanalysis*, 11 (1999) 839–844.
53 A. Sargent, T. Loi, S. Gal and O.A. Sadik, *J. Electroanal. Chem.*, 470 (1999) 144–156.
54 A. Malinauskas, *Polymer*, 42 (2001) 3957–3972.
55 (a) O.A. Sadik and D.M. Witt, *Environ. Sci. Technol.*, 33(17) (1999) 368; (b) W. Land, O.A. Sadik, D. Leibensperger and M. Breimer, Using Support Vector Machines for Development and Testing Intelligent Sensors to Combat Terrorism. In: C.H. Dagli, A.L. Buczak, J. Ghosh, M.J. Embrechts, O. Ersoy and S.L. Kernel, *ANNIE*, 12 (2002) 811.
56 (a) S. Cosnier, *Biosens. Bioelectron.*, 14 (1999) 443–456; (b) P.N. Bartlett and P.R. Birkin, *Synth. Met.*, 61 (1993) 15–21.
57 (a) O.A. Sadik and J.M. Van Emon, *Chem Tech*, 27 (1997) 38; (b) A. Mulchandani and O.A. Sadik, Environmental Chemical Sensors & Biosensors, *ACS Symposium Series*, 762 (2000); (c) W. Schuhmann, *Mikrochim. Acta*, 121 (1995) 1–29.
58 M. Trojanowicz and T. Krawczynski vel Krawczyk, *Mikrochim. Acta*, 121 (1995) 167–181.
59 P.N. Bartlett and J.M. Cooper, *J. Electroanal. Chem.*, 362 (1993) 1–12.
60 M.E. Tess and J.A. Cox, *J. Pharm. Biomed. Anal.*, 19 (1998) 55–68.
61 M.A. Breimer, G. Yevgeny, S. Sy and O.A. Sadik, *Nano Lett.*, 1(6) (2001) 305–308.
62 (a) D. Sazou and C. Georgolios, *J. Electroanal. Chem.*, 429 (1997) 81–93; (b) R. Mazeikiene and A. Malinauskas, *Eur. Polym. J.*, 36 (2000) 1347–1355.
63 J.C. Lacroix, J.L. Camalet, S. Aeiyach, K. Chane-Ching, J. Petitjean, E. Chauveau and P.C. Lacaze, *J. Electroanal. Chem.*, 481 (2000) 76–81.
64 L.C. Lacroix, F. Maurel and P.C. Lacaze, *J. Am. Chem. Soc.*, 123 (2001) 1989–1996.
65 D.E. Tallman, C. Vang, G.G. Wallace and G.P. Bierwagen, *J. Electrochem. Soc.*, 149 (2002) 173–180.
66 H.S.O. Chan and S.C. Ng, *Prog. Polym. Sci.*, 23 (1998) 1167–1231.
67 N.C. Foulds and C.R. Lowe, *J. Chem. Soc., Faraday Trans. 1*, 82 (1986) 1259–1264.

68 N.F. Almeida, E.J. Beckman and M.M. Ataai, *Biotechnol. Bioeng.*, 42 (1993) 1037–1045.
69 S. Sadki, P. Schottland, N. Brodie and G. Sabouraud, *Chem. Soc. Rev.*, 29 (2000) 283–293.
70 O.A. Sadik and G.G. Wallace, *Anal. Chim. Acta*, 279 (1993) 209–212.
71 O.A. Sadik, M. John and G.G. Wallace, *Analyst*, 119 (1994) 1997–2000.
72 A. Sargent and O.A. Sadik, *Anal. Chim. Acta*, 376 (1998) 125–131.
73 G.G. Wallace, M. Smyth and H. Zhao, *Trends Anal. Chem.*, 18(4) (1999) 245–251.
74 E.C. Hernandez, A. Witkowski, S. Daunert and L.G. Bachas, *Mikrochim. Acta*, 121 (1995) 63–72.
75 H. Rockel, J. Hubner, R. Gleiter and W. Schuhmann, *Adv. Mater.*, 6 (1994) 567–571.
76 W. Schuhmann, *Synth. Met.*, 41–43 (1991) 429–432.
77 A. Dupont-Filliard, A. Roget, T. Livache and M. Billon, *Anal. Chim. Acta*, 449 (2001) 45–50.
78 S. Cosnier and A. Le Pellec, *Electrochim. Acta*, 44(11) (1999) 1833–1836.
79 T.W. Lewis, G.G. Wallace and M.R. Smith, *Analyst*, 124 (1999) 213–219.
80 O. Oureghi, A. Senillou, N. Jaffrezic-Renault, C. Martelet, H. Ben Ouada and S. Cosnier, *J. Electroanal. Chem.*, 501 (2001) 62–69.
81 K. Haupt, *Analyst*, 126 (2001) 747–756.
82 R.J. Ansell, O. Ramstrom and K. Mosbach, *Clin. Chem.*, 42 (1996) 1506–1512.
83 F.L. Dickert, M. Tortschanoff, W.E. Bulst and G. Fischerauer, *Anal. Chem.*, 71 (1999) 4559–4563.
84 B. Jakoby, G.M. Ismail, M.P. Byfield and M.J. Vollekop, *Sens. Actuators A*, 76 (1999) 93–97.
85 S.P. Sakti, S. Rösler, R. Lucklum, P. Hauptmann, F. Bühling and S. Ansorge, *Sens. Actuators A*, 76 (1999) 98–102.
86 H.S. Ji, S. McNiven, K.H. Lee, T. Saito, K. Ikebukuro and I. Karube, *Biosens. Bioelectron.*, 15 (2000) 403–409.
87 T.A. Sergeyeva, S.A. Piletsky, A.A. Brovko, E.A. Slinchenko, L.M. Sergeeva, T.L. Panasyuk and A.V. El'skaya, *Analyst*, 124 (1999) 331–334.
88 M. Jakusch, M. Janotta, B. Mizaikoff, K. Mosbach and K. Haupt, *Anal. Chem.*, 71 (1999) 4786–4791.
89 Z. Gao, G. Binyamin, H.H. Kim, S.C. Barton, Y. Zhang and A. Heller, *Angew. Chem. Int. Ed.*, 41(5) (2002) 810–813.
90 S. Cosnier, B. Galland, C. Gondran and A. Le Pellec, *Electroanalysis*, 10 (1998) 808–813.
91 (a) S. Andreescu, D. Fournier and J.L. Marty, *Anal. Lett.*, 39(9) (2003) 1865–1885; (b) S. Andreescu, V. Magearu, A. Lougarre, D. Fournier and J.L. Marty, *Anal. Lett.*, 34(4) (2001) 171–180.
92 M.S. Nieuwenhuizen and J.L.N. Harteveld, *Sens. Actuators*, B40 (1997) 167–173.
93 A.L. Jenkins, O.M. Uy and G.M. Murray, *Anal. Chem.*, 71 (1999) 373–378.

94 L.L. Hench and J.K. West, *Chem. Rev.*, 90 (1990) 33–79.
95 I. Gill and A. Ballesteros, *Trends Biotechnol.*, 18 (2000) 282–296.
96 B.C. Dave, B. Dunn, J.S. Valentine and J.I. Zink, *Anal. Chem.*, 66 (1994) 1120A–1127A.
97 D. Avnir, S. Braun, O. Lev and M. Ottolenghi, *Chem. Mater.*, 6 (1994) 1605–1614.
98 B. Dunn and J.I. Zink, *Chem. Mater.*, 9 (1997) 1431–1434.
99 B. Li, L. Niu, W. Kou, Q. Deng, G. Cheng and S. Dong, *Anal. Biochem.*, 256 (1998) 130–132.
100 X. Chen and S. Dong, *Biosens. Bioelectron.*, 18 (2003) 999–1004.
101 B. Wang, B. Li, Q. Deng and S. Dong, *Anal. Chem.*, 70 (1998) 3170–3174.
102 S. Dong and X. Chen, *Rev. Mol. Biotechnol.*, 82 (2002) 303–323.
103 M. Albareda-Sirvent, A. Merkoci and S. Alegret, *Sens. Actuators*, B69 (2000) 153–163.
104 I. Gill and A. Ballesteros, *Trends Biotechnol.*, 18 (2000) 282–296.
105 J. Wang, P.V.A. Pamidi and D. Su Park, *Anal. Chem.*, 68 (1996) 2705–2708.
106 S. Andreescu, L. Barthelmebs and J.L. Marty, *Anal. Chim. Acta*, 464 (2002) 171–180.
107 V. Glezer and O. Lev, *J. Am. Chem. Soc.*, 115 (1993) 2533–2534.
108 S. Yang, Y. Lu, P. Atanasov, E. Wilkins and X. Long, *Talanta*, 47 (1998) 735–743.
109 U. Narang, P.N. Prasad, F. Bright, K. Ramanathan, N. Kumar, B. Malhorta, M. Kamalasana and S. Chandra, *Anal. Chem.*, 66 (1994) 3139–3144.
110 J. Wang, *Anal. Chim. Acta*, 399 (1999) 21–27.
111 S. Iijima, *Nature*, 354 (1991) 56–58.
112 P.M. Ajayan, *Chem. Rev.*, 99 (1999) 1787–1799.
113 L. Vayssieres, A. Hagfelt and S.E. Lindquist, *Pure Appl. Chem.*, 72(1) (2000) 47–52.
114 W.C.W. Chan and S.M. Nie, *Science*, 281 (1998) 2016–2021.
115 R.H. Baughman, A.A. Zakhidov and W.A. de Heer, *Science*, 297 (2002) 787–792.
116 M.D. Rubianes and G.A. Rivas, *Electrochem. Commun.*, 5 (2003) 689–694.
117 J.J. Davis, K.S. Coleman, B.R. Azamian, C.B. Bagshaw and M.L.H. Green, *Chem. Eur. J.*, 9 (2003) 3732–3739.
118 G. De la Torre, W. Blau and T. Torres, *Nanotechnology*, 14 (2003) 765–771.
119 M. Shim, A. Javey, N. Wong, S. Kam and H. Dai, *J. Am. Chem. Soc.*, 123 (2001) 11512–11513.
120 J. Kong and H. Dai, *J. Phys. Chem.*, B105 (2001) 2890–2893.
121 R.J. Chen, Y. Zhang, D. Wang and H. Dai, *J. Am. Chem. Soc.*, 123 (2001) 3838–3839.
122 S. Banerjee and S.S. Wong, *Nano Lett.*, 2(1) (2002) 49–53.
123 Q. Chen, L. Dai, M. Gao, S. Huang and A. Mau, *J. Phys. Chem.*, B105 (2001) 618–622.
124 G.S. Duesberg, M. Burghard, J. Muster, G. Philipp and S. Roth, *Chem. Commun.*, 17 (1998) 435–436.

125 V. Krstic, G.S. Duesberg, J. Muster, M. Burghard and S. Roth, *Chem. Mater.*, 10 (1998) 2338–2340.
126 R.J. Chen, Y. Zhang, D. Wang and H. Dai, *J. Am. Chem. Soc.*, 123 (2001) 3838–3839.
127 J.N. Coleman, A.B. Dalton, S. Curran, A. Rubio, A.P. Davey, A. Drury, B. McCarthy, B. Lahr, P.M. Ajayan, S. Roth, R.C. Barklie and W.J. Blau, *Adv. Mater.*, 12 (2000) 213–216.
128 T.W. Ebbesen, H. Hiura, M.E. Bisher, M.M.J. Treacy, J.L. Shreeve-Keyer and R.C. Haushalter, *Adv. Mater.*, 8 (1996) 155–159.
129 F. Balvoine, P. Schultz, C. Richard, V. Mallouh, T.W. Ebbesen and C. Miokowski, *Angew. Chem. Int. Ed.*, 38 (1999) 1912–1915.
130 R.J. Chen, Y. Zhang, D. Wang and H. Dai, *J. Am. Chem. Soc.*, 123 (2001) 3838–3839.
131 Z. Guo, P.J. Sadler and S.C. Tsang, *Adv. Mater.*, 10 (1998) 701–703.
132 J. Wang and M. Musameth, *Anal. Chem.*, 75(9) (2003) 2075–2079.
133 M. Shim, N.W.S. Kam, R.J. Chen, Y. Li and H. Dai, *Nano Lett.*, 2(4) (2002) 258–288.
134 S.C. Tsang, Z. Guo, Y.K. Chen, M.L.H. Green, H.A.O. Hill, T.W. Hambley and P.J. Sadler, *Angew. Chem. Int. Ed.*, 36 (1998) 22198.
135 M. Shim, N.W.S. Kam, R.J. Chen, Y. Li and H. Dai, *Nano Lett.*, 2 (2002) 285–288.
136 Z.J. Guo, P.J. Sadler and S.C. Tsang, *Adv. Mater.*, 10 (1998) 701–703.
137 F. Balavoine, P. Schultz, C. Richard, V. Mallouh, T.W. Ebbesen and C. Mioskowski, *Angew. Chem. Int. Ed.*, 38 (1999) 1912–1915.
138 P.G. Collins, K. Bradley, M. Ishigami and A. Zettl, *Science*, 287 (2000) 1801–1804.
139 R.J. Chen, Y.G. Zhan, D.W. Wang and H.J. Dai, *J. Am. Chem. Soc.*, 123 (2001) 3838–3839.
140 B.F. Erlanger, B.-X. Chen, M. Zhu and L. Brus, *Nano Lett.*, 1 (2001) 465–467.
141 J.L. Bahr and J.M. Tour, *J. Mater. Chem.*, 12 (2002) 1952–1958.
142 S.S. Wong, E. Joselevich, A.T. Woolley, C.L. Cheung and C.M. Lieber, *Nature*, 394 (1998) 52–55.
143 J. Liu, A.G. Rinzler, H.J. Dai, J.H. Hafner, R.K. Bradley, P.J. Boul, A. Lu, T. Iverson, K. Shelimov, C.B. Huffman, F.J. Rodriguez-Macias, Y.S. Shon, T.R. Lee, D.T. Colbert and R.E. Smalley, *Science*, 280 (1998) 1253–1256.
144 D.B. Mawhinney, V. Naumenko, A. Kuznetsova and J.T. Yates, *J. Am. Chem. Soc.*, 122 (2000) 2383–2384.
145 E.T. Mikelson, C.B. Huffman, A.G. Rinzler, R.E. Smalley, R.H. Hauge and L.J. Margrave, *Chem. Phys. Lett.*, 296 (1998) 188–194.
146 N. Tagmatarchis, V. Georgakilas, M. Prato and H. Shinohara, *Chem. Commun.*, 18 (2002) 2010–2011.
147 M.A. Hamon, H. Hui, P. Bhowmik, H.M.E. Itkis and R.C. Haddon, *Appl. Phys. A*, 74 (2002) 333–338.
148 F. Pompeo and D.E. Resasco, *Nano Lett.*, 2 (2002) 369–373.

149 Y. Chen, R.C. Haddon, S. Fang, A.M. Rao, P.C. Eklund, W.H. Lee, E.C. Dickey, E.A. Grulke, J.P. Pendergrass, A. Chavan, B.E. Haley and R.E. Smalley, *J. Mater. Res.*, 13 (1998) 2423–2431.

150 V. Georgakilas, K. Kordatos, M. Prato, D.M. Guldi, M. Holzinger and A. Hirsch, *J. Am. Chem. Soc.*, 124 (2002) 760–761.

151 C.V. Nguyen, L. Delzeit, A.M. Cassell, J. Li, J. Han and M. Meyyappan, *Nano Lett.*, 2 (2002) 1079–1081.

152 J.L. Bahr, J. Yang, D.V. Kosynkin, M.J. Bronikowski, R.E. Smalley and J.M. Tour, *J. Am. Chem. Soc.*, 123 (2001) 6536–6542.

153 S.E. Kooi, U. Schelecht, M. Burghard and K. Kern, *Angew. Chem. Int. Ed.*, 41 (2002) 1353–1355.

154 B.R. Azamian, K.S. Coleman, J.J. Davis, N. Hanson and M.L.H. Green, *Chem. Commun.*, 4 (2002) 366–367.

155 P.J. Britto, K.S.V. Santhanam and P.M. Ayajan, *Bioelectrochem. Bioenerg.*, 41 (1996) 121–125.

156 J. Wang, M. Li, Z. Shi, N. Li and Z. Gu, *Anal. Chem.*, 74 (2002) 1993–1997.

157 M. Musameh, J. Wang, A. Merkoci and Y. Lin, *Electrochem. Commun.*, 4 (2002) 743–746.

158 Y. Zhao, W.D. Zhang, H. Chen and Q.M. Luo, *Talanta*, 58 (2002) 529–534.

159 Z.H. Wang, J. Liu, Q.L. Liang, Y.M. Wang and G. Luo, *Analyst*, 127 (2002) 653–658.

160 J.N. Wohlstadter, J.L. Wilbur, G.B. Sigal, H.A. Biebuyck, M.A. Billadeau, L. Dong, A.B. Fischer, S.R. Gudibande, S.H. Jameison, J.H. Kenten, J. Leginus, J.K. Leland, R.J. Massey and S.J. Wohlstadter, *Adv. Mater.*, 15(14) (2003) 1184–1187.

161 S.E. Baker, W. Cai, T.L. Lasseter, K.P. Weidkamp and R.J. Hamers, *Nano Lett.*, 2(12) (2002) 1413–1417.

162 G. Decher, B. Lehr, K. Lowak, Y. Lvov and J. Schmitt, *Biosens. Bioelectron.*, 9 (1994) 677–684.

163 H. Schmidt, *Mater. Sci. Technol.*, 16 (2000) 1356–1358.

164 E.W. Jager, E. Smela and O. Inganas, *Science*, 290 (2000) 1540–1545.

165 A.J. Haes and R.P. Van Duyne, *J. Am. Chem. Soc.*, 124 (2002) 10596–10604.

166 D. Andreescu and O.A. Sadik, 2003 Eastern Analytical Symposium, 17–20 November (2003) Somerset, NJ, USA.

167 O.D. Velev and E.W. Kaler, *Langmuir*, 15(11) (1999) 3693–3698.

168 J.R. Durrant, J. Nelson and D.R. Klug, *Mater. Sci. Technol.*, 16 (2000) 1345–1348.

169 E. Topoglidis, T. Lutz, R.L. Willis, C.J. Barnett, A.E.G. Cass and J.R. Durrant, *Faraday Discuss.*, 116 (2000) 35–46.

170 M. Albrecht, G. Rodriguez, J. Schoenmaker and G. van Koten, *Org. Lett.*, 2(22) (2000) 3461–3464.

171 K.J. Franz, N. Singh and S.J. Lippard, *Angew. Chem. Int. Ed.*, 39 (2000) 2120–2122.

172 W. Laureyn, D. Nelis, P. Van Gerwen, K. Baert, L. Hermans, R. Magnee, J.J. Pireaux and G. Maes, *Sens. Actuators B*, 68 (2000) 360–370.
173 P. Van Gerwen, W. Laureyn, W. Laureys, G. Huyberechts, M. De Beeck, K. Baert, J. Suls, W. Sansen, P. Jacobs, L. Hermans and R. Martens, *Sens. Actuators B*, 49 (1999) 73–80.
174 S. Sole, A. Merkoci and S. Alegret, *Trends Anal. Chem.*, 20 (2001) 102–110.
175 F. Cespedes and S. Alegret, *Trends Anal. Chem.*, 19 (2000) 276–284.
176 K.R. Rogers, J.Y. Becker and J. Cembrano, *Electrochim. Acta*, 45 (2000) 4373–4379.
177 R.N. Adams, *Anal. Chem.*, 30 (1958) 1576.
178 L. Gorton, *Electroanalysis*, 7 (1995) 23–45.
179 S. Andreescu and O.A. Sadik, *Anal. Chem.*, 76 (2003) 552–560.
180 S. Alegret, F. Cespedes, E. Martinez-Fabregas, D. Martorell, A. Morales, E. Centelles and J. Munoz, *Biosens. Bioelectron.*, 11 (1996) 35–44.
181 J.M. Zen, H.H. Chung and A.S. Kumar, *Anal. Chem.*, 74 (2002) 1202–1206.
182 J. Wang, O. Rincon, R. Polsky and E. Domínguez, *Electrochem. Commun.*, 5 (2003) 83–86.
183 S. Kelley, E. Boon, J. Barton, J. Jackson and M. Hill, *Nucleic Acids Res.*, 27(24) (1999) 4830–4837.
184 J. Wang, *Chem. Eur. J.*, 5 (1999) 1681–1685.
185 I. K'Owino, R. Agarwal and O.A. Sadik, *Langmuir*, 19 (2003) 4344–4350.
186 M. Islam, M. Khanin and O.A. Sadik, *Biomacromolecules*, 4 (2003) 114–121.
187 H. Xu, E. Lee, O.A. Sadik, R. Bakhtiar, J. Drader and C. Hendrikson, *Anal. Chem.*, 71 (1999) 5271–5278.
188 C.J. Picket, S.K. Ibrahim and D.L. Hughes, *Faraday Discuss.*, 116 (2000) 235–244.
189 K.J. Albert, N.S. Lewis, C.L. Schauer, G.A. Sotzing, S.E. Stitzel, T.P. Vaid and D.R. Walt, *Chem. Rev.*, 100 (2000) 2595–2626.
190 I. Lundström, M.S. Shivaraman, C.S. Svenson and L. Lundkvist, *Appl. Phys. Lett.*, 26 (1975) 55–57.
191 P. Bergveld, *Sens. Actuators*, 8 (1985) 109–127.
192 M. Stutzmann, G. Steinhoff, M. Eickhoff, O. Ambacher, C.E. Nebel, J. Schalwig, R. Neuberger and G. Muller, *Diamond Relat. Mater.*, 11 (2002) 886–891.
193 R.W. Boyd and J.E. Heebner, *Appl. Opt.*, 40(31) (2001) 5742–5747.
194 B.J. Luff, R.D. Harris, J.S. Wilkinson, R. Wilson and D.J. Schiffrin, *Opt. Lett.*, 21 (1996) 618–620.
195 B.J. Luff, J.S. Wilkinson, J. Piehler, U. Hollenbach, J. Ingenhoff and N. Fabricius, *J. Lightwave Technol.*, 16 (1998) 583–592.
196 S. Blair and Y. Chen, *Appl. Opt.*, 40 (2001) 570–582.
197 W. Lukosz, *Biosens. Bioelectron.*, 6 (1991) 215–225.
198 S. Sampath and O. Lev, *Anal. Chem.*, 68 (1996) 2015–2021.
199 H. Notsu, T. Tatsuma and A. Fujishima, *J. Electroanal. Chem.*, 523 (2002) 86–92.

200 T.N. Rao, I. Yagi, T. Miwa, D.A. Tryk and A. Fujishima, *Anal. Chem.*, 71 (1999) 2506–2512.
201 C.E. Troup, I.C. Drummond, C. Graham, J. Grice, P. John, J.I.B. Wilson, N.G. Jubber and N.A. Morrison, *Diamond Relat. Mater.*, 7 (1998) 575–580.
202 R.W. Murray, A.G. Ewing and R.A. Durst, *Anal. Chem.*, 59 (1987) 379A–385A.
203 A. Walcarius, *Anal. Chim. Acta*, 384 (1999) 1–16.
204 A. Walcarius, S. Rozanska, J. Bessiere and J. Wang, *Analyst*, 124 (1999) 1185–1190.
205 H. Kotte, B. Grundig, K.D. Vorlop, B. Stehlitz and U. Stottmeister, *Anal. Chem.*, 67 (1995) 65–70.
206 J. Wang and A. Walcarius, *J. Electroanal. Chem.*, 407 (1996) 183–187.
207 J. Wang and A. Walcarius, *J. Electroanal. Chem.*, 404 (1996) 237–242.
208 R.R. Richardson, J.A. Miller Jr. and W.M. Reichert, *Biomaterials*, 14(8) (1993) 627–635.
209 A. Merkoci and S. Alegret, *Trends Anal. Chem.*, 21 (2002) 717–725.
210 S.A. Piletsky and A.P.F. Turner, *Electroanalysis*, 14 (2002) 317–323.
211 R. Langer, *Chem. Eng. Sci.*, 50(24) (1995) 4109–4121.
212 R. Lewis, *Bioscience*, 39(5) (1989) 288–291.
213 A.J. Wang and G.A. Rechnitz, *Anal. Chem.*, 65 (1993) 3067–3070.
214 J.D. Newman, S.F. White, I.E. Tothill and A.P.F. Turner, *Anal. Chem.*, 67 (1995) 4594–4599.
215 K. Dill, D.D. Montgomery, W. Wang and J.C. Tsai, *Anal. Chim. Acta*, 444 (2001) 69–78.
216 V.M. Mirsky, *Trends Anal. Chem.*, 21 (2002) 439–450.
217 C.B. Kriz, K. Radevik and D. Kriz, *Anal. Chem.*, 68 (1996) 1966–1970.
218 J. Jia, B. Wang, A. Wu, G. Cheng, Z. Li and S. Dong, *Anal. Chem.*, 74 (2002) 2217–2223.
219 D. Shan, S. Cosnier and C. Mousty, *Anal. Chem.*, 75 (2003) 3872–3879.
220 T. Cassagneau and F. Caruso, *Adv. Mater.*, 14 (2002) 1837–1841.
221 T. Cassagneau and F. Caruso, *Adv. Mater.*, 14 (2002) 1629–1633.
222 P.R. Coulet, *J. Membr. Sci.*, 68 (1992) 217–228.
223 C.M.A. Brett, *Pure Appl. Chem.*, 73(12) (2001) 1969–1977.
224 S. Bender and O.A. Sadik, *Environ. Sci. Technol.*, 32 (1998) 788–797.
225 M.P. Marco and D. Barcelo, *Meas. Sci. Technol.*, 7 (1996) 1547–1562.
226 P.B. Luppa, L.J. Sokoll and D.W. Chan, *Clin. Chim. Acta*, 314 (2001) 1–26.
227 M. Keugsen, *Naturwissenchaften*, 89 (2002) 433–444.
228 M. Minunni, S. Tombelli, E. Mariotti, M. Mascini and M. Mascini, *Fresenius J. Anal. Chem.*, 369 (2001) 589–593.
229 S.S. Iqbal, M.W. Mayo, J.G. Bruno, B.V. Bronk, C.A. Batt and J.P. Chambers, *Biosens. Bioelectron.*, 15 (2000) 459–478.
230 O.A. Sadik, W.H. Land and J. Wang, *Electroanalysis*, 15(14) (2003) 1149–1159.

Chapter 8

Electrochemical antibody-based sensors

Judith Rishpon and Virginia Buchner

8.1 ELECTROCHEMISTRY MERGES WITH IMMUNOLOGY

Antibody-based biosensors are based on the principle that antigen–antibody interactions can be transduced directly into a measurable physical signal. Scientists have long endeavoured to combine antibody molecules as the biological sensing element with electrochemical detection systems for sensing applications. In the early 1970s, researchers began exploring the possibility of building direct immunosensors by fixing antibodies to piezoelectric or potentiometric transducers, but a report by Aizawa et al. in 1979 [1] was the first to pave the way for the merger of immunology with electrochemistry. The authors described the principle of amperometric immunosensors whose selectivity depends on the immunochemical affinity of an antigen for its corresponding antibody bound to a membrane or electrode. The concentration of the antigen is converted to an electrical signal that allows for both quantitative and qualitative measurements of the analyte in real time. Warsinke et al. [2] defined immunosensors as devices in which the immunoreactant is intimately associated with or within a physicochemical transducer or transducing microsystem.

Antibody-based electrochemical systems allow the rapid and continuous analysis of a binding event without requiring added reagents or separation/washing steps. Combining the sensitivity and specificity of the antigen–antibody binding interaction with the high sensitivity and relative simplicity of modern electroanalytical techniques has produced immunoelectrochemical sensors that have a potential in such areas as clinical diagnostics, environmental monitoring, food and water quality control, and bioprocess analysis.

In this chapter, we will describe the most frequently applied principles in electrochemical immunoassays. We first consider the principles of antigen–antibody reactions and their incorporation into various conventional immunoassay systems. Next, we discuss the basic principles of electrochemical detection and then examine the merging of electronics and immunology into antibody-based electrochemical analytical systems.

8.2 PRINCIPLES OF CONVENTIONAL IMMUNOASSAYS

8.2.1 Brief historical synopsis

Immunoassays are based on the capture of an antigen (analyte) or a group of related target analytes by a selective antibody. The hallmark of the antibody-mediated immune response is specificity, first characterized as early as 1928 by Landsteiner and van der Scheer [3], who showed that an immunologic method can select between the optical isomers of phenyl-(*p*-aminobenzoylamino) acetate. When an analyte is present in complex matrices in concentrations $<10^{-6}$ M, the method of choice with respect to sensitivity, cost, time, and handling is the antibody-based immunoassay. The field of immunochemistry evolved from clinical serology, which for decades had been using various forms of antibody-binding reactions for medical diagnostic tests and epidemiologic studies. Subsequently, immunoassays found use in environmental studies as well. The results of immunochemical methods usually agree with those of traditional chemical methods (like gas chromatography–mass spectrometry), especially when the efficiency of sample preparation steps is comparable. The emergence of monoclonal antibody production and recombinant antibody-fragment technology has further extended the sensitivity of the immunoassay.

8.2.2 The antigen–antibody reaction

By definition, a molecule cannot be an antigen (derived from "*anti*body *gen*erator") without a complementary antibody produced by lymphocytes bearing specific receptors for that antigen. Thus, an antigen is concomitantly a molecule that evokes an antibody response in animals or humans and one that reacts with antibodies, irrespective of its ability to generate them. An antigen *selects* a complementary *specific* antibody from a pool of cells producing antibodies that were formed before the antigen was ever seen. This notion was first put forth by Erlich at the end of the 19th century [4], but began to attract worldwide interest only 70 years later when the Nobel Prize laureate, Sir Mcfarlane Burnet, published his theory of clonal selection [5].

The portion of the antigen (several amino acids long) that is in contact with the antibody is called an *epitope*. Because the immune system specifically differentiates between different organisms or compounds, antigen–antibody complexes show a high specificity, namely, antibodies generated against a given immunogen will not react with unrelated antigens.

The strength of binding reactions (affinity) is important in antibody-based sensors. The term affinity, which describes the binding of an antibody to a single epitope of an antigen, is defined as the calculation of attraction and energy forces between the molecules. Antibodies that bind strongly to an antigen are called *high affinity antibodies*, and the higher the affinity of the antibody that is used, the greater the amount of antigen that will be bound for a given concentration. The reaction between an analyte (Ag) and its complementary antibody (Ab) is reversible, and the complex so formed can easily dissociate, depending on the affinity. Thus,

$$Ag + Ab \rightleftharpoons AgAb. \tag{8.1}$$

The reaction of a high affinity Ab with one epitope of the Ag favours the complex formation.

Because an antigen contains many different immunogenic sites (epitopes), the serum from an immunized animal can contain many structurally different antibodies, each directed toward a different epitope. The serum thus contains *polyclonal* antibodies. The possession of more than one antibody-combining site can cause difficulties with quantitative binding studies because a polyclonal antibody preparation can bind to more than one epitope. The term *avidity* is a measure of the functional affinity of a polyclonal antiserum for a polyvalent antigen. For this reason, haptens are often used for affinity-discriminating work because they contain only one epitope and therefore, they are monovalent with respect to binding to polyclonal antibodies. A hapten is a well-defined low-molecular-weight chemical grouping that is not immunogenic on its own but will react with preformed antibodies elicited by injecting the hapten linked to an immunogenic, high-molecular-weight protein "carrier" molecule, such as keyhole limpet haemocyanin or bovine serum albumin. Chemically coupling the hapten to the carrier produces an immunogenic conjugate that incorporates the structure of the non-immunogenic hapten, so that the antibodies formed will bind uncoupled analyte as well. Thus, antibodies can be generated against almost any analyte, even if the analyte alone is not immunogenic.

The problem of multiple epitopes in an antigen can be overcome by using monoclonal antibodies produced in vitro, a technique first described in 1975 by Milstein and Kohler [6] who received the Nobel Prize for their revolutionary achievement. Antibody-secreting hybridoma cell colonies are cloned by means of limiting dilution, so that only one antibody-producing cell (clone) remains in each well. From this single parental cell, all the progeny secrete the same monoclonal antibody to a single antigenic epitope, even if the antigen has many epitopes. The high specificity and affinity of a monoclonal antibody for its epitope results in specific binding of the analyte. The target epitope can be present in the nano- to picomolar range among hundreds of other substances, even if such substances exceed the concentration of the target by 2 to 3 orders of magnitude.

8.2.3 The antibody (immunoglobulin) molecule

The immunoglobulin (Ig) molecule is a four-polypeptide unit, constructed of two identical heavy (large) and two identical light (small) chains joined by inter-chain disulfide (S–S) linkages to form a flexible "Y"-shape. Before the advent of monoclonal antibody technology, monoclonal antibody-producing clones could be obtained from patients with malignant multiple-myeloma tumours because each tumour produces a different homogeneous Ig molecule. Comparing the amino acid sequences of such proteins revealed variations in structure at the N-terminal portions of their heavy and light chains (termed V_H and V_L, respectively, which are located at the ends of the arms in the "Y", whereas the respective sequences of the other half of the light chain (C_L) and the rest of the heavy chain (C_H) remained relatively constant. The constant region of the heavy chains designates the antibody class (IgG, IgM, IgA, IgD, IgE), and a small, hypervariable region in the V_H and V_L domains contains the antigen-binding site.

The Ig molecule can be split enzymatically into two identical fragments (Fab—antigen binding fragment, located in the variable region), each containing a single antigen-combing site, and a third fragment (Fc—crystallizable fragment) lacking the ability to bind antigen. Antibodies have been reduced in size, dissected into minimal binding fragments, and rebuilt into multivalent high-affinity reagents. The Fv (variable fragment) module, comprising the aligned antibody V_H and V_L domains, is the smallest fragment that retains the full monovalent binding affinity of the intact parent antibody.

8.2.4 Recombinant antibody technology

Because producing Fvs by the proteolysis of intact immunoglobulins is difficult, recombinant single-chain Fv (scFv) molecules, in which a polypeptide linker secures the V_H and V_L domains, are used to imitate Fv fragments. The in vitro method of choice for selecting high-affinity antibody fragments is bacteriophage (phage) display technology. One advantage of this method is avoiding the use of animals for antibody production. Antibody-phage display libraries have been used for generating specific binders of proteins that recognize membrane receptors, continuous epitopes, and soluble proteins. Consequently, recombinant scFvs can be used both for ligand capture/immobilization and for ligand detection—two essential constituents of an immunosensor. Recombinant antibody fragments have been fused to radioisotopes, drugs, toxins, enzymes, and biosensor surfaces.

An efficient system for the production of recombinant antibodies is cellulose-assisted refolding technology, as described by Berdichevsky et al. [7]. The expressed scFvs were fused to a cellulose-binding domain (CBD) from the bacterium *Clostridium thermocellum* in the format scFv–CBD. The resulting fusion proteins were obtained in high yield from bacterially produced inclusion bodies that become solubilized and then refold while immobilized on cellulose. The refolded and purified scFv–CBD fusion proteins can be used to form cellulose-based affinity matrices or, as described herein, can be immobilized on a cellulose matrix that makes up part of the immunoelectrochemical sensor device.

8.2.5 Immunoassay techniques

In immunoassay techniques, antibody–antigen reactions are detected by a reagent conjugated with an appropriate label. Two basic immunoassay schemes are used: homogeneous and heterogeneous. Homogeneous immunoassays, characterized as reactions occurring in solution between a free antigen and a free antibody, can be performed without separation or washing steps. If the capture antibody is bound to a solid surface, an increase in the labelled antigen in solution is measured in the presence of an antigen. Heterogeneous immunoassay refers to the specific adsorption of an antigen onto an antibody that is attached to a solid support. This technique usually necessitates the labelling of the antibody and, in general, a separation after the reaction of the complex antigen–antibody from the free antibody. Association and dissociation constants in heterogeneous immunoassays, however, can significantly differ from the values found in reactions taking place in solution. In solid-phase reactions with high solid-phase binding capacities and low analyte concentrations, the solution in the direct vicinity of the solid phase depletes, thus decreasing the binding rate because of diffusional limitations.

In principle, immunoassays with labelled compounds can be carried out in the following ways. (a) In competitive assays, the labelled analyte competes with the unlabelled analyte for the epitope-binding site of the antibody. The key feature of a competitive assay is that maximal assay sensitivity is attained using an amount of antibody tending to zero. The sensitivity of a competitive immunoassay is determined by the affinity of the antibody for its antigen. As the affinity of an antibody can be in the range of 10^5–10^{12} M^{-1} [8], competitive immunoassays using antibodies with a Kd = 10^{-12} M reach their highest sensitivity in the picomolar range. Varying the assay conditions can decrease the lower detection limit even further. (b) In non-competitive assays (e.g., "sandwich" or sequential saturation assays), a significant excess of antibodies over the antigen is used. Here, diffusion processes are much more important than antibody affinity. The lowest analyte concentration detectable using a competitive design is in the order of 10^7 molecules ml^{-1}, whereas a non-competitive method is potentially capable of measuring concentrations that are lower by several orders of magnitude [9]. Additionally, because of the excess of one immunoreactant, non-competitive assays are usually more rapid than competitive assays.

In heterogeneous assays, the detection is usually performed in a competitive format, in which labelled antigen is added to the sample solution, or in a sandwich format, in which after washing the surface, secondary labelled antibodies are specifically adsorbed onto the immobilized antigens. For the latter format, the antigen must have two different epitopes—one for adsorption onto the capture antibody and one for the secondary antibody adsorption:

$$\boxed{Ab_1\text{—}\langle\bullet Ag\circ\rangle\text{—}{}^*Ab_2} \qquad (8.2)$$

where Ab_1 is the capture antibody to epitope 1 ($\bullet$) on the antigen, and $^*Ab_2$ is the labelled antibody to epitope 2 ($\circ$).

In the competitive format, the signal is *inversely* proportional to the antigen concentration, whereas in the sandwich format, the signal is *directly* proportional to the antigen concentration.

Immunoassays are used to identify various compounds and to detect microorganisms in clinical and environmental specimens. The utility of antibodies for diagnostic and therapeutic applications depends primarily on their affinity and on their kinetic and stability properties. The binding of an antigen to the appropriate antibody is accompanied by small physicochemical changes. In most immunoassays, the binding event is visualized via an auxiliary reaction, in which one immunoreactant is labelled with a substance that can easily be measured by spectrophotometric methods. In 1959, Yalow and Berson [10] were the first to use a radioactive compound as label in an immunoassay, yielding the term "radioimmunoassay" (RIA). Many different labels—enzymes, fluorophors, redox compounds, cofactors, fluorescence quenchers, chemiluminescence metals, latex particles, and liposomes—have been applied in immunoassays. For solid-phase immunoassays, the reactants are immobilized on a solid support, such as plastic or nitrocellulose paper.

8.2.6 Enzyme immunoassays with and without separation

Enzyme immunoassays (EIA) play an important role in clinical diagnostics, veterinary medicine, environmental control, and bioprocess analysis. Antibodies are coupled to enzymes like peroxidase or phosphatase, whose products can be measured after the degradation of a substrate. Because of its high selectivity and sensitivity, EIA enables the detection of a broad spectrum of analytes in complex samples. A solid phase EIA performed in a plastic microtitre plate is called an *enzyme-linked immunosorbent assay* (ELISA). The coloured products produced in the ELISA can be measured spectrophotometrically rather than in a scintillation counter as for the RIA.

Because heterogeneous EIAs usually require multiple incubation and washing steps, automatization and miniaturization are difficult. The ELISA, the most successful standard heterogeneous EIA, is usually performed in a 96-well microtitre plate containing immobilized antibody or antigen, thus allowing the concurrent determination of multiple samples. After sequential incubations of sample, antibody– or antigen–enzyme conjugate, and enzyme–substrate reagents with washing steps in between, the readout is performed by measuring the absorbance or fluorescence of each well with a standard microtitre plate reader.

A sandwich ELISA is used to search for a desired analyte in a test solution, as follows: the solid phase is coated with analyte-specific "capture" antibodies to pull the analyte out of the test sample. After washing, the amount of analyte bound to the solid phase can be determined by adding an excess of enzyme-labelled analyte-specific antibody. The specificity of the method can be improved by using a sandwich-type, two-site assay, in which the capture and labelled antibodies have specificities for different parts of the analyte, as mentioned above. The performance of an immunoassay in a standard microtitre plate requires several hours. Such long incubation times are mostly linked to inefficient mass transport from the solution to the surface, whereas the immunocapture itself is a rapid process.

Antibody-based, fully automated immunosensors for small molecules have been used to detect explosives (see Refs. [11,12]) and biological warfare agents (see Refs. [13,14]), and can be used to analyse drinking water and extracts at hazardous waste sites (for examples, see Refs. [15–17]). In the following section, we will discuss how the addition of electrochemical detection has broadened the capabilities of such devices.

8.3 PRINCIPLES OF ELECTROCHEMICAL ANALYSIS

8.3.1 Electroanalytical measurements

An electrochemical measurement is performed in an electrochemical cell containing at least two electrodes: one is the working electrode and the second a reference electrode. Redox of the analyte takes place on the working electrode, which can be constructed from a variety of materials, including platinum, gold, silver, glassy carbon, nickel, and palladium. The reference electrode is typically made from silver/silver chloride (Ag/AgCl) or saturated calomel (SCE), and the potential of this electrode is known and constant. In amperometric detection, a third auxiliary electrode is often used. The purpose of the auxiliary electrode (counter electrode) is to carry current flow away from the reference electrode.

8.3.2 Screen-printed electrodes

Screen-printing techniques began in Egypt about 3000 years ago for patterned textile production, as well as for building the Great Wall of China. The technique has been applied to the microelectronics industry in the manufacture of printed circuit boards. The advent of screen-printing techniques using metal-loaded inks for the deposition of metal layers onto planar backing supports opens up the possibility of mass production of electrodes. Much of the materials science development required for this task is in place, and a variety of ceramic and plastic strip structures are possible. Because individual electrode costs are kept low by using screen-printing techniques, the electrodes can be discarded after a single use, hence eliminating cross-contamination from reuse and facilitating calibration-free use as dipsticks. Screen-printing technology also provides a high level of precision and reproducibility during bulk manufacture.

Two different types of substrates are commonly employed in screen-printing, ceramic and plastic-based materials. Grennan et al. [18] have characterized the production of a screen-printed electrode for use in an amperometric biosensor system.

8.3.3 Transducers

Electrochemical transducers usually use enzymes or redox compounds to generate an electrochemical signal, either amperometric or potentiometric. The transducer system must acquire signals that are unique to the probe system and generate low noise signals that can be further processed without degradation to provide a human observer with an indication of probe system activity. Potentiometric, amperometric, conductive, and impedance transducers have been applied in direct and indirect electrochemical immunoassays.

The most widely used biosensors measure electrochemical effects. Regardless of the transducer or application involved, Gizeli and Lowe [19] proposed that an ideal system should have the following capabilities:

- to detect and quantify the analyte within the required concentration range and a reasonable time (preferably a few seconds);
- to transduce the binding event without externally added reagents;
- to repeat the measurement on the same device, due to a reversible immunochemical reaction; and
- to detect specific binding of the analyte in real samples.

Currently, immunosensors provide a powerful tool for monitoring biospecific interactions in real time or for deriving information about the binding kinetics of an immunoreaction or the structure/function relationships of molecules. Amperometric, potentiometric, conductive, and impedimetric transducers have been applied in direct and indirect electrochemical immunoassays.

8.3.3.1 Potentiometric transducers
Potentiometric transducers measure the potential under conditions of constant current. This device can be used to determine the analytical quantity of interest, generally the concentration of a certain analyte. The potential that develops in the electrochemical cell is the result of the free-energy change that would occur if the chemical phenomena were to proceed until the equilibrium condition is satisfied. For electrochemical cells containing an anode and a cathode, the potential difference between the cathode electrode potential and the anode electrode potential is the potential of the electrochemical cell. If the reaction is conducted under standard-state conditions, then this equation allows the calculation of the standard cell potential. When the reaction conditions are not standard state, however, one must use the Nernst equation to determine the cell potential. Physical phenomena that do not involve explicit redox reactions, but whose initial conditions have a non-zero free energy, also will generate a potential. An example of this would be ion-concentration gradients across a semi-permeable membrane; this can also be a potentiometric phenomenon and is the basis of measurements that use ion-selective electrodes (ISEs).

An ISE produces a potential that is proportional to the concentration of an analyte. Making measurements with an ISE is therefore a form of potentiometry. The most common ISE is the pH electrode, which contains a thin glass membrane that responds to the hydrogen ion concentration in a solution. The potential difference across an ion-sensitive membrane is as follows:

$$E = K - (2.303RT/nF)\log(a) \tag{8.3}$$

where K is a constant to account for all other potentials, R, the gas constant, T, the temperature, n, the number of electrons transferred, F, Faraday's constant, and a is the activity of an analyte ion.

A plot of measured potential vs. $\log(a)$ will therefore give a straight line.

Because ISEs are susceptible to several interferences, samples and standards are diluted 1:1 with total ionic strength adjuster and buffer (TISAB). The TISAB consists of 1 M sodium chloride to adjust the ionic strength, acetic acid/acetate buffer to control pH, and a metal-complexing agent. The ISEs consist of an ion-selective membrane, an internal reference electrode, an external reference electrode, and a voltmeter.

Direct electrochemical immunoassays use changes in charge densities or conductivities for transduction and do not require any auxiliary reaction. The ability to work label-free is very attractive, especially for the development of in vivo immunosensors because if low-affinity antibodies are applied, then real-time measurement can be accomplished without adding a hazardous reagent. Over two decades ago, Janata [20] observed a potential change when he incubated mannan with a potentiometric electrode covered with PVC membrane-immobilized concanavalin A (ConA). Thereby, the binding process could be followed directly in real time without a labelling reagent. In 1984, Keating and Rechnitz [21] used a potassium-sensitive electrode with a dioxin–ionophore conjugate within a PVC membrane for the determination of anti-dioxin-antibodies. Although many publications (see, for example, Refs. [22–24]) have described similar effects, whether the potential change is a result of disturbed ion transport or of charge changes at the protein surface during binding is not yet completely clear (for a critical review see Ref. [25]).

Constant-current chronopotentiometry has been used for the detection of immunoprecipitation. In an EIA, the biocatalysed precipitation of an insoluble product produced on an electrode was used as an amplification path for biosensing. Chronopotentiometry was suggested by the authors as a rapid transduction means (a few seconds) (see Refs. [26,27]).

8.3.3.2 Amperometric transducers

Amperometric transducers are used more frequently than other transducers in electrochemical immunoassays because they allow the rapid detection of currents over a broad linear range, with detection limits as low as 10^{-10} M. As shown by Weber and Long [28], commercial amperometric biosensors measure currents generated when electrons are exchanged between a biological system and an electrode. Because amperometric detection is a heterogeneous process of electron transfer, electrochemical measurements at electrode interfaces are easier than spectrophotometric measurements to execute when working with very small volumes. In most cases, amperometric devices detect changes in current under conditions of constant potential. Nevertheless, measuring the current resulting from various forms of potential changes has led to a higher sensitivity and to a better understanding of the processes occurring in the vicinity of the electrode.

Cyclic voltammogrammes trace the transfer of electrons during a redox reaction. The reaction begins at a certain potential. As the potential changes, it controls the point at which the redox reaction will occur. Electrodes are placed in an electrolyte solution containing an analyte that will undergo a redox reaction. The system starts off with an initial potential at which no redox reaction can take place. If the potential change starts in the cathodic direction, at a critical potential during the forward scan the electroactive species will begin to be reduced. After the reversal of potential scan direction and depletion of the oxidized species, oxidation takes place. Pemberton et al. [29], Sargent et al. [30], and Anne et al. [31] demonstrated the use of cyclic voltammetry in electrochemical immunosensors. Pulses, differential pulses, and square wave voltammetry are as used for higher sensitivity. Pulsed amperometric techniques can be used to monitor antigen–antibody binding in real time without using a labelled compound. Pulsed amperometric detection (PAD) was used by Sargent and Sadik [32] in combination with a polymer-modified antibody electrode on which an antibody or antigen that binds the analyte of interest is immobilized on the surface of a membrane electrode. The sample is applied in flow injection mode.

In indirect electrochemical immunoassays, the binding reaction is visualized indirectly via an auxiliary reaction by a labelling compound. Because antibodies and antigens are usually not electrochemically active within the desired potential range, redox-active compounds can be applied as labels for indication. Although the ability to work reagent-free (which can be accomplished with potentiometric and capacitive immunoassays) is not possible with indirect amperometric immunoassays, non-specific binding of molecules other than the labelled immunoreactant will not contribute to the signal, which is an advantage. In amperometric immunoassays, the labelling redox compound should have the following properties (from Ref. [2]):

- is electroactive in a potential range of 0 to -200 mV [33] does not cause electrode fouling,
- has no side reactions with the matrix,
- is stable in buffer, and
- chemical groups are available for coupling.

To develop amperometric immunosensor devices, Blonder et al. [34] insulated the electrical contact between a redox protein and an electrode surface upon association of an antibody to an antigen monolayer assembled on the electrode. In one configuration, a mixed monolayer consisting of the N-ε-(2,4-dinitrophenyl (DNP))lysine antigen and ferrocene units acting as electron transfer mediators was applied to sense dinitrophenyl antibody (DNP-Ab) in the presence of glucose oxidase (GOx) and glucose. In the absence of DNP-Ab, the mixed monolayer electrode stimulated the electrocatalysed oxidation of glucose, resulting in an amplified amperometric response. Association of the DNP-Ab to the modified electrode blocked the electrocatalytic transformation. The extent of the electrode insulation by the DNP-Ab is controlled by the Ab concentration in the sample. In the second configuration, an N-ε-(2,4-DNP) lysine antigen monolayer assembled on a gold electrode was applied to sense the DNP-Ab in the presence of a redox-modified GOx, exhibiting electrical communication with the electrode surface. Two kinds of redox-modified "electrically wired" GOx were applied: GOx modified by N-(ferrocenylmethyl)-caproic acid, Fc-GOx, and a novel electrobiocatalyst generated by reconstitution of apo-GOx with a ferrocene-modified FAD semi-synthetic cofactor. Electrocatalytic oxidation of glucose by the electrically wired biocatalysts proceeds in the presence of the antigen monolayer electrode, giving rise to an amplified amperometric signal. The electrocatalytic transformation is blocked by the association of the DNP-Ab to the monolayer electrode. The extent of electrode insulation toward the bioelectrocatalytic oxidation of glucose is controlled by the concentration of DNP-Ab in the samples. The application of biocatalysts for the amperometric sensing of antigen–antibody interactions at the electrode surface makes the electrode insensitive to microscopic pinhole defects in the monolayer assembly. The antigen monolayer electrode is applied to sense the DNP-Ab in a concentration range of 1–50 μg ml^{-1}.

Tiefenauer et al. [35] used a nanostructured gold electrode with an interelectrode space of about 30 nm for an amperometric immunosensor. A hapten was immobilized in that space on SiO_2. In combination with ferrocene- or heme-modified antibodies, this concept can be used to increase the efficiency of redox-labelled immunoassays. One immunoreactant must be immobilized, however. A principle in which antigen, labelled antigen, and antibody can be applied in homogeneous phase is appealing because of the fast complexation rate in solution and the circumvention of the immobilization procedure.

Calvo et al. [36] recently achieved the multilayer immobilization of antibody and redox polymer molecules on a gold electrode as a strategy for the potential development of an amperometric immunosensor. The step-by-step assembly of antibiotin IgG on an Os(bpy)(2)ClPyCH(2)NH poly(allylamine) redox polymer (PAH-Os) adsorbed on thiolated gold electrodes was proved by quartz crystal microbalance (QCM) and atomic force microscopy (AFM) experiments, confirming the electrochemical evidence. The increase of redox charge during the layer-by-layer deposition demonstrated that charge propagation within the layers is feasible. The multilayer structure proved to be effective for the molecular recognition of horseradish peroxidase–biotin conjugate (HRP–biotin), as confirmed by the QCM measurements and the electrocatalytic reduction current obtained upon hydrogen peroxide addition. The catalytic current resulting from PAH-Os mediation increased with the number of assembled layers. Furthermore, the inventory of IgG molecules on the supramolecular self-assembled structure and the specific and non-specific binding of HRP–biotin conjugate were confirmed by the QCM transient studies, giving information on the kinetics of IgG deposition and HRP–biotin conjugate binding to the IgG.

8.3.3.3 Impedimetric transducers
Besides potentiometric and amperometric transducers, impedimetric transducers have been used for the real-time and label-free measurement of the antigen–antibody reaction. Impedimetric devices are based on the principle that for electrolytic capacitors, the capacitance depends on the thickness and the dielectric behaviour of a dielectric layer on the surface of a metal plate. The first impedimetric immunosensor was developed by Newman et al. [37], who used interdigitated copper electrodes on a glass surface isolated by a silicium oxide-coated parylene layer. The use of a silicium oxide-immobilized antigen enables the determination of the antibodies via capacity decrease. Gebert et al. [38,39] described a tantalum strip onto which tantalum oxide was grown electrochemically and used for the immobilization of mouse IgG. Determining anti-mouse-IgG was possible with a measuring range of $0.2-20$ ng ml^{-1}, which corresponds to $1.3 \times 10^{-12}-1.3 \times 10^{-10}$ M. The authors stated that this principle is limited to the determination of relatively large molecules, and that the non-specific binding and noises related to the

fluid or to its movement are problematic. Riepl et al. [40] used a similar approach, but this group used a gold electrode incubated with 16-mercaptohexadecanoic acid to immobilize an anti-HSA-antibody via carbodiimide. Although the stable gold–sulphur binding with this long-chain thiol resulted in low drifting, the sensitivity for the determination of HSA was lower than that for the sensor described by Gebert for IgG. Berggren and Johansson [41] worked on a gold electrode, using thioctic acid as a linker for the immobilization of anti-human chorionic gonadotropin hormone (HCG)-antibodies. For a direct immunosensor, the calculated detection limit for HCG 624 of 15×10^{-15} M (0.5 pg ml^{-1}) is remarkably sensitive. Ameur et al. [42] described direct immunodetection through impedance measurements on gold functionalized electrodes. As gold has the ability to react with sulfhydryl groups, the authors explored three protocols for the fixation of antibodies, based either on the self-assembling properties of functional thiols bearing long alkyl chains or on the possibility of a direct coupling of antibody moieties. The analysis of the electrochemical impedance behaviour of such layers offers a sensitive method for the direct detection of the antibody–antigen interaction. The addition of a redox couple in the tested solution, acting as an amplifier, enables analyte detection as low as a few picograms ml^{-1}. Alfonta and colleagues [26,43] reported the use of impedance measurements in conjunction with enzyme amplification to detect immunoprecipitation. The biocatalysed precipitation of an insoluble product produced on electrode support was used as an amplification path. An antigen monolayer electrode was used to sense the DNP-Ab applying an anti-antibody–HRP conjugate as a biocatalyst for the oxidative precipitation by hydrogen peroxide to yield the insoluble product. When common problems for label-free affinity sensors (including optical), such as limitations for determining low-molecular-weight analytes and matrix effects, have been solved, impedimetric electrochemical immunoassays could have a great potential for direct immunosensing.

8.3.3.4 Conductimetric transducers
Conductimetric devices detect changes in conductivity between two electrodes. Conductivity simply provides a measure of the ionic concentration and mobility in a solution but can be exploited for biosensing by using an enzyme to drive ionic speciation changes. Any enzymatic reaction can change solution conductivity if it generates a net change in the concentration of an ionized species. Substances that can be assayed in this manner include paracetamol, creatinine, L-asparagine, penicillin G, and glucose, as shown by Thompson et al. [44]. Yet measuring conductivity is difficult because of the variable ionic background of clinical samples and the relatively small

conductivity changes that occur in such high ionic strength solutions. A second, comparative "blank" electrode must be used, but variable drift at two separate electrodes poses a universal drawback. An immunosensor for urine methamphetamine (MA) was described by Yagiuda et al. [45], who also used surface conductivity. Anti-MA antibody was immobilized onto platinum electrodes (surface area 0.95 cm^2, 0.5 mm apart) by crosslinking with glutaraldehyde. Following incubation with the sample, the subsequent binding reaction between anti-MA and MA led to a decreased conductivity of the immobilized anti-MA layer between the electrodes. The sensor had a linear range of 1 ± 10 mg l^{-1} MA in urine, with a detection limit of 0.1 mg l^{-1}.

8.3.4 The recognition element

The antibody is an ideal recognition element for bioassays and biosensors. Like enzyme–substrate or hormone–receptor interactions, epitope–antibody recognition usually occurs through a fairly accurate complementarily in shape, allowing the ligands to come close to each other so that normal intermolecular forces are permitted to become relatively strong. On the basis of the immunoassay format used, the assay can be either direct (the immunochemical reaction is directly determined by measuring the physical changes induced by the formation of the complex) or indirect (a sensitively detectable label is combined with the antibody or antigen of interest, as in the ELISA technique). A sufficiently high selectivity can be obtained by using non-labelled immunosensors, although non-specific adsorption onto the biorecognition surface remains a major problem in direct systems.

8.3.5 Construction of an electroimmunosensor

Combining ELISA techniques with biosensors to form electrochemical immunosensors has increased the range, speed, and sensitivity of immunoassays. For antigen detection, the electrode is coated with (immobilized) antibody, specific for the antigen of interest, and immersed in a solution containing an unknown concentration of the antigen. In a competition format, the solution contains a mixture of a known amount of antigen–enzyme conjugate plus an unknown concentration of sample antigen. After a suitable period, the antigen and antigen–enzyme conjugate will be distributed between the bound and free states, dependent upon their relative concentrations. The unbound material is washed off, discarded, and the amount of antigen–enzyme conjugate bound is determined directly from the transduced signal. More advanced immunosensors use the direct detection of antigen bound to the antibody-coated surface of the biosensor. Figures 8.1–8.4 depict in graphic form simple configurations for four different electrochemical immunoassay techniques, in which the biosensor merely replaces the traditional colorimetric detection system (see Section 8.2.5 for immunoassay details).

Figure 8.1a depicts the simple, direct electrochemical detection of antigen or antibody, in which analyte capture by the antibody-coated electrode is directly measured by the changes in charge density or conductivity. Figure 8.1b shows the opposite approach, using an antigen-coated electrode. The scheme of an electrochemical competition immunoassay is shown in Fig. 8.2, in which the labelled analyte competes with the unlabeled analyte for the epitope-binding site of the antibody.

In a sandwich format, the antibody coated electrode is first immersed in a solution containing the unknown concentration of the antigen, the solution is then washed and the electrode is immersed in a solution containing a second antibody labelled with an enzyme or a redox couple that will be captured by the antigen on the electrode. Figure 8.3 shows a scheme of a heterogeneous, non-competitive electrochemical sandwich-type immunoassay, in which the Ab-coated electrode binds the analyte and secondary labelled Abs specifically adsorb onto the antigens immobilized by the first Ab.

The general principle of enzyme-channelling immunoassays, in which Ab–Ag binding brings two enzyme labels into close proximity to speed sequential reactions, is described in Fig. 8.4. One enzyme conjugate is

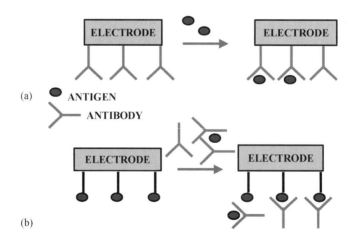

Fig. 8.1. Direct electrochemical detection of antigen or antibody.

Electrochemical antibody-based sensors

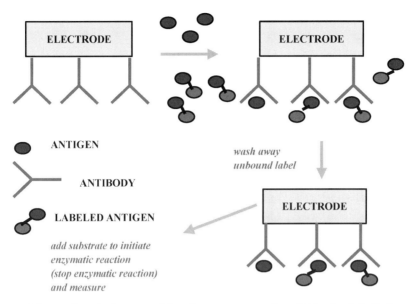

Fig. 8.2. Electrochemical competition immunoassay, in which the labelled analyte competes with the unlabelled analyte for the epitope-binding site of the antibody.

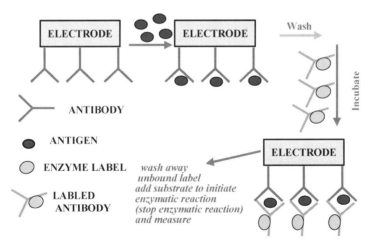

Fig. 8.3. Electrochemical sandwich immunoassay, in which the Ab-coated electrode binds the analyte and secondary labelled Abs specifically adsorb onto the antigens immobilized by the first Ab.

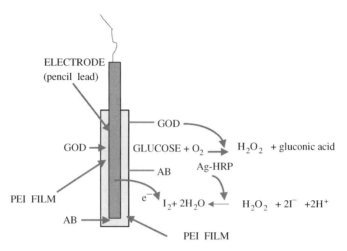

Fig. 8.4. Schematic illustration of the principle of the separation-free, amperometric enzyme channelling immunoassay with immobilized glucose oxidase and antibody on the PEI-modified electrode surface.

immobilized on the surface of an electrochemical detector, while the other is added, along with an appropriate substrate, as a reagent to the sample phase.

8.3.5.1 Effect of ionic strength

Hardeman et al. [46] measured the impedance response of an uncoated sensor in air and buffer solutions of changing ionic strength and pH to investigate the effect of the electrical properties of a buffer on the detection of an immunochemical reaction. The authors found that ionic strength, rather than pH, has a dominant effect on the sensor impedance. Schyberg et al. [47] performed an impedance analysis of an amino-silane grafted Si/SiO_2 structure. The hypothesis that each molecular layer is a perfect dielectric has not yet been confirmed, not only the capacitative part of the grafted structure must be studied but also the total impedance. This approach becomes even more obvious in the case of an antibody layer for which the size of the molecule enables the formation of a dense, well-structured layer.

8.3.5.2 Redox cycling

When redox recycling is applied, higher amplifications are achieved because the redox compound is oxidized and reduced in a cyclic manner, thus the indication of one labelled antigen or antibody molecule generates multiple signal equivalents. Sanderson and Anderson [48] first studied redox cycling in 1985. For redox cycling, microelectrodes offer higher sensitivity than macroelectrodes of conventional size because an electroactive molecule can approach the microelectrode from every direction (spherical diffusion). Therefore, the flux of electroactive molecules toward the electrode is much greater for a microelectrode than for a macroelectrode, for which the diffusion is planar. Sanderson and Anderson pioneered the interdigitated array (IDA) microelectrode, which consists of a pair of microband array electrodes that mesh with each other. Because each set of microband electrodes in the IDA can be potentiostated individually, a reduced species generated at one microband electrode diffuses across a small gap to the adjacent electrode and then is oxidized and diffuses back to the original electrode. Redox cycling increases the currents at both electrodes. Several different approaches to redox recycling can be applied, using electrode–electrode, electrode–enzyme, or enzyme–enzyme couples.

Niwa et al. [49], Wollenberger et al. [50], and Paeschke et al. [51,52] ascertained that the principle of electrode coupling can be used to enhance the sensitivity of an electroactive compound down to the nanomolar range, which is equivalent to an amplification factor of about 20. Redox recycling with an IDA electrode having an electrode width of 1.5 μm and an interelectrode space of 0.8 μm was used to design an electrode-supported ELISA or for a separation-free displacement assay. Anti-hapten antibodies coupled to small polymeric beads are saturated with hapten–ferrocene or hapten–peptide–ferrocene conjugates, and the analyte (free hapten) displaces the conjugates, which are then detected by the IDA electrode. The advantage of this kind of redox recycling is that a sensitive amperometric determination is reached without the use of a transducer-immobilized biomolecule, which should make the device more stable.

Thomas et al. [53] produced an electrochemical sandwich ELISA on paramagnetic microbeads using microfabricated platinum IDAs with 25 pairs of electrodes having 1.6-μm gaps and 2.4-μm widths. The enzyme reaction was measured continuously by positioning the microbeads near the electrode surface with a magnet. With mouse IgG as the analyte and β-galactosidase as the enzyme label, electrochemical recycling occurred via the oxidation of p-aminophenol (PAP), the product of the enzyme–substrate reaction, to p-quinone imine (PQI) at $+290$ mV followed by PQI reduction to PAP at -300 mV vs. Ag/AgCl. Dual-electrode detection amplifies the signal fourfold compared with single-electrode detection.

Di Gleria et al. [54] proposed the use of electrode–enzyme coupling for the amplified readout of a redox label, using a lidocaine–ferrocene conjugate in combination with GOx. The use of ferrocene as a mediator for glucose determination within GOx enzyme electrodes provided the rationale for using it as the labelling compound for a homogeneous competitive immunoassay. In that study, the ferrocene label was oxidized at a gold working electrode at a potential of +380 mV, and the reaction of GOx in the presence of glucose reduced the label back. A standard calibration curve for lidocaine in sevenfold diluted plasma over the concentration range 5–50 μM, showing a linear relation between the catalytic current and the concentration of lidocaine, agreed with the results of the standard enzyme multiplied immunoassay technique (EMIT) immunoassay.

The last type of redox recycling is substrate recycling rather than true redox recycling. Two enzymes are used, with the product of the first enzyme reaction serving as substrate for the second enzyme. In turn, the product of the second reaction acts as the substrate for the first enzyme. Both the degradation of a co-substrate and the production of a product like hydrogen peroxide can be quantified electrochemically. This concept can be applied to other electrochemical transducers, such as ion selective electrodes, because label conversion is not accomplished with the electrode. Amplification factors of 3–48,000 were reported by Scheller and colleagues [55–58] for amperometric multienzyme electrodes with the appropriate substrates.

Antigen-coupled derivatives are applied in redox-labelled immunoassays. Because such derivatives are not accepted as strongly as the natural substrate, the amplification is less effective than that obtained with the natural substrate. For the quantification of ferrocene derivatives, Ghindilis et al. [59] used as a couple laccase and pyrroloquinoline quinone (PQQ)-dependent glucose dehydrogenase. After the enzymes were immobilized and sandwiched between a dialysis membrane and a polyethylene membrane, the enzyme membrane was fixed on the top of a Clark electrode, which indicated the oxygen consumption resulting from the laccase reaction. In the presence or absence of glucose, respective lower detection limits of 0.2 or 10,000 nM for ferrocene benzoylisothiocyanate were reached.

Bier et al. [60] used a similar approach to detect the herbicide 2,4-dichlorophenoxyacetic acid (2,4-D), with NADH-independent oligosaccharide dehydrogenase and tyrosinase serving as the enzyme couple and tyrosine as the redox (substrate) label. Anti-2,4-D-antibodies were immobilized within the enzyme layer and loaded with 2,4-D-Tyr conjugates. The 2,4-D diffuses into the enzyme/antibody layer and displaces the 2,4-D-Tyr conjugate. In the presence of glucose, the conjugate is recycled by the enzyme couple and oxygen consumption is measured. The detection limit reported for 2,4-D was 25 nM.

In non-amplified redox-labelled electrochemical immunoassays, the indication of one antigen or one antibody molecule generates one signal equivalent. Performing this type of assay is feasible only if the concentration of the analyte to be determined is in the lower micromolar range, which lies in the sensitivity range of an amperometric sensor for the redox compound. More sensitive immunoassays require amplification principles, as shown by Bordes et al. [61,62]. For a multi-analyte immunoassay, the use of a 5 min preconcentration step of the cationic labels (ferrocene ammonium salt with $E_p = +260$ mV, cobaltocenium salt with $E_p = -1.05$ V) at a negatively charged nafion-loaded carbon paste electrode resulted in a linear response down to 75 nM. The nafion-loaded carbon paste surface must be renewed after each measurement to regenerate the sensor.

8.3.5.3 Flow-through and flow-injection systems
A microflow cell ensuring the continuous flux of new substrate prevents the accumulation or adsorption of products in the vicinity of the working electrode. The miniature size of an electrochemical cell combined with screen-print electrodes offers the advantage of working with low volumes, which is important when dealing with hazardous materials. Kaneki et al. [63] described an electrochemical enzyme immunoreactor with sample and reagents flowing through 20 µl, antibody-coated capillaries.

Heterogeneous-phase immunosensors working in flow conditions (flow-through sensors) combine the sensitivity and selectivity of immunoassays with the accuracy and ease of automation of flow techniques. Moreover, flow techniques offer the advantage of reusing the immunosorbent for a large number of assays, thereby lowering the analysis cost and raising the reproducibility of the assays. Detection occurs at an unmodified electrode that dips into the incubation solution or is an integrated part of an electrochemical flow-through cell. Such assays can be carried out in test tubes or in microtitre plates or, for automatic performance, in flow systems with reactor-immobilized antibodies. The ability to break up the Ab–Ag complex without affecting the binding activity of the immobilized reagent is an essential property for such techniques. Current regeneration techniques are based on the exposure of immune complexes to a reagent that weakens the Ab–Ag interaction, thereby allowing the physical separation of both species. The immunoreactor can be reused only if its activity remains intact after the regeneration process. Two different alternatives should be studied in the development of flow-based immunosensors: (1) antibody or (2) antigen/hapten immobilization. The first choice is the most useful format according to Puchades and Maquieira [64]. The main advantage of this option is that reusing the immunoreagent not only saves expensive antibody preparations but also reduces assay steps. Any denaturation of the antibody during the regeneration step would preclude the possibility of obtaining successful results. On the other hand, the immobilization of the antigen or hapten conjugate is advantageous regarding its reusability. In this case, regeneration can be achieved without a loss of activity of the immobilized reagent, which can still be recognized by the antibody. According to Thompson et al. [65], in both formats a solid phase is used to immobilize either the antibody or the antigen, thus allowing the separation of free fractions from bound immunocomplexes. Hence, selecting the correct immobilization support is another important issue. Because they minimize sample dispersion, packed bed reactors offer, as pointed out by Ho [66], good performance for enhancing sensitivity. Moreover, the solid phase should not only permit the attachment of immunoreagents in an oriented way but also be reusable, as recommended by Morais et al. [67]. Moreover, immunosorbents should tolerate organic solvents and have good hydrodynamic properties, as shown by Gonzalez-Martinez et al. [68].

Ghindilis et al. [69] used a flow-through immunosensor, based on a high surface area carbon immunoelectrode. Dispersed carbon material served as a carrier for immobilized antibodies and as electrode material. A disposable sensing element (immunocolumn) containing dispersed carbon material with immobilized antibodies also served as a working electrode, and a current collector connected the working electrode to the measuring device.

Abad-Villar et al. [70] used an immunoreactor and an electrochemical detector based on gold bands in the design of a flow system. The solid support on which either the immune interaction or the electronic transference occurs was obtained by sputtering gold over Kapton (a polyimide). Alkaline phosphatase (AP)-labelled IgG was physically adsorbed onto the gold surface of the reactor, with two eight-way valves allowing an independent treatment of reactor and detector. The enzymatic reaction (hydrolysis of naphthylphosphate) and the immune interaction (between IgG–AP and anti-AP) were monitored by the amount of naphthol produced.

8.3.5.4 Enzyme-tagged immunoelectrodes
For an electrochemical enzyme immunoassay (ECEIA), the labelling enzyme should have the following properties: (i) high kcat, (ii) electrochemically active products, (iii) enzyme substrates and enzymes stable in buffer, and (iv) a low side reaction of the enzyme products. In ECEIAs, the assay principles (competitive, non-competitive) are similar to those used in conventional electrochemical immunoassays. As the detection of the enzyme products finally takes place at an interface (electrode), additional features can be brought into play, especially for developing separation-free assays.

For example, for the detection of bacteria, Rishpon et al. [71] used antibodies labelled with alkaline phosphatase in a sandwich-type electroimmunoassay. Carbon-felt electrodes containing immobilized antibodies successfully detected a few organisms from the test solution and their antibody probe. Dai and Meyerhoff [72] described two homogeneous, non-separation, binding immunoassays with polycation-sensitive membrane electrode detection. In the first, the authors used a synthetic polycationic-analyte conjugates (polycation labelled analyte) and a limited amount of binding protein sites (antibodies, etc.) to modulate the electrochemical response toward the conjugate. Upon conjugate binding to large binding proteins/antibodies in solution, the polycation label could not be efficiently extracted into the transduction membrane of the polycation-sensitive device, greatly diminishing the observed EMF response. In the presence of analyte molecules, competitive binding frees up more of the conjugate, resulting in an increase in EMF response relative to the analyte concentration. Sub-micromolar levels of a model analyte (theophylline) were measured without any discrete washing or separation steps. In the second, trypsin was the enzymatic label in an EIA, using conjugates of trypsin–biotin as a model analyte. A polycation-sensitive membrane electrode detected the decrease in the EMF response to the polycationic protein protamine as it was cleaved into smaller fragments by the enzyme, responding to the analyte at 10 nM levels.

8.4 APPLICATIONS

8.4.1 Overview

Antibody-based electrochemical detection has been used in many applications for many years. In this section, we present a representative selection of useful applications using different types of antibody-based electrodes and detection principles. In particular, the type of antibody–antigen interaction, the surface modification, and the sensor performance will be addressed.

The properties of high specificity and a wide applicability with many analytes have led to the widespread use of immunoanalytical techniques. The benefits of electrochemical sensors include technical simplicity, speed, and convenience via direct transduction to electronic equipment. Combining these two systems offers the possibility of a convenient assay technique with high selectivity. Because of the complexity of immunoassay methods, such devices have not yet found widespread use. Nevertheless, electrochemical immunosensors offer the potential for fast, simple, cost-effective analysis of many substances of environmental, biomedical, pharmaceutical, and agricultural interest.

8.4.2 Diagnostic, biomedical, and veterinary applications

Progress has been slow in the development of biomedical applications for electrochemical immunosensors. The first application of amperometric immunosensors in medical diagnostics was reported by Aizawa in 1979 [1], who described the detection of human chorionic gonadotropin (HCG) by a membrane-bound antibody. Five years later Boitieux et al. [73] published their method for the amperometric determination of antigens in serum, based on the biological model of "hepatitis B surface-antigen antibodies". Specific antibodies labelled with GOx were immobilized onto a gelatine membrane and applied in a solid phase sandwich procedure. The active membrane was then fixed onto a pO_2 electrode, and the sensor was immersed in a standard glucose solution. A signal correlating to the antigen concentration of the sample was obtained by measuring the consumption of oxygen following enzymatic degradation of the substrate. The authors foresaw that "the use of a computerized enzyme immunosensor could easily be extended to assays of other antigens and haptens that are usually determined by radioimmunoassay". Another 3 years passed, before Yao and Rechnitz [132] described an amperometric enzyme-immunosensor in which co-immobilized GOx and anti-T4 antiserum were used to measure thyroxine at a detection limit of 15 nM.

In 1989 Rishpon [74] developed an immunosensor for the electrochemical determination of the isoenzyme lactic dehydrogenase isoenzyme 5 (LDH-5) comprising a monoclonal antibody to LDH-5 covalently bound to a glassy carbon electrode (GCE) and a computerized electrochemical system. This combination formed the basis of a highly sensitive fast and convenient technique for the determination of LDH isoenzymes in human sera. Over a decade later, Kelly et al. [75] described an amperometric immunosensor for the detection of the LDH isoenzyme LD-1. The assay is based on the covalent immobilization of polyclonal antibodies onto a pre-activated Immunodyne ABC membrane, which is then reacted with standard LD-1 solutions and placed onto a platinum working electrode polarized at +600 mV vs. Ag/AgCl. Detection of the enzymatic oxidation of NADH at the electrode surface yielded a calibration curve for LDH in the range of 0.005–0.12 U.

For in vitro cortisol detection, Xu and Suleiman [76] co-immobilized horseradish peroxidase and cortisol-antibody on a chemically activated affinity membrane mounted over the tip of an oxygen electrode to produce a reusable amperometric immunosensor. The association between the analyte and the co-immobilized antibody inhibits the electrocatalytic current response to the substrate. The calibration curve for cortisol was linear in the concentration range of 1×10^{-7}–1×10^{-5} M.

Cook [77] used an antibody-based electrode and a corticosteroid–peroxidase conjugate to measure the competitive binding of endogenous corticosteroids in conscious animals in real time. A dialysate membrane enclosing the electrode so that the corticosteroids could equilibrate across the membrane, as well as the small size of the probe (350 μm Da), allowed its implantation into circulatory systems or tissue for in vivo measurements. The authors reported a detection sensitivity of 0.2–0.6 μg, 100 ml^{-1} of cortisol or corticosterone. Subsequently, the author [78] was the first to report the rapid on-line measurement of corticotropin-releasing hormone (CRH) and cortisol from a discrete brain area within or in the dialysate of a microdialysis probe. Polyclonal antibodies for either cortisol or CRH affixed on a platinum electrode within the probe were used in a competitive immunoassay, measured as current change. These probes were used for extracellular measurements in the amygdala, a limbic brain region, of sheep, but could be used directly in vivo. They are stable in vivo (>72 h) and regenerable, offering on-line measurement, with rapid time resolution, of neurohumoral substances.

A novel amperometric immunosensor, based on graphite paste (graphite powder and paraffin oil) was developed by Stefan and Aboul-Enein [79] for detecting the thyroid hormone (+)-3,3′,5,5′-tetraiodo-L-thyronine (L-T-4). The graphite paste is impregnated with mouse monoclonal anti-(+)-3,3′,5,5′-tetraiodo-L-thyronine (anti-L-T-4). The immunosensor is reliable for the assay of L-T-4 in thyroid and in drugs, using a chronoamperometry technique, at ppt up to ppb concentration levels. The potential used for the L-T-4 assay is +450 mV vs. Ag/AgCl electrode. The surface of the immunosensor can be regenerated by simply polishing, thereby obtaining a fresh immunocomposite ready to be used in a new assay.

In 2003, Dai et al. [80] reported a strategy for a reagentless immunosensor for the determination of carcinoma antigen-125 (CA-125) based on the immobilization of antigen and the direct electrochemistry of HRP bound to an antibody. The immunosensor was prepared by immobilizing CA-125 with titania sol–gel on a GCE by the vapour deposition method. The incubation of the immunosensor in phosphate buffer solution containing HRP-labelled CA 125 antibody leads to the formation of an HRP-modified surface. The immobilized HRP displays its direct electrochemistry with a rate constant of 3.04 ± 1.21 s^{-1}. With a competition mechanism, a differential pulse voltammetric determination method for CA-125 was established by the peak current decrease of the immobilized HRP. The current decrease resulted from the competitive binding of the CA-125 in sample solution and the immobilized CA-125 to the limited amount of HRP-labelled CA-125 antibody. Under optimal conditions, the current decrease was proportional to the CA-125 concentration, ranging from 2 to 14 U ml^{-1} with a detection limit of 1.29 U ml^{-1} at a current decrease by 10%. The CA-125 immunosensor showed good accuracy and acceptable precision and fabrication reproducibility with intra-assay CVs of 8.7 and 5.5% at 8 and 14 U ml^{-1} CA-125 concentrations, respectively, and inter-assay CV of 19.8% at 8 U ml^{-1}.

The following year, the same laboratory [81] prepared a novel immunosensor for the rapid separation-free determination of carcinoembryonic antigen (CEA) in human serum by co-immobilizing thionine and HRP-labelled CEA antibody on a GCE through covalently binding them to GCE with a glutaraldehyde linkage. The electrochemical behaviour of the immobilized thionine displays a surface-controlled electrode process with an average electron transfer rate constant of 4.74 ± 2.99 s^{-1}. It can be used as an electron transfer mediator for detecting the enzymatic activity of HRP-labelled anti-CEA antibody. After the immunosensor is incubated with the CEA solution for 40 min at 23°C, access of the activity centre of the HRP to thionine is partly inhibited, leading to a linear decrease in the catalytic efficiency of HRP in the oxidation of immobilized thionine by H_2O_2 at -300 mV. Under optimal conditions, the detection limit for the CEA immunoassay is 0.1 ng ml^{-1}. The immunosensor shows good accuracy and acceptable storage stability, precision, and reproducibility with intra-assay CVs of 6.1 and 5.8% at 2.5 and 50 ng ml^{-1} CEA, respectively, and an inter-assay CV of 6.3% at 50 ng ml^{-1}.

Du et al. [82] described a novel electrochemical immunosensor for carbohydrate antigen 19-9 (CA19-9), a marker of pancreatic cancer, based on the immobilization of CA19-9 with titania sol–gel on a graphite electrode (GE) by vapour deposition. Binding of the HRP-labelled CA19-9 antibody with the immobilized antigen catalyses the oxidation of catechol by hydrogen peroxide, providing a competitive method for measuring serum CA19-9. The response current decreased with increasing CA19-9 concentration in the incubation solution. Under optimal conditions, the current decrease of the immunosensor was proportional to CA19-9 concentrations in the range of 3–20 U ml^{-1}, with a detection limit of 2.68 U ml^{-1} at a current decrease of 10%. The detection of CA19-9 in serum samples obtained from clinically diagnosed patients with pancreatic carcinoma showed acceptable accuracy.

Prostate specific antigen (PSA) is a reliable clinical tool for diagnosing and monitoring prostate cancer. Sarkar et al. [83] developed an amperometric detection procedure for detecting total PSA in human serum using a three-electrode system, with hydroxyethyl cellulose and rhodinized carbon as the working electrode. The electrochemical response is monitored by a sandwich immunoassay on the working electrode. The monoclonal capture antibody to PSA is immobilized on the working electrode with a HRP-labelled monoclonal antibody as the tracer antibody.

Stefan and Bokretsion [84] constructed lately an amperometric immunosensor based on diamond paste (diamond powder and paraffin oil) impregnated with anti-azidothymidine antibody for the assay of azidothymidine (AZT, an approved and widely used antiretroviral drug for the treatment of human immunodeficiency virus infection) in its pharmaceutical formulations. The potential used for azidothymidine assay was $+240$ mV vs. Ag/AgCl electrode.

A novel quantitative polypyrrole-based potentiometric biosensor using as model analytes hepatitis B surface antigen, troponin I, digoxin, and tumour necrosis factor was recently described by Purvis et al. [85]. The potentiometric biosensor detects enzyme labelled immunocomplexes formed at the surface of a polypyrrole coated, screen-printed gold electrode. Detection is mediated by a secondary reaction that produces charged products (a "charge-step" procedure). A shift in potential is measured at the sensor surface, caused by local changes in redox state, pH and/or ionic strength. The magnitude of the difference in potential is related to the concentration of the formed receptor–target complex. The technology is rapid (<15 min), ultrasensitive (<50 fM), and reproducible (CV $< 5\%$ at 0.1 ng ml^{-1}).

Alanine aminotransferase (ALT) is regarded as one of the most sensitive indicators of hepatocellular damage. Xuan et al. [86] generated monoclonal antibodies to human recombinant ALT and fabricated them for use in an electrochemical immunosensor for detecting the enzyme in human samples. The ALT immunosensor comprises the following: (1) anti-ALT antibody-immobilized outer membrane; (2) pyruvate oxidase-absorbed inner membrane; (3) a self-assembled monolayer mediator-coated gold working electrode and an Ag/AgCl reference electrode. The chronoamperometric measurement of the immunosensor was performed in less than 5 min with 40 μl of PBS containing substrates and ALT without a washing step. The dynamic range of the ALT immunosensor was between 10 pg ml^{-1} and 1 μg ml^{-1}, at the respective detection limits and sensitivity of 10 pg ml^{-1} and 26.3 nA/(ng ml^{-1}).

Scheller's group [87] generated creatinine-specific antibodies for sensitive and specific immunochemical creatinine determinations in a reusable sensor. The creatinine sensor was constructed by fixing a creatinine-modified membrane on top of a platinum working electrode, which was then incorporated into a stirred electrochemical measuring cell. For creatinine determination, the authors incubated the test sample with a monoclonal antibody and a mouse anti-IgG–GOx conjugate and then applied it to the measuring cell, which was then washed. Glucose was added, and the hydrogen peroxide produced was registered at $E_{appl} = +600$ mV vs. Ag/AgCl. The measuring range was 0.01–10 μg ml^{-1}. The highest sensitivity for creatinine was achieved at 330 ng ml^{-1} (3 μM), with a lower detection limit of 4.5 ng ml^{-1} (40 nM). This value is far below the relevant clinical range of 5–17 μg ml^{-1} (44–150 μM) and enables the reliable determination of very low creatinine concentrations in serum, for which standard methods cannot be applied.

A high level of glycosylated haemoglobin A(1c) [HbA(1c)], an indicator of glucose intolerance, is related to cardiovascular risk in the general population without diabetes mellitus. To detect HbA(1c) in human serum, Scheller's group [88] used membrane-immobilized haptoglobin as affinity matrix fixed in front of a Pt-working electrode for an amperometric immunosensor. The HbA1c assay was carried out in a two-step procedure including selective haemoglobin enrichment on the sensor surface and specific HbA1c detection by a GOx labelled anti-HbA1c antibody. Hydrogen peroxide generated by the enzyme label was oxidized at +600 mV vs. Ag/AgCl. A standard curve for HbA1c was obtained with a linear range between 0 and 25% HbA1c of total haemoglobin, which correspond to 7.8–39 nM. ELISA studies confirmed the advantage of a sandwich-type format with haptoglobin as capture molecule for selective haemoglobin binding over the direct adsorption method. Results by the sandwich immunoassay showed a linear correlation within the clinically relevant range 5–20% (CV < 3).

Two immunosensors developed by O'Regan et al. [89,90] have demonstrated their usefulness for the early assessment of acute myocardial infarction (AMI). Human heart fatty-acid binding protein (H-FABP) is a biochemical marker for the early assessment of AMI. The authors constructed an amperometric immunosensor for the rapid detection of H-FABP in whole blood. The sensor is based on a one-step, direct sandwich assay in which the analyte and an alkaline phosphatase (AP) labelled antibody are simultaneously added to the immobilized primary antibody, using two distinct monoclonal mouse anti-human H-FABP antibodies. The substrate p-aminophenyl phosphate is converted to p-aminophenol by AP, and the current generated by its subsequent oxidation at +300 mV vs. Ag/AgCl is measured. The total assay time is 50 min, and the standard curve was linear between 4 and 250 ng ml^{-1}. The intra- and inter-assay coefficients of variation were below 9%. No cross-reactivity of the antibodies was found with other early cardiac markers, and endogenous substances in whole blood did not have an interfering effect. Myoglobin is another biochemical marker for the early assessment of AMI. The same group developed an amperometric immunosensor for the rapid detection of myoglobin in whole blood based on a one-step indirect sandwich assay using a polyclonal goat anti-human cardiac myoglobin antibody with monoclonal mouse anti-myoglobin and goat anti-mouse IgG conjugated to AP as the detecting antibody. The standard curve was linear between 85 and 925 ng ml^{-1}. The intra- and inter-assay coefficients of variation were below 8%. No cross-reactivity of the antibodies was found with other cardiac proteins.

Coumarin, the anticoagulant of choice for the treatment of thromboembolic disorders in the outpatient setting, requires frequent monitoring of blood levels of its metabolite 7-hydroxycoumarin. An antibody-based amperometric biosensor was developed by Dempsey et al. [91] to study the reaction of 7-hydroxycoumarin with an anti-7-hydroxycoumarin (7-OH-coumarin) antibody immobilized at the surface of a GCE contained behind a cellulose dialysis membrane. Upon binding of the antibody to the antigen, a 90% decrease in peak height was well defined using differential-pulse voltammetry. This system provides an ideal method for studying both antibody specificity and the kinetics of antibody–antigen interactions. The same group [92] also developed a size exclusion redox-labelled immunoassay (SERI) for homogeneous amperometric creatinine determination.

Dijksma et al. [93] developed a reusable electrochemical immunosensor for the direct detection of the 15.5 kDa protein interferon-gamma (IFN-gamma) at the attomolar level. The IFN-gamma was adsorbed physically to self-assembled monolayers of cysteine or acetylcysteine on electropolished polycrystalline gold electrodes. A specific antibody (MD-2) against IFN-gamma was covalently attached, following the carbodiimide/succinimide activation of the monolayer. Activation of the carboxylic groups, attachment of MD-2, and deactivation of the remaining succinimide groups with ethanolamine were monitored impedimetrically at a frequency of 113 Hz, a potential of +0.2 V vs. SCE and an AC modulation amplitude of 10 mV. In the thermostatted set-up (23.0°C), samples of unlabelled IFN-gamma were injected and the binding with immobilized MD-2 was monitored with AC impedance or potential-step methods. This AC impedance approach provides detection limits as low as 0.02 fg ml^{-1} (approximately 1 aM) IFN-gamma.

Ivnitski and Rishpon [94] developed a one-step, separation-free, amperometric enzyme immunosensor, consisting of an antibody electrode that is low cost, disposable, and operates without washing or separation steps, using human luteinizing hormone (hLH) as a model analyte. Pencil lead was used as the working electrode as shown in Fig. 8.4. The immunosensor combines the following signal-amplification systems: enzyme-channelling immunoassay, accumulation of the redox mediators (I2/I-); cyclic regeneration of an enzyme (peroxidase) substrate at the (polyethylenimine) polymer/electrode interface; and control of the hydrodynamic conditions at the interface of the antibody electrode. The immune reactions were monitored electrochemically in situ, and the binding curves were directly visualized on a computer screen. In model systems using a sandwich immunoassay, the immunosensor detected concentrations of rabbit IgG and hLH as low as 1 ng ml^{-1}.

Darain et al. [95] constructed a disposable and mediatorless immunosensor based on a conducting polymer (5,2′:5′2″-terthiophene-3′-carboxylic acid)-coated screen-printed carbon electrode using a separation-free homogeneous technique for the detection of rabbit IgG. Horseradish peroxidase and streptavidin were covalently bonded with the polymer on the electrode and a biotinylated antibody was immobilized on the electrode surface using avidin–biotin coupling. This sensor is based on a competitive assay between free and labelled antigen for the available binding sites of antibody. Glucose oxidase is used as a label and in the presence of glucose, the hydrogen peroxide formed by the analyte–enzyme conjugate was reduced by the enzyme channelling via HRP bonded on the electrode. Amperometric monitoring of the catalytic current at -0.35 V vs. Ag/AgCl resulted in a detection limit of 0.33 μg ml^{-1}. This method showed a linear range of RIgG concentrations from 0.5 to 2 μg ml^{-1}.

A differential pulse voltammetry transducer with mouse anti-erythromycin (and anti-tylosin) monoclonal antibodies serving as molecular recognition elements was developed by Ammida et al. [96] for the detection of erythromycin and tylosin in bovine muscle, using a disposable electrochemical ELISA in a screen-printed electrode system. The immunochemical system makes use of the competition assay principle, using an erythromycin (or tylosin)–BSA conjugate as coating molecule. After competition between free and coated analyte for the antibodies, the activity of AP-labelled antiglobulins is measured electrochemically using 1-naphthylphosphate as substrate. The detection limit of the assay was 0.2 ng ml^{-1} for erythromycin and 2.0 ng ml^{-1} for tylosin, whereas the sensitivity (25% inhibition concentration) was 1.0 ng ml^{-1} for erythromycin and 3.0 ng ml^{-1} for tylosin.

An indirect amperometric immunosensor for the detection of HCG was described by Chetcuti et al. [97]. The assay consisted of an anti-HCG monoclonal antibody immobilized on a GCE and the use of a sandwich assay with HRP conjugated to a second anti-HCG monoclonal antibody. Electrochemical detection of the enzymatic reduction of benzoquinone to hydroquinone permitted the quantification of HCG with a limit of linearity at 200 mIU ml^{-1} and a detection limit of 11.2 mIU ml^{-1}.

The integration of a system into a flow injection system allows the monitoring of real-time antibody–antigen interactions using electrochemical methods. Killard et al. [98] developed an electrochemical flow injection immunoassay for the real-time monitoring of biospecific interactions. The authors used screen-printed electrodes incorporating the electroactive polymer, polyaniline, and enzyme-labelled antibodies as the biomolecular recognition component. A potentiometric immunosensor using covalently immobilized anti-IgG on a silver electrode for detecting IgG was developed by Feng et al. [99]. Detection was based on the change in the electric potential before and after the antigen–antibody reaction.

Pemberton et al. [29,100,101] developed a single-use electrochemical biosensor for cow's milk progesterone, based on a competitive immunoassay under thin-layer, continuous-flow conditions. An anti-progesterone monoclonal antibody was deposited onto screen-printed carbon electrodes. The rationale for determining progesterone in milk was to have a rapid means of determining the onset of oestrus in dairy cattle.

8.4.3 Environmental monitoring and food analysis

An amperometric immunosensor using sequential injection analysis techniques to detect the herbicide 2,4-D in water was described by Wilmer and Trau [102,103]. This rapid competitive EIA used an alkaline phosphatase-labelled monoclonal antibody directed against the herbicide and an immunoreactor with 2,4-D immobilized via BSA, either to Eupergit in a column or directly to the surface of a glass capillary. A detection limit of the immunosensor at 0.1 μg 2,4-D l^{-1}, without enrichment of the analyte, pointed to the feasibility of making automatic measurements of 2,4-D in drinking and ground water.

Several years later, Zeravik and Skladal [104] described the construction of a regenerable flow-through immunosensor for detecting 2,4-D, using a screen-printed electrochemical system. The gold-based composite electrode was modified with a self-assembled monolayer of cysteamine; the amino group obtained was then linked to 2,4-D using the common N-hydroxysuccinimide/carbodiimide reaction. A peroxidase-conjugated monoclonal antibody was the tracer. The degradation of the substrates by the enzyme at the electrode surface was quantified amperometrically at 50 mV vs. the screen-printed silver pseudo-reference electrode. The limit of detection (based on the 20% decrease of relative binding of the tracer) was 0.12 mg l^{-1}.

Disposable screen-printed electrodes were used by Delcarlo et al. [105] for the immunochemical detection of polychlorinated biphenyls (PCBs). In the following year, a direct electrochemical immunosensor for monitoring PCBs was developed by Bender and Sadik [106]. Their assay for PCB determination is based on the measurement of the current following the specific binding between PCB and an anti-PCB antibody-immobilized conducting polymer matrix. The linear dynamic range of the immunosensor was between 0.3 and 100 ng ml^{-1}, with a correlation coefficient of 0.997 for Aroclor 1242. The method detection limits for Aroclor 1242, 1248, 1254, and 1016 were 3.3, 1.56, 0.39, and 1.66 ng ml^{-1}, respectively, at a signal-to-noise (S/N) ratio of 3. Recoveries of spiked Aroclor 1242 and 1254 from industrial effluent water, rolling mill, and seafood plant pre-treated water at 0.5 and 50 ng ml^{-1} ranged from 103 to 106%. This method can be applied for the continuous monitoring of effluents like waste streams and groundwater.

An amperometric immunosensor for detecting polycyclic aromatic hydrocarbons (PAHs), based on disposable screen-printed carbon electrodes, was developed by Fahnrich et al. [107]. Phenanthrene-9-carboxaldehyde coupled to bovine serum albumin via adipic acid dihydrazide was the coating antigen, detected by monoclonal mouse anti-phenanthrene. Alkaline phosphatase was used in combination with the substrate p-aminophenyl phosphate (p-APP) for detection at +300 mV (vs. Ag/AgCl). Good results were achieved with an indirect co-exposure competition assay with a limit of detection of 0.8 ng ml^{-1} (800 ppt) and an IC(50) of 7.1 ng ml^{-1} (7.1 ppb) for phenanthrene. An indirect competition assay detected phenanthrene with a limit of detection of 2 ng ml^{-1} (IC(50): 15 ng ml^{-1}) and an indirect displacement assay with a limit of detection of 2 ng ml^{-1} (IC(50): 11 ng l^{-1}) at a 5 μl surface coating of 8.8 μg ml^{-1} phenanthrene–BSA conjugate. A coating concentration of 2.2 μg ml^{-1} enabled detection with a limit of detection of 0.25 ng ml^{-1} (250 ppt) with the indirect competition assay. Anthracene and chrysene, but not benzo[g,h,i]perylene and dibenzo[a,h] anthracene, showed strong cross-reactivity.

Genetically engineered biomolecules provide an attractive approach for the development of electrochemical immunosensors. Zeravik et al. [108] described a concept based on the Peroxidase-chip (P-chip)-antibody co-immobilization for the detection of environmental pollutants like s-triazine herbicides. In this concept, recombinant peroxidase is immobilized on the gold electrode (P-chip) in such a way that direct electron transfer is achieved. The recognition and quantitation of the target analyte is realized through the competition between the simazine–GOx conjugate and free simazine for the binding sites of the monoclonal antibody co-immobilized with peroxidase on the gold electrode. The immunosensor exhibited a 50% signal decrease

(IC50 value) at approximately 0.02 μg l^{-1}. A concentration of 0.1 ng l^{-1} gave a signal clearly distinguishable from the blank, whereas the ELISA using the same antibody had a typical detection limit of about 1 μg l^{-1}.

Restricted requirements for nitrogen reduction at wastewater treatment plants have increased the need for assays determining the inhibition of nitrification, which can result from the inhibition of *Nitrobacter* species. Sandén et al. [109] used an electrode made of glassy carbon for the immunologic reaction in an amperometric enzyme-linked immunoassay designed to specifically quantify *Nitrobacter*. The method uses a monoclonal capture antibody and an AP-labelled secondary antibody. The enzyme substrate 5-bromo-4-chloro-3-indolyl phosphate generates an electroactive product that is detected amperometrically. This principle was later used by the authors for quantifying *Nitrobacter* in activated sludge from several wastewater treatment plants in Sweden [110]. All strains of the nitrite-oxidizing genera *Nitrobacter* and *Nitrococcus*, but not other bacterial strains and isolates, reacted strongly with the monoclonal antibody.

Bäumner and Schmid [111] described a disposable amperometric immunomigration sensor for the detection of triazine pesticides in water samples. The authors used thick film electrodes printed on PVC as strip-type transducers, with monoclonal antibodies against atrazine and terbutylazine serving as biorecognition elements. Hapten-tagged liposomes entrapping ascorbic acid as a marker molecule were used to generate and amplify the signal, using a capillary gap and a wicking filter membrane strip as the migration zone. For detecting the signal on a GE, liposomes were lysed with Triton X-100, and the released ascorbic acid was quantified at a potential of +300 mV vs. printed Ag/AgCl. The sensitivity of measurements in tap water was below 1 μg l^{-1} for atrazine and terbutylazine. Another approach to the detection of atrazine was reported by Keay and McNeil [112], who developed a separation-free, electrochemical competitive ELISA technique. The assay incorporates disposable screen-HRP modified electrodes in conjunction with single-use atrazine immune membranes. A monoclonal antibody specific for atrazine was immobilized onto Biodyne C membranes and then placed over the electrode surface. The assay was based on competition for available binding sites between free atrazine and an atrazine–GOx conjugate. Enzyme channelling via the HRP electrode reduced the peroxide formed by the conjugate in the presence of glucose. The HRP was, in turn, re-reduced by direct electron transfer at a potential of +50 mV vs. Ag/AgCl.

The application of a potentiometric immunosensor for detecting the herbicide simazine in food was described by Yulaev et al. [113]. After a competitive immune reaction on the gold planar electrode surface, the peroxidase label was detected by a potentiometric transducer, with a detection limit of 3 ng ml^{-1} simazine. The technique was applied to detect the herbicide in meat extracts, milk, tomatoes, cucumbers, and potatoes. For the first three substances, simazine was detected quantitatively without pre-treatment. The results confirmed the efficacy of the sensor for the food quality control.

Rapid methods to detect pathogenic bacteria in food products are important alternatives to laborious and time-consuming culture procedures. Many investigators have published papers on the use of electrochemical immunosensors for the identification and quantification of bacterial pathogens *Escherichia coli* O157, *Staphylococcus aureus*, and *Salmonella* spp., which are common contaminants of food or water.

Mirhabibollahi et al. [114–117] published a series of papers describing an amperometric enzyme-linked immunosensor for the electrochemical detection and quantification of bacteria in foods. A catalase-labelled anti-protein-A antibody, specific for the detection of *S. aureus*, was used in a modified sandwich ELISA. The catalase-released O_2 was monitored using an amperometric oxygen electrode. The rate of current increase was proportional to the antigen concentration (protein A or *S. aureus*). Protein A was detected reliably at 0.1 ng ml^{-1}, representing a 20-fold increase in sensitivity over the conventional ELISA using horseradish peroxidase. The amperometric electrode detected *S. aureus* in inoculated food samples at concentrations of 10^{-3}–10^{-4} cfu ml^{-1}. A similar sensitivity of detection was obtained with samples of beef and milk. One disadvantage of this approach is that natural variations in the protein-A content of different strains of the pathogen could make the system unreliable for quantifying absolute cell numbers.

Rishpon's group [118] described the development of heterogeneous, enzyme-tagged immunoelectrochemical assays for detecting small numbers of bacteria, using disposable carbon felt discs that served both as electrodes and as the heterogeneous solid phase. Antibodies raised against the microbe were immobilized on the carbon felt via a diaminoalkane–biotin–avidin–biotin bridge, using AP as a label. The bound antibodies were monitored by following the electro-oxidation of aminophenol, produced enzymatically from *p*-PAP by the immobilized AP at the electrode surface. A small number (approximately 80 microorganisms ml^{-1}) of bacteria could be quantified using this system. The same group [119] later developed an amperometric enzyme-channelling immunosensor for detecting *S. aureus*. The sensor consists of a disposable, polymer-modified, carbon (pencil lead) electrode on which enzyme 1 is co-immobilized with a specific antibody that binds the corresponding bacterial antigen in a test solution. An immunologic reaction brings enzyme 1 and a conjugate of enzyme 2 into close proximity at the electrode surface, and the signal is amplified through enzyme channelling. The localization of both enzymes on the electrode surface limits the enzymatic reactions to the polymer/membrane/electrode interface. The sensor overcomes the problem of discriminating between the signal that is produced by the immune-bound enzyme label on the electrode surface and the background level of signal that emerges from the bulk solution. The combination of enzyme-channelling reactions, optimizing hydrodynamic conditions, and electrochemically regenerating mediators within the membrane layer of the antibody electrode significantly increases the signal-to-noise ratio of the sensor. In pure culture, *S. aureus* cells can be detected at concentrations as low as 1000 cells ml^{-1}. The antibody-captured bacteria on the electrode are visualized by electron microscopy, as shown in Fig. 8.5.

The quantitative determination of total and faecal coliform bacteria is essential for water quality control. Mittelmann et al. [120] used a sensitive inexpensive amperometric enzyme biosensor based on the electrochemical detection of β-galactosidase activity, for determining the density of total coliforms, represented by *E. coli* and *Klebsiella pneumoniae*. After degradation of the substrate *p*-amino-phenyl-β-D-galactopyranoside, the direct determination of β-D-galactosidase activity in the electrochemical cell yielded a detectable signal for both coliform types in concentrations ranging from about 10^3 to 10^7 cfu ml^{-1}. The results showed a good linear correlation with the number of bacteria measured by the plate counting method. Brewster and Mazenko [121] used filtration capture and immunoelectrochemical detection for the rapid assay of enzyme-labelled *E. coli*. After filtration of the sample, the labelled cells were detected by placing the filter on the surface of an electrode, incubating the electrode with the enzyme substrate, and measuring

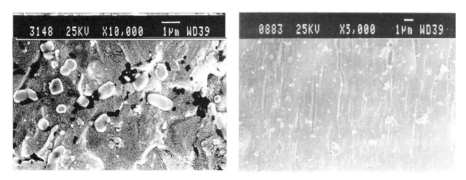

Fig. 8.5. Adsorption of *S. aureus* cells to a RbIgG modified electrode surface observed by electron microscopy. Left: electrode surface with no bacteria. Right: *S. aureus* cells adsorbed to the electrode surface. Magnification, 10,000 ×.

the current produced by oxidation of the electroactive enzyme product. A detection limit of 5000 cells ml^{-1} was obtained for *E. coli* O157:H7. Fewer than 50 cells could be detected when very small sample volumes (10 µl) were used to minimize background current.

Abdel-Hamid et al. [122] used a flow-injection amperometric immunofiltration assay system for the rapid detection of total *E. coli* and *Salmonella*. Disposable porous nylon membranes served as a support for the immobilization of anti-*E. coli* or anti-*Salmonella* antibodies. The assay system consists of a flow-injection system, a disposable filter-membrane, and an amperometric sensor. A sandwich immunoassay specifically and directly detected 50 cells ml^{-1} total *E. coli* or 50 cells ml^{-1} *Salmonella*. The immunosensor can be used as a highly sensitive and automated bioanalytical device for the rapid quantitative detection of bacteria in food and water.

A label-free electrochemical impedance immunosensor for the rapid detection of *E. coli* O157:H7 consists of immobilized anti-*E. coli* antibodies on an indium–tin oxide IDA microelectrode [123]. The binding of *E. coli* cells to the IDA microelectrode surface increases the electron-transfer resistance, which is directly measured with electrochemical impedance spectroscopy in the presence of $[Fe(CN)_{(6)}]^{(3-/4-)}$ as a redox probe. The electron-transfer resistance correlates with the concentration of *E. coli* cells in a range from 4.36×10^5 to 4.36×10^8 cfu ml^{-1} with a detection limit of 10^6 cfu ml^{-1}.

Dong et al. [124] described an immunosensor for staphylococcal enterotoxin C1 (SEC1), using an electrochemical self-assembled gold electrode. The set-up comprises a three-electrode system and a layer of antibody to SEC1, which is detected electrochemically at a constant antibody concentration. Fourier transform spectroscopy was used to confirm the adsorption of the SEC1 antibody onto the gold electrode and the association of SEC1 antigen and antibody.

A major innovation is the use of variable-region F_V antibody fragments to improve the antigen-binding surface of immunosensors. Benhar et al. [125] designed an electrochemical immunosensor for detecting pathogenic bacteria in foods, using recombinant scFvs on screen-printed carbon electrodes to capture and quantify the target antigen. To capture the enzyme β-galactosidase, an anti-β-galactosidase scFv–CBD fusion protein was immobilized onto a cellulose filter and placed on a screen-print carbon electrode; the enzyme activity was measured by lactose injection. A second electroimmunosensor was constructed using two isolated scFvs prepared against *Listeria monocytogenes*, a common bacterial contaminant of cooked ready-to-eat meats and seafoods, smoked fish, and dairy products, for detecting the microorganisms in a sandwich immunoassay. One antibody expressing a CBD-containing fusion protein was isolated from a phage library, and an identical antibody fused to alkaline phosphatases served as the second antibody. A graphic depiction of the phage set-up is presented in Fig. 8.6.

The detection of *L. monocytogenes* and another food pathogen *Bacillus cereus* was achieved using an immunosensor without a label developed by Susmel et al. [126]. Gold electrodes were produced using screen-printing and the gold surfaces were modified by thiol-based self-assembled monolayers based on the use of thioctic acid, mercaptopropionic acid, and mercaptoundecanoic acid. Following antibody immobilization via the optimum monolayer, the redox behaviour and diffusion coefficient of a potassium hexacyanoferrate(II) probe was monitored in the absence and presence of an analyte. In the presence of analyte, impedance of the diffusion of redox electrons to the electrode surface following the formation of the antibody–bacteria immunocomplex caused a change in the apparent diffusion coefficient of the redox probe.

8.4.4 Biological warfare agents

The development of advanced techniques is important for the sensitive and effective detection of biological threat agents, particularly in cases where the analysis includes complex samples containing multiple contaminating factors.

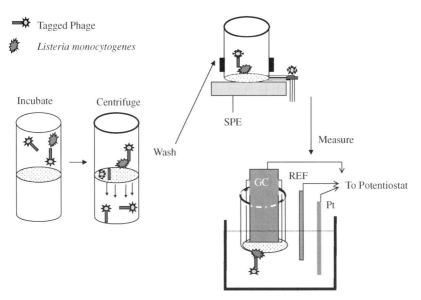

Fig. 8.6. The experimental set-up for *listeria* capture and detection by scFv fragments.

Uithoven et al. [127] reported the use of an instrument employing a light addressable potentiometric sensor and a flow-through immunofiltration-enzyme assay system for the rapid and specific identification of biological warfare agents. The system was designed to assay up to eight agents simultaneously and provides an indication of the absence or presence of a threat within 15 min. Measurements of blank samples and samples containing *Bacillus subtilis* spores in the concentration range of 104–106 cfu ml^{-1} indicated a limit of detection of 3×10^3 cfu ml^{-1} for *B. subtilus*. This instrument offers a potential interim approach for addressing the need for the rapid detection of biological warfare agents in the field.

8.4.5 Illegal drug screening

Suleiman and Xu [128] described a novel reusable amperometric immunosensor for the determination of cocaine. Horseradish peroxidase and benzoylecgonine-antibody were co-immobilized on a chemically activated affinity membrane, which was then mounted over the tip of an oxygen electrode. The enzymatic electrocatalytic current response to the substrate is inhibited by the association of the antigen to the co-immobilized antibody. The calibration plot for cocaine was linear in the concentration range of $1 \times 10^{-7} - 1 \times 10^{-5}$ M.

Another drug detection system was described by Ivison et al. [129], who used a redox mediated amperometric detection system in a screening immunoassay for urinary amphetamines. The EMIT assay with optical detection is well established as a method for screening for amphetamines in urine. Here the authors adapted EMIT reagents for simplified amperometric detection, using a water-soluble redox mediator, 1,2-naphthoquinone sulfonic acid (NQSA). This reagent allowed measurement at a low electrode potential, minimizing interference by electroactive species. A working potential of +100 mV (vs. Ag/AgCl) was used, resulting in a negligible background interference response to a 1:10 urine dilution, which allowed ready detection of 1 mM NADH. Glucose-6-phosphate dehydrogenase (G6PDH) activity was measured by an initial rate method, from 0.2 to 10 U l^{-1}. With optimal dilutions of G6PDH–amphetamine conjugate and amphetamine antibody (1:30 and 1:50, respectively), a standard plot of initial rate of oxidation vs. D-amphetamine concentration was obtained over the range 0.74–14.8 μmol l^{-1} (0.1–2.0 μg ml^{-1}). The plot provided a demonstrable limit of detection of 2.5 μmol l^{-1} (0.3 μg ml^{-1}).

An amperometric immunosensor in the competitive format was developed by Luangaram et al. [130] for the detection of MA in urine. The electrodes consist of carbon paste and Ag/AgCl screen-printed on heat-sealing film, with a monoclonal anti-methamphetamine antibody as the biorecognition element. Methamphetamine, but not amphetamine, was detected by the conversion of P-aminophenyl phosphate to electroactive *p*-aminophenol in the range of 200 ng ml^{-1} (lower detection limit) to 1500 ng ml^{-1} MA in a nearly linear dose response curve. In urine samples spiked with MA, the respective accuracy and precision of the assay were 91.5–104.4% and 15.8–24.4%.

8.5 CONCLUSIONS

The incorporation of antibody-based electrodes as the recognition element in electrochemical biosensing has facilitated the quantification of minute amounts of an analyte in real time. Immunoelectrochemical sensors can turn centralized quantitative immunoassays into portable biosensors by miniaturizing and placing the means of signal detection adjacent to the biocomponent. Hence, a tray or test tube containing the assay components moving through a huge, cumbersome instrument (the sensing element) can now be replaced by a combination of the two, namely, an electrode covered with specific antibodies to the target analyte (see Ref. [131]). In the medical area, this approach means point-of-care monitoring, with an enormous reduction in time from taking the sample to analysis to making a therapeutic decision, as well as less reliance on delicate equipment and highly trained operators.

Current research efforts are focusing on devising rapid, non-separation electrochemically based immunoassay methods that quantify a wide range of species, ranging from small environmental analytes like herbicides to intact bacteria and viruses. Separation-free immunoassays are very attractive for use under field conditions because no washing procedures are necessary. For medical diagnostics, miniaturized hand-held immunoassay devices will enable the measurement of point-of-care diagnostic testing at the bedside, in critical-care units, and in physicians' offices; the test results can be acted upon immediately. Such an apparatus should be easy to handle, reliable, inexpensive, and produce results within minutes. Similar criteria apply to areas that are not directly related to medical diagnosis, such as environmental analysis or hazardous materials control, for which the release of toxic substances requires immediate identification. Conventional detection methods are time-consuming, usually taking from 7 to 10 days to produce reliable measurements, which is far too long for perishable food or a hazardous chemical spill, for example. For the 21st century, immunoelectrochemical techniques should prove to be a powerful tool not only for biomedical needs but also for the environmental monitoring of PCBs, heavy metals, and pesticide/herbicide residues, as well as for the continuous monitoring of waste streams and groundwater to comply with government regulations.

REFERENCES

1. M. Aizawa, A. Morioka, S. Suzuki and Y. Nagamura, *Anal. Biochem.*, 94 (1979) 22–28.
2. A. Warsinke, A. Benkert and F.W. Scheller, *Fresenius J. Anal. Chem.*, 366 (2000) 622–634.
3. K. Landsteiner and J. van der Scheer, *The Specificity of Serological Reactions.* Harvard University Press, Cambridge, MA, 1936, reprinted 1962 by Dover Publications, New York.
4. P. Erlich, In: Melchers.al (Ed.), *Progress in Immunology.* Springer, Berlin, 1890.
5. M. Burnet, *The Clonal Selection Theory of Acquired Immunity*, Cambridge University Press, Cambridge, UK, 1959.
6. G. Kohler and C. Milstein, *Nature*, 256 (1975) 495–497.
7. Y. Berdichevsky, R. Lamed, D. Frenkel, U. Gophna, E.A. Bayer, S. Yaron, Y. Shoham and I. Benhar, *Protein Expression Purif.*, 17 (1999) 249–259.
8. E. Harlow and D. Lane, *Antibodies: A Laboratory Manual.* Cold Spring Harbor Laboratory, Cold Spring Harbor, NY, 1988.
9. R. Ekins, *Nature*, 340 (1989) 256–258.
10. R.S. Yalow and S.A. Berson, *Nature*, 184 (1959) 1648–1649.
11. J.C. Bart, L.L. Judd, K.E. Hoffman, A.M. Wilkins and A.W. Kusterbeck, *Environ. Sci. Technol.*, 31 (1997) 1505–1511.
12. J.C. Bart, L.L. Judd and A.W. Kusterbeck, *Sens. Actuators, B*, 39 (1997) 411–418.
13. S. Koch, H. Wolf, C. Danapel and K.A. Feller, *Biosens. Bioelectron.*, 14 (2000) 779–784.
14. T. O'Brien, L.H. Johnson, J.L. Aldrich, S.G. Allen, L.T. Liang, A.L. Plummer, S.J. Krak and A.A. Boiarski, *Biosens. Bioelectron.*, 14 (2000) 815–828.
15. E. Mallat, D. Barceló, C. Barzen, G. Gauglitz and R. Abuknesha, *Trends Anal. Chem.*, 20 (2001) 124–132.
16. E. Turiel, P. Fernandez, C. Perez-Conde, A.M. Gutierrez and C. Camara, *Talanta*, 47 (1998) 1255–1261.
17. C. Wittmann, F.F. Bier, S.A. Eremin and R.D. Schmid, *J. Agric. Food Chem.*, 44 (1996) 343–350.
18. K. Grennan, A.J. Killard and M.R. Smyth, *Electroanalysis*, 13 (2001) 745–750.
19. E. Gizeli and C.R. Lowe, *Curr. Opin. Biotechnol.*, 7 (1996) 66–71.
20. J. Janata, *J. Am. Chem. Soc.*, 97 (1975) 2914–2916.

21. M.Y. Keating and G.A. Rechnitz, *Anal. Chem.*, 56 (1984) 801–806.
22. M. Gothoh, E. Tamiya, M. Momoi, Y. Kagawa and I. Karube, *Anal. Lett.*, 20 (1987) 857–870.
23. U. Pfeifer and W. Baumann, *Fresenius J. Anal. Chem.*, 343 (1992) 541–549.
24. L. Engel and W. Baumann, *Fresenius J. Anal. Chem.*, 349 (1994) 447–450.
25. P. Bergveld, *Biosens. Bioelectron.*, 6 (1991) 55–72.
26. L. Alfonta, A. Bardea, O. Khersonsky, E. Katz and I. Willner, *Biosens. Bioelectron.*, 16 (2001) 675–687.
27. L. Alfonta, A.K. Singh and I. Willner, *Anal. Chem.*, 73 (2001) 91–102.
28. S.G. Weber and J.T. Long, *Anal. Chem.*, 60 (1988) A903–908.
29. R.M. Pemberton, J.P. Hart, P. Stoddard and J.A. Foulkes, *Biosens. Bioelectron.*, 14 (1999) 495–503.
30. A. Sargent, T. Loi, S. Gal and O.A. Sadik, *J. Electroanal. Chem.*, 470 (1999) 144–156.
31. A. Anne, C. Demaille and J. Moiroux, *J. Am. Chem. Soc.*, 121 (1999) 10379–10388.
32. A. Sargent and O.A. Sadik, *Anal. Chim. Acta*, 376 (1998) 125–131.
33. G. Marko-Varga, J. Emnéus, L. Gorton and T. Ruzgas, *Trends Anal. Chem.*, 14 (1995) 319–328.
34. R. Blonder, E. Katz, Y. Cohen, N. Itzhak, A. Riklin and I. Willner, *Anal. Chem.*, 68 (1996) 3151–3157.
35. L.X. Tiefenauer, S. Kossek, C. Padeste and P. Thiebaud, *Biosens. Bioelectron.*, 12 (1997) 213–223.
36. E.J. Calvo, C. Danilowicz, C.M. Lagier, J.M. Manrique and M. Otero, *Biosens. Bioelectron.*, 19 (2004) 1219–1228.
37. A.L. Newman, K.W. Hunter and W.D. Stanbro, In: *2nd International Meeting on Chemical Sensors, Bordeaux*, pp. 596–598, 1986
38. A. Gebbert, M. Alvarezicaza, H. Peters, V. Jager, U. Bilitewski and R.D. Schmid, *J. Biotechnol.*, 32 (1994) 213–220.
39. A. Gebbert, M. Alvarezicaza, W. Stocklein and R.D. Schmid, *Anal. Chem.*, 64 (1992) 997–1003.
40. M. Riepl, V.M. Mirsky, I. Novotny, V. Tvarozek, V. Rehacek and O.S. Wolfbeis, *Anal. Chim. Acta*, 392 (1999) 77–84.
41. C. Berggren and G. Johansson, *Anal. Chem.*, 69 (1997) 3651–3657.
42. S. Ameur, C. Martelet, N. Jaffrezic-Renault and J.M. Chovelon, *Appl. Biochem. Biotechnol.*, 89 (2000) 161–170.
43. E. Katz, L. Alfonta and I. Willner, *Sens. Actuators, B*, 76 (2001) 134–141.
44. J.C. Thompson, J.A. Mazoh, A. Hochberg, S.Y. Tseng and J.L. Seago, *Anal. Biochem.*, 194 (1991) 295–301.
45. K. Yagiuda, A. Hemmi, S. Ito, Y. Asano, Y. Fushinuki, C.Y. Chen and I. Karube, *Biosens. Bioelectron.*, 11 (1996) 703–707.
46. S. Hardeman, T. Nelson, D. Beirne, M. Desilva, P.J. Hesketh, G. Jordan-Maclay and S.M. Gendel, *Sens. Actuators, B*, 24–25 (1995) 98–102.

47 C. Schyberg, C. Plossu, D. Barbier, N. Jaffrezic-Renault, C. Martelet, H. Maupas, E. Souteyrand, M.H. Charles, T. Delair and B. Mandrand, *Sens. Actuators, B*, 26–27 (1995) 457–460.
48 D.G. Sanderson and L.B. Anderson, *Anal. Chem.*, 57 (1985) 2388–2393.
49 O. Niwa, M. Morita and H. Tabei, *Electroanalysis*, 3 (1991) 163–168.
50 U. Wollenberger, M. Paeschke and R. Hintsche, *Analyst*, 119 (1994) 1245–1249.
51 M. Paeschke, U. Wollenberger, C. Kohler, T. Lisec, U. Schnakenberg and R. Hintsche, *Anal. Chim. Acta*, 305 (1995) 126–136.
52 M. Paeschke, U. Wollenberger, T. Lisec, U. Schnakenberg and R. Hintsche, *Sens. Actuators, B*, 27 (1995) 394–397.
53 J.H. Thomas, S.K. Kim, P.J. Hesketh, H.B. Halsall and W.R. Heineman, *Anal. Biochem.*, 328 (2004) 113–122.
54 K. Di Gleria, H.A. Hill, C.J. McNeil and M.J. Green, *Anal. Chem.*, 58 (1986) 1203–1205.
55 U. Wollenberger, F. Schubert, D. Pfeiffer and F.W. Scheller, *Trends Biotechnol.*, 11 (1993) 255–262.
56 J. Raba and H.A. Mottola, *Anal. Biochem.*, 220 (1994) 297–302.
57 C.G. Bauer, A.V. Eremenko, E. Ehrentreich-Forster, F.F. Bier, A. Makower, H.B. Halsall, W.R. Heineman and F.W. Scheller, *Anal. Chem.*, 68 (1996) 2453–2458.
58 C.G. Bauer, A.V. Eremenko, A. Kuhn, K. Kurzinger, A. Makower and F.W. Scheller, *Anal. Chem.*, 70 (1998) 4624–4630.
59 A.L. Ghindilis, A. Makower and F.W. Scheller, *Anal. Lett.*, 28 (1995) 1–11.
60 F.F. Bier, E. Ehrentreich-Forster, R. Dolling, A.V. Eremenko and F.W. Scheller, *Anal. Chim. Acta*, 344 (1997) 119–124.
61 A.L. Bordes, B. Limoges, P. Brossier and C. Degrand, *Anal. Chim. Acta*, 356 (1997) 195–203.
62 A.L. Bordes, B. Schollhorn, B. Limoges and C. Degrand, *Talanta*, 48 (1999) 201–208.
63 N. Kaneki, Y. Xu, A. Kumari, H.B. Halsall, W.R. Heineman and P.T. Kissinger, *Anal. Chim. Acta*, 287 (1994) 253–258.
64 R. Puchades and A. Maquieira, *Crit. Rev. Anal. Chem.*, 26 (1996) 195–218.
65 R.Q. Thompson, H. Kim and C.E. Miller, *Anal. Chim. Acta*, 198 (1987) 165–172.
66 M.H. Ho, *Methods Enzymol.*, 137 (1988) 271–287.
67 S. Morais, M.A. Gonzalez-Martinez, A. Abad, A. Montoya, A. Maquieira and R. Puchades, *J. Immunol. Methods*, 208 (1997) 75–83.
68 M.A. Gonzalez-Martinez, R. Puchades and A. Maquieira, *Trends Anal. Chem.*, 18 (1999) 204–218.
69 A.L. Ghindilis, R. Krishnan, P. Atanasov and E. Wilkins, *Biosens. Bioelectron.*, 12 (1997) 415–423.
70 E.M. Abad-Villar, M.T. Fernandez-Abedul and A. Costa-Garcia, *Anal. Chim. Acta*, 409 (2000) 149–158.
71 J. Rishpon, Y. Gezundhajt, L. Soussan, I. Rosenmargalit and E. Hadas, In: P.R. Mathewson and J.W. Finley (eds.), *Biosensor Design and Application*, ACS

Symposium Series No. 511. American Chemical Society, Washington, DC, 1992, pp. 59–72.
72 S. Dai and M.E. Meyerhoff, *Electroanalysis*, 13 (2001) 276–283.
73 J.L. Boitieux, J.L. Romette, N. Aubry and D. Thomas, *Clin. Chim. Acta*, 136 (1984) 19–28.
74 J. Rishpon and I. Rosen, *Biosensors*, 4 (1989) 61–74.
75 S. Kelly, D. Compagnone and G. Guilbault, *Biosens. Bioelectron.*, 13 (1998) 173–179.
76 Y.H. Xu and A.A. Suleiman, *Anal. Lett.*, 30 (1997) 2675–2689.
77 C.J. Cook, *Nat. Biotechnol.*, 15 (1997) 467–471.
78 C.J. Cook, *J. Neurosci. Methods*, 110 (2001) 95–101.
79 R.I. Stefan and H.Y. Aboul-Enein, *J. Immunoassay Immunochem.*, 23 (2002) 429–437.
80 Z. Dai, F. Yan, J. Chen and H. Ju, *Anal. Chem.*, 75 (2003) 5429–5434.
81 Z. Dai, F. Yan, H. Yu, X. Hu and H. Ju, *J. Immunol. Methods*, 287 (2004) 13–20.
82 D. Du, F. Yan, S. Liu and H. Ju, *J. Immunol. Methods*, 283 (2003) 67–75.
83 P. Sarkar, P.S. Pal, D. Ghosh, S.J. Setford and I.E. Tothill, *Int. J. Pharm.*, 238 (2002) 1–9.
84 R. Stefan and R.G. Bokretsion, *J. Immunoassay Immunochem.*, 24 (2003) 319–324.
85 D. Purvis, O. Leonardova, D. Farmakovsky and V. Cherkasov, *Biosens. Bioelectron.*, 18 (2003) 1385–1390.
86 G.S. Xuan, S.W. Oh and E.Y. Choi, *Biosens. Bioelectron.*, 19 (2003) 365–371.
87 A. Benkert, F. Scheller, W. Schossler, C. Hentschel, B. Micheel, O. Behrsing, G. Scharte, W. Stocklein and A. Warsinke, *Anal. Chem.*, 72 (2000) 916–921.
88 D. Stollner, W. Stocklein, F. Scheller and A. Warsinke, *Anal. Chim. Acta*, 470 (2002) 111–119.
89 T.M. O'Regan, M. Pravda, C.K. O'Sullivan and G.G. Guilbault, *Talanta*, 57 (2002) 501–510.
90 T.M. O'Regan, L.J. O'Riordan, M. Pravda, C.K. O'Sullivan and G.G. Guilbault, *Anal. Chim. Acta*, 460 (2002) 141–150.
91 E. Dempsey, C. O'Sullivan, M.R. Smyth, D. Egan, R. Okennedy and J. Wang, *Analyst*, 118 (1993) 411–413.
92 A. Benkert, F.W. Scheller, W. Schoessler, B. Micheel and A. Warsinke, *Electroanalysis*, 12 (2000) 1318–1321.
93 M. Dijksma, B. Kamp, J.C. Hoogvliet and W.P. van Bennekom, *Anal. Chem.*, 73 (2001) 901–907.
94 D. Ivnitski and J. Rishpon, *Biosens. Bioelectron.*, 11 (1996) 409–417.
95 F. Darain, S.U. Park and Y.B. Shim, *Biosens. Bioelectron.*, 18 (2003) 773–780.
96 N.H. Ammida, G. Volpe, R. Draisci, F. delli Quadri, L. Palleschi and G. Palleschi, *Analyst*, 129 (2004) 15–19.

97 A.F. Chetcuti, D.K.Y. Wong and M.C. Stuart, *Anal. Chem.*, 71 (1999) 4088–4094.
98 A.J. Killard, S.Q. Zhang, H.J. Zhao, R. John, E.I. Iwuoha and M.R. Smyth, *Anal. Chim. Acta*, 400 (1999) 109–119.
99 C.L. Feng, Y.H. Xu and L.M. Song, *Sens. Actuators, B*, 66 (2000) 190–192.
100 R.M. Pemberton, J.P. Hart and J.A. Foulkes, *Electrochim. Acta*, 43 (1998) 3567–3574.
101 R.M. Pemberton, J.P. Hart and T.T. Mottram, *Biosens. Bioelectron.*, 16 (2001) 715–723.
102 M. Wilmer, D. Trau, R. Renneberg and F. Spener, *Anal. Lett.*, 30 (1997) 515–525.
103 D. Trau, T. Theuerl, M. Wilmer, M. Meusel and F. Spener, *Biosens. Bioelectron.*, 12 (1997) 499–510.
104 J. Zeravik and P. Skladal, *Electroanalysis*, 11 (1999) 851–856.
105 M. Delcarlo, I. Lionti, M. Taccini, A. Cagnini and M. Mascini, *Anal. Chim. Acta*, 342 (1997) 189–197.
106 S. Bender and O.A. Sadik, *Environ. Sci. Technol.*, 32 (1998) 788–797.
107 K.A. Fahnrich, M. Pravda and G.G. Guilbault, *Biosens. Bioelectron.*, 18 (2003) 73–82.
108 J. Zeravik, T. Ruzgas and M. Franek, *Biosens. Bioelectron.*, 18 (2003) 1321–1327.
109 B. Sanden, L.H. Eng and G. Dalhammar, *Appl. Microbiol. Biotechnol.*, 50 (1998) 710–716.
110 B. Sandén and G. Dalhammar, *Appl. Microbiol. Biotechnol.*, 54 (2000) 413–417.
111 A.J. Bäumner and R.D. Schmid, *Biosens. Bioelectron.*, 13 (1998) 519–529.
112 R.W. Keay and C.J. McNeil, *Biosens. Bioelectron.*, 13 (1998) 963–970.
113 M.F. Yulaev, R.A. Sitdikov, N.M. Dmitrieva, E.V. Yazynina, A.V. Zherdev and B.B. Dzantiev, *Sens. Actuators, B*, 75 (2001) 129–135.
114 B. Mirhabibollahi, J.L. Brooks and R.G. Kroll, *Appl. Microbiol. Biotechnol.*, 34 (1990) 242–247.
115 B. Mirhabibollahi, J.L. Brooks and R.G. Kroll, *Lett. Appl. Microbiol.*, 11 (1990) 119–122.
116 B. Mirhabibollahi, J.L. Brooks and R.G. Kroll, *J. Appl. Bacteriol.*, 68 (1990) 577–585.
117 J.L. Brooks, B. Mirhabibollahi and R.G. Kroll, *Appl. Environ. Microbiol.*, 56 (1990) 3278–3284.
118 E. Hadas, L. Soussan, I. Rosenmargalit, A. Farkash and J. Rishpon, *J. Immunoassay*, 13 (1992) 231–252.
119 J. Rishpon and D. Ivnitski, *Biosens. Bioelectron.*, 12 (1997) 195–204.
120 A.S. Mittelmann, E.Z. Ron and J. Rishpon, *Anal. Chem.*, 74 (2002) 903–907.
121 J.D. Brewster and R.S. Mazenko, *J. Immunol. Methods*, 211 (1998) 1–8.
122 I. Abdel-Hamid, D. Ivnitski, P. Atanasov and E. Wilkins, *Anal. Chim. Acta*, 399 (1999) 99–108.
123 L. Yang, Y. Li and G.F. Erf, *Anal. Chem.*, 76 (2004) 1107–1113.

124 S.Y. Dong, G. Luo, J. Feng, Q.W. Li and H. Gao, *Electroanalysis*, 13 (2001) 30–33.
125 I. Benhar, I. Eshkenazi, T. Neufeld, J. Opatowsky, S. Shaky and J. Rishpon, *Talanta*, 55 (2001) 899–907.
126 S. Susmel, G.G. Guilbault and C.K. O'Sullivan, *Biosens. Bioelectron.*, 18 (2003) 881–889.
127 K.A. Uithoven, J.C. Schmidt and M.E. Ballman, *Biosens. Bioelectron.*, 14 (2000) 761–770.
128 A.A. Suleiman and Y.H. Xu, *Electroanalysis*, 10 (1998) 240–243.
129 F.M. Ivison, J.W. Kane, J.E. Pearson, K. Jenny and P. Vadgama, *Electroanalysis*, 12 (2000) 778–785.
130 K. Luangaram, D. Boonsua, S. Soontornchai and C. Promptmas, *Biocatal. Biotransform.*, 20 (2002) 397–403.
131 D. Monroe, *Crit. Rev. Clin. Lab. Sci.*, 28 (1990) 1–18.
132 T. Yao and G.A. Rechnitz, *Biosensors*, 3 (1987/88) 307–312.

Chapter 9

Immunoassay: potentials and limitations

Catalin Nistor and Jenny Emnéus

9.1 INTRODUCTION

Bioassays can be classified according to two categories: comparative (or "functionally specific") and analytical (or "structurally specific") [1]. The former compares relative effects of different substances or a mixture of substances on a biological system (e.g., whole animal, tissue, cell culture, receptor, etc.) and the results are represented as units of effect, which differ depending on the target biological system. By contrast, analytical bioassays measure the amount of a certain analyte in the test sample and are normally unable to be employed for the measurement of mixtures of substances without their prior separation. Results of analytical assays are represented in units of mass and should invariably be independent of the assay system used. Immunoassay can be considered as a "hybrid technique", since its working principle is based on the molecular recognition specific for comparative methods, but is very often applied for quantification of analytes, as the analytical assays are.

Since the immunoassay principle was introduced by Yalow and Berson [2] for the determination of insulin and by Ekins [3] for the determination of thyroxine, there has been an exponential growth in immunoassay development, not only in the range of applications, but also in the number of novel and ingenious designs. This enormous scientific interest is clearly reflected by the number of scientific publications that have appeared in the last four decades in the immunoassay field (Fig. 9.1).

The immunoassay is today a relatively well-established technique, with numerous applications in clinical analysis, but also with a considerable growing interest in other fields, such as the food industry or environmental monitoring. Benefits such as high selectivity and sensitivity (for certain analytes not attainable by any alternative methods), speed and relatively low cost of analysis have also led to rapid commercialization of immunoassay kits and automated analysers [4].

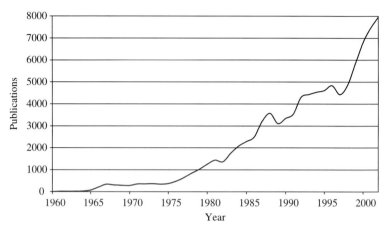

Fig. 9.1. The number of scientific articles published in the immunoassay field has an increasing trend.

Since immunoassays are primarily analytical techniques, in addition to studies for a better understanding of the nature of antibody–antigen interaction, there are continuous efforts to improve immunoassay performance (e.g., sensitivity, selectivity, precision and accuracy) in terms of robustness and reliability when analysing complex samples. The present chapter attempts to summarize the most commonly used immunoassay concepts, as well as the main approaches employed for the improvement of immunoassay sensitivity, selectivity and precision. The discussion is focussed around the main thermodynamic and kinetic principles governing the antibody–antigen interaction, and the effect of diverse factors, such as assay design, concentration of reactants, incubation time, temperature and sample matrix, is reviewed in relation to these principles. Finally, particular aspects on immunoassay standardization are discussed as well as the main benefits and limitations on screening vs. quantification of analytes in real samples.

9.2 BASIC ASPECTS OF IMMUNOASSAY

9.2.1 Antibody–antigen interactions—basic concepts

Immunoassays are based on the employment of antibodies as analytical reagents. *Antibodies* are glycoproteins from the immunoglobulin family, which are involved in the immune recognition and self-defence systems. *Immunoglobulin G* (IgG), the most abundant antibody in serum, is most often used for immunoassay development. An IgG has a tetrameric polypeptide structure, consisting of two identical heavy–light chain hetero-dimers, linked by disulphide bridges to form the characteristic Y-shape of the molecule [5]. Each of the heavy (H) and light (L) chains consists of two domains of amino acid sequences: one that shows very little variation among immunoglobulin molecules (i.e., the constant region, F_c) and another that can vary among different antibody molecules (i.e., the variable region or antigen-binding fragment, F_{ab}). Five different F_c region isotypes, IgG, IgM, IgA, IgD and IgE, have been identified, which differ from the structural point of view, as well as by carbohydrate content.

Immunoassays derive their unique characteristics from two important properties of antibodies: the strength of the antibody–antigen binding force and the narrow selectivity within a class of compounds, sometimes even with absolute specificity for the analyte they were designed for. These two properties allow selective measurement of very low concentrations of analyte even in very complex matrices (e.g., biological fluids, residual waters, etc.). For clarity of presentation, the main terms and concepts used in immunoassays are listed in "Glossary".

The essential step in an immunoassay is the molecular recognition of the antigen by the antibody due to stereochemical complementarity. This recognition is mediated by the antibody-binding site (known as the paratope) composed of six loops of hypervariable peptide sequences, which bind to a specific structure in the antigen molecule (the epitope) [6]. Four types of forces promote the binding: (1) electrostatic interactions, (2) dispersion forces, (3) hydrogen bonds and (4) hydrophobic interactions. The high selectivity of an antibody for its corresponding antigen is governed by sterical factors especially. A lack of complementarity between the binding sites causes repulsive forces and prevents a very close proximity at an intermolecular distance where the weak, attractive forces become effective [7]. The role of hydrophobic interactions, based on the repulsion of water by non-polar groups in the structure of the two molecules, is essential. The change in the microenvironment between the paratope and epitope interfaces thus increases the strength of the electrostatic forces between the antibody and the antigen, since the water molecules no longer compete with the charged groups involved in the interaction. Electrostatic interactions enhance significantly the affinity (i.e., 100–10,000 times for one bond). However, they are also relatively more effective at long distances, since their decrease is proportional to the inverse square root of the distance between the interacting groups, which can be compared with, e.g., dispersion (van der Waals) interactions, which decrease with the inverse sixth power of the intermolecular distance (see Fig. 9.2) [7]. This is why non-specific interactions of antibodies with other molecules are often electrostatic.

The complex nature of the antibody–antigen interactions, as well as the large diversity of antigens/epitopes, makes it almost impossible to predict in advance the optimum conditions for an immunoassay. Variations in pH and ionic strength of the reaction environment usually have an important role in changing the antibody affinity because of modification in the polar forces. Organic solvents may also disrupt hydrophobic interactions and are sometimes used to dissociate antigen–antibody complexes. Nevertheless, the main factors to take into consideration when optimizing an immunoassay are the assay format, as well as the nature and concentration of the immunoreactants.

9.2.2 Immunoassay sensitivity

Appreciation of the principles governing assay design is obviously an essential requirement for its application. Surprisingly for immunoassays, very little attention has been paid to these principles for many years following the initial description of the technique. To some extent, this is understandable due to many practical problems limiting the performance of immunoassays, such as

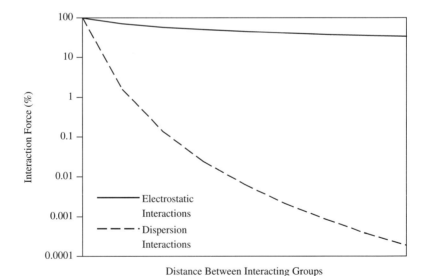

Fig. 9.2. Relative decrease of the attractive force with distance between interacting groups.

Immunoassay: potentials and limitations

difficulties in isolation, purification and labelling of macromolecular antigens, and in obtaining antisera with sufficient sensitivity and selectivity.

9.2.2.1 Precision vs. accuracy

Since immunoassay sensitivity is intimately related to assay precision, it is important to define and distinguish between some commonly used terms.

- *Precision* of an assay describes how well an assay result can be reproduced. *Within-run precision* is defined as the precision of the same sample run on several occasions with the same assay under the same assay conditions. *Between-run precision* is an index of the ability of the assay to reproduce the same result on the same sample at different runs, time moments, by different operators or in different labs. Precision is described by the variance (s^2), standard deviation (s) and coefficient of variation (CV).

  Precision estimates from several sources (different assays, different operators, etc.) may be combined by calculating the root mean square (RMS) CV as described by Eq. (9.1), in which N is the number of data points in each assay from 1 to n:

$$\text{RMS} = \frac{\sqrt{\text{CV}_1 N_1 \text{CV}_2 N_2 \cdots \text{CV}_n N_n}}{N_1 + N_2 + \cdots + N_n}. \tag{9.1}$$

- *Accuracy* shows how close the measured value is to the "real value". *Bias* is a measure of inaccuracy and is usually reported as a relative bias (%RE); it might be *constant*, having the same value regardless of the signal/analyte concentration, or *proportional*, where the assay reads a higher or lower constant percentage than the true value.
- *Error* encompasses both the imprecision of data and its inaccuracies. High precision does not indicate high accuracy or vice versa.

9.2.2.2 Expressions of immunoassay sensitivity

For most immunoassays, the mean response is a non-linear function of the analyte concentration (the dose–response curve) and the variance of replicate measurements is a non-constant function of the mean response, as seen in Fig. 9.3.

High sensitivity is an intrinsically desirable property of any analytical technique. The definition of sensitivity has, however, become a subject of considerable controversy particularly for immunoassays, being considered by different scientific groups as the slope in the middle of the dose–response curve [8,9], the midpoint of the dose–response curve (IC_{50}) [10] or the minimal

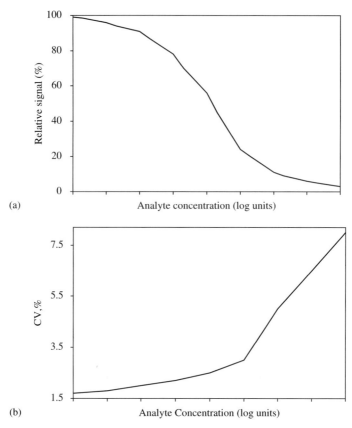

Fig. 9.3. A typical immunoassay calibration curve (a) and the variation of the signal CV (%) with concentration (b).

detectable concentration (MDC) for the analyte [11,12]. Since the most common way of plotting the immunoassay calibration for an analyte is in relative units, the slope of the dose–response curve becomes irrelevant for comparison between sensitivities of different immunoassays (see Fig. 9.4).

The analytical sensitivity (AS) in immunoassay is defined as the analyte concentration that gives a signal significantly different from the blank, i.e., two or three times the standard deviation (s). High assay sensitivity can thus be obtained by increasing the "signal-to-noise ratio", either by an increase of signal in the presence of the analyte or by decreasing the "noise" (i.e., the errors in measurement). For a given system, assay optimization may therefore be regarded simply as an exercise to identify the sources of errors, and determination of the appropriate concentrations of reagents that will minimize these. Such models consider that the error in measurement is nearly constant

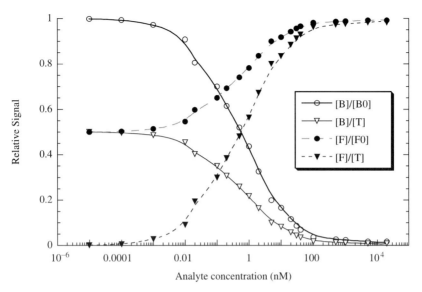

Fig. 9.4. Different ways of plotting immunoassay data for the same analyte calibration lead to different shapes/slopes of the curve. The plotted parameters stand for different fractions of the analyte; [B], bound; [F], free; [B0], bound at zero dose; [F0], free at zero dose; T, total.

over the region between the signal in the absence of the analyte ("zero dose") and the MDC of the analyte, and that there is a linear correlation of the signal with the concentration within this region [13]. However, the values calculated in this way are not necessarily reliable. One reason for this is that the calculated AS very often lies below the lowest concentration for the dynamic range of the response, sometimes by one order of magnitude. One assumes, for the purpose of calculating AS, that the *fitted calibration curve* (see Section 9.2.2.3) is a close reflection of the dose–response curve of an immunoassays and this is most often not the case, as shown in Fig. 9.5a where the calculated sensitivity (S1) is not reflecting the true value (S2).

Another reason is that there is a high variation in assay precision close to zero dose, i.e., it usually "drops" asymptotically parallel with the Y-axis reaching a minimum value around the middle of the calibration curve (see Fig. 9.5b). Thus, a considerable difference in CV in the absence of the analyte and at the MDC is expected. This is why the concept of functional sensitivity (FS) was introduced [14], defined as the lowest concentration of analyte for which the CV falls below 20%. FS is a parameter that is far more practical than AS and it is derived from the knowledge of the precision profile (PP), i.e., the changes in precision with the concentration of the analyte (Fig. 9.5b) [14].

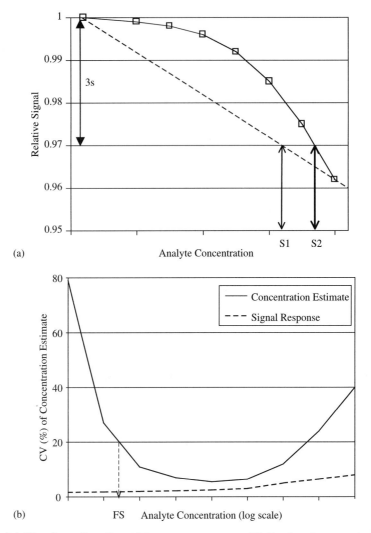

Fig. 9.5. (a) Bias in estimation of immunoassay sensitivity due to erroneous curve fitting: S1, calculated sensitivity; S2, true sensitivity. (b) Precision profile (PP): variation of precision (CV in %) for the signal response and estimated concentration (translated from the signal response) within the whole immunoassay range.

Assessment of precision in immunoassay requires the measurement of variability at defined concentrations using solutions containing standard analyte. There are several problems associated with such an approach. First, the precision in between the discrete levels measured must be interpolated. Second, the precision is estimated in standard solutions, thus the matrix effects or differences between samples are not taken into consideration. There is, however, no perfect solution for any of these problems. Precision of a high number of calibrating solutions has to be estimated and a sample matrix as similar as possible to the one to be analysed has to be evaluated in order to avoid wrong interpretation of assay results. The total error can then be plotted against a wide range of concentrations to obtain the PP of the assay (Fig. 9.5b) [14], from which the FS can be determined.

Immunoassays are usually applied for one of the following purposes.

- *Detection*. Identification of the presence or absence of a particular substance, i.e., disease marker, drug of abuse or environmental pollutant.
- *Quantification*. Accurate determination of the level of a particular substance, within a range of interest.
- *Monitoring*. Assessment of changes in concentration of a specific analyte in time.

Optimization of an assay should be performed in accordance with its specific application, as listed above. The minimal distinguishable difference in concentration (MDDC) at a certain concentration level with a certain probability depends on the precision at that specific level. In terms of clinical utility, the assay should thus be optimized so that the precision is best at the clinical decision cut-off level for a specific analyte. In environmental monitoring this would correlate to optimizing the assay so that the precision is best at the lowest allowable level for the presence of a pollutant in environmental waters. Thus, the MDDC will be at its minimum at this point and hence give the least errors in clinical classification or environmental screening. In practice, this means that assays should be optimized so that the clinical or environmental decision level appears at the mid-point of the curve, i.e., at the IC_{50}, since here the error is at its minimum (Fig. 9.5b).

The analytical range of immunoassays is defined as the range between the lower limit of quantification (LLQ) and upper limit of quantification (ULQ) for which interpolated results have acceptable levels of accuracy and precision [15]. This "acceptable level" has been recommended to be at 10% CV [16] or at 20% total error (e.g., sum between relative accuracy and precision). The MDC is similarly recommended as to be the lowest concentration of the analyte for which the CV drops under 20% (i.e., FS) [16] or when the total error is below 25% [15].

9.2.2.3 Non-linearity of the standard curve
Even though numerous articles have been published on mathematical modelling for fitting immunoassay data [17–19], interpretation of the results

from the non-linear response–concentration and error–concentration relationships is not always well understood and often wrongly applied in practice. Computational errors caused by inappropriate fitting models to analyse experimental data are of a pure technical nature and are common to all analytical methods, even though they might be more accentuated in the case of immunoassays, which frequently operate close to their limits of sensitivity [1].

A fitting equation model should have the following features: (1) the number of fixed numerical parameters that characterize the curve should be minimal (preferably four); (2) the shape, slope and position of the curve described by the equation should be flexible; (3) the form should be monotonic (i.e., no reversals of slope) [18]. The variation of the total error over the analytical range of response should be characterized as a second non-linear equation [15].

A commonly accepted fitting approach is the "four-parameter logistic" equation (4PL) used to fit the mean concentration–response relationship, and the power-of-the-mean equation to fit the response–error relationship [15,18]. In some applications a fifth parameter describing the asymmetry of the curve might be incorporated, ensuring a higher versatility of the model [18].

The 4PL (Eq. (9.2)) provides an accurate representation of the sigmoidal relationship between the analytical response and the logarithm of the analyte concentration in immunoassays:

$$Y = \frac{(A - D)}{1 + \left(\frac{X}{C}\right)^B} + D. \qquad (9.2)$$

Y is the expected response and X is the corresponding concentration. A, B, C and D are the four parameters of the equation, where A gives an estimate of expected response at zero dose, B is the slope (e.g., response/concentration) in the middle of the calibration curve, C is the IC_{50} and D is the expected response at "infinite" dose. This curve satisfies all the conditions specified for the response–concentration relationship and also closely approximates the mass-action equations [20]. Weighting of the results is recommended for fitting dose–response data from immunoassay, in order to compensate the heterogeneity of response variances in the response–error relationship [17,18].

Acceptability criteria for a fitting model should always be verified for a particular application by evaluating the %RE between back calculated and nominal concentrations of the calibration samples [15,21]. The recommendations state a maximal acceptable limit for the %RE to 10% and that standard curves from a minimum of three analytical runs should be evaluated when selecting a model [15]. It should, however, be realized that with all precautions taken, the model will never lead to a "perfect fitting", since the bias is always higher at the two extremes of the curve than in its middle region. Thus, care should be taken when analysing samples containing very low or very high concentrations of analyte.

A minimum of six non-zero concentrations of the analyte should be tested when fitting a calibration equation to a non-linear response–concentration relationship. One calibrator should be around the IC_{50} value and all the other should be spaced to the left and right of this value, based on logarithmic progression with each concentration tested at least in duplicate [22]. "Anchoring" calibration points at the extremes of the calibration curve are beneficial in curve fitting and lead to improved accuracy and precision [23].

The error–response relationship is usually expressed as a product of two terms, as shown in the following equation:

$$Y = \alpha E(Y)^\theta, \qquad (9.3)$$

where α is a proportionality constant, $E(Y)$ is the error associated to the response Y and θ is a shaping parameter, which is constant across runs. It is recommended that the parameters α and θ be experimentally estimated from multiple runs because of the limited degree of replication in a single run [18,22].

9.2.3 The principle of antibody occupancy: classification of immunoassay

It is essential to be capable of predicting the optimum reactant concentrations in order to design sensitive immunoassays. Depending on the purpose, one can choose different working concentrations of immunoreactants. If, for instance, a certain decision concentration level is critical, as, for example, in diagnosis or for environmental screening of pollutants/toxicants, the precision of the assay around that specific concentration level should be improved as much as possible. On the other hand, if the analyte needs to be measured over a wide concentration range, the assay should be designed accordingly. By modulating the reactant concentrations the assay characteristics can be completely changed, e.g., measurement of the analyte in a particular range of concentrations requires particular concentrations of immunoreactants.

Although mainly mathematical models can give quantitative estimates of the optimal concentrations of reagents, Ekins [24] has demonstrated that qualitative conclusions, regarding mainly the relative sensitivity of a certain assay format, may be easily drawn by examining the assay design in terms of the fraction of antibody-binding sites that are measured, referred to as *the antibody occupancy principle*. All assays can essentially be described in one of the two ways: either they are based on the measurement of antibody sites occupied by analyte/antigen or by the measurement of unoccupied binding sites. Immunoassays are commonly divided into two major classes, competitive immunoassays and non-competitive immunoassays, as will be described below.

9.2.3.1 Competitive (reagent limited) immunoassays

Competitive immunoassays are based on the measurement of antibody-binding sites unoccupied by the analyte. An increase in the concentration of analyte causes a reduction of unoccupied sites. The measurement is performed indirectly by means of competition between the analyte and a competitor/tracer[1] molecule (usually a labelled analyte derivative) for a limited number of antibody-binding sites. In principle, such assays can be described by Eqs. (9.4) and (9.5), together with their corresponding equilibrium constants K and K^*:

$$\text{Ag} + \text{Ab} \underset{k_2}{\overset{k_1}{\Leftrightarrow}} \text{Ab-Ag}, \qquad K = \frac{k_1}{k_2} = \frac{[\text{Ab-Ag}]}{[\text{Ab}][\text{Ag}]}, \qquad (9.4)$$

$$\text{Ag}^* + \text{Ab} \underset{k_4}{\overset{k_3}{\Leftrightarrow}} \text{Ab-Ag}^*, \qquad K^* = \frac{k_3}{k_4} = \frac{[\text{Ab-Ag}^*]}{[\text{Ab}][\text{Ag}^*]}, \qquad (9.5)$$

where Ab, Ag, Ag*, Ab–Ag, Ab–Ag* are the antibody, antigen/analyte, competitor, antibody–analyte and antibody–competitor, respectively; the values between square brackets are the corresponding molar concentrations, k_1 and k_3 are the association rate constants, k_2 and k_4 are the dissociation rate constants and K and K^* are the equilibrium- or affinity constants for the Ab–Ag and Ab–Ag* complexes, respectively.

The first immunoassay developed was based on the principle described above, with an insulin competitor, i.e., an analyte derivative labelled with a radioactive isotope (^{125}I) and the analyte (insulin) competing for a limited amount of immobilized anti-insulin antibodies. After the equilibrium was reached and the unbound competitor was removed, the residual radioactivity was correlated to the concentration of insulin [2]. Since then, numerous other variants of competitive immunoassay formats have been described, either in a

---

[1] Here we use the term competitor for any compound used for indirect detection of the analyte by competing with it for a limited number of antibody-binding sites. A tracer is a competitor that is labelled with a detectable group.

similar format, i.e., by immobilization of the antibody on a solid support ("direct competitive assays") or by immobilization of an analyte derivative on the solid support ("indirect competitive assays"). The two different assay formats will be discussed below in more detail.

9.2.3.1.1 Competitive immunoassay formats

In a *direct competitive heterogeneous immunoassay*, the competitor and the analyte present in solution compete for a limited amount of antibody-binding sites. A typical way of performing such an assay is shown in Fig. 9.6. In the most frequently employed format, the labelled competitor is mixed with the sample containing the analyte, and the formed mixture is added over a solid phase with immobilized antibody. When the equilibrium is reached, the unbound analyte- and labelled competitor are removed from the solution and the labelled fraction bound to the solid phase is quantified. The obtained signal is inversely proportional to the analyte concentration.

Usually, immobilization of antibodies can be performed either directly by passive adsorption, covalent coupling, or indirectly by using a solid surface pre-coated with protein A or G [25] or a secondary immunoglobulin (Ab_2, anti-Ab_1) [26] for a favourable orientation of the antibody on the solid surface with the binding sites towards the sample solution. This orientated immobilization of antibodies is important for the assay performance, taking into account the special characteristics of solid-phase immunoassays (see Section 9.3.4.4).

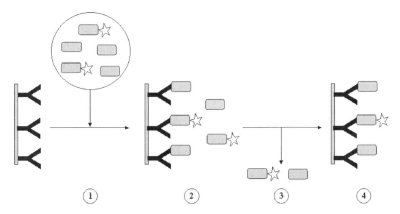

Fig. 9.6. *Direct competitive immunoassay with immobilized antibodies*: The analyte competes with a labelled analyte derivative ("tracer") for a limited amount of antibody-binding sites (1) until equilibrium (2) is established. Then, the excess of analyte and tracer (3) are removed by washing of the solid support, and the fraction bound to the antibody is detected (4).

In an *indirect competitive immunoassay*, the competitor is commonly immobilized on a solid surface while the antibody and the analyte are added in the adjacent solution. After the competition, the fractions that are unbound to the solid surface are removed and the bound *primary* antibody is usually measured by the addition of a labelled *secondary* antibody (see Fig. 9.7). Even though this assay format is not as frequently used as the direct competitive immunoassay, it provides an advantage when analysing complex samples, i.e., the label (usually an enzyme) does not come into direct contact with the sample matrix, so that interferences with the detection step are minimized.

9.2.3.1.2 Limitations of competitive immunoassay
For a label of infinite specific activity and in the absence of unspecific binding, the concentrations of both antibody and competitor tend to go towards zero. The theoretical limit of sensitivity (the MDC) is given by the following equation [27]:

$$\text{MDC} = 2\frac{CV_0}{K}, \tag{9.6}$$

where CV_0 is the relative measurement error at zero dose (experimental error). For antibodies with equilibrium constants (K) of 10^9-10^{11} l mol^{-1}, high

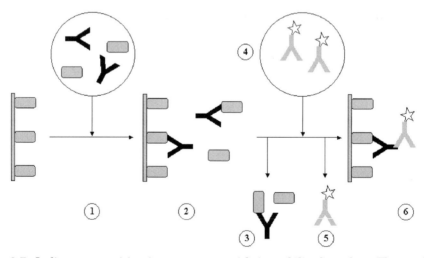

Fig. 9.7. *Indirect competitive immunoassay with immobilized analyte*: The analyte competes with an immobilized analyte derivative for a limited amount of antibody-binding sites (1) until equilibrium (2) is established. The surplus of analyte and antibody (3) are removed by washing of the solid support and an excess of labelled secondary anti-immunoglobulin antibody (4) is added onto the solid support. After the excess of secondary antibody is removed (5), the fraction of bound label is detected (6).

specific activity of the label and 1% experimental error, Eq. (9.6) predicts sensitivities between 0.2 and 20 pmol l^{-1}.[2] For competitive immunoassays the sensitivity is mainly limited by the ratio between the affinity constants of the competitor vs. analyte and their absolute values, but other factors, such as the error in signal measurement, the level of non-specific binding and the residual errors, have also to be taken into consideration.

9.2.3.2 Non-competitive (reagent excess) immunoassays
Non-competitive (or immunometric) immunoassays are based on the measurement of antibody-binding sites occupied by the analyte. The signal therefore is expected to rise with increasing analyte concentration. In contrast to competitive assays, the best sensitivity in non-competitive assays is attained when the maximal concentration of antibody is used, to ensure that the highest possible proportion of analyte is bound. However, increasing the amount of labelled antibody also increases the non-specific binding, until a point where the level of non-specific binding increases faster than the rate of specific binding. This point of maximum sensitivity depends on the value of non-specific binding, its associated error, as well as the errors in signal measurement [5].

9.2.3.2.1 Non-competitive immunoassay formats
In *single-site non-competitive assays* (Fig. 9.8), first described by Miles [28], an excess of labelled antibodies is incubated with the analyte. This reaction is followed by the removal of the unreacted labelled antibodies with the help of a large excess of solid-phase bound antigen. The amount of labelled antibody used dictates the assay format. Thus, if the antibody concentration is low, the assay approaches that of the indirect competitive format. On the other hand, if the concentration of labelled antibody is high enough, the equilibrium reaction in relation (4) should be driven towards a higher degree of completion, so that some of the limitations of sensitivity dictated by affinity (e.g., a major factor restricting the sensitivity of competitive assays) are avoided.

In *double-site immunometric assays* (i.e., "sandwich assays" [29], see Fig. 9.9) the sample containing the analyte is incubated with both a *capture antibody*, usually immobilized on a solid phase, and a labelled *detecting antibody*. After the excess of labelled antibody is removed, the measured signal can be correlated directly to the analyte concentration.

---

[2] The model considers that the equilibrium constants for the analyte and competitor are of the same order of magnitude, which often is not the case for small hapten analytes due to difference in structure of tracer and analyte.

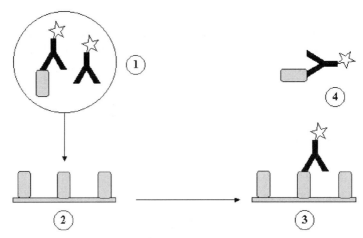

Fig. 9.8. *Single-site immunometric assay*: The sample containing the analyte, previously incubated with a surplus of labelled anti-analyte antibodies (1), is passed through a column with immobilized antigen (2), able to trap the excess of antibodies (3). The analyte–antibody complex is eluted from the column (4) and can be detected downstream.

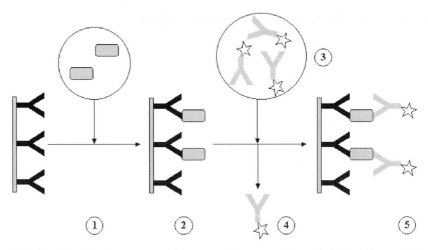

Fig. 9.9. *Double-site immunometric assay*: The sample containing the analyte is incubated with the capture antibodies immobilized on a solid support (1), until equilibrium (2) is established. Then, labelled detection antibodies (3), recognizing a different epitope on the antigen than capture antibodies, are added onto the support, and the excess of detection antibodies is removed by washing (4). The amount of the label bound to the support is detected (5).

Since this assay format involves a double molecular recognition, the selectivity is considerably improved. This format is, however, limited for use with larger antigens that have the intrinsic ability to bind two different antibodies to two different epitopes and is thus not applicable to small haptens.

9.2.3.2.2 Limitations of non-competitive immunoassay

In contrast to competitive immunoassays, the number of theoretical models describing the kinetics and thermodynamics of non-competitive immunoassays is considerably lower and most of them are derived from double-site immunometric configurations. The dose–response curve for such assays is sigmoid; the signal increasing with the analyte concentration with a plateau value reached when the capture antibody becomes saturated.

The limiting sensitivity (MDC) of immunometric assays at infinite specific activity of the label is described by the following equation [27]:

$$\text{MDC} = \frac{\text{NSB} \times \text{CV}_0}{K}, \tag{9.7}$$

where NSB is the level of non-specific binding and the CV_0 is the relative error of the response at zero dose. According to Eq. (9.7), the maximum sensitivity should be of the order $0.2-20$ fmol l^{-1} for a system with 0.1% NSB, 1% error of response at zero dose, equilibrium constants (K) between 10^9 and 10^{11} l mol^{-1} and infinitely high specific activity of the label. This means three orders of magnitude better sensitivity than for competitive methods.

9.2.4 Immunoassay selectivity

Selectivity describes the ability of an assay to produce a measurable response only for the analyte of interest. *Interferences* are considered to be any factor causing a bias in the assay result and can be divided in two major groups: *specific* and *non-specific*, as will be described below.

9.2.4.1 Specific interferences: cross-reactivity

A specific interference is caused by the heterogeneity of the antigenic response (e.g., for polyclonal antibodies) or by compounds having epitope(s) similar in structure to the one of the analyte, and which can directly compete with the basic immunoreaction ("true cross-reactants"). It is well recognized that such interferences are derived from the selectivity of the antibody–ligand interaction [6,30]. As pointed out by Berzofsky and Schechter [31], the concept of antibody specificity is complementary to the concept of cross-reactivity of antigens. Typical cases of cross-reactivity are, e.g., when compounds from the same chemical class are present simultaneously in the same sample [32], or when metabolites of the main analyte that also contain cross-reactive epitopes are produced in situ.

There are several ways to express *cross-reactivity* (CR) between specific interfering compounds and the analyte of interest, but the most common one is the ratio between the concentration of standard analyte and that of the cross-reactant, each leading to a 50% change in signal vs. the signal in the absence of the analyte, see Eq. (9.8):

$$\%\mathrm{CR} = \frac{\mathrm{IC}_{50,\mathrm{Ag}}}{\mathrm{IC}_{50,\mathrm{CR}}}, \tag{9.8}$$

where $\mathrm{IC}_{50,\mathrm{Ag}}$ and $\mathrm{IC}_{50,\mathrm{CR}}$ are the concentrations of analyte and cross-reactant, respectively, which produce a change in signal of 50% vs. the signal at zero dose.

The CR might, however, vary across the assay concentration range, thus verification of CR at different concentration ranges is recommended [33]. For example, when polyclonal antibodies are used, the observed CR reflects the influence of a particular antibody clone having the largest influence in the binding reaction at a particular analyte concentration level. At low analyte concentration this will predominantly be for high-affinity antibodies, while at high concentration, the weaker-affinity clones will have the largest influence. One useful technique in this relation was proposed for the development of polyclonal steroid assays in which a small quantity of a cross-reactive steroid was added deliberately in order to block a minor, but particular cross-reactive antibody clone [16].

Relative potency (RP, Eq. (9.9)) is another way of describing CR and gives a better estimate of the CR at different positions of the dose–response curve:

$$\mathrm{RP} = \left(F \frac{K_{\mathrm{Ag}}}{K_{\mathrm{CR}}} \right) + B, \tag{9.9}$$

where B and F are the bound and free fractions of the analyte and K_{Ag} and K_{CR} are the equilibrium constants for the analyte and the cross-reactant, respectively [34]. Another way of calculating the CR within the whole assay range was recently proposed [35].

Regardless of the assay design, the best way of estimating CR (and *interference* in general) is, however, by spiking the compound under study into samples containing the endogenous analyte at a fixed concentration, e.g., at the IC_{50} level. Then, the CR may be defined as the concentration of interfering species that produces a significant change of the reference signal [16]. Complete characterization of CR with all the possible chemicals structurally related to the analyte is, however, technically impossible.

Depending on the degree of reactivity with structurally related compounds, antibodies used in immunoassay development can be classified as: (1) highly specific, recognizing basically only the compound of interest from a whole chemical class; (2) cross-reactive within a chemical class, with more or less the same binding affinity for a set of compounds; and (3) cross-reactive, but with different degrees of binding affinity for compounds from the target chemical class.

Antibodies from class (1) and (2) can be used for the development of analyte-specific and group-selective immunoassay, respectively. However, production of class (3) antibodies often occurs and they are difficult to use for quantification of analyte(s) in unknown samples. In such situation, incomplete or incorrect characterization of the antibody CR in the specific assay format might lead to erroneous quantification of the analyte. In general, only the target analyte is used for calibration and the results are often expressed in "equivalents of target analyte concentration", which is a more or less accurate estimation of all cross-reactants present in the sample [33,36].

Qualitative conclusions of the presence or absence of cross-reactants in the sample matrix can easily be drawn by measuring the analytical response at different dilutions of the sample: non-parallel curves for the standard and sample demonstrate a lack in structural identity [1]. This simple result is based on the principle that differences in molecular structure are likely to lead to differences in reaction thermodynamics and kinetics, even when identical epitopes are involved in the antibody–antigen reaction. Such differences are generally manifested as differences in the apparent affinity constants governing the reactions, which can be revealed as alterations of assay characteristics [1,16].

The mentioned tests do not provide a general solution, but only an apparent one, since CR also depends on a set of other factors: heterogeneity of antibody population (e.g., polyclonal vs. monoclonal), relative affinity for the analyte/cross-reactant of each antibody, recognition of several antigenic determinants, physical state of the molecule (e.g., immobilized or free in solution) or matrix effects (e.g., protein content of the sample, pH, ionic strength, etc.) [37], as well as synergistic effects [38]. It is thus suggested for CR to be tested in the presence of different concentrations of the analyte and in a matrix as close to the sample to be analysed as possible.

9.2.4.2 Non-specific interferences: matrix effects (other than cross-reactivity)
Non-specific interferences are due to factors that alter the final result in a different way than competition with the antibody binding of the analyte. These effects are classified as *matrix effects*, which is why they often are referred to as "false cross-reactants". Incorrect evaluation of such effects might lead to wrong decisions/diagnostics, sometimes with disastrous consequences [39,40]. In principle, the sample, buffer, label, antibodies, separation and wash steps can all contribute with unspecific components in terms of the signal measured. Some mechanisms of non-specific interference are: (1) alteration of the effective analyte concentration; (2) interferences with antibody binding; (3) interferences with the separation of bound and free tracer; (4) interferences with the detection.

Taking into consideration the complex composition of many real samples, as well as the basic protein structure of antibodies, non-specific response changes due to other components than the analyte to determine are very likely to occur. Even though many possible alternatives for matrix effects are described here, there are numerous situations when the factors that cause the effect are practically unknown. Matrix effects are strongly dependent on the assay configuration: thus, in a competitive assay the error is higher if the free tracer fraction is measured and the effect is less pronounced in an indirect format using secondary-labelled antibodies than in a direct format with a labelled analyte derivative as the tracer [41].

Matrix effects can be quantified by comparing a standard calibration curve in standard buffer and in the sample/sample extract [38]. A common effect is a reduction in the response intensity in that particular sample (see Fig. 9.10): the stronger the matrix, the lower the response.

Matrix effects have traditionally been dealt with by standard addition, i.e., addition of known amounts of standard analyte in the sample and extrapolating the result to zero added standard. However, this method is difficult to apply for immunoassay because of the non-linearity of the concentration–response relationship. An alternative approach is presented as a modality to optimize the standards and buffer, so that it reproduces as closely as possible the effects of a specific matrix on immunoassay performances [42].

Obviously, the easiest way to circumvent a matrix effect is to perform the calibration in analyte-free samples [15,44,45] or to correct the data with blanks prepared from the same matrix subjected to the entire protocol of extraction and analysis [46,47]. The availability of analyte-free samples is limited, however, especially in environmental monitoring, due to the great variability between sample compositions. Moreover, preparation of

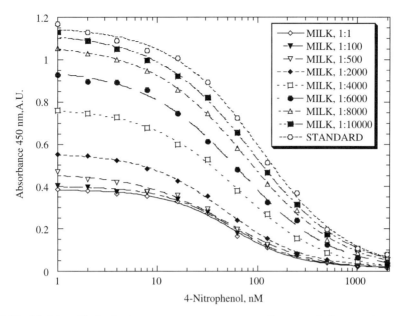

Fig. 9.10. Matrix effects in immunoassay are generally revealed as a decrease in absolute signal at any level of spiked analyte concentration. This effect decreases with the dilution of the sample. Example presented: an indirect competitive ELISA for 4-nitrophenol in milk at different dilutions. A 4-nitrophenol derivative coupled to bovine serum albumin was used as the competitor, an anti-4-nitrophenol (polyclonal, from rabbit) as the antibody and goat anti-rabbit immunoglobulin G coupled to peroxidase as the detection system [43]. (Reprinted with permission.)

analyte-free samples is not always without consequences, since some of the matrix components might be removed together with the analyte [15].

Depending on the complexity of the sample matrix, a pre-cleaning step is a possible solution to remove interfering effects; nevertheless, the choice should be a good compromise in relation to the total time and cost of analysis [33]. Supercritical fluid- [48,49], microwave-assisted- [50] or solid-phase extraction (SPE) [51–53] have been used as pre-cleaning steps prior to immunoassay. However, in certain situations, the extraction step led to a decrease in accuracy and recovery, probably due to analyte loss during clean-up/evaporation/re-dissolution steps [53].

Since the sample matrix generally gives the same effect on the analytical response, as if the analyte would be present in the sample, a positive bias when measured by immunoassay vs. a reference method is often observed [54,55]. Simple mathematical models that parallel the strategy of sample addition for correcting these negative effects have been proposed [56,57].

These models rely on an observed constant relationship between the absorbance at zero dose in the standard buffer ($OD_{0,standard}$) and the sample matrix ($OD_{0,sample}$) [56], between the two calibration slopes [58], or between relative absorbance and analyte concentration in the sample [57]. In one such model [56], an index of matrix interference (I_m) and a correction factor (N) have been defined according to Eqs. (9.10) and (9.11):

$$I_m = \frac{OD_{0,standard} - OD_{0,sample}}{OD_{0,standard}} \times 100, \tag{9.10}$$

$$N = \frac{100 - I_m}{100}. \tag{9.11}$$

By using N as a correction coefficient in an ELISA for the fungicide tetraconazole in fruit extracts [56] or for the insecticide diazinon in surface water samples [58], a considerable improvement in assay accuracy could be attained. A similar constant relationship has been observed between the relative signals given by two spiked analyte concentrations in various matrices (see Eq. (9.12)):

$$\frac{OD_{C1,sample}}{OD_{C2,sample}} = c, \tag{9.12}$$

where c is a constant, $C1$ and $C2$ are the two spiked analyte concentrations, with $C1 > C2$ and both of them in the dynamic response range [57]. The value c was shown to be valid for different sample matrices, with a CV of below 5%. Taking into account the continuous decrease in absolute slope with the increase in analyte concentration in immunoassay, the value c could be set as an indicator in a semi-quantitative test: all the samples with a ratio c smaller than the set value are positive, while all the others are negative. The new method allowed a considerable reduction in the number of 2,4-dichlorophenoxiacetic acid false positives identified in a set of apple, peach, orange and potato samples at a decision level of $5\ \mu g\,l^{-1}$ [57]. A recent approach combining sequential dilution of the sample and standard addition allowed circumvention of matrix effects and simultaneous quantification of two cross-reactants in an ELISA combined with multivariate analysis [43].

9.3 DETERMINANTS OF SENSITIVITY AND SELECTIVITY

Taking into consideration the great complexity of assay formats described in the literature, the following discussion will be limited to competitive immunoassays. The relevance of such limitation is supported by the data

Immunoassay: potentials and limitations

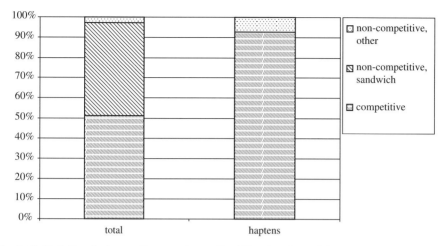

Fig. 9.11. Relation between the numbers of publication describing competitive vs. non-competitive immunoassay.

in Fig. 9.11: more than 50% of the total number of published assays and more than 90% of the assays for small molecules (haptens) are competitive.[3] If one excludes the sandwich-type assays, which are generally not suitable for haptens, the number of competitive immunoassays is in great majority.

Below follows a discussion around the effect of diverse factors on immunoassay characteristics, addressing basic aspects such as immunoreagents (i.e., antibody and antigens) quality, immunoreagent concentrations, incubation time and temperature, as well as other factors influencing each of the main steps performed in a competitive heterogeneous immunoassay: *labelling, hapten structure and design, effective analyte concentration, separation of antibody-bound from free antigen fractions* and *detection*.

9.3.1 Immunoreagent quality

Polyclonal antibodies can be used as whole serum or in purified form, the latter either as the total immunoglobulin fraction or affinity purified. Due to the heterogeneity of the antibody-binding sites of polyclonals, different selectivity for different antigens will appear and this problem is simply avoided by the use of monoclonal antibodies. However, even though it is often

---

[3] Numerical data are a result of a literature search performed in the CAS database using the software SciFinder Scholar 2000. This search could not cover all immunoassay publications, but was limited only to the ones in which the assay format (competitive or non-competitive) was specified. We consider, however, that the probability for the assay format to be mentioned is the same for both competitive and non-competitive immunoassays.

claimed that monoclonal antibodies lead to assays with lower cross-reactivity, this is not always true. In an immunoassay for atrazine using two monoclonal antibodies, the %CR for simazine was 61 and 67%, respectively, whereas it was found to be 12 and 20% for two polyclonal antisera developed against the same hapten derivative [59].

Although high cross-reactivity is sometimes seen as a problem in immunoassay development, antibodies with a high CR for compounds within the same chemical class can favourably be used either for post-column detection of the cross-reactants separated by HPLC [60–64] or for "selective" SPE in columns with immobilized antibodies (immunoaffinity SPE), followed by HPLC separation of the trapped cross-reactants (see Fig. 9.12) [65].

Highly specific antibodies have been used for a similar purpose as above in immunoextraction recycling as shown in Fig. 9.13 [66]. This technique is based on the passage of a fluorochrome-labelled sample through a battery of small immunoaffinity columns; each column extracting a single specific analyte. Detection is realized by sequential elution of bound analytes via a valve system and laser-induced fluorescence. This system was applied for selective detection of different cytokines [66].

Besides monoclonal and polyclonal antibodies, Fab or F(ab)$_2$ fragments can be employed for immunoassay development. Their use is of considerable

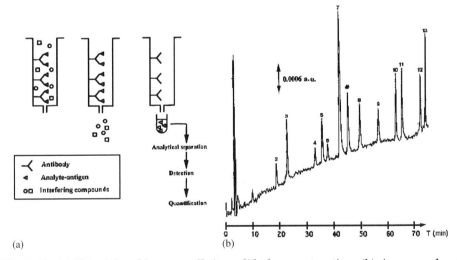

Fig. 9.12. (a) Principle of immunoaffinity solid-phase extraction. (b) An example of on-line immunotrapping and RP-HPLC of 13 phenylurea herbicides spiked into water. Analytes: 1, phenuron; 2, metoxuron; 3, monuron; 4, methabenzthiazuron; 5, chlortoluron; 6, fluomethuron; 7, isoproturon; 8, difenoxuron; 9, buturon; 10, linuron; 11, chlorbromuron; 12, difluzbenzuron; 13, neburon [65]. (Reprinted with permission.)

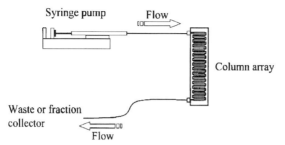

Fig. 9.13. The basic set-up for a immunoextraction recycling procedure. Each column in the array contains a highly specific antibody, able to trap only one specific analyte from the sample. The sequential elution is performed with the help of three-way switching valves, connected after each antibody column [66]. (Reprinted with permission.)

importance especially when immunoassays are applied for analysis of complex samples (i.e., serum, soil extracts, etc.) in which the probability of unspecific binding by using whole IgG molecules is high [67]. Another important aspect under these circumstances is that Fab fragments improve the detection of the analyte in microcavities due to their smaller molecular size.

9.3.2 Immunoreagent concentration

For competitive immunoassays, the rate of signal change with the concentration of analyte is greatest when the amount of competitor is small, with the best sensitivity obtained for a minimal concentration of competitor $[Ag^*]$. However, such reduction to infinite small concentrations of competitor is limited by the errors in signal measurement, and considering the variation in non-specific binding, which is related to the total concentration of the competitor, this becomes significant especially at low levels of specific binding. As a direct consequence, the total concentration of antibody also needs to be considerably reduced, in order to minimize the errors in estimating the unoccupied binding sites.

In order to decide the lowest possible working concentrations of competitor and antibody, the development of a competitive immunoassay usually starts by performing a two-dimensional dilution experiment, i.e., the antibody concentration, $[Ab]$, is varied in one "dimension" and the competitor concentration, $[Ag^*]$, is varied in the other. In this way the optimal $[Ab]$ and $[Ag^*]$ can be determined experimentally and the lowest possible $[Ag^*]$ and $[Ab]$ are usually selected, when 30–70% of the tracer is bound to the antibody ($[Ab-Ag^*]$). However, the contradiction in defining immunoassay sensitivity, as mentioned above, has led to different values of the theoretically predicted optimal concentrations of antibody and competitor. For example, Berson and Yalow [68] define sensitivity as the slope at the mid-point of the calibration curve and the predicted optimal values were obtained for a $[Ab] = 0.5/K$ and a relative tracer binding at $[Ag^*-Ab]/[Ag^*]$ of 0.33, considering that the precision of the assay is the same over the whole dynamic range. Ekins [69], on the other hand, defines sensitivity as the MDC, taking into account that the precision of the assay varies within the assay range, i.e., it is the lowest in the area of the mid-point of the curve and the highest at the extremes (see Fig. 9.5b), as well as with changes in the assay design. Under these conditions, the predicted reactant concentrations to obtain the lowest MDC were $[Ab] = 2.25/K$ and $[Ag^*] = 1.25/K$.

Alteration of the effective analyte concentration can occur due to blocking, binding, complexation, or changes in the structural conformation of the analyte. Numerous analytes exist in nature bound to macromolecules or proteins; a good example is given by the thyroid hormones thyroxine (T_4) and triiodothyronine (T_3), which are bound to proteins in the blood in proportions of 99.98 and 99.7%, respectively [70]. Only the free fraction is biologically active, since it can pass through biological membranes to the target tissues. The concentration of T_3 and T_4 in plasma is an important indicator of the thyroid status, while their total concentration shows the potential thyroid hormone activity [70].

Atrazine has been found to form non-extractable residues of up to 50% of the amount originally applied in soil, while other pesticides such as 3,4-dichloroaniline and trifluralin produce even higher amounts of non-extractible residues [67]. These residues are mainly bound to the organic matrix of the soil, especially to humic acids, and this fraction of analyte cannot be assessed by normal extraction procedures. Therefore, specific techniques have been developed for monitoring of non-extractible residues of pesticides. One of the most common approaches is immunolabelling, where (fluorescent) labelled anti-pesticide residue antibodies are incubated with soil samples [71, 72]. After incubation, samples are transferred to cover slips and the binding visualized by a photomicroscope. By the use of antibodies with different selectivity, this technique facilitated the identification of atrazine groups involved in interaction with soil matrix [72,73]. Similar immunolabelling model studies were carried out for the identification of atrazine in plant tissues [74–76]. Quantification of matrix-bound residues of analytes by immunolabelling is however rather difficult, thus alternative immunoassay formats have been developed. Competitive enzyme-based, direct non-competitive enzyme-based [72,73] and sandwich-type assays [77] were developed and applied for the determination of atrazine-bound residues in soil.

The mentioned assays led to atrazine residue recoveries between 25 and 1000%, which is a variation that might be due to the high complexity of the sample matrix and the non-homogeneity of the standards used [77]. Improved characteristics were obtained after a better characterization of the used standards [78]. A further improvement was achieved by replacing the anti-humic acid antibodies, used for capturing the analyte, with basic proteins [79,80]. These are easily available and have the advantage that a broader variety of humic acids can be bound; therefore, the assay should be independent of the composition of a particular soil. The use of this assay format improved both the sensitivity and signal intensity.

The phenomenon of unspecific analyte blocking can be beneficial in terms of assay selectivity in certain situations. If the affinity of the analyte and cross-reactant(s) for binding to large molecules (other than an antibody) in the sample are sufficiently different, their separation based on size exclusion becomes possible. For example, digoxin-like immunoreactive factors (DLIF) are tightly (non-heat releasable) bound to serum proteins [81], while digoxin is bound only 23 and 6% at 37 and 1°C, respectively [82]. Cooling down the serum sample proved to be sufficient for an effective separation of DLIF from free digoxin by ultrafiltration with a 10 KDa cut-off membrane [81]. An alternative method to size exclusion is protein precipitation by various agents, i.e., trichloroacetic acid, salicylic acid or ammonium sulphate [83].

9.3.3 Incubation time and temperature

An important aspect in the development of competitive immunoassays is to decide how long to allow the incubation of reactants to progress in order to attain equilibrium. Motulsky [84] demonstrated that the incubation time until attaining equilibrium depends on the relative values of the dissociation rate constants of the competitor and analyte from their corresponding antibody complexes, as well as on their concentration. When the dissociation of the antibody–competitor complex is much faster than that of the antibody–analyte complex, the predicted equilibration time is strongly dependent on the analyte concentration, whereas the kinetic constants and the concentration of the competitor do not matter. The time necessary to reach the equilibrium at IC_{50} is equal to $1.75/k_2$, whereas full equilibrium at maximum analyte binding is reached at $3.5/k_2$ (i.e., the lower the analyte concentration, the faster the reaction kinetics). On the other hand, when the dissociation of the antibody–analyte complex is much faster than that of the antibody–competitor complex, the time required to reach equilibrium at IC_{50} depends mainly on the competitor. If the competitor concentration is very low (i.e., $[Ag^*]_0 \ll (1/K^*)$), the concentration of the analyte does not affect the time necessary to reach the equilibrium, e.g., this is $3.5/k_4$. As the competitor concentration is increased infinitely, the equilibration time starts changing at different levels of analyte concentration (i.e., it is halved to $1.75/k_4$ at IC_{50}, but it remains the same for reaching equilibrium at high concentrations of analyte, having a value of $3.5/k_4$). As shown in Fig. 9.14, the kinetic profile before reaching equilibrium also differs, depending on which of the dissociation rate constant is limiting the whole process. However, regardless of the kinetic constants, the IC_{50} changes very slowly during the final phases of the experiment and an acceptable approximation of the equilibrium IC_{50} may be attained in less than half the time necessary for the equilibrium to be established (see Fig. 9.15).

As a direct consequence, if the equilibration time is not correctly estimated, the immunoassay performances are expected to depend on the order of mixing the immunoreagents [85,86]. Practically, three mixing orders are possible: (1) pre-mixing of analyte and competitor, followed by the addition of antibody (e.g., "true competition"); (2) pre-mixing of the analyte and antibody, followed by the addition of the competitor (e.g., "analyte displacement"); (3) pre-incubation of the competitor with the antibody, then addition of the analyte (e.g., "competitor displacement") [85].

Depending on the dissociation rate constants of the analyte and competitor, the distribution of the fractions bound to the antibody will vary in time and the sensitivity of the assay will depend on how far from

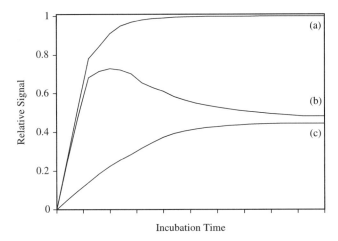

Fig. 9.14. Differences in the kinetic profile before the establishment of equilibrium in a reagent-limited immunoassay depending on which dissociation constant (e.g., for analyte or competitor) limits the process: (a) no analyte; (b) slowly dissociating analyte; (c) rapidly dissociating analyte. (Adapted from Ref. [84].)

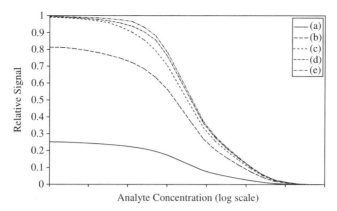

Fig. 9.15. Evolution of the signal at different moments (a–e) before equilibrium at different levels of the calibration curve in a competitive immunoassay. (Adapted from Ref. [84].)

equilibrium the measurement is performed [84]. Generally, a pre-incubation of the antibody with the sample containing the analyte prior to the competition step improves the immunoassay characteristics [87,88]; however, this is not a necessary condition [89]. If the time necessary to reach equilibrium is relatively small, there are certain practical advantages of using the "competitor displacement" mode, e.g., sometimes a higher stability of antibody–competitor mixtures as compared with separate solutions [90].

The differences in kinetic properties, i.e., association and dissociation rate constants (see relations (9.4) and (9.5)) allow optimization of the incubation time so that the influence of the cross-reacting species on the analytical signal is minimal [84]. A study on the CR of an anti-testosterone antiserum [91] showed that even though the association half-time (which is inversely proportional to the association rate constant) for testosterone, dihydrotestosterone, androst-4-ene-3,17-dione and estradiol is similar, the dissociation half-time for testosterone is much longer than for the other three steroids. In the beginning of the incubation step, the amount of any ligand bound to the antibody is thus dictated by the corresponding association constant (k_1), which means a high CR in the above example. As equilibrium is approached, the dissociation rates k_2 contribute substantially to the binding dynamics; therefore, the cross-reactivity for testosterone decreases with time. In general terms, this means that the more rapid the equilibrium is established, the faster the cross-reactivity is reduced. Association rate constants have been found to be comparable (i.e., 10^7–$10^8 \, \mathrm{l \, mol^{-1} \, s^{-1}}$) for a large variety of antibody–antigen interactions, with even smaller differences when comparing two antigens binding to the same antibody [92,93]. The CR must thus be carefully evaluated before deciding a reduction in the immunoreaction incubation time. In accordance with the above discussion, any factor that influences the time after which the equilibrium of the antibody–antigen reaction is completed will affect the cross-reactivity as well.

Temperature is a factor often used to modulate the affinity of antibody–antigen interactions; the dependence between the equilibrium (affinity) constant and temperature is given by the following equation:

$$K = A\, e^{-\Delta G/RT}, \tag{9.13}$$

where R is the gas constant (1.93 cal mol^{-1} K^{-1}), K is the equilibrium constant, ΔG is the free energy change and A is a constant for each specific reaction.

The dependence of K on temperature plays an important role in establishing the relationship between the relative roles of enthalpy- (ΔH) and entropy (ΔS) change in the antibody–antigen interaction. This relationship is expressed by the van't Hoff equation (Eq. (9.14)), which is obtained by differentiating Eq. (9.13) (taking into account that $\Delta G = \Delta H - T\Delta S$):

$$\frac{d \ln K}{dT} = \frac{\Delta H}{RT^2}. \tag{9.14}$$

Equation (9.14) reflects the dependence of K only on ΔH and the non-dependence on ΔS. Thus, if ΔH is negative, an increase in temperature will decrease the equilibrium constant and vice versa. Table 9.1 shows some

TABLE 9.1

Thermodynamic constants for some antibody–antigen interactions. (Reprinted with permission.)

Antigen/class	ΔH^0 (kcal mol^{-1})	ΔS^0 (kcal mol^{-1})	$T\,°C$	$K \times 10^{-9}\, l/m$
Aldosterone/A	-3	42	5/37	5.0/2.9
Digoxin/A	$+1.6$	55	5/37	0.88/1.2
Estradiol/A	-4.6	40	6/25	34/20
Estriol/A	0	51	4/37	2.1/2.1
Testosterone/A	-1	50	4/37	8.9/7.3
Thyroxin/A	0	48	4/36	0.69/0.65
Angiotensin I/B	-9	22	5/37	14/2.5
Angiotensin II/B	-9	27	5/37	90/16
Gastrin/B	-8	32	5/37	590/120
Insulin/B	-7	34	5/37	120/31

ΔH^0 – standard enthalpy change of interaction, ΔS^0 – standard entropy change, and K – equilibrium constants at two different temperatures (T), respectively [94].

examples of the dependence of the equilibrium constants for different antibody–antigen interactions [94]. From the relative contributions of the enthalpy and entropy to the free energy of binding, and thus to the equilibrium constant, different "classes" of interactions have been described. Group A includes antibodies against small and relatively hydrophobic molecules (e.g., steroids), which are characterized by small, near zero ΔH values. For this group, the temperature dependence of the equilibrium constant is relatively small. Group B includes antibodies against peptides, with moderately negative ΔH values and for which the temperature dependence of binding is substantial (with higher equilibrium constants at lower temperatures). These data indicate the importance of interpreting the temperature dependence of cross-reactivity in immunoassays.

According to the classification presented above, the equilibrium constants in immunoassays for environmental pollutants, most of which are relatively small hydrophobic molecules (potentially group A compounds), should not be considerably affected by the incubation temperature; however, supporting data for such an assumption have, at least to our knowledge, not been presented in the literature yet.

9.3.4 Other factors

9.3.4.1 Labelling
The influence of immunoreactant concentrations on the immunoassay sensitivity is relatively easy to predict for any assay design. In particular for the measurement of the unoccupied antibody sites, the concentration of the antibody and tracer is directly dependent on the label and detection mode employed, as well as on the conjugation procedure used for the labelling reaction [89,95]. Even though special precautions are taken, labelling effects can be observed for the antibody or the analyte/hapten conjugated to marker molecules, when either the structure of the antibody-binding sites or essential epitopes are structurally changed or hindered. Sometimes, labelling can cause an enhancement in recognition, as in *bridge recognition* (see below). The nature of the labelling reagent or method is important, especially for haptens, as shown in the following examples.

Radiolabelling is often performed by oxidation of iodide by different agents for iodination of tyrosine and histidine residues in peptides. However, unfavourable oxidation reactions can occur simultaneously, e.g., cysteine and methionine can be oxidized to their corresponding sulphone or sulphonic acid, so that neutral amino acids are converted to highly acidic groups, thus changing the electrostatic distribution within the structure of the labelled peptide. Exchanging chloramine T as the oxidation reagent with lactoperoxidase/glucose oxidase can minimize this effect [96].

Labelling with enzymes drastically alters the properties of the labelled compounds, because of the large sizes of the enzyme labels, and the effect is more pronounced for small haptens than for larger compounds (e.g., peptides and proteins). A practice often used is to optimize the length of the "spacer arm" between the structure containing the antigenic determinants and the enzyme label, so that hindering effects are avoided [97,98].

If the labelling reaction is incomplete, special care has to be taken so that all the uncoupled immunoreagent- and label molecules are separated from the reaction mixture. Depending on the assay format, the presence of unlabelled antibody/antigen in the tracer mixture leads to false results, while unconjugated label molecules generate an unspecific background signal. A particular case is represented by the *bridge recognition* effect that appears when the same conjugation chemistry/spacer is used for the labelling reaction as is used for the synthesis of immunogens from haptens conjugated to large carrier proteins. In this case, some of the antibodies can recognize the spacer arm between the hapten and the label.

9.3.4.2 Hapten/competitor structure and design
Theoretically, any change in the structure of the competitor used for assay development will alter the antibody recognition, thus affecting the relative affinity for the competitor vs. the analyte. It has frequently been reported that the immunoassay sensitivity is improved by introducing a certain degree of heterology in the chemical structure of the competitor [99–101], and some authors suggested that different competitor/antibody combinations modify the assay selectivity [102,103]. One example of the latter is found in a recent work for the development of an indirect competitive immunoassay for 2,4,6-trichlorophenol, in which five different starting haptens for synthesis of the competitor were tested (see the structures in Fig. 9.16) [89].

As can be seen, hapten **1** has nearly the same structure as the analyte and hapten **5**, except that the spacer arm blocks the hydroxyl group, so that it cannot participate in proton transfer reactions. Hapten **2** also has the hydroxyl group blocked and, moreover, the heterology vs. the analyte structure is enhanced by the change in position of where one of the chlorine atoms binds to the aromatic ring. Haptens **3** and **4** have the phenolic group free, the spacer arm has the same structure and is placed at the same position as in hapten **5**, but one chlorine atom is missing in each case. Molecular modelling to evaluate the degree of heterology between each hapten and the target analyte was performed. The differences existing between the haptens

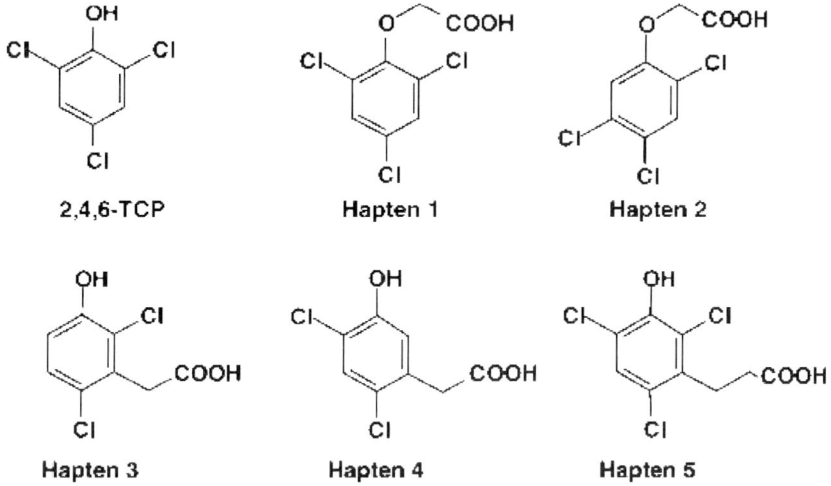

Immunoassay: potentials and limitations

Fig. 9.16. Chemical structures of the analyte (2,4,6-TCP) and five different potential haptens to be used for synthesis of the competitor [89]. (Reprinted with permission.)

and the analyte with regard to atomic distances, punctual atomic charges, pK_a, relative affinity constants of the antibody–hapten–BSA conjugates and IC_{50} of the final immunoassay (e.g., using the BSA–hapten conjugate) are presented in Table 9.2 [89].

According to the data in Table 9.2, the most different hapten compared to the analyte (TCP) is hapten **2** while the most closely related in geometrical

TABLE 9.2

Relative dissimilarities[a] between the target analyte and haptens, relative affinity constant of the antibody–hapten–BSA conjugate and IC_{50} of the final immunoassay. (Reprinted with permission.)

Hapten	Geometry	Punctual charge distribution	pK_a	RO^- (%)	K_{aff}^{rel} (%)	IC_{50} (nM)
1	0.057	0.554	–	–	0.75	20.7
2	0.352	0.556	–	–	0.41	11.1
3	0.255	0.468	7.99	24.5	3.70	43.2
4	0.278	0.286	6.94	78.4	0.50	189.7
5	0.169	0.036	6.63	88.1	1	n.a.
TCP	–	–	6.59	89.0	–	–

Conditions and abbreviations: pH 7.5; antiserum, As43; RO^-, percentage of phenoxide species; K_{aff}^{rel}, affinity constant of the antibody–hapten–BSA interaction relative to the affinity of the antibody for hapten **5**–BSA conjugate; n.a., not available [89].
[a]This has been expressed as the root mean square of the addition of errors of the distances and the punctual charges, respectively, between the corresponding atoms of both molecules (analyte vs. each hapten).

and electronic properties is hapten **5**. Hapten **1** shows expectable geometrical similarities, while the charge distribution is very different. This trend was correctly followed by the relative affinity constant (see Table 9.2). As can be seen, IC_{50} values for the developed assays also correlated well (e.g., inverse proportionality) with the estimated relative affinity and the heterology of the competitors.

A review by Hennion and Barceló [33] describes the influence of the hapten structure design on the selectivity within the triazine class of immunoassays (for chemical structures see Fig. 9.17). If the immunogen was synthesized via a spacer group that replaces the chlorine atom (hapten type **1a**), the developed antiserum showed a high cross-reactivity (i.e., %CR = 89%) for ametryn [104] and, in a different assay configuration, a very high CR for propazine (%CR = 250%) and a relatively low CR for simazine (%CR = 6.4%) [85,86]. If the ethyl group was replaced with an isopropyl (hapten type **1b**), the CR for propazine, prometryn and prometron was relatively high (i.e., %CR = 80%),

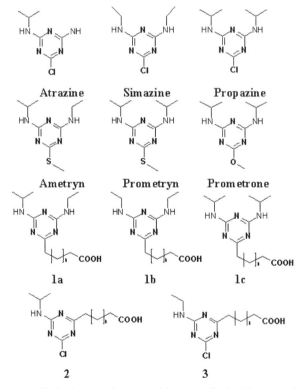

Fig. 9.17. Structure of triazine analytes and hapten derivatives used for immunogen synthesis.

while the replacement of the isopropyl group with an ethyl (hapten type **1c**) led to a low affinity for simazine (i.e., %CR = 16%) [104].

If the spacer arm for immunogen synthesis was placed instead of the ethyl group (hapten type **2**), only the isopropyl and chlorine can serve as antigenic determinants, and thus the relatively high CR for propazine vs. simazine was not surprising (i.e., %$CR_{propazine}$/%$CR_{simazine}$ = 33) [86]. Monoclonal antibodies developed using the same type of hapten with a spacer containing four carbon atoms allowed reduction of the cross-reactivity for simazine from 10 to 2.5% [105]. By contrast, when the immunogen synthesis was performed by replacing the isopropyl group with the spacer arm (hapten type **3**), the relative %$CR_{propazine}$/%$CR_{simazine}$ was considerably reduced (i.e., 6.52) [86].

9.3.4.3 Hapten density

Hapten molecules are often coupled to carrier proteins (e.g., coating antigens used for the development of indirect competitive assays) or to enzyme labels (e.g., tracers used for detection in competitive assays). A protein has usually many different sites to which the hapten can be coupled and the chosen chemistry should obviously be one that leaves the active site as little affected as possible. The number of haptens that can be coupled to one protein molecule (hapten density) by a particular chemistry is essential for the final assay outcome [106–110].

The increase in the hapten to protein molar ratio leads to competitor molecules carrying more and more antigen derivatives. Considering the antibody–competitor interaction at the molecular level (e.g., the affinity is dictated by the interaction of one antibody and one competitor molecule), this increase in hapten density causes an apparent increase in the affinity constant for the competitor (K_{app}^*) at a more or less constant value for the analyte affinity K_{app}. Moreover, the use of whole antibody molecules, containing two antigen-binding sites, and the presence of several antigen derivative residues on the same molecule, can cause strong avidity effect, with a considerable decrease in assay sensitivity [106–110].

Table 9.3 shows the effect of increase in molar ratio of an atrazine hapten coupled to the enzyme β-galactosidase (β-GAL) in a flow immunoassay for the determination of atrazine. A high hapten density had a negative impact on the assay sensitivity, i.e., the lowest MDC and IC_{50} values were obtained for the lowest molar ratio of hapten to enzyme.

9.3.4.4 Immobilization

Special effects due to immobilization of either the antibody or the antigen/competitor on solid surfaces are discussed below in more detail,

TABLE 9.3

Effect of the hapten/enzyme label ratio on the immunoassay MDC and IC_{50}

Initial molar ratio hapten/label	Assay characteristics	
	MDC (nM)	IC_{50} (nM)
10	0.70	26.46
33	0.82	30.82
66	1.59	31.43

See the text for details.

since heterogeneous assays nowadays are the most often employed, both in research and commercial assays.

Solid-phase immunoreagents can be immobilized by four general procedures: (a) passive adsorption; (b) covalent attachment to functionalized solid phases; (c) affinity-based immobilization; and (d) other non-adsorbent, non-covalent methods (e.g., avidin–biotin interaction) [16]. Although chemical coupling of proteins to solid phases is gaining in popularity, many immunoassays rely on non-covalent adsorption. The process by which the latter occurs is poorly understood, but it seems certain that binding essentially involves hydrophobic interactions. Knowing that most proteins tend to adopt a conformation in which hydrophilic groups are located on the outside of the protein and hydrophobic groups inside it, it seems inevitable that hydrophobic binding to polymer surfaces will cause conformational changes in the adsorbed protein. Thus, adsorbed antibodies might undergo conformational changes, affecting not only the number of active binding sites, but also their affinity for the antigen. This is reflected by the fact that it is quite common that antibodies which work well in solution show little or no antigen recognition when bound to a solid phase via non-covalent adsorption: monoclonal antibodies are more affected than polyclonal antibodies, most probably due to the diversity of antibody molecules in the latter [111].

One important aspect to take into consideration for interface reactions involving antigen–antibody recognition is "non-specific interactions" between the species in solutions and the solid phase. Thus, "blocking" of unspecific sites is almost always required, with inert proteins (e.g., serum albumin, casein, non-fat milk) or non-ionic detergents being the most employed blocking agents. The choice of a certain blocking agent depends mainly on the solid phase used, but the particular antigen–antibody reaction also plays an important role. The mechanism of "the blocking" process is not very well known; however, it is believed that the protein blocking agents "fill in the gaps" of the solid phase unoccupied by the immobilized immunoreagent. Detergents probably act in various ways, but in contrast to protein blockers they generally prevent hydrophobic aggregation of macromolecules and non-specific adsorption on the solid surface, rather than filling the unoccupied sites on the solid phase. Their structure determines which effect is predominant: for example, Tween 20 (i.e., the commercial name of polyoxyethylene sorbitan monolaureate) has proven to prevent both protein–protein interactions and non-specific adsorption on solid surfaces, CHAPS (i.e., 3-[(cholamidopropyl)dimethylammonio]-1-propane sulphonate) blocks mainly macromolecular aggregation, while SDS (i.e., sodium dodecyl sulphate) may interfere with the specific antibody–antigen binding [16].

Diffusion is another limiting factor in most heterogeneous immunoassays. The association rate constant (k_1 and k_3 in Eqs. (9.4) and (9.5)) is dependent on the viscosity of the surrounding environment and a geometric factor called the *sticking coefficient* (SC), where the latter has a maximum value for reactants of equal size and uniform reactivity. It has been demonstrated that a considerable increase in diffusion was obtained by increasing the shacking rate of a microtitre plate with immobilized antibodies, with a direct effect on the time required for the antibody–antigen equilibrium to be established [111]. In the absence of constant stirring, when the mass transfer is a function only of the molecular size and the viscosity, the association rate constant for an antibody–hapten reaction should be approximately $10^9 \, \text{l mol}^{-1} \, \text{s}^{-1}$ [112]. Most association rate constants are, however, substantially lower than $10^9 \, \text{l mol}^{-1} \, \text{s}^{-1}$ due to a low SC, which reflects that antibody and antigen can only bind when they have a correct alignment. The SC is high when the degree of rotational freedom of the two reactants is high, but is strongly affected when one of the components is bound, due to lowering of the association rate constant.

For direct competitive assays, it has been reported that a *boundary layer* (BL) is formed at the surface–solution interface [111], due to the establishment of a fast pseudo-equilibrium between the surface-bound antigen and its concentration in close vicinity to the surface (within the BL), the real equilibrium reaction being limited by the mass transfer of the antigen from the solution bulk. In the BL the local concentration of antibody-binding sites is very high and the dissociation rate constants of the complex are thus much lower than for antibody–antigen complexes in solution, sometimes by two orders of magnitude [111]. It has been suggested that the binding reaction on solid-phase immobilized antibodies is essentially irreversible. The organization of antibodies on the surface is not homogeneous, but rather non-uniformly distributed. There is some evidence that antibodies might form fractal clusters [113], i.e., the modified solid phase consists of islands of highly organized antibodies, a process similar to the formation of crystalline structures. These clusters may lead to apparent positive allosteric effects; there is even some evidence that the association rate constant in solid-phase immunoassay is variable.

9.3.4.5 Interferences with antibody binding
Non-specific interferences with antibody binding can occur due to inactivation of the binding sites, unspecific binding or incorrect orientation on a solid support. Several examples are presented below.

- *Physical masking of the binding sites*, i.e., by non-oriented immobilization on solid supports.
- *Alteration of the conformation of the binding sites*, i.e., by changes in pH, ionic strength, organic modifiers, etc. Some anti-microbial agents, such as sodium azide, are known to affect the antibody binding. Some detergents, used for blocking of unspecific binding, might inhibit the antigen–antibody reaction, especially if they are not highly purified and contain impurities (e.g., Tween 20 might contain free peroxides [96]).
- *The low dose hook effect* is a small change in signal in the opposite direction as compared to the normal response–concentration relationship at very low concentrations of analyte [114,115]. In some cases this is attributed to positive allosteric effects for antibody–tracer binding at low concentrations of the analyte.
- *The high dose hook effect* is a phenomenon observed for one-step "sandwich" assays at high concentrations of analytes, when different analyte molecules saturate the capture and detection antibodies, respectively [16].
- *Complement- and rheumatoid factors* [116,117], *heterophile and anti-animal antibodies* [118,119] are often present in serum, plasma and whole blood samples. These generally bind to the F_c regions of immunoglobulins, tending to form cross-linked structures of non-reactive antibodies. A good general solution to overcome this type of interference is to use only Fab or $F(ab)_2$ fragments for immunoassay development.

Covinsky et al. [120] describe a case of a male patient having increased immunoassay response to cardiac troponin I, thyrotropin, human chorionic gonadotropin, α-fetoprotein and CA-125 simultaneously. This response was consistent with myocardial infarction, hyperthyroidism, pregnancy (!), as well as hepatic or ovarian cancer! This somewhat strange response in a male patient was later found to be due to an interference caused by a body infection of Gram-negative bacteria that induced broadly reactive anti-immunoglobulin antibodies. Incubation of the patient's plasma with *E. coli* prior to immunoassay led to normal results due to removal of an interfering IgM fraction.

9.3.4.6 Separation of fractions in heterogeneous immunoassay
Separation of free from antibody-bound analyte/competitor fractions in heterogeneous immunoassays is very seldom complete and the efficiency of fraction separation has a direct effect on the quality of the assay. For example, in competitive immunoassays that employ a labelled analyte derivative as tracer and measurement of the antibody-bound fraction, interferences with the separation step can be due to:

- *failure to remove all the free tracer* ("non-specific binding") which can result in a residual background signal;
- *partial "displacement"* of the tracer bound to the antibody;
- *inconsistency in separation* due to the variability of the sample matrix.

Fraction-separation methods used in immunoassay can be divided into two major groups: *liquid-phase* and *immunoreagent modified solid-phase separations*.

The two main principles governing the separation of free from antibody-bound fractions in *liquid phase* are size exclusion and selective precipitation. Some specific examples follow.

- *Electrophoresis, gel filtration and gel chromatography* have been used for fraction separation in immunoassay, but they are too expensive and time consuming for routine use [121].
- *Charcoal or silicate adsorption* is often used to remove free hapten molecules, thus leaving the antibody-bound label in the liquid phase [96].
- *Restricted-access-based materials (RAM)* can trap small hapten-based tracers, where the fraction separation is governed both by size exclusion and hydrophobic interactions [60,62,86,122,123]. There are indications that the RAM phase "competes" with the antibody-bound tracer and analyte [124], as shown in Fig. 9.18.
- *Precipitation with salts, organic solvents and polyethylene glycol (PEG)* has often been employed as an approach for fraction separation, which usually attracts the water essential for maintaining the 3D structure of antibody molecules, so that they precipitate [5]. Ammonium sulphate is a common salt used for antibody precipitation [125], but since it results in high levels of unspecific binding (i.e., 5–20%), it is seldom used nowadays for fraction separation purposes.

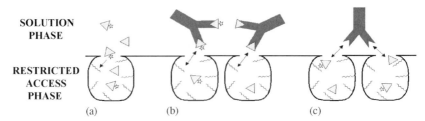

Fig. 9.18. Effects due to the interaction between a RAM phase and a mixture an antibody (Y shaped), antigen (triangle) and labelled antigen (triangle + star): (a) the phase traps small analyte and tracer molecules by mixed effects, i.e., size exclusion and hydrophobic interactions; (b) if the hydrophobic forces are strong enough, the phase "removes" the tracer/analyte from their antibody complexes; (c) the free antibody binds to free analyte/tracer previously trapped by the RAM phase.

- *Secondary antibody precipitation* is sometimes used for separation of large analytes, possessing at least two epitopes that can be bound simultaneously by two different antibodies [126]. The divalent nature of antibodies is responsible for the formation of a lattice consisting of cross-linked antibodies and analyte, a lattice that can be easily separated by filtration or decantation.
- *Mixed precipitation methods*, i.e., PEG-assisted second antibody precipitation, are sometimes used for fraction separation, combining the specificity and low level of unspecific binding of the second-antibody method with the speed of PEG-based precipitation [127].

The simplest way of fraction separation, however, is when one of the immunoreactants (i.e., either the antigen or the antibody) is immobilized on a solid phase, i.e., *immunoreagent modified solid-phase separations*, such as plastic tubes, microtitre plate, latex-, glass- or magnetic beads, dipsticks or nitrocellulose membranes [121]. The efficiency of separation is determined by the nature of the solid surface. Unspecific binding of the label to the solid phase can occur, especially when the latter has hydrophobic properties [96]. The adsorption of the antibody to the solid surface causes its partial inactivation, which can explain some unexpected effects when comparing the same antibody used in liquid and solid-phase systems (see also Section 9.3.4.4).

9.3.4.7 Detection

Factors influencing the detection step are generally specific for each type of label, i.e., the presence of endogenous signal generating substances in the sample (i.e., membrane-bound or soluble peroxidases in HRP-based assays), enzyme effectors (inhibitors or activators), fluorescence quenching agents, etc., in the sample matrix.

Enzymes are nowadays the labels that are most used in immunoassay. Their main disadvantage is their susceptibility to interferences due to changes in assay conditions, e.g., pH, ionic strength, content of organic solvents. The addition of anti-microbial agents such as azides or mercury derivatives also can affect the enzyme activity when using peroxidase labels [96]. The presence of other catalysts in the sample can also have an effect on the enzyme immunoassay result, e.g., Cu(II) ions promote luminol chemiluminescence in the presence of hydrogen peroxide [16].

9.4 PARTICULAR ASPECTS OF IMMUNOASSAY STANDARDIZATION

Obviously, the universal goal of any measurement technique is to obtain reproducible results regardless whether the samples come from different sources with different matrix effects, are run by different operators, in different laboratories, on different occasions and using different lots of reagents. This is usually accomplished by characterization of the newly developed assay in terms of sensitivity, selectivity, robustness and correctness (i.e., accuracy and precision). Evaluation of sensitivity and precision does not normally constitute a problem for a newly developed immunoassay, and accuracy can be attained by comparison with the results obtained by a reference ("standard") method. However, the evolution of assay standardization from this point on is much more difficult. The following section deals with application-specific problems related to validation and standardization of immunoassay.

9.4.1 Availability of standard materials

There are many factors to be taken into consideration for assay standardization. The first factor is the immunoreagents themselves. It is known that the antibodies are pleomorphic over much of their molecular structure even in uniform populations [128]. The analyte should be stable, available in a pure form, the structure should be well understood and invariable with time, and this is generally the case for hapten analytes [129]. More problematic are analytes where the molecular size is large and biologically variable (e.g., certain hormones and enzymes) [128]. Standardization is thus almost impossible because the analyte itself cannot be obtained in an homogeneous, pure form, or the standard is not identical to the corresponding substance in the sample to be measured [130–132].

This is a general problem in clinical, pharmaceutical and, in some situations, in food analysis (e.g., "enhanced crops", "genetically modified plants") [1,15,133,134]. When a reference standard does not have well-defined physical characteristics, its biological activity (e.g., affinity binding or enzymatic activity per unit or weight or volume) might be used and documented as the standard [15].

The availability of secondary standards (i.e., sample matrix-based standards) has the advantage that the calibration and analysis can be performed in a similar matrix, however, without complete elimination of the matrix effects [133]. For many analytes that are structurally heterogeneous (e.g., some hormones and tumour markers), even the development of reference methods constitutes a serious problem [133].

Problems related to the misuse of standards are less evident to the user. The most obvious example is represented by inter-laboratory tests where the same samples are analysed, but where the result from different laboratories differs dramatically, sometimes by several orders of magnitude, because different "standard materials" are employed [131]. Small structural differences between standards might be totally "invisible" to an antibody, which might lead to different results due to differences in CR, even though no internal test will show the assay system to be invalid! Another important source of errors in immunoassay is the lot-to-lot variability and stability of reagents (antibodies, enzyme conjugates, etc.) [15].

9.4.2 Immunoassay validation with emphasis on applications for environmental monitoring

As has increasingly been highlighted in recent years, immunoassays can offer simple, rapid and relatively cheap ways for the determination of pollutants in environmental samples, especially for those that are relatively difficult to measure by conventional analytical methods and require frequent and rapid monitoring [33,45,135–138]. However, in contrast to clinical and pharmaceutical applications, the development of commercial immunoassays for environmental monitoring and their acceptance by registration agencies are still in their early stages. This might be due to the very few efforts to establish guidelines for standardization and quality assurance of immunoassays with application to environmental samples [139]. Immunoassay can, nevertheless, find excellent applicability as pre-screening methods for identification of positive samples and quantification during environmental-fate and residue studies becomes possible after a good characterization of the assay in terms of selectivity, robustness, accuracy and precision. Incompletely characterized assays/commercial assay kits often lead to incorrect results [140–142].

For any analytical method, the goal of the validation is to document that the new procedure generates reliable analytical results, which meet the requirements of the laboratory and of the intended user [15]. Thus, a necessary condition prior to validation is to set "theoretical limits" for detection of the analyte. These cut-off levels are decided depending on the existing regulations, the compound under study and the sample matrix to be analysed [33]. Moreover, the complete design of the validation experiment, the statistical procedures for data analysis and statistical acceptance criteria that define the rules for accepting/rejecting a result should be specified. US-EPA has published a complete guide to validate and use immunoassays in environmental analysis [143].

The following example of validation is constructed around the EC Drinking Water Directive that sets the maximum admissible concentration for many pollutants, including pesticides, at 100 ng l^{-1}: the concentration of pesticides in other matrices (i.e., wastewater, fruits or vegetables) might reach concentrations that are higher by several orders of magnitude [53,56,57]. Taking into consideration the existing regulations, the UK Department Drinking Water Inspectorate sets the accuracy targets for any method to be used for analysis of drinking water at the following values: limit of detection 10 ng l^{-1}, total standard deviation 2.5 ng l^{-1} or 5% of the nominal concentration, bias not higher than 5 ng l^{-1} or 10% of the analyte concentration, total error of 10 ng l^{-1} or 20%, whichever is greater [144]. However, these limits are too restrictive for immunoassays applied in environmental monitoring. Firstly, there are very few immunoassays for environmental pollutants that can reach a detection limit of 10 ng l^{-1}. The main reason is that most of the pollutants are relatively small molecules, possessing a limited number of possible epitopes, so that production of high-affinity antibodies is extremely difficult. Instead, a good necessary condition for newly developed assays would be that the IC_{50} value (e.g., where the assay displays the best precision) should have a value as close as possible to 100 ng l^{-1}. Moreover, the precision and accuracy values recommended for pharmaceutical and clinical analysis (e.g., not more than 20% total error from the nominal value, except around the limits of quantification where it should be 25% [15]) are also applicable for immunoassay applications to environmental analysis.

Each immunoassay should also be characterized in terms of cross-reactivity and possible matrix effects before analysis. In particular, the influence of the compounds frequently present in water samples (see Table 9.4 [144])

TABLE 9.4

Possible interfering compounds in water samples [144]. (Reprinted with permission.)

Possible interferent	Range	Possible interferent	Range
Fulvic/humic acids	10 mg l^{-1}	Calcium	200 mg l^{-1}
Iron	200 µg l^{-1}	Magnesium	50 mg l^{-1}
Copper	5 µg l^{-1}	Free chlorine	0.5 mg l^{-1}
Lead	50 µg l^{-1}	Chloramine	0.5 mg l^{-1}
Nickel	50 µg l^{-1}	Trihalomethanes	0.1 mg l^{-1}
Zinc	5 mg l^{-1}	Chloroform	50 µg l^{-1}
Manganese	50 µg l^{-1}	Trichloroacteic acid	50 µg l^{-1}
Chloride	400 mg l^{-1}	Linear alkylbenzene sulphonates	10 µg l^{-1}
Nitrate	100 mg l^{-1}	Alkyl ethoxylates	10 µg l^{-1}
Sulphate	250 mg l^{-1}	Di-hardened tallow cations	10 µg l^{-1}
Bicarbonate	200 mg l^{-1}	pH	6–8.5

and some of their mixtures should be studied prior to comparison with a standard method. It is recommended that at least six standard concentrations are used for immunoassay calibration, and at least three validation samples (e.g., two close to each limit of quantification and one around the midpoint of the calibration). In particular, for water samples, a standard containing a low concentration of analyte (i.e., 10 ng l^{-1}), one at 100 ng l^{-1} and another at a higher concentration (i.e., 800 ng l^{-1}) need to be tested [144]. Each sample should be run at least in duplicate since immunoassays tend to have relatively high inter-assay imprecision. At least six validation tests should be performed, but no more than two per day [15].

The best statistical acceptance/rejection criteria for immunoassay results have proved to be the use of confidence intervals for total measurement error (e.g., including both accuracy and precision) and equivalence testing procedures [15], and a 90% two-sided confidence interval has been proposed for the acceptance of immunoassay data [145]. Recoveries of 70–120% are considered acceptable according to US-EPA guidelines and the maximum permissible level of false negatives is 5% [38].

Because standardization of immunoassays requires so many compromises, much work remains to be done, not only to accurately define the analytical goals for immunoassays, but also to develop the necessary technology to achieve them. The whole process of immunoassay development and standardization is laborious, costly and very time consuming, so that priorities need to be set depending on the values and consequences of the output information obtained.

9.5 CONCLUSIONS

Immunoassays, along with all other methods based on biorecognition, are a great achievement for the field of analytical chemistry. The first area to benefit from the advantages of this technique is probably clinical analysis, where selective and sensitive determination of macromolecules is often necessary, as it is very difficult, or sometimes impossible to identify and quantify macromolecules by any alternative method. Recently, the immunoassay application field was extended, more or less successfully, to the determination of small molecules (below 1000 Da) as well. The acceptance of immunoassays is, however, relatively limited in some fields, probably due to the differences between the classical analytical methods and immunoassays, the latter requiring special conditions of operation, characterization and data interpretation, due to the extraordinary nature of the antibody–antigen interaction, as well as that of many possible interfering reactions.

This chapter covers the theoretical grounds, the possibilities to rationally design the "perfect" assay in terms of sensitivity for any specific application, as well as the limitations of immunoassays, mainly in relation to operation close to the sensitivity limit, susceptibility to saturation kinetics and interferences due to sample matrix components.

REFERENCES

1. R. Ekins, *Scand. J. Clin. Lab. Invest.*, 51 (1991) 33–46.
2. R.S. Yalow and S.A. Berson, *Nature*, 184 (1960) 1648–1649.
3. R.P. Ekins, *Clin. Chim. Acta*, 5 (1960) 453–459.
4. L.J. Sokoll and D.W. Chan, *Anal. Chem.*, 71 (1999) 356R–362R.
5. C. Davies, In: D. Wild (Ed.), *The Immunoassay Handbook*. Stockton Press, New York, 1994, pp. 8–38.
6. M.H.V. van Regenmortel, *J. Immunol. Methods*, 216 (1998) 37–48.
7. P. Tijssen, *Practice and Theory of Enzyme Immunoassays*, 7th ed., Elsevier, Amsterdam, 1992.
8. R.S. Yalow and S.A. Berson, In: R.L. Hayes, F.A. Goswitz and B.E.P. Murphy (Eds.), *Radioisotopes in Medicine: in vitro Studies*. US Atomic Energy Commission, Oak Ridge, TN, 1968, pp. 7–39.
9. H.L. Pardue, *Clin. Chem.*, 43 (1997) 1831–1837.
10. D. Rodbard, In: W.D. Odell and W.H. Daughaday (Eds.), *Principles of Competitive Protein Binding*. J.P. Lippincott, Philadelphia, PA, 1971, pp. 204–259.
11. R. Ekins and P. Edwards, *Clin. Chem.*, 43 (1997) 1824–1831.

12 R. Ekins and P. Edwards, *Clin. Chem.*, 44 (1998) 1773–1775.
13 D. Rodbard, *Anal. Biochem.*, 90 (1978) 1–12.
14 R.P. Ekins, In: W.M. Hunter and J.E.T. Corrie (Eds.), *Immunoassay for Clinical Chemistry*. Churchill Livingstone, Edinburgh, 1983, pp. 76–105.
15 J.W.A. Findlay, W.C. Smith, J.W. Lee, G.D. Nordblom, I. Das, B.S. DaSilva, M.N. Khan and R.R. Bowsher, *J. Pharm. Biomed. Anal.*, 21 (2000) 1249–1273.
16 C. Davies, In: D. Wild (Ed.), *The Immunoassay Handbook*. Stockton Press, New York, 1994, pp. 83–115.
17 D. Rodbard and H.A. Feldman, *Methods Enzymol.*, 36 (1975) 3–22.
18 R.A. Dudley, P. Edwrds, R.P. Ekins, D.J. Finney, I.G.M. McKenzie, G.M. Raab, D. Rodbard and R.P.C. Rodgers, *Clin. Chem.*, 31 (1985) 1264–1271.
19 B. Nix, In: D. Wild (Ed.), *The Immunoassay Handbook*. Stockton Press, New York, 1994, pp. 117–123.
20 D.J. Finney, *Clin. Chem.*, 29 (1983) 1762–1766.
21 H.T. Karnes and C. March, *J. Pharm. Biomed. Anal.*, 9 (1991) 911–918.
22 D.J. Finney and P. Philips, *Appl. Stat.*, 26 (1978) 312–320.
23 J.W.A. Findlay and I. Das, *J. Clin. Ligand Assay*, 20 (1997) 49–55.
24 R. Ekins, In: C.P. Price and D.J. Newman (Eds.), *Principle and Practice of Immunoassay*. Macmillan, London, 1991, pp. 96–153.
25 B.B. Dzantiev, A.V. Zherdev, I.Y. Moreva, O.G. Romanenko, L.A. Sapegova and S.A. Eremin, *Appl. Biochem. Microbiol.*, 30 (1994) 752–759.
26 J. Gascón, A. Oubina, I. Ferrer, P. Önnerfjord, G. Marko-Varga, B.D. Hammock, M.-P. Marco and D. Barceló, *Anal. Chim. Acta*, 330 (1996) 41–51.
27 T.M. Jackson and R.P. Ekins, *J. Immunol. Methods*, 87 (1983) 13–20.
28 L.E.M. Miles and C.N. Hales, *Nature*, 219 (1968) 186–189.
29 G.M. Addison and C.N. Hales, In: K.E. Kirkham and W.M. Hunter (Eds.), *Radioimmunoassay Methods*. Churchill Livingstone, Edinburgh, 1970, pp. 447–461, 481–487.
30 J.J. Miller and R.J. Valdes, *Clin. Chem.*, 37 (1991) 144–153.
31 J.A. Berzofsky and A.N. Schechter, *Mol. Immunol.*, 18 (1981) 751–763.
32 M. Nording and P. Häglund, *Anal. Chim. Acta*, 487 (2003) 43–50.
33 M.C. Hennion and D. Barceló, *Anal. Chim. Acta*, 362 (1998) 3–34.
34 R.P. Ekins, *Br. Med. Bull.*, 30 (1974) 3–11.
35 C. Nistor and J. Emnéus, *Anal. Bioanal. Chem.*, 375 (2003) 1076–1081.
36 M. Badea, C. Nistor, Y. Goda, S. Fujimoto, S. Dosho, A. Danet, D. Barceló, F. Ventura and J. Emnéus, *Analyst*, 128 (2003) 849–856.
37 J. Büttner, *Scand. J. Clin. Lab. Invest.*, 51 (1991) 11–20.
38 IUPAC; Agrochemicals, A.C.D.O., *Pure Appl. Chem.*, 67 (1995) 2065–2088.
39 R.A. Cole, K.M. Rinne, S. Shahabi and A. Omarani, *Clin. Chem.*, 45 (1999) 313–314.
40 S. Rotmensch and L.A. Cole, *Lancet*, 355 (2000) 712–715.
41 C. Nistor and J. Emnéus, unpublished results, 2002

42   R.J. Bushway, B. Perkins, S.A. Savage, S.J. Lekousi and B.S. Ferguson, *Bull. Environ. Contam. Toxicol.*, 42 (1989) 899–904.
43   C. Nistor, J. Christensen, N. Ocio, L. Nørgaard and J. Emnéus, *Anal. Bioanal. Chem.*, published on-line on Nov. 26, 2004, ISSN: 1618-2642 (Paper), 1618-2650 (Online).
44   P. Rauch, I. Hochel, E. Berankova and J. Kas, *J. Dairy Res.*, 56 (1989) 793–797.
45   M.C. Hennion, *Analusis*, 26 (1998) M149–M155.
46   W.H. Newsome, J.M. Young and P.J. Collins, *J. Assoc. Off. Anal. Chem.*, 76 (1993) 381–386.
47   B.S. Ferguson, D.E. Kelsey, T.S. Fan and R.J. Bushway, *Sci. Total Environ.*, 132 (1993) 415–428.
48   J.M. Wong, Q.X. Li, B.D. Hammock and J.N. Seiber, *J. Agric. Food Chem.*, 39 (1991) 1802–1807.
49   V. Lopez-Avila, C. Charan and W.F. Beckert, *Trends Anal. Chem.*, 13 (1994) 118–126.
50   G. Xiong, J. Liang, S. Zou and Z. Zhang, *Anal. Chim. Acta*, 371 (1998) 97–103.
51   D.S. Aga and E.M. Thurman, *Anal. Chem.*, 65 (1993) 2894–2898.
52   D.S. Aga, E.M. Thurman and M.L. Pomes, *Anal. Chem.*, 66 (1994) 1495–1499.
53   A. Abad, M.J. Moreno, R. Pelegri, M.I. Martinez, A. Saez, M. Gamon and A. Montoya, *J. Chromatogr. A*, 833 (1999) 3–12.
54   M.-P. Marco, S. Chiron, J. Gascón, B.D. Hammock and D. Barceló, *Anal. Chim. Acta*, 311 (1995) 319–329.
55   G. Shan, W.R. Leeman, S.J. Gee, J.R. Sanborn, A.D. Jones, D.P.Y. Chang and B.D. Hammock, *Anal. Chim. Acta*, 444 (2001) 169–178.
56   S. Cairoli, A. Arnoldi and S. Pagani, *J. Agric. Food Chem.*, 44 (1996) 3849–3854.
57   S.J. Richman, S. Karthikeyan, D.A. Bennett, A.C. Chung and S.M. Lee, *J. Agric. Food Chem.*, 44 (1996) 2924–2929.
58   J.J. Sullivan and K.S. Goh, *J. Agric. Food Chem.*, 48 (2000) 4071–4078.
59   A.D. Lucas, P. Schneider, R.O. Harrison, J.N. Seiber, B.D. Hammock, J.W. Biggar and D.E. Rolston, *Food Agric. Immunol.*, 3 (1991) 155–167.
60   A.J. Oosterkamp, H. Irth, U.R. Tjaden and J. van der Greef, *Anal. Chem.*, 66 (1994) 4295–4301.
61   A.J. Oosterkamp, H. Irth, M. Beth, K.K. Unger, U.R. Tjaden and J. van der Greef, *J. Chromatogr. B: Biomed. Appl.*, 653 (1994) 55–61.
62   A.J. Oosterkamp, H. Irth, L. Heintz, G. Marko-Varga, U.R. Tjaden and J. van der Greef, *Anal. Chem.*, 68 (1996) 4101–4106.
63   E.S.M. Lutz, A.J. Oosterkamp and H. Irth, *Chim. Oggi*, 15 (1997) 11–15.
64   H. Irth, A.J. Oosterkamp, U.R. Tjaden and J. van der Greef, *Trends Anal. Chem.*, 14 (1995) 355–361.
65   N. Delaunay, V. Pichon and M.C. Hennion, *J. Chromatogr. B*, 745 (2000) 15–37.
66   T.M. Phillips, *J. Biochem. Biophys. Methods*, 49 (2001) 253–262.
67   A. Dankwardt and B. Hock, *Chemosphere*, 45 (2001) 523–533.
68   S.A. Berson and R.S. Yalow, *Clin. Chim. Acta*, 22 (1968) 51–69.

69 R.P. Ekins, G.B. Newman, R. Piyasa, P. Banks and J.D.H. Slater, *J. Steroid Biochem.*, 3 (1972) 289–304.
70 D. Wild, In: D. Wild (Ed.), *The Immunoassay Handbook*. Stockton Press, New York, 1994, pp. 323–344.
71 A. Hahn, F. Frimmel, A. Haisch, G. Henkelmann and B. Hock, *Z. Pflanzenernähr. Bodenk.*, 155 (1992) 203–208.
72 A. Dankwardt and B. Hock, *GIT Fachz. Lab.*, 39 (1995) 721–722.
73 A. Dankwardt, B. Hock, R. Simon, D. Freitag and A. Kettrup, *Environ. Sci. Technol.*, 30 (1996) 3493–3500.
74 S.J. Huber and C. Sautter, *Z. Planzenernähr. Bodenk.*, 93 (1986) 608–613.
75 G. Sohn, C. Sautter and B. Hock, *Planta*, 181 (1990) 199–203.
76 T. Giersch, G. Sohn and B. Hock, *Anal. Lett.*, 26 (1993) 1831–1845.
77 P. Ulrich, M.G. Weller and R. Niessner, *Fresenius J. Anal. Chem.*, 354 (1996) 352–358.
78 E. Simon, M.G. Weller and R. Niessner, *Fresenius J. Anal. Chem.*, 360 (1998) 824–826.
79 P. Pfortner, M.G. Weller and R. Niessner, *Fresenius J. Anal. Chem.*, 360 (1998) 781–783.
80 P. Pfortner, M.G. Weller and R. Niessner, *Fresenius J. Anal. Chem.*, 360 (1998) 192–198.
81 R.J. Valdes and S.W. Graves, *J. Clin. Endocrinol. Metab.*, 60 (1985) 1135–1143.
82 D.S. Lukas and A.G. de Martino, *J. Clin. Invest.*, 48 (1969) 1041–1053.
83 S.J. Soldin, C. Stephey, E. Giesbrecht and R. Harding, *Clin. Chem.*, 32 (1986) 1591.
84 H.J. Motulsky and L.C. Mahan, *Mol. Pharmacol.*, 25 (1983) 1–9.
85 P. Önnerfjord, S.A. Eremin, J. Emnéus and G. Marko-Varga, *J. Immunol. Methods*, 213 (1998) 31–39.
86 P. Önnerfjord, S.A. Eremin, J. Emnéus and G. Marko-Varga, *J. Chromatogr. A*, 800 (1998) 219–230.
87 M.G. Weller, L. Weil and R. Niessner, *Mikrochim. Acta*, 108 (1992) 29–40.
88 J. Gascón, A. Oubina, B. Ballesteros, D. Barceló, F. Camps, M.-P. Marco, M.A. González-Martínez, S. Morais, R. Puchades and A. Maquieira, *Anal. Chim. Acta*, 347 (1997) 149–162.
89 R. Galve, F. Sanchez-Baeza, F. Camps and M.P. Marco, *Anal. Chim. Acta*, 452 (2002) 191–206.
90 O.A. Melnichenko, S.A. Eremin and A.M. Egorov, *J. Anal. Chem.*, 51 (1996) 512.
91 R.F. Vining, P. Compton and R. McGinley, *Clin. Chem.*, 27 (1981) 910–913.
92 A.H. Sehon, *Ann. NY Acad. Sci.*, 103 (1963) 626–629.
93 T.W. Smith and K.M. Skubitz, *Biochemistry*, 14 (1975) 1496–1502.
94 P.M. Keane, W.H.C. Walker, J. Gauldie and G.E. Abraham, *Clin. Chem.*, 22 (1976) 70–73.
95 E.H. Gendlof, W.L. Casale, B.P. Ram, J.H. Tai, J.J. Pestka and L.P. Hart, *J. Immunol. Methods*, 92 (1986) 15–24.

96 G.W. Wood, *Scand. J. Clin. Lab. Invest.*, 51 (1991) 105–112.
97 J.F. Brady, J.R. Fleeker, R.A. Wilson and R.O. Mumma, In: R.G.M. Wang, C.A. Franklin, R.C. Honeycutt and J.C. Reinhert (Eds.), *Biological Monitoring of Pesticide Exposure, Measurement, Estimation, and Risk Exposure*. ACS, Washington, DC, 1989, pp. 262–284.
98 A.S. Hill, D.P. McAdam, S.L. Edward and J.H. Skerritt, *J. Agric. Food Chem.*, 41 (1993) 2011–2018.
99 A. Abad, M.J. Moreno and A. Montoya, *J. Agric. Food Chem.*, 46 (1998) 2417–2426.
100 B. Ballesteros, D. Barceló, F. Sanchez-Baeza, F. Camps and M.P. Marco, *Anal. Chem.*, 70 (1998) 4004–4014.
101 Y.J. Kim, Y.A. Cho, H.-S.L. Lee and T. Yong, *Anal. Chim. Acta*, 494 (2003) 29–40.
102 S.I. Wie and B.D. Hammock, *J. Agric. Food Chem.*, 32 (1984) 1294–1301.
103 A. Oubiña, D. Barceló and M.-P. Marco, *Anal. Chim. Acta*, 387 (1999) 267–279.
104 R.J. Bushway, B. Perkins, S.A. Savage, S.J. Lekousi and B.S. Ferguson, *Bull. Environ. Contam. Toxicol.*, 40 (1988) 647.
105 J.M. Schlaeppi, W. Föry and K.A. Ramsteiner, *J. Agric. Food Chem.*, 37 (1989) 1532.
106 E. Burestedt, C. Nistor, U. Schagerlöf and J. Emnéus, *Anal. Chem.*, 72 (2000) 4171–4177.
107 R.Z. Dintzis, M.H. Middleton and H.M. Dintzis, *J. Immunol.*, 135 (1985) 423–427.
108 K.R. Luehrsen, S. Davidson, Y.J. Lee, R. Rouhani, A. Soleimani, T. Raich, C.A. Cain, E.J. Collarini, D.T. Yamanishi, J. Pearson, K. Magee, M.R. Madlansacay, V. Bodepudi, D. Davoudzadeh, P.A. Schueler and W. Mahoney, *J. Histochem. Cytochem.*, 48 (2000) 133–145.
109 S.W. Garden and P. Sporns, *J. Agric. Food Chem.*, 42 (1994) 1379–1391.
110 R. Galve, M. Nichkova, F. Camps, F. Sanchez-Baeza and M.P. Marco, *Anal. Chem.*, 74 (2002) 468–478.
111 J.E. Butler, In: E.P. Diamandis and T.K. Christopulos (Eds.), *Immunoassay*. Academic Press, San Diego, CA, 1996, pp. 205–225.
112 M. Stenberg and H. Nygren, *J. Immunol. Methods*, 113 (1988) 3–8.
113 M. Werthén, M. Stenber and H. Nygren, *Prog. Colloid Polym. Sci.*, 82 (1990) 349.
114 M.S. Barbarakis, W.G. Qaisi, S. Daunert and L.G. Bachas, *Anal. Chem.*, 65 (1993) 457–460.
115 L.G. Bachas, P.F. Lewis and M.E. Meyerhoff, *Anal. Chem.*, 56 (1984) 1723–1726.
116 J. Krahn, D.M. Parry, M. Leroux and J. Dalton, *Clin. Biochem.*, 32 (1999) 477–480.
117 G. Katwa, G. Komatireddy and S.E. Walker, *J. Rheumatol.*, 28 (2001) 2750–2751.
118 L.J. Kricka, *Clin. Chem.*, 45 (1999) 942–956.
119 C. Hennig, L. Rink, U. Fagin, W.L. Jabs and H. Kirchner, *J. Immunol. Methods*, 235 (2000) 71–80.

120 M. Covinsky, O. Laterza, J.D. Pfeifer, T. Farkas-Szallasi and M.G. Scott, *Clin. Chem.*, 46 (2000) 1157–1161.
121 D. Wild and C. Davies, In: D. Wild (Ed.), *The Immunoassay Handbook*. Stockton Press, New York, 1994, pp. 49–82.
122 P. Önnerfjord and G. Marko-Varga, *Chromatographia*, 51 (2000) 199–204.
123 C. Nistor, A. Oubiña, M.-P. Marco, D. Barceló and J. Emnéus, *Anal. Chim. Acta*, 426 (2001) 185–195.
124 C. Nistor, M.-P. Marco, M. Tudorache and J. Emnéus, unpublished results.
125 H.G. Dahlen, K.H. Voigt and H.P. Schneider, *Horm. Metab. Res.*, 8 (1976) 61–66.
126 R.D. Utiger, M.L. Parker and W.H. Daughaday, *J. Clin. Invest.*, 41 (1962) 254–261.
127 W.G. Wood, G. Stalla, O.A. Müller and P.C. Scriba, *J. Clin. Chem. Clin. Biochem.*, 17 (1979) 111–114.
128 R.F. Ritchie, *Scand. J. Clin. Lab. Invest.*, 51 (1991) 3–10.
129 J. Gosling, *Scand. J. Clin. Lab. Invest.*, 51 (1991) 95–104.
130 R. Rej and P. Drake, *Scand. J. Clin. Lab. Invest.*, 51 (1991) 47–54.
131 R. Thorpe, *J. Immunol. Methods*, 216 (1998) 93–101.
132 U.-H. Stenman, *Clin. Chem.*, 47 (2001) 815–820.
133 J.T. Witcher, *Scand. J. Clin. Lab. Invest.*, 51 (1991) 21–32.
134 C.R. Lipton, J.X. Dautlick, G.D. Grothaus, P.L. Hunst, K.M. Magin, C.A. Mihaliak, F.M. Rubio and J.W. Stave, *Food Agric. Immunol.*, 12 (2000) 153–164.
135 J.P. Sherry, *Crit. Rev. Anal. Chem.*, 23 (1992) 217–300.
136 E.P. Meulenberg, W.H. Mulder and P.G. Stoks, *Environ. Sci. Technol.*, 29 (1995) 553–561.
137 J. Sherry, *Chemosphere*, 34 (1997) 1011–1025.
138 P. Behnish, *Environ. Int.*, 27 (2001) 441–442.
139 P. Krämer, *Anal. Chim. Acta*, 376 (1998) 3–11.
140 S.M. Lee and S. Richman, *J. Assoc. Off. Anal. Chem.*, 74 (1991) 893.
141 D.B. Baker, R.J. Bushway, S.A. Adams and C. Macomber, *Environ. Sci. Technol.*, 27 (1993) 562–564.
142 D. Zaruk, M. Comba, J. Struger and S. Young, *Anal. Chim. Acta*, 444 (2001) 163–168.
143 US-EPA, Quality Assurance Unit Staff, *Office of Environmental Measurement and Evaluation*, Immunoassay Guidelines for Planning Environmental Projects, US-EPA, New England, 1996, pp. 45.
144 IUPAC; Commission of Water Chemistry, *Pure Appl. Chem.*, 67 (1995) 1533–1548.
145 P. Hubert, P. Chiap, J. Crommen, B. Boulanger, E. Chapuzet, N. Mercier, S. Bervoas-Martin, P. Chevalier, D. Grandjean, P. Lagorce, M. Lallier, M.C. Laparra, M. Laurentie and J.C. Nivet, *Anal. Chim. Acta*, 391 (1999) 135–148.

GLOSSARY

Accuracy: Degree to which the value obtained in the assay corresponds to the true value

Affinity: The strength of binding between antibody and antigen

Analyte: A compound or family of compounds in a sample to be determined in an assay

Antibody (Ab): Protein from Ig class produced in an animal in response to an antigen, which can specifically react to form an antigen–antibody complex

Antibody-binding site: Region of the antibody molecule that can react with antigenic determinants of antigens = paratope

Antigen (Ag): An entity (chemical compound, virus, bacteria) that specifically reacts with an antibody to form the antigen–antibody complex

Antigenic determinant: Region of the antigen molecule recognized and bound by an antibody = epitope

Antigenicity: The ability of a molecule to bind to an antibody

Antiserum: Antibody-containing serum

Bound fraction: Antigen/competitor present as a complex, attached to the corresponding antibody; the fraction of the reaction mixture which contains the antigen–antibody complex

Carrier protein: Large non-immunogenic protein that, when coupled to a hapten, induces production of anti-hapten antibodies in an animal

Competitive assay: Immunoassay based on the principle of competition between the analyte in the sample ("unknown") and competitor

Competitor: Antigen derivative that is modified (i.e., labelled, coupled to a large protein) and serves for indirect detection of the analyte in competitive immunoassays

Conjugate: Antigen/antigen derivative or antibody covalently coupled to a label (= tracer) or hapten coupled to a carrier protein molecule (= immunogen)

Control: Sample-like preparation containing a known amount of analyte

Cross-reaction: Antibody recognition and binding of more than one compound

ELISA: Enzyme-linked immunosorbent assay—heterogeneous immunoassay using enzyme-labelled antigens or antibodies

Enzyme: Protein capable of catalysing specific reactions of substrates into products

False negative: A negative interpretation of a sample containing the target analyte(s) at or above the detection level

False positive: A positive response for a sample containing the analyte(s) under the detection level

Free fraction: Antigen/competitor that is not attached to the antibody; the fraction of the reaction mixture that contains the free antigen/competitor

Hapten: Small molecule, not immunogenic by itself, but which can react with specific antibodies

Heterogeneous immunoassay: Type of immunoassay that requires separation of bound and free fractions

Homogeneous immunoassay: Type of immunoassay that does not require separation of bound and free fractions

Immunoassay: Analytical technique that uses antibodies as reagents for binding of analyte(s) in the sample

Immunogen: Compound able to induce an immune response (i.e., production of antibodies) in an animal

Immunoglobulin: A class of globular proteins = antibody

Ligand: A member of a binding pair, generally the smaller member

Matrix effect: Any bias in the assay result

Monoclonal antibodies: Homogeneous antibody populations directed at against a single antigenic determinant, produced by a single clone of cells

Polyclonal antibodies: Heterogeneous antibody populations that exhibit different properties and binding characteristics

Precision: Extent to which certain measurements agree to each other, usually stated as coefficients of variation, standard deviation or confidence limits

Replicates: Repeated independent determinations on the same sample by the same analyst at essentially the same time under the same conditions

Response: Measured characteristics when a sample is analysed

Sandwich immunoassay: Immunoassay variant in which the analyte is detected by binding to two different antibodies

Selectivity: Characteristic of a method to distinguish between true/specific and inaccurate/non-specific results

Sensitivity: Ability of an analytical method to detect a response variation due to changes in analyte concentration

Solid-phase separation: Method of fraction separation in which the binding reagent is immobilized on an insoluble material (magnetic particles, polymers, etc.)

Standard curve: Relationship between the analytical response and known concentrations of standard analyte = calibration curve = dose–response curve

Substrate: Substance that can be transformed under the action of enzymes into a product. This leads to a detectable change in the concentration of either the substrate or the product
Tracer: Antigen or antibody covalently coupled to a label used for the indirect detection of the analyte in the sample
Zero dose: Standard/sample in which the analyte is absent

Chapter 10

Non-affinity sensing technology: the exploitation of biocatalytic events for environmental analysis

Elena Domínguez and Arántzazu Narváez

10.1 INTRODUCTION

Catalytic reactions are commonly used in environmental technologies with applications ranging from pollution abatement to pollution prevention. Examples include complex biocatalytic pathways embedded into natural microbial communities for the bioremediation of toxic compounds into harmless end products, and new and cleaner catalytic synthetic routes for the production of bulk and fine chemicals [1]. The Organization for Economic Cooperation and Development (OECD) has recognized enzyme technology as an important component of sustainable industrial development [2,3].

In addition to safer and sustainable environmental technologies, the abatement and prevention of pollution also require control and warning systems for the diagnosis and provision of feedback for achieving operating conditions within predefined limits and minimum acceptable risk exposures (i.e., environmental quality standards, EQS). Real time, cost-effective and field-deployable analytical systems that provide temporal and spatial chemical and biological information are thus needed for sound environmental management decisions. This includes rapid diagnostic tools as well as long-term monitoring devices.

The 2001 ATSDR's priority hazardous substances list includes 275 compounds ranked on a combination of criteria including their frequency, toxicity, and potential for human exposure at American priority list sites. The European list of priority substances (Water Framework Directive, 2000/60/EC) categorizes 33 substances or groups of substances, all of which pose a major threat to environmental waters. At present, 11 substances are identified as hazardous and, thus, their emission, discharges and losses should cease not later than in 20 years. For another 14 substances, eight of which are pesticides, the Commission has to make proposals on their status. However, most of these 14 substances are already internationally recognized as hazardous, and should also be eliminated from European waters. The remaining substances are considered not to be hazardous. The complete list is summarized in Table 10.1.

Of all the pollutants released into the environment every year by human activity, persistent organic pollutants (POPs) are among the most dangerous. These organic substances possess toxic characteristics, are persistent, bioaccumulative, prone to long-range transboundary atmospheric transport and deposition, and are likely to cause significant adverse human health or environmental effects near to and distant from their sources [4]. These internationally recognized POPs include eight pesticides, such as Aldrin, Chlordane, Dieldrin, Endrin, Heptachlor, Mirex, Toxaphene, and DDT, two industrial chemicals (PCB and hexachlorobenzene) and two groups of by-products (i.e., dioxins and furans). The structural formulae of these compounds are depicted in Fig. 10.1.

In addition to these "conventional" priority pollutants, there are many other substances that, although not being traditionally regarded as environmental threats, are emerging in the environment as relevant pollutants as a consequence of their intensive use or after the recognition of unexpected biological effects [5]. For instance, antibiotics and, in general, human and veterinary drugs or the large and diverse group of chemical compounds categorized as endocrine disrupting compounds (e.g., phthalates, bisphenol A, nonylphenol) are presently well recognized as emerging pollutants [6,7]. Some of these compounds and their metabolites are not efficiently degraded in treatment plants and are very persistent in the environment, causing an actual threat to ecosystems [8].

Environmental monitoring should also include many other compounds that, although not produced or discharged by human activities (i.e., xenobiotics), are nonetheless relevant for the assessment of EQS. For instance, the concentration of nutrients or naturally occurring substances cannot exceed the levels appropriate for the normal functioning of ecosystems.

It is obvious that the large variety of chemical species that have to be monitored and controlled prohibits approaching environmental analysis by a single means or technique. This is further complicated by the complexity and very different nature of environmental samples: water, waste, soil, and air. Furthermore, water matrix effects will vary depending on whether surface, ground- or wastewater is tackled. The wide range of concentrations encountered in these applications also challenge the suitability of any single analytical technique. For example, superfund sites typically require

TABLE 10.1
List of priority substances in the field of water policy

Name	CAS number	Major uses or emissions sources
Priority hazardous substances		
Brominated diphenylethers[a]	Not applicable	Flame retardants
Cadmium and its compounds	7440-43-9	Batteries, pigments
C$_{10-13}$-chloroalkanes	85535-84-8	Metal working fluids, flame retardant
Hexachlorobenzene	118-74-1	No use in EU but unintentional by-product, e.g., in PVC
Hexachlorobutadiene	87-68-3	No use in EU but unintentional by-product
Hexachlorocyclohexane[b]	608-73-1	Plant protection product (insecticide)
Mercury and its compounds	7439-97-6	Batteries, thermometers, tooth filling, chlor-alkali industry
Nonylphenols[c]	25154-52-3	Chemical intermediate, industrial detergent and others
Pentachlorobenzene	608-93-5	Intermediate in the production of quintozene (plant protection product)
Polyaromatic hydrocarbons	Not applicable	Combustion by-products, metal treatment, wood treatment (creosote) and others
Tributyltin compounds[d]	688-73-3	Antifouling paints of ships
Priority substances under review as possible priority hazardous substances		
Anthracene	120-12-7	Chemical intermediate, wood preservative (creosote), combustion by-product
Atrazine	1912-24-9	Plant protection product (herbicide)
Chlorpyrifos	2921-88-2	Plant protection product (insecticide)

continued

TABLE 10.1 (Continuation)

Name	CAS number	Major uses or emissions sources
Di(2-ethylhexyl)phthalate (DEHP)	117-81-7	Plasticizer in soft-PVC
Diuron	330-54-1	Plant protection product (herbicide)
Endosulfan[e]	115-29-7	Plant protection product (insecticide)
Isoproturon	34123-59-6	Plant protection product (herbicide)
Lead and its compounds	7439-92-1	Batteries and many other products
Naphthalene	91-20-3	Chemical intermediate, wood preservative (creosote), combustion by-product
Octylphenols[f]	1806-26-4	Similar to nonylphenol
Pentachlorophenol	87-86-5	Biocide in wood or textiles
Simazine	122-34-9	Plant protection product (herbicide)
Trichlorobenzenes[g]	12002-48-1	Chemical intermediate, process solvent and others
Trifluralin	1582-09-8	Plant protection product (herbicide)
Priority substances		
Alachlor	15972-60-8	Plant protection product (herbicide)
Benzene	71-43-2	Synthesis of other chemicals
Chlorfenvinphos	470-90-6	Plant protection product (insecticide)
1,2-Dichloroethane	107-06-2	Production of vinyl chloride monomer for PVC production
Dichloromethane	75-09-2	Solvent, aerosol, foam blowing agent
Fluoroanthene[h]	206-44-0	

Non-affinity sensing technology

Nickel and its compounds	7440-02-0	More than 300,000 products mainly as alloys, e.g., stainless steel
Trichloromethane (chloroform)	67-66-3	Chemical intermediate, e.g., production of HCFC (blowing agent and refrigerant)

Decision No. 2455/2001/EC of the European Parliament and of the Council of 20th November 2001 establishing the list of priority substances in the field of water policy and amending Directive 2000/60/EC.
[a]These group of substances normally include a considerable number of individual compounds. At present, appropriate indicative parameters are not given. Only pentabromobiphenylether is identified as priority hazardous substance.
[b]The gamma-isomer of HCH (CAS No. 58-89-9), also known as lindane, is also used separately as an insecticide and it is categorized as priority substance.
[c]In the framework of Directive 91/414/EC, it is foreseen that nonylphenol and nonylphenol ethoxylates as active substances in pesticides have to be withdrawn from the market as from July 2003.
[d]Tributyltin-cation (CAS No. 36643-28-4) is categorized as priority substance.
[e]Alpha-endosulfan (CAS No. 959-98-8) is categorized as priority substance.
[f]p-tert-octylphenol (CAS No. 140-66-9) is categorized as priority substance.
[g]1,2,4-Trichlorobenzene (CAS No. 120-82-1) is categorized as priority substance.
[h]Fluoroanthene is on the list as an indicator of other, more dangerous polyaromatic hydrocarbons.

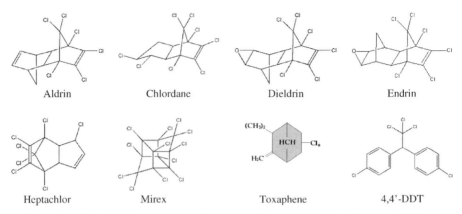

Fig. 10.1. Chemical structures of some persistent organic pollutants (POPs).

monitoring in the mg g^{-1} to μg g^{-1} (ppm) range, whereas agricultural run-off monitoring may require measurements in the μg g^{-1} to ng g^{-1} (ppb) range [9]. This variety can also be found in waters intended for human consumption with parametric values in the mg l^{-1} (e.g., nitrates) or in the μg l^{-1} range (e.g., pesticides, heavy metals).

Bioanalytical and biomimetic principles and, particularly, mechanism-based bioanalytical tools, conveniently integrated into transducing systems, are specially suited for environmental analysis for several reasons:

- Effect related, rather than chemical-based responses, can be obtained thereby bridging exposure and effect assessment and helping in a faster identification of the causative agent(s).
- The exquisite selectivity of biological receptors can simplify the overall analytical process, with minimum treatment and handling of samples, facilitating on-line and continuous measurements.
- Real-time measurements are obtained, enabling fast decision-making and prevention of risk exposures.
- The nanoscale dimensions of the biorecognition receptor allow for miniaturized systems with decreased reagent and power consumption and with the concomitant increase of spatial and temporal information per unit area of device enabling simultaneous multiparametric information.
- Miniaturization also leads to in situ devices with minimal disturbance of hydrogeologic parameters enabling information concerning the concentration profile and movement of the compound(s) of interest.
- New parameters based on chemical imaging rather than on chemical identification can be obtained through sensor arrays that, although beyond

legislative compliance, may permit site characterization and temporal or spatial evolution trends.

These considerations, in addition to the high sensitivity of bioanalytical tools, may result in over-reactions by authorities who could initiate bioremedial procedures while neglecting the ability of the species ecoweb to return to steady-state ecostasis. Therefore, environmental management should always be based a priori on a case-by-case basis [10].

10.2 FOCUS AND SCOPE

This chapter focuses on well-recognized and emerging biocatalytic approaches that can be specifically applied to the screening or monitoring of environmentally relevant chemical species for the assessment of good ecological and chemical status. This chapter covers classical catalytic approaches, and well-characterized enzyme reactions, as well as other catalytic events in living systems that through catabolic activities elicit useful (bio)analytical information. Accordingly, this chapter does not include biochemical sensing technologies that are not based on both biorecognition and catalytic events. The exploitation of affinity interactions—particularly immunoreactions—for environmental applications is presented in Chapters 8 and 9. It should be remembered that chemical sensors, with no participation of any biorecognition, have significantly contributed to the reduction of emissions in industrial processes or during the operation of motor vehicles. The implementation of solid electrolytes, conductive polymers and/or semiconductors and microelectronics sensor manufacture for atmospheric environmental monitoring of oxygen, ozone, carbon dioxide, volatile organic compounds (VOCs), toxic, combustible and refrigerant gases is having a great impact in the abatement of hazardous gases. Any reader interested in this technology can find a comprehensive overview in the book by Campbell [11]. Additionally, there are commercially available borehole sensors for the simultaneous measurement of water quality parameters: conductivity, pH, redox potential (ORP), dissolved oxygen, nitrate, chloride, etc. A list of these devices as well as an information exchange web site can be found on the Internet at http://www.sentix.org.

Also, in line with the authors' view, this chapter does not intend to exhaustively reflect the broad and diverse literature in this field. There are several books [12–16] and reviews fully devoted to specific aspects of biosensors for environmental analysis [17–26]. The reader is advised to consult these references for more specific insight within this area. Instead, environmental relevant analytes whose detection has been extensively approached by catalytic pathways are critically reviewed. The content of this chapter focuses on the description of biosensors within the environmental area with as criterion not only the selection of specific parameters but also the evaluation of the new monitoring or screening challenges that will have to be fulfilled in the near future in order to preserve the environment. The (bio)engineering tools that are presently expanding and improving the biocatalytic approach in environmental applications are treated in a separate section. In this way, we hope that the future of biosensors will be more clearly illustrated because, irrespective of how much work has been done, there is still much to explore and exploit.

10.3 THE UPCOMING PHASE IN ENVIRONMENTAL ANALYSIS: FROM RESEARCH TO MANDATORY MONITORING

The European Environment Agency defines screening as a process used to determine whether a project requires an environmental assessment and what type and level of assessment would be necessary. In contrast, monitoring is defined as a combination of observation and measurement for the performance of a plan, programme or measure and its compliance with environmental policy and legislation. In practical terms, these definitions seem to overlap if screening is considered, as it traditionally is, as the provision of fast and cost-effective "qualitative" information enabling the discrimination of samples according to quality standards, admissible concentrations or threshold levels. The characterization of water bodies or specific sites can then be achieved through extensive screening that renders useful information on the spatial distribution of the targeted pollutants. This information, primarily a binary or dichotomous response, has to be reliable and, consequently, the uncertainty [27] or unreliability [28] for indicating the interval in which the errors can occur (i.e., false positives and false negatives) has to be assessed for sound decisions. Temporal or long-term evolution of these pollutants requires quantitative information which is traditionally reserved for monitoring tasks.

As concerns the protection of water bodies in the Member States of the European Union (EU) the Water Framework Directive distinguishes between three types of monitoring for surface waters: surveillance, operational and investigative monitoring. In terms of groundwater chemical status, surveillance and operational monitoring are indicated for providing information that can be used in the assessment of long-term trends in chemical status. All these monitoring programs will be mandatory by the end of 2006. Surveillance monitoring aims at the provision of information for use in the assessment of long-term trends both as a result of changes in natural conditions and through anthropogenic activity. The objectives of operational monitoring are to establish the chemical status of water bodies that are at risk and the presence of any long-term anthropogenically induced upward trend in the concentration of any pollutant. Investigative monitoring of surface waters may also be required in specified cases. For instance, when surveillance monitoring indicates that the objectives set are not likely to be achieved and operational monitoring has not already been established or if the magnitude and impact of accidental pollution have to be ascertained. A key issue in all these monitoring programmes is the level of precision and confidence required in order to avoid misclassification and wrong assessment of a water body's status. Although the Directive has not specified these levels, they should always enable meaningful assessments of status in time and space. Obviously, the levels of confidence and precision achievable must be balanced against the cost of the monitoring programmes. The frequency of monitoring is also very much determined by the amount of information necessary to provide sufficient data for reliable assessment and to detect the impact of relevant pressures.

Overall, bioanalytical tools enabling the assessment of "good water status" may find a niche of applications within these monitoring programmes. In other words, the protection of European water resources demands the provision of cost-effective analytical tools to assess pollution trends. The demonstration of reliable biosensing principles can be expected to start in the immediate future with a new phase of implementing bioanalytical principles in field applications emerging.

10.4 CATALYSIS IN LIVING SYSTEMS: FROM ESSENTIAL NATURAL PROCESSES TO SENSING DEVICES FOR ENVIRONMENTAL ANALYSIS

Living systems rely on kinetic stability for their existence. They are all thermodynamically unstable, i.e., they would combust to carbon dioxide and water if the system were to reach thermodynamic equilibrium. Consequently, life processes depend upon the ability to restrict these thermodynamic tendencies by controlled kinetics to produce energy as needed. A key feature of living systems is the use of catalysts for the controlled release of that energy. Examples of such catalysts are enzymes which control the synthesis and degradation of biologically important molecules. A sound approach to understanding the reaction rate enhancement by enzymes consists of a mixture of thermodynamic principles, i.e., by lowering the free energy of activation, and mechanistic descriptions by aligning substrates (orientation and proximity effects) through adding or withdrawing protons, electrons, etc. As catalysts, enzymes participate in each reaction cycle and are returned to their original state after each cycle; in other words, one biocatalyst, many reaction cycles.

The key role that enzymes play in living systems are incorporated into preliminary approaches for the assessment of soil degradation [29] and quality [30]. Soil enzyme activities are closely correlated to soil chemical parameters and biological parameters. Since soil enzyme activity is a much more sensitive measurement than organic carbon data, it is being considered as an early indicator of changes in soil organic matter and soil health [31]. Soil enzymes are mainly produced by plants and microorganisms with some of their functions associated with the microbes themselves, such as dehydrogenase activity, which is mainly localized in the plasma membrane of bacteria or in mitochondrial membranes of fungi. Other enzymes are synthesized and secreted extracellularly by bacteria or fungi (e.g., phosphatases, urease, cellulases, pectinases) becoming part of the soil matrix as extracellular enzymes [31–33]. Although this approach cannot be exactly classified as a true sensing technology, the rationale behind it is based on the in vitro measurement of particular enzymes' activities as indicators of microbial metabolism. These data, conveniently processed in combination with other physico-chemical parameters and microbial properties, may contribute to the assessment of soil stress. In this line, enzyme activities could also be contemplated as bioindicators of detrimental effects in ecosystems. For instance, the inhibition of cholinesterase activity has been proposed for the evaluation of exposure and effects of organophosphorous and carbamate pesticides to crustaceans [34] and molluscs [35]. Due regard, as will be discussed further, must be given to the fact that in vivo activities may substantially differ from those found in laboratory bioassays and thus, evolution trends—spatial and/or temporal—ought to be the only feasible approach for obtaining relevant information.

Almost 3200 different enzymes have been listed and categorized by the International Union of Biochemistry and Molecular Biology in its last report in 1992. An encyclopaedic description of more than 7000 commercially available enzymes can be found in Ref. [36]. Table 10.2 collects some industrial enzymes' suppliers. Enzymes exhibiting the same catalytic function are known as homologous enzymes and they fall into two classes: heteroenzymes and isoenzymes. The first group includes enzymes derived from different sources but which catalyse identical reactions, yet show different chemical and kinetic characteristics. A comprehensive enzyme information system, termed BRENDA, is available via the Internet (http://www.brenda.uni-koeln.de).

TABLE 10.2

Some commercial suppliers of enzymes

Suppliers	URL
Sigma-Aldrich-Fluka, St. Louis, MO, USA	http://www.sigma-aldrich.com
Worthington Biochemical, Lakewood, NJ, USA	http://www.worthington-biochem.com
EMD Biosciences/Merck Biosciences (Germany): Calbiochem, Novabiochem, Novagen	http://www.merckbiosciences.co.uk
Roche Diagnostics, Mannheim, Germany	http://www.roche-applied-science.com
Novozymes, Bagsvaerd, Denmark	http://www.novozymes.com
Diversa, San Diego, CA, USA	http://www.diversa.com
DSM Food Specialties, Delft, The Netherlands	http://www.dsm.nl/dfs
Biozyme Laboratories, UK	http://www.biozyme.com
BioCatalytics, Pasadena, CA, USA	http://www.biocatalytics.com
Amano Enzyme, Nagoya, Japan	http://www.amano-enzyme.co.jp
Biocon India, Bangalore, India	http://www.biocon.com
ThermoGen, Woodridge, IL, USA	http://www.thermogen.com
Serva, Germany (Invitrogen group)	http://www.serva.de
Genencor International, Rochester, NY, USA	http://www.genencor.com
Toyobo, Tokyo, Japan	http://www.toyobo.co.jp
New England Biolabs, Beverly, MA, USA	http://www.neb.com

It contains the main collection of enzyme functional data. Isoenzymes are different molecular forms of the same enzyme that are found in the same animal or organism, although they show a pattern of distribution between tissues. A priori, the use of enzymes as biorecognition elements for sensing devices seems to be guaranteed, not only by their large variety, availability and different reaction mechanisms, but also because they are selective catalysts thus opening up easy and direct means of transduction. Since enzymes spontaneously regenerate over the catalytic cycle, they render reversible systems. Consequently, and in line with IUPAC terminology, they are inherently chemical sensors in contrast to chemical probes [37]. Moreover, if the catalytic recognition reaction is not prone to direct transduction and/or is thermodynamically unfavourable (having a low equilibrium constant, K_e), coupled catalytic reactions result in new products or cofactors and in shifted overall reactions that permit the generation of a quantitative signal. Biocatalytic events are also used as selective detection units in flow injection analysis and many other techniques [38,39]. In these applications, enzymes are usually immobilized in packed-bed reactors, enabling a large surplus of catalyst in order to avoid response decay by enzyme inactivation [40].

Although catalysis is featured as the distinct property of enzymes for bioanalytical purposes, the effect of activators or inhibitors on the reaction rate, the intrinsic nature of allosteric enzymes as well as the enzyme's conformational changes upon binding of the substrate substantially increase the possibilities of sensor designs and target analytes. The fundamentals of enzyme kinetics in homogeneous and heterogeneous phases can be found in a multitude of textbooks and thus this chapter will not cover these aspects. Table 10.3 summarizes the main parameters to be considered in enzyme-catalysed reactions obeying a Michaelis–Menten behaviour. In absolute terms, enzymes have turnover numbers from as slow as one catalytic event per minute to 10^5 per second as in the case of catalase reaching the diffusional limit in aqueous solutions [41]. The relative values over non-enzymatic rates of transformation—acceleration rates—are often 10^{10} as for protease-mediated hydrolysis of peptide bonds, and can reach 10^{23} in the case of orotidine-5′-phospate decarboxylase in the pyrimidine biosynthetic pathway. Sensitivity can be further improved by amplification cycles either by coupling

TABLE 10.3

Kinetic terms used in enzyme catalysed reactions: $E + S \underset{k_{-1}}{\overset{k_1}{\rightleftharpoons}} ES \overset{k_2}{\rightarrow} E + P$

Definition	Symbols	Mathematical expression	Units
Maximum rate of reaction	V_{max}	$k_2[E]_t$	$M\,s^{-1}$
Rate of reaction	v	$\dfrac{V_{max}[S]}{[S] + K_m}$	$M\,s^{-1}$
Initial rate of reaction	v_0	$\dfrac{V_{max}[S]}{K_m}$	$M\,s^{-1}$
Michaelis constant	K_m	$\dfrac{k_{-1} + k_2}{k_1}$	M
Catalytic rate constant or turnover number	$k_2 (k_{cat})$	$\dfrac{V_{max}}{[E]_t}$	s^{-1}
Specificity constant	$\dfrac{k_{cat}}{K_m}$		$M^{-1}\,s^{-1}$
Acceleration rate	$\dfrac{k_{cat}}{k_{uncat}}$		
Equilibrium constant of the reaction	K_e	$\dfrac{k_1 k_2}{k_{-1}}$	

other catalytic reactions [42,43] or by efficient transduction chemistry as will be discussed in the following sections. In addition to the use of enzymes as bio-recognition elements in biocatalytic sensors, their inherent properties make them well suited as reporters or labels of other sensing principles with no catalytic events, enabling a multitude of detection schemes (see Chapters 8 and 9).

Despite all these advantages commercialization—not only of biocatalytic sensors but of any kind of biosensors—has been rather scarce within the field of environmental applications [20,44]. Global sales of biosensors in 2002 are estimated in the order of US$2500 million, being dominated by one application, the glucose sensor [45]. Within the biosensors' market, the sales of environmental biosensors range between 1 and 8%, depending on the source of information although there is a general agreement that this market sector shows one of the highest growth rates [45–47]. Rogers has addressed this issue, categorizing the limiting factors for commercialization as [20]:

- Technical issues associated with measurement of pollutants in environmental samples, and
- Practical issues associated with development and marketing within the highly regulated and diverse environmental monitoring area.

Any analytical technique faces a similar situation though bioanalytical systems appear to be particularly susceptible to these obstacles, most likely due to the relatively high development cost as well as the limited experience in the commercialization of biosensors for other markets [20]. Industry estimates of a realistic cost to market for chemical sensor or biosensor systems development range from US$10 million to US$20 million [48]. The high development cost should be considered under the umbrella of the environmental market size that, although wide and in increasing demand, is highly fragmented and constrained by legal compliance.

Overall, technical issues seem to be solved more rapidly than market obstacles can be overcome. In fact, the operational characteristics of biosensors have greatly improved with the advances in microelectronics, resulting in microminiaturized systems capable of dispensing, mixing, separating, and detecting (see Chapters 6 and 12). Equally important, novel signal transducer technology and array systems in combination with new data management and data processing are also contributing to the design of new and fit-for-purpose sensing devices [49]. The difficulties related to environmental samples are not the unique limitations as concerns bioanalytical approaches and it seems that the twin hallmarks of enzymes, selectivity and rates of acceleration, are not sufficiently fitted for diagnostics purposes, requiring an amalgam of additional properties or new biocatalysts not found in nature because they confer no evolutionary advantage. These limitations are not exclusive within diagnostics applications and are common for all industrial applications. Among them are stability, better tolerance to organic solvents, and tuned selectivity with higher specificity constants for non-natural substrates. As will be further discussed, advances in protein engineering and biomimetics, although not fully implemented into bioanalytical devices, offer immense possibilities for new technical developments.

10.5 BIOCATALYTIC PROCESSES FOR ENVIRONMENTAL SENSING PURPOSES

10.5.1 Complex catalytic pathways or catabolic approaches

Despite the vast number of available enzymes, the exploitation of enzymatic selectivity in environmental analysis seems, a priori, a paradox. In general, if living systems had efficient catalysts to metabolize the large variety of organic and inorganic compounds that are environmentally relevant, pollution would not be a threat to ecosystems and human health. This is not in contradiction with the existence of catabolic pathways in some microorganisms that can break down organometallic, aliphatic, aromatic or heteroaromatic pollutants by oxidative, reductive and hydrolytic transformations [50]. Indeed, this is the basis for natural attenuation and bioremediation processes. Nonetheless, sorption of pollutants to soil components, toxic pollutant concentrations, and physico-chemical conditions may limit efficient biodegradation. Usually a cocktail of microorganisms is necessary to convert the pollutants into less toxic compounds or even to attain their mineralization. Theoretically, and if these catabolic pathways are to be used for biorecognition or remediation purposes, the right choice of microbial species should be dictated by the natural microbial population at the contaminated site. A British company is using this approach not only for remediation purposes but also for the design of biosensors based on the transduction of the specific catabolic events (Toxmonitor™) [www.remedios.com.uk]. Two difficulties are inherent to this approach. First, the transduction of the catabolic breakdown is not straightforward and secondly, if using isolates from contaminated sites, the enzymes may not be expressed under ex vivo conditions.

It is clear from the above discussion that the myriad chemical transformations carried out by living organisms can be considered as catalytic systems for bioanalytical purposes. Natural evolution has resulted in adequate coupling of multienzymatic steps, optimizing catalytic efficiency (k_{cat}/K_m) by finding their precise value to maximize enzymatic conversions [51]. However, as many waste sites have a witches' brew of non-biogenic compounds, it is somewhat restrictive to seek stable and multiple pathways within the bacterial communities. Presently, the use of biocatalytic events embedded in whole cells for environmental analysis falls into two categories:

- Deleterious physiological response of viable intact cells by foreign compounds, and,
- Promoter recognition by a specific pollutant followed by gene expression, enzyme synthesis and catalytic activity.

10.5.1.1 Enzyme networks in intact cells: biosensors for acute toxicity and BOD

10.5.1.1.1 Bioanalytical tools for evaluation of acute toxicity

This group includes rather non-selective sensors and they mainly tackle toxicity information. The obvious advantage of using intact cells is that they can provide information about the bioavailability of the analyte because the pollutant must be taken up by the cell before a response is produced. Based on this principle, simple bioanalytical approaches have been developed for determining the toxicity of environmental samples. The commercially available Microtox® and DeltaTox®, both manufactured by Azur Environmental (Strategic Diagnostics Inc.) and LUMIStox from Dr. Bruno Lange GmbH & Co. KG are based on the inhibition of the bioluminescent response of the bacterium *Vibrio fischeri* when exposed to toxic substances, according to the following reaction:

$$FMNH_2 + RCHO + O_2 \xrightarrow{\text{Bacterial luciferase}} FMN + RCOOH + H_2O + h\nu(490 \text{ nm})$$
(10.1)

$FMNH_2$ and FMN are the reduced and oxidized form of riboflavin phosphate (flavin mononucleotide), respectively; RCHO and RCOOH, long-chain aldehyde and acid (e.g., decanal and decanoic acid, respectively); and $h\nu$, emitted light. This reaction is catalysed by bacterial luciferase, offering direct means of optical transduction. The assay has been internationally adopted as a rapid screening test of acute toxicity enabling NOEC (i.e., no observable effect concentration or highest concentration with no adverse effects) and EC_{10}–EC_{50} values to be determined after a 15–30 min exposure period and reconstitution of the freeze-dried cells [52]. This assay has been interpreted as a surrogate for some of the much more costly and time-consuming fish tests [53,54]. A comparison of the acute toxicity values found with different commercial instruments can be found in Ref. [55]. DeltaTox® is a truly portable instrument enabling a qualitative identification of polluted areas whilst Microtox® quantifies acute toxicity through EC values. Toxicity results from both assays agree well but their correlation with concentration measurements obtained with classical techniques depends on the sample fraction [56].

Despite the sensitivity of this approach, one of its major limitations is that it is not ecologically representative of freshwater systems since a saltwater bacterium is used. Risk-assessment and protection of man-made environments should optimally be performed with organisms relevant to the particular environment. For instance, toxicity evaluation of the water inlets in a wastewater treatment plant (WWTP) seems better characterized by the monitoring of the oxygen uptake inhibition of activated sludge than by Microtox assays [57]. The higher sensitivity of the latter (lower EC_{50}) seems to be less relevant than the respirometry values obtained with the heterogeneous microbial population of activated sludge for the performance and management of the WWTP.

Other intact bacterial species, *Escherichia coli* or *Pseudomonas putida*, constitute the basis of the CellSense™ mediated amperometric sensor manufactured by Euroclone Ltd. *E. coli* may show high sensitivity to a range of organoxenobiotics [58] while *P. putida* is a natural soil bacteria possessing relative tolerance and, in some cases, degradative capabilities to a wide range of organic molecules [59]. Table 10.4 summarizes the EC_{50} values obtained for several pollutants determined with *E. coli* or *P. putida* and compared with the values obtained with the bioluminescent assay. Toxicity units (TU) can be defined, in order to establish a direct parameter, as the reciprocal of the EC_{50} values [61].

The cells are immobilized on the surface of screen-printed disposable and single-use electrodes. In the presence of a cocktail of respiratory substrates and upon addition of a redox mediator, potassium hexacyanoferrate (III) or *p*-benzoquinone, the metabolic status of the cells is disturbed by the presence of pollutants. Electrochemical oxidation of the mediator gives a signal that inversely correlates with the concentration of the pollutant. In principle, this sensor's design has the inherent advantage of enabling a wider and more strategic choice of organisms or cells. For instance, cyanobacteria [62], activated sludge [63], eukaryotic algae [64], or cultured fish cells [65] have also been used for the measurement of acute toxicity by electrochemical means of transduction. Ideally, any biosensor approach to toxicity assessment should be capable of incorporating a wide range of species and cell types if it is to be suitable for use in different environmental scenarios.

Ecotoxicity is by no means soundly assessed by these simple devices, yet they offer a primary and simple approach to the toxicity imposed by

TABLE 10.4

Reported toxicity values using the *Vibrio fisheri* and the Cellsense biosensor either with *P. putida* or *E. coli* expressed in 50% inhibition (EC_{50}) and toxicity units (TUs) for some pollutants [60]

Compounds	*Vibrio fisheri*		*P. putida*		*E. coli*	
	EC_{50} (µg ml^{-1})	TUs	EC_{50} (µg ml^{-1})	TUs	EC_{50} (µg ml^{-1})	TUs
Alcohol ethoxylate: C10EOx	3.2	30.8	75.0	1.3	92.0	1.1
Alcohol ethoxylate: C12EOx	0.6	182.0	69.0	1.4	596.0	0.2
Ametryn	18.6	5.4	2.2	45.3	ni	ni
2-Chlorophenol	34.8	2.9	296.0	0.3	250.0	0.4
4-Chlorophenol	21.2	4.7	239.0	0.4	201.0	0.5
Cypermethrin	110.4	0.9	36.1	2.8	ni	ni
Deltamethrin	100.8	1.0	28.0	3.6	ni	ni
2,4-Dichlorophenol	2.8	35.1	247.0	0.4	393.0	0.2
Dimethoate	ni	ni	0.9	113.0	ni	ni
Endosulfan	5.6	17.8	3.4	29.6	ni	ni
Fenamiphos	31.6	3.2	0.9	109.0	ni	ni
2-Naphthalene sulfonate 2-NPS	1580.0	0.1	nd	nd	7100.0	<0.1
Nonylphenol	0.4	277.8	ni	ni	500.0	0.2
Nonylphenol polyethoxylates: NPEOs	2.7	37.0	7.0	14.3	328.0	0.3
Pentachlorophenol	ni		320.0	0.3	0.04	2703.0
Polyethylene glycol: PEG4	127.4	0.8	33.0	3.0	400.1	0.2
2,4,6-Trichlorophenol	ni	ni	256.0	0.4	0.7	149.2

ni: Not investigated, nd: Not detected.

combinations of pollutants, where synergetic or antagonic effects are essential for interpreting the fate of pollutants [66]. A battery of microscale toxicity tests involving different levels of cellular organization is then necessary [67]. For instance, mitochondrial enzyme reactions have the potential to be sensitive and efficient screening tests for general toxicity [68]. The clear advantage derives from the possibility of using this approach as a surrogate for higher animals. Mitochondrial enzymes and enzyme reactions, including mitochondrial electron transport, are well-understood processes in living organisms. Thus, reverse electron transport (RET) assays based on the effects of toxic compounds on NAD^+ reduction by submitochondrial particles have been suggested as an alternative to Microtox for screening water-soluble chemicals [69]. In practice, the reaction conditions are arranged so that the enzyme $NADH_2$ dehydrogenase (ubiquinone) is operating backwards and the effect of pollutants is related to the rate of NAD^+ reduction induced by ATP and succinate at the first level of the respiratory chain [70–72]. Inhibition of RET (i.e., production of NADH) has been found to correlate well with whole-animal toxicity in aquatic organisms [73] and humans [74]. The vast variety of electrochemical transduction for NAD^+/NADH [75,76] opens up the road for the construction of true biosensors based on RET.

10.5.1.1.2 Bioanalytical tools for the determination of biochemical oxygen demand

Intact microbial cells are also used for the measurement of biochemical oxygen demand (BOD), mainly in WWTP. Wastewaters loaded with biodegradable organic compounds increase the metabolism of immobilized microorganisms with a concomitant depletion of molecular oxygen [77,78]. Since the first report by Karube in 1977, the development of this sensor has been extensively investigated with well-focussed research in order to replace the time-consuming BOD_5 assay. As a result, several companies manufacture automatic biosensing instruments and, equally important, the use of these devices has been incorporated into industrial standard methods in Japan [20]. Commercial BOD-sensor systems are being or have been produced by Nisshin Electric Co. Ltd, DKK TOA Corp., Aucoteam GmbH, Dr. Bruno Lange GmbH, and Medingen GmbH [25,79]. Market research indicates that there are approximately 55 million BOD tests performed annually in OECD countries, with an unknown but significant number being performed in non-OECD countries. This represents an extraordinary market for this kind of sensor.

Generally, budding yeast cells are immobilized on the surface of a Clark electrode. Figure 10.2 shows the scheme of a BOD sensor with the classical approach of monitoring molecular oxygen depletion or the more recent

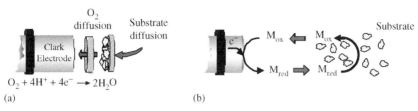

Fig. 10.2. Sensing principles for the determination of the biochemical oxygen demand (BOD). (a) Immobilized cells on the surface of a Clark electrode and (b) batch assay with cells in suspension and with a redox mediator.

approach that makes use of a redox mediator (vide infra). Yeast cells are especially suitable because they are able to use a wide substrate spectrum combined with a wide measuring range. The determination of BOD with a microbial sensor is a quick test for biological activity by a selected microbial species, whereas the BOD_5 measures the sum of various biochemical processes in a biosludge during a period of 5 days. Consequently, the BOD sensor values may differ from standard BOD_5 measurements, depending on the kind of wastewater (e.g., industrial or domestic) [79,80]. The choice of microorganisms is crucial if strict comparison between official and sensor values is pursued. In order to achieve better correlations, the sensor can be constructed with activated sludge [81] or, in the case of samples containing high molecular weight compounds, these can be previously hydrolysed with α-amylase, cellulase and protease [82]. Placing these enzymes on the surface of microorganisms by means of recombinant DNA technology offers novel tailored microorganisms with wider applicability [83]. The use of dimorphic yeast (i.e., budding and mycelial) such as *Arxula adeninivorans* LS3 enables BOD measurements in salty waters of coastal and island regions [84]. Alternatively, reproducing the assay format of the BOD_5 (i.e., batch cells) the extended 5 days of incubation is shortened to 1 h by using a redox mediator, potassium hexacyanoferrate (III), that acts as an electron acceptor in the absence of molecular oxygen. In the presence of organic matter, the respiratory activity of *E. coli* cells is then increased, transferring the electrons to the mediator, which is reoxidized at 450 mV vs. Ag/AgCl as summarized in the following reactions:

$$CH_2O + H_2O + 4\left[Fe(CN)_6^{3-}\right] \xrightarrow{\text{microorganisms}} CO_2 + 4H^+ + 4\left[Fe(CN)_6^{4-}\right] \quad (10.2)$$

$$Fe(CN)_6^{4-} \xrightarrow{E_{app}} Fe(CN)_6^{3-} + 1e^- \quad (10.3)$$

This approach enables a high surplus of electron donor, in contrast to the limited dissolved oxygen, and thus higher rates and complete degradation can be achieved in shorter times [85–88]. This approach has resulted in the development of MICREDOX® by Lincoln Ventures Ltd and is intended to be manufactured and commercialized by Aotea Bio Ltd, New Zealand.

10.5.1.1.3 Other analytes determined through catabolic pathways
Many other biosensors based on intact cells have been reported that aim at more specific parameters [89–93]. However, the approach of using one particular type of cell for a target analyte may be less competitive than the use of non-selective biological components combined with the analysis of response patterns. This approach results in an increased number of detectable targets without concomitant increase of biological components. Using this principle, several toxicants have been detected with a luminescent bacterial biosensor through the analysis of luminescence fingerprints [94]. The system is calibrated to seven known pollutants and, with the use of a computer algorithm, the comparison of unknown samples with the calibration response-curves permits the prediction of the pollutant at 95% confidence level [94]. Microalgae of genus *Klebsormidium* and *Chlorella* have been used for the construction of an algae-based biochip enabling the specific discrimination of formaldehyde and methanol in a mixture of VOCs at concentrations below the relevant values to human health (10 ppb–10 ppm) [95]. The response is based on the specific rate of decrease of quantum efficiency of electron transport of photosystem II. Overall, this technology seems particularly attractive in cases where target analytes impose serious restrictions on other more specific catalytic or affinity approaches, or require complicated biorecognition pathways.

Alternatively and in contrast with the previous approaches targeting toxicity or specific parameters, the transduction of microbial metabolism can also be applied for the detection of bacteria. However, this sensing principle shows poor selectivity and long response times. This type of biosensor can only be used for well-defined samples because of the possible presence of enzymes from sources other than the tackled species [96].

10.5.1.2 Expression of selective enzymes in engineered bacteria: biosensors for specific pollutants
As previously mentioned, many different species of soil- and water-borne bacteria have evolved a variety of mechanisms that allow them to survive and grow in non-benign environments. Although the mechanism for achieving such resistance may be much more intriguing than a biocatalytic pathway, as in the case of metals [97,98], some enzymes are known to play a key role in this mechanism thus permitting the detection of specific analytes. For instance, haloalkane dehalogenases and haloacid dehalogenases are involved in the

cleavage of carbon–halogen bonds for the aerobic degradation of haloaliphatics. The transcription of the genes encoding these and other enzymes that participate in the degradation pathways is regulated, meaning that expression of these catabolic enzymes is induced by the presence of the compounds that they degrade. This type of transcriptional control is achieved by the interaction of transcriptional activator protein with specific gene promoters. These proteins contain a DNA binding domain, a transcriptional activation domain, and a recognition domain [99]. Organic molecules bind to the recognition domain and induce a conformational change in the protein that results in enhanced interaction of the DNA binding domain with specific promoter sequences. This interaction triggers the formation of a complex on the promoter DNA consisting of the transcriptional activator, σ-factor 54, integration host factor, and the RNA polymerase. The complex efficiently initiates transcription of the genes encoding the catabolic enzymes that lie directly downstream of the promoter sequence [100]. A schematic representation of the biosensing systems based on promoter recognition and gene expression is depicted in Fig. 10.3.

The specific promoter gene, including the analyte-binding regulatory protein gene and its upstream promoter, are usually fused to a structural gene reporter inside a host cell (e.g., $E.\ coli$). The reporter gene produces a protein, in most cases an enzyme, enabling easy transduction of the promoter activation. These reporter genes include those that code for bioluminescent proteins (i.e., lux or luc from bacterial or firefly luciferase, respectively), green fluorescent protein (GFP) from marine jellyfish ($Aequorea\ victoria$) and β-galactosidase. These reporter genes enable optical and electrochemical transduction according to Eqs. (10.1), (10.4), and (10.5), respectively:

$$\text{Luciferin} + \text{ATP-Mg}^{2+} + \text{O}_2 \xrightarrow{\text{Firefly luciferase}} \text{Oxyluciferin} + \text{AMP} + \text{PPi} + h\nu(560\text{ nm}) \tag{10.4}$$

$$p\text{-aminophenyl-}\beta\text{-galactopyranoside} \xrightarrow{\beta\text{-galactosidase}} p\text{-aminophenol} + \text{galactopyranoside} \tag{10.5}$$

The discovery of transcriptional activators and their corresponding promoter sequences—offering the exquisite specificity of this regulation—together with the fusion of the well-known and well-behaved reporter genes, has made possible the development of bioengineered bacteria for specific pollutants. Since the pioneering work of King at the beginning of the 1990s using the promoter of a gene ($nahG$) coding for naphthalene degrading enzyme [101], many other set of genes have been identified.

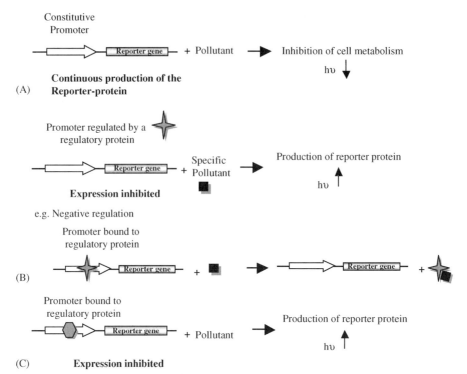

Fig. 10.3. Different analytical approaches based on bacterial systems. (a) In the presence of toxic substances, the protein production is suppressed as a consequence of cell death or inhibition of its metabolism. (b) The reporter gene is controlled by a promoter that will respond to a specific pollutant. In this case, the pollutant binds specifically to the regulatory protein being released from the promoter and allowing the transcription of the reporter gene. (c) The reporter gene is expressed under the control of a promoter that is responsive to a wide range of conditions that are potentially harmful to the cells, thus the presence of a very broad spectrum of pollutants induces the expression of the reporter gene.

Examples include genes coding for heavy metals [102–112], naphthalene [101,113,114], fluorene [115], halo-organic [116–123] and aliphatic compounds [124], biphenyls [125], and BTX (benzene, toluene and xylene) [100, 126,127]. This bioanalytical approach has been reviewed by Daunert et al. [128] and also by Kohler and Belkin [119,129], focussing on environmental analysis.

Commercialization of this approach has been realized for the measurement of the bioavailability of heavy metals (Biomet®). This test uses a relevant soil bacterium (*Ralstonia metallidurans*) that when adequately engineered produces light when a specific heavy metal enters the cell.

The fact that engineered microorganisms are used for the control and monitoring of specific pollutants has not stopped the US Environmental Protection Agency from approving, in October 1996, the release into contaminated soil ecosystems of *Pseudomonas fluorescens* HK44 for the monitoring of PAH bioavailability during a long-term bioremediation process [130].

Legal acceptance, though rather restricted up to now, commercially available tests, and intensive research reinforce this strategy for environmental monitoring, but technically the use of these biological constructs is not problem free. In fact, few applications of these sensing devices to real sample analysis are reported [131] and procedures to enumerate HK44 cell numbers in soil under long-term (over 2 years) field conditions are being developed [132]. The major limitation derives from the confounding variability in the luminescent responses due to different growth stage and cell numbers. This phenomenon of cell-to-cell variability, even in cells with the same genetic background, has been described [133,134] and, a possible explanation for this "genetic noise" has been proposed [135]. In addition, and as for any catalytic process, pH and T are also critical parameters for data reliability. For instance, the minimum parameter deviation required to alter lux gene expression during naphthalene catabolism in *P. putida* RB1353 was a change in temperature of 1°C, a change in pH of 0.2, or a change in initial cell number of one order of magnitude [114]. This calls for very strict control of experimental conditions to attain reproducible results. To overcome this problem and that of interfering sample matrix effects (e.g., presence of toxic compounds or cell viability), a second reporter protein whose signal is not coupled to the specific regulatory protein of the target analyte has been successfully integrated into the plasmid [136]. As a result, an internal correction signal is obtained, counting for the non-specific stimuli, and thus enabling signal correction of the specific first reporter. It seems that the technical difficulties hampering the operational performance of these cells for field applications can be resolved by elegant strategies that recall the well-known standard addition method in instrumental analysis. The idea of introducing a second reporter has also been adapted for the design of an optical imaging fibre-based cell array biosensor that can accommodate a variety of sensing strains. The rationale here is to set a second reporter to visualize the cell locations in the microwell array before beginning the assay [134].

Irrespective of these advances, the limited understanding of the genetic systems that control bacterial responses to pollutant species and the specificity of the interaction between the transcription activator protein and its chemical effector intrinsically restrict the construction and use of these bacterial biosensors [121]. This limitation has been tackled by the modification of the regulatory protein that detects the presence of the pollutant. Some of these proteins are well known and characterized and thus their chemical sensing domain can be altered through mutagenesis, resulting in a new potential for the detection of chemical species that are not effectors of the wild protein [121].

Other promoters of genes belonging to global regulatory circuits, which are thus induced by a broad spectrum of chemicals and environmental conditions, are used with a similar strategy as described above for the specific promoters. The lack of selectivity in these cell constructs is exploited for the monitoring of toxicity, including genotoxicity [94,137–142].

10.5.2 Simple catalytic reaction: isolated enzymes

Since the seminal work of Clark and Lyons in the early 1960s on the first enzyme electrode for glucose, a multitude of enzyme sensors have been described in the literature exploiting different means of transduction and covering almost the whole range of commercially available enzymes. Although there is still a gap between this research and the delivery of a product to the market, some of this research has substantially contributed to a better understanding of biocatalysts. For instance, the tailoring of electrical contact between redox enzymes and electrodes has resulted in multidisciplinary efforts attempting to comprehend the kinetics of redox reactions associated with biocatalysts at the electrode/solution interface.

Basically, the use of single or several enzymes for environmental applications falls into three main categories that dictate the main characteristics and drawbacks of the corresponding enzyme sensors. Accordingly, this is the scheme used in the following description that is further classified by relevant environmental analytes. Obviously, a particular compound can be detected by different catalytic approaches (e.g., phosphates, pesticides) and, vice versa, a single enzyme can be used for detecting different analytes through different mechanisms (e.g., tyrosinase).

10.5.2.1 Catalytic transformation of the analyte
10.5.2.1.1 General considerations
In principle, this is the best—or at least the simplest—choice for the design of catalytic sensors. They can be directly designed as reagentless devices and no regeneration steps are needed if continuous monitoring is intended. A priori, the catalytic conversion of the analyte is inherently more selective than any other approach. Optimally, the sensor response should be limited by mass

transfer and thus its performance will be independent of any kinetic step with the concomitant benefits of stability and sensitivity. This is not a simple task for the design of enzyme sensors because different processes control the response of the sensor. In an immobilized enzyme film, the analyte diffuses first to the film interface and then partitions into the film. There it forms the Michaelis complex with the active form of the enzyme with an equilibrium constant ($K = k_1/k_{-1}$, see Table 10.2). This complex dissociates at a rate constant k_2, which, as previously mentioned, is usually termed the turnover number of the enzyme, and forms a product with the concomitant regeneration of the biocatalyst in its active form.

The above sequence is by far the simplest reaction scheme and only valid for a certain class of enzymes (e.g., hydrolases) with no need of co-substrates or cofactors. As will be described further, oxidoreductases are very useful in environmental applications but they impose additional requirements on redox cofactors, such as flavin nucleotides (FAD or FMN), heme, pyrroloquinoline, iron–sulfur clusters or pyridine nucleotides as in the case of NAD^+/NADH dependent dehydrogenases. In the last mentioned case, the fact that the cofactor is not permanently bound within the enzyme means that the design of reagentless configurations has to be really precise and alternatives to solve this obstacle have to be specifically addressed. Suitable electron donors, for instance oxygen in the case of oxidases, or non-natural redox partners, and optimally the electrode as in the case of direct electron transfer in peroxidases (see Chapters 2 and 3), are responsible for the regeneration of the active redox form of the enzyme. This is an additional redox reaction within the catalytic pathway that has to be considered because its reaction rate may become, in many instances, the limiting step of the enzyme sensor.

In amperometric enzyme electrodes, if the rate of direct electron transfer is slow compared with the turnover number of the enzyme or does not occur at all, enhanced sensitivity is most commonly accomplished by the design of efficient redox mediators, i.e., transducing chemistry that enables communication between the enzyme's redox centres and the electrode's surface. In other words, the choice of this chemistry, sometimes other redox proteins, is of vital importance in order to achieve the required sensitivity. A scheme of the general reactions in an enzyme-based sensor is depicted in Fig. 10.4. The methodology for the calculation of all these rate constants has been debated and widely described in the literature for amperometric enzyme sensors. Smyth and co-workers [143,144] have used the ratio between the turnover number and the Michaelis constant, i.e., the specificity constant, to assess the efficiency of organic phase enzyme electrodes. However, this constant neglects the interplay of mass transport, charge propagation across the enzyme film,

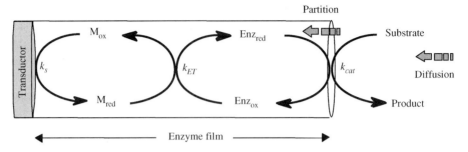

Fig. 10.4. Schematic representation of the different processes that control the response of an enzyme sensor. The scheme exemplifies an amperometric sensor with an oxidoreductase electrochemically connected to the electrode through a redox mediator.

and non-linear Michaelis–Menten kinetics [145,146]. In general, analytical solutions are not attainable, and two approaches have been adopted to cope with this problem: digital simulation and approximate analytical solutions for a number of limiting cases [147]. Most of the published work in this area deals with systems in which either the enzyme, or both enzyme and mediator, are free diffusing species, with less attention to the case where enzyme and mediator are coimmobilized in the adsorbed film. The approach based on the search of approximate analytical solutions was pioneered by Savéant and co-workers [148,149] and Albery and Hillman [150]. They have reported a kinetic analysis of the reaction between a free diffusing substrate and a mediator bound to a polymer-modified electrode. Later incorporation of non-linear enzyme kinetics results in a more involved mathematical problem, and approximate analytical solutions have been derived for amperometric biosensors by considering separately the cases of saturated and unsaturated kinetics [146,151]. Bartlett and co-workers [152,153] later developed a rather extensive approach, focusing on the diffusion–reaction coupling within a tridimensional biocatalytic film. They have obtained approximate analytical solutions for a number of limiting cases, whereas numerical solutions were obtained for intermediate situations.

The kinetic behaviour of electrochemical biosensors is most commonly characterized using the dependence of the steady-state amperometric current on the substrate concentration. This type of analysis has some limitations because it does not allow for a decoupling of the enzyme–mediator and enzyme–substrate reaction rates. The additional information required to complete the kinetic analysis can be extracted either from the potential dependence of the steady-state catalytic current or from the shift of the half-wave potential with substrate concentration [154]. Savéant and co-workers [155] have presented the theoretical analysis of an electrocatalytic system including an immobilized redox enzyme connected to the electrode by a freely diffusing mediator. They discussed the shape of the transient and steady-state voltammetric waves in terms of the kinetic competition between substrate and co-substrate and also considered the influence of substrate diffusion on the current. A further implementation of this electrode characterization includes a completely reagentless enzyme electrode that incorporates a redox polymer into the film and also taking into account substrate depletion outside the bioelectrocatalytic film [154]. This work permits an easy and direct study of the overall processes that determine the response of the enzyme electrode. The accurate and direct estimation of the active enzyme concentration in the film still remains, in practice, the most difficult problem for a real determination of the transfer rate between the enzyme and the transducing chemistry (k_{ET}).

A classical way to circumvent poor communication of the enzyme with the chosen transducing chemistry and to increase stability has been to increase the thickness of the bioelectrocatalytic film. The concomitant diffusion problems within the film may ruin this approach. A rational approach to deliver efficient enzyme electrodes demands: (i) the strategic immobilization in highly permeable films or thin films where mass transport is favoured (e.g., self-assembled monolayers) (see Chapter 1), (ii) the choice of suitable electrode material (see Chapter 7), (iii) the design of efficient redox chemistry (if necessary for electrical communication), (iv) the tuning of the intrinsic characteristics of the enzymes by protein engineering, and, if necessary, (v) the concurrence of other approaches (e.g., photochemical) for alternative means of enzyme regeneration. In the following section, the state of the art of enzyme-based biosensors for environmental applications will be described.

10.5.2.1.2 Phenolic compounds
The detection of phenolic compounds through their catalytic conversion by tyrosinase or laccases appears to be the main application within this category of enzyme sensors. Tyrosinase (EC 1.14.18.1) or polyphenol oxidase is a copper-containing monooxygenase enzyme that catalyses the oxidation of phenolic substrates to the corresponding quinone species, with the concomitant reduction of oxygen to water. Its active site contains a coupled binuclear copper complex. Similarly, laccase (EC 1.10.3.2) is also a copper oxygenase but containing four copper atoms [156]. In common with tyrosinase, laccases are characterized by low substrate selectivity although their catalytic competence varies widely depending on the origin of the enzyme. Simple diphenols such as hydroquinone and catechols are good substrates for the majority of laccases as

they are able to catalyse the oxidation of other substituted polyphenols, aromatic amines, or benzenethiols although unable, in contrast to tyrosinase, to oxidize monophenols at significant reaction rates. Accordingly, laccases and tyrosines share identical possibilities for transduction.

There are two main reasons for the wide use of these enzymes in environmental analysis. First, they show a broad selectivity enabling the determination of a "phenol index" as contemplated by legislation to assess the quality of environmental waters. It is not surprising then that since the 1970s there have been efforts to replace or complement the time consuming and laborious 4-aminoantipyrine method with enzymatic sensors. Secondly, the catalytic reaction opens up different means of transduction. In most cases the produced quinones are the transducing parent species because they readily polymerize, forming coloured compounds [157], or because they can be electrochemically reduced either directly in a large variety of electrode materials [158–163] or indirectly with the incorporation of more reversible redox couples [164–166]. The use of surface-modified electrodes with redox mediators circumvents the poor electrochemical reversibility of the quinone reduction at non-acidic pH compatible with the oxygenase activity, which is the main cause of the poor stability and the high limits of detection of many configurations. If reversibility of the system, both chemical and electrochemical, is achieved, then the amplification cycle renders limits of detection for phenol in the order of 10 nM (0.9 μg l^{-1}) [166–169]. Figure 10.5 shows the catalytic and electrochemical reactions in mediated tyrosinase amperometric sensors.

Mechanistic aspects of the action of tyrosinase and the usual transduction schemes have been summarized on several occasions [166,170–173]. In short, this copper enzyme possesses two activities, mono- and di-phenolase. Due to the predominant presence of the mono phenolase inactive form (met-form), the enzyme is inherently inefficient for the catalysis of these monophenol derivatives. However, in the presence of a diphenol, the catalytic cycle is activated to produce quinones and the scheme results in an efficient biorecognition cascade. This activation is achieved more efficiently when combined with electrochemical detection through the reduction of the produced quinones [166], as illustrated in Fig. 10.5. Consequently, a change in the rate-limiting step can be observed through kinetic to diffusion controlled sensors with a concomitant increase in stability and sensitivity, as depicted in Fig. 10.6.

The relative sensitivity of tyrosinase-based sensors for different phenolic compounds very much depends on the electrode material, the immobilization matrix or procedure, whether a transducing chemistry is incorporated, and

Non-affinity sensing technology

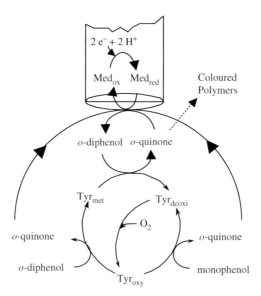

Fig. 10.5. Catalytic and electrochemical reactions in mediated tyrosinase amperometric sensors.

the nature of the solvent. For instance, Bisphenol-A and 17β-estradiol, rather unusual compounds for catalytic transformation, have been detected with boron-doped diamond electrodes coated with tyrosinase [174]. Accordingly, very different trends in sensitivities have been reported when a large repertoire of phenols is tested [158,159,165,175,176]. Nevertheless, this has not limited the use of these sensors for the determination of the global phenol content in different types of waters and to correlate the "biosensor index" with the phenol index obtained with the official colorimetric method [165,177]. The reported results show the validity of this approach for the delivery of fast, cost-effective and reliable information on phenol content. In general, this biosensor index results in lower values than those obtained by chromatographic methods, but it is a useful approach to monitor, for instance, the efficiency of a WWTP or bioremediation processes [178]. Table 10.5 illustrates the evaluation of the degradation of these pollutants depending on the different stage of the water treatment in two WWTPs in the area of Catalonia (Spain). Although the results obtained with tyrosinase sensors are lower than those obtained with solid-phase extraction and LC–MS for these samples, as reported by Barceló and co-workers [179], the efficiency of the treatment plant for these compounds can be soundly assessed by the biosensor response.

A multitude of publications describe enzyme sensors for the detection of phenolic compounds, reflecting the large variety of possible configurations.

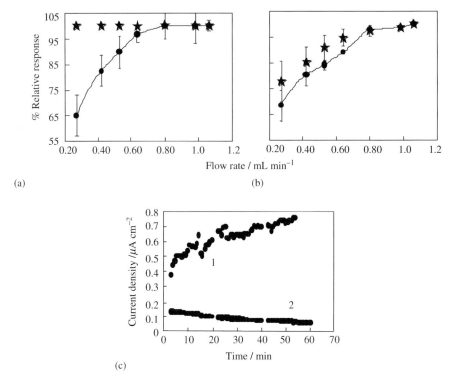

Fig. 10.6. Change of the rate limiting step in tyrosinase carbon paste electrodes. (a) and (b) show the flow rate dependence of steady-state currents for unmediated (a) and mediated (b) enzyme electrodes. Applied potential: -0.05 V vs. Ag/AgCl for (★) 1 μM phenol and (●) 0.5 mM ferrocyanide as control of diffusion limited response. Error bars show the standard deviation of six electrodes. The change from kinetic to diffusional control in mediated electrodes results in higher sensitivity and improved operational stability as demonstrated in (c) where (1) represents the FIA response of mediated electrodes and (2) unmediated with consecutive 20 μl injections of 10 μM phenol in a thin-layer cell. Applied potential: -0.05 V vs. Ag/AgCl, mobile phase: 0.25 M phosphate buffer pH 6.0 and flow rate 0.7 ml min^{-1}.

This, in addition to the less common detection techniques for these oxygenases (e.g., ENFET [180] or calorimetric [181]), the search for new laccases, the degree of maturity that these sensors have generally achieved and, mainly, the search for better or fit-for-purpose performance requires a focus on the following features:

(a) Extended selectivity. Despite the broad selectivity that different laccases and tyrosinases show, other catalytic schemes have also been investigated. These include the coupling of tyrosinase and laccase [182, 183], tyrosinase and peroxidase [184,185] and the direct exploitation of

TABLE 10.5

Measured phenolic content normalized to catechol concentration in six different wastewater samples

Wastewater treatment plant	Effluent	Primary settlement	Raw influent	Phenolic removal
Llagosta	3.7 ppb	308 ppb	451 ppb	99.2%
Igualada	30 ppb	1.2 ppm	2.3 ppm	98.7%

Data obtained by the standard addition method with a tyrosinase sensor in flow injection analysis. Experimental conditions: 0.1 M phosphate buffer pH 6.0, injection volume: 500 μl, flow rate: 0.5 ml min^{-1}. ppm (mg l^{-1}), ppb (μg l^{-1}).

peroxidase for the detection of chlorophenols [186]. Wild type laccase from *Trametes versicolor* has demonstrated activity for a large variety of *para* and *ortho* substituted monophenols and diphenols [187] while recombinant laccases have also demonstrated activity for 1-naphtol, benzidine (4,4′-diaminobiphenyl) and *o*-anisidine (2-methoxyaniline) [188]. The combination of tyrosinase and laccases broadens the spectrum of biocatalysed phenolic compounds. In contrast to this approach, the use of both enzymes in separate electrodes but operating simultaneously has also been explored for the discrimination of binary phenol mixtures by multivariate analysis [189]. The promiscuity of peroxidases in accepting electron donors in the presence of hydrogen peroxide has resulted in enzyme electrodes for the detection of phenols that usually are not catalysed by the monooxygenases, such as chlorinated derivatives and, mainly 2-amino-4-chlorophenol [190–192]. A less explored combination of enzymes, tyrosinase and peroxidase, has also been reported in pursuing signal amplification as compared with monoenzyme electrodes [193–195]. Irrespective of the enhanced sensitivity, any configuration based on peroxidase ruins the reagentless design of these sensors. This has been circumvented by a cooperative mechanism between both enzymes that electrogenerates hydrogen peroxide from dissolved oxygen at -50 mV vs. Ag/AgCl [184]. The limit of detection achieved with carbon screen-printed electrodes modified with both enzymes and entrapped in poly(carbamoylsulfonate) hydrogels is 2.5 nM (0.2 μg l^{-1}) for phenol. High amplification cycles have also been attained by combining tyrosinase or laccase with either quinoprotein glucose dehydrogenase or oligosaccharide dehydrogenase on a Clark electrode [196,197].

(b) Wang et al. have implemented tyrosinase-based sensors into a compact and rugged design for on-site and in situ determination of phenol pollution. The sensor consists of a three-electrode assembly (in a

PVC housing tube) connected through a three-pin environmentally sealed rubber connector, to a long shielded cable [198]. This remote biosensor has also been tested with tyrosinase and laccase for broader selectivity of the device [199]. Further implementation of this concept has resulted in the so-called "lab-on-a-cable" based on scaling down an electrochemical flow system to a cable platform [200]. In contrast to the previous realization and aiming at a more versatile system for pollutant detection, the cable platform consists of a microdialysis sampler, reservoirs (for reagents and waste solutions), a miniaturized pump, a fluidic network and an amperometric flow detector. The clear advantage of this design, compared with the remote sensor, for direct and in situ application is its robustness against ionic strength, pH and hydrodynamic fluctuations [200].

(c) Other transducing schemes, different to those described with monooxygenases or peroxidases, have also been tested making use of dehydrogenases or oxidases. For instance, cellobiose dehydrogenase or quinoprotein glucose dehydrogenase, both non-dependent NAD^+/NADH dehydrogenases generate an amplification cycle upon direct electrochemical oxidation of the phenolic compounds. In the presence of the corresponding substrates, cellobiose or glucose, respectively, the dehydrogenase then reduces the electrochemically-generated quinone, resulting in a closed electrochemical and catalytic cycle [201]. These sensors have been applied in WWTPs for on-site monitoring of the efficiency of the water treatment [179]. In contrast to other sensors based on tyrosinase, these devices show higher responses than those reported from chromatographic analysis but the wider selectivity of these devices and the ability to be specifically inhibited open up a route for their use in toxicity evaluation [179]. Similarly, the use of this transducing scheme is also applicable to glucose oxidase because the flavin adenine dinucleotide (FAD) can be continuously reoxidized by the transient quinone. The combination of this scheme with a chemical treatment with bis(trifluoroacetoxy)iodobenzene enables the detection in the nM range of all chlorophenols, including pentachlorophenol [202]. The use of bilirubin oxidase in the presence of NADH has also demonstrated its ability to detect pentachlorophenol, measuring the fluorescence rate in a simple assay format in solution [203].

10.5.2.1.3 Inorganic nitrogen: nitrate, nitrite, and ammonium ions
These ions are relevant parameters in environmental analysis because they are involved in important biological processes such as nitrate reduction, nitrite reduction and denitrification. Nitrate and nitrite reductases play a key

role in the assimilatory and dissimilatory pathway of nitrate. In the assimilatory pathway, nitrogen is assimilated into the cell as fully reduced nitrogen (ammonium, NH_4^+). When other nitrogen sources are limited, nitrate becomes the predominant nitrogen source in terrestrial, aquatic and marine systems. The first step of biological nitrate assimilation is the reduction of nitrate to nitrite, which is catalysed by multidomain redox enzymes, nitrate reductases (NR), as follows:

$$NO_3^- + 2e^- + 2H^+ \xrightarrow{\text{Nitrate reductase}} NO_2^- + H_2O \tag{10.6}$$

This two-electron reduction uses NADH or NADPH as electron donor depending on the particular nitrate reductase (EC 1.7.1.1 or EC 1.7.1.2, respectively). Nitrite is then reduced to ammonium in a six-electron process that involves the transfer of three electron pairs from NAD(P)H:

$$NO_2^- + 6e^- + 8H^+ \xrightarrow{\text{Nitrite reductase}} NH_4^+ + 2H_2O \tag{10.7}$$

In contrast, many gram-negative bacteria contain a nitrate reductase (EC 1.7.99.4 and/or 1.9.6.1) that also reduces nitrate to nitrite although under anaerobic conditions. The dissimilatory nitrite reduction leading to denitrification encompasses then the reduction of nitrite to nitric oxide by dissimilatory nitrite reductases (NiR, EC 1.7.2.1) that, in combination with nitric oxide reductases (NOR) and nitrous oxide reductases (N_2OR), transform nitrite into nitrogen:

$$NO_2^- \rightarrow NO \rightarrow N_2O \rightarrow N_2 \tag{10.8}$$

Nitrate levels have increased substantially over the last 50 years because of the increased use of nitrogen-based fertilizers. The environmental relevance of nitrates has been addressed by a European Nitrates Directive (1991/676/EC) for the establishment of good farming practice and action programmes in nitrate vulnerable zones, including provisions to improve soil condition. In addition, the presence of acidifying contaminants in soil favours the leaching out of nutrients with subsequent loss of soil fertility and possible eutrophication problems in water and an excess of nitrates in drinking water. The implementation of this directive requires the capacity of laboratories for measuring nitrates as well as setting up monitoring programmes, over at least 1 year, to obtain nitrate levels and the degree of eutrophication of surface and ground waters. Nitrate and ammonium are also contemplated in the Water Framework Directive (2000/60/EC) as core chemical parameters to be monitored in all selected groundwater bodies. Nitrate concentration is monitored in municipal water supplies worldwide and in foodstuffs to prevent exposure of populations to harmful levels. The US Public Health Service has established allowable limits for nitrate and nitrite in potable water at 10 and 0.6 ppm (as N), respectively. In Europe, Directive 98/83/EC has defined these maximum values for nitrate, nitrite and ammonium ions as 50, 0.1, and 0.5 mg l^{-1} (ppm), respectively. In air, the recommended threshold limit value for human exposure to ammonia is 25 ppm (18 mg m^{-3}).

In spite of the fact that there is a multitude of analytical methods for these analytes, including ion-selective electrodes for nitrate and ammonium, attempts to produce catalytic sensors have been extensively reported in the literature since the pioneering work of Guilbault in the mid-1970s [204–206]. The lack of selectivity of nitrate ion-selective electrodes is responsible for most of these efforts. In the case of amperometric detection there has been a great interest in the electrochemical communication between reductases and electrode surfaces that, if sufficiently efficient, will result in sensitive devices. Different nitrate and nitrite reductases have been isolated, characterized [207–209] and used for the design of catalytic amperometric sensors for nitrate and nitrite. The NAD(P)H-dependent nitrate reductases can communicate directly with bipyridinium radical cations, which substitute the native NADH cofactor as the electron donor [210]. Consequently, the immobilization of nitrate reductase in a bipyridinium-functionalized polythiophene film results in electrically contacted enzyme-electrodes that enable the electrochemical detection of NO_3^-. The dissimilatory nitrate reductase from *E. coli* is dependent on the cytochrome *c* cofactor. This enzyme can be communicated through microperoxidase-11 resulting in electrobiocatalysed reduction of nitrate to nitrite [211–213].

Although conducting polymers have demonstrated direct electrochemical communication with nitrate reductases, the incorporation of electron relay groups within the polymer matrix provides a more efficient pathway for electron hopping between the enzyme and the electrode surface. Several artificial electron donors can shuttle electrons to the oxidized form of nitrate reductase with methyl viologen being the best choice due to its very negative redox potential [214–216]. The electron transfer reactions can be represented as follows:

$$MV^{2+} + e^- \Leftrightarrow MV^{+\bullet} \tag{10.9}$$

$$NR_{red} + NO_3^- \rightarrow NR_{ox} + NO_2^- \tag{10.10}$$

$$NR_{ox} + MV^{+\bullet} \rightarrow NR_{red} + MV^{2+} \tag{10.11}$$

where MV^{2+} and $MV^{+\bullet}$ represent the oxidized methyl viologen and the viologen radical cation, respectively; NR_{red} and NR_{ox} are the reduced and oxidized forms of the enzyme. Initally, the reduction of methyl viologen was only achieved by either chemical or photochemical means [214,217].

Truly amperometric enzyme sensors for nitrate and nitrite were first described in 1994 [216,218]. Efficient electrocatalytic currents were demonstrated by using electropolymerization of highly purified nitrate reductase (not commercially available) within an amphiphilic polypyrrole-viologen derivative. This results in reagentless nitrate sensors with a limit of detection in the sub-ppb range although the residual enzyme activity in the electrode was very low compared with the initial enzyme loading [218]. This transducing chemistry has been further applied to commercially available nitrate reductase from *Aspergillus niger*, evaluating the loss of enzyme activity during the immobilization process [219]. Nitrate reductase from *Pseudomonas stutzeri* has been investigated for its electrochemical communication with 19 different redox mediators [220]. The best mediators were those of the phenothiazine group although the limit of detection remains above the maximum admissible concentration for environmental applications. Nitrate reductases are generally adsorbed on the electrode surface or co-entrapped with the redox polymers. Other strategies have also been reported, including either the organization of an electrically-contacted electrode through the crosslinking of an affinity complex generated between a de novo four-helix bundle hemoprotein and nitrate reductase on the electrode support [221], or the construction of self-assembled multilayers of nitrate reductase intercalated with cationic viologen-functionalized polyvinylpyridinium [222].

Nitrite reductase has been mediated with 1-methoxy-N-methylphenazonium (1-methoxy-NMP$^+$) [216] or 1-methoxy-N-methylphenazonium methyl sulfate (1-methoxy-PMS) [223] or phenosafranin [224], all resulting in limits of detection enabling direct detection of nitrite although few environmental applications are reported. The enzyme has also been connected to gold electrodes in the presence of *Achromobacter cycloclastes* apopseudoazurin [225]. The electrochemical characterization of a fusion protein consisting of nitrite reductase and a maltose-binding domain has been described on electrodes modified with electropolymerized films coated with the fusion protein and poly(benzyl viologen) [226]. The fact that the counterpart nitrite reductase without the binding domain did not show electrocatalytic currents under similar immobilization conditions suggests certain recognition of the electropolymerized film for the maltose-binding domain of the catalytic protein.

The required working potential (≈ -700 mV and ≈ -200 mV for nitrate and nitrite detection, respectively), which requires sample purging, is not the only problem confronting these sensors. When developed with commercially available reductases, the low purity of the enzyme preparations becomes the limiting step for successful results. It is not surprising then that whole cell sensors for nitrate and nitrite have also been intensively investigated [227]. An elegant alternative to overcome this problem has been described, based on the use of ultrathin-film composite membrane (UTFCM) [228]. An UTFCM consists of an ultrathin-film of the desired chemically-selective material (typically a polymer) that has been coated across the surface of a microporous support membrane. The UTFCM separates the sample solution from an aqueous internal sensing solution that contains the enzyme and the redox mediator. The ultrathin film prevents leaching of these materials from the internal solution but allows the analyte species to permeate from the sample into the sensing solution. This concept has been applied to nitrate sensors with UTFCM permeable to anions and no leaching of cations, as required in the case of nitrate reductase mediated with methyl viologen [229]. Films of 1 μm thickness are necessary for preventing the leaching of MV^{2+} while allowing the permeation of anions. The response time remains in the order of 10 s with no interference of nitrite (an electrochemical interference of nitrite with the viologen radical cation has been reported [230]) and a limit of detection much lower than the maximum allowable nitrate concentration in waters. Although this limit of detection remains one order of magnitude higher than that reported by Cosnier et al. [218], the application of this sensor for the analysis of well and ditch waters has been demonstrated with no need of enzyme purification. After 3 and 6 days at room temperature, the sensor retained 94 and 85% of its initial sensitivity, respectively. Other nitrate sensor based on nitrate reductase from corn seedlings and using Nafion to ensure a reservoir of methyl viologen has proven its applicability for the analysis of nitrates in fertilizers and drinking water [231].

Other electrochemical and optical transduction techniques have also been used for the construction of nitrate and nitrite enzyme sensors. They all suffer from the necessity of regenerating the oxidation state of the reductases upon catalytic conversion of the analyte thus complicating the design of reagentless devices. An enzyme field effect transistor for nitrate (ENFET) based on the control of the gate potential through the activation of reductive enzyme transformations has been described by Willner and co-workers [232]. N,N'-Dialkyl-4,4'-bipyridinium relay units are covalently linked to the gate surface, enabling the formation of a complex between nitrate reductase and the bipyridinium unit, which is further crosslinked with glutaraldehyde to yield a stable relay–enzyme layer on the gate interface. In the presence of dithionite as an electron donor, the biocatalysed reduction of NO_3^- to NO_2^- is stimulated, the potential of the gate being controlled by the concentration of nitrate. The limit of detection of this ENFET is very much dependent on the chain length that ties the bipyridinium units to the gate interface and, if conveniently designed, remains at least one order of magnitude below the admissible concentration for nitrate. Irrespective of the analytical properties of this sensor, the major merit of this work is the demonstration that the gate potential can be controlled by a redox label through the activation of a biocatalytic reduction [232]. Optical biosensors for nitrate [233] and nitrite [234] using the corresponding reductases have been described, both based on the measurement of the spectroscopic changes that occur during the reductase catalytic cycle. These systems achieve limits of detection for nitrate and nitrite that are perfectly comparable with those reported with amperometry and thus are good prospects for environmental applications. A colorimetric nitrate enzyme test kit is commercially available from The Nitrate Elimination Co., Inc (http://www.nitrate.com). It uses highly purified nitrate reductase from corn leaves (EC 1.7.1.1) enabling the discrimination of water samples between safe and non-admissible levels according to the intensity of the produced pink colour. The assay is also amenable to automated analysers [235].

The catalytic detection of ammonium ions has not been extensively investigated in contrast with the large variety of potentiometric and amperometric chemical sensors and optical sensors described in the literature [236]. Similarly, the detection of ammonia in air has merited different approaches in the field of chemical sensors. Screen-printed electrodes modified with Meldola's Blue and covered with a polycarbonate membrane constitute the basis of the catalytic detection of NH_4^+. The measurement is based on the electrocatalytic reduction of NADH upon addition of glutamate dehydrogenase to a stirred solution containing NADH, 2-oxoglutarate and ammonium ions. The rate of current decrease (nA s^{-1}), measured at 50 mV, correlates to the concentration of ammonium ions in the sample. Recoveries of ammonium ions in spiked pond and tap waters at the level of 0.1 ppm are close to 100%, which demonstrates the feasibility of this assay for the detection of ammonium ions in waters [237].

10.5.2.1.4 Phosphate ions
Like nitrates, phosphates are included in the indicative list of main pollutants (Water Framework Directive, 2000/60/EC) because of their contribution to eutrophication. Phosphorus is an essential nutrient for all plants and in modern agriculture this element has to be supplied to the crops as fertilizer. Wind erosion, surface runoff and leaching constitute the main pathways for transport of phosphorus from terrestrial to aquatic ecosystems. The process is accelerated by agriculture, animal husbandry and anthropogenic discharges.

The result will often be the eutrophication of lakes and coastal waters [238]. The area of phosphate sensors, including both chemical and biosensors, has been reviewed in 1998 by Engblom [239]. Phosphate biosensors have been reviewed by Amine and Palleschi [240].

Enzyme electrodes for the detection of phosphate fall into two categories: either phosphate acts as inhibitor or phosphate is the co-substrate of a catalytic system. In principle, both cases are problematic for the design of reagentless systems. In the second approach, the easiest scheme includes the use of pyruvate oxidase that catalyses the oxidation of pyruvate in the presence of phosphate and oxygen with the production of acetylphosphate, carbon dioxide and hydrogen peroxide [241]. The enzyme requires the presence of FAD, thiamine pyrophosphate (TPP), and magnesium ions for full activity. A limit of detection of 74 nM ($\approx$ 10 ppb) with this oxidase has been reported by chemiluminescence detection of H_2O_2 through luminol, p-iodophenol and peroxidase in an FIA system with the enzymes immobilized in small reactors [242]. This overcomes the lack of stability of the pyruvate oxidase, enabling the detection of phosphate for 2 weeks. This system has been further improved [243] for automatic operation and extensively tested in river waters of Japan [244]. It was found that pretreatment of the samples with activated carbon was necessary to remove interferences, thereby obtaining an acceptable correlation ($r = 0.94$) of the sensor response with the modified molybdenum blue method [245]. Amperometric transduction of pyruvate oxidase has been reported using platinized glassy carbon electrodes modified with a conductive redox polymer obtained by potentiostatic polymerization of Os(bpy)$_2$pyCl, pyrrole and thiophene [246]. This permits the electrochemical communication of the bound FAD enzyme redox centre when adsorbed on top of the electrode, resulting in a limit of quantification of 2 ppm for phosphate at anodic potentials (350–500 mV) and in the absence of oxygen. These sensors are stable with a loss of sensitivity of 12% after 10 days and they have been successfully applied to the measurement of pyruvate in biological fluids whilst application for phosphate in environmental samples is still to be demonstrated [247]. A recombinant pyruvate oxidase has been used, enabling the detection of 1.5 ppb (15 nM) phosphate in a reagentless configuration [248].

Other approaches include an initial enzyme, usually a phosphorylase, that uses phosphate as co-substrate giving a product that is the substrate for a second enzyme, generally an oxidase. The use of phosphorylase A was first proposed by Guilbault and Cserfalvi in 1976 using phosphoglucomutase and NAD(P)$^+$/NAD(P)H dependent glucose-6-phosphate dehydrogenase for amperometric detection through NAD(P)H [249]. This preliminary system did not work with immobilized enzymes and starch in an attempt to design a reagentless biocatalytic sensor. A similar reagentless configuration has been reported with electrocatalytic oxidation of NADH at 200 mV with Os(1,10-phenanthroline-5,6-dione)$_2$Cl$_2$ as redox mediator [250]. A scheme of the most widely reported enzymatic cascades is given in Fig. 10.7. The incorporation of a phosphatase enables the recycling of phosphate, giving rise to an amplification scheme with improved limits of detection. The sensor response is generally obtained with a Clark-type electrode or by direct oxidation of the produced hydrogen peroxide. Table 10.6 summarizes the main features of these sensors. Although remarkably low limits of detection are obtained, environmental applications are not generally reported with the remarkable exception of the group of Karube in Japan.

The detection scheme based on maltose phosphorylase, mutarotase and glucose oxidase with chemiluminescence detection and implemented into an FIA system has been thoroughly investigated on freshwater samples [264]. Interferences were eliminated with a cation-exchange resin as the only sample pretreatment and the results of 31 river and pond waters correlate fairly well with the molybdenum assay ($r = 0.96$) [264]. A laboratory based multichannel FIA equipment has the enzymes immobilized in a packed bed

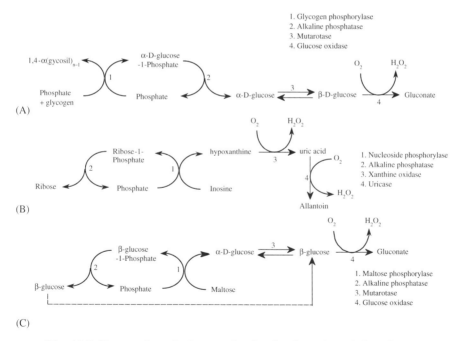

Fig. 10.7. Reported catalytic cascades for the detection of phosphates.

TABLE 10.6
Characteristics of selected enzyme sensors for phosphate determination

Catalytic scheme[a]	Detection	Limit of detection	Stability	Application	Reagents	Reference
A	O_2 depletion	10 μM	–	–	None	[251]
A[b]	NADH 200 mV	1 mM	Storage 21 days	–	None	[250]
B[c]	O_2 depletion	0.3 mM	>70 assays/30 days	–	Inosine	[252]
B[c]	H_2O_2 and uric acid 700 mV	1.25 μM	160 assays/30% loss after 3 weeks	Food products and blood plasma	Inosine	[253]
B[c]	H_2O_2 and uric acid 700 mV	0.1 μM	30% loss after 20 h of operation	–	Inosine	[254]
B[c]	TCNQ 100 mV	0.22 mM	–	–	Inosine	[255]
B[d]	O_2	25 nM	>8 days; 300 assays	–	Inosine	[256]
B	Peroxidase fluorescence	50 nM	–	–	Inosine, Amplex Red	[257]
B[e]	F sensitive semiconductor		Storage 21 days	–	Inosine, p-fluoroaniline	[258]
C	H_2O_2 600 mV	10 nM			Maltose	[259]
C[f]	H_2O_2 600 mV	0.1 μM			Maltose	[260]
C[g]	H_2O_2 600 mV	1 μM	30% loss after 42 days/>8 h of operation	–	Maltose	[261]
C[g]	Peroxidase chemiluminescence	10 nM	>2 weeks	River waters	Maltose	[244]
Pyruvate oxidase	Amperometric, not detailed	0.2 μM	50% loss after 7 days of operation	–	Pyruvate, FAD, TPP, Mg^{2+}	[262]
Pyruvate oxidase		20 μM	12% loss after 10 days	–		[246]

Analyte	Enzyme	Transduction	Detection limit	Lifetime	Samples	Interferences	Ref
Pyruvate	Pyruvate oxidase[h]	H_2O_2, 800 mV	4.8 μM	50% loss after 6 months/33% loss after 220 h of operation	Serum samples	Pyruvate	[263]
Pyruvate	Pyruvate oxidase	Chemiluminescence ARP-luminol	0.16 μM	37 days	Spiked samples	Pyruvate, FAD, TPP, Mg^{2+}, luminol, ARP	[243]
Pyruvate	Pyruvate oxidase	Peroxidase chemiluminescence	96 nM	>2 weeks			[244]

TCNQ: 7,7,8,8-tetracyanoquinodimethane, TPP: tyramine pyrophosphate, ARP: peroxidase from *Arthromyces ramosus*, FAD: flavin adenine dinucleotide.
[a]See Fig. 10.7.
[b]No recycling with phosphatase and NAD/NADH dependent dehydrogenase.
[c]Neither phosphatase nor uricase was included.
[d]Catalase is included to prevent enzyme inactivation.
[e]Neither phosphatase nor mutarotase was included. Peroxidase is included.
[f]Neither phosphatase nor mutarotase was included.
[g]No phosphatase is included.
[h]Recombinant pyruvate oxidase from *Lactobacillus plantarum* containing the cofactors FAD, TPP, and Mg^{2+} bound to the enzyme active centre.

reactor and the transducing peroxidase reaction is incorporated in solution to a mixing cell.

With the exception of the sensor based on pyruvate oxidase, most of the catalytic systems for phosphate include an enzymatic cascade that in some cases has as many as five enzymes. This complicates the manufacture of these biosensors as well its basic design because the enzyme ratio and the absolute amount of enzymes will determine their analytical properties. In principle, the kinetic behaviour of this multienzymatic system could be modelled with adequate software. In any case, any intermediate and/or transducing reaction should not limit the overall conversion and particular attention should be given to the specificity constants (k_{cat}/K_m) of each individual enzyme. However, the possible lack of information about these constants, the equilibrium constant of each chemical reaction, in addition to possible diffusional restrictions and different relative inactivation of each enzyme upon immobilization, may complicate this approach. An empirical approach for this optimization has been undertaken by Cosnier et al. in the system based on maltose phosphorylase [261] whilst the optimum ratio for the couple nucleoside phosphorylase and xanthine oxidase has been studied by D'Urso and Coulet [254]. In this case, it was found that the optimum ratio of enzymes in solution prior to immobilization was five, i.e., a higher surplus of the phosphorylase enzyme was needed for maximum response of the biosensor. This contrasts with the optimum ratio found when these enzymes are assayed in homogeneous systems where the optimum ratio follows the trend of the relative specificity constants [257]. In other words, a higher surplus of xanthine oxidase vs. phosphorylase is required to keep the ratio k_{cat}/K_m of the first reaction lower than that of the oxidase, according to their respective kinetic constants.

10.5.2.1.5 Organophosphorous pesticides and chemical warfare agents
A notable exception to the most common approach for the catalytic detection of pesticides (i.e., enzyme inhibition) is the use of organophosphorous hydrolase (OPH, aryltriphosphate dialkylphosphohydrolase from *Pseudomonas diminuta*, EC 3.1.8.1). This enzyme permits the direct reagentless detection of organophosphate compounds in one-step detection with no need of previous incubation and reactivation/regeneration processes. Additionally, OPH shows a narrower selectivity than acetyl cholinesterase with no activity for carbamates. In principle, enzyme sensors based on OPH are less prone to false positives caused by any external factor diminishing enzyme activity as in the case of inhibition sensors. OPH catalyses the hydrolysis of a wide range of organophosphorous pesticides (e.g., parathion, paraoxon, diazinon, dursban,

methyl parathion, coumaphos and acephate) and chemical warfare agents (e.g., soman, sarin, VX, and tabun) [265,266]. Some of the chemical structures of these pesticides and warfare agents are depicted in Figs. 10.8 and 10.9, respectively.

As shown in Fig. 10.10, this enzyme catalyses the hydrolysis of organophosphorous compounds with the production of two protons as a result of the cleavage of the O–P, P–F, P–S or P–CN bonds and an alcohol, which in many cases is chromophoric and/or electroactive. This opens up various transduction schemes that show different analytical properties depending on the immobilization procedure. Muchaldani et al. have reviewed the different transduction schemes of this enzyme [267]. The influence of different chemistries and immobilization methods on the loading, activity and stability of this enzyme has also been described [268]. The authors found that the use of a sol–gel process to form a structurally stable porous matrix followed by functionalization with 3-aminopropyltrimethoxysilane and protein immobilization with glutaraldehyde resulted in the best procedure in terms of enzyme loading, activity and stability [268]. Some of the catalytic systems based on OPH (EC 3.1.8.1) and reported in the literature are summarized in Table 10.7. The combination of a dual transduction scheme, amperometric and a silicon-based-pH-sensitive electrolyte-insulator-semiconductor, has successfully increased the spectrum of detectable compounds. Both enzyme sensors were coupled in an FIA system enabling the simultaneous detection of all organophosphorous compounds by the potentiometric sensor while the amperometric only responded with a very high sensitivity to organophosphorous pesticides [278]. The amperometric enzyme electrode has been implemented to a remote configuration for the in situ monitoring of surveillance and detoxification activities [279]. Amperometric detection of OPH results in limits of detection in the nM range, offering the advantage of removing interferences of oxidizable compounds by a previous pulse during 2 s at 700 mV vs. Ag/AgCl and enabling direct detection of parathion in river water [277]. Alternatively, Mulchandani et al. have used a recombinant *E. coli* with surface-expressed OPH activity and immobilized cells coupled to optical [280], potentiometric [281] or amperometric transduction [267]. The analytical properties of these cell sensors do not significantly differ from those obtained with the counterpart enzyme sensors although slightly higher stability could be envisaged.

Another hydrolytic enzyme (OPAA, organophosphorous acid anhydrolase, EC 3.1.8.2) has been used to unambiguously discriminate between chemical warfare agents (fluorine-containing organophosphorous compounds) and the chemically-related pesticides [282,283]. The ability of this enzyme, derived

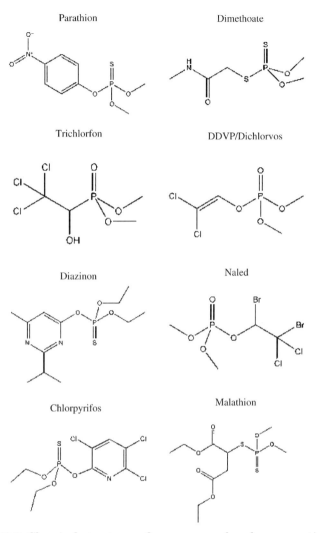

Fig. 10.8. Chemical structures of some organophosphorous pesticides.

from an extremophile (see Section 10.6.1), to distinguish between fluorine compounds and pesticides has been demonstrated with a full characterization of the kinetic constants for a wide range of compounds [282]. Maximum activity of the OPAA is attained in the presence of 50 mM NH_4Cl, pH 8.5, with 0.1 mM $MnCl_2$, and 100 mM NaCl. The transduction of OPAA with pH-FET results in similar detectability for diisopropyl fluorophosphate (DFP) as obtained with the OPH from *Pseudomonas* and using a pF sensitive

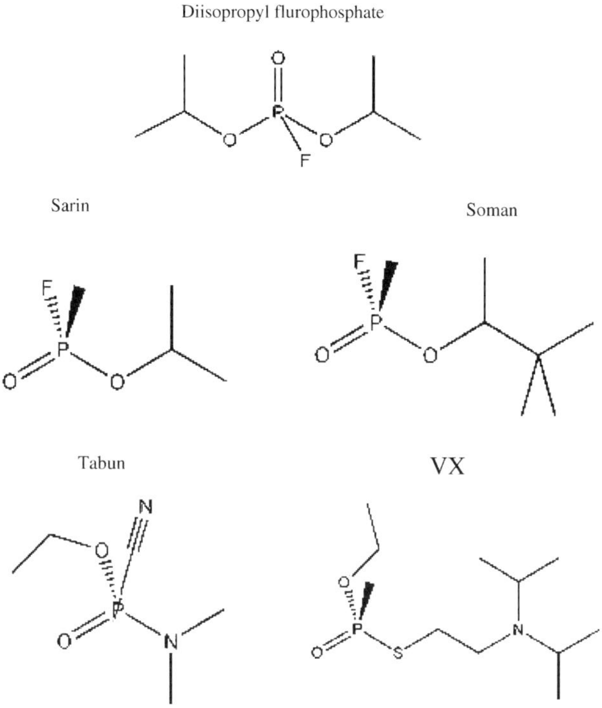

Non-affinity sensing technology

Fig. 10.9. Chemical structures of nerve agents.

sensor [272]. The transduction of OPAA with a pH-FET and a pF sensor has also been reported with limit of detection for DFP of 25 μM (≈ 4.7 ppm) and 20 μM (≈ 3.7 ppm), respectively [282].

10.5.2.1.6 Heavy metals

Like pesticides, heavy metals are traditionally tested by enzyme inhibition or modulation of catalytic activity. Several metalloproteins behave as chelators for specific metals with no known catalytic reactions. Such heavy metal binding sites exist in metallothioneins and in various protein elements of bacterial heavy metal mechanisms and have been exploited for specific detection through affinity events. Nevertheless and as previously mentioned, bacterial resistance mechanisms can also be linked to catalytic pathways. For instance, cytochromes c3 and hydrogenases from sulfate and sulfur reducing bacteria [284,285] are well suited for bioremediation purposes because they can reduce various metals such as U(V) and Cr(VI) [286,287]. Cytochrome c3 has been reported to catalyse Cr(VI) and U(VI) reduction in *Desulfovibrio vulgaris* [288,289], suggesting

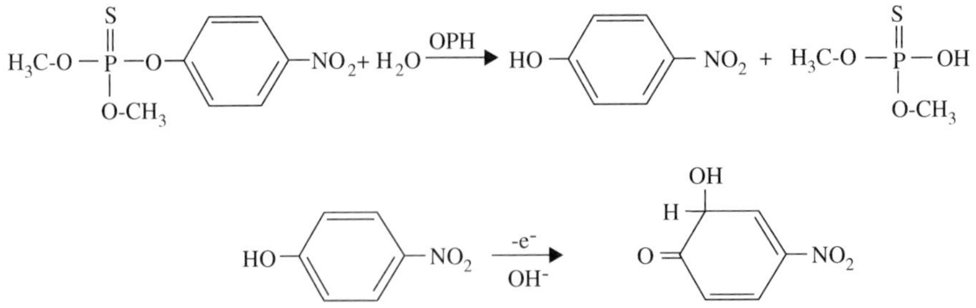

Fig. 10.10. Reaction scheme of the organophosphorus hydrolase catalysed reaction of parathion and the electrochemical detection of the product: *p*-nitrophenol.

that this cytochrome may function as both U(VI) and Cr(VI) reductase. Cytochromes *c*3 are multiheme redox proteins with very negative redox potentials that have also been isolated and characterized from *Desulfovibrio norvegicum* [287]. A preliminary study for the demonstration of cytochromes *c*3 as catalytic proteins for the detection of Cr(VI) has been reported [290]. The authors describe different means of protein immobilization. Adsorption on glassy carbon electrodes resulted in a very narrow dynamic range for Cr(VI) (0.2–0.8 mg l^{-1}), which was extended by incorporation of a dialysis membrane (0.20–6.84 mg l^{-1}). Cathodic currents are obtained at -570 mV vs. SCE enabling direct regeneration of the reduced cyt *c*3 in deoxygenated solutions [290].

10.5.2.1.7 Other analytes measured in aqueous phases
The "drug metabolizing enzymes" include a diverse group of biocatalysts responsible for the metabolism of a vast variety of xenobiotic compounds. From an enzymological point of view, they are most noted for their broad substrate selectivity. Some members of the cytochrome P450 (P450 or CYP) and flavin monooxygenase (FMO) families are known to metabolize more than 50 structurally diverse compounds. Both type of enzymes are categorized as Phase I in drug metabolism because they catalyse oxidative reactions like hydroxylations and demethylations. Phase II enzymes catalyse conjugative and/or other oxidative reactions that usually result in the formation of soluble compounds that are more readily excreted (e.g., glutathione transferases, sulfotransferases, etc.) [291].

Oxidation of organic molecules by P450s is rather complex, but the overall reaction can be simplified as:

$$RH + O_2 + NADPH + H^+ \rightarrow ROH + H_2O + NADP^+ \tag{10.12}$$

TABLE 10.7
Enzyme sensors for direct detection of organophosphate pesticides based on organophosphate hydrolase (EC 3.1.8.1)

Transduction	Immobilization	LOD	Response time	Stability	Reference
pH-FET	Covalent link to silica microspheres dip-coated on the FET gate	1 μM paraoxon	<10 s	2.5–5 months	[269]
Optical (400 nm)	Cross-linking with BSA and GA in a Byodine nylon membrane	10 μM paraoxon and parathion	10 min steady response	>10 days	[270]
pH-capacitive EIS	Adsorption with GA	≈2 μM paraoxon	3 min (steady response)	10% loss after 2 months operation	[271]
F selective electrodes (batch mode)	Covalent link to silica beads with GA	25 μM diisopropyl fluorophosphate	—	57% loss after 6 months	[272]
pH-glass electrode	Cross-linking with BSA and GA and dialysis membrane	2 μM paraoxon, ethyl parathion, methyl parathion; 2 μm diazinon	2–10 min	>1 month 20 assays	[273]
Amperometric/850 mV	Adsorption with Nafion on screen-printed electrodes	90 nM paraoxon, 70 nM methyl parathion	10 s	—	[274]
Amperometric 850 mV	Co-immobilization with BSA to a nylon net attached to carbon paste electrodes	15 nM parathion, 20 nM paraoxon	—	—	[275]
Amperometric 900 mV	Glass bead reactor (630 μl) with GA	20 nM paraoxon, 20 nM methyl parathion	FIA 30 h^{-1}	>1 month	[276]

continued

TABLE 10.7 (Continuation)

Transduction	Immobilization	LOD	Response time	Stability	Reference
Amperometric 850 mV	Adsorption with PEI on screen-printed electrodes	10 nM parathion	FIA < 1 min	–	[277]
Dual Amperometric	Cross-linking with GA on gold with SAM	≈70 nM paraoxon	FIA 20 h^{-1}	Several hours in continuous operation	[278]
Potentiometric (capacitive field-effect sensor)	Cross-linking with GA	2 μM paraoxon, 6 μM dichlorvos			

GA: glutaraldehyde, BSA: bovine serum albumin, FIA: flow injection analysis. Potentials vs. Ag/AgCl.

The electron pathway of this reaction depends on the type of P450. Type I can be found in mitochondrial and bacteria and consists of an iron–sulfur protein, NADPH or NADH-dependent reductase (a FAD-type flavoprotein) and P450 enzyme. The electrons are supplied by NADH and transferred first to putidaredoxin reductase. They are then transferred to the heme of P450 through the iron–sulfur protein (putidaredoxin) which is the direct redox partner of P450. Type II are microsomal P450 systems and comprise NADPH:P450 reductase (FAD- and FMN-containing flavoprotein), and P450 enzyme [292]. The large number of identified P450s have a specific nomenclature that is currently available at http://drnelson.utmem.edu/cytochromeP450.html [293,294].

These enzymes are good candidates for environmental applications targeting a broad spectrum of compounds although the difficulties in their isolation, reconstitution of activity and stabilization have hampered most of their utility. Their application in the pharmaceutical industry for both drug discovery and development has resulted in methods for heterologous expression of recombinant P450s using *E. coli* and baculovirus. Biosensing applications are also limited by the inherent difficulties in their transduction. While in some reactions such as *o*-deethylation of 7-ethoxycoumarin, hydroxylation of aniline, N-demethylation of aminopyrine, or hydroxylation of hexobarbital, progress of the reaction can be easily followed by absorbance or fluorescence [295], in many other reactions, e.g., epoxidation of aromatic compounds or hydroxylation of aliphatics, this becomes more difficult. Most of the latest reactions are particularly relevant in environmental applications because they involve alkanes, and polyaromatic and polychlorinated hydrocarbons.

Generally accepted and widely applicable methods to measure the enzymatic activity of various P450s are the measurement of oxygen consumption by an oxygen electrode [296,297] and competition assay between the standard substrate and various other chemicals as putative substrates [298]. Aiming at more specific and practical approaches, there has been intensive research on non-enzymatic, particularly electrochemical, electron transfer pathways on P450 that while retaining its catalytic properties permit easy and direct transduction for sensing systems [299–302]. Many of these studies have been undertaken with P450$_{cam}$, which catalyses the hydroxylation of camphor through a two electron process in a Type I pathway. The 3D structure of a cytochrome P450$_{cam}$ is shown in Fig. 10.11. Most of this research aims at the elimination of the needed cofactors and electron mediating proteins and at improved substrate specificity [303–305]. For instance, oxidation of styrene has been achieved using both wild type and the Y96F

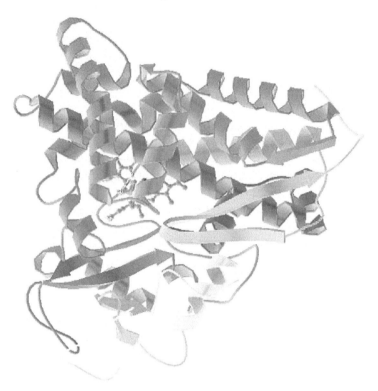

Fig. 10.11. 3D view of a cytochrome P450$_{cam}$ (Camphor Monooxygenase) (EC 1.14.15.1) with bound camphor. This structure has been obtained from the Protein Data Bank and displayed with MolScript program.

mutated form of the enzyme [306]; naphthalene, pyrene, butane, and propane have also been reported as substrates for P450$_{cam}$ [307,308]. Furthermore, the dehalogenation reactions of hexachloroethane, carbon tetrachloride, and other polyhalomethanes by P450$_{cam}$ and some of its mutants have also been reported [309,310]. Direct electrochemistry of myoglobin and cytochrome P450$_{cam}$ in alternate layer-by-layer films with DNA and other polyions has shown enzyme-like catalysis of styrene epoxidation in the presence of oxygen [311].

A notable exception to the voltammetric transduction of P450 has been reported by Hara et al. [312]. The authors used a fusion enzyme between rat P450 1A1 monooxygenase and yeast NADPH-cytochrome P450 oxidoreductase expressed in *Saccharomyces cerevisiae* that is immobilized in an agarose layer on the gate of an ISFET for the monitoring of the *o*-deethylation of 7-ethoxycoumarin to 7-hydroxycoumarin. As depicted in Eq. (10.12), the consumption of protons can be followed through a potential shift. Because the enzyme is entrapped in a gel matrix and immobilized on the gate, and a non-modified gate is also connected in the circuit, the difference in the voltage between the modified enzyme and the control can be correlated to the concentration of substrate [312]. This system seems applicable to other compounds, e.g., 3,4-dichlorophenol and 2,6-dichlorophenol, but further studies are required to elucidate the response mechanisms.

There is still a substantial gap between this research and analytical performance and more research is still required for a better understanding of the complex catalytic P450 systems (see Section 10.6). Presently, screening of planar PAH, PCB, PCCDs/PCDFs is carried out using the induction of P450 in a cell-based microtitre assay [313,314]. If xenobiotics are present, the catalytic activity increases with concomitant emission of light through a gene reporter. It would be desirable for on-site and faster screening to implement this bioassay—which already saves time and cost compared with classical techniques—in a more efficient biosensing approach.

10.5.2.1.8 Other analytes directly measured in gas phase
Formaldehyde is a ubiquitous airborne pollutant and exits as a dilute vapour at room temperature. Annual global production of formaldehyde is some 10 million tons, half of this volume being utilized for the production of phenol, urea and melamine–formaldehyde resins, which are widely used in the manufacturing of building plates, plywood and lacquer materials [315]. Its widespread use in the chemical industry poses risk exposure to a large proportion of workers [316]. Indoor contamination, in the order of 0.1–0.6 ppm, has also been reported as a consequence of its use in building materials in new homes [317] while the highest concentrations have been measured in cigarette smoke with concentrations of 30–40 ppm [318]. A formaldehyde vapour concentration of 0.5 ppm is the German threshold limit in a workplace environment [319]. The first biosensor was described by Guilbault and was based on formaldehyde dehydrogenase (FDH from *Candida boidinii*, EC 1.2.1.1) that, in contrast to the extensively used FDH in aqueous phase (from *P. putida*, EC 1.2.1.46), uses glutathione as well as NAD^+ as cofactor [320]. A piezoelectric crystal was modified with an enzyme solution containing both cofactors and allowed to dry. On exposure to formaldehyde, a change in frequency was correlated to the amount of analyte in the 10 ppb–10 ppm range at 50% relative humidity, this parameter being of crucial importance for the performance of the sensor [321]. Because low humidity is detrimental to enzyme activity, Turner et al. immobilized FDH from *P. putida* in a reverse micelle medium which did not dehydrate

significantly over time and allowed direct-gas phase monitoring [322]. The partition of formaldehyde vapour into the reverse micelle results in the catalytic formation of NADH that is directly oxidized at 800 mV vs. Ag/AgCl using screen-printed electrodes, according to the following reaction:

$$HCHO + NAD^+ + H_2O \xrightarrow{FDH} HCOOH + NADH + H^+ \qquad (10.13)$$

Under these conditions a dynamic range between 1 ppb and 1.3 ppm can be obtained [322]. Hämmerle et al. have described an electrochemical cell containing the enzyme electrode that can directly operate in the gas phase with a limit of detection of 0.3 ppm and no loss of response during 7 h of operation [323]. Formaldehyde dehydrogenase has also been incorporated into a field-effect transistor for the measurement of catalytically-produced protons and the ENFET incorporated to a sampling system that strips the aldehyde into an aqueous phase [324].

In spite of the fact that formaldehyde is traditionally considered a major threat in air pollution, a new risk factor associated with formaldehyde has been revealed. Some advanced technologies of potable water pretreatment include the ozonation process, which not only disinfects the water, but also removes iron and manganese and eliminates unpleasant flavours and odours. However, as a result of the ozonation of some organic substances, for example humus traces, several kinds of toxic and mutagenic aldehydes are formed particularly acetaldehyde, benzaldehyde, propanal, glyoxal, methylglyoxal and, mostly, formaldehyde [325]. The use of formaldehyde dehydrogenase for the quantification of aqueous formaldehyde has been extensively exploited in the literature making use of amperometric [75,76], potentiometric [326] or optical [92] transduction.

As mentioned above, the need to use the soluble cofactor NAD^+ complicates the use of this enzyme and many other dehydrogenases in a reagentless configuration. The use of large molecular weight NAD^+ derivatives (i.e., NAD^+-dextran or PEG-NAD^+) has been intensively investigated [327–329] including catalytic recycling of the reduced PEG-NADH by salicylate hydrolase (EC 1.14.13.1) [330] (see Section 10.6). Alternatively, nanocrystalline CdS self-assembled on gold surfaces and combined with formaldehyde dehydrogenase results in a biological–inorganic hybrid, enabling the catalytic oxidation of the aldehyde in the absence of the natural cofactor [331]. The semiconductor nanoparticles have to be included in reverse micelles to prevent enzyme inactivation. Despite the rather poor values of the reported FDH photocurrents in this work, the replacement of NAD^+ by the semiconductor photoactive material acting as a synthetic charge transfer system opens up new routes for the redesign of photoelectrochemical sensing systems [332]. In such systems, the electrons and holes generated in the conduction and valence bands of the semiconductor, respectively, can be used by the enzyme for the reduction and oxidation of the substrate. Other approaches to avoid the soluble cofactor for the determination of formaldehyde include the use of highly or partially purified alcohol oxidase from *Hansenula polymorpha* (EC 1.1.3.13) immobilised on a pH field-effect transistor [333]. The authors describe in detail the unusual catalytic activity of the oxidase ENFET that does not respond to methanol, demonstrating a working range for aqueous formaldehyde of $g\,l^{-1}$ (5–200 mM) [333].

Unfortunately, on-site demonstrations of these devices is not generally considered. An exception to this rather academic trend has been recently reported. A field test in a factory using formic acid-containing glue for glulam products has been performed with an enzyme sensor that is selective for formic acid [334,335]. In parallel to the measurements with the biosensor, a well-defined reference method was used for sampling and analysing formic acid. It was found that the biosensor worked satisfactorily in this environment, achieving detectability at the ppt level (0.03 ppb, mg m^{-3}) [335].

10.5.2.2 Enzyme sensors based on the inhibition of the catalytic activity
10.5.2.2.1 General considerations
Many hazardous substances with a negative impact on human health and the environment are known to inhibit the catalytic activity of enzymes. Among them, pesticides and heavy metals are well-known inhibitors of some hydrolases and oxidoreductases. Accordingly, this has been widely exploited for the design of enzyme sensors. Several recent reviews have covered this issue, focussing on a particular class of enzymes [21,336], or the detection technique [92,337–339]. In principle, the choice of a particular enzyme would be determined by the specific mode of action of a targeted pollutant. Cholinesterases from insects would be the enzymes of choice for the detection of insecticides such as organophosphate and carbamate pesticides [340]. Dithiocarbamate fungicides, which act on living systems by inhibition of aldehyde dehydrogenases involved in amino acid biosynthesis, are preferentially measured through the redox enzyme [341], whilst other enzymes from the same class, e.g., tyrosinase and peroxidases, are better options for the detection of herbicides that disrupt photosynthetic reactions. However, the reality of farming practices, which are not always openly declared, and the use of binary mixtures of pesticides, plus the fact that the modes of action of more than 300 pesticides used in agriculture are not completely understood

challenge not only this bioanalytical approach but also any other means of analysis. Heavy metals may be an easier target for a bioanalytical technique because their toxicity in biological systems is better understood, including the metal form that is particularly more toxic or predominant in a specific environment [342] and the source of pollution, the manufacturing industry, is also well recognized. The problem here is the broad spectrum of selectivity that heavy metals present when inhibiting catalytic activities, including both classes of enzymes, and thus poor selectivity is likely when this approach is used for a single metal species.

In contrast to other bioanalytical approaches that offer a better and well-defined chemical selectivity (e.g., immunodetection), enzyme sensors based on inhibition are inherently and a priori less selective. The potential revenue of most of these—in many cases somewhat formal or academic—developments should be contemplated from a wider perspective than merely environmental implementation. First, the mechanisms of enzyme inhibition and its translation into any analytical configuration are nowadays straightforward. Initially, enzyme assays based on catalytic inhibition were restricted to either reversible inhibitors or to one single assay as in the case of irreversible mechanisms [343]. The discovery that powerful nucleophilic compounds such as 2-pyridinealdoxime methiodide (2-PAM) or [1,1'-trimethylenebis-4-(hydroxyiminomethy)-pyridimium bromide] (TMB-4) could reactivate the catalytic activity of phosphorylated cholinesterases widened the scope of inhibition assays. Intensive research on cholinesterases has enabled the preparation of (bio)engineered variants of these and related enzymes (see Section 10.6). Part of this work was realized after in depth knowledge of the kinetic characteristics of the inhibition pathway giving indirect information on the chemical nature and the conformational characteristics of the active site(s) of the enzyme(s). As a result, this research has provided useful bioanalytical tools where the combination of several enzymes with adequate data processing permits the discrimination of a single pesticide in a complex mixture [344] or the progress of a global pesticide parameter can be monitored and used as an early-warning system. The maximum European level for total pesticide concentration is $0.5 \ \mu g \ l^{-1}$ whilst $0.1 \ \mu g \ l^{-1}$ is the threshold concentration for a single compound (Directive 98/83/EC).

The mode of action of inhibitors can be divided into two groups: reversible or irreversible. A detailed explanation can be found in textbooks of enzyme kinetics and assays [345]. The critical aspects and kinetics of both mechanisms are summarized in Table 10.8. In brief, in all cases the presence of a substrate is a prerequisite and it has to be added at a concentration that does not limit the reaction rate of the non-inhibited reaction which can be kept as a reference value. The different reactions occurring within the different mechanisms of inhibition are summarized in Fig. 10.12. It should be kept in mind that these generic kinetic formalisms refer to equilibrium reactions in homogeneous phases whilst the signal of an enzyme sensor implies heterogeneous phases and transient and/or local equilibria that may not be traced to either Michaelis–Menten (rapid equilibrium) or Briggs–Haldane (stationary state) models. In addition, for any biocatalytic sensor, the nature of the heterogeneous phase where the enzyme is retained or immobilized may also affect the catalytic activity, the sensitivity towards a particular inhibitor or the mechanism of inhibition [347–349].

Unlike the previously described group of enzyme sensors that rely on catalytic transformation and that optimally should work under diffusional control, these sensors are obliged to perform under kinetic control. Basically, the measurement of the inhibition can be obtained through steady-state (Fig. 10.13a) or kinetic (Fig. 10.13b) measurements. Figure 10.13 exemplifies the main features of both approaches. The kinetic measurement is obviously preferred from a practical point of view [350] whilst better sensitivity can be achieved when inhibition is calculated from two separate measurements [351].

In search of both sensitivity and practicability of the assay, the reaction pathway and the particular mechanism of inhibition may preliminarily determine the particular detection principle. Without exception, most common detection techniques have been implemented with this biorecognition principle, employing varied assay formats using either batch mode or any modality of flow injection analysis.

10.5.2.2.2 Organophosphorous and carbamate pesticides
These compounds represent a large percentage of currently used pesticides including herbicides, insecticides, and fungicides with harmful effects on other animals and humans. Figure 10.14 shows some of the chemical structures of carbamate pesticides, organophosphorus pesticides are depicted in Fig. 10.8. Their biological effect is related to the ability to inhibit cholinesterase, which plays a key role in the cholinergic transmission by catalysing the rapid hydrolysis of the neurotransmitter acetylcholine into acetate and choline. Acetylcholinesterase effectively terminates the chemical impulse at rates that are similar to a diffusion-controlled process, allowing a rapid and repetitive response [352]. Accordingly, the analysis of organophosphorous and carbamate pesticides based on their ability to inhibit acetylcholinesterase activity gives an indication of the toxicity of a particular

TABLE 10.8

Kinetics and characteristics of enzyme's inhibition mechanisms (after Ref. [346])

	Reversible	Irreversible
Reactions	$E + I \overset{K_i}{\rightleftharpoons} EI$	$E + I \rightleftharpoons [EI] \overset{k_i}{\longrightarrow} EI'$
Equations	Dissociation constant K_i $$v' = \frac{V'_{max}[S]}{K'_m + [S]}$$ • Competitive: $V'_{max} = V_{max}$ $$K'_m = K_m\left(1 + \frac{[I]}{K_i}\right)$$ • Non-competitive: $K'_m = K_m$ $$V'_{max} = \frac{V_{max}}{1 + \frac{[I]}{K_i}}$$ • Uncompetitive: $$K'_m = \frac{K_m}{1 + \frac{[I]}{K_i}}$$ $$V'_{max} = \frac{V_{max}}{1 + \frac{[I]}{K_i}}$$	Bimolecular rate constant k_i $\ln(v/v') = k_i[I]t$, $[I] \gg [E]$, $t < 10$–15 min
Binding to the inhibitor	Non-covalent	Covalent
Need of preincubation	No	Yes

sample. Both type of pesticides differ in their mode of action: carbamate pesticides are competitive, reversible inhibitors whilst organophosphorous pesticides are irreversible inhibitors forming a highly stable complex with the enzyme, resulting in a comparably higher toxicity of these compounds (see Section 10.6). Thus, the inhibition of organophosphorous compounds is not dependent on the presence of the substrate and in this respect it may be viewed as non-competitive, as follows:

$$\text{Enz–Ser–OH} + [R(R')O]_2\text{–PO–H} \overset{K_i}{\rightarrow} \text{Enz–Ser–O–PO–[OR(R')]}_2 \qquad (10.14)$$

The enzyme complex is impervious to a further nucleophilic attack of water (see Eq. (10.15)), and unless very strong nucleophilic agents are

Non-affinity sensing technology

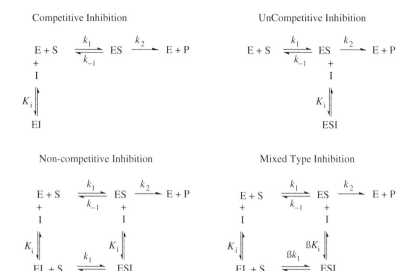

Fig. 10.12. Illustration of the reactions occurring in the different mechanisms of reversible enzyme inhibition.

added (vide supra) no reversibility of this inhibition is observed. In contrast, the compctitive mechanism of carbamate compounds, as a truly reversible reaction, becomes non-detectable when the substrate concentration is high. However, as previously mentioned, all these features can be tuned by

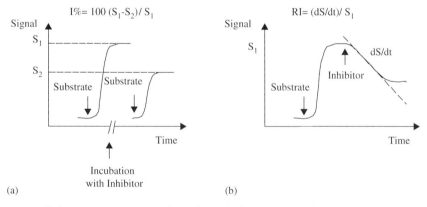

Fig. 10.13. Schematic representation of the different methodologies for estimation of the enzyme inhibition. (a) Indirect determination: the biosensor response is measured before and after its incubation with the inhibitor for a given time. (b) Direct determination: the inhibitor is added during the steady-state reaction.

485

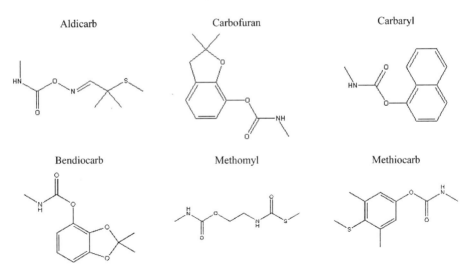

Fig. 10.14. Chemical structures of carbamates all having in common: $CH_3-NH-COO-R$.

adjusting the concentration of the substrate and the enzyme immobilization procedure [349].

Acetyl-cholinesterases (AChE, EC 3.1.1.7, also called true cholinesterase) and acylcholine acylhydrolases (or pseudocholinesterases including butyrylcholinesterase, BuChE, EC 3.1.1.8) are commercially available enzymes from different biological sources that catalyse the hydrolysis of acetyl- or butyrylcholine into choline and acetate or butyrate, respectively, according to the following reaction:

$$R-COO(CH_2)_2N^+(CH_3)_3 + H_2O \xrightarrow{\text{Cholinesterase}} R-COOH + HO(CH_2)N^+(CH_3)$$
(10.15)

Unless synthetic substrates are used, biosensors based on a single enzyme are required to detect the changes in pH, directly or indirectly, by potentiometric or optical indicators, respectively. This is the only parameter that can be easily correlated to the concentration of the pesticide. A portable fibre-optic biosensor for the determination of carbaryl and dichlorvos, using acylcholine acylhydrolase from horse serum immobilized in a Durapore membrane and bromcresol purple as indicator in a sol–gel layer, has been described and applied to different waters with evaluation of interferences through recovery experiments [353]. Pesticide recoveries ranged between 85 and 110%, which can be considered good—the acceptable recovery range for drinking water varies between 70 and 130% [354]. Groundwater samples had the lowest recoveries among all samples that were directly analysed with no pretreatment. After the system is calibrated, the biosensing probe is incubated with the sample for 10 min and after addition of substrate the rate of absorbance decrease is measured for 2 min, enabling the calculation of the percentage of inhibition. The operational stability differs depending on the pesticide, with 25 cycles of analysis for carbaryl and 15 cycles in the case of dichlorvos, which shows higher toxicity as it corresponds to its irreversible mechanism. The limits of detection, calculated as the concentration causing 10% of inhibition, were 108 and 5.2 $\mu g\, l^{-1}$ for carbaryl and dichlorvos, respectively [353]. Other immobilization procedures and pH-sensitive dyes have also been described always introducing the sensing layer(s) at the tip of a bifurcated fibre-optic head, which can be integrated in a flow-through cell [92].

Cholinesterase field-effect transistors have also been reported for a large variety of pesticides [338,339,355]. Organophosphorous pesticides can be easily determined at the μM ($\approx$ppm) level using 2-PMA for reactivation of cholinesterase activity with reproducible values for at least 30 cycles [355]. This technique has been applied to monitor the photodegradation rate of methyl-parathion evaluating the reaction products with HPLC and inhibition of *Vibrio fischeri* luminescence [356]. From these results, it can be seen that toxicity values from *V. fischeri* diminish concomitantly with the photodegradation of methyl-parathion while the concentrations measured with the enzyme sensor follow the trend of the appearance of reaction products. However, higher values are reported if strict comparison with the formation of methyl-paraoxon and 4-nitrophenol, as main reaction products, is taken into consideration [356]. A toxicity screening of wastewaters using a *Paramecium* assay and a cholinesterase potentiometric sensor has also been described [349]. The oxidation products seem to show higher inhibition, which has been further exploited as a tool for improved sensitivity in cholinesterase biosensors using glass electrodes [357,358]. Light addressable potentiometric cholinesterase sensor has been demonstrated for the nM ($\approx$ppb) detection of paraoxon and bendiocarb [359]. Conductometric detection has also been demonstrated and compared with field-effect transistors using AChE, BuChE and urease aiming at specific detection of single species by multivariate analysis [360].

Non-natural substrates broaden the scope of detection techniques when using one single enzyme. For instance, by using a non-fluorescent synthetic substrate, indoxyl-acetate, the highly fluorescent enzyme product indoxyl can be correlated to ppb ($\mu g\, l^{-1}$) levels of methidation or naled although poor stability was observed for continuous monitoring [361]. The use of *o*-nitrophenyl acetate enables the quantification of the inhibition through the formation of the yellow product o-nitrophenol [362]. Amperometric transduction becomes feasible if acetylthiocholine or butyrylthiocholine is used because acetate or butyrate is produced with thiocholine that can be further oxidized according to the following reaction:

$$2HS-(CH_2)_2-N^+(CH_3)_3 \xrightarrow{\text{anodic oxidation}} (CH_3)_3N^+-(CH_2)_2-S-S-(CH_2)_2-N^+(CH_3)_3 + 2H^+ + 2e^- \quad (10.16)$$

There is a large variety of electrode materials as well as immobilization procedures with very varied limits of detection and dynamic ranges for very different organophosphorous and carbamate pesticides (see reviews in Refs. [21,337–339,363]. Importantly, amperometry opens up strategic ways of electrode modification conveying electrocatalytic properties that permit operation at lower potentials thereby diminishing the risk of non-selective redox processes. The use of 7,7,8,8,-tetracyanoquinodimethane-modified screen printed electrodes was pioneered by Kulys in 1991 who reported sensitivities for different organophosphates at 100 mV vs. Ag/AgCl between 0.06 and 8 ppm (mg l^{-1}) [364]. In 1992, Sklàdal reported a limit of detection for paraoxon of 80 ppt (ng l^{-1}) by crosslinking BuChE with glutaraldehyde on a carbon electrode modified with cobalt phthalocyanine [365]. Marty has approached the validation of sensor data with HPLC results in a variety of water samples [366]. His group also carried out a series of studies investigating different enzymes and immobilization procedures [367–370] including a His-tag modified recombinant AchE for oriented immobilization on nickel-modified thick film electrodes [371], as well as its characterization in water miscible organic solvents [372,373] and the study of different metabolites of chlorpyrifos [374]. The use of a soluble mediator, hexacyanoferrate, has also been tested and incorporated into a microflow system [375]. The reported limits of detection were at the nM level, enabling the detection of most pesticides at low ppb levels in the presence of 5% acetonitrile. The working potential can be lowered by using 4-aminophenyl acetate as substrate and oxidizing the resulting 4-aminophenol at 200 mV vs. SCE [376,377]. Detection by crystal quartz microbalance with 3-indolyl acetate upon hydrolysis produces an insoluble 3-hydroxyindole that precipitates at the gold surface, giving a limit of detection of 50 nM for paraoxon and 0.1 μM for carbaryl [378]. These limits are, in general, slightly higher than those routinely reported with amperometric sensors.

Obviously, the incorporation of a second or even a third enzyme complicates the manufacture of the biosensor and could compromise its stability if continuous operation is required. On the other hand, the possibilities for detection are clearly expanded because more products are formed and sensitivity may be enhanced if a recycling pathway is included. The general scheme for the incorporation of one or two more enzymes is as follows:

$$\text{HO(CH}_2)_2\text{N}^+(\text{CH}_3)_3 + \text{O}_2 \xrightarrow{\text{choline oxidase}} \text{HOOC–CH}_2\text{N}^+(\text{CH}_3)_3 + \text{H}_2\text{O}_2 \quad (10.17)$$

The formation of betaine and hydrogen peroxide with consumption of oxygen permits detection with a Clark electrode [379] or through direct oxidation of the hydrogen peroxide at ≈ 700 mV vs. Ag/AgCl. Using the latter, the determination of organophosphorous pesticides in soil extracts has been described by Campanella et al. [380]. Soil extracts were obtained by placing 145 g of soil in a beaker containing 500 ml of either distilled water or ethanol/water mixture (60/40), and stirring for 1 h at room temperature. The suspension was then centrifuged and the supernatant directly measured with the bienzymatic sensor. The authors concluded that ethanol did not interfere and they did not observe any adverse matrix effects from the extracted spiked samples [380]. The use of both enzymes avoids the regeneration step by immobilizing the cholinesterase in a large reactor coupled to an electrode modified with choline oxidase [381]. The large ratio between the amount of ChE in the reactor and the concentration of the inhibitor is the basis of this assay format [381].

Reduction of hydrogen peroxide can be further achieved with peroxidase and suitable electron donors:

$$\text{H}_2\text{O}_2 + 2\text{RH}_2 \xrightarrow{\text{peroxidase}} 2\text{H}_2\text{O} + 2\text{RH} \quad (10.18)$$

Depending on the characteristics of the electron donor, optical techniques and amperometry can be used. The use of the three enzymes, in homogeneous phase, has been applied in combination with pervaporation as sample pretreatment, for the determination of dichlorvos in soil and water samples [382]. Acceptable recoveries for dichlorvos (90–110%) in natural waters and spiked soil samples were reported. The whole procedure offers a sample frequency of $2\,\text{h}^{-1}$ [382]. Microgravimetric measurements of this multienzymatic system have also been reported using benzidines (3,3'- diaminobenzidine) as substrate of peroxidase, which is oxidized to an insoluble product [383]. Similarly to the previous example using this technique, the limits of detection achieved for carbaryl and dichlorvos are not as good as those obtained amperometrically.

A strict comparison between the sensitivities achieved with mono or multienzymatic systems using the same or a different detection principle is somewhat difficult because very different pesticides are tested, the assay formats also differ, the limits of detection may be calculated (if declared) considering different percentages of inhibition (i.e., 5 or 10%), and only rarely is the 50% inhibition reported as a possible reference value for direct comparison with immunosensors. However, taking paraoxon as the most assayed compound (the oxidation product of parathion), it does not seem that the introduction of new enzymes significantly and systematically enhances any analytical parameter. With regard to the detection technique, it seems that amperometry and fibre-optic biosensors offer higher sensitivity than the ENFETs and these are more sensitive than systems using the crystal quartz microbalance. However, the real comparison should be made with field samples where interferences may challenge the reliability of any configuration. Then, the competitiveness of any particular enzyme configuration and detection principle for the determination of organophosphorous and carbamate pesticides could be shown on the basis of other reference techniques, preferably combining chemical analysis and toxicity evaluation. Strictly speaking, toxicity identification evaluation (TIE) protocols demand combined chemical and bioanalytical approaches for identification of compounds responsible for the toxicity of the samples. In addition, it should always be considered that the global effect of each pesticide in a sample may not correspond to the additive response of individual compounds [283].

Other catalytic pathways have also been exploited for the indirect biorecognition of these compounds. The inhibition of alkaline phosphatase [384], acid phosphatase and glucose oxidase [385] or tyrosinase [386] has also been investigated as possible routes for the detection of organophosphorous and carbamate pesticides. Dithiocarbamate fungicides strongly and irreversibly inhibit the catalysis of aldehyde dehydrogenase, enabling the design of single-use modified electrodes that detect amperometrically the oxidation of NADH [387,388]. These compounds have also been approached through the inhibition of tyrosinase [389] or peroxidase [390].

As previously mentioned, most of the work related to the detection of these pesticides aims at the provision of a general or global parameter that is indicative of the toxicity of a sample. In principle, the lack of selectivity of cholinesterases is well suited for this task. Most of the reported approaches may fail if a particular pesticide is targeted, unless a very specific sample pretreatment permits a more selective isolation and preconcentration of the analyte. However, this contradicts the need for a simple and time-efficient device. It is not surprising then, that aiming at a more selective detection of pesticides, new tools have to be explored. The use of other catalytic routes for organophosphorous pesticides has been discussed (see "Organophosphorous pesticides and chemical warfare agents"). The combination of organophosphorous hydrolase and acetylcholinesterase has been investigated for the discrimination of organophosphorous neurotoxins and non-neurotoxic counterparts [283]. Alternatively and aiming at a more selective biorecognition (i.e., compound oriented), the group of Schmid is investigating genetically engineered cholinesterases that, combined in sensor arrays with data processing of artificial neural networks, may provide the route for the discrimination of a particular analyte [336]. Bachmann et al. have combined three AchE mutants of *Drosophila melanogaster* (DmAChE) and wild type DmAChE in a multielectrode configuration for the multianalyte detection of organophosphates and carbamates [344,391]. This type of array enabled the quantitative discrimination of paraoxon and carbofuran with a resolution error of 0.4 μg l^{-1} for paraoxon and 0.5 μg l^{-1} for carbofuran. A limit of determination 0.5 μg l^{-1} for each analyte could be achieved by incorporating mutant DmY408F, which was specifically designed by site-directed mutagenesis. Selective sensitivity for methamidophos has also been accomplished by using DmAChE variants [392]. One of the seven mutants tested displayed a 40 times higher sensitivity than the wild type electric eel AChE with a limit of detection of 1.4 μg l^{-1}. This limit was also lower than that achieved with the wild type DmAChE: 4.8 μg l^{-1} [392].

10.5.2.2.3 Heavy metals
In general, heavy metals are detrimental to enzyme activity, causing adverse effects in ecosystems. Table 10.9 summarizes the main characteristics of heavy metals related to environmental analysis [342]. The biological effects of heavy metals are mainly attributed to strong interaction with essential thiol groups like cysteine residues near the binding or active sites of proteins. This constitutes the basis of many enzyme electrodes that, indirectly and in a rather non-specific manner, can determine heavy metals. This lack of selectivity has been approached as a general indication of toxicity or for pollution monitoring using an array of enzymes and processing the responses through pattern recognition algorithms [393–395]. The validity of this approach remains to be demonstrated in the field in complex environmental matrices [396]. Generally, urease or other enzymatic systems are used for the detection of specific metals. Some of these examples are presented in Table 10.10.

The detectability of these sensors very much depends on the particular enzyme(s), the detection used and the assay format. In general, as previously mentioned, higher sensitivity is obtained by performing the measurement in two steps, first incubation with the substrate and then addition of the sample containing the target metal. Although some of these sensors have limits of detection significantly below the mandatory EC values, their utility for specific metals has to be further investigated in environmental samples in order to evaluate the efficiency of the somewhat tedious and complicated protocols for suppression of the interferences from other metals. For instance, urease can be inhibited by ion metals in the following decreasing order: Ag(I) > Hg(II) > Cu(II) $\gg$ Ni(II) > Co(II) > Cd(II) > Fe(III) > Zn(II) > Pb(II). Contrary to other groups of hazardous compound (e.g., PAHs, pesticides), legislation imposes restricted values for each single metal in waters intended for human consumption (98/83/EC) (see Table 10.9). This needs to be considered when assessing the analytical characteristics of all these sensors. Special precaution should also be taken in the assessment of bioavailability from data obtained with these sensors. The concept of bioavailability has been defined differently within various disciplines and its extended use may be misleading. The concept of contaminant bioavailability in soil and sediment has been revised by the US EPA [404].

10.5.2.2.4 Other analytes

Phosphates have been detected through several catalytic cascades using the inhibition of alkaline phosphatase and coupling glucose oxidase [405] or

TABLE 10.9

Main characteristics of heavy metals relevant in the analysis of waters intended for human consumption (after Ref. [342])

Metal	Main soluble metal ions	Other relevant forms	Most toxic forms	Threshold concentration[a]
Mercury	Hg^{2+}	Alkyls, phenyls	Methyl mercury	$1\ \mu g\ l^{-1}$ (ppb)
Cadmium	Cd^{2+} only	–	Species dependent; Cd^{2+}	$5\ \mu g\ l^{-1}$
Selenium	Se^{4+}, Se^{6+}	H_2Se, $(CH_3)_2Se$	Elemental	$10\ \mu g\ l^{-1}$
Arsenic	As^{3+}, As^{5+}	–	AsH_3, As^{3+}	$10\ \mu g\ l^{-1}$
Lead	Pb^{2+}	Alkyls, carbonyls	Tetraethyl and methyl lead	$10\ \mu g\ l^{-1}$
Nickel	Ni^{2+}	$Ni(H_2O_6)^{2+}$	Elemental, carbonyls	$20\ \mu g\ l^{-1}$
Chromium	$Cr^{6+} > Cr^{3+}$	–	Cr^{6+}	$50\ \mu g\ l^{-1}$
Copper	$Cu^{2+} > Cu^{+}$	Organo complexes	Elemental, Cu^{2+}	$2\ mg\ l^{-1}$ (ppm)

[a]Maximum admissible concentrations according to the European Directive 98/83/EEC.

TABLE 10.10

Examples of enzyme inhibition biosensors for the detection of heavy metals

Analyte(s)	Enzyme(s)	Detection	LOD	Reference
Hg^{2+}	Urease (solution)	SAW/impedance	50 ppb	[397]
Hg^{2+}, Cu^{2+}, Cd^{2+}	Urease	Conductimetric	0.5 ppm Cd	[398]
Hg^{2+}	Urease	pH-FET	0.2 ppm	[399]
Hg^{2+}, Hg^{+}, phenyl-Hg	Urease	pH-FET	10 ppb Hg^{2+} or Hg^{+}, 20 ppb phenyl-Hg	[400]
Hg^{2+}	Invertase (solution) and GOx	Amperometric 650 mV vs. SCE	1 ppb	[401]
Cd^{2+}, Zn^{2+}, Ni^{2+}, Pb^{2+}, Sn^{2+}, Co^{2+}, Cu^{2+}	Cholinesterases, urease and GOx	pH-sensitive EIS	≈19 ppb (Cu), ≈0.2 ppm (Cd, Co, Ni), ≈6 ppm (Pd, Zn, Sr)	[402]
Hg^{2+}, Ag^{+}, Pb^{2+}, Cu^{2+}, Zn^{2+}	LDH and LOx	Clark electrode	0.4 ppb (200 ppb)[a] Hg, 31 ppb (0.6 ppm) Cu, 0.3 ppm (1.6 ppm) Zn, 2 ppb Ag, 40 ppb Pb	[403]

SAW, surface acoustic wave (impedance); FET, field-effect transistor; EIS, electrolyte-insulator-semiconductor; GOx, glucose oxidase; LDH, lactate dehydrogenase; LOx, lactate oxidase.
[a]Values obtained with soluble or (immobilized) LDH.

tyrosinase [406] for amperometric detection of reaction products. Tyrosinase has been widely exploited for the detection of different compounds [407]. Examples include: atrazine [408], cocktails of chlorophenols [409], benzoic acid [410], azide [411], and cyanide [412]. Glutathione-S-transferase has been used for the detection of captan [413] or atrazine [414] and acetolactate synthase for the detection of trifensulfuron methyl and sulfuron methyl [415]. The validity of all these approaches for sound environmental applications should follow the same guidelines as for the inhibition of acetyl cholinesterase.

10.5.2.3 Enzyme-based sensors without catalytic transformation of the analyte
According to the classification presented in this chapter, this group of enzyme sensors exploits the interaction between the analyte and any domain of the protein with a concomitant "change" that can be detected, in most cases, with no need of catalytic transformation. Although a rather formal approach and scarcely implemented in sensor configurations, this route may open up new ways of exploiting biorecognition events by enzymes. They include very different interactions and means of detection that can be categorized in three major groups: (i) detection through reactivation of the apoenzyme, (ii) direct or indirect detection of the interaction with the holoenzyme, and (iii) rybozymes or non-protein enzymes. All these approaches differ in their degree of maturity. While the use of apoenzymes may not be competitive as a further catalytic step is required for detection, the use of holoenzymes and rybozymes may expand significantly in the near future concomitantly with the technical development of simple instruments measuring polarization, anisotropy or life-time fluorescence. The scope of analytes to be transduced within these approaches remains, presently, rather narrow.

10.5.2.3.1 Use of apoenzymes for the detection of metal ions
Generally, apoenzymes of metalloenzymes can be used for the detection of the corresponding metal ion. Restoration of enzyme activity obtained in the presence of the metal ion can be correlated to its concentration. This principle has been demonstrated in the detection of copper while evaluating reconstituted catalytic activities in galactose oxidase and ascorbate oxidase and also in the detection of zinc since this ion is essential for the activity of carbonic anhydrase and alkaline phosphatase [416]. The need of stripping the metal for the preparation of the apoenzyme may demand tedious procedures and a catalytic assay with the addition of the substrate is always required for detection.

10.5.2.3.2 Labelled or label-free detection of the interaction between the analyte and the holoenzyme
The use of the holoenzyme could be an easier approach although means of detection have to be provided. Indirect optical transduction through the labelling of the holoenzyme is generally used. Site-specific labelling ensures the reporting of conformational changes that the protein undergoes upon analyte binding. This approach is extensively used with non-catalytic proteins but it has also been demonstrated with the metalloenzyme carbonic anhydrase for the detection of zinc in a sensor configuration using fibre optic [417]. The basis of this sensor is the association of the zinc holoenzyme with the inhibitor 5-(dimethylamino)-1-naphthalenesulfonamide (dansylamide), leading to a titrable change in the fluorescence emission wavelength and intensity. Sub-ppb levels of zinc can be quantified by ratiometric methods by measuring the emission of free dansylamide at 580 nm and the emission of bound dansylamide at 470 nm when excited by 326 nm light, or by changes in the apparent fluorescence lifetime as the fraction of holoenzyme and thus bound dansylamide changes [418]. In contrast to the detection of the dansylamide, which relies upon zinc-dependent changes in the wavelength and intensity of the fluorescence emission, the change in anisotropy of the fluorophore between the free and bound enzyme states has also been correlated to ppb levels of zinc in solution by redesigning the chemistry of the sulfonamide probe [419]. By judicious design of the fluorescent labels and producing variants of the enzyme by site-directed mutagenesis, the selectivity and sensitivity performance of this sensor can be tuned. The rational design of these sensors involves selecting amino acids for modification that are in close proximity to the binding cleft by looking at either the NMR or the 3D crystal structure of the protein (see "Site-directed mutagenesis"). For instance, sensors based on a variant of human carbonic anhydrase II respond to Cu(II) and Zn(II) concentrations in the picomolar range and Cd(II), Ni(II), and Co(II) concentrations in the nanomolar range, even in the presence of millimolar concentrations of calcium and magnesium. By the use of different fluorescent labels, sensors responding to Cu(II), Co(II) and Ni(II) but not to Zn(II) and Cd(II) have also been reported [420,421].

Similarly to the detection of affinity events in immunosensors, the direct detection of the interactions between analyte and proteins generally requires highly sensitive techniques that may not offer the required robustness for routine analysis, in situ or on-site environmental practice. Surface plasmon resonance (SPR) is currently used for the real-time detection of a multitude of biological binding events. This technique has been applied to the detection of cadmium due to its interaction with the binding site of urease [422]. The low mass of the analyte does not seem to prohibit its detection, up to 10 mg l^{-1}, through SPR measurements since possible structural changes in the adsorbed protein layer can occur upon binding of the metal [422].

10.5.2.3.3 Non-protein enzymes: catalytic nucleic acids
In 1982 it was recognized that nucleic acids and particularly RNA may catalyse a broad range of reactions such as formation, cleavage and rearrangements of some chemical bonds. Catalytic RNAs are known as ribozymes and they can be found in organelles of plants and lower eukaryotes, in amphibians, in prokaryotes, in bacteriophages and in viroids or satellite viruses. Arguably, they may not be true biocatalysts since they suffer a net change in the catalysed reaction. The exception is the RNase P which is an endoribonuclease responsible for the removal of $5'$ leader sequences from transfer RNA precursors. Engineered catalytic RNA is able to accelerate a myriad of chemical transformations with acceleration rates in the order of 10^{11}, specificity constants k_{cat}/K_m up to $10^8 \text{ M}^{-1} \text{ min}^{-1}$, although protein enzymes are still a 1000 times more powerful catalysts than ribozymes for similar reactions [423].

There is interest in the use of catalytic nucleic acids as biorecognition receptors in sensors as they offer inherent advantages compared with catalytic proteins. Among them: simplicity, low cost, easy synthesis, and more importantly, the possibility to be designed and/or evolved through an iterative selection in a combinatorial approach. Since nucleic acids can be easily replicated, they offer a unique advantage in comparison to combinatorial selection of organic ligands or peptides (see Section 10.6.3.3).

Most of the known ribozymes require divalent cations, usually Mg^{2+}, which is used for proper assembly of a complex 3D structure that provides an appropriate environment for catalysis. Metal ions can also act as cofactors in the chemical reaction enabling catalysis by activation of the nucleophile, stabilization of the transition state and protonation of the leaving group in general acid catalysis [424]. Accordingly, ribozymes could be considered as metalloenzymes. This fact has suggested the use of ribozymes as sensing elements for the detection of metal ions.

An important input for sensing applications of engineered nucleic acids has been their ability to show allosteric properties and thus perform as precision molecular switches. Allosteric ribozymes are frequently termed as aptazymes since they combine nucleic acid binding properties (aptamers) and catalytic activity like their protein enzyme counterparts. They possess an effector binding site that is distinct from the active site of the catalyst and it modulates its activity by inducing conformational changes. The possibility to design and/or evolve the existing aptazymes enables the tuning of their analytical properties, in terms of effector specificity, affinity, kinetics and even the type of enzymatic activity [425]. Consequently, a repertoire of aptazymes has been developed for the recognition and quantification of organic molecules including, among many others, ATP, theophiline, FMN and proteins [425,426]. The recognition is based on the binding of the analytes that are specific effectors of self-nucleic acid cleavage reactions that concomitantly releases the label (e.g., ^{32}P-labeled aptazyme) [427]. Alternatively, the use of aptazyme ligases simplifies the detection procedure. Since aptazyme ligases directly transduce analyte recognition by the formation of a covalent bond, they can be immediately adapted for use in biosensors [426]. One component, either the aptazyme or the substrate, can be immobilized and the second, complementary component (substrate or analyte) can be appropriately labelled. Addition of the analyte will instigate the formation of a covalent bond and therefore the coimmobilization of the labelled second component. Aptazyme ligase arrays are shown to apprehend and quantitate a wider range of analytes than would be possible with typical ELISA assays [426]. Presently, most of the technical interest for aptazymes has been limited to pharmaceutical or therapeutic applications.

In contrast of the applications of aptazymes, the use of non-allosteric and metal ion-dependent catalytic nucleic acids in environmental analysis has been reported [428,429]. It has been demonstrated that in vitro selection of catalytic nucleic acids can change the metal specificity or binding activity of existing ribozymes, resulting in catalytic nucleic acids highly specific for Pb^{2+} [428], or Zn^{2+}/Cu^{2+} [429]. In this line, Lu has developed a new class of detection system based on deoxyribozymes for the detection of lead ions [430]. He has developed an in vitro selected metal ion-dependent deoxyribozyme capable of cleaving a single RNA linkage within a DNA substrate (i.e., DNA/RNA chimera) with high activity and specificity for lead. The assay is based on the labelling of the 5′-end of the substrate with a fluorophore (TAMRA) and the 3′-end of the deoxyribozyme with a corresponding quencher (Dabcyl) that effectively quenches fluorescence upon hybridization. The addition of Pb^{2+} induces the catalytic cleavage of the substrate and concomitant suppression of the quenching with a fluorescence signal at 580 nm. The dynamic range of this assay is between 10 nM and 4 μM with a selectivity coefficient for most other metal ions in the order of 80. Spiked water samples from Lake Michigan have not shown any matrix interference [431]. An elegant refinement of this assay was the introduction of gold nanoparticles for detection [432] and performing a second hybridization with a specific 12-mer oligonucleotide carrying

Au nanoparticles [433]. Briefly, the assay consists of a first hybridization of the substrate with the DNAzyme and then a new hybridization with the oligonucleotide functionalized gold nanoparticles rendering a dendritic structure. This structure results in the aggregation of gold particles with the appearance of blue colour. In the presence of Pb^{2+}, the DNAzyme catalyses the hydrolytic cleavage of the substrate, prohibiting the formation of nanoparticle aggregates and thus a change from blue to red colour denotes the presence of lead. The ratio of absorbance at 522 and 700 nm is used for the quantification of Pb^{2+}, resulting in a dynamic range very similar to that obtained by quenching [433]. It is expected that in the near future more of these applications will be reported enlarging, with no apparent limits, the possibilities for customized and highly selective biorecognition with catalytic RNA and DNA.

10.6 EMERGING CATALYTIC MOLECULES FOR ENVIRONMENTAL APPLICATIONS

Enzymes have had billions of years of evolution in order to sustain living systems as needed. That these naturally evolved molecules can then accomplish any other task outside their intrinsic requirements is simply fortuitous. It is thus not surprising that the use of enzymes for environmental, industrial or biotechnological applications is not easily reconcilable with their intrinsic characteristics.

Enzyme-based applications for environmental screening or monitoring demand tailored biocatalysts performing catalysis in non-natural substrates and/or in non-usual or hostile media. Moreover, the increased complexity of contaminated environmental sites also demands efficient biodegradation of xenobiotic compounds through new and multiple engineered pathways where the tailored biocatalysts should perform in their host microbial cells [434].

Throughout this chapter the use of enzymes that are mainly obtained from mesophilic organisms has been described in biosensing applications. Most of these enzymes are obtained either from cultured microorganisms or expressed in recombinant organisms for efficient production. The narrow set of conditions for optimal catalysis, the possible deactivation and the capricious stability of enzymes somewhat hinders the complete exploitation of the exquisite capabilities of these biocatalysts. Many different and diverse tools are nowadays available for finding or designing new enzymes exhibiting tailored abilities for ad hoc applications.

10.6.1 New biocatalysts from natural environments: extremozymes

For many years, one way to solve these drawbacks has been the identification and isolation of new biocatalysts with the desired characteristics. The most common procedure for isolation and identification of biocatalysts has also been applied to microbial populations living in harsh environments such as hot springs, abyssal hot vents or geothermal power plants. These microorganisms, known as extremophiles can be found in environments of extreme temperature, ionic strength, pH, pressure, metal concentrations or radiation levels. These natural reservoirs of extremozymes directly offer new activities and extreme stabilities but these microorganisms also offer novel metabolic pathways. Table 10.11 summarizes the identified extremozymes found in some extremophiles.

The adaptation of these enzymes to extreme environments is related to different factors that influence their 3D structure. Among them: changes in specific amino acid residues (mainly glycine), increased number of ion pairs formed, increase in the extent of secondary structure formation and truncated amino and carboxyl termini, increased hydrophobic interaction at subunit interfaces, reduction in the size of loops and in the number of cavities, and an excess of negatively charged amino acids on the protein surface [435,436]

Since the conditions required for the growth of extremophiles include extreme conditions and an anaerobic environment, less than 1% of these strains are currently amenable to growth in culture [437]. The large-scale culture of extremozymes relies on the expression of the corresponding genes in a non-native mesophilic environment (*E. coli*). Indeed, the heterologous expression of thermophile and psychrophile enzymes generally does not compromise their inherent characteristics [438], the thermophiles being easily purified through a simple heat treatment [439]. In contrast, the expression of halophilic and acidophilic or alkalinophilic enzymes presents some difficulties for preserving activity. In the case of halophilic enzymes their reactivation is possible by a denaturation step with guanidine hydrochloride or urea followed by renaturation in the presence of NaCl [436].

These extremozymes not only serve as excellent models for understanding protein stability but also have significant biotechnological potential. To date, most of the technical interest of these enzymes derives from the food and chemical industry [440]. Equally important, the finding of the thermostable DNA polymerase of *Thermus aquaticus* [441] has facilitated and expanded many methodologies in biochemistry and molecular biology. As concerns environmental applications, the use of extremozymes has been restricted to whole organisms, particularly moderate thermophilic anaerobes (75°C) for the

TABLE 10.11
Different kinds of enzymes found in extreme environments

Phenotype	Habitat	Enzymes	Typical genera
Thermophile	High temperature		
	Moderate thermophiles (45–65°C)	Amylases, xylanases	*Methanobacterium, Thermoplasma, Thermus*[a], some *Bacillus*[a] species, *Aquifes*[a], *Archaeoglobus, Hydrogenobacter*[a], *Methanothermus, Pyrococcus, Pyrodictium, Pyrolobus, Sulfolobus, Thermococcus, Thermoproteus, Thermotoga*[a]
	Thermophiles (65–85°C)	Proteases, DNA polymerases	
	Hyperthermophiles (>85°C)		
Psychrophile	−2–20°C	Proteases, dehydrogenases, amylases	*Alteromonas*[a], *Psychrobacter*[a]
Acidophile	pH < 4	Sulfur oxidation, chalcopyrite concentrate	*Acidianus, Desulfurolobus, Sulfolobus, Thiobacillus*[a]
Alkalophile	pH > 9	Cellulases	*Natronobacterium, Natronococcus*, some *Bacillus*[a] species
Halophile	2–5 M NaCl	Malate dehydrogenase, *p*-nitrophenylphosphate phosphatase	*Haloarcula, Halobacterium, Haloferax, Halorubrum*
Piezophile	High pressure	Whole microorganism	
Metalophile	High metal concentration	Whole microorganism	
Radiophile	High radiation levels	Whole microorganism	
Microaerophile	Growth in <21% O_2		

[a]Genera of the domain Bacteria; all others are Archaea.

degradation of different contaminants such as synthetic pyrothroid insecticides [442], 3-chlorobenzoate [443] or halophenols including 2-chlorophenol, 2-bromophenol [444]. The waste problem of cheese whey in the dairy industry has been alleviated by the addition of thermophilic mixed populations of *Bacillus* sp., thereby doubling the degradation rates of lactose, proteins and COD as compared with equivalent treatment with mesophilic microorganisms [445].

Irrespective of the above applications, the enormous potential of these biocatalysts remains to be fully exploited. For instance, new perspectives in bioremediation processes have been envisaged, focusing on the use of (i) radiophile microorganisms in radionucleotide contaminated sites [446], (ii) thermophilic enzymes performing at high temperatures with enhanced yields (i.e., higher availability of hydrophobic compounds), reactivity (i.e., higher transfer rates and lower diffusional restrictions) thus reducing the risk of contamination [447], (iii) psychrophilic polymer-degrading enzymes in the paper industry [448] and (iv) halophilic enzymes that possess salt-enriched solvation shells, retaining their catalytic activity in environments with low water activity [449].

A more specific example has been previously illustrated through an OPAA that was originally isolated from a halophilic bacterial isolate of the genus *Altermonas* [450] (see "Organophosphorous pesticides and chemical warfare agents"). The gene responsible for the DFP hydrolysing capability of the original strain has been isolated and cloned into *E. coli* and sequenced. This enzyme efficiently hydrolyses high levels of fluorine-containing organophosphates with a unique ability to discriminate chemical warfare agents and organophosphorous pesticides [282]. It is worth mentioning the period between this discovery and its demonstration of suitability for analytical purposes. The extremozyme was isolated in 1991, 5 years later the cloned enzyme was reported and in 2001 the biosensor was characterized. This should be taken into consideration for further realizations.

The biotechnological potential of extremozymes is exploited by emerging companies, for instance, Diversa (http://www.diversa.com) that combines natural and directed evolution to deliver tailored products with enhanced properties. The company extracts the genetic material from non-cultivable organisms of almost all ecosystems, including harsh environments, providing a resource of genetic diversity and, through evolution methods (vide infra), it develops enzymes for pharmaceutical, agriculture, chemical and industrial markets.

10.6.2 Catalytic antibodies: abzymes

Another approach in searching for new biocatalysts has been the exploitation of the mammalian immune system by the effective production of antibodies with tailored catalytic properties. Basically, the design of haptens, mimicking the proper transition state (TS) of the catalysed reaction, induces the production of these catalytic antibodies, which also called abzymes. Consequently, one single protein assembles the specificity of antibodies and the catalytic activity of enzymes. This is achieved by changing their ability to use binding energies, i.e., their affinity for the transition state of the reaction should increase in detriment to the affinity for the antigen in the ground state which is thermodynamically favourable to product release. Although this idea was formulated in 1969 by Jencks and demonstrated for the fist time in 1986 by Schultz [451] and Lerner [452], humans already produce natural catalytic antibodies as a consequence of some auto-immune disorders.

In general, these molecules base their recognition on a specialized binding site and, similarly to enzymes, they catalyse a large variety of reactions, displaying a similar specificity, stereospecificity, kinetics and competitive inhibition, than their enzyme counterparts [453]. However, the rate of accelerations obtained by transition state stabilization still remains lower than that obtained with enzymes, i.e., 10^3-10^4 fold [454]. In search of improved efficiency and since no stable transition state analogues can reproduce all the characteristics of the transition state analogues, new and more sophisticated alternatives have been developed to elicit catalytic antibodies [455]. These include

(a) Reactive immunization or inducing catalytic residues such as acid/base groups or nucleophiles in the antibody-binding sites through a tailored chemically reactive hapten. This approach changes the selection criterion from simple binding to chemical reactivity and, although restricted to some reactions, it has served to commercialize the first catalytic antibody (38C2) that catalyses aldol and retro-aldol reactions from a wider range of substrates with high catalytic activity [456].
(b) An entropic trap that is based on proximity effects using the antibody-binding energy to constrain the molecules into a reactive conformation, and the antibody behaving as "entropy traps".
(c) Bait and switch that exploit charge complementarity [457]. This approach is limited to one or two catalytic groups.
(d) Anti-idiotypic antibody characterized by the use of a complex antigen in order of increasing the structural features responsible for catalysis.

This approach is based on the idea that the immune system can be regarded as a network of interacting idiotopes [458]. A major postulate is that for each immunoglobulin generated against an antigenic determinant (e.g., enzyme active site), there exists a complementary antibody directed against the idiotopic determinants of the immunoglobulin. When idiotopes are superimposed upon the complementary region of the binding sites (anti-idiotopes) molecules can mimic the antigen structures and are designed as internal images of the original antigens (i.e., enzymes).

Other strategies have focused on refinements of immunization and screening protocols, improved transition state analogues, site directed mutagenesis, directed evolution or the incorporation of cofactors (e.g., flavins, heme and metal ions) to the abzyme by using analogues containing both the cofactor and substrate, or by chemical semisynthesis. This improves the efficiency of abzymes as well as eliciting catalytic antibodies with non-classical catalytic activities. Frequently, several strategies are combined. For instance, the in vitro abzyme evolution (vide infra) of a hydrolytic abzyme generated by the classical in vivo method with immunization of the transition-state analogue, has resulted in optimizing the differential affinity for the transition state relative to the ground state, giving rise to molecules with a 6- to 20-fold increase in turnover [459].

The rational design of these catalytic antibodies enables the creation of new active sites and the evaluation of possible alternative pathways for a particular reaction through the comparison with naturally occurring enzymes. Nevertheless, since their large-scale production can be performed via bacterial expression systems, an almost endless array of new biocatalysts can be produced [454]. Actually, more than 100 reactions have been accelerated by antibodies, including reactions for which no enzyme exists. Consequently, many applications can be envisaged, mainly targeting therapeutics and the pharmaceutical industry [455]. In therapeutics, since abzymes are biocompatible with long serum half-lives, they can destroy peptides or carbohydrates associated with viral particles, or tumour cells or in cancer treatment by the strategy called antibody directed abzyme pro-drug therapy [460]. Abzymes in organic syntheses could be used for the controlled release and activation of specific molecules attached to biodegradable polymeric material [461]. One interesting application of these molecules is the possibility to use them for neutralizing organophosphorus chemical warfare agents, being the first step towards the production of therapeutically active abzymes to treat poisoning by nerve agents [462].

10.6.3 Redesign of catalytic activities

Despite the possibilities created by the incorporation of extremozymes and abzymes, the natural diversity found in living systems cannot address all practical requirements for biocatalytic applications. As de novo design approach of an enzyme is time consuming and risky, since a tiny miscalculation could be catastrophic [453], presently the practical approach consists of the redesign of known biocatalysts by protein engineering in order to enhance their overall performance. Redesign should be understood as, optimally, an alteration of the enzyme structure that causes a predicted change in its function, i.e., changes in substrate specificity, inversion of stereochemistry, engineering new reaction mechanisms into the same active site, and conversion of ligand-binding sites into catalytic centres [463].

Additionally, the new enzymes discovered by classical screening methods could provide new starting-points for possible improvements of such new catalysts through chemical modifications or protein engineering. The difficulty, in case of a large library, is to screen many reactions for catalysis, which requires the unambiguous detection of small catalytic enhancements. Furthermore, this screening should also be related to the intended application [464].

The different but complementary strategies for redesigning natural enzyme can be approached from the perspective of how rational the strategy could be. Thus, the existing biocatalyst can be fine-tuned by rational design, being information intensive and computer assisted. In other words, the more rational the desired approach, the more knowledge is necessary to exactly interplay protein sequence and function.

10.6.3.1 Rational redesign of biocatalysts
The active site's structures are sufficiently conserved during the evolution process of natural enzymes, and thus the identification of the target residues controlling the particular enzyme behaviour, can in some cases be performed by structural and sequence comparison. The aim is to assign the enzyme to a protein superfamily in which scaffolding and structural architecture is predictable [434]. The selection of the best blueprint for a given protein family and the prediction for the best modification of these catalytic residues or structural alterations can be performed by an iterative computer design using available databases, e.g., PROCAT, for deciphering the appropriate geometries of active site residues for different reactions, GenBank, EMBL, SWISS-PROT, and PDB [463]. Once the target region has been identified, multiple available tools of different complexity could be used, including site-directed mutagenesis, specific chemical modification or molecular lego.

10.6.3.1.1 Site-directed mutagenesis

Site-directed mutagenesis is a powerful molecular technique for introducing point mutations, through two main methods: two-round or single-round PCR. Site-directed mutagenesis has been successfully used to redesign enzyme targeting, specifically, the following alterations [463]:

(a) Substrate specificity without altering the chemical mechanism of the reaction. This has been pursued in different oxidoreductases (e.g., dehydrogenases), hydrolases (e.g., acetylcholinesterases, lactamases, proteases), transferases (e.g., aminotransferases, restriction enzymes). More specifically, targeted mutations in the substrate-binding pocket of the cytochrome P450 BM-3, a double hydrophobic substitution at the entrance of the substrate channel for promoting the initial recognition and subsequent entry of hydrophobic PAHs, combined with mutations (Ala264Gly and/or Phe87Ala) in the active centre for better access of PAHs to the heme, have resulted in an improvement of the activity for the oxidation of certain PAHs (i.e., phenanthrene, fluoranthene and pyrene). The mutant enzyme exhibits 40–200 fold enhanced activity compared with the wild type enzyme and with simultaneous enhancement of NADPH oxidation. The Ala264Gly mutation causes the coupling efficiency between NADPH oxidation and substrate hydroxylation, while the Phe87Ala mutation changed the substrate specificity [465].

(b) Cofactor requirements, specifically focusing on changes of selectivity of NADH vs. NADPH [466] or substitution of ATP for non-natural ATP analogues [467]. The modification of the residue involved in the interaction with the 2′-phosphate group of the AMP moiety of NADPH, together with additional mutations, has resulted in most cases in a successful change in affinity of this redox pair. Therefore, the introduction of seven mutations renders, for instance, a glutathione reductase with an affinity 18,000-fold higher for NADH than the natural NADPH cofactor [468].

(c) Stability. There are no strict routes for the design of mutants with improved stability since the rules for stabilization have not been established because hundreds of amino acids can be involved. In some cases, higher thermal stability has been achieved through an increase of the conformational rigidity and thus the introduction of glycine to alanine and alanine to proline substitutions and a disulfide bridge in a thermolysin-like protease, has allowed a 340-fold increase in thermal stability at 100°C [469].

(d) Stereospecificity. The enantioselectivity of target enzymes for substrates or products has been successfully modified, and attempts have been made to alter the stereochemistry of hydride transfer reaction [470].

(e) Reaction mechanisms. The search for new catalytic reactions in wild biocatalysts has been one of the major goals in this approach. The notion that a family of closely related enzymes possess different catalytic activities that arose from a common progenitor enzyme as a consequence of some modifications of its active site opens up the ability to redirect a common intermediate reaction into different end-products [471]. From a practical point of view, this activity transfer can be done by control and optimization of side or intermediate reactions through specific alterations of the active site. This is the case for the carboxyl/cholinesterases family, which is inhibited by organophosphorus insecticides and nerve agents, through the formation of a covalent intermediate analogous to the acylated enzyme intermediates, leading to a phosphorylated enzyme (see Eq. (10.16)) [472]. Given that the second hydrolytic step (i.e., a nucleophilic attack) to regenerate the free enzyme is rate limiting (probably due to different stereochemistries of the phosphoryl and carbonyl moieties) [473] this inhibition is essentially irreversible. The substitution of a glycine (Gly^{117}) in BuChE with a histidine residue (Gly117His) near to the oxyanion hole of the active site, confers to butyrylcholinesterase an organophosphorous hydrolase activity that preserves its carboxylesterase activity [474,475]. Similarly, the Gly137Asp substitution converts a carboxylesterase to an organophosphorous hydrolase and confers insecticide resistance on a blowfly [473]. However, this mechanism of reaction differs from the one-step described in the organophosphorus hydrolase which has been isolated and purified from several microorganisms (e.g., *Pseudomona diminuta*) in soils and activated sludges [266].

10.6.3.1.2 Specific chemical modifications
A different but complementary tool for the rational redesign of biocatalysts is their chemical modification. In principle, this approach permits the incorporation of an almost unlimited range of functionalities including non-natural amino acids. Essentially, anything that can be incorporated into a peptide can be installed in a larger protein. The most relevant advantage is that chemical modification can be coupled to engineered proteins expanding the possibilities to redesign or create new catalysts.

It is illustrative to mention that the first chemical modification of enzymes was earlier than the discovery of site-directed mutagenesis. In 1966 a chemical approach for converting the active site serine-derived hydroxyl group present in subtilizin to a thiol was reported [476,477]. Since then, chemical modification has been demonstrated to be a rapid and inexpensive approach for enhancing and modifying the analytical properties of biocatalysts for many specific applications. For instance, using rather random modifications the stability and chemical properties of enzymes can be improved although, in general, to the detriment of catalytic activity. More specifically, the introduction of hydrophilic groups on the surface of the protein usually prevents incorrect refolding after reversible denaturation [478] and crosslinking of enzymes by different agents [479] or the binding to polymers usually renders enhanced stability [480]. These and many other chemistries are currently used in the design of biosensors.

Irrespective of the potential of the random chemical modification, this tool usually renders a mixture of semisynthetic enzymes as consequence of their inherent low selectivity and, in some cases, low efficiency. Additionally, the yield and accuracy of these modifications in limited chemical functional groups is not always predictable. Consequently, and in order to exploit all the possibilities of the chemical modification of enzymes, it is necessary to achieve specific chemical reactions of the target enzyme with high specificity and efficiency.

This "specific" chemical modification permits the redesign of enzymes by (i) alteration of their catalytic activity, including changes in substrate specificity and the introduction of functionalities that allow the regulation of the enzyme by exogenous ligands, (ii) the incorporation of new enzyme activities in both metal and non-metal containing systems, and (iii) enzyme printing [481]. Different strategies for the modification of an existing enzyme activity can be applied, including [481]:

(a) Atom replacement as described for the production of the thiosubtilizin described above.
(b) Segment reassembly, i.e., the replacement of large portions of the protein via selective proteolysis or chemical cleavage. For instance, since the S-peptide of RNase S is required for its catalytic activity, modifications of this peptide can be used for the modulation of the rate of the enzymatic reaction. Hamachi et al. have prepared different forms of this enzyme whose activity is modulated by metal ions. The introduction of negatively charged moieties into the S-peptide in positions where the enzyme is usually occupied by cationic residues leads to a decrease in the activity. Thus, the addition of metal ions inverts the charge at the peptide restoring the wild type activity [482].
(c) Specific modification exploiting the distinctive reactivity of the thiol group present in cysteine residues. This approach is generally more useful when combined with site-directed mutagenesis. Accordingly, the incorporation of new cysteine residues at specific positions of the target enzyme permits

its specific modification by chemical means. This approach is been extensively used to decipher the reaction mechanism of enzymes, particularly the redox class of enzymes. In the case of a heme monooxygenase P450$_{cam}$, a cysteine-free mutant has been prepared by substitution of the cysteines for alanines. Then, the further production of single cysteine mutants permits the specific covalent linking of N-ferrocenylmaleimide for its electrochemical communication with the electrode surface and offering a well-characterized pathway for electron transfer [483]. A more sophisticated strategy has been reported by Shumyantseva et al. using various P450 isoforms. They prepared semi-artificial flavohaemoproteins containing, in one polypeptide chain, both the heme and flavin nucleotides [484]. These flavohaemoproteins, obtained through chemical modification of CytP450 2B4 and 1A2 by the use of reactive derivatives of flavin nucleotides, proved to be adequate functional models of the microsomal monooxygenase system [484]. Importantly, CytP450 2B4 is responsible for the metabolism of substituted amines, and CytP450 1A2 catalyses the oxygenation of aromatic compounds (e.g., PAHs) [485]. In addition to these mechanistic studies, the modification of specific cysteine irrespective of the number of internal cysteine residues in the target enzyme [486], or the labelling of proteins within living cells by chemical reaction of specific cysteine residues [487, 488], significantly expands the possibilities and areas of applications of this synthetic route.

(d) The covalent cofactor attachment can be used for the modification of catalytic activity or for the introduction of new catalytic reactions but overall, and equally important, it permits reagentless configurations of many enzyme based sensors. Stocker et al. have redesigned an FAD-dependent amino acid oxidase by chemically attaching an FAD analogue. The main feature of the resulting semisynthetic enzyme is its higher stability with similar kinetic constants than the native holoenzyme [489]. Similarly, NAD(P)/NAD(P)H dependent dehydrogenases have been bound to NAD analogues through carbodiimide reaction [490]. The semisynthetic dehydrogenases show lower specific activity than the native enzyme but, importantly, they do not require addition of the soluble cofactor and they can also be regenerated by a coupled enzyme reaction [491].

Chemical modification also opens up the possibility of transforming a non-catalytic protein into an enzyme-like protein. This allows old catalytic reactions to be performed by new proteins that have been modified with catalytic moieties (metal and non-metal complexes) [492,493]. For instance, human serum albumin has acquired catalytic abilities by complex formation with iron protoporphyrin IX [494].

A final possibility of this synthetic route is the creation of catalytic bioprints that were demonstrated 5 years earlier, in 1984, than the well-known molecularly imprinted polymers (MIPs) [495]. The synthetic route is also similar to that of MIPs and it basically consists of the partial or complete denaturation of the protein, its renaturalization in the presence of the template (a substrate analogue), the chemical cross-linking of the resulting complex and, finally, the removal of the template [495]. This strategy has been used to prepare fluorohydrolase in testing several proteins (e.g., BSA, egg albumin, casein) and using diisopropyl phosphoric acid as template. The best bioprint, using dimethyl pimelimidate as cross-linker agent, showed fluorohydrolase activity for phenylmethylsulfonylfluoride and for the potent acetylcholinesterase inhibitor, diisopropylphosphorofluoridate, with turnover numbers in the order of 125 min^{-1} [496].

10.6.3.1.3 Molecular lego
All previous strategies rely on the alteration(s) of the target residues responsible of a particular enzyme behaviour for the preparation of customized enzymes. Recently, Gilardi et al. have demonstrated a new and sophisticated strategy: molecular LEGO [497]. Basically, this approach is based on the assembly of simple modules or domains with the desired proprieties for building up chimeric proteins at will. The availability of the 3D structures of the different domains allows the use of computational methods for generating 3D models of the possible complexes required for the rational design of the assembly [497]. Based on this strategy, these authors succeeded in building up an artificial multidomain CytP450, targeting some of the major drawbacks of this family: their low interaction with the electrode surface and solubility (the human P450 is poorly soluble). An additional advantage of this approach is the creation of a library of new P450 random variants. The chimeric protein is successfully expressed in its active form, displaying improved electrochemical properties [498].

10.6.3.2 Directed evolution of enzymes
In spite of the fact that biocatalysts with improved properties and notable successes have been reported by a rational approach, it is precisely this rationalism what constrains the possibilities of all these strategies. In fact, the redesign of a new biocatalyst may also require a myriad new capabilities that

are by far too ambitious to be approached by specific alterations of an enzyme's structure. But even in the case of pursuing one single target (e.g., a higher turnover number), it would be rather naive to think that this can be achieved by a point or a few specific modifications, neglecting the complex structure that an enzyme possesses to perform at a certain rate with a particular substrate. Similarly, and as previously mentioned, the stability of enzymes can hardly be featured or traced to a limited number of amino acids. In addition, most of the rational approaches are costly requiring considerable investments and knowledge and are slow to deliver.

In contrast to rational approaches, the directed evolution of enzymes is based on the search of useful functionalities in libraries randomly generated and on improvement by suitable and proper selection. The directed evolution combines two powerful and independent technologies: methods for the generation of random genetic libraries and strategies for the selection of variant enzymes with the specific capabilities [499–503]. This process can result in biocatalysts with non-natural proprieties, since the proteins are expressed in recombinant cells decoupled from its biological functions and evolved under unusual conditions. One additional advantage is the possibility to tailor not only individual proteins, but also the whole biosynthetic and catabolic pathways [471].

A scheme with the key steps in a typical enzyme evolution experiment is illustrated in Fig. 10.15. After the generation of the library of mutant genes (targeted into a specific region or distributed throughout the gene), the gene library/vector is inserted and expressed in an appropriate microorganism, which is individually screened for the specific propriety. The obtained mutant gene is again subjected to subsequent evolution cycles, in order to create a defined pressure leading to the formation and identification of the desired enzyme.

This evolutionary approach can be performed following two strategies. The first is based on an iterative process, where large changes in function are obtained by accumulating small changes over many generations; either sequentially or by recombination to minimize the number of silent or deleterious mutations [504]. In other words, mimicking the natural evolution process in the laboratory to acquire the desired function. The second approach is to direct a much higher level of random mutations to a relative small region of the gene [505], resulting in novel combinations with the desired functions, since the unfavourable effect of intermediates can be non-possible from single steps [471].

In order to generate molecular diversity, directed evolution makes use of random mutagenesis applying standard molecular tools (such as, error prone

Non-affinity sensing technology

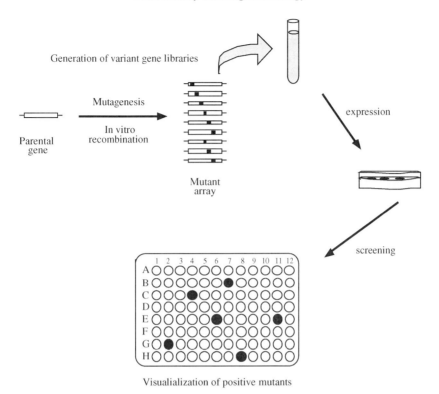

Fig. 10.15. Key steps of a typical directed enzyme evolution experiment.

PCR, phage display, or saturation mutagenesis), or in vitro recombination. In the second case, DNA shuffling, pioneered by Stemmer, has emerged as one of the most powerful techniques for in vitro protein evolution [504]. This technique creates hybrid gene libraries by homologous recombination of related gene parents. This "molecular sex" creates new genes that code for proteins with sequence information from any or all parents [471]. Then, genes from multiple parents and even from different species can be shuffled in a single step [506]—operations that are forbidden in nature but may be very useful for rapid adaptation [471]. This tool could be applied in families such as cytochrome P450 or estradiol dioxygenases, which are involved in many bacterial pathways for the degradation of aromatic compounds, as starting points to search for enhanced activity, stability and selectivity, particularly when introducing non-natural substrates or in non-natural conditions.

Directed evolution has emerged, within a short time, as a powerful strategy. The list of enzymes subjected to a targeted improvement is continuously increasing, due to the short time required to achieve the desired capability in comparison with other approaches. Some examples of environmental relevance include haloalkane dehalogenase [507] and the P450 family that have been exposed to directed evolution in order to enhance their stability and activity for relevant pollutants. Complex enzyme pathways have also been subjected to directed evolution [508]. The advantage of applying directed evolution to metabolic pathways is that it can optimize not only single enzymes but also interactions among them and, thus, the overall metabolic flux of the system [509–511]. Some examples in which directed evolution has been performed to tailor chemical transformations of environmental relevance are shown in Table 10.12.

Directed evolution requires both the efficient generation of molecular diversity and the effective selection of the variant enzymes. In this sense, the difficulties of generating sufficient diversity and the lack of universal selection procedures hamper the incorporation of new reactions in the evolved enzymes. Furthermore, since true selection based on the essential character of the desired capability or such capability carried on a clone that survives or grows faster is rare, one of the major challenges is the screening process for identifying the best mutants. Consequently, these mutants have to be retrieved on the basis of detecting the enzymatic activity. This process may be directed either towards a very few active mutants, in the simplest case, or the most active mutants from a moderately active background population [525]. Several techniques are used for this screening: pH change, fluorescence, heat evolution, or substrate depletion, etc. However, they all generally offer poor throughputs, particularly when high sensitivity is mandatory for accurate screening. Alternatively, new approaches are being developed. Some examples include confining the evolution process within a water–oil emulsion [526] or in a physical support [525], or by using substrate arrays as enzyme fingerprinting tools [527]. The importance of this step can be shown by the fact that a company, Maxygen, claims as its unique technological advantage, the high percentage of functional variants that can be found based on more definitive screening techniques (http://www.maxygen.com).

Despite the significant achievements already attained, this technique is still in its infancy. New experimental methods for creating genetic diversity, and advances in screening and selection technologies will reduce time and cost, making it possible to tackle more difficult problems, including complex systems [528].

10.6.3.3 Artificial enzymes

The whole range of technologies for the design of biocatalysts concludes with a completely synthetic approach escaping from the limitations posed by the

TABLE 10.12
Selected examples of enzymes of environmental relevance that have been redesigned by directed evolution

Biocatalyst	Environmental relevance	Target function	Enhancement factor k_{cat}/K_m	Amino acid substitutions	Reference
Cyt P450 family	Detoxification of xenobiotics				
Cyt P450$_{cam}$		Increased activity in peroxide shunt pathway, towards napthalene	5–20-fold	1–2	[512]
P450 BM3		Highly efficient alkane hydroxylase	Butane 108-fold (k_{cat}: 1800 min^{-1}), octane 38-fold, propane 57-fold	11	[513]
			Alkane epoxidation: benzene: new activity. styrene, 1-hexene, cyclohexene, and propylene: 17–100-fold		[514]
		High activity in organic solvents THF DMSO	10-fold in 2% THF 6-fold in 25% DMSO		[515]
Haloalkane dehalogenase	Detoxification effect: halogenated aliphatic family compounds	Selectivity 1,2,3-trichloropropane (TCP)	8	2	[516]
		Thermostability 55°C	30,000-fold	8 (mutations)	[517]

continued

TABLE 10.12 (Continuation)

Biocatalyst	Environmental relevance	Target function	Enhancement factor k_{cat}/K_m		Amino acid substitutions	Reference
Biphenyl dioxygenase	PCB biodegradation	Enhanced degradation of PCBs		Expanded selectivity for benzene and toluene	3–4	[518]
				Enhanced activity for 2,6-dichlorobiphenyl		[519]
Catechol 2,3-dioxygenase	Member of the superfamily of extradiol dioxygenase, key enzyme for dihydroxylated intermediates	Thermostability 65°C	66	13–26-fold	—	[520]
Toluene ortho-monooxygenase	Slow degradation of trichloroethylene and other toxic intermediates	Enhanced activity for chlorinated aliphatics: Trichloroethylene Naphthalene		2-fold 6-fold	1	[521]
Atrazine chlorohydrolase	Degradation of pesticides: atrazine	Enhanced activity and substrate range			11	[522]
Triazine hydrolases				150-fold new activities for: Prometron and Prometryn	9	[523]

Laccase	Use for pulp bleaching and delignification for the paper industry and degradation of polycyclic aromatic hydrocarbons	Enhanced expression in *Saccharomyces cerevisiae*. Improved activity and stability	10-fold expression 22-fold turnover number ABTS	13	[524]
Arsenate detoxification pathway		Arsenic resistance	12-fold (rate of arsenate reduction)	(3 mutations)	[508]

intrinsic chemistry of proteins. By doing so, it is anticipated that some of the inherent drawbacks of catalytic proteins can be suppressed. The rationale behind the design of enzyme mimics is then to prepare new molecules that, being inherently stable, could rival rates of acceleration and specificity.

These enzyme mimics, sometimes termed synzymes, can be constructed using different synthetic routes mainly based on the previously mentioned strategies: (i) the design approach, (ii) the transition state analogues or (iii) the catalytic activity-selection approach [453]. In the first case, a host molecule is modified with a functionality that is present in the natural counterpart and is known to be involved in the catalysed reaction. This is the case of some catalytic cyclodextrines, where a manganese porphyrin carrying four β-cyclodextrines results in a P450 mimic. When used with iodosobenzene as co-oxidant, the β-cyclodextrin tetramer is capable of a selective epoxidation of stilbene substrates as well as the hydroxylation of substrates with appropriate size containing *tert*-butylphenyl groups that can hydrophobically bind into the cyclodextrin cavities, with reaction rates comparable to its natural counterpart [529]. Catalytic MIPs are a younger field than catalytic antibodies but the basic concepts remain very similar [530]. Molecular imprinting is a polymerization process resulting in a porous polymer with binding sites, which can selectively bind the molecule with which they were imprinted. If the imprinting molecule is a transition state analogue, it should then show catalytic proprieties. The present scope of catalytic MIPs is still relatively limited. One of the most successful catalytic MIPs has been described by Wulff et al. who reported 588- and 1435-fold rates of acceleration for the hydrolysis of aryl carbonate and aryl carbamate, respectively [531].

These two strategies have the obvious weakness that the desired catalytic property is not in the screening criteria and thus, effective catalysts may often fail. The third approach substantially differs from the previous ones. A combinatorial library of polymers or peptides is created and then the pursued catalytic activity is screened. There are different methods to generate the chemical library and diversity here does not seem to be a problem, but, in common with directed evolution, screening in a parallel combinatorial fashion is not yet possible. An example of a combinatorial polymer capable of catalysing the biologically relevant dehydration of β-hydroxy ketone has been reported [532]. Presently, this approach is by far the least investigated but its capabilities may be further exploited along with improvements in combinatorial methods and high throughput techniques [453].

Finally, a completely different way for mimicking the catalytic reaction of enzymes consists in the use of artificial active sites. The use of a tris(2-2'-bipyridil) copper(II) chloride complex as a biomimetic tyrosinase catalyst allows the construction of a biosensor for several phenolic compounds with a good sensitivity and selectivity [533]. Although the synthetic enzyme field has been developing for many years, with some impressive advances, still this approach cannot be compared with natural enzymes or engineered proteins.

10.7 CONCLUDING REMARKS

The different routes for the exploitation of catalytic events in environmental analysis have been presented and discussed, reviewing available and demonstrated technology for a large range of chemicals. This area is obviously alive and well with good perspectives. The maturity of several bioanalytical approaches allows them to enter the stage of full demonstration. A more focussed approach is then necessary in order to avoid reiterations of already demonstrated technologies in favour of targeting the specific and relevant problems that the new European legislation is dictating. In this effort, a more active participation of stakeholders is necessary with the support of national and European resources to consolidate these initiatives. Legal compliance is not the unique goal, particularly for the research community which has the responsibility of offering scientific criteria to tackle the problems of new and emerging contaminants. This requires the exploration and exploitation of new technologies from protein engineering as described here but also of many others presented in this book. For most of them, the driving force for new research and development is not environmental applications although it is clear that this field also demands the exploitation of these technologies. Bioanalytical screening and monitoring, in addition to bioremediation techniques, are key applications for the preservation and maintenance of the environment.

Acknowledgements

The authors wish to acknowledge financial support from the EU in several RTD projects (EV5V-CT92-0109; QLK3-CT-2000-01481) as well as the European Thematic Networks BIOSET and SENPOL (EVK1-1999-20001). The Community of Madrid is acknowledged for project (07M/0071/2001). A. N. acknowledges the Spanish Ministry of Science and Technology for a research contract under the Ramón y Cajal program.

REFERENCES

1. R.J. Farrauto and R.M. Heck, *Catal. Today*, 55 (2000) 179.
2. OECD, *Biotechnology for Clean Industrial Products and Processes*. Organization for Economic Cooperation and Development, Paris, 1998.
3. OECD, *The Application of Biotechnology to Industrial Sustainability*. Organization for Economic Cooperation and Development, Paris, 2001.
4. E. Eljarrat and D. Barceló, *Trends Anal. Chem.*, 22 (2003) 655.
5. C.G. Daughton and T.A. Ternes, *Environ. Health Perspect.*, 107 (1999) 907.
6. O.A. Jones, N. Voulvoulis and J.N. Lester, *Environ. Toxicol.*, 22 (2001) 1383.
7. M. Petrovic, M. Solé, M.J. López de Alda and D. Barceló, *Environ. Toxicol. Chem.*, 21 (2002) 2146.
8. M.S. Díaz-Cruz, M.J. López de Alda and D. Barceló, *Trends Anal. Chem.*, 22 (2003) 340.
9. K.R. Rogers, *Biosens. Bioelectron.*, 10 (1995) 533.
10. R. Atlas, In: J.M. Lynch and A. Wiseman (Eds.), *Environmental Biomonitoring: The Biotechnology Ecotoxicology Interface*. Cambridge University Press, New York, 1998, pp. 147.
11. M. Campbell (Ed.), *Sensor Systems for Environmental Monitoring*. Kluwer Academic, Dordrecht, 1998.
12. U. Bilitewski and A.P.F. Turner (Eds.), *Biosensors for Environmental Monitoring*. Harwood Academic, Amsterdam, 2000.
13. D.P. Nikolelis, U.J. Krull, J. Wang and M. Mascini (Eds.), *Biosensors for Direct Monitoring of Environmental Pollutants in Field*, NATO ASI Series. Partnership Sub-Series 2, Environment, Vol. 38. Kluwer Academic, Dordrecht, 1997.
14. B. Hock (Ed.), *Biosensors for Environmental Diagnostics*. B.G. Teubner, Stuttgart, 1998.
15. A. Mulchandani and O.A. Sadik (Eds.), *Chemical and Biological Sensors for Environmental Monitoring*, ACS Symposium Series 762, American Chemical Society, Washington, DC, 2000.
16. E. Domínguez and S. Alcock (Eds.), *Proceedings of the 1st SENSPOL Workshop: Sensing Technologies for Contaminated Sites and Groundwater*. University of Alcalá, Alcalá de Henares, 2001.
17. D. Van der Lelie, P. Corbisier, W. Baeyens, S. Wuertz, L. Diels and M. Mergeay, *Res. Microbiol.*, 145 (1994) 67.
18. M.P. Marco and D. Barceló, *Meas. Sci. Technol.*, 7 (1996) 1547.
19. J.M. Van Emon, C.L. Gerlach and K. Bowman, *J. Chromatogr. B*, 715 (1998) 211.
20. K.R. Rogers and C.L. Gerlach, *Environ. Sci. Technol.*, 33 (1999) 500A.
21. C. Nistor and J. Emnéus, *Waste Manage.*, 19 (1999) 147.
22. I. Karube and Y. Nomura, *J. Mol. Catal. B- Enzym.*, 10 (2000) 177.
23. J. Rishpon, *Rev. Environ. Health*, 17 (2002) 219.
24. S. Kroger, S. Piletsky and A.P.F. Turner, *Mar. Pollut. Bull.*, 45 (2002) 24.
25. H. Nakamura and I. Karube, *Anal. Bioanal. Chem.*, 377 (2003) 446.

26 S. Rodríguez-Mozaz, M.P. Marco, M.J. López de Alda and D. Barceló, *Pure Appl. Chem.*, 76 (2004) 723.
27 A. Pulido, I. Ruisánchez, R. Boqué and F.X. Rius, *Trends Anal. Chem.*, 22 (2003) 647.
28 A. Ríos, B. Barceló, L. Buydens, S. Cárdenas, K. Heydorn, B. Karlberg, K. Klemm, B. Lendl, B. Milman, B. Neidhart, R.W. Stephany, A. Townshend, A. Zschunke and M. Valcárcel, *Accred. Qual. Assur.*, 8 (2003) 68.
29 M.A. Aon and A.C. Colaneri, *Appl. Soil Ecol.*, 18 (2001) 255.
30 M.C. Lobo and J.J. Ibáñez (Eds.), *Preserving Soil Quality and Soil Biodiversity. The Role of Surrogate Indicators*. IMIA/CCM–CSIC, Madrid, 2003.
31 R.P. Dick, In: C.E. Pankhurst, B.M. Doube and V.V.S.R. Gupta (Eds.), *Biological Indicators of Soil Health*. CAB International, Wallingford, NY, 1997, pp. 121–156.
32 R.L. Sinsabaugh, *Biol. Fertil. Soils*, 17 (1994) 69.
33 R.P. Dick, D.P. Breakwell and R.F. Turco, In: J.W. Doran and A.J. Jones (Eds.), *Methods for Assessing Soil Quality*, Special Publication, Vol. 49. Soil Science Society of America, Madison, WI, 1996, pp. 247–271.
34 P.G. Wells, M.H. Depledge, J.N. Butler, J.J. Manock and A.H. Knap, *Mar. Pollut. Bull.*, 42 (2001) 799.
35 T.S. Galloway, N. Millward, M.A. Browne and M.H. Depledge, *Aquat. Toxicol.*, 61 (2002) 169.
36 J.S. White and D.C. White, *Sourcebook of Enzymes*. CRC Press, Boca Raton, FL, 1997.
37 D. Thévenot, K. Toth, R.A. Durst and G.S. Wilson, *Biosens. Bioelectron.*, 16 (2001) 121.
38 G. Marko-Varga and E. Domínguez, *Trends Anal. Chem.*, 10 (1991) 290.
39 R. Freitag, *J. Chromatogr. B*, 722 (1999) 279.
40 L. Gorton, G. Marko-Varga, E. Domínguez and J. Emnéus, In: S. Lam and G. Malikin (Eds.), *Analytical Applications of Immobilised Enzyme Reactors*. Blackie Academic & Professional, London, 1994, pp. 1–21.
41 G. Pettersson, *Eur. J. Biochem.*, 184 (1989) 561.
42 F.W. Scheller, R. Renneberg and F. Schubert, *Methods Enzymol.*, 137 (1988) 29.
43 F.W. Scheller and U. Wollenberger, In: A.J. Bard and M. Stratmann (Eds.), *Encyclopedia of Electrochemistry*. Wiley-VCH, Amsterdam, 2002, pp. 431.
44 G. Ramsay (Ed.), *Commercial Biosensors: Applications to Clinical, Bioprocess, and Environmental Samples*. Wiley, London, 1998.
45 J.D. Newman, P.J. Warner, L.J. Tigwell and A.P.F. Turner, *Biosensors: An Inside View*. Cranfield Biotechnology Centre, Cranfield, 2002.
46 Business Communications Company, Inc., *Biosensors and Chemical Sensors (RC-053X)*, Norwalk, 1999.
47 Frost & Sullivan, *Biosensors: Emerging Technologies and Growth Opportunities (D247)*, 2002.

48 H.H. Weetall, *Biosens. Bioelectron.*, 14 (1999) 237.
49 W. Göpel, *Sens. Actuators B*, 52 (1998) 125.
50 S.N. Agathos and W. Reineke (Eds.), *Biotechnology for the Environment: Soil Remediation, Focus on Biotechnology*, Biotechnology Series, Vol. 3B. Kluwer Academic, Dordrecht, 2002.
51 R. Heinrich, E. Meléndez-Hevia and H. Cabezas, *Biochem. Mol. Biol. Educ.*, 30 (2002) 184.
52 S.M. Steinberg, E.J. Poziomek, W.H. Englemann and K.R. Rogers, *Chemosphere*, 30 (1995) 2155.
53 A.A. Bulich, K.K. Tung and G. Sheibner, *J. Biolumin. Chemilumin.*, 5 (1990) 71.
54 ISO, Water Quality, ISO 11348-1, 2, and 3, Geneva, 1998.
55 V.L.K. Jennings, M.H. Rayner-Brandes and D.J. Bird, *Water Res.*, 35 (2001) 3448.
56 F.S. Mowat and K.J. Bundy, *Environ. Int.*, 27 (2001) 479.
57 M. Gutiérrez, J. Etxebarria and L. de las Fuentes, *Water Res.*, 36 (2002) 919.
58 G.M. Sinclair, G.I. Paton, A.A. Meharg and K. Killham, *FEMS Microbiol. Lett.*, 174 (1999) 273.
59 B.J. Reid, K.T. Semple, C.J. MacLeod, H.J. Weitz and G.I. Paton, *FEMS Microbiol. Lett.*, 169 (1998) 227.
60 M. Farré and D. Barceló, *Trends Anal. Chem.*, 22 (2003) 299.
61 M. Farré, O. Pasini, M.C. Alonso, M. Castillo and D. Barceló, *Anal. Chim. Acta*, 426 (2001) 155.
62 H.G. Preuss and E.A.H. Hall, *Anal. Chem.*, 67 (1995) 1940.
63 M.R. Evans, G.M. Jordinson, D.M. Rawson and J.G. Rogerson, *Pestic. Sci.*, 54 (1998) 447.
64 P. Pandard, P. Vasseur and D.M. Rawson, *Water Res.*, 27 (1993) 427.
65 M.E. Polak, D.M. Rawson and B.G.D. Haggett, *Biosens. Bioelectron.*, 11 (1996) 1253.
66 S. Preston, N. Coad, J. Townend, K. Killham and G.I. Paton, *Environ. Toxicol. Chem.*, 19 (2000) 775.
67 G. Persoone, C. Jansen and W. De Coen (Eds.), *New Microbiotests for Routine Toxicity Screening and Biomonitoring*. Kluwer Academic/Plenum Publishers, New York, 2000.
68 H.W. Read, J.M. Harkin and K.E. Gustavson, In: P.W. Wells, K. Lee and C. Blaise (Eds.), *Microscale Testing in Aquatic Toxicology: Advances, Techniques and Practice*. CRC Press, Boca Raton, FL, 1998, 31–52.
69 M. Weideborg, E.A. Vik, G.D. Öfjord and O. Kjönnö, *Environ. Toxicol. Chem.*, 16 (1997) 384.
70 E. Argese, C. Bettiol, G. Giurin and P. Miana, *Chemosphere*, 38 (1999) 2281.
71 L. Knobeloch, G. Blondin and J. Harkin, *Environ. Toxicol. Water Qual.*, 9 (1994) 231.
72 J.K. Pollak and D.J. Oakes, *Toxicology*, 136 (1999) 41.

73 A.D. Betterman, J.M. Lazorchak and J.C. Dorofi, *Environ. Toxicol. Chem.*, 15 (1996) 319.
74 L. Knobeloch, G. Blondin and J. Harkin, *Bull. Contam. Toxicol.*, 44 (1990) 661.
75 I. Katakis and E. Domínguez, *Mikrochim. Acta*, 126 (1997) 11.
76 L. Gorton and E. Domínguez, In: A.J. Bard and M. Stratmann (Eds.), *Encyclopedia of Electrochemistry*. Wiley-VCH, Amsterdam, 2002, pp. 67–143.
77 I. Karube, T. Matsugana, S. Mitsuda and S. Suzuki, *Biotechnol. Bioeng.*, 19 (1977) 1535.
78 K. Riedel, In: A. Mulchandani and K.R. Rogers (Eds.), *Enzyme and Microbial Biosensors: Techniques and Protocols*. Humana Press, Totowa, NJ, 1998, pp. 199–224.
79 K. Riedel, G. Kunze and A. König, *Adv. Biochem. Eng. Biotechnol.*, 75 (2001) 81.
80 S. Rastogi, P. Rathee, T.K. Saxena, N.K. Mehra and R. Kumar, *Curr. Appl. Phys.*, 3 (2003) 191.
81 J. Liu, L. Björnsson and B. Mattiasson, *Biosens. Bioelectron.*, 14 (2000) 883.
82 K. Tag, A.W.K. Kwong, M. Lehmann, C.Y. Chan, R. Renneberg, K. Riedel and G. Kunze, *J. Chem. Technol. Biotechnol.*, 75 (2000) 1080.
83 P. Samuelson, E. Gunneriusson, P.A. Nygren and S. Ståhl, *J. Biotechnol.*, 96 (2002) 129.
84 M. Lehmann, C. Chan, A. Lo, M. Lung, K. Tag, G. Kunze, K. Riedel, B. Grundig and R. Renneberg, *Biosens. Bioelectron.*, 14 (1999) 295.
85 N. Pasco, K. Baronian, C. Jeffries and J. Hay, *Appl. Microbiol. Biotechnol.*, 53 (2000) 613.
86 K. Morris, K. Catterall, H. Zhao, N. Pasco and R. John, *Anal. Chim. Acta*, 442 (2001) 129.
87 N.F. Pasco, J.M. Hay and J. Webber, *Biomarkers*, 6 (2001) 83.
88 K. Catterall, H. Zhao, N. Pasco and R. John, *Anal. Chem.*, 75 (2003) 2584.
89 Y. Nomura, K. Ikebukuro, K. Yokoyama, T. Takeuchi, Y. Arikawa, S. Ohno and I. Karube, *Biosens. Bioelectron.*, 13 (1998) 1047.
90 J.J. Pancrazio, J.P. Whelan, D.A. Borkholder, W. Ma and D.A. Stenger, *Ann. Biomed. Eng.*, 27 (1999) 697.
91 S.F. D'Souza, *Biosens. Bioelectron.*, 16 (2001) 337.
92 M.D. Marazuela and M.C. Moreno-Bondi, *Anal. Bioanal. Chem.*, 372 (2002) 664.
93 T.S. Han, S. Sasaki, K. Yano, K. Ikebukuro, K. Atsushi, T. Nagamune and I. Karube, *Anal. Lett.*, 36 (2003) 539.
94 N.L. Turner, A. Horsburgh, G.I. Paton, K. Killham, A. Meharg, S. Primrose and N.J.C. Strachan, *Environ. Toxicol. Chem.*, 20 (2001) 2456.
95 B. Podola, E.C.M. Nowack and M. Melkonian, *Biosens. Bioelectron.*, 19 (2004) 1253.
96 D. Ivnitski, I. Abdel-Hamid, P. Aranasov and E. Wilkins, *Biosens. Bioelectron.*, 14 (1999) 599.
97 S. Silver and G. Ji, *Environ. Health Perspect.*, 102 (1994) 107.
98 D.A. Rouch, B.T.O. Lee and A.P. Morby, *J. Ind. Microbiol.*, 14 (1995) 132.

99   S. Kustu, A.K. North and D.S. Weiss, *Trends Biochem. Sci.*, 6 (1991) 397.
100 B.M. Willardson, J.F. Wilkins, T.A. Rand, J.M. Schupp, K.K. Hill, P. Keim and P.J. Jackson, *Appl. Environ. Microbiol.*, 64 (1998) 1006.
101 J.M.H. King, P.M. DiGrazia, B. Applegate, R. Burlage, J. Sanseverino, P. Dunbar, F. Larimer and G.S. Sayler, *Science*, 249 (1991) 778.
102 M. Virta, J. Lampinem and M. Karp, *Anal. Chem.*, 67 (1995) 667.
103 N. Peitzsch, G. Eberz and D.H. Nies, *Appl. Environ. Microbiol.*, 64 (1998) 453.
104 P. Corbisier, D. van der Lelie, B. Borremans, A. Provoost, V. de Lorenzo, N.L. Brown, J.R. Lloyd, J.L. Hobman, E. Csöregi, G. Johansson and B. Mattiasson, *Anal. Chim. Acta*, 387 (1999) 235.
105 J.O. Lappalainen, M.T. Karp, J. Nurmi, R. Juvonen and M.P.J. Virta, *Environ. Toxicol.*, 15 (2000) 443.
106 S. Rossbach, M.L. Kukuk, T.L. Wilson, S.F. Feng, M.M. Pearson and M.A. Fisher, *Environ. Microbiol.*, 2 (2000) 373.
107 I. Biran, R. Babai, K. Levcov, J. Rishpon and E.Z. Ron, *Environ. Microbiol.*, 2 (2000) 285.
108 A. Roda, P. Pasini, N. Mirasoli, M. Guardigli, C. Russo, M. Musiani and M. Baraldini, *Anal. Lett.*, 34 (2001) 29.
109 A. Ivask, K. Hakkila and M. Virta, *Anal. Chem.*, 73 (2001) 5168.
110 T. Petanen, M. Virta, M. Karp and M. Romantschuk, *Microb. Ecol.*, 41 (2001) 360.
111 F.F. Roberto, J.M. Barnes and F. Bruhn, *Talanta*, 58 (2002) 181.
112 T. Yamagata, M. Ishii, M. Narita, G.C. Huang and G. Endo, *Water Sci. Technol.*, 46 (2002) 253.
113 A. Heitzer, K. Malachowsky, J.E. Thonnard, P.R. Bienkowski, D.C. White and G.S. Sayler, *Appl. Environ. Microbiol.*, 60 (1994) 1487.
114 J.G. Dorn, R.J. Frye and R.M. Maier, *Appl. Environ. Microbiol.*, 69 (2003) 2209.
115 L. Bastiaens, D. Springael, W. Dejonghe, P. Wattiau, H. Verachtert and L. Diels, *Res. Microbiol.*, 152 (2001) 849.
116 V. de Lorenzo, S. Fernández, M. Herrero, U. Jakubzik and K.N. Timmis, *Gene*, 130 (1993) 41.
117 Y. Rozen, A. Nejidat, K.H. Gartemann and S. Belkin, *Chemosphere*, 38 (1998) 633.
118 S. Köhler, T.T. Bachmann, J. Schmitt, S. Belkin and R.D. Schmid, *Sens. Actuators B*, 70 (2000) 139.
119 S. Köhler, S. Belkin and R.D. Schmid, *Fresenius J. Anal. Chem.*, 366 (2000) 769.
120 X. Guan, S. Ramanathan, J.P. Garris, R.S. Shetty, M. Ensor, L.G. Bachas and S. Daunert, *Anal. Chem.*, 72 (2000) 2423.
121 A.A. Wise and C.R. Kuske, *Appl. Environ. Microbiol.*, 66 (2000) 163.
122 M. Tauber, R. Rosen and S. Belkin, *Talanta*, 55 (2001) 959.
123 S.M. Park, H.H. Park, W.K. Lim and H.J. Shin, *J. Biotechnol.*, 103 (2003) 227.
124 P. Sticher, M.C.M. Jaspers, H. Stemmler, H. Harms, A.J.B. Zehnder and J.R. van der Meer, *Appl. Environ. Microbiol.*, 63 (1997) 4053.

125 A.C. Layton, M. Muccini, M.M. Ghosh and G.S. Sayler, *Appl. Environ. Microbiol.*, 64 (1998) 5023.
126 Y. Ikariyama, S. Nishiguchi, T. Koyama, E. Kobatake, M. Aizawa, M. Tsuda and T. Nakazawa, *Anal. Chem.*, 69 (1997) 2600.
127 L. Stiner and L.J. Halverson, *Appl. Environ. Microbiol.*, 68 (2002) 1962.
128 S. Daunert, G. Barett, J.S. Feliciano, R.J. Shetty, S. Shrestha and W. Smith-Spencer, *Chem. Rev.*, 100 (2000) 2705.
129 S. Belkin, *Curr. Opin. Microbiol.*, 6 (2003) 206.
130 G.S. Sayler, C.D. Cox, R. Burlage, S. Ripp, D.E. Nivens, C. Werner, Y. Ahn and U. Matrubutham, In: R. Fass, Y. Flashner and S. Reuveny (Eds.), *Novel Approaches for Bioremediation of Organic Pollution.* Kluwer Academic/Plenum Press, New York, 1999, pp. 241.
131 G. Palmer, R. McFadzean, K. Kilham, A. Sinclair and G. Paton, *Chemosphere*, 36 (1998) 2683.
132 S. Ripp, D.E. Nivens, C. Werner and G.S. Sayler, *Appl. Microbiol. Biotechnol.*, 53 (2000) 736.
133 P.V. Attfield, H.Y. Choi, D.A. Veal and P.J.L. Bell, *Mol. Microbiol.*, 40 (2001) 1000.
134 I. Biran, D.M. Rissin, E.Z. Ron and D.R. Walt, *Anal. Biochem.*, 315 (2003) 106.
135 M.B. Elowitz, A.J. Levine, E.D. Siggia and P.S. Swain, *Science*, 297 (2002) 1183.
136 M. Mirasoli, J. Feliciano, E. Michelini, S. Daunert and A. Roda, *Anal. Chem.*, 74 (2002) 5948.
137 Y. Davidod, R. Rozen, D.R. Smulski, T.K. Van Dyk, A.C. Vollmetr, D.A. Elsmore, R.A. LaRossa and S. Belkin, *Mutat. Res.*, 466 (2000) 97.
138 M.B. Gu and G.C. Gil, *Biosens. Bioelectron.*, 16 (2001) 661.
139 M.B. Gu and S.H. Choi, *Water Sci. Technol.*, 43 (2001) 147.
140 A.M. Horsburgh, D.P. Mardlin, N.L. Turner, R. Henkler, N. Strachan, L.A. Glover, G.I. Paton and K. Killham, *Biosens. Bioelectron.*, 17 (2002) 495.
141 E.J. Kim, Y. Lee, J.E. Lee and M.B. Gu, *Water Sci. Technol.*, 46 (2002) 51.
142 O. Bechor, D.R. Smulski, T.K. Van Dyk, R.A. LaRossa and S. Belkin, *J. Biotechnol.*, 94 (2002) 125.
143 E. Iwuoha, I. Leister, E. Miland, M. Smyth and C. Fágáin, *Anal. Chem.*, 69 (1997) 1674.
144 E. Iwuoha, M. Smyth and M. Lyons, *Biosens. Bioelectron.*, 12 (1997) 53.
145 P.N. Bartlett, P. Tebbutt and R.G. Whitaker, *Prog. React. Kinet.*, 16 (1991) 55.
146 P. Bartlett and R.G. Whitaker, *J. Electroanal. Chem.*, 224 (1987) 27.
147 P. Bartlett and K. Pratt, *Biosens. Bioelectron.*, 8 (1993) 451.
148 C. Andrieux, J. Dumas-Bouchiat and J. Savéant, *J. Electroanal. Chem.*, 131 (1982) 1.
149 J. Leddy, A. Bard, J. Maloy and J. Savéant, *J. Electroanal. Chem.*, 187 (1985) 205.
150 W. Albery and A. Hillman, *J. Electroanal. Chem.*, 170 (1984) 27.
151 J. Kulys, V. Sorochinskii and R. Vidziunaite, *Biosensors*, 2 (1986) 135.
152 P. Bartlett and K. Pratt, *J. Electroanal. Chem.*, 397 (1995) 61.

153 M.E.G. Lyons, J.C. Greer, C.A. Fitzgerald, T. Bannon and P.N. Bartlett, *Analyst*, 121 (1996) 715.
154 J. Calvente, A. Narváez, E. Domínguez and R. Andreu, *J. Phys. Chem. B*, 107 (2003) 6629.
155 B. Limoges, J. Moiroux and J. Savéant, *J. Electroanal. Chem.*, 521 (2002) 8.
156 N. Durán, M.A. Rosa, A. D'Annibale and L. Gianfreda, *Enzyme Microb. Technol.*, 31 (2002) 907.
157 P. Paranjpe, S. Dutta, M. Karve, S. Padhye and R. Narayanaswamy, *Anal. Biochem.*, 294 (2001) 102.
158 F. Ortega, E. Domínguez, G. Jönsson-Pettersson and L. Gorton, *J. Biotechnol.*, 31 (1993) 289.
159 J. Wang, L. Fang and D. López, *Analyst*, 119 (1994) 455.
160 P. Önnerfjord, J. Emnéus, G. Marko-Varga, L. Gorton, F. Ortega and E. Domínguez, *Biosens. Bioelectron.*, 10 (1995) 607.
161 E. Burestedt, A. Narváez, T. Ruzgas, L. Gorton, J. Emnéus, E. Domínguez and G. Marko-Varga, *Anal. Chem.*, 68 (1996) 1605.
162 E.S.M. Lutz and E. Domínguez, *Electroanalysis*, 8 (1996) 117.
163 N. Pena, A.J. Reviejo and J.M. Pingarrón, *Talanta*, 55 (2001) 179.
164 J. Kulys and R.D. Schmid, *Anal. Lett.*, 23 (1990) 589.
165 H. Kotte, B. Gründig, K.-D. Vorlop, B. Strehlitz and U. Stottmeister, *Anal. Chem.*, 67 (1995) 65.
166 M. Hedenmo, A. Narváez, E. Domínguez and I. Katakis, *J. Electroanal. Chem.*, 425 (1997) 1.
167 S. Cosnier, J.J. Fombon, P. Labbe and D. Limosin, *Sens. Actuators B*, 59 (1999) 134.
168 S. Zhang, H. Zhao and R. John, *Anal. Chim. Acta*, 441 (2001) 95.
169 S.Q. Liu, J.H. Yu and H.X. Ju, *J. Electroanal. Chem.*, 540 (2003) 61.
170 A. Sánchez-Ferrer, J. Rodríguez-López, F. García-Cánovas and F. García-Carmona, *Biochim. Biophys. Acta*, 1247 (1995) 1.
171 M.G. Peter and U. Wollenberger, In: F.W. Scheller, F. Schubert and J. Fedrowitz (Eds.), *Frontiers in Biosensorics*, Vol. I. Birkhäuser, Basel, 1997, pp. 63.
172 C. Wittman, K. Riedel and R.D. Schmid, In: E. Kress-Rogers (Ed.), *Handbook of Biosensors and Electronic Noses*. CRC Press, Frankfurt, 1997, pp. 299.
173 L. Coche-Guerente, P. Labbé and V. Mengeaud, *Anal. Chem.*, 73 (2001) 3206.
174 H. Notsu, T. Tatsuma and A. Fujishima, *J. Electroanal. Chem.*, 523 (2004) 86.
175 C. Nistor, J. Emnéus, L. Gorton and A. Ciucu, *Anal. Chim. Acta*, 387 (1999) 309.
176 K. Streffer, E. Vijgenboom, A. Tepper, A. Makower, F.W. Scheller, G.W. Canters and U. Wollenberger, *Anal. Chim. Acta*, 427 (2001) 201.
177 J. Parellada, A. Narváez, M.A. López, E. Domínguez, J.J. Fernández, V. Pavlov and I. Katakis, *Anal. Chim. Acta*, 362 (1998) 47.
178 J. Svitel and S. Miertus, *Environ. Sci. Technol.*, 32 (1998) 828.

179 C. Nistor, A. Rose, M. Farré, L. Stoica, U. Wollenberger, T. Ruzgas, D. Pfeiffer, D. Barceló, L. Gorton and J. Emnéus, *Anal. Chim. Acta*, 456 (2002) 3.
180 T.M. Anh, S.V. Dzyadevych, A.P. Soldatkin, N.D. Chien, N. Jaffrezic-Renault and J.M. Chovelon, *Talanta*, 56 (2002) 627.
181 A. Wolf, R. Huttl and G. Wolf, *J. Therm. Anal. Calorim.*, 61 (2000) 37.
182 A.I. Yaropolov, A.N. Kharybin, J. Emnéus, G. Marko-Varga and L. Gorton, *Anal. Chim. Acta*, 308 (1995) 137.
183 G. Marko-Varga, J. Emnéus, L. Gorton and T. Ruzgas, *Trends Anal. Chem.*, 14 (1996) 319.
184 S.C. Chang, K. Rawson and C.J. McNeil, *Biosens. Bioelectron.*, 17 (2002) 1015.
185 B. Serra, B. Benito, L. Agui, A.J. Reviejo and J.M. Pingarrón, *Electroanalysis*, 13 (2001) 693.
186 T. Ruzgas, J. Emnéus, L. Gorton and G. Marko-Varga, *Anal. Chim. Acta*, 311 (1995) 245.
187 B. Haghighi, L. Gorton, T. Ruzgas and L.J. Jönsson, *Anal. Chim. Acta*, 487 (2003) 3.
188 J. Kulys and R. Vidziunaite, *Biosens. Bioelectron.*, 18 (2003) 319.
189 R.S. Freire, M.M.C. Ferreira, N. Duran and L.T. Kubota, *Anal. Chim. Acta*, 485 (2003) 263.
190 T. Ruzgas, E. Csöregi, J. Emnéus, L. Gorton and G. Marko-Varga, *Anal. Chim. Acta*, 330 (1996) 123.
191 F.D. Munteanu, A. Lindgren, J. Emnéus, L. Gorton, T. Ruzgas, E. Csöregi, A. Ciucu, R.B. van Huystee, I.G. Gazaryan and L.M. Lagrimini, *Anal. Chem.*, 70 (1998) 2596.
192 S.S. Rosatto, G.D. Neto and L.T. Kubota, *Electroanalysis*, 13 (2001) 445.
193 S. Cosnier and I.C. Popescu, *Anal. Chim. Acta*, 319 (1996) 145.
194 S.E. Stanca, I.C. Popescu and L. Oniciu, *Talanta*, 61 (2003) 501.
195 H. Notsu and T. Tatsuma, *J. Electroanal. Chem.*, 566 (2004) 379.
196 A.L. Ghindilis, A. Makower, C.G. Bauer, F.F. Bier and F.W. Scheller, *Anal. Chim. Acta*, 304 (1995) 25.
197 F.F. Bier, E. Ehrentreich-Förster, F.W. Scheller, A. Makower, A.V. Ermenko, U. Wollenberger, C.G. Bauer, D. Pfeiffer and D. Michael, *Sens. Actuators B*, 33 (1996) 5.
198 J. Wang and Q. Chen, *Anal. Chim. Acta*, 312 (1995) 39.
199 R.S. Freire, S. Thongngamdee, N. Durán, J. Wang and L.T. Kubota, *Analyst*, 127 (2002) 258.
200 J. Wang, J. Lu, S.Y. Ly, M. Vuki, B. Tian, W.K. Adeniyi and R.A. Armendariz, *Anal. Chem.*, 72 (2000) 2659.
201 U. Wollenberger and B. Neumann, *Electroanalysis*, 9 (1997) 366.
202 C. Saby, K.B. Male and J.H.T. Luong, *Anal. Chem.*, 69 (1997) 4324.
203 D. Cybulski, K.B. Male, J.M. Scharer, M. Moo-Young and J.H.T. Luong, *Environ. Sci. Technol.*, 33 (1999) 796.
204 W.R. Hussein and G.G. Guilbault, *Anal. Chim. Acta*, 72 (1974) 381.

205 W.R. Hussein and G.G. Guilbault, *Anal. Chim. Acta*, 76 (1975) 183.
206 C.-H. Kiang, S.S. Kuan and G.G. Guilbault, *Anal. Chim. Acta*, 80 (1975) 209.
207 C.J. Kay, L.P. Solomonson and M.J. Barber, *Biochemistry*, 30 (1991) 11445.
208 B.A. Averill, *Chem. Rev.*, 96 (1996) 2951.
209 W.G. Zumft, *Microbiol. Mol. Biol. Rev.*, 61 (1997) 553.
210 I. Willner, E. Katz and N. Lapidot, *Bioelectrochem. Bioenerg.*, 29 (1992) 29.
211 A. Narváez, E. Domínguez, I. Katakis, E. Katz, K.T. Ranjit, I. Ben Dov and I. Willner, *J. Electroanal. Chem.*, 430 (1997) 227.
212 F. Patolsky, E. Katz, V. Heleg-Shabtai and I. Willner, *Chem. Eur. J.*, 4 (1998) 1068.
213 E. Katz, S.V. Heleg, A. Bardea, I. Willner, H.K. Rau and W. Haehnel, *Biosens. Bioelectron.*, 13 (1998) 741.
214 C.H. Kiang, S.S. Kuan and G.G. Guilbault, *Anal. Chem.*, 50 (1978) 1319.
215 R.B. Mellor, J. Ronnenberg, W.H. Campbell and S. Diekmann, *Nature*, 355 (1992) 717.
216 B. Strehlitz, B. Gründing, K.D. Vorlop, P. Bartholmes, H. Kotte and U. Strottmeister, *Fresenius J. Anal. Chem.*, 349 (1994) 676.
217 I. Willner, A. Riklin and N. Lapidot, *J. Am. Chem. Soc.*, 112 (1990) 6438.
218 S. Cosnier, C. Innocent and Y. Jouanneau, *Anal. Chem.*, 66 (1994) 3198.
219 G. Ramsay and S.M. Wolpert, *Anal. Chem.*, 71 (1999) 504.
220 D. Kirstein, L. Kirstein, F. Scheller, H. Borcherding, J. Ronnenberg, S. Diekmann and P. Steinrücke, *J. Electroanal. Chem.*, 474 (1999) 43.
221 I. Willner, V. Heleg-Shabtai, E. Katz, H.K. Rau and W. Haehnel, *J. Am. Chem. Soc.*, 121 (1999) 6455.
222 N.F. Ferreyra, L. Coche-Guerente, P. Labbe, E.J. Calvo and V.M. Solis, *Langmuir*, 19 (2003) 3864.
223 S. Sasaki, I. Karube, N. Hirota, Y. Arikawa, M. Nishiyama, M. Kukimoto, S. Horinouchi and T. Beppu, *Biosens. Bioelectron.*, 13 (1998) 1.
224 B. Strehlitz, B. Gründig, W. Schumacher, P.M.H. Kroneck, K.-D. Vorlop and H. Kotte, *Anal. Chem.*, 68 (1996) 807.
225 T. Kohzuma, S. Shidara and S. Suzuki, *Bull. Chem. Soc.*, 67 (1994) 138.
226 Q. Wu, G.D. Storrier, F. Pariente, Y. Wang, J.P. Shapleigh and H.D. Abruña, *Anal. Chem.*, 69 (1997) 4856.
227 L.H. Larsen, T. Kjaer and N.P. Revsbech, *Anal. Chem.*, 69 (1997) 3527.
228 B. Ballarin, C. Brumlik, D. Lawson, W. Liang, L. Van Dyke and C. Martin, *Anal. Chem.*, 64 (1992) 2647.
229 L.M. Moretto, P. Ugo, M. Zanata, P. Guerriero and C.R. Martin, *Anal. Chem.*, 70 (1998) 2163.
230 N.F. Ferreyra, S.A. Dassie and V.M. Solis, *J. Electroanal. Chem.*, 486 (2000) 126.
231 S.A. Glazier, E.R. Campbell and W.H. Campbell, *Anal. Chem.*, 70 (1998) 1511.
232 M. Zayats, A.B. Kharitonov, E. Katz and I. Willner, *Analyst*, 126 (2001) 652.
233 J.W. Aylott, D.J. Richardson and D.A. Russell, *Analyst*, 122 (1997) 77.

234 C.C. Rosa, H.J. Cruz, M. Vidal and A.G. Oliva, *Biosens. Bioelectron.*, 17 (2002) 45.
235 C.J. Patton, A.E. Fischer, W.H. Campbell and E.R. Campbell, *Environ. Sci. Technol.*, 36 (2002) 729.
236 B. Strehlitz, B. Gründig and H. Kopinke, *Anal. Chim. Acta*, 403 (2000) 11.
237 A.K. Abass, J.P. Hart, D.C. Cowell and A. Chappell, *Anal. Chim. Acta*, 373 (1998) 1.
238 H. Tiessen (Ed.), *Phosphorous in the Global Environment: Transfers, Cycles and Management*. Wiley, Chichester, 1995.
239 S.O. Engblom, *Biosens. Bioelectron.*, 13 (1998) 981.
240 A. Amine and G. Palleschi, *Anal. Lett.*, 37 (2004) 1.
241 T. Kubo, M. Inagawa, T. Sugawara, Y. Arikawa and I. Karube, *Anal. Lett.*, 24 (1991) 1711.
242 K. Ikebukoro, H. Wakamaru, I. Karube, I. Kubo, M. Inagawa, T. Sugawara, Y. Arikawa, M. Suzuki and T. Takeuchi, *Biosens. Bioelectron.*, 10/11 (1996) 959.
243 H. Nakamura, K. Ikebukuro, S. McNiven, I. Karube, H. Yamamoto, K. Hayashi, M. Suzuki and I. Kubo, *Biosens. Bioelectron.*, 12 (1997) 959.
244 H. Nakamura, H. Tanaka, M. Hasegawa, Y. Masuda, Y. Arikawa, Y. Nomura, K. Ikebukuro and I. Karube, *Talanta*, 50 (1999) 799.
245 S. Saheki, A. Takeda and T. Shimazu, *Anal. Biochem.*, 148 (1985) 277.
246 N. Gajovic, K. Habermüller, A. Warsinke, W. Schuhmann and F.W. Scheller, *Electroanalysis*, 11 (1999) 1377.
247 N. Gajovic, G. Binyamin, A. Warsinke, F.W. Scheller and A. Heller, *Anal. Chem.*, 72 (2000) 2963.
248 M. Suzuki, H. Kurata, Y. Inoue, Y. Shin, I. Kubo, H. Nakamura, K. Ikebukuro and I. Karube, *Denki Kagaku*, 66 (1998) 579.
249 G.G. Guilbault and T. Cserfalvi, *Anal. Lett.*, 9 (1976) 277.
250 J.J. Fernández, J.R. López, X. Correig and I. Katakis, *Sens. Actuators B*, 47 (1998) 13.
251 U. Wollenberger and F.W. Scheller, *Biosens. Bioelectron.*, 8 (1993) 291.
252 E. Watanabe, H. Endo and K. Toyama, *Biosens. Bioelectron.*, 3 (1988) 297.
253 K.B. Male and J.H.T. Luong, *Biosens. Bioelectron.*, 6 (1991) 581.
254 E.M. D'Urso and R.P. Coulet, *Anal. Chim. Acta*, 281 (1993) 535.
255 J. Kulys, I.J. Higgings and J.V. Bannister, *Biosens. Bioelectron.*, 6 (1992) 581.
256 U. Wollenberger, F. Schubert and F.W. Scheller, *Sens. Actuators B*, 7 (1992) 412.
257 M.J. Vazquez, B. Rodriguez, C. Zapatero and D.G. Tew, *Anal. Biochem.*, 320 (2003) 292.
258 C. Menzel, T. Lerch, T. Scheper and K. Schügerl, *Anal. Chim. Acta*, 317 (1995) 259.
259 N. Conrath, B. Gründing, S. Hüwel and K. Cammann, *Anal. Chim. Acta*, 309 (1995) 47.
260 S. Hüwel, L. Haalck, N. Conrath and F. Spener, *Enzyme Microb. Technol.*, 21 (1996) 413.

261 C. Mousty, S. Cosnier, D. Shan and S.L. Mu, *Anal. Chim. Acta*, 443 (2001) 1.
262 F. Mizutani, S. Yabuki, Y. Sato, T. Sawaguchi and S. Iijima, *Electrochim. Acta*, 45 (2000) 2945.
263 V.G. Gavalas and N.A. Chaniotakis, *Anal. Chim. Acta*, 427 (2001) 271.
264 H. Nakamura, M. Hasegawa, Y. Nomura, K. Ikebukuro, Y. Arikawa and I. Karube, *Anal. Lett.*, 36 (2003) 1805.
265 D.M. Munnecke, *J. Agric. Food Chem.*, 28 (1980) 105.
266 V. Lewis, W. Donarski, J. Wild and F. Raushel, *Biochemistry*, 27 (1988) 1591.
267 A. Mulchandani, W. Chen, P. Mulchandani, J. Wang and K.R. Rogers, *Biosens. Bioelectron.*, 16 (2001) 225.
268 A.K. Singh, A.W. Flounders, J.V. Volponi, C.S. Ashley, K. Wally and J.S. Schoeniger, *Biosens. Bioelectron.*, 14 (1999) 703.
269 A.W. Flounders, A.K. Singh, J.V. Volponi, S.C. Carichner, K. Wally, A.S. Simonian, J.R. Wild and J.S. Schoeniger, *Biosens. Bioelectron.*, 14 (1999) 715.
270 A. Mulchandani, S. Pan and W. Chen, *Biotechnol. Prog.*, 15 (1999) 130.
271 M.J. Schöning, M. Arzdorf, P. Mulchandani, W. Chen and A. Mulchandani, *Sens. Actuators B*, 91 (2003) 92.
272 A.L. Simonian, B.D. diSioudi and J.R. Wild, *Anal. Chim. Acta*, 389 (1999) 189.
273 P. Mulchandani, A. Mulchandani, I. Kaneva and W. Chen, *Biosens. Bioelectron.*, 14 (1999) 77.
274 A. Mulchandani, P. Mulchandani and W. Chen, *Anal. Chem.*, 71 (1999) 2246.
275 S.H. Chough, A. Mulchandani, P. Mulchandani, W. Chen, J. Wang and K.R. Rogers, *Electroanalysis*, 14 (2002) 273.
276 P. Mulchandani, W. Chen and A. Mulchandani, *Environ. Sci. Technol.*, 35 (2001) 2562.
277 V. Sacks, I. Eshkenazi, T. Neufeld, C. Dosoretz and J. Rishpon, *Anal. Chem.*, 72 (2000) 2055.
278 J. Wang, R. Krause, K. Block, M. Musameh, A. Mulchandani, P. Mulchandani, W. Chen and M.J. Schoning, *Anal. Chim. Acta*, 469 (2002) 197.
279 J. Wang, L. Chen, A. Mulchandani, P. Mulchandani and W. Chen, *Electroanalysis*, 11 (1999) 866.
280 A. Mulchandani, I. Kaneva and W. Chen, *Anal. Chem.*, 70 (1998) 5042.
281 A. Mulchandani, P. Mulchandani, I. Kaneva and W. Chen, *Anal. Chem.*, 70 (1998) 4140.
282 A.L. Simonian, J.K. Grimsley, A.W. Flounders, J.S. Schoeniger, T.C. Cheng, J.J. DeFrank and J.R. Wild, *Anal. Chim. Acta*, 442 (2001) 15.
283 A.L. Simonian, E.N. Efremenko and J.R. Wild, *Anal. Chim. Acta*, 444 (2001) 179.
284 F. Battaglia-Brunet, S. Foucher, A. Denamur, I. Ignatiadis, C. Michel and D. Morin, *J. Ind. Microbiol. Biotechnol.*, 28 (2002) 154.
285 B. Chardin, A. Dolla, F. Chaspoul, M.L. Fardeau, P. Gallice and M. Bruschi, *Appl. Microbiol. Biotechnol.*, 60 (2002) 352.

286 M. Assfalg, I. Bertini, M. Bruschi, C. Michel and P. Turano, *Proc. Natl Acad. Sci. USA*, 99 (2002) 9750.
287 C. Michel, M. Brugna, C. Aubert, A. Bernadac and M. Bruschi, *Appl. Microbiol. Biotechnol.*, 55 (2001) 95.
288 D.R. Lovle, P.K. Widman, J.C. Woodward and J.P. Phillis, *Appl. Environ. Microbiol.*, 59 (1993) 3572.
289 C. Cervantes, J. Campos-Garcia, S. Devars, F. Gutiérrez-Corona, H. Loza-Tavera, J. Torres-Guzman and R. Moreno-Sanchez, *FEMS Microbiol. Rev.*, 25 (2001) 335.
290 C. Michel, F. Battaglia-Brunet, C. Tran Minh, M. Bruschi and I. Ignatiadis, *Biosens. Bioelectron.*, 19 (2003) 345.
291 F.P. Guengerich, In: P.R. Ortiz de Montellano (Ed.), *Cytochrome P450*. Plenum Press, New York, 1995, pp. 473.
292 K.N. Degtyarenko, *Protein Eng.*, 8 (1995) 737.
293 F.P. Guengerich, *Chem.-Biol. Interact.*, 106 (1997) 161.
294 F.P. Guengerich, *Drug Dev. Res.*, 49 (2000) 4.
295 K. Nakamura, I.H. Hanna, H. Cai, Y. Nishimura, K.M. Williams and F.P. Guengerich, *Anal. Biochem.*, 192 (2001) 280.
296 F. Schubert and F.W. Scheller, *Methods Enzymol.*, 137 (1988) 152.
297 N. Sugihara, Y. Ogona, K. Abe, Y. Murakami, Y. Kondo and T. Akaike, *Polym. Adv. Technol.*, 9 (1998) 858.
298 C.R. Crespi, V.P. Miller and B.W. Penman, *Anal. Biochem.*, 248 (1997) 188.
299 R.W. Estabrook, K.M. Faulker, M.S. Shet and C.W. Fischer, *Methods Enzymol.*, 272 (1996) 44.
300 K.M. Faulkner, M.S. Shet, C.W. Fisher and R.W. Estabrook, *Proc. Natl Acad. Sci. USA*, 92 (1995) 7705.
301 V.V. Shumyantseva, T.V. Bulko, S.A. Alexandrova, N.N. Sokolov, R.D. Schmid, T.T. Bachmann and A.I. Archakov, *Biochem. Biophys. Res. Commun.*, 263 (1999) 678.
302 V.V. Shumyantseva, T.V. Bulko, T.T. Bachmann, U. Bilitewski, R.D. Schmid and A.I. Archakov, *Arch. Biochem. Biophys.*, 377 (2000) 43.
303 E.I. Iwuoha and M.R. Smyth, *Biosens. Bioelectron.*, 18 (2003) 237.
304 N. Bistolas, A. Christenson, T. Ruzgas, C. Jung, F.W. Scheller and U. Wollenberger, *Biochem. Biophys. Res. Commun.*, 314 (2004) 810.
305 V.V. Shumyantseva, T.V. Bulko, R.D. Schmid and A.I. Archakov, *Biosens. Bioelectron.*, 17 (2002) 233.
306 M.P. Mayhew, V. Reipa, M.J. Holden and V.L. Vilker, *Biotechnol. Prog.*, 16 (2000) 610.
307 P.A. England, C.F. Harford-Cross, J.A. Stevenson, D.A. Rouch and L.L. Wong, *FEBS Lett.*, 424 (1998) 271.
308 S.G. Bell, J.A. Stevenson, H.D. Boyd, S. Campbell, A.D. Riddle, E.L. Orton and L.L. Wong, *Chem. Commun.*, (2002) 490.
309 C.E. Castro, R.S. Wade and N.O. Belser, *Biochemistry*, 24 (1985) 204.

310 M.E. Walsh, P. Kytitsis, N.A. Eady and L.L. Wong, *Eur. J. Biochem.*, 267 (2000) 5815.
311 Y.M. Lvov, Z. Lu, J.B. Schenkman, X. Zu and J.F. Rusling, *J. Am. Chem. Soc.*, 120 (1998) 4073.
312 M. Hara, Y. Yasuda, H. Toyotama, H. Ohkawa, T. Nozawa and J. Miyake, *Biosens. Bioelectron.*, 17 (2002) 173.
313 J.W. Anderson, S.S. Rossi, R.H. Tukey, T. Vu and L.C. Quattrochi, *Environ. Toxicol. Chem.*, 14 (1995) 1159.
314 V.A. McFarland, C.V. Ang, D.D. McCant, L.S. Inouye and A.S. Jarvis, US-Army Engineer Research and Development Center (TN-DOER-C8), Vicksburg, 1999.
315 P. Patnaik, *Handbook of Environmental Analysis. Chemical Pollutants in Air, Water and Soil and Soil Wastes*. CRC Press, Boca Raton, FL, 1997.
316 J.S. Noble, C.R. Strang and P.R. Michael, *Am. Ind. Hyg. Assoc.*, 54 (1993) 723.
317 E.P. Agency, *Fed. Reg.*, 49 (1984) 21870.
318 T. Higgenbottam, C. Feyerdbank and T.J.H. Clark, *Thorax*, 35 (1980) 246.
319 K. Schmid, K.H. Schaller, J. Angerer and G. Lehnert, *Zent. Bl. Hyg. Umweltmed.*, 196 (1994) 139.
320 G.G. Guilbault, *Anal. Chem.*, 55 (1983) 1682.
321 J.H.T. Luong and G.G. Guilbault, In: L.J. Blum and P.R. Coulet (Eds.), *Biosensor Principles and Practice*. Marcel Dekker, New York, 1991, pp. 107–138.
322 M.J. Dennison, J.M. Hall and A.P.F. Turner, *Analyst*, 121 (1996) 1769.
323 M. Hämmerle, E.A.H. Hall, N. Cade and D. Hodgins, *Biosens. Bioelectron.*, 11 (1996) 239.
324 F. Vianello, A. Stefani, M.L. Di Paolo, A. Rigo, A. Lui, B. Margesin, M. Zen, M. Scarpa and G. Soncini, *Sens. Actuators B*, 37 (1996) 49.
325 D.S. Schechter and P.C. Singer, *Ozone Sci. Eng.*, 17 (1995) 53.
326 Y.I. Korpan, A.P. Soldatkin, M.V. Gonchar, A.A. Sibirny, T.D. Gibson and A.V. El'skaya, *J. Chem. Technol. Biotechnol.*, 68 (1997) 209.
327 R. Lammert, I. Ogbomo, T. Baumeister, J. Danzer, R. Kittsteiner-Eberle and H.L. Schmidt, *GBE Monogr.*, 13 (1989) 93.
328 B. Leca and J.L. Marty, *Biosens. Bioelectron.*, 12 (1997) 1083.
329 K. Délécouls-Servat, A. Bergel and R. Basséguy, *J. Appl. Electrochem.*, 31 (2001) 1095.
330 K.K.W. Mak, U. Wollenberger, F.W. Scheller and R. Renneberg, *Biosens. Bioelectron.*, 18 (2003) 1095.
331 M.L. Curri, A. Agostiano, G. Leo, A. Mallardi, P. Cosma and M. Della Monica, *Mater. Sci. Eng. C*, 22 (2002) 449.
332 V. Pardo-Yissar, E. Katz, J. Wasserman and I. Willner, *J. Am. Chem. Soc.*, 125 (2003) 622.
333 Y.I. Korpan, M.V. Gonchar, A.A. Sibirny, C. Martelet, A.V. El'skaya, T.D. Gibson and A.P. Soldatkin, *Biosens. Bioelectron.*, 15 (2000) 77.

334 K.J.M. Sandström, J. Newman, A.L. Sunesson, J.O. Levin and A.P.F. Turner, *Sens. Actuators B*, 70 (2000) 182.

335 K.J.M. Sandström, A.L. Sunesson, J.O. Levin and A.P.F. Turner, *J. Environ. Monit.*, 5 (2003) 477.

336 H. Schulze, S. Vorlova, F. Villatte, T.T. Bachmann and R.D. Schmid, *Biosens. Bioelectron.*, 18 (2003) 201.

337 M. Trojanowicz, *Electroanalysis*, 14 (2002) 1311.

338 S. Sole, A. Merkoci and S. Alegret, *Crit. Rev. Anal. Chem.*, 33 (2003) 127.

339 S. Sole, A. Merkoci and S. Alegret, *Crit. Rev. Anal. Chem.*, 33 (2003) 89.

340 E. Usden, *Anticholinesterases Agents*. Pergamon Press, London, 1970.

341 M. Wiegand-Rosinus, U. Obst, K. Haberer and A. Wild, *Vom Wasser*, 71 (1990) 225.

342 E. Merian, *Metals and Their Compounds in the Environment*. VCH, Weinheim, 1991.

343 G.A. Evtugyn, H.C. Budnikov and E.B. Nikolska, *Talanta*, 46 (1998) 465.

344 T.T. Bachmann, B. Leca, F. Vilatte, J.L. Marty, D. Fournier and R.D. Schmid, *Biosens. Bioelectron.*, 15 (2000) 193.

345 R.A. Copeland, *Enzymes: A Practical Introduction to Structure, Mechanisms, and Data Analysis*. Wiley-VCH, New York, 2000.

346 B. Leca, T. Noguer and J.L. Marty, In: U. Bilitewski and A.P.F. Turner (Eds.), *Biosensors for Environmental Monitoring*. Harwood Academic, Amsterdam, 2000, pp. 76–86.

347 J.-C. Gayet, A. Haouz, A. Geloso-Meyer and C. Burstein, *Biosens. Bioelectron.*, 8 (1993) 117.

348 G.A. Evtugyn, H.C. Budnikov and E.B. Nokolskaya, *Analyst*, 121 (1996) 1911.

349 G.A. Evtugyn, E.P. Rizaeva, E. Stoikova, V.Z. Latipova and H. Budnikov, *Electroanalysis*, 9 (1997) 1124.

350 P. Skládal, *Anal. Chim. Acta*, 269 (1992) 281.

351 J.L. Marty, K. Sode and I. Karube, *Electroanalysis*, 4 (1992) 249.

352 M. Bazelyansky, E. Robey and J.F. Kirsch, *Biochemistry*, 25 (1986) 125.

353 V.G. Andreou and Y.D. Clonis, *Biosens. Bioelectron.*, 17 (2002) 61.

354 P. Cunniff (Ed.), *Official methods of analysis, Pesticide and Industrial Chemical Residues*, 16th ed., 5th rev. ed., Vol. I. AOAC International, Gaithersburg, MD, 1999.

355 S.V. Dzyadevych, A.P. Soldatkin, Y.I. Korpan, V.N. Arkhypova, A.V. El'skaya, J.M. Chovelon, C. Martelet and N. Jaffrezic-Renault, *Anal. Bioanal. Chem.*, 377 (2003) 496.

356 S.V. Dzyadevych, A.P. Soldatkin, V.N. Arkhypova, A.V. El'skaya, J.M. Chovelon, C.A. Georgiou, C. Martelet and N. Jaffrezic-Renault, *Sens. Actuators B*, (2004) in press.

357 H.S. Lee, Y.A. Kim, D.H. Chung and Y.T. Lee, *Int. J. Food Sci. Technol.*, 36 (2001) 263.

358 H.S. Lee, Y.A. Kim, Y.A. Cho and Y.T. Lee, *Chemosphere*, 46 (2002) 571.

359 J.C. Fernando, K.R. Rogers, N.A. Anis, J.J. Valdes, R.G. Thompson, A.T. Eldefrawi and M.E. Eldefrawi, *J. Agric. Food Chem.*, 41 (1993) 511.
360 V.N. Arkhypova, S.V. Dzyadevych, A.P. Soldatkin, A.V. El'skaya, N. Jaffrezic-Renault, H. Jaffrezic and C. Martelet, *Talanta*, 55 (2001) 919.
361 A. Navas Díaz and M.C. Ramos Peinado, *Sens. Actuators B*, 38–39 (1997) 426.
362 J.W. Choi, Y.K. Kim, I.H. Lee, J.H. Min and W.H. Lee, *Biosens. Bioelectron.*, 16 (2001) 937.
363 N. Jaffrezic-Renault, *Sensors*, 1 (2001) 60.
364 J. Kulys and E.J. D'Costa, *Biosens. Bioelectron.*, 6 (1991) 109.
365 P. Sklàdal and M. Mascini, *Biosens. Bioelectron.*, 7 (1992) 335.
366 J.L. Marty, N. Mionetto, S. Lacorte and D. Barceló, *Anal. Chim. Acta*, 311 (1995) 265.
367 E.V. Gogol, G.A. Evtugyn, J.L. Marty, H.C. Budnikov and V.G. Winter, *Talanta*, 53 (2000) 379.
368 S. Andreescu, L. Barthelmebs and J.L. Marty, *Anal. Chim. Acta*, 464 (2002) 171.
369 S. Andreescu, A. Avramescu, C. Bala, V. Magearu and J.L. Marty, *Anal. Bioanal. Chem.*, 374 (2002) 39.
370 C. Bonnet, S. Andreescu and J.L. Marty, *Anal. Chim. Acta*, 481 (2003) 209.
371 S. Andreescu, V. Magearu, A. Lougarre, D. Fournier and J.L. Marty, *Anal. Lett.*, 34 (2001) 529.
372 S. Andreescu, T. Noguer, V. Magearu and J.L. Marty, *Talanta*, 57 (2002) 169.
373 T. Montesinos, S. Perez Munguia, F. Valdez and J.L. Marty, *Anal. Chim. Acta*, 431 (2001) 231.
374 G. Jeanty, C. Ghommidh and J.L. Marty, *Anal. Chim. Acta*, 436 (2001) 119.
375 T. Neufeld, I. Eshkenazi, E. Cohen and J. Rishpon, *Biosens. Bioelectron.*, 15 (2000) 323.
376 C. La Rosa, F. Pariente, L. Hernández and E. Lorenzo, *Anal. Chim. Acta*, 295 (1994) 273.
377 C. La Rosa, F. Pariente, L. Hernández and E. Lorenzo, *Anal. Chim. Acta*, 308 (1995) 129.
378 J.M. Abad, F. Pariente, L. Hernández, H.D. Abruña and E. Lorenzo, *Anal. Chem.*, 70 (1998) 2848.
379 F.N. Kok, F. Bozoglu and V. Hasirci, *Biosens. Bioelectron.*, 17 (2002) 531.
380 L. Campanella, R. Cocco, M.P. Sammartino and M. Tomassetti, *Sci. Total Environ.*, 123/124 (1992) 1.
381 F. Botrè, G. Lorenti, F. Mazzei, G. Simonetti, F. Porcelli, C. Botrè and G. Scibona, *Sens. Actuators B*, 18–19 (1994) 689.
382 F.D. Reyes, J.M.F. Romero and M.D.L. de Castro, *Anal. Chim. Acta*, 408 (2000) 209.
383 N.G. Karousos, S. Aouabdi, A.S. Way and S.M. Reddy, *Anal. Chim. Acta*, 469 (2002) 189.
384 M.S. Ayyagari, S. Kamtekar, R. Pande, K.A. Marx, J. Kummar, S.K. Tripathy and D.L. Kaplan, *Mater. Sci. Eng. C*, 2 (1995) 191.

385 F. Mazzei, F. Botrè and C. Botrè, *Anal. Chim. Acta*, 347 (1996) 63.
386 W.R. Everett and G.A. Rechnitz, *Anal. Chem.*, 70 (1998) 807.
387 R. Rouillon, J.-J. Mestres and J.L. Marty, *Anal. Chim. Acta*, 347 (1997) 63.
388 A. Avramescu, S. Andreescu, T. Noguer, C. Bala, D. Andreescu and J.L. Marty, *Anal. Bioanal. Chem.*, 374 (2002) 25.
389 M.T. Pérez Pita, A.J. Reviejo, M.F.J. de Villena and J.M. Pingarrón, *Anal. Chim. Acta*, 340 (1997) 89.
390 J. Wang, E. Dempsey, A. Eremenko and M.R. Smyth, *Anal. Chim. Acta*, 279 (1993) 203.
391 T.T. Bachmann and R.D. Schmid, *Anal. Chim. Acta*, 401 (1999) 95.
392 T. Montesinos, S. Perez Munguia, J.L. Marty and K. Mitsubayashi, *J. Adv. Sci.*, 12 (2000) 217.
393 D.C. Cowell, A.A. Dowman and T. Ashcroft, *Biosens. Bioelectron.*, 10 (1995) 509.
394 T. Danzer and G. Schwedt, *Anal. Chim. Acta*, 318 (1996) 275.
395 S.J. Young, J.P. Hart, A.A. Dowman and D.C. Cowell, *Biosens. Bioelectron.*, 16 (2001) 887.
396 K. Jung, G. Bitton and B. Koopman, *Water Res.*, 29 (1995) 1929.
397 D.H. Liu, A.F. Yin, K. Chen, K. Ge, L.H. Nie and S.Z. Yao, *Anal. Lett.*, 28 (1995) 1323.
398 G.A. Zhylyak, S.V. Dzyadevich, Y.I. Korpan, A.P. Soldatkin and A.V. El'skaya, *Sens. Actuators B*, 24–25 (1995) 145.
399 V. Volotovsky, Y. Jung Nam and N. Kim, *Sens. Actuators B*, 42 (1997) 233.
400 T. Krawczynski vel Krawczyk, M. Moszczynska and M. Trojanowicz, *Biosens. Bioelectron.*, 15 (2000) 681.
401 H. Mohammadi, M. El Rhazi, A. Amine, A.M. Oliveira Brett and C.M.A. Brett, *Analyst*, 202 (2002) 1088.
402 A.L. Kukla, N.I. Kanjuk, N.F. Starodub and Y.M. Shirshova, *Sens. Actuators B*, 57 (1999) 213.
403 S. Fennouh, V. Casimiri, A. Geloso-Meyer and C. Burstein, *Biosens. Bioelectron.*, 13 (1998) 903.
404 L.J. Ehlers and R.G. Luthy, *Environ. Sci. Technol.*, 302 (2003) 296A.
405 Y. Su and M. Mascini, *Anal. Lett.*, 28 (1995) 1359.
406 S. Cosnier, C. Gondran, J.C. Watelet, W. De Giovani, R.P.M. Furriel and F.A. Leone, *Anal. Chem.*, 70 (1998) 3952.
407 A. Rescigno, F. Sollai, B. Pisu, A. Rinaldi and E. Sanjust, *J. Enzyme Inhib. Med. Chem.*, 17 (2002) 207.
408 F.A. McArdle and K.C. Persaud, *Analyst*, 118 (1993) 419.
409 J.L. Besombes, S. Cosnier, P. Labbé and G. Reverdy, *Anal. Lett.*, 28 (1995) 255.
410 M.D. Morales, S. Morante, A. Escarpa, M.C. González, A.J. Reviejo and J.M. Pingarrón, *Talanta*, 57 (2002) 1189.
411 F. Daigle, F. Trudeau, G. Robinson, M.R. Smyth and D. Leech, *Biosens. Bioelectron.*, 13 (1998) 417.

412 T. Tatsuma, K. Komori, H.H. Yeoh and N. Oyama, *Anal. Chim. Acta*, 408 (2000) 233.
413 J.-W. Choi, Y.-K. Kim, S.-Y. Song, I. Lee and W.H. Lee, *Biosens. Bioelectron.*, 18 (2003) 1461.
414 V.G. Andreou and Y.D. Clonis, *Anal. Chim. Acta*, 460 (2002) 151.
415 J.L. Marty, N. Mionetto, T. Noguer, F. Ortega and C. Roux, *Biosens. Bioelectron.*, 8 (1993) 273.
416 I. Satoh, *Ann. NY Acad. Sci.*, 613 (1990) 401.
417 R.B. Thompson and E.R. Jones, *Anal. Chem.*, 65 (1993) 730.
418 R.B. Thompson and M.W. Patchan, *Anal. Biochem.*, 227 (1995) 123.
419 D. Elbaum, S.K. Nair, M.W. Patchan, R.B. Thompson and D.W. Christianson, *J. Am. Chem. Soc.*, 118 (1996) 8381.
420 R.B. Thompson, B.P. Maliwal and C.A. Fierke, *Anal. Biochem.*, 267 (1999) 185.
421 C.A. Fierke and R.B. Thompson, *Biometals*, 14 (2001) 205.
422 L.M. May and D.A. Russel, *Anal. Chim. Acta*, 500 (2003) 119.
423 N.K. Tanner, *FEMS Microbiol. Rev.*, 23 (1999) 257.
424 V.J. DeRose, *Curr. Opin. Struct. Biol.*, 13 (2003) 317.
425 R. Breaker, *Curr. Opin. Biotechnol.*, 13 (2002) 31.
426 J. Hesselberth, M. Robertson, S. Knudsen and A. Ellintong, *Anal. Biochem.*, 312 (2003) 106.
427 S. Seetharaman, M. Zivarts, N. Sudarsan and R. Breaker, *Nat. Biotechnol.*, 19 (2001) 336.
428 T. Pan and O. Uhlenbeck, *Nature*, 358 (1992) 560.
429 B. Cuenoud and J. Szostak, *Nature*, 375 (1995) 611.
430 J. Li and Y. Lu, *J. Am. Chem. Soc.*, 122 (2000) 10466.
431 Y. Lu, J. Liu, J. Li, P. Bruesehoff, C.-B. Pavot and K. Brown, *Biosens. Bioelectron.*, 18 (2003) 529.
432 C.A. Mirkin, R.L. Letsinger, R.C. Mucic and J.J. Storhoff, *Nature*, 382 (1996) 607.
433 J.W. Liu and Y. Lu, *J. Am. Chem. Soc.*, 125 (2003) 6642.
434 C. Walsh, *Nature*, 409 (2001) 226.
435 G.A. Sellek and J.B. Chaudhuri, *Enzyme Microb. Technol.*, 25 (1999) 471.
436 D.W. Hough and M.J. Danson, *Curr. Opin. Chem. Biol.*, 3 (1999) 39.
437 J. Eichler, *Biotechnol. Adv.*, 19 (2001) 261.
438 F. Niehaus, C. Bertoldo, M. Kahler and G. Antranikian, *Appl. Microbiol. Biotechnol.*, 51 (1999) 711.
439 M.E. Bruins, A.E.M. Janssen and R.M. Boom, *Appl. Biochem. Biotechnol.*, 90 (2001) 155.
440 H. Huber and K.O. Stetter, *J. Biotechnol.*, 64 (1998) 39.
441 R. Saiki, D. Gelfand, S. Stoffel, S. Scharf, R. Higuchi, G. Horn, K. Mullis and H. Erlich, *Science*, 239 (1988) 487.
442 S. Maloney, T. Marks and R. Sharp, *J. Chem. Technol. Biotechnol.*, 68 (1998) 357.
443 S. Maloney, T. Marks and R. Sharp, *Lett. Appl. Microbiol.*, 24 (1997) 441.
444 U. Reinscheid, M. Bauer and R. Muller, *Biodegradation*, 7 (1997) 445.

445 M.R. Kosseva, C.A. Kent and D.R. Lloyd, *Biochem. Eng. J.*, 15 (2003) 125.
446 D.C. Demirjian, F. Morís-Varas and S.C. Cassidy, *Curr. Opin. Chem. Biol.*, 5 (2001) 144.
447 M.E. Bruins, A.J.H. Thewessen, A.E.M. Janssen and R.M. Boom, *J. Mol. Catal. B*, 21 (2003) 31.
448 S. Cummings and G. Black, *Extremophiles*, 3 (1999) 81.
449 D. Madern, C. Ebel and G. Zaccai, *Extremophiles*, 4 (2000) 91.
450 J. DeFrank and T. Cheng, *J. Bacteriol.*, 173 (1991) 1938.
451 S. Pollack, J. Jacobs and P. Schultz, *Science*, 234 (1986) 1570.
452 A. Tramontano, K. Janda and R. Lerner, *Science*, 234 (1986) 1566.
453 W.B. Motherwell, M.J. Bingham and Y. Six, *Tetrahedron*, 57 (2001) 4663.
454 T. McCormack, G. Keating, A. Killard, B.M. Manning and R. O'Kennedy, In: D. Diamond (Ed.), *Principles of Chemical and Biological Sensors.* Wiley, New York, 1998.
455 C. Tellier, *Transfus. Clin. Biol.*, 9 (2002) 1.
456 G. Zhong, T. Hoffmann, R. Lerner, S. Danishefsky and C. Barbas III, *J. Am. Chem. Soc.*, 119 (1997) 8131.
457 M. Barbany, H. Gutierrez-de-Teran, F. Sanz, J. Villa-Freixa and A. Warshel, *ChemBioChem*, 4 (2003) 277.
458 N. Jerne, *Ann. Immunol. (Paris)*, 125C (1974) 373.
459 N. Takahashi, H. Kakinuma, L.D. Liu, Y. Nishi and I. Fujii, *Nat. Biotechnol.*, 19 (2001) 563.
460 P. Wentworth, A. Datta, D. Blakey, T. Boyle, L. Partridge and G. Blackburn, *Proc. Natl Acad. Sci. USA*, 93 (1996) 799.
461 A. Cordova and K.D. Janda, *J. Am. Chem. Soc.*, 123 (2001) 8248.
462 P. Vayron, P.-Y. Renard, T. Taran, C. Créminon, Y. Frobert, J. Grassi and C. Mioskowski, *Proc. Natl Acad. Sci. USA*, 97 (2000) 7058.
463 T.M. Penning and J.M. Jez, *Chem. Rev.*, 101 (2001) 3027.
464 J.R. Cherry and A.L. Fidantsef, *Curr. Opin. Biotechnol.*, 14 (2003) 438.
465 A.B. Carmichael and L.L. Wong, *Eur. J. Biochem.*, 268 (2001) 3117.
466 N. Scrutton, A. Berry and R. Perham, *Nature*, 343 (1990) 38.
467 K. Shah, Y. Liu, C. Deirmengian and K.M. Shokat, *Proc. Natl Acad. Sci. USA*, 94 (1997) 3565.
468 P. Mittl, A. Berry, N. Scrutton, R. Perham and G. Schulz, *Protein Sci.*, 3 (1994) 1504.
469 B. Van den Burg, G. Vriend, O. Veltman, G. Venema and V. Eijsink, *Proc. Natl Acad. Sci. USA*, 95 (1998) 2056.
470 K. Nakajima, H. Kato, J. Oda, Y. Yamada and T. Hashimoto, *J. Biol. Chem.*, 274 (1999) 16563.
471 F.H. Arnold, *Nature*, 409 (2001) 253.
472 M. Brufani, M. Marta and M. Pomponi, *Eur. J. Biochem.*, 157 (1986) 115.
473 R. Newcomb, P. Campbell, D. Ollis, E. Cheah, R. Russell and J. Oakeshott, *Proc. Natl Acad. Sci. USA*, 94 (1997) 7464.

474 O. Lockridge, R. Blong, P. Masson, M. Froment, C. Millard and C. Broomfield, *Biochemistry*, 36 (1997) 786.
475 C. Millard, O. Lockridge and C. Broomfield, *Biochemistry*, 6 (1998) 237.
476 L. Polgar and M. Bender, *J. Am. Chem. Soc.*, 88 (1966) 3153.
477 K. Neet and D. Koshland, *Proc. Natl Acad. Sci. USA*, 56 (1966) 1606.
478 V. Mozhaev, *Trends Biotechnol.*, 11 (1993) 88.
479 N. Clair and M. Navia, *J. Am. Chem. Soc.*, 114 (1992) 7314.
480 P. Sundaram and R. Venkatesh, *Protein Eng.*, 11 (1998) 691.
481 D. Qi, C.-M. Tann, D. Haring and M. Distefano, *Chem. Rev.*, 101 (2001) 3081.
482 I. Hamachi, R. Eboshi, J. Watanabe and S. Shinkai, *J. Am. Chem. Soc.*, 122 (2000) 4530.
483 K.K.-W. Lo, L.-L. Wong and H.A.O. Hill, *FEBS Lett.*, 451 (1999) 342.
484 V. Shumyantseva, V. Uvarov, O. Byakova and A. Archakov, *Biochem. Mol. Biol. Int.*, 38 (1996) 829.
485 A.I. Archakov, V.G. Zgoda and I.I. Karuzina, *Vop. Med. Khim.*, 44 (1998) 3.
486 P. Dawson, T. Muir, I. Clark-Lewis and S. Kent, *Science*, 266 (1994) 776.
487 S. Adams, R. Campbell, L. Gross, B. Martin, G. Walkup, Y. Yao, J. Llopis and R. Tsien, *J. Am. Chem. Soc.*, 124 (2002) 6063.
488 B. Griffin, S. Adams, J. Jones and R. Tsien, *Methods Enzymol.*, 327 (2000) 565.
489 A. Stocker, H. Hecht and A. Buckmann, *Eur. J. Biochem.*, 238 (1996) 519.
490 M.-O. Månson, P.O. Larsson and K. Mosbach, *Eur. J. Biochem.*, 86 (1978) 455.
491 M.-O. Månson, P.O. Larsson and K. Mosbach, *FEBS Lett.*, 98 (1979) 309.
492 H. Kuang, M. Brown, R. Davies, E. Young and M. Distefano, *J. Am. Chem. Soc.*, 118 (1996) 10702.
493 T. Rana and C. Meares, *Proc. Natl Acad. Sci. USA*, 88 (1991) 10578.
494 V. Shumyantseva, V. Uvarov, O. Byakova and A. Archakov, *Arch. Biochem. Biophys.*, 354 (1998) 133.
495 S. Saraswathi and M. Keyes, *Enzyme Microb. Technol.*, 6 (1984) 98.
496 D.E. Albert, M.B. Douglas, M.A. Hintz, C.A. Youngen and M.H. Keyes, *Enzyme Microb. Technol.*, 14 (1992) 885.
497 S. Sadeghi, Y. Meharenna, A. Fantuzzi, F. Valetti and G. Gilardi, *Faraday Discuss.*, 116 (2000) 135–153.
498 G. Gilardi, Y.T. Meharenna, G.E. Tsotsou, S.J. Sadeghi, M. Fairhead and S. Giannini, *Biosens. Bioelectron.*, 17 (2002) 133.
499 F. Arnold, *Protein Eng.*, 2 (1988) 21.
500 F. Arnold, *Curr. Opin. Biotechnol.*, 4 (1993) 450.
501 F.H. Arnold, *Acc. Chem. Res.*, 31 (1998) 125.
502 I.P. Petrounia and F.H. Arnold, *Curr. Opin. Biotechnol.*, 11 (2000) 325.
503 F.H. Arnold, P.L. Wintrode, K. Miyazaki and A. Gershenson, *Trends Biochem. Sci.*, 26 (2001) 100.
504 W. Stemmer, *Nature*, 370 (1994) 389.
505 M. Black, T. Newcomb, H.-M. Wilson and L. Loeb, *Proc. Natl Acad. Sci. USA*, 93 (1996) 3525.

506 M. Kikuchi, K. Ohnishi and S. Harayama, *Gene*, 243 (2000) 133.
507 D. Janssen, *Curr. Opin. Chem. Biol.*, 8 (2004) 150.
508 A. Crameri, G. Dawes, E. Rodriguez, S. Silver and W. Stemmer, *Nat. Biotechnol.*, 15 (1997) 436.
509 D. Umeno, A.V. Tobias and F.H. Arnold, *J. Bacteriol.*, 184 (2002) 6690.
510 D. Umeno and F.H. Arnold, *Appl. Environ. Microbiol.*, 69 (2003) 3573.
511 D. Umeno and F.H. Arnold, *J. Bacteriol.*, 186 (2004) 1531.
512 H. Joo, Z.L. Lin and F.H. Arnold, *Nature*, 399 (1999) 670.
513 A. Glieder, E.T. Farinas and F.H. Arnold, *Nat. Biotechnol.*, 20 (2002) 1135–1139.
514 E.T. Farinas, M. Alcalde and F.H. Arnold, *Tetrahedron*, 60 (2004) 525.
515 T.S. Wong, F.H. Arnold and U. Schwaneberg, *Biotechnol. Bioeng.*, 85 (2004) 351.
516 T. Bosma, J. Damborsky, G. Stucki and D. Janssen, *Appl. Environ. Microbiol.*, 68 (2002) 3582.
517 K.A. Gray, T.H. Richardson, K. Kretz, J.M. Short, F. Bartnek, R. Knowles, L. Kan, P.E. Swanson and D.E. Robertson, *Adv. Synth. Catal.*, 343 (2001) 607.
518 T. Kumamaru, H. Suenaga, M. Mitsuoka, T. Watanabe and K. Furukawa, *Nat. Biotechnol.*, 16 (1998) 663.
519 F. Bruhlmann and W. Chen, *Biotechnol. Bioeng.*, 63 (1999) 544–551.
520 M. Kikuchi, K. Ohnishi and S. Harayama, *Gene*, 236 (1999) 159–167.
521 K.A. Canada, S. Iwashita, H. Shim and T.K. Wood, *J. Bacteriol.*, 184 (2002) 344.
522 L. Wackett, *Ann. NY Acad. Sci.*, 864 (1998) 142.
523 S. Raillard, A. Krebber, Y. Chen, J. Ness, E. Bermúdez, R. Trinidad, R. Fullem, C. Davis, M. Welch, J. Seffernick, P. Wakett, W. Stemmer and J. Minshl, *Chem. Biol.*, 8 (2001) 891.
524 T. Bulter, M. Alcalde, V. Sieber, P. Meinhold, C. Schlachtbauer and F.H. Arnold, *Appl. Environ. Microbiol.*, 69 (2003) 5037.
525 N. Cohen, S. Abramov, Y. Dror and A. Freeman, *Trends Biotechnol.*, 19 (2001) 507.
526 D.S. Tawfik and A. Griffiths, *Nat. Biotechnol.*, 16 (1998) 652.
527 J. Reymond and D. Wahler, *ChemBioChem*, 3 (2002) 701.
528 F.H. Arnold and A. Volkov, *Curr. Opin. Chem. Biol.*, 3 (1999) 54.
529 R. Breslow, Y. Huang, X. Zhang and J. Yang, *Proc. Natl Acad. Sci. USA*, 94 (1997) 11156.
530 D. Robinson and K. Mosbach, *J. Chem. Soc., Chem. Commun.*, 14 (1989) 969.
531 A. Strikovsky, D. Kasper, M. Grün, B. Green, J. Hradil and G. Wulff, *J. Am. Chem. Soc.*, 122 (2000) 6295.
532 F. Menger, J. Ding and V. Barragan, *J. Org. Chem.*, 63 (1998) 7578.
533 M. Sotomayor, A. Tanaka and L. Kubota, *Electrochim. Acta*, 48 (2003) 855.

Chapter 11

Biosensors for bioprocess monitoring

Ursula Bilitewski

11.1 INTRODUCTION

The traditional application area of biotechnological products is the food industry, which is still an important market with the production of beer and wine and also food additives. However, medical applications are receiving increasing attention, for example, antibiotics, vaccines, therapeutic proteins or proteins used in diagnostic assays are increasingly produced by (recombinant) microorganisms. The most important producing organisms are bacteria, such as *Escherichia coli*, *Bacillus subtilis*, *Corynebacterium glutamicum* or *Streptomyces spec.*, yeasts, such as *Saccharomyces cerevisiae* and *Pichia pastoris*, fungi, such as *Aspergillus nidulans* or *Penicillium notatum* and animal cell lines, such as hybridoma cells for the production of monoclonal antibodies, chinese hamster ovary (CHO) and baby hamster kidney (BHK) cells. Additionally, cells or cell systems are developed as therapeutic means, e.g., stem cells, dendritic cells or artificial tissues or organs based on the cultivation of hepatocytes, keratinocytes, etc. Each of these organisms has its own requirements on the composition of the medium and cultivation strategies for optimal growth and productivity [1,2]. The development of recombinant organisms additionally requires specific strategies for genetic modification, as metabolic pathways, secretion pathways for proteins or suitable promoters for the induction of the biosynthesis are specific properties of an organism [3–10].

The originally applied empirical strategies for the optimization of processes turned out to be of only limited value in the long term and for products of high economic value, such as therapeutic proteins, and the number of monitoring systems and control strategies continuously increased. The so-called "low-level control" based on temperature, pressure, pH and pO_2 [11]

E-mail address: ubi@gbf.de (U. Bilitewski).

is still the predominant strategy in routine production processes, as suitable in situ sensors exist for a long time. On-line determination of nutrients, metabolic products, secreted recombinant proteins can be done by HPLC or GC [12,13]. However, due to the complexity of the instruments and requirements on sample purity and analysis time, these devices are often restricted to off-line analysis, whereas on-line analysis of specific components is done by systems based on biochemical analysis or biosensor principles. They comprise a suitable biochemical receptor, usually an enzyme or an antibody, in immobilized form, and the biochemical reaction proceeding in the presence of the analyte is monitored by an appropriate transducer (an amperometric or potentiometric electrode, a thermistor, fluorimeter, photometer, etc.). As these systems are not sterilizable due to the heat sensitivity of the biochemical receptor, they have to be coupled to the process via suitable interfaces maintaining the sterile integrity of the process. Several attempts to design in situ probes did not result in routine devices, and at present flow-through or flow-injection analysis (FIA) devices are the most popular. They are coupled to the process via sampling systems allowing automated removal of samples without affecting the sterile integrity of the process [13–17].

The application of these systems, in particular for the determination of glucose, and to a minor degree of lactate and methanol, becomes increasingly routine, as after a long time of research applications suitable sampling and monitoring systems are commercially available (for example, www.ysi.com; www.trace.de; www.flownamics.com). This lead to further stimulation of the field and accordingly the number of reports is increasing describing not only monitoring but also control of cultivations using the concentration of the carbon source as variable [12,18–20]. In recent years there appeared a number of books and review articles describing the state-of-the-art of these systems [11,14,15,17,21], and thus they will not be considered in detail in this contribution. Exceptions are the determination of "new" carbon sources, approaches for the improvement of the stability and sensitivity of sensors, new sampling devices or the application of the system to control the bioprocess.

Nowadays, there are additional requirements on bioprocess monitoring systems. These are partly due to governmental regulations on the production of pharmaceuticals, requiring validated processes delivering, for example, a constant yield of cells, product and product quality. The latter includes not only the quality of the product itself, such as the biological activity of a therapeutic protein, but also its purity, i.e., the presence or absence of interfering compounds [22,23].

In addition, new strategies are under development for the optimization of bioprocesses, in particular, for the recombinant production of therapeutic proteins. It is increasingly recognized that the success of a bioprocess, i.e., the yield of the desired product, is not only influenced by the choice of the producing strain or cell line, of the gene copy number (plasmid copy number) or the promoter in a recombinant strain [6,24] or the induction, fermentation and feeding strategy [25,26], but also by the efficient use of the metabolic potential of the host. The production of a heterologous protein can exert such a significant burden on the cellular metabolism that vital functions are impaired [27–29]. That is why new parameters are evaluated which should give physiologically relevant information, derived not only indirectly from parameters outside the cells [27,30–33], but also directly from the cells and the internal variables. The physiologic state of cells is traditionally judged from several parameters: microscopic investigations show the cell morphology and can give first information on the viability and metabolic state of the cells. A more detailed picture, however, is obtained from chemical analysis. The application of isotopically labelled compounds as substrates and the analysis of the distribution of these isotopes on metabolic intermediates and products [34–39], the determination of enzyme activities in different cell extracts, the analysis of the intracellular protein composition [28,37,40,41] of the energy charge and of important precursors or intermediates [9,42–44] give a picture of the existence and efficiency of metabolic pathways [45]. Observed limitations within a pathway leading directly to the desired product were used to optimize the strain further by appropriate genetic modifications (metabolic engineering) [30,46–48]. However, these modifications often did not lead to the expected success [46]. This is at least partly due to the connections of different pathways through common substrates, intermediates, precursors or regulating compounds requiring more complex analysis of the cells. That is why at present, new methods for the analysis of cells are under investigation [45]. They range from the evaluation of new on-line systems indicating the cell physiology through key parameters to off-line analysis of the global cell physiology. Among these new approaches are the application of on-line microscopy [49,50], near-infrared (NIR) spectroscopy [51,52], nuclear magnetic resonance (NMR) spectroscopy [21,36], HPLC analysis [8,42] and enzymatic sensor systems for key compounds of cell metabolism, gel electrophoresis for protein analysis [23,28,40] and DNA chips for gene expression (mRNA) analysis. Microscopic and spectroscopic methods are attractive, because they are applicable as in vivo methods allowing the investigation of cells without the need for sampling or cell disruption. However, until now microscopy and NIR-spectroscopy were only occasionally

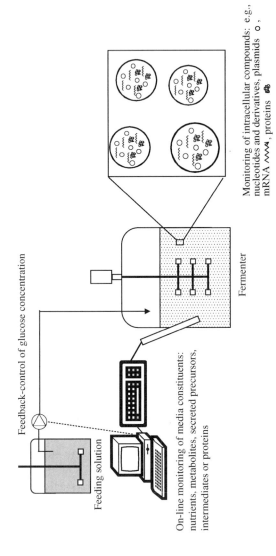

Fig. 11.1. Position of biochemical analytical systems for bioprocess monitoring.

used for on-line bioprocess monitoring [53]. Other spectroscopic methods (in particular NMR) still need further improvements with respect to sensitivity and signal processing before being applicable as routine methods [34,51].

Considering the above mentioned trends in bioprocess analysis the following article will present and discuss

(a) approaches for on-line monitoring and control of substrate and product concentrations with enzyme electrodes,
(b) contributions of biosensors for the determination of product quality,
(c) possibilities for quantification of plasmid DNA,
(d) enzyme-based on-line analysis of indicator compounds for cell physiology or metabolic stress,
(e) arrays for gene expression analysis.

An overview of the different aspects covered by these topics is given in Fig. 11.1.

11.2 MONITORING AND CONTROL OF SUBSTRATE AND PRODUCT CONCENTRATIONS WITH ENZYME ELECTRODES

The most important carbon source in biochemical engineering is glucose, as it is used in most microbial fermentations as well as in animal cell cultivations. Hence, glucose sensor systems are applied to bioprocess monitoring for almost 20 years [14,15] and reached a reliability which allowed their application even to process control (see Section 11.2.5). However, in particular, some microbial processes require alternative substrates. Examples are fermentations of recombinant strains of the yeast *P. pastoris* which are grown initially on glycerol and during the production phase on methanol [19,26,53,54]. Glycerol is also a by-product of alcoholic fermentations of yeasts, e.g., during the production of wine. Fuel ethanol is the product of yeast fermentations with the fermentation medium containing sugars such as glucose, xylose and galactose resulting from the hydrolysis of lignocellulose. That is why enzyme-based systems were developed not only for ethanol [55,56], but also for glycerol [57], and various monosaccharides [56] (Section 11.2.1). Of major concern for sensors to be applied to bioprocess monitoring is the stability of the sensors during application (Section 11.2.2) and the specificity for the target analytes (Section 11.2.3). The latter one can be improved, for example, by the combination with chromatography or with suitable sampling devices (Section 11.2.4).

11.2.1 Enzyme electrodes for alcohols and saccharides

Ethanol was determined using alcohol oxidase isolated either from *P. pastoris* [55] or *Hansenula polymorpha* [56] and immobilized on glass beads entrapped in an enzyme reactor or adsorbed on graphite powder of a carbon paste electrode. It is well known for oxidase-based sensor systems that the determination of the analyte can be based on the determination of hydrogen peroxide formed from the oxidase reaction. Again, several approaches were used: hydrogen peroxide was oxidised directly at potentials of $+700$ mV using platinum electrodes made, for example, from screen-printing technology [55] or it was reduced in a bi-enzymatic approach by horseradish peroxidase which was re-reduced via direct electron transfer from a carbon paste electrode [56] or via mediators.

A similar approach was used for the detection of monosaccharides [56]: pyranose oxidase delivered a suitable oxidase-based sensor with the hydrogen peroxide being determined again via the direct electron transfer from the carbon paste electrode to the enzyme horseradish peroxidase. In all bi-enzyme systems, both the enzymes were entrapped within the carbon paste. These electrodes were used as detectors in a liquid chromatography system allowing the simultaneous determination not only of several carbohydrates but also of ethanol within approximately 20 min. Ethanol production and carbohydrate consumption were monitored on-line during a fermentation of *P. stipitis* for 16 h (see also Section 11.2.3).

For the determination of glycerol no suitable oxidase is readily available. That is why glycerol kinase was used to convert glycerol to glycerol-3-phosphate, which was oxidised in the following by the corresponding oxidase to glycerone-3-phosphate and hydrogen peroxide [57] (Eqs. (11.1) and (11.2)).

Glycerokinase :   Glycerol + ATP → Glycerol-3-phosphate + ADP   (11.1)

Glycerol-3-phosphate oxidase :

Glycerol-3-phosphate + O_2 + H_2O → Gycerone-3-phosphate + H_2O_2 (11.2)

For maximum sensitivity and stability of the system both enzymes had to be used in different compartments and, hence, the glycerol kinase was immobilized on glass beads and kept in an enzyme reactor whereas the glycerol-3-phosphate oxidase was immobilized on a pre-activated membrane and placed directly in front of the electrode. The co-substrate of the kinase reaction, ATP, was added together with Mg^{2+} to the working buffer. The system was applied to off-line monitoring of glycerol production during the fermentation of *S. cerevisiae* in must from Italian Trebbiano grapes. Applications to cultivations of *P. pastoris* are not yet mentioned.

11.2.2 Stability of enzyme sensors

The stability of enzyme sensors is dependent on the stability of the particular enzyme structure and can be influenced by conditions chosen during immobilization and storage [58]. Already the immobilization of the protein may improve its stability compared to the corresponding enzyme solution, because the enzyme structure loses flexibility due to the coupling to a solid support and thus, movements of protein domains leading to denaturation are decreased and the stability is increased [59,60]. Additionally, the amount of immobilized enzyme influences the signal height and the stability [61,62]. Starting with low amounts of immobilized enzyme, for example, with less than a monolayer coverage of the electrode, the signals will increase with the amount of the enzyme, because the enzymatic turnover rate limits the sensor signals (kinetically controlled regime). When a certain enzyme loading on the electrode is achieved, substrate molecules reaching the electrode surface will be converted instantaneously and the signal is limited by the diffusion rate of the substrate to the electrode surface (diffusion controlled regime). As the enzyme activity does not limit the signal, it can decrease due to denaturation without affecting the signal height and thus, the sensors have a high apparent stability. Hence, already the choice of immobilization methods allowing to increase the amount of enzyme, such as the entrapment in carbon paste, in membranes or in enzyme reactors, can improve the sensor stability.

Moreover, it is known that certain additives have a beneficial effect on the enzyme stability [60], in some cases this is combined with an improvement of the sensitivity and lower detection limits. Sugars, salts and polyols are known to stabilize proteins in solutions or during lyophilization [63,64]. For sensor development, in particular, the effects of polymeric additives were investigated, because they could be entrapped together with the enzyme in membranes or pastes. It was observed that polyelectrolytes, such as polyethylenimine, diethylaminoethyl (DEAE)-dextran, polylysine and Gafquat (copolymer from vinylpyrollidone and dimethylaminoethylmethacrylate) improve the stability [65] and also the sensitivity [66,67] of sensors, just as additional proteins, for example, bovine serum albumine (BSA) or lysozyme [68] and sugar alcohols, mainly lactitol. The reason for the stabilizing effect is still not completely understood and for each enzyme the particular optimal mixture of additives has to be found [56,66,69]. It is often assumed that the polymers interact with the protein through electrostatic or hydrophobic interactions [70] forming a kind of cage and thus protecting it from unfolding and denaturation. However, this simple model cannot explain the observed dependencies on the concentrations of the additives, the need for a specific optimization for each protein and the observation that in some cases ternary mixtures are beneficial. Moreover, computer modelling of the interaction of horseradish peroxidase with the monomers of Gafquat, DEAE-dextran and polyethylenimine showed specific binding sites only for polyethylenimine [71], which was not reported to have significant stabilizing effects on the protein. Nevertheless, the addition of these compounds to the immobilization matrix often improves the stability of the resulting sensor significantly, so that they are applicable to on-line monitoring of cultivations even for periods of several weeks [20,69]. Even if it is not possible to include these additives in the immobilization matrix their addition to the storage solution proved to be beneficial, extending the lifetime of the glycerol-3-phosphate oxidase from 6 to 30 days [57].

11.2.3 Specificity of enzyme sensors

The specificity of an enzyme electrode is mainly given by the specificity of the enzyme. Enzymes such as amino acid oxidase [72], alcohol oxidase [73] or pyranose oxidase [56] catalyse the oxidation of several substrates and, hence, the corresponding sensor systems also respond to a range of compounds. This was utilized for the development of a sensor for the detection of the total amount of protein in a sample [74]. A protease was used for the degradation of the proteins to amino acids, which could be determined subsequently by the amino acid oxidase. However, usually a low specificity of an enzyme prevents the specific determination of analytes, unless it is coupled with the separation of compounds. Thus, Lidén et al. [56] coupled the sensor for monosaccharides, which was based on pyranose oxidase, to liquid chromatography. This allowed the simultaneous specific determination of glucose, galactose and xylose, as all the three compounds were separated on the chromatographic column and were detected post-column by the pyranose oxidase modified electrode. Analysis times of approximately 20 min allowed the on-line determination of these compounds during the fermentation of *P. stipitis*.

The application of chromatographic or electrophoretic systems leads to the separation of not only several substrates but also of sample constituents, which could influence the enzyme activity or lead to enzyme-independent signals [75]. An alternative approach to eliminate these interferences is the restriction of the access of interfering compounds to the enzyme and the transducer. This can be achieved by adapting the permeability of the enzyme membrane via a suitable porosity or via appropriate hydrophilic and hydrophobic properties. These parameters can be changed by changing the composition of the pre-polymer mixture which is used for enzyme membrane formation [76]. Alternatively, additional membranes can be used, which could either cover directly the electrode, such as Nafion membranes [77], or be inserted in the flow system as membrane modules [55,69] or even as sampling devices [56,78]. Membrane modules, which are inserted in a flow system, utilize the principles of dialysis [55,66,69,73] or pervaporation [79]. These modules comprise the donor channel through which the sample is pumped and the acceptor channel containing only the buffer. Both the channels are separated by a membrane through which the analyte has to diffuse (Fig. 11.2). Depending on the membrane material and membrane properties, such as thickness and pore size, the permeability for compounds is influenced [73]. Thus, interfering compounds are separated by the use of hydrophobic membranes, which allow the permeation of only volatile compounds such as alcohols [55,73]. This separation of volatile and non-volatile compounds is still amplified in pervaporation modules, as they have an air gap between donor stream and membrane, which can only be passed by volatile sample constituents. Membrane modules always influence the linear range of the system, as only a fraction of the total amount of analyte permeates the membrane and the analytical ranges are usually shifted to higher concentrations. Thus, the introduction of membrane modules is a mean

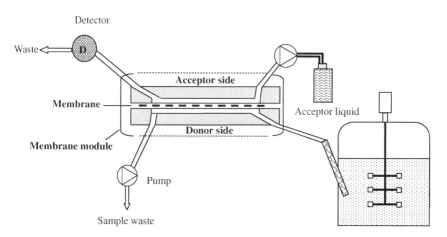

Fig. 11.2. Membrane module as interface between fermenter and detector systems.

to reduce the influence of possible interferents and to adjust the linear range of the system to the concentration range of the analyte in the real samples avoiding any further dilution. Appropriately designed modules could be used as interfaces to the bioprocess medium, i.e., as sampling modules. Corresponding approaches are described in Section 11.2.4.

11.2.4 Sampling devices

As enzyme electrodes cannot be sterilized, suitable sampling systems are a major issue in all reviews dealing with the application of biosensors for bioprocess monitoring. These sampling systems maintain the sterile integrity of the bioprocess and allow at the same time continuous access to analytes present in the medium. Yellow Springs Instruments offers a system, which pulls samples directly from a bioreactor to the analysis system and flushes the sampling line afterwards with an aseptic solution (www.ysi.com). They suggest the combination of this system with a filtration module for bioprocesses, in which loss of cells is a concern.

The application of membranes as interface between the bioprocess and the analytical system, generally a flow-through system such as a FIA-device, is favoured in most of the approaches. The driving force for the analyte to pass the membrane is either a pressure gradient, i.e., the samples are filtered, or a concentration gradient, i.e., the analyte passes the membrane by diffusion. In earlier reviews, mainly filtration modules were described, and some are commercially available for several years (e.g., www.applikon.com; www.flownamics.com; www.trace.com). However, filtration of cultivation media through membranes can cause membrane fouling and membrane blocking due to precipitated cells. That is why, nowadays, diffusion modules are receiving increasing interest and first systems are also commercially available [78]. Dialysis probes are equipped with a membrane of known molecular weight cut-off and perfused with a perfusion liquid, e.g., water or buffer. The analyte passes the membrane driven by the concentration gradient and the permeability of the membrane. Thus, the total liquid volume in the reactor is not affected and dialysis probes can also be applied to small-scale fermenters. Moreover, the efficiency of transfer through the membrane is dependent on the flow rate of the perfusion liquid and can thus be adapted to the linear range of the analytical system [78]. Suitable sampling modules exist in different designs depending on the application. For a more extended discussion the reader is referred to Chapter 12 on microdialysis modules.

11.2.5 Control of substrate concentrations

The carbon source is one of the most important medium components in fermentations, as its availability influences cell growth, cell metabolism, cell viability and productivity. Hence, a control of its concentration is highly desirable. However, a lack of reliable real-time information on the substrate concentration prevented its application in advanced control strategies. Nowadays, glucose and methanol analysers together with suitable sampling strategies were improved to a degree that the control of cultivation via the substrate concentration is possible.

Already in 1987, Mizutani et al. [18] described the control of glucose concentration in fed-batch cultures of *S. cerevisiae* to two different levels (10 and 0.3 g/l) using a glucose analyser, which was coupled to the fermenter using a ceramic or a membrane filter. Glucose concentrations were measured at intervals of 1 min and were controlled using an on/off control scheme for a peristaltic pump by feeding a glucose solution into the fermenter. The glucose concentration was kept to the desired value within 10% for the higher glucose concentration. Controlling the glucose at the lower level of 0.3 g/l led to glucose concentrations in the range from 0.08 to 0.54 g/l. For a cultivation of *M. ruteus*, the set-point for glucose was 2.5 g/l and the range of variation was about 20%, which was probably due to the longer measurement intervals of 7 min.

The influence of the glucose concentration on the expression of a recombinant protein by *S. cerevisiae* was investigated by Iijima et al. [80]. With an adaptive control strategy and measurements for every 5 min they observed a fairly good cell growth but a reduced gene expression when the glucose concentration was 10 g/l, whereas gene expression was improved and growth rate was decreased at low glucose levels of 0.15 g/l. Suitable sampling, monitoring and control systems are now commercially available (www.ysi.com).

P. pastoris is another yeast frequently used for recombinant protein production. Usually, the recombinant gene is located behind the gene of the alcohol oxidase (*AOX1*), and hence, expression is induced by methanol. In fed-batch cultivations the organism was grown to high cell density (80 g/l) on glycerol [19]. In the following fed-batch phase glycerol was replaced by methanol, the concentration of which was controlled to 1 g/l using a flow-through device with a diffusion membrane module as interface to the process (www.trace.de).

Because of their long duration (10–60 days) and high cost of media, maintenance of sterility is a major concern when monitoring and controlling perfusion animal cell cultivations. That is why biochemical analysers or FIA-devices were connected to the fermenter via cross-flow filtration modules in combination with a sterile barrier flushed with a bactericidal agent (1 M NaOH) between sample measurements [12,81]. Almost constant glucose concentrations at the chosen set-points were obtained for 42–70 days in cultivations of CHO cell cultivations using a PI control algorithm for the perfusion pump [12,20]. This led to cell viabilities of more than 90% throughout the whole cultivation time illustrating the successful maintenance of stable cultivation conditions.

11.3 BIOSENSOR SYSTEMS FOR THE EVALUATION OF PRODUCT QUALITY

Progress in recombinant DNA-technology made possible the large-scale production of recombinant proteins of diagnostic or therapeutic interest. Due to the economic value of these proteins efficient production strategies are required which are established by monitoring not only the standard parameters of cultivations, but also the concentration of the target protein. Thus, a number of methods are described for off-line or on-line protein determination during bioprocesses ranging from NIR spectroscopy [82], gel electrophoresis with visual inspection or image analysis of protein band intensities [28,54,83] to ELISA techniques [26] and antibody-based FIA-systems [14,15,17]. These methods give quantitative information on the total amount of the target protein with different degrees of accuracy.

However, in particular, for medical applications the structural integrity and biological activity of recombinant proteins have to be guaranteed and thus, also the amount of active protein has to be determined. Depending on the type of protein (enzyme or receptor ligand) different assay formats were established, most of which were not yet automated and applied to on-line monitoring of cultivations. The activity of enzymes is usually determined by suitable enzyme assays, i.e., the sample under investigation is incubated with an enzyme substrate and product formation is recorded. The activity of glucose oxidase was monitored in lysates of *Aspergillus niger* using 1,1'-dimethylferricinium as mediator in a deaerated solution by Luong et al. [84]. The detection of enzyme activity was based on the re-oxidation of the mediator at a platinum working electrode at a potential of 250 mV vs. Ag/AgCl. This amperometric assay was as sensitive as the standard photometric assay based on the reduction of dichlorophenol–indophenol (DCPIP) and was successfully applied to off-line monitoring of an *A. niger* cultivation without being affected by the turbidity of the samples.

Automation of an enzyme assay by FIA allowed van Putten et al. [85] on-line monitoring of the production of alkaline serine protease by *B. licheniformis*. An enzyme substrate was used, from which *p*-nitrophenol was cleaved by the enzyme and thus enzyme activity was determined photometrically (340 nm). The corresponding FIA-set-up was integrated in a six-channel FIA-system, which also allowed on-line monitoring of other medium components. Though the enzyme assay performed in the FIA-system was identical to the assay used for off-line analysis, enzyme activities determined on-line were always significantly lower than the off-line data, but the general dependence of the enzyme activity on the cultivation time (total ca. 45 h) was the same. The differences were attributed to fouling of the membrane sampling unit.

Combination of automated sterile cell sampling and cell disintegration by discontinuous treatment with ultrasound allowed on-line monitoring of even intracellular enzyme activities, as demonstrated by Kracke-Helm et al. [86]. β-galactosidase was determined directly in the unpurified cell extract, which was injected as sample in a FIA-system. The substrate was *o*-nitrophenyl-β-D-galactopyranoside (ONPG), which is hydrolysed to produce *o*-nitrophenol, which was determined photometrically. Total assay time including cell disintegration was 18 min, which allowed on-line determination of the protein during an *E. coli* cultivation. In this set-up, the on-line system delivered higher values than off-line analysis (9%), which may be due to the loss of enzyme activity during sample storage and handling.

Without the need for cell disintegration β-galactosidase activity was determined by Biran et al. [87], as they used the *lacZ* gene as reporter gene for the induction of gene expression after isopropyl β-D-thiogalactopyranoside (IPTG) addition or after entry into the stationary phase. The assay relied on the electrochemical detection of *p*-aminophenol, which is produced from the substrate *p*-aminophenyl-β-D-galactopyranoside by β-galactosidase. Screen-printed electrodes were used as detectors and placed directly in the Erlenmeyer flask used for *E. coli* cultivation. The onset of gene expression could be followed without any further sample treatment.

Another example of on-line monitoring of enzyme activities was given by Künnecke et al. [88], when a FIA-system was used for the determination of enzyme activities during protein purification by fast protein liquid chromatography (FPLC). Photometric assays for four different oxidases were established in a FIA-system extending the linear range by the so-called zone sampling method. The FIA-device was coupled to the FPLC unit behind a photometer monitoring protein absorbance at 280 nm. Thus, the parallel monitoring of the enzyme activities allowed the identification of the enzyme in the various protein containing fractions.

A similar idea was followed by Bracewell et al. [89]. However, their target analyte was an antibody fragment specific for hen egg lysozyme. This protein was produced by an *E. coli* strain which harboured a corresponding high copy plasmid. The protein was found in the culture supernatant and purified by affinity chromatography. A FIA-system was established which contained the optical affinity sensor system IAsys as detector. The sensor cuvette was replaced by a flow-through cell allowing automated delivery of all reagents to the sensor surface. The sensor comprised two measuring cells, one of which contained immobilized hen egg lysozyme (measuring cell) and the other, immobilized turkey egg lysozyme (reference cell) (Fig. 11.3). The sample simultaneously addressed both the surfaces and the response from the reference surface was subtracted from that of the measuring surface. The surfaces had to be regenerated by an acid pulse. The whole measuring time was 30 s allowing on-line monitoring of loading of the column and of the elution profile. The breakthrough of the protein indicating the saturation of

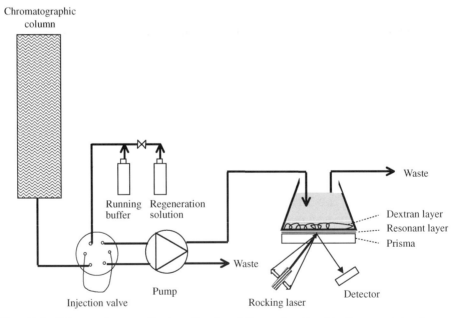

Fig. 11.3. Scheme for monitoring the breakthrough of a binding protein from a chromatographic column using the resonant mirror as affinity sensor system.

the column could be monitored only by the optical biosensor, because the target protein was only a minor fraction from the total protein in the feed of the column and thus it was not detected at 280 nm.

The application of optical affinity sensor technologies is the method of choice, when the biological activity of binding proteins, such as antibodies, receptors, receptor ligands, is to be determined. Standard assays indicating the biological activity of receptor ligands are cell-based assays with the cells being stimulated to protein expression or proliferation by the bioactive protein, and cell stimulation is determined either radioactively [90] or by protein expression analysis [83,91]. Alternatively, receptor binding assays are performed in microtiter plates, in which the analyte displaces a radioactively labelled ligand [90]. As the basis of the biological activity of these proteins is their binding to a suitable receptor, optical affinity sensor systems are ideal devices for the quantitative determination of the biological activity of the recombinant protein, as the receptor can be immobilized on the sensor surface and binding of the analyte followed in real-time. This approach was followed for the characterization and process validation of a chimeric antibody for the human Interleukin-2 receptor [92] and for the characterization of an artificial receptor for a human insulin analogue [93]. Additionally, off-line process monitoring was achieved during the production of the antibody fragment D1.3 Fv, which is specific for hen egg lysozyme, by *E. coli* [94] and of the vascular endothelial growth factor by a recombinant *P. pastoris* strain [95]. The importance of these investigations is evident, as particularly recombinant proteins are often not secreted in the medium by the producing bacterial strain (e.g., *E. coli*) and aggregate forming intracellular inclusion bodies containing incorrectly folded, not active proteins. Thus, refolding strategies have to be developed and the activity of the resulting product has to be demonstrated [83,90]. Using a receptor sensor based on the surface plasmon resonance (SPR) the success of the refolding procedures, i.e., the formation of biologically active proteins, could be followed directly during the refolding process without any sample pre-treatment besides dilution using the example of recombinant human bone morphogenetic protein-2 (rhBMP-2) production [96]. It was shown that in some cases the amount of active protein was only a fraction of the total amount of the produced target protein (less than 30%) [97]. Most of the reports mentioned above utilize SPR as the measuring principle, but other affinity sensing systems are also applicable [89,94,98].

Compared to previous investigations on protein determinations by immunosensors [15,81,99], binding of the recombinant protein to the sensor surface with the immobilized receptor only occurs when the binding site of the recombinant protein is correctly folded, i.e., the recombinant protein is present in its biologically active form. Binding to an antibody occurs, if the epitope recognized by the antibody is present and accessible, which does not necessarily relate to the biological activity of the protein.

11.4 QUANTIFICATION OF PLASMID DNA

The recombinant production of proteins usually follows a two-step strategy: in the first step, biomass is generated and in the second step, the expression of the target protein is induced. This is done, because heterologous protein production interferes with the host metabolism (see following sections) and usually leads to a reduction of cell growth. A prerequisite for this strategy are high-quality plasmids in which the protein gene is placed. One target for optimization of protein production is gene dosage, i.e., the number of recombinant gene copies per host genome which can be increased by increasing the plasmid copy number [100]. The replication of plasmids is dependent on the plasmid sequence itself but also on growth conditions. Thus, the necessity for validated and controlled processes [22] includes the control over the plasmid copy number leading to the need for suitable assays [101].

DNA concentrations are usually determined by spectrophotometric methods or by gel or capillary electrophoresis. The absorbance at 254 nm [102] or 260 nm [103] allows the reliable determination of the concentration for purified plasmids (50 μg/ml gives an $OD_{260} = 1$ [104]). The purity of the plasmid is controlled either by the ratio of the absorbances at 260 and 280 nm ($OD_{260}/OD_{280} = 1.8$ for purified DNA, decreases with protein contamination [104]) or by gel or capillary electrophoresis. One drawback of this method is the need for rather high DNA-concentrations (>0.25 μg/ml [104]) so that fluorescence methods were developed as alternatives. They rely on changes of spectral properties of DNA-binding compounds, such as PicoGreen [103] or bisbenzimide (Hoechst dye 33258) [102].

A prerequisite for the determination of plasmid concentrations is the lysis of the cells, which is usually done by mixing the cells with an alkaline SDS-solution, followed by plasmid purification by ion exchange chromatography after neutralization with potassium acetate. Incubation of the obtained plasmid with PicoGreen allowed the determination of plasmid concentrations in the range from 15 to 280 ng/ml [103]. This method was also applied to process samples showing that no interferences were observed, if the samples were diluted by a factor of 1/400. Whereas Noites et al. [103] performed plasmid isolation manually, this procedure was automated using a flow injection system by Nandakumar et al. [102]. Not only cell sampling and mixing with the lysis solution was done automatically, but also binding to a miniaturized ion exchange column followed by elution. A whole cycle required approximately 20–25 min. With a fluorescence detector, determination was possible from 1 to 40 µg/ml, the range was shifted to 20–100 µg/ml, when the UV-detector was used. The system was applied to monitor the plasmid concentration on-line during an *E. coli* cultivation. It was observed that off-line analysis always gave higher values than on-line analysis. The reason for this deviation was not elucidated.

These analytical methods are unspecific methods as they rely on spectral properties of DNA or of compounds binding to DNA. The identity of the isolated plasmid DNA had to be confirmed by digestion using restriction enzymes [103]. An alternative could be the application of DNA-sensors, i.e., the detection of the hybridization of specific polymerase chain reaction (PCR) products to immobilized oligonucleotides. The PCR leads to an amplification of a DNA-sequence enclosed by the sequences of the two PCR-primers. If both primers are used in almost equal concentrations double-stranded DNA-products are obtained, which, however, cannot hybridize to oligonucleotides without denaturation. Thus, variants of the conventional PCR were established, e.g., by the choice of primer sequences leading to two complementary DNA-strands of different lengths [105] or by at least a 10-fold excess of one primer leading to the preferred synthesis of the corresponding sequence and thus to a significant amount of single-stranded products [106]. This strategy was also followed by Kuhlmeier [107] who amplified a specific sequence of the plasmid pUC 19 by an asymmetric PCR and could follow the increasing amount of PCR product with increasing PCR-cycle numbers using a SPR instrument (BIAcore) as detector. Without any further plasmid purification he performed the PCR directly in the cell lysate, which he obtained from thermal lysis of *E. coli*. Thus, he combined the first step of the PCR cycle with the thermal lysis of the cells. In Fig. 11.4a the success of the PCR on these plasmid samples is shown, and in Fig. 11.4b the signals obtained with the nucleic acid sensor based on SPR using samples from the asymmetric PCR [107]. As expected signals increased linearly with increasing PCR cycle numbers.

The combination of automated cell sampling, thermal cell lysis, on-line asymmetric PCR (as suggested by microsystem technology [108]) and specific detection of the PCR products could allow in future an improved control of plasmid copy numbers during fermentation. If, however, the final product of the process is a pure plasmid for use in gene therapy, additional investigations are required to guarantee the absence of contaminating genomic DNA, proteins or chemicals used during plasmid isolation [109].

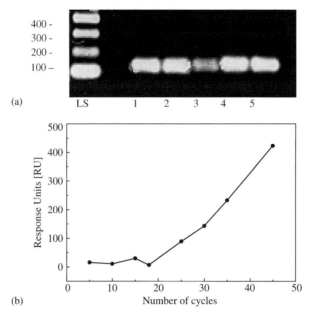

Fig. 11.4. (a) Agarose-gel electrophoresis of the 125 bp PCR product resulting from amplification of a specific sequence on the pUC plasmid. For the PCR a range of different template concentrations were used. The lanes correspond to *Escherichia coli* cells grown to different $OD_{600\ nm}$ values: lane 1, purified pUC19 (1 ng); lane 2, OD at 0.028; lane 3, OD at 0.078; lane 4, OD at 0.23 and lane 5, OD at 0.63. LS contains a molecular weight marker (100 bp ladder). (b) Dependency of sensor signals on the number of asymmetric PCR cycles, after 25 cycles of a conventional, symmetric PCR in a first step. After hybridization of the PCR product to the immobilized plasmid-specific oligonucleotide, the sensor surface was regenerated by NaOH.

11.5 ENZYME-BASED SYSTEMS FOR INDICATORS OF CELL PHYSIOLOGY AND METABOLIC STRESS

Analysis of regulatory networks involved in the adaptation of the metabolism of microorganisms to various environmental conditions, such as starvation, revealed the particular importance of nucleotides and amino acids. These compounds function as signals for starvation, precursors in metabolic pathways, energy sources or are involved in enzyme activity regulation [10,32,33,42,44,110–112]. They are usually determined off-line by HPLC [8,10,32,33,42,44,112], as chromatographic or electrophoretic determinations allow the simultaneous determination of all compounds of interest. However, for some of them also enzymatic methods are described, which may also be applied in future to bioprocess monitoring.

11.5.1 Amino acid analysis

Enzyme electrodes in particular for glutamate and glutamine are known for a number of years and were mainly applied to monitoring of animal cell cultivations [14,15,112–115]. They are usually based on glutamate oxidase and/or glutaminase. Glutaminase converts glutamine to glutamate and ammonia (Eq. (11.3)):

Glutaminase: glutamine → NH_3 + L-glutamate (11.3)

Thus, the enzyme can be coupled to ammonia electrodes, field effect transistors or chemical methods of ammonia determination [115]. Alternatively, a bi-enzyme system can be applied combining glutaminase with glutamate oxidase (Eq. (11.4)):

Glutamate oxidase:

L-glutamate + O_2 + H_2O → α-ketoglutarate + NH_3 + H_2O_2 (11.4)

This allows the combination with the amperometric detection of hydrogen peroxide. However, if glutamine and glutamate are simultaneously present in a sample, a differential measurement is required.

Less wide-spread than the enzymatic determination of these amino acids is the application of enzyme sensor systems for precursors or other amino acids. Collins et al. [116] described a flow injection system for the determination of α-ketoglutarate during an industrial fermentation. The system was based on glutamate dehydrogenase (Eq. (11.5)) and glutamate oxidase (Eq. (11.4))

Glutamate dehydrogenase:

α-ketoglutarate + NH_3 → α-iminoglutarate → L-glutamate (11.5)

Both enzymes were immobilized on beads (125–315 μm) and serially used in enzyme reactors either as packed beds or as expanded beds. Again detection was based on amperometric H_2O_2 detection. Co-reagents, such as NADH and NH_3, were added to the carrier. The expanded bed reactors showed a lower sensitivity and longer response times compared to the packed bed system. However, the risk of clogging due to sample constituents was minimal, moreover, cellular samples could also be analysed whereas for the packed-bed system cells had to be removed by centrifugation.

Gilis et al. [117] developed biosensors for the determination of L-alanine and its precursor pyruvate. They adapted the principle of a photometric assay, which was based on alanine dehydrogenase (Eq. (11.6))

Alanine dehydrogenase:

$$\text{L-alanine} + \text{NAD}^+ + \text{H}_2\text{O} \rightarrow \text{pyruvate} + \text{NADH} + \text{NH}_4^+ \qquad (11.6)$$

This reaction was coupled to the oxidation of NADH by hexacyanoferrate(II) catalysed by diaphorase (Eq. (11.7)).

$$\text{Diaphorase:} \quad \text{NADH} + 2\text{Fe(CN)}_6^{3-} \rightarrow \text{NAD}^+ + 2\text{Fe(CN)}_6^{4-} + \text{H}^+ \qquad (11.7)$$

The electrochemical detection utilized the re-oxidation of hexacyanoferrate(II) on a platinum electrode. For pyruvate determination this assay was extended to a 3-enzyme system by the addition of glutamate pyruvate transaminase, which produces alanine from pyruvate. All enzymes were used in solution in a reaction chamber of approximately 2 μl directly in front of the electrode. The cofactor NAD^+ was coupled to dextran with a molecular weight of 40,000 to avoid its replacement for each assay. As the sensor responded to L-alanine and pyruvate again a differential measurement was required when a sample contained both compounds. It was applied to off-line monitoring of a cultivation of *S. cerevisiae* and data showed good correlation to the photometric assays.

11.5.2 Nucleotide analysis

Enzyme sensors or assays exist for the determination of adenosine triphosphate (ATP) and its degradation products inosine 5'-monophosphate (IMP), inosine (HxR), hypoxanthine (Hx) and xanthine (X) as ATP is used as an indicator of the presence of microorganisms and the concentrations of its degradation products are used as indicators for fish and meat freshness in food industry [118].

Enzyme systems for the determination of ATP differ in the achievable lower detection limits depending on the enzymes and detection principles. The enzymatic system derived from the light-generating mechanism in fireflies utilizes the generation of light by the luciferin–luciferase system (Eq. (11.8))

Luciferase:

$$\text{ATP} + \text{D-Luciferin} + O_2 \rightarrow \text{Oxyluciferin} + PP_i + \text{AMP} + CO_2 + \text{light} \quad (11.8)$$

Gamborg and Hansen [119] suggested a flow-injection system, in which they mixed enzyme, reagent and sample solution directly in front of a photomultiplier and achieved a lower detection limit of 10^{-10} M. They faced the problem to obtain stable standard solutions in this low concentration range, which made further improvements impossible.

Less sensitive systems are obtained when enzyme reactions are used, in which ATP is one of the cofactors. For example, the phosphorylation of glucose by hexokinase requires ATP as cosubstrate (Eq. (11.9)). Thus, a glucose electrode was combined with hexokinase and signal reduction was observed when ATP was present [120].

Hexokinase: $\quad$ Glucose + ATP $\rightarrow$ Glucose-6-phosphate + ADP $\quad (11.9)$

Coupling this reaction to additional enzyme reactions converting the ADP (Eq. (11.10)) and the glucose-6-phosphate (Eq. (11.11)) lead to recycling of ATP and signal amplification by factors from 15 to 1000 [121,122]. The lower detection limit was in the range of 10^{-9} M.

Pyruvate kinase:

$$\text{ADP} + \text{phosphoenolpyruvate} \rightarrow \text{pyruvate} + \text{ATP} \quad (11.10)$$

Glucose-6-phosphate dehydrogenase:

$$\text{Glucose-6-phosphate} + NAD^+ \rightarrow NADH + \text{gluconate-6-phosphate} \quad (11.11)$$

No application to real samples, in particular, to fermentation monitoring, was reported.

The determination of the nucleobases IMP, HxR, Hx and X follows the natural degradation pathway of ATP (Eq. (11.12)).

5′-nucleotidase: $\quad\quad\quad\quad$ IMP $\rightarrow$ HxR

Nucleoside phosphorylase: $\quad$ HxR + P_i $\rightarrow$ Hx

Xanthine oxidase: $\quad\quad\quad\quad$ Hx + O_2 $\rightarrow$ X $\rightarrow$ uric acid + H_2O_2 $\quad (11.12)$

Thus, differential measurements are required to determine all compounds separately. However, as xanthine and hypoxanthine are both substrates for xanthine oxidase each of them can only be determined, if the other is not present. For food analysis, hypoxanthine usually is of major importance [118], and in fermentation monitoring it was observed that hypoxanthine accumulated intracellular, whereas xanthine was found in the medium [44]. Thus, it was possible to establish a FIA-system with immobilized xanthine oxidase and a screen-printed platinum electrode as detector to determine xanthine in samples from an *E. coli* cultivation [123].

11.6 DNA ARRAYS

Microorganisms achieve the synthesis of precursors or building blocks required to adapt to the conditions in the fermenter not only by the regulation of enzyme activities but also by the regulation of the expression of genes leading to changes in protein concentrations and allowing the synthesis of proteins required only occasionally. For example, growth rate, cell density and the composition of media influence the requirement for enzymes involved in respiration [124–126], the synthesis of building blocks [127,128], central metabolic pathways [124,129,130] or protein translation [127]. In addition, a temperature increase leads to the induction of the so-called stress proteins or heat-shock proteins [40,131,132], and similar proteins are induced as response to the production of heterologous proteins [28,40,133,134] and other process-related stresses [135]. A number of these investigations were based on the analysis of the cytosolic protein fraction by 2D-gel electrophoresis [28,40, 128,129], others, however, analyse the degree of gene transcription directly on the mRNA level. For protein analysis increasingly protein arrays become available, which are usually based on the recognition of the protein by antibodies [136,137]. However, these arrays were not yet applied to bioprocess analysis and thus they are not considered in this contribution though they utilize biosensor principles.

The transcription of specific genes (detection of specific mRNAs) is monitored for the optimization of recombinant overexpression of heterologous proteins already for a number of years [100]. The availability of global methods, such as DNA chips or membranes covering a whole genome or parts of a genome, allows a more general molecular analysis of cell physiologies, completing available biochemical knowledge and data from protein analysis. Several methods for global gene expression analysis are available, most of which do not utilize strict biosensor principles, as they include a number of separate manual handling procedures, signal recording after drying and some even radioactive labelling [138] (Section 11.6.1). However, it was shown that the use of fluorescent labels together with glass chips as support for immobilized probes gives essentially the same results as radioactive labels

Biosensors for bioprocess monitoring

in combination with membrane supports [139]. Moreover, there are first suggestions to couple fluorescence scanners to fluidic systems [140] and to expand SPR technology to arrays [141]. Thus, in the following reports on the biotechnological application of gene expression monitoring are mentioned irrespective of the detection principle and assay format (Section 11.6.2).

11.6.1 Fundamentals

11.6.1.1 Influences on the hybridization reaction
DNA arrays are based on the specific base-pairing of complementary nucleotides leading to double-stranded sequences of nucleic acids. Unlike traditional Northern Blot analysis, the strand being specific for the gene under investigation is immobilized as capture probe and the corresponding counterpart, the target, is isolated from the sample and present in the solution. For gene expression monitoring chips are used with either immobilized oligonucleotides (length < 70 nucleotides) (e.g., www.amershambiosciences.com; www.mwg-biotech.com; www.chem.agilent.com; www.affymetrix.com) or immobilized and denatured PCR products (length 400–1000 bp) (e.g., www.eurogentec.com; www.sigma-genosys.com; www.corning.com) representing all open reading frames (ORFs) of a genome or only a set of genes [142,143].

As other affinity reactions, the hybridization reaction of nucleic acids or oligonucleotides is characterized by the affinity constant and by the kinetic constants of the association and dissociation reaction. The affinity is influenced firstly by the sequences of the nucleic acids, and secondly by experimental conditions. Binding of the guanine–cytosine (G–C) base pair is stronger than that of the adenine–thymine (A–T) base pair leading to higher affinities of G–C rich sequences. Moreover, as each base pair contributes to the strength of the overall binding, the affinity increases with the length of the interacting sequences, i.e., the number of matching nucleotides [144]. If the sequence of matching base pairs is interrupted by a mismatch, this does not totally prevent hybridization of both strands but leads to a reduced stability [145]. The degree of destabilization is dependent on the lengths of the remaining perfectly matching sequences and, thus, on the position of the mismatch. Usually, double-strands are less stable, when the mismatch is localized in an internal position compared to mismatches at the end of the sequence [145,146]. Thus, signal intensities increase with the length of the probe and decrease, if mismatches are present [146].

Even for given nucleic acid strands the affinity can be influenced by additives to the hybridization solution [104,147]. It was found that formamide inhibits the formation of hydrogen bonds and thus reduces the stability of the double-stranded helix. Moreover, nucleic acids are negatively charged due to the phosphate groups of the nucleotide backbone. Without the compensation of this charge by counterions, even complementary single-strands are electrostatically repelled leading to the stabilization of double-strands with increasing concentrations of monovalent cations (Na^+). Another important experimental parameter is the hybridization temperature [146,147] as increasing temperature leads to increasing dissociation rates and a separation of double-strands into single-strands, a reaction called "melting of DNA". The temperature leading to a dissociation of 50% of the double-strands is called the melting temperature T_m of the sequence and is used for the characterization of the stability of the double-strand. For oligonucleotides it can roughly be calculated from the sequence by using the approximation

$$T_m = 2°C \times (A-T) + 4°C \times (G-C), \qquad (11.13)$$

with (A–T) being the number of A–T pairs and (G–C) the number of G–C pairs in the sequence.

For longer hybrids the approximation

$$T_m = 81.5°C + 16.6 \log[c_{Na^+}] + 0.41(\%G + C) - 500/n \qquad (11.14)$$

is used [104] with n being the length of the hybridising strand and c_{Na^+} the concentration of Na^+ ions.

If T_m exceeds the hybridization temperature by 10–15°C, efficient hybridization is observed. If, however, the hybridization temperature is much lower than T_m, hybridization of strands containing mismatches or being only partly complementary will occur.

The influences of oligonucleotide sequences on the specificities of the hybridization reactions have to be considered when DNA chips are designed. The length and sequence of the capture probe has to be chosen so that it is specific for a single gene. Additionally, sequences of oligonucleotides to be combined on an array should not differ too much in the G–C-content to obtain similar melting temperatures of the corresponding double-strands. Kane et al. [148] investigated the sensitivity and specificity of a chip containing oligonucleotides with a length of 50 bases (50-mers). They could identify experimental conditions so that only sequences with more than 75% sequence similarity contribute to the signal by cross-hybridization, and sequences which were only marginally similar should not contain a stretch of complementary sequence of >15 contiguous bases. If the oligonucleotides were shorter than 50 bases, they found too many non-target sequences with a similarity of >75%, if they were longer, e.g., PCR products with >100 bases, the probability for a sufficient number of complementary bases also increases.

These observations are compatible with observations from northern blot analysis, where occasionally multiple, non-target genes are indicated by the chosen labelled probe. Thus, if oligonucleotides with 50–60 nt are used, it is assumed that a single probe per gene allows specific and reliable monitoring of the expression of this gene. If shorter oligonucleotides (20–25 nt) are used, usually several probes are to be designed for each gene [147].

However, the length of the probe is important not only for the stability of the double-strand and the specificity of the hybridization, but also for the kinetics of the reaction. The hybridization rate is mainly determined by the access of the targets to the immobilized probes, which is influenced by the density of the immobilized probes [149], the length and complexity of the sequence, and by the diffusion rate, which depends on temperature, concentrations and viscosity of the target solution. Additionally, experimental conditions leading to a destabilization of the double-strand, such as low ionic strength or the presence of mismatches, reduce the hybridization rate, as the dissociation is accelerated as could be observed with a SPR device [107]. Increasing the length of the probe increases the stability of the double-strand. On the other hand, secondary structures may be formed, preventing hybridization with the target, and the complexity of the sequence increases, i.e., the degree of non-repetitive sequences, leading also to a reduction of the hybridization rate. For example, binding of mRNA to a 24-mer comprising only thymidine (polyT) resulted in a flow-through system in significantly higher signals compared to mRNA captured via gene specific capture probes [150]. It was found that hybridization equilibrium is reached only after at least 24 h hybridization time, even if oligonucleotide probes are used [146,151].

11.6.1.2 Immobilization
Nucleic acids (oligonucleotides or PCR products) are immobilized either by physical forces or by (bio)chemical reactions [143]. Major substrates are glass slides or nylon membranes, depending on the detection method chosen. However, assays in microtiterplates were also described [152].

As nucleotides are negatively charged, they interact by electrostatic attraction with positively charged surfaces as delivered by nylon membranes [131] or glass slides pre-treated with poly-L-lysine [129,153] (http://cmgm.Stanford.edu/pbrown/protocols) or aminopropyltri-ethoxysilane (APTS) [154] (www.corning.com). Capture probes are spotted on the surface and dried. If PCR products are used, they are denatured either prior [154] or after [124] spotting. Usually, UV-irradiation is recommended as an additional cross-linking step, but heating to 60 and 120°C is also possible [147]. As each nucleotide contributes with an additional charge, the electrostatic forces increase with increasing length of the nucleic acids, which makes this immobilization method applicable mainly for PCR products. Zammatteo et al. [154] showed that already for a 255 bp capture probe the efficiency of this electrostatic attraction exceeded the efficiency of even covalent attachment.

The alternative to immobilization via physical interactions is covalent binding. This requires the availability of suitable functional groups on both the probe to be immobilized and the immobilization substrate, usually glass. Aminated glass chips were used, which are further activated by carbodiimide to allow coupling of carboxylated or phosphorylated nucleic acids [154] or further modification prior to nucleic acid immobilization [155]. Additionally, amino-functionalized oligonucleotides or PCR products are coupled via bifunctional reagents, such as diisothiocyanates, disuccinimidyl derivatives, or trichlorotriazine [155–157]. Amino-functionalized nucleic acids, however, can be coupled easily to amine-reactive groups present in a long-chain hydrophilic polymer [148,158] or to epoxy- [159] or aldehyde-modified glass surfaces. Resulting Schiff bases can be reduced by sodium borohydride. This method is proved to be highly effective and specific and suitable for the application of very small volumes of liquid, though the binding kinetics showed an increasing efficiency with increasing time of contact from the DNA-solution to the glass surface [154].

As known from bioconjugate chemistry additional functional groups can be used for attachment of biomolecules. Thus, coupling via SH-groups [149,155,160], a benzaldehyde-semicarbazide-reaction [161] or the reaction between a streptavidin-coated surface and a biotinylated oligonucleotide [150,162,163] are only some examples.

Most of these methods are applied to either PCR products or pre-synthesized oligonucleotides in combination with liquid-spotting systems. This allows control and optimization of the amount of the immobilized single-strand [149]. SH-coupling or benzaldehyde-semicarbazide-coupling delivered maximal densities of approximately 10–100 pmol/cm^2 [160,161], whereas binding of amino-functionalized PCR products to an aldehyde glass surface resulted in only 600 fmol/cm^2 [154]. However, only 10–30% of the immobilized probes took part in the hybridization reaction so that the optimal probe density was even further reduced [149]. The density of spots in these arrays is limited by mechanical constraints of liquid delivery or surface manipulation.

As solid phase synthesis of oligonucleotides is well established, oligonucleotide capture probes can also be synthesized directly on the chip surface. This was described by Pease et al. [164], and later by Weiler and Hoheisel [165] and Blanchard et al. [166]. The basic reaction is the reaction of phosphoramidite-activated deoxynucleosides with suitable functional groups, usually hydroxyl groups, on the glass or polypropylene surface. At Affymetrix these hydroxyl groups are generated at selected sites by illumination of the chip through an appropriate mask [164,167]. As a first step the solid support is derivatized with a covalent linker molecule terminated with a special, photolabile-protecting group. Illumination leads to deprotection and the formation of hydroxyl groups. In the next step the nucleoside derivative to be coupled is added, being the corresponding 3′-phosphoramidite and 5′-photoprotected. Illumination with another mask generates a different pattern of hydroxyl groups allowing each desirable sequence to be synthesized. As the number of probes in one array is limited by the physical size of the array and the achievable photolithographic resolution, approximately 400,000 oligonucleotides were synthesized on $1.28 \times 1.28 \text{ cm}^2$ chips. Originally, these high-density arrays were designed for sequencing by hybridization [164], but they are also used for gene expression monitoring [125,130]. A sequence of 200–300 bases of the gene of interest is chosen and a number of non-overlapping 25-mer probes is designed and synthesized on the chip together with mismatch control probes containing a single base difference in the central position. This redundancy should improve accuracy and improve the signal-to-noise-ratio.

Hughes et al. [168], however, found a single 60-mer oligonucleotide per gene adequate when transcript abundance ratios were determined, which limits the number of probes required to cover all genes within a whole genome to some tens of thousands and makes array fabrication by ink-jet printing feasible.

11.6.1.3 Detection principles
Hybridization of a target nucleic acid to the immobilized probe is an affinity reaction between two complementary reaction partners. Hence, this reaction was followed in real-time by affinity sensor systems, such as SPR devices [145,149,150,169], resonant mirrors [144] or grating coupler systems [170]. Some systems offer sensor chips covered with a carboxymethylated dextran layer, which prevents the direct covalent immobilization of the probe due to electrostatic repulsion of the negatively charged oligonucleotide from the negatively charged polymer. The preferred immobilization method is the biochemical immobilization using streptavidin immobilized in the polymer and biotinylated probes. Binding efficiencies up to 83% were obtained [162]. Real-time monitoring of the hybridization allowed the analysis of the influence of probe and target length, probe and target concentration and the position of mismatches not only on the resulting steady state signal but also on the association and dissociation rate. It was shown that the affinity constants

determined by SPR correlated well with melting temperature and decreasing affinities as decreasing lengths of the target influenced mainly the dissociation rate [145]. Most of these investigations were done with oligonucleotides. However, the determination of mRNA was also possible though rather high concentrations were required [150].

Usually, the detection of mRNAs utilize specific features of nucleic acids, i.e., the possibility to synthesize a copy DNA-strand (cDNA) by a reverse transcriptase reaction which allows the integration of labelled compounds. This is done either by the use of labelled primers or labelled nucleotides as components of the reaction mixture. Suitable labels are radioactive isotopes, such as ^{33}P [128] or ^{32}P, fluorescent dyes, such as fluorescein [171], Cy3 or Cy5 [148,172,173], biotin [125,158] or micro- and nanoparticles [174,175]. Labelling with biotin requires additional staining, e.g., with streptavidin-phycoerythrin [125], streptavidin-Alexa647 [158] or antibiotin–antibody- or streptavidin-horseradish peroxidase-conjugates [173,176]. The latter one is the basis of the so-called tyramide signal amplification (TSA) technology, also called *c*atalysed *r*eporter *d*eposition (CARD), which leads to the deposition of fluorescent dyes as a result of the peroxidase reaction. Thiol group-modified oligonucleotides were attached to gold nanoparticles via the SH-groups [174]. Hybridization to complementary target oligonucleotides triggered a red to purple colour change in solution [174] and light pink spots on arrays [175] due to the local increases in particle concentrations. However, sandwich-type assays were required with one strand being immobilized and the other being labelled and both cross-linked by the target strand. Signals were amplified by a factor of 10^5 using reduction of silver ions by hydroquinone on the surface of gold nanoparticles. This allowed the visualization and even quantification of the target hybridization by a simple flatbed scanner [175].

Consideration of the above mentioned labels usually integrated in cDNAs (Cy3, Cy5, fluorescein, Alexa647, phycoerythrin, biotin) show that the most often used detectors are fluorescence detectors allowing the analysis of chip surfaces. Light sources are preferably lasers with the appropriate wavelengths, nowadays systems comprising more than one light source and thus allowing the excitation of different dyes are available (e.g., www.mwg-biotech.com). The emitted fluorescence is captured either by CCD-cameras or by photomultipliers with the latter showing the higher sensitivity requiring, however, the movement of either the chip or the detector to obtain an image of the chip. CCD-cameras allow the analysis of larger areas of the chip at a time, but due to the reduced sensitivity extended integration times may be required. Though planar chips dominate gene expression analysis, systems using bundles of optical fibres were also described [171].

Electrochemical detection principles, as an alternative to optical detection, are also described for DNA-analysis [163,177,178]. They utilize the direct electrochemical oxidation of guanine, electron transfer properties of DNA double-strands or redox-active markers, such as daunomycin or ferrocene derivatives. However, these approaches did not yet develop into complex array systems to be used for gene expression monitoring. An interesting combination of electrochemical and optical properties of nucleic acids was presented by Heller et al. [179]. They generated electric fields by polarization of single electrodes for an active transport of negatively charged nucleic acids. Target nucleic acids were concentrated by the application of a positive bias and an acceleration of the hybridization reaction was observed. Reversal of the polarity removed the not-bound targets and the application of a current pulse increased stringency of the hybridization, improving specificity of the system. Detection, however, was based on fluorescence monitoring of Bodipy Texas Red, which was used for probe labelling. The device was called an "active microelectronic DNA chip device" and contained 100 test sites.

11.6.1.4 Data analysis
With the complete elucidation of genomic sequences of organisms and the appearance of first DNA chips tools were available for global analysis of organisms and thus for an increased understanding of biological systems. The expectations were high, in particular, as handling of DNA chips and generating data seemed to be rather easy [180]. However, the application of whole-genome arrays raised such a huge amount of data points, e.g., data for the expression of approximately 6200 genes for a single sample from a yeast cultivation, or of 4200 genes for an *E. coli* cultivation, that they can only be analysed, if suitable computational methods are available [181]. They have to cover aspects such as image analysis, storage and organization of results, comparison of expression profiles and functional interpretation [182,183]. Moreover, increasing the application of arrays often showed a limited reproducibility of raw data, even when experiments were performed on the same system in the same laboratory [184]. Thus, normalization strategies were developed and statistical methods tested to improve the quality of data [181,185–187]. Considering these aspects already during the design of experiments is the prerequisite for reliable data. Arfin et al. [187] showed that statistically significant changes of gene expression were only detected with a multiplication of cultivations, and the well-designed application of different arrays to the different mRNA pools to be compared.

11.6.2 Gene expression analysis in biochemical engineering

Most biotechnological applications of methods for gene expression analysis were described for *E. coli* strains [124,127,129,133–135,139,187–189] due to their importance for heterologous protein production [6] and the resulting available wealth of information on cultivation conditions [2], cell physiology, genetics and metabolism [190]. However, applications to *B. subtilis* [128,191, 192], *C. glutamicum* [126,153], *Streptomyces coelicolor* [172] and the yeast *S. cerevisiae* [125,130] were also described.

Various aspects of the influence of cultivation conditions on *E. coli* were investigated. One of the major concerns in optimization of bioprocesses is the adaptation of the microorganisms to the various aspects of stress imposed on the cells during cultivation. Thus, monitoring of the expression of stress genes was the major focus, in particular, in these investigations which considered only a subset of genes [28,124,134,135]. It was observed that genes coding for heat-shock proteins and being members of the σ^{32}-regulon (σ^{32}: *rpoH* gene product), such as dnaK, clpB, GroEL, IbpA, were induced at least transiently not only due to temperature increase [139] but also due to scale-up [135], high-cell density cultivation [129,132] or IPTG-induced expression of heterologous proteins [28,124]. The expression of these genes was not affected when different carbon sources (glucose, glycerol, acetate) were used [124], and a down-regulation was observed for cells cultivated in a 2 l-fermenter at constant temperature and pH reaching the stationary phase or being in growth arrest [188]. However, the general stress response regulated by the sigma factor σ^s (*rpoS* gene product) was induced under these conditions and was suppressed, when a heterologous protein was produced (6 l-fermenter, fed-batch) [134]. This regulon was also induced when short-chain fatty acids, such as acetate, were present in the medium, e.g., due to the cultivation in minimal medium, thus contributing to a higher stability of the cells towards acid stress [126,189]. Global analysis of gene expression highlighted the influence on the expression of additional genes, for example, of those involved in glycolysis, TCA-cycles, respiration or biosynthesis of amino acids and of those with yet unknown function [129]. Some results correlated well with previous knowledge, others are not yet understood, in particular, when genes belonging to the same regulon were found to be differently regulated [124,133, 188]. On the basis of the available information, a genome-scale computational model of the transcriptional regulatory and metabolic network was constructed for *E. coli* [193] to allow not only the description but also the prediction of cellular reactions.

Comparable investigations, though in lesser detail, were performed for other strains of biotechnological relevance. Thus, the influences of anaerobic conditions [126] or nitrogen limitation [192] on gene expression in *B. subtilis* and in *S. cerevisiae* [125,130] were investigated. Modifications in the expression of genes involved in carbon metabolism, electron transfer, iron uptake, stress response or production of antibiotics were identified, of which the physiological relevance was partly obvious whereas other phenomena cannot yet be explained [125]. Similar trends were also stated when *Streptomyces ceolicolor* [172] or *C. glutamicum* [126,153] were analysed. Thus, DNA-arrays were considered to be an excellent tool to study the physiological state of cells [126] and may complement or even replace proteomic studies relying on 2D-gel electrophoresis as they are easier to handle and allow the detection of transcripts of proteins which are hardly to be detected by electrophoresis (e.g., membrane bound proteins) [191].

A number of experiments were performed in shaking flasks, others in fermenters. Consideration of the results obtained from the detailed molecular analysis, which showed influences of even slight modifications of experimental conditions on gene expression, makes the careful control of experimental conditions indispensable [184]. This may be of importance even for non-standard biotechnological investigations, e.g., toxicity tests [194] or fundamental investigations of cell physiology. It was already observed that cultivation of cells in a fermenter lead to higher reproducibilities of gene expression data than reported for cultivations in shaking flasks [184,188].

11.7 TRENDS

In the past, biosensor development for bioprocess monitoring focussed on the development of reliable and stable automated systems for the determination of nutrients (e.g., glucose) and metabolic products (e.g., lactate, ethanol, glutamine). A prerequisite was suitable sampling systems allowing automated withdrawal of cell-free samples without affecting the sterile integrity of the process. After a long period of research and development, now these systems are commercially available and are applied even in industrial production processes. Moreover, monitoring of concentrations of the carbon source can be extended to control these concentrations to constant levels.

Thus, bioprocess-related biosensor development today has to deal with new challenges in biochemical engineering.

Regulations due to the pharmaceutical application of biotechnological products require additional analysis to guarantee the quality of products. Biosensors were shown to be useful instruments to quantify the biological activity of proteins and may also be the valuable tools for the determination of the gene copy number, thus allowing a more efficient control of the process. However, they cannot be used to monitor the absence of interfering compounds without a clear description of the analyte, i.e., the contaminating compound.

In addition to "classical" microorganisms, cultivation of cells derived from various tissues, such as liver, skin, heart or bone, becomes increasingly important. They can not only be used for therapeutic purposes (e.g., transplants of tissues), but also as test systems for toxicity or drug activity. Monitoring the cultivation of these cells does not necessarily need the determination of "new" analytes, but analytical systems have to consider special requirements of these cells, i.e., small cultivation volumes, 3D structure of cell layers or co-cultivation of several cell types. Thus, conventional sampling and analytical systems are not ideally suited as required sample volumes are too large and give more an integral than a detailed picture. Imaging methods for in vivo analysis or sampling via microsystems, such as microdialysis devices, in combination with biochemical analysis may allow the extension of analytical methods known from "conventional" cultivations to these new cell systems.

Rational improvements of cultivation efficiencies, measured as productivity, cell viability or functionality, are expected from a more detailed understanding of cell physiologies. That is why global analytical methods, e.g., DNA chips, protein gel electrophoresis, metabolite determination by chromatography, NMR or mass spectrometry, find application in bioprocess development units. Of these methods, only DNA chips and future protein chips are related to biosensor principles. A stringent combination of several already existing technologies, such as cell sampling, cell disintegration, automation by fluid handling, may allow the automated, quasi-continuous determination of important intracellular parameters, such as gene expression analysis, synthesis of selected metabolites or precursors. This will lead to a more detailed understanding of cellular reactions as the dynamic behaviour of intracellular parameters can be followed. However, whereas gene expression may be followed by the development of array-based affinity sensor systems, only selected metabolites or precursors can be detected by biosensor principles, because no common biochemical analytical principle exists. Nevertheless, suitable computing methods will increasingly become a limiting issue for the full exploitation of achievable data.

REFERENCES

1. A.L. Demain and J.E. Davies, *Manual of Industrial Microbiology and Biotechnology*, 2nd ed., American Society for Microbiology Press, Washington, DC, 1999.
2. D. Riesenberg and R. Guthke, *Appl. Microbiol. Biotechnol.*, 51 (1999) 422.
3. B.R. Glick and J.J. Pasternak, *Molecular Biotechnology*, 2nd ed., American Society for Microbiology Press, Washington, DC, 1998.
4. H. Hauser and R. Wagner, *Mammalian Cell Biotechnology in Protein Production*. Walter de Gruyter, Berlin, 1997.
5. H. Hauser and G.J. Zylstra, *Curr. Opin. Biotechnol.*, 12 (2001) 437.
6. S.C. Makrides, *Microbiol. Rev.*, 60 (1996) 512.
7. R. Korke, A. Rink, T.K. Seow, M.C.M. Chung, C.W. Beattie and W.-S. Hu, *J. Biotechnol.*, 94 (2002) 73.
8. J.H. Choi and S.Y. Lee, *Appl. Microbiol. Biotechnol.*, 64 (2004) 625.
9. G.B. Nyberg, R.R. Balcarcel, B.D. Follstad, G. Stephanopoulos and D.I.C. Wang, *Biotechnol. Bioeng.*, 62 (1999) 336.
10. H.J. Cruz, J.L. Moreira and M.J.T. Carrondo, *Biotechnol. Bioeng.*, 66 (1999) 104.
11. K. Schügerl, *J. Biotechnol.*, 85 (2001) 149.
12. K.B. Konstantinov, Y. Tsai, D. Moles and R. Matanguihan, *Biotechnol. Prog.*, 12 (1996) 100.
13. B. Danielsson, *Curr. Opin. Biotechnol.*, 2 (1991) 17.
14. J. Bradley, W. Stöcklein and R.D. Schmid, *Process Control Quality*, 1 (1991) 157.
15. U. Bilitewski and I. Rohm, *Handbook of Biosensors and Electronic Noses*. CRC Press, Boca Raton, 1997, pp. 435.
16. J.R. Woodward and R.B. Spokane, *Commercial Biosensors: Applications to Clinical, Bioprocess and Environmental Samples*. John Wiley & Sons, New York, 1998, pp. 227.
17. R. Freitag, *Biosensors in Analytical Biotechnology*. Academic Press, San Diego, 1996.
18. S. Mizutani, S. Iijima, M. Morikawa, K. Shimizu, M. Matsubara, Y. Ogawa, R. Izumi, K. Matsumoto and T. Kobayashi, *J. Ferment. Technol.*, 56 (1987) 325.
19. G. Cornelissen, H. Leptien, D. Pump, U. Scheffler, E. Sowa, H.H. Radeke and R. Luttmann, *Proc. Comput. Appl. Biotechnol.*, (2001) 8.
20. L.R. Castilho, J.T. Schumacher, J. Nothnagel, U. Bilitewski, F.B. Anspach, W.-D. Deckwer, Proceedings of Biotecnologia Industrial, Brasil, November, 2001.
21. T. Scheper, *Bioanalytik*. Vieweg Verlagsgesellschaft mbH, Braunschweig, 1991.
22. F. Bylund, E. Collet, S.-O. Enfors and G. Larsson, *Bioprocess Eng.*, 18 (1998) 171.
23. S. Müllner, *Adv. Biochem. Eng./Biotechnol.*, 83 (2003) 1.
24. A. Seeger, B. Schneppe, J.E.G. McCarthy, W.-D. Deckwer and U. Rinas, *Enzyme Microb. Technol.*, 17 (1995) 947.
25. D. Mattanovich, W. Kramer, C. Lüttich, R. Weik, K. Bayer and H. Katinger, *Biotechnol. Bioeng.*, 58 (1998) 296.

26 A.K. Chauhan, D. Arora and N. Khanna, *Process Biochem.*, 34 (1999) 139.
27 F. Hoffmann and U. Rinas, *Biotechnol. Bioeng.*, 76 (2001) 333.
28 B. Jürgen, H.Y. Lin, S. Riemschneider, C. Scharf, P. Neubauer, R. Schmid, M. Hecker and T. Schweder, *Biotechnol. Bioeng.*, 70 (2000) 217.
29 M. Cserjan-Puschmann, W. Kramer, E. Duerrschmid, G. Striedner and K. Bayer, *Appl. Microbiol. Biotechnol.*, 53 (1999) 43.
30 H. Shi and K. Shimizu, *Biotechnol. Bioeng.*, 58 (1998) 139.
31 M. Carlsen, J. Nielsen and J. Villadsen, *J. Biotechnol.*, 45 (1996) 81.
32 S. Takac, G. Calik, F. Mavituna and G. Dervakos, *Enzyme Microb. Technol.*, 23 (1998) 286.
33 G.B. Nyberg, R.R. Balcarcel, B.D. Follstad, G. Stephanopoulos and D.I.C. Wang, *Biotechnol. Bioeng.*, 62 (1999) 324.
34 K. Lee, F. Berthiaume, G.N. Stephanopoulos and M.L. Yarmush, *Tissue Eng.*, 5 (1999) 347.
35 K. Schmidt, A. Marx, A.A. de Graaf, W. Wiechert, H. Sahm, J. Nielsen and J. Villadsen, *Biotechnol. Bioeng.*, 58 (1998) 254.
36 H.P.J. Bonarius, B. Timmerarends, C.D. de Gooijer and J. Tramper, *Biotechnol. Bioeng.*, 58 (1998) 258.
37 J. Zhao, T. Baba, H. Mori and K. Shimizu, *Appl. Microbiol. Biotechnol.*, 64 (2004) 91.
38 H. Zhang, K. Shimizu and S. Yao, *Biochem. Eng. J.*, 16 (2003) 211.
39 J.-N. Phue and J. Shiloach, *J. Biotechnol.*, 109 (2004) 21.
40 F. Hoffmann and U. Rinas, *Biotechnol. Prog.*, 16 (2000) 1000.
41 L. Peng and K. Shimizu, *Appl. Microbiol. Biotechnol.*, 61 (2003) 163.
42 T. Ryll and R. Wagner, *J. Chromatogr.*, 570 (1991) 77.
43 P. Calik, G. Calik, S. Takac and T.H. Özdamar, *Biotechnol. Bioeng.*, 64 (1999) 151.
44 U. Rinas, K. Hellmuth, R. Kang, A. Seeger and H. Schlieker, *Appl. Environ. Microbiol.*, 61 (1995) 4147.
45 K.C. Schuster, *Adv. Biochem. Eng.*, 66 (1999) 185.
46 R. Krämer, *J. Biotechnol.*, 45 (1996) 1.
47 J. Nielsen, *Biotechnol. Bioeng.*, 58 (1998) 125.
48 C. Khosla and J.D. Keasling, *Nat. Rev. Drug Discov.*, 2 (2003) 1019.
49 A. Spohr, C. Dam-Mikkelsen, M. Carlsen, J. Nielsen and J. Villadsen, *Biotechnol. Bioeng.*, 58 (1998) 541.
50 C. Bittner, G. Wehnert and T. Scheper, *Biotechnol. Bioeng.*, 60 (1998) 24.
51 A.G. Cavinato, D.M. Mayes, Z. Ge and J.B. Callis, *Anal. Chem.*, 62 (1990) 1977.
52 G. Vaccari, E. Dosi, A.L. Campi, A. Gonzalez-Varay, R.D. Matteuzzi and G. Mantovani, *Biotechnol. Bioeng.*, 43 (1994) 913.
53 T. Yano and M. Harata, *J. Ferment. Bioeng.*, 77 (1994) 659.
54 M.C. d'Anjou and A.J. Daugulis, *Biotechnol. Bioeng.*, 72 (2001) 1.
55 J. Mohns and W. Künnecke, *Anal. Chim. Acta*, 305 (1995) 241.

56 H. Lidén, T. Buttler, H. Jeppsson, G. Marko-Varga, J. Volc and L. Gorton, *Chromatographia*, 47 (1998) 501.
57 D. Compagnone, M. Esti, M.C. Messia, E. Peluso and G. Palleschi, *Biosens. Bioelectron.*, 13 (1998) 875.
58 C.O. Fágáin, *Stabilizing Protein Function*. Springer, Heidelberg, 1997.
59 Q. Chen, G.L. Kenausis and A. Heller, *J. Am. Chem. Soc.*, 120 (1998) 4582.
60 J. Heller and A. Heller, *J. Am. Chem. Soc.*, 120 (1998) 4586.
61 D. Pfeiffer, E.V. Ralis, A. Makower and F.W. Scheller, *J. Chem. Technol. Biotechnol.*, 49 (1990) 255.
62 D. Pfeiffer, K. Setz, T. Schulmeister, F.W. Scheller, H.B. Lück and D. Pfeiffer, *Biosens. Bioelectron.*, 7 (1992) 661.
63 R. Wimmer, M. Olsson, M.T. Neves Petersen, R. Hatti-Kaul, S.B. Petersen and N. Müller, *J. Biotechnol.*, 55 (1997) 85.
64 J.F. Carpenter, K.-i. Izutsu and T.W. Randolph, *Drugs Pharm. Sci.*, 96 (1999) 123.
65 T.D. Gibson, J.N. Hulbert and J.R. Woodward, *Anal. Chim. Acta*, 279 (1993) 185.
66 I. Rohm, M. Genrich, W. Collier and U. Bilitewski, *Analyst*, 121 (1996) 877.
67 V.G. Gavalas, N.A. Chaniotakis and T.D. Gibson, *Biosens. Bioelectron.*, 13 (1998) 1205.
68 M.D. Gouda, M.A. Kumar, M.S. Thakur and N.G. Karanth, *Biosens. Bioelectron.*, 17 (2002) 503.
69 J.T. Schumacher, I. Münch, T. Richter, I. Rohm and U. Bilitewski, *J. Mol. Catal. B Enzym.*, 7 (1999) 67.
70 T.D. Gibson, B.L.J. Pierce, J.N. Hulbert and S. Gillespie, *Sens. Actuators B*, 33 (1996) 13.
71 A. Schmidt, J.T. Schumacher, J. Reichelt, H.-J. Hecht and U. Bilitewski, *Anal. Chem.*, 74 (2002) 3037.
72 P. Sarkar, I.E. Tothill, S.J. Setford and A.P.F. Turner, *Analyst*, 124 (1999) 856.
73 W. Künnecke and R.D. Schmid, *J. Biotechnol.*, 14 (1990) 127.
74 P. Sarkar, *Microchem. J.*, 64 (2000) 283.
75 J. Wang, M.P. Chatrathi, B. Tian and R. Polsky, *Anal. Chem.*, 72 (2000) 2514.
76 G.A.M. Mersal, M. Khodari and U. Bilitewski, *Biosens. Bioelectron.*, 20 (2004) 305.
77 G. Fortier, M. Vaillancourt and D. Belanger, *Electroanalysis*, 4 (1992) 275.
78 N. Torto, T. Laurell, L. Gorton and G. Marko-Varga, *Anal. Chim. Acta*, 379 (1999) 281.
79 I. Papefstathiou, U. Bilitewski and M.D. Luque de Castro, *Anal. Chim. Acta*, 330 (1996) 265.
80 S. Iijima, Y. Soo Park and T. Kobayashi, *Proceedings of Asia-Pacific Biochemical Engineering Conference*, 1992, pp. 39.
81 A. Gebbert, M. Alvarez-Icaza, H. Peters, V. Jäger, U. Bilitewski and R.D. Schmid, *J. Biotechnol.*, 32 (1994) 213.
82 S. Harthun, K. Matischak and P. Friedl, *Anal. Biochem.*, 251 (1997) 73.

83 L.F. Vallejo, M. Brokelmann, S. Marten, S. Trappe, J. Cabrera-Crespo, A. Hoffmann, G. Gross, H.A. Weich and U. Rinas, *J. Biotechnol.*, 94 (2002) 185.
84 J.H.T. Luong, C. Masson, R.S. Brown, K.B. Male and A.-L. Nguyen, *Biosens. Bioelectron.*, 9 (1994) 577.
85 A.B. van Putten, F. Spitzenberger, G. Kretzmer, B. Hitzmann and K. Schügerl, *Anal. Chim. Acta*, 317 (1995) 247.
86 H.-A. Kracke-Helm, L. Brandes, B. Hitzmann, U. Rinas and K. Schügerl, *J. Biotechnol.*, 20 (1991) 95.
87 I. Biran, L. Klimentiy, R. Hengge-Aronis, E.Z. Ron and J. Rishpon, *Microbiology*, 145 (1999) 2129.
88 W. Künnecke, H.M. Kalisz and R.D. Schmid, *Anal. Lett.*, 22 (1989) 1471.
89 D.G. Bracewell, A. Gill, M. Hoare, P.A. Lowe and C.H. Maule, *Biosens. Bioelectron.*, 13 (1998) 847.
90 G. Siemeister, B. Schnurr, K. Mohrs, C. Schächtele, D. Marmé and G. Martiny-Baron, *Biochem. Biophys. Res. Commun.*, 222 (1996) 249.
91 T. Kirsch, J. Nickel and W. Sebald, *EMBO J.*, 19 (2000) 3314.
92 F. Deckert and F. Legay, *Anal. Biochem.*, 274 (1999) 81.
93 D.M. Disley, P.R. Morrill, K. Sproule and C.R. Lowe, *Biosens. Bioelectron.*, 14 (1999) 481.
94 A. Gill, D.G. Bracewell, C.H. Maule, P.A. Lowe and M. Hoare, *J. Biotechnol.*, 65 (1998) 69.
95 B.v. Tiedemann and U. Bilitewski, *Biosens. Bioelectron.*, 17 (2002) 983.
96 J. Wendler, L.F. Vallejo, U. Rinas and U. Bilitewski, *Anal. Bioanal. Chem*, Submitted for publication.
97 G. Zeder-Lutz, A. Benito and M.H.V. van Regenmortel, *J. Mol. Recognit.*, 12 (1999) 300.
98 M. Michalzik, J. Wendler, J. Rabe, S. Büttgenbach and U. Bilitewski, *Sens. Actuators B*, in press.
99 R. Polzius, F.F. Bier, U. Bilitewski, V. Jäger and R.D. Schmid, *Biotechnol. Bioeng.*, 42 (1993) 1287.
100 D. Mattanovich, R. Weik, S. Thim, W. Kramer, K. Bayer and H. Katinger, *Ann. NY Acad. Sci.*, 782 (1996) 182.
101 K. Friehs, *Adv. Biochem. Bioeng./Biotechnol.*, 86 (2004) 47.
102 M.P. Nandakumar, E. Nordberg Karlsson and B. Mattiasson, *Biotechnol. Bioeng.*, 73 (2001) 406.
103 I.S. Noites, R.D. O'Kennedy, M.S. Levy, N. Abidi and E. Keshavarz-Moore, *Biotechnol. Bioeng.*, 66 (1999) 195.
104 F. Lottspeich and H. Zorbas, *Bioanalytik*. Spektrum Akademischer Verlag, Heidelberg, 1998.
105 E. Kai, S. Sawata, K. Ikebuqro, T. Iida, T. Honda and I. Karube, *Anal. Chem.*, 71 (1999) 796.

106 N. Bianchi, C. Rutigliano, M. Tomasetti, G. Feriotto, F. Zorzato and R. Gambari, *Clin. Diagn. Virol.*, 8 (1997) 199.
107 D. Kuhlmeier, *Nachweis von Nukeinsäure-Wechselwirkungen mit Hilfe optischer Biosensoren*, 2000, opus.tu-bs.de/opus/volltexte/2000/95/pdf/20000224a.pdf
108 G.H.W. Sanders and A. Manz, *Trends Anal. Chem.*, 19 (2000) 364.
109 D.M.F. Prazeres, G.N.M. Ferreira, G.A. Monteiro, C.L. Cooney and J.M.S. Cabral, *Trends Biotechnol.*, 17 (1999) 169.
110 D.R. Gentry, V.J. Hernandez, L.H. Hguyen, D.B. Jensen and M. Cashel, *J. Bacteriol.*, 175 (1993) 7982.
111 A. Konopka, *Curr. Opin. Microbiol.*, 3 (2000) 244.
112 J. Ljunggren and L. Häggström, *Biotechnol. Bioeng.*, 44 (1994) 808.
113 R. Renneberg, G. Trott-Kriegeskorte, M. Lietz, V. Jäger, M. Pawlowa, G. Kaiser, U. Wollenberger, F. Schubert, R. Wagner, R.D. Schmid and F.W. Scheller, *J. Biotechnol.*, 21 (1991) 173.
114 W. Vahjen, J. Bradley, U. Bilitewski and R.D. Schmid, *Anal. Lett.*, 24 (1991) 1445.
115 T.S. Stoll, P.-A. Ruffieux, M. Schneider, U. von Stockar and I.W. Marison, *J. Biotechnol.*, 51 (1996) 27.
116 A. Collins, M.P. Nandakumar, E. Csöregi and B. Mattiasson, *Biosens. Bioelectron.*, 16 (2001) 765.
117 M. Gilis, H. Durliat and M. Comtat, *Anal. Chim. Acta*, 355 (1997) 235.
118 I.-S. Park, Y.-J. Cho and N. Kim, *Anal. Chim. Acta*, 404 (2000) 75.
119 G. Gamborg and E.H. Hansen, *Anal. Chim. Acta*, 285 (1994) 321.
120 G. Davis, M.J. Green and H.A.O. Hill, *Enzyme Microb. Technol.*, 8 (1986) 349.
121 X. Yang, G. Johansson, D. Pfeiffer and F. Scheller, *Electroanalysis*, 3 (1991) 659.
122 E.H. Hansen, M. Gundstrup and H.S. Mikkelsen, *J. Biotechnol.*, 31 (1993) 369.
123 T. Richter, *Entwicklung alternativer Methoden zur Nukleotidanalytik in der Bioprozessüberwachung*, 2000, www.biblio.tu-bs.de/ediss/data/20000803a/20000803a.pdf
124 M.-K. Oh and J.C. Liao, *Biotechnol. Prog.*, 16 (2000) 278.
125 J.J.M. Ter Linde, H. Liang, R.W. Davies, H.Y. Steensma, J.P. van Dijken and J.T. Pronk, *J. Bacteriol.*, 181 (1999) 7409.
126 R.W. Ye, W. Tao, L. Bedzyk, T. Young, M. Chen and L. Li, *J. Bacteriol.*, 182 (2000) 4458.
127 H. Tao, C. Bausch, C. Richmond, F.R. Blattner and T. Conway, *J. Bacteriol.*, 181 (1999) 6425.
128 B. Jürgen, R. Hanschke, M. Sarvas, M. Hecker and T. Schweder, *Appl. Microbiol. Biotechnol.*, 55 (2001) 326.
129 S.H. Yoon, M.-J. Han, S.Y. Lee, K.J. Jeong and J.-S. Yoo, *Biotechnol. Bioeng.*, 81 (2003) 753.
130 V.M. Boer, J.H. de Winde, J.T. Pronk and M.D.W. Piper, *J. Biol. Chem.*, 278 (2003) 3265.

131 A. Muffler, S. Bettermann, M. Haushalter, A. Hörlein, U. Neveling, M. Schramm and O. Sorgenfrei, *J. Biotechnol.*, 98 (2002) 255.
132 R.T. Gill, M.P. deLisa, J.J. Valdes and W.E. Bentley, *Biotechnol. Bioeng.*, 72 (2001) 85.
133 M.-K. Oh and J.C. Liao, *Metabol. Eng.*, 2 (2000) 201.
134 T. Schweder, H.Y. Lin, B. Jürgen, A. Breitenstein, S. Riemschneider, V. Khalameyzer, A. Gupta, K. Büttner and P. Neubauer, *Appl. Microbiol. Biotechnol.*, 58 (2002) 330.
135 T. Schweder, E. Krüger, B. Xu, B. Jürgen, G. Blomsten, S.-O. Enfors and M. Hecker, *Biotechnol. Bioeng.*, 65 (1999) 151.
136 D.S. Wilson and S. Nock, *Curr. Opin. Chem. Biol.*, 6 (2001) 81.
137 S.R. Weinberger, E.A. Dalmasso and E.T. Fung, *Curr. Opin. Chem. Biol.*, 6 (2001) 86.
138 B.R. Jordan, *DNA Microarrays: Gene Expression Applications*. Springer, Berlin, 2001.
139 C.S. Richmond, J.D. Glasner, R. Mau, H. Jin and F.R. Blattner, *Nucl. Acids Res.*, 27 (1999) 3821.
140 F.F. Bier and F. Kleinjung, *Fresenius J. Anal. Chem.*, 371 (2001) 151.
141 A.J. Thiel, A.G. Frutos, C.E. Jordan, R.M. Corn and L.M. Smith, *Anal. Chem.*, 69 (1997) 4948.
142 M. ElAtifi, I. Duprè, B. Rostaing, E.M. Chambaz, A.L. Benabid and F. Berger, *BioTechniques*, 33 (2002) 612.
143 M. Campas and I. Katakis, *Trends Anal. Chem.*, 23 (2004) 49.
144 H.J. Watts, D. Yeung and H. Parkes, *Anal. Chem.*, 67 (1995) 4283.
145 B. Persson, K. Stenhag, P. Nilsson, A. Larsson, M. Uhlén and P.-A. Nygren, *Anal. Biochem.*, 246 (1997) 34.
146 D.R. Dorris, A. Nguyen, L. Gieser, R. Lockner, A. Lublinsky, M. Patterson, E. Touma, T.J. Sendera, R. Elghanian and A. Mazumder, *BMC Biotechnol.*, 3 (2003) 6.
147 A. Relogio, C. Schwager, A. Richter, W. Ansorge and J. Valcarcel, *Nucl. Acids Res.*, 30 (2002) e51.
148 M.D. Kane, T.A. Jatkoe, C.R. Stumpf, J. Lu, J.D. Thomas and S.J. Madore, *Nucl. Acids Res.*, 28 (2000) 4552.
149 A.W. Peterson, R.J. Heaton and R.M. Georgiadis, *Nucl. Acids Res.*, 29 (2001) 5163.
150 B. Henze, C. Bebber, J.J. van den Heuvel and U. Bilitewski, *Detection of mRNA Using the BIAcore,* 2002, http://w210.ub.uni-tuebingen.de/dbt/volltexte/2002/454/
151 H. Dai, M. Meyer, S. Stepaniants, M. Ziman and R. Stoughton, *Nucl. Acids Res.*, 30 (2002) e86.
152 T.T. Nikiforov and Y.-H. Rogers, *Anal. Biochem.*, 227 (1995) 201.
153 A. Loos, C. Glanemann, L.B. Willis, X.M. O'Brian, P.A. Lessard, R. Gerstmeir, S. Guillouet and A.J. Sinskey, *Appl. Environ. Microbiol.*, 67 (2001) 2310.

154 N. Zammatteo, L. Jeanmart, S. Hamels, S. Courtois, P. Louette, L. Hevesi and J. Remacle, *Anal. Biochem.*, 280 (2000) 143.
155 H. Hakala and H. Lönnberg, *Bioconjugate Chem.*, 8 (1997) 232.
156 P. Wilkins Stevens, M.R. Henry and D.M. Kelso, *Nucl. Acids Res.*, 27 (1999) 1719.
157 M. Beier and J.D. Hoheisel, *Nucl. Acids Res.*, 27 (1999) 1970.
158 R. Ramakrishnan, D. Dorris, A. Lublinsky, A. Nguyen, M. Domanus, A. Prokhorova, L. Gieser, E. Touma, R. Lockner, M. Tata, X. Zhu, M. Patterson, R. Shippy, T.J. Sendera and A. Mazumder, *Nucl. Acids Res.*, 30 (2002) e30.
159 W.G. Beattie, L. Meng, S.L. Turner, R.S. Varma, D.D. Dao and K.L. Beattie, *Mol. Biotechnol.*, 4 (1995) 213.
160 Y.-H. Rogers, P. Jiang-Baucom, Z.-J. Huang, V. Bogdanov, S. Anderson and M.T. Boyce-Jacino, *Anal. Biochem.*, 266 (1999) 23.
161 M.A. Podyminogin, E.A. Lukhtanov and M.W. Reed, *Nucl. Acids Res.*, 29 (2001) 5090.
162 D. Kuhlmeier, E. Rodda, L.O. Kolarik, D.N. Furlong and U. Bilitewski, *Biosens. Bioelectron.*, 18 (2003) 925.
163 G. Marrazza, I. Chianella and M. Mascini, *Biosens. Bioelectron.*, 14 (1999) 43.
164 A.C. Pease, D. Solas, E.J. Sullivan, M.T. Cronin, C.P. Holmes and S.P.A. Fodor, *Proc. Natl. Acad. Sci.*, 91 (1994) 5022.
165 J. Weiler and J.D. Hoheisel, *Anal. Biochem.*, 243 (1996) 218.
166 A.P. Blanchard, R.J. Kaiser and L.E. Hood, *Biosens. Bioelectron.*, 11 (1996) 687.
167 G.H. McGall and F.C. Christians, *Adv. Biochem. Eng.*, 77 (2002) 21.
168 T.R. Hughes, M. Mao, A.R. Jones, J. Burchard, M.J. Marton, K.W. Shannon, S.M. Lefkowitz, M. Ziman, J.M. Schelter, M.R. Meyer, S. Kobayashi, C. Davis, H. Dai, Y.D. He, S.B. Stephaniants, G. Cavet, W.L. Walker, A. West, E. Coffey, D.D. Shoemaker, R. Stoughton, A.P. Blanchard, S.H. Friend and P.S. Linsley, *Nat. Biotechnol.*, 19 (2001) 342.
169 P. Nilsson, B. Persson, M. Uhlén and P.-A. Nygren, *Anal. Biochem.*, 224 (1995) 400.
170 F.F. Bier, F. Kleinjung and F.W. Scheller, *Sens. Actuators B*, 38/39 (1997) 78.
171 J.A. Ferguson, T.C. Boles, C.P. Adams and D.R. Walt, *Nat. Biotechnol.*, 14 (1996) 1681.
172 J. Huang, C.-J. Lih, K.-H. Pan and S.N. Cohen, *Genes Dev.*, 15 (2001) 3183.
173 A. Badiee, H.G. Eiken, V.M. Stehen and R. Lovlie, *BMC Biotechnol.*, 3 (2003) 23.
174 J.J. Storhoff, R. Elghanian, R.C. Mucie, C.A. Mirkin and R.L. Letsinger, *J. Am. Chem. Soc.*, 120 (1998) 1959.
175 T.A. Taton, C.A. Mirkin and R.L. Letsinger, *Science*, 289 (2000) 1757.
176 E.J.M. Speel, A.H.N. Hopman and P. Komminoth, *J. Histochem. Cytochem.*, 47 (1999) 281.
177 J. Wang, *Chem. Eur. J.*, 5 (1999) 1681.
178 S.O. Kelley, N.M. Jackson, M.G. Hill and J.K. Barton, *Angew. Chem. Int. Ed.*, 38 (1999) 941.

179 M.J. Heller, A.H. Forster and E. Tu, *Electrophoresis*, 21 (2000) 157.
180 P.O. Brown and D. Botstein, *Nat. Genet. Suppl.*, 21 (1999) 33.
181 A. Butte, *Nat. Rev. Drug Discovery*, 1 (2002) 951.
182 J. Tamames, D. Clark, J. Herrero, J. Dopazo, C. Blaschke, J.M. Fernández, J.C. Oliveros and A. Valencia, *J. Biotechnol.*, 98 (2002) 269.
183 A. Brazma, U. Sarkans, A. Robinson, J. Vilo, M. Vingron, J. Hoheisel and K. Fellenberg, *Adv. Biochem. Eng./Biotechnol.*, 77 (2002) 113.
184 M.D.W. Piper, P. Daran-Lapujade, C. Bro, B. Regenberg, S. Knudsen, J. Nielsen and J.T. Pronk, *J. Biol. Chem.*, 277 (2002) 37001.
185 G.C. Tseng, M.-K. Oh, L. Rohlin, J.C. Liao and W.H. Wong, *Nucl. Acids Res.*, 29 (2001) 2549.
186 K.A. Baggerly, K.R. Coombes, K.R. Hess, D.N. Stivers, L.V. Abruzzo and W. Zhang, *J. Comput. Biol.*, 8 (2001) 639.
187 S.M. Arfin, A.D. Long, E.T. Ito, L. Tolleri, M.M. Riehle, E.S. Paegle and G.W. Hatfield, *J. Biol. Chem.*, 275 (2000) 29672.
188 D.-E. Chang, D.J. Smalley and T. Conway, *Mol. Microbiol.*, 45 (2002) 289.
189 C.N. Arnold, J. McElhanon, A. Lee, R. Leonhart and D.A. Siegele, *J. Bacteriol.*, 183 (2001) 2178.
190 F.C. Neidhardt, *Escherichia coli and Salmonella, Cellular and Molecular Biology*, 2nd ed., American Society of Microbiology, Washington, DC, 1996.
191 M. Hecker and S. Engelmann, *Int. J. Med. Microbiol.*, 290 (2000) 123.
192 H. Jarmer, R. Berka, S. Knudsen and H.H. Saxild, *FEMS Microbiol. Lett.*, 206 (2002) 197.
193 M.W. Covert, E.M. Knight, J.L. Reed, M.J. Herrgard and B.O. Palsson, *Nature*, 429 (2004) 92.
194 T.K. Baker, M.A. Carfagna, H. Gao, E.R. Dow, Q. Li, G.H. Searfoss and T.P. Ryan, *Chem. Res. Toxicol.*, 14 (2001) 1218.

Chapter 12

Coupling of microdialysis sampling with biosensing detection modes

Danila Moscone

12.1 INTRODUCTION

The sampling process can be the most critical aspect of an analysis, the step where sample loss and contamination errors are most likely to occur. This is especially true when the "material" to be analysed is a very complex medium, such as a biological fluid, a tissue, an organ, a plant, or when sampling takes place in a particular environment, as is the case for evaluating wastewater or bioreactor content during a bioprocess. Sometimes, the sampling site may be hard to access, may require special care so as to avoid contamination, and/or the sample could be available in a very low quantity. Also the sampling time could be an important parameter in cases where the sample composition can vary considerably in response to a different stimulus, physiological status, or, in case of a living being, simply because the analyte of interest could have a circadian profile.

In all these cases, the microdialysis sampling technique has great potential. This technique arose historically in the field of neuroscience, having been introduced in 1972 by Delgado et al. [1] for use in brain research as an evolution of the push–pull technique devised by Gaddum back in 1961 [2] and of the long-term dialysis sac implantation explored by Bito since 1966 [3].

Bito introduced the concept of a dialysis membrane that separates two fluid compartments, where the external one usually represents the site to be sampled. The other compartment contains a liquid whose composition (ionic strength and pH, for example) tentatively matches that of the external fluid being sampled.

Permeable substances present in one of the compartments, and absent in the other, will diffuse down their concentration gradient across the membrane. Usually, the membrane is hydrophilic; hence, this passive transport principally concerns hydrophilic molecules of low molecular weight. Large non-permeable molecules such as proteins, enzymes and peptides present on one side of the membrane are excluded from this transport, but sometimes an osmotic pressure can develop causing a flux of water across the membrane towards the high molecular weight molecules. The minimization of this latter effect has been accomplished by using membranes with very small surface areas and by flowing fresh isotonic solutions at very low flow rates on one side of the membrane in order to avoid any net exchange of ions and loss of fluid through the membrane. Moreover, the continuous perfusion of a fresh sampling fluid contributes to maintain a concentration gradient for a high rate of transport of the analytes in the sampling stream, and this fact constitutes the major difference between a simple equilibrium dialysis and the microdialysis technique.

The availability of membranes of very small cylindrical size, the so-called hollow fibres used in artificial kidney cartridges, thus provided a great impulse to the development of this technique, and much of the original research work was done by Ungerstedt. He implanted "hollow fibres" into rat brains in order to mimic the function of blood vessels [4]; since then, this simple "microdialysis probe" has been markedly improved [5–8]. A vast number of different designs have appeared [9–14] resulting in probes that, because of their narrow cylindrical shape, can be handled like a needle and easily implanted into brain and many other tissues.

The basic principles of this technique have been extensively reviewed by different authors who offered theoretical and physicochemical considerations [15–20]. By the first half of 2004, a Medline search using "microdialysis" as keyword reveals the presence of a little less than 9000 articles, attesting to the great interest of this technique. It should be noted that, if the great majority of these papers deal with the biomedical field, there is a growing application in other areas such as the sampling and control of enzymatic bioprocesses or fermentation as well as the analysis of environmental samples. Gorton and Torto pioneered these fields of application and have written a very interesting review in 1999 [20].

12.2 THE PROBE

The central component of any microdialysis system is the probe itself. Nowadays, many scientists design and create their own probes depending on the purpose of the study. However, an increasing number of companies supply probes ready to use and made in different shapes, dimensions, shafts materials, fibres, etc.

Table 12.1 shows a list of suppliers (probably incomplete) of probes and the equipment necessary to carry out this technique.

TABLE 12.1
List of supplier of microdialysis equipment

Company	Probes	Pumps	Micro-samplers	Swivels	Stereotaxic equipments	Software	Website
CMA/Microdialysis	✓	✓	✓	✓	✓	✓	www.microdialysis.se
Bioanalytical Systems Inc.	✓	✓	✓	✓	✓	✓	www.bioanalytical.com
Harvard Apparatus	✓	✓		✓	✓		www.harvardapparatus.com
Metalant	✓	✓	✓	✓	✓		www.microbiotech.se
Eicom	✓	✓	✓	✓	✓		www.eicom.co.jp
Marsil Scientific	✓	✓	✓	✓	✓		www.marsilsci.com
Agnthos	✓	✓	✓	✓			www.agnthos.se
Applied Neurosciences	✓						www.appliedneuroscience.co.uk

Typically, a microdialysis probe consists of a small piece (2–10 mm) of a cylindrical dialysis membrane (the hollow fibre), connected to an inlet and an outlet tube of suitable dimensions. Despite the differences in design, what essentially typifies the probe is the position of the inlet and outlet tubes. Basically, two types of probes can be distinguished: in the first one, the two tubes are positioned in a serial arrangement; in the other type, the inlet and outlet tubes are concentric or parallel. Consequently, in the first type, which is normally termed a "linear probe", the hollow fibre is glued in between. In the second one, it is affixed at the tip, and this is usually called a "concentric probe". In addition to these two basic designs, other probes are available such as the "loop probe", the "shunt probe", and the "flexible probe". Figure 12.1 shows some schematic representations of these probes.

The linear probe is the simplest design and is the most adapted for sampling of soft tissues such as muscle [21,22], adipose and subcutaneous tissue [23–25], liver [26], and tumours [27,28–30]. The advantages of this kind of probe are the simplicity of construction, the small dimensions and high flexibility (all of which minimize the tissue damage) and the ruggedness, which is sometimes achieved by inserting into the lumen of the hollow fibre

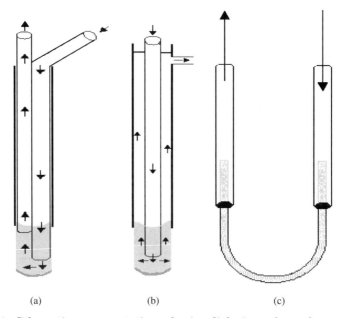

Fig. 12.1. Schematic representation of microdialysis probes where the inlet and outlet tubes are in a parallel (a), concentric (b) and serial (c) configuration. The grey part represents the hollow fibre, while the arrows indicate the direction of the flowing perfusate.

one or two twisted wires, in order to prevent the folding of the fibre or its collapse when implanted in muscles of free moving animals. One disadvantage is that, when implanted in vivo, two holes are necessary, one for the entrance and one for the exit, increasing the risk of microbial attacks.

The loop probes are linear probes with a longer membrane (10–50 mm), often folded to form a loop. Their large surface area increases the amount of analyte retrieved by the probe and therefore they are a good choice for in vitro microdialysis and for sample cleanup.

The shunt probe can be considered as a variation because the linear probe is contained inside another plastic tube (the shunt) where the fluid to be sampled passes, this arrangement allowing the continuous sampling from moving fluids. First described by Scott and Lunte [31] it is usually applied for bile sampling [32,33] but in vitro applications are also described such as that for milk analysis [34].

The concentric probe, patented by Ungerstedt (USA Patent 1987) is the most used, although it is also the most difficult to construct. It can be made very thin (around 300 μm in diameter); the material of the shaft may be rigid (metal) or flexible (fused silica or plastic); the length of the membrane at the tip can be varied from a mm to a cm; in some configurations it can be autoclavable, and it is the most available of the commercial probes (see Table 12.1).

Because of its small dimensions and conformation, it is easy to handle as a needle. It also presents the lowest degree of invasiveness when implanted in vivo, since it requires only a single insertion point. When the shaft is rigid, the probe can be easily glued to the skull of animals for intra-cerebral investigations. By contrast, the flexible conformation is the most suitable for implantation in blood vessels and in soft tissue because it permits freedom of movement without injury. This kind of probe can be assembled in larger and more robust designs that are useful for sampling from bioreactors and fermenters, where stirring may be continuous.

12.3 THE MEMBRANE

The hollow fibre is the most crucial part of the microdialysis probe. It acts as a membrane, and its characteristics affect performance in the sampling step as well as the probe's suitability for the selected application. Hollow fibres are commercially available in different materials, the most common being polycarbonate (PC), regenerated cellulose (Cuprophan, CU), cellulose acetate (CA), polyacrylonitrile (PAN), polyethersulphone (PES), polysulphone (PE), and polyamide (PA). Generally, the fibres have an outer diameter between 200 and 500 μm, a wall thickness between 9 and 100 μm (some of these membranes having a supporting layer of the same material as the fibre), and a molecular weight cut-off (MWCO) from 3000 to 20,000 Da, although larger cut-off limits are available. The MWCO characterizes the dialysis performance by controlling the size of the molecules that will be retained by the membrane. This parameter is related to the size of the pores in the membrane and to the pore distribution. However, other characteristics of a molecule such as its shape, charge, degree of hydration, the nature of the solvent, pH and ionic strength, in addition to the molecular weight, are important parameters to be considered.

For in vivo implantation and for bioprocess application, both sterilizability and biocompatibility are required, although the latter is a vague concept for which a variety of definitions exist. The biocompatibility aspect is very important both in vivo and in vitro when working with microorganisms or enzymes, which might absorb onto the membranes. In fact, when exposed to complex media, a membrane–protein interaction can occur, leading to an undesirable fouling driven by the adsorptive and adhesive interaction of proteins and cells to the outer surface of the probe. Unfortunately, especially in the biomedical field, the choice of the dialysis membrane seems to be based primarily on the commercial availability with little attention being paid to membrane optimization. In addition, a well-reasoned choice of the most suitable membrane for the selected application is made more difficult by the fact that often membranes made of the same material and/or with similar MWCO, but supplied by different companies (which do not give precise information about), can have different performance. For example, the degree of rejection of large molecules depends not only on the magnitude of the MWCO, but also on its sharpness. In addition, the term "polysulphone" membrane can be misleading if no specification of the polymer structure is given [35].

Recently, studies on biofouling have appeared in the literature [36–38] while membrane optimization has been more extensively studied in the field of bioprocess [39–45].

12.4 SAMPLING CONSIDERATIONS

As mentioned before, microdialysis is a sampling technique controlled by diffusion. Because a fluid continuously perfuses the probe, equilibrium conditions will not be reached, and only a fraction of the actual concentration of the analytes present in the medium surrounding the probe will be collected. The amount of analyte collected is normally defined as "recovery".

The "relative recovery" is defined as the ratio between the concentration in the dialysate (C_{in}) and the concentration outside the probe (C_{out}):

$$\text{Recovery} = C_{in}/C_{out} \tag{12.1}$$

and is the value obtained if volume concentrations are used, while the absolute amount of mass removed from the medium of interest per time interval is defined as "absolute recovery". The relative recovery is also described as the "extraction fraction" (EF).

A number of variables, such as the perfusion flow rate, the membrane surface area and geometry, the MWCO of the membrane, the diffusion characteristics of the collected analyte, the composition of the perfusion medium, and the temperature, influence recovery parameters. When the microdialysis is carried out in vivo, the recovery can also be affected by some tissue properties, including tissue tortuosity, the extracellular space volume, the tissue blood fluid and the tissue metabolism of the substance.

Nevertheless, under a given set of experimental conditions, most of these parameters remain constant and a calibration can be performed. Recovery is initially determined in vitro with standard solutions or in a more complex medium that approaches the one to be sampled. However, it is rare that these calibrations completely match the results obtained during real experiments. This is especially true when the experiments are performed in vivo.

Mathematical models that take into consideration the effects of both in vitro and in vivo parameters have been proposed by Jacobson [46], who described the recovery as a relationship between the flow rate, the membrane area and the mass exchange:

$$R = \text{EF} = C_{in}/C_{out} = 1 - \exp(-K_m S/Q) \tag{12.2}$$

where Q is the flow rate used, S the membrane surface area, and K_m an overall mass transfer coefficient comprehensive of dialysate, membrane and tissue coefficients.

Bungay [47] attempted to further specify the mass transfer coefficient:

$$\text{EF} = C_{in}/C_{out} = 1 - \exp[-1/Q(R_m + R_d + R_e)] \tag{12.3}$$

where R_m, R_d, and R_e represent, respectively, the mass transfer resistance in the dialysis membrane, in the dialysate, and in the external medium.

As pointed out previously, the EF obtained during the real experiments seldom equals that observed during the in vitro calibration, due to different diffusion coefficients in different media. Processes such as adsorption, complexation, and precipitation can trap analytes, leading to the underestimation of available concentrations. Nevertheless, the data obtained always represent a good approximation of the process under investigation, which can in any case be treated quantitatively [20].

As has been well illustrated by Torto et al. [20], Eq. (12.3) can play a significant role in determining the value of the extraction fraction. For example, for very soluble low molecular weight analytes, if pure water or a buffering reagent is used as perfusion liquid, the term R_e in Eq. (12.3) can be ignored ($R_e = 0$). Also with $R_d \ll R_m$, the EF for an analyte crossing a given membrane is dependent principally on R_m.

The latter can be evaluated by the linearization of Eq. (12.3), plotting $-\ln(1 - EF)$ vs. $1/Q$. From the slope of the linear regression, the factor $[1/(R_m)]$ (assuming R_d and R_e as zero) can be obtained [48], and it can be considered as a permeability factor, PF (PF = $1/R_m$).

R_m is also useful for selecting more suitable membranes. For different membranes with equivalent geometry, at a constant perfusion rate, the relationship is

$$R_m = \ln(r_0/r_1)/2\pi L D_m \Phi_m + A \tag{12.4}$$

where r_0 and r_1 are the membrane's outer and inner radii, respectively, L the membrane's effective dialysis length, D_m the effective diffusion coefficient in the aqueous phase of the membrane, and Φ_m the aqueous phase volume fraction of the membrane [48].

For membranes with the same L and r_0/r_1, the membrane's porosity and tortuosity will contribute to $D_m \Phi_m$, and consequently will determine the value of EF [48].

The factor A, called "Andrade effect", introduced into Eq. (12.4) accounts for the initial decrease in EF due, amongst other factors, to polymer (membrane)–protein or polymer–analyte interactions. The interaction of the membrane with the matrix will evidently affect Eq. (12.4), resulting in an initial decrease in EF [49–52].

By optimizing the probe's inner geometry, it is possible to maximize the performance of a microdialysis probe, as demonstrated by Torto et al. [53,54]. In the cited papers, they used an in situ *tunable probe*, i.e., a concentric probe where the inner tube of the probe is moveable lengthwise (tunable), and any effective dialysis length can be achieved. Moreover, in the case of particle blockage of the inner tube, this can be simply replaced by another tube while the probe body is still in position, without perturbing the on-going monitoring.

By using inner cannulae having different inner and outer diameters, they reported how the microdialysis EF may be enhanced by increasing the outer diameter of the inner cannula [53,54]. While this model has been well characterized in vitro, its in vivo application is more difficult because physiological processes can also affect the recovery.

When the exact concentration of an analyte is needed, an in vivo calibration is required. Several techniques based on different approaches have been developed, such as the "no net flux" [55], the "internal reference" [56–58] and the "retrodialysis" [59] methods. All of these methods are complex, difficult and time-consuming and none of them can be considered completely satisfactory for all applications. Methods for probe calibration have recently been reviewed in depth and can be found in the literature [60–63].

In many cases, however, if only the monitoring of changes in concentration of a given analyte is of interest, determination of the absolute concentration may not be necessary and simpler calibration methods can be applied. For example, when microdialysis is used for monitoring the concentration of a substance in blood, the dialysate concentration may be calibrated against blood concentrations, assuming that there is a linear relationship between the dialysate and blood concentration. This approach has been used for subcutaneous glucose sensors [64,65].

Often, after sampling, a successful separation of the analyte of interest from a complex matrix is the first goal of an analysis. One of the key points of microdialysis lies precisely in the fact that these two steps, the sampling and the separation, at least from macromolecules, are obtained at the same time. The sample clean-up step, which would normally be needed before the application of most analytical techniques, is generally avoided and, because enzymes are excluded as well, no further degradation of metabolites occurs after sample collection.

The advantageous features of microdialysis techniques have been clearly illustrated by Ungerstedt since 1991 [16] and are by now well known, but it is also necessary to point out some problems. One of these, the difficulty in achieving a proper calibration, has already been mentioned. In addition, microdialysis probe design can have stringent requirements. Small probes with a high extraction fraction are necessary if users must sample from restricted areas such as organ or plant tissues. Here, it is essential that the sampling site not be disturbed. For example, in neurochemistry [66] a small and effectively designed probe can enable the measurement of neurotransmitters directly at sites of release and avoids diffusion effects that could affect data interpretation. A small probe design is also important for monitoring mineral uptake by plants when the probe is placed directly into the stem tissue [67]. However, the probes need to achieve a high extraction fraction to reduce demands on the sensitivity of the detection system.

Another critical factor can be, in fact, the size of the sample. The flow rates used in this technique are usually very low, between 1 and 10 μl/min, and recently ultraslow flow rates (less than 0.3 μl/min) [68] have also been proposed in order to increase the recovery of a selected analyte. This approach results in very small sample volumes (a few microlitres) often with a very low analyte concentration. At this level, evaporation is also a problem to be considered. To obtain larger samples, both the collection time and/or the flow rate can be increased, but in this case, the temporal resolution and the recovery will also decrease. The analytical techniques used in the analysis of the dialysate hence have to be adequate for these volumes and concentrations, and very sensitive detection systems have been developed.

Since microdialysis sampling is a continuous process, coupling it to an on-line analysis system could be a solution to these problems. From this point of view, microdialysis could come close to meeting the requirements of an ideal sampling system, providing a representative sample, a sterile barrier, and usually not disturbing the sampling site [69]. At the same time, continuous monitoring can provide real-time information about relevant variables in the system under study. Researchers have not overlooked this opportunity, and many examples of this strategy can be found in the literature; some of them will be illustrated in the following sections.

12.5 ON-LINE COUPLING TO BIOSENSING DEVICES

Historically microdialysis samples, collected in a discontinuous way, have been analysed by high-performance liquid chromatography using various types of detectors, the most common types being UV, electrochemical, fluorescence and mass spectrometry detectors.

By now, chromatographic techniques have been successfully coupled with microdialysis for on-line analysis [70–74]. The advantage of this approach is the amount of data available since several analytes can be measured almost simultaneously. However, the low time resolution, due to the time necessary for the elution of all the compounds from the chromatographic column, and hence the low analysis frequency, presents a problem. The relatively high cost and the bulkiness of the equipment as well as more particular problems such as loss of sample and carry-over in the injection valve can also negatively affect the analysis [75].

A different approach has been the coupling of microdialysis with biosensing techniques. These latter are "non-separation" techniques based on molecular recognition of the target analyte, which may form a complex with a bioreceptor, allowing its selective identification even in a complex mixture.

The reaction between the analyte and the bioreceptor produces a physical or chemical output signal normally relayed to a transducer, which then generally converts it into an electrical signal, providing quantitative information of analytical interest. The transducers can be classified based on the technique utilized for measurement, being optical (absorption, luminescence, surface plasmon resonance), electrochemical, calorimetric, or mass sensitive measurements (microbalance, surface acoustic wave), etc. If the molecular recognition system and the physicochemical transducer are in direct spatial contact, the system can be defined as a biosensor [76]. A number of books have been published on this subject and they provide details concerning definitions, properties, and construction of these devices [77–82].

The selectivity characteristic of biosensors allows one to avoid separation procedures after the microdialysis collection, thus greatly improving the overall analysis, in terms of both instrumentation and the time required. Moreover, since microdialysis and chromatography entail dilutions, the avoidance of chromatographic separation also allows one to obtain a more concentrated sample which can be analysed more easily and accurately.

The above-mentioned bioreceptor is a molecule that recognizes the analyte by means of a biochemical mechanism, such as enzyme–substrate, antibody–antigen interactions, nucleic acid base pairing, or cellular interactions. Sometimes, this biorecognition step can be different from the detection step, as in the case of flow systems where an enzyme or an antibody is contained in a reactor and the detector is located downstream. Although such a system is not a biosensor, it can be considered as a "biosensing device" and can offer several advantages over conventional biosensors because it is not strongly dependent on enzyme reaction kinetics, as long as the 100% of substrate conversion has been achieved in the reactor. Thus, this approach can be used to overcome problems such as enzyme activity loss over time, variations in local oxygen concentration, local temperature, etc. [83]. In a reactor, for instance, the enzyme loading can be increased if an enzyme is not very active so that an amount greater than what could be immobilized onto a transducer can be used in order to achieve an adequate signal. In other cases, for example, when two enzymes of different stability have to be used sequentially for obtaining the analytical signal, it could be convenient to separate them in order to not waste the still active enzyme while the other one is losing its activity.

When such a reactor device has to be coupled to a microdialysis probe, the dialysate residence time in the reactor must be considered. It has to be sufficient for the complete conversion of the analyte, but at the same time, this delay, when added to the delays resulting from the residence time in the microdialysis probe and in the connecting tubes, should ideally not increase the response time of the system very much, thus decreasing the time resolution of the analysis.

In any case, both biosensors and biosensing devices have been coupled to microdialysis and are considered among the "non-separation-based methods" [83]. The drawback of biosensing approaches is that they are usually able to measure just one analyte at a time, in contrast with separation-based methods such as chromatography and electrophoresis, which allow the detection of several analytes. However, if the primary interest is not the identification of unknown compounds, but, for example, the monitoring of variations in a single metabolite or drug, the optimization of therapeutic responses, or the control of a bioprocess via a marker analyte, the use of a specific sensor, which can be employed in a continuous manner, can provide useful information, and can also help to avoid the analysis of hundreds of samples or to reduce the number of animals necessary for a study.

Another drawback of this kind of sensor is its sensitivity to other active compounds (interferents) that can be present in the medium to be analysed, giving rise to an additional unspecific signal. Several strategies have been developed to solve this problem, but sometimes more than one approach is necessary, especially when the concentration of these interferents is quite high in comparison to those of the metabolites of interest.

However, despite these drawbacks, one of the greatest advantages of the coupling of microdialysis with biosensing techniques is the unique opportunity offered for continuous monitoring of a given analyte *in real time*, both in vivo and in vitro.

12.6 THE ISSUE OF CONTINUOUS GLUCOSE MONITORING

Perhaps, the earliest example of coupling of a dialysis probe with a glucose biosensing device was reported by Mandenius in 1984 [84] and this approach was applied to bioprocess control in a fermenter broth. The authors determined the glucose concentration at the outlet of a very large dialysis probe through calorimetric measurement of the heat produced by the reaction of glucose oxidase (GOD) with its substrate inside an enzyme thermistor. In 1988 [85], the same authors coupled the dialysis probe, reduced to 1 mm in diameter, to an amperometric oxygen electrode, measuring the O_2 decrease after the reaction of glucose with GOD and catalase co-immobilized in a packed-bed enzymatic reactor.

The clinical importance of the glucose measurements in diabetic patients and the commercial potentialities of a system able to continuously monitor this metabolite have continued to stimulate efforts by many scientists to develop such a device.

It is not surprising that the most successful commercial biosensor is precisely that for glucose. Apart from the particular stability and low cost of GOD, which renders this enzyme very "user friendly" for researchers, there is also considerable interest on the part of companies in financing projects related to the clinical measurement of this metabolite.

The size of the overall market for glucose measurements can be deduced from a 2001 report about a market analysis, summarized as follows on Internet [86]:

> Glucose monitoring continues to be one of fastest-growing segments of the $21 billion in vitro diagnostics market. In 2000, the U.S. diabetic population was approximately 16 million people, with 800,000 new cases diagnosed and 200,000 deaths. The diabetic population is estimated to continue to grow twice as fast as the general population. Contributing factors include unhealthy lifestyles, increased obesity, aging, a 10–12% increase in type II diabetes among children and minority groups such as African-Americans and Hispanics, and an increased number of insulin-pump users who monitor their blood levels up to six times a day. Education and awareness campaigns should also lead to increased monitoring, which will in turn lead to better glucose control and reduced complications from diabetes. Finally, industry players enjoy abnormal margins, since manufacturers distribute 70% of their products directly to retail channels. All told, it is estimated that the U.S. market will grow 15% to $2.6 billion, whereas the worldwide market will grow 12% to $4 billion, approaching $6 billion by 2004.

While some large companies such as Roche (Basel, Switzerland), Johnson & Johnson (New Brunswick, NJ), Bayer (Leverkusen, Germany), and Abbott (Illinois, USA) are contending in the market for fingertip disposable glucose strips, the next step in blood-sugar control is real-time glucose monitoring, with alarms that warn the diabetic of an impending high- or low-blood-sugar episode. Once realized, it would be possible to combine this system with automatic insulin pumps that could be the platform for an implantable artificial pancreas. Continuous glucose-monitoring systems are intended to replace the point-in-time measurements of traditional glucose meters and strips, providing physicians with glucose-level trend information that can be critical in making therapeutic decisions. Moreover, they offer the additional advantage of reduced pain. A successful system of this type could significantly improve patient compliance and facilitate better disease management.

Considering the importance of continuous glucose monitoring, it is useful to make some general observations: in order to succeed in long-term in vivo monitoring, the ideal method for whichever analysis system should be non-invasive. As a matter of fact, some non-invasive approaches to continuous glucose monitoring have been reported in the literature. Mainly, these methods have been based on optical techniques such as near-infrared spectroscopy or light scattering [87–89]. However, even when elaborate mathematical models have been developed to filter out interfering molecules, these methods have still not satisfactorily met the criteria for reliability, reproducibility, sensitivity, calibration, miniaturization, and reasonable cost.

Because of the objective difficulties of the non-invasive approaches, another way to perform in vivo continuous monitoring has been to adopt the so-called "minimally invasive" methods, which measure the glucose concentration in the interstitial fluid of the skin. One of these is the insertion of a needle-shaped glucose biosensor directly into the subcutaneous tissue; another is to insert just the microdialysis probe under the skin, while the determination of the metabolite occurs just outside the body.

The first of these methods was the first explored by researchers [64, 90–94], and after 30 years of effort, there is finally a "continuous glucose monitoring system" (CGMS), which is currently marketed by MiniMed (Silmar, CA, USA). Up to now, it has only been used by healthcare professionals [95]. The system employs an approach similar to a Holter monitor, 72-h implantation, automatic glucose measurement every 5 min, and a 2-week data-storage capability. The sensor is a tiny electrode inside a small needle that is introduced under the skin. Once inserted, only the sensor remains under the skin and a thin cable connects the sensor to a pager-sized glucose monitor.

The major problem that has plagued these kinds of implantable biosensors is the gradual decrease in sensitivity and in some cases a complete loss of function within just hours of implantation. Biofouling, oxygen limitation, electrochemical interference and GOD inactivation have been considered as explanations of this behaviour. For instance, a tissue reaction to the sensor implantation may result in a limitation in the blood supply to the tissue surrounding the probe and thus in a lower availability of glucose and oxygen.

Therefore, a reliable calibration method becomes critical. Because of both the altered in vivo sensitivity and the accuracy demanded for continuous monitoring systems, the use of an in vitro pre-calibration is not feasible.

The use of the blood glucose value measured with conventional clinical methods or a finger-stick system in order to make a correlation with the subcutaneous glucose concentration is possible, even though this assumes that blood and interstitial fluid concentrations are equal or at least constantly proportional [96]. Nevertheless, the calibration procedures needed to obtain reliable results remain a matter of debate: whether a calibration process based on one blood glucose level (single-point calibration), or a two-point in vivo calibration, with each one related to a different steady-state level of blood glucose [97], is necessary. However, the two-point calibration is only feasible in a clinical setting and is not very convenient for diabetics. Otherwise, frequent recalibration of the sensors might be necessary. Also the CGMS currently on the market requires an initial calibration-meter finger-stick measurement and four finger-stick measurements from midnight to midnight each day. Moreover, a considerable drift of the sensor signal due to the foreign body reaction limits the number of recalibrations. For this reason, the use of the CGMS is not allowed for more than 3 days [95].

12.6.1 In vivo glucose monitoring through microdialysis coupled to electrochemical biosensors

The other approach taken for continuous in vivo monitoring has been the use of the microdialysis sampling method coupled to a glucose biosensor usually located on-line in a flow cell. This approach has several advantages: the glucose biosensor is protected against protein biofouling or foreign body attack by the tissue; the sensitivity of the sensor can be checked often and the sensor recalibrated, if necessary; the hollow fibre can be easily sterilized, avoiding damage to the biosensing element; finally, the overall process of in vivo implantation can be better managed and processes such as inflammation and clotting can be easily monitored.

The possibility of monitoring changes in subcutaneous glucose concentration by the microdialysis method was first demonstrated by Lönnroth [98]. Other reports followed [99–104], but in all of these, measurements were made in a discontinuous way using traditional techniques.

The microdialysis technique was subsequently adopted by the same groups (such as Pfeiffer and Shichiri) which had initially developed the in vivo needle sensors, now employing these same sensors in a flow system connected to a microdialysis probe for sampling [105–108].

The Pfeiffer group was very active in the first half of the 1990s. They coupled a commercially available concentric microdialysis probe to a modified "Glucosensor Unitec Ulm", the apparatus that had already been developed in their laboratory [109]. They then followed the glucose content of the extracellular subcutaneous fluid in rats [106], in healthy humans, and in diabetic patients for up to 48 h [107,110,111]. The calibration was performed by the one-point method using the actual blood glucose level after a breaking-in period of 30 min following the implantation of the microdialysis probe. The subcutaneous sensor signal was measured and compared to the glucose level independently determined on the dialysate and these values correlated well. In 1994, they introduced the "Ulm Zucher Uhr", a portable apparatus in which the microdialysis probe was connected to a pocket-sized device which contained the glucose sensor, a micropump, a data storage system, the software needed to process the data, and a liquid crystal display. Moreover, the measured blood glucose concentrations were telemetrically transferred once a minute to an indicator, the "Sugar Watch", worn by the patient [112]. This prototype, however, never reached the market.

Shichiri also made use of a previously developed needle glucose sensor, combining it with a microdialysis probe [108] to create a miniaturized glucose monitoring system. Working at a very low flow rate of 2 μl/min, and utilizing a hollow fibre 15 mm long, the combined system showed a linear range for glucose determination from 0 to 22 mM, and a constant and almost complete in vivo glucose recovery. The delay in response was about 6–8 min. The system was assessed in five healthy subjects and eight diabetic patients and there was a good correlation between the subcutaneous glucose concentration and the venous values. Their results also demonstrated the feasibility of continuous glucose monitoring up to 7/8 days. The system in fact functioned for 4 days without any in vivo calibrations, and then for 7 days with the introduction of in vivo calibration. However, up to now, this system has not been further developed.

Schoonen and Schmidt have combined the microdialysis method with a miniaturized Clark-type oxygen electrode [113]. In this system the enzymes GOD and catalase were contained in solution in a reservoir and continuously perfused at a fixed rate into the microdialysis probe and then to the sensor cell. All the glucose diffusing into the probe was catalytically converted and the difference between the basal level of oxygen and the value after its depletion by the enzymatic reaction was correlated with the glucose concentration. In subsequent studies, a successful implant functioning up to 9 days was reported [114], but a significant day-by-day variation of in vivo calibration factors was also noted. Moreover, the possibility of enzyme leakage toxic to humans could not be excluded.

The system has been improved through redesigning the "closed-flow" microdialysis perfusion system of the sensor to minimize the risk of enzyme leakage by immobilizing the enzymes GOD and catalase in an enzyme reactor.

On the basis of the results from this group, the Roche Diagnostic company has recently produced a wearable apparatus that has been tested in 23 diabetic patients for up to 72 h. The extra-corporeal unit displays a glucose value every minute. The signal is corrected for the time needed for fluid transport from the microdialysis probe to the sensor (31 min). The one-point calibration mode was used after an initial equilibration time of about 4.7 h. Continuous monitoring was feasible for at least 3 days with no time-dependent decline in sensitivity to glucose being noted [115].

In the early 1990s, our group also developed a system for continuous glucose monitoring, coupling microdialysis with a flow cell containing an electrochemical glucose biosensor [116,117]. The initial experiments were carried out with bench-sized instrumentation and involved studies both in vitro and in vivo on animals and humans. This relatively new system, for its time, was first fully characterized for in vitro samples. The microdialysis probe was a home-built linear device made by gluing a single hollow fibre, 20 mm long, 200 μm of diameter, 20,000 MWCO, between two pieces of nylon tubing of suitable dimensions. A tungsten wire was also inserted in the lumen of the fibre to increase the strength of the probe. The glucose biosensor was assembled by positioning two membranes, an acetate cellulose membrane with a nominal cut-off of 100 Da (for the exclusion of the electrochemical interferents) and a nylon net with immobilized GOD, on top of a platinum electrode contained in a thin-layer transducer cell for liquid chromatography (BAS, IN, USA). Once the probe output was connected to the inlet of the cell, the in vitro influence of flow rate, probe length and temperature were evaluated. Figure 12.2 shows a diagram of the flow system.

After the in vitro characterization, the system was tested in vivo in rabbits and in humans: the first unexpected, and not previously noted, problem was a sharp decrease in the sensitivity after about 1 h following both an intravenous (in the rabbit) and an oral (in the human) glucose load.

Through the use of an injection valve inserted into the flow system just after the microdialysis probe, it was possible to send standard glucose solutions directly to the measuring cell and demonstrate that the observed decrease in sensitivity was to be attributed to the glucose biosensor and not to an eventual blockage of the microdialysis probe. This inhibition was reversible, with the biosensor recovering its initial sensitivity 15–30 min after washing with buffer. Efforts were then directed at characterizing and solving this problem. The hypothesis was that endogenous substances could

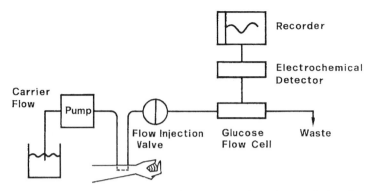

Fig. 12.2. Diagram of the flow system. After an initial trial period, the flow injection valve was excluded from the system.

dialyse through the fibre and could interfere with the enzymatic reaction or with the electrodic reaction.

Because the microdialysis probe is supposed to provide for both a sampling and a "clean-up" step, efforts were made to improve this latter aspect. We tested several hollow fibres made of different materials and with decreasing MWCO from 20,000 to 6000 Da. Using probes with lower cut-offs only delayed the onset of the phenomenon, which set in 4–5 h from the beginning of the in vivo experiment. The decrease in sensitivity was smaller, but was not eliminated, suggesting the passage of small proteins and/or peptides through even the 6000 Da cut-off fibre; these subsequently adhered to the glucose membrane.

We also succeeded in obtaining substantial improvement using a different flow cell (wall-jet), which has a different geometry and a different flow-pathway relative to the thin-layer cell used previously. While in the thin-layer cell the flow is parallel to the working electrode, in the wall-jet cell it is perpendicular. In such a way, the washing effect of the nozzle, located just in front of the glucose biosensor, was able to clean the surface of the sensor, eliminating the deposits (sometimes visible after long-term in vivo experiments) that had eventually formed on the biosensor in the thin-layer cell. With the wall-jet cell, a rather stable signal and good sensitivity were obtained, even when used with probes having higher MWCO (Fig. 12.3) [117].

Subsequent experiments were all performed using the wall-jet cell. In these studies, a number of hospitalized patients with different pathologies (but all subjected to an oral glucose load) were monitored. The one-point in vivo calibration method was adopted, assuming that the constant value of current observed before the glucose load was proportional to the concentration measured in the blood. This implies a linear relation between the biosensor

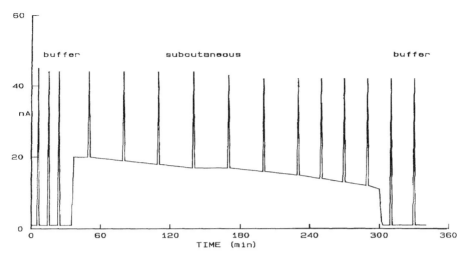

Fig. 12.3. In vivo experiment with 10-mm hollow fibre with 20,000 MWCO inserted subcutaneously in a rat and a wall-jet cell as detector. The peaks correspond to spikes of glucose standard solution sent directly into the flow cell by the injection valve, while the continuous line represents the subcutaneous glucose monitoring through the microdialysis probe.

output (current) and glucose concentration in the range of interest. The subcutaneous glucose values were compared with blood samples that were taken every 30 min during 3 h of monitoring and analysed for glucose by standard laboratory procedures. The subcutaneous values measured with the sensor correlated well with the blood glucose levels. In 15 experiments, there were only three cases in which a delay was observed between changes in blood glucose and related sensor measurements. The delay was calculated to be about 30 min [118].

By exploiting the microdialysis technique, the miniaturization of the whole instrumentation was achieved, also through the use of a peristaltic pump which turned out to be the only type of pump suitable for coupling to an electrochemical cell. The high sensitivity of the latter, in fact, revealed that all the syringe pumps tested were not really "pulse-free".

In Fig. 12.4 an interesting experiment performed with one of these prototypes to explore the effects of the location of the subcutaneous implants is reported. Two microdialysis devices were implanted in the same patient: one in the forearm and the second in the skin of the abdomen. The results were impressive in that despite the differing locations and lengths of the probes (1 cm in the forearm, 2 cm in the abdomen), several fluctuations in glucose level, having the same time course, were observed from each of the two probes.

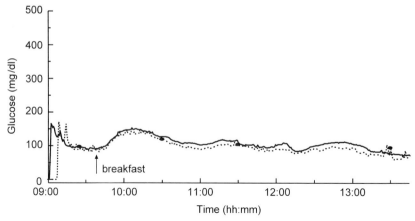

Fig. 12.4. In vivo monitoring of subcutaneous glucose on the same patient by two independent apparatuses. Two microdialysis probes of different length were inserted in the paraumbilical skin (2 cm, continuous line) and in the forearm (1 cm, dashed line). The dots correspond to blood glucose measurements.

Since the devices were completely independent, we could exclude variations caused by pump rate or voltage fluctuations. From such experiments, we could demonstrate how the glucose concentrations in subcutaneous tissue vary continuously and consider their significance. While such fluctuations have often been ascribed to external or local factors, it seems they may instead be related to glucose metabolism which was causing such changes as detected in different subcutaneous tissue sites.

However, these same studies showed that muscular contractions could often affect the signal output. In several cases, we observed transient collapses of the microdialysis hollow fibre as a consequence of involuntary contractions. Therefore, we found that the umbilical region is in general more suitable as a position for insertion of the microdialysis device since fewer muscular contractions occur there.

Several prototypes of the portable device described above have been produced, including the instrument, GlucoDay®, now commercialized by A. Menarini Diagnostics (Florence, Italy).

This device is composed of a linear subcutaneous probe that is connected to a wearable unit about the size of a walkman. The latter is comprised of a programmable micropump providing flow rates from 10 to 100 μl/min; a wall-jet cell with a glucose biosensor; a disposable fluid line which includes a pressure sensor; a microcontroller for pump speed programming, signal acquisition and data storage; an LC display which shows glucose values every second; a keyboard; an RS232 interface and a 9 V battery. Data can also be visualized continuously on a computer through an infrared communicating port. The system takes a glucose value every second and stores an averaged value every 3 min, for a total of 480 measurements per day. Two plastic bags, one for the buffer reservoir and one for the waste complete the apparatus for a total weight of about 250 g. Alarms are also present to warn of hypo- or hyperglycaemia. A picture of the instrument is shown in Fig. 12.5.

The actual glucose sensor (a platinum electrode covered by three membranes: cellulose acetate, nylon net with covalently linked GOD, and a polycarbonate protective membrane) is located in a miniaturized wall-jet cell. The sensor exhibits excellent performance, with a linear range extending up to 27 mM, thanks to the microdialysis dilution effect which was estimated to be 1:10 for the probe length used and for the flow rate set by the instrumentation. Long-term stability tests revealed that the biosensor still maintains its initial activity after incubations of 4 weeks at 45°C, 11 weeks at 37°C, and 32 weeks at room temperature (see Table 12.2). From these results, a shelf life of more than 2 years at 2–8°C can be extrapolated [119].

The system was first tested on rabbits, then on human volunteers, and in Figs. 12.6–12.8 profiles of subcutaneous glucose monitoring up to 24 and 48 h are shown.

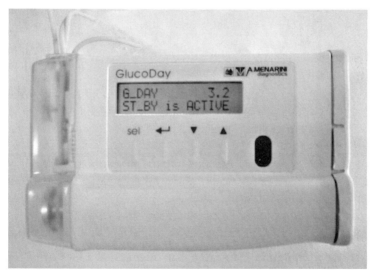

Fig. 12.5. Picture of the instrument GlucoDay® (with the permission of A. Menarini Diagnostic, Florence, Italy).

TABLE 12.2

Lifetime of glucose biosensors

Temperature	Sensor	Weeks					
		0	3	10	16	24	32
29 ± 2°C	A	13.4	12.8	13.0	11.2	10.0	12.0
29 ± 2°C	B	13.5	11.0	12.4	12.0	10.6	10.0
29 ± 2°C	C	13.0	10.5	11.2	10.2	11.0	10.5

In these trials, a one-point in vivo calibration of the instrument was performed, testing the blood and/or capillary glucose levels and correlating this value with the current output obtained by the instrument. Studies undertaken to optimize the calibration procedures showed that the best accuracy was obtained performing the calibration at 60 min, or better yet 120 min, after the insertion of the fibre. This is the time necessary for the microdialysis probe to reach equilibrium under the skin and for the tissue to recover from the insult of fibre insertion. In Figs. 12.6 and 12.7, two subsequent 24-h runs on the same patient are displayed. During the first 24 h the glucose was monitored by the GlucoDay® and verified by

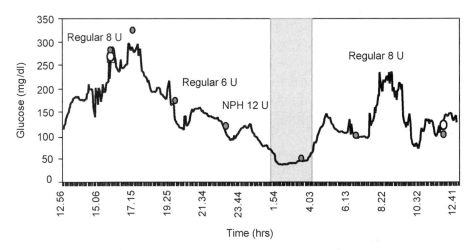

Fig. 12.6. Twelve hours in vivo monitoring with the GlucoDay® on a diabetic patient on day 1. When indicated, the patient assumed some amounts of insulin, both in regular and fast (NPH) formulation. The grey box highlights a nocturnal episode of hypoglycaemia. The grey dots represent the finger-stick controls, while the black dots correspond to blood measurements (with the permission of A. Menarini Diagnostic, Florence, Italy).

Coupling of microdialysis sampling with biosensing detection modes

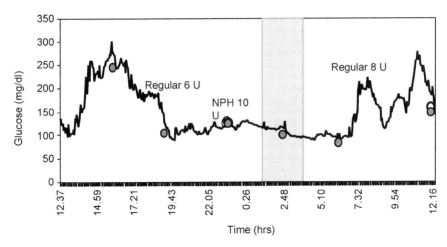

Fig. 12.7. Glucose monitoring on day 2 on the same patient of Fig. 12.6. Conditions as in Fig. 12.6 (with the permission of A. Menarini Diagnostic, Florence, Italy).

seven finger-stick measurements. Moreover, two venous blood checks were also performed, at the beginning and at the end of the experiment. The patient was able to proceed with his normal life during the trials.

It is interesting to note how the nocturnal hypoglycaemia seen on the first day, as evidenced by the grey box in the figures, was completely remedied the night after by the adjustment of the dosage of a quick formulation of insulin. Figure 12.9 presents the results of 48 h of monitoring in which the validity of

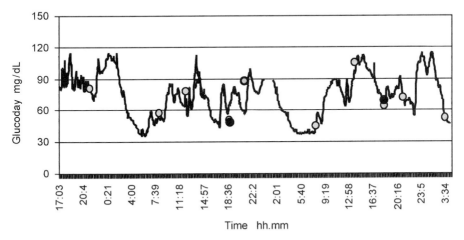

Fig. 12.8. Continuous glucose monitoring for 58 h. Dots as in Fig. 12.6 (with the permission of A. Menarini SFR, Florence, Italy).

the calibration obtained by the single-point method was verified by venous checks at 24 and 48 h, as well as by comparison with intermittent capillary blood measurements. The bias observed against capillary samples was -6.3%, and from venous blood samples was around -3% [120].

To evaluate the accuracy of the GlucoDay®, a multi-centre study was performed on 70 diabetic patients in seven different Italian centres [121]. Briefly, 24-h monitoring was performed and the results were correlated with nine venous blood samples collected throughout the day. The patients were also asked to complete a questionnaire ranking discomfort and pain during the monitoring period. Results of the statistical analysis showed that subcutaneous glucose levels, determined with a single-point calibration over 24 h, correlated well with the venous values ($r = 0.9, P < 0.001$) over a range of 40–400 mg/dl. An error-grid analysis of 391 data points revealed that 97% of the values ($n = 381$) were in the zone A and B (acceptable), 3% ($n = 9$) in the zone C, only one in the zone D, and none in the zone E (C–E zones = potentially dangerous). Both the insertion of the fibre and any discomfort during normal daily activities were well tolerated by the great majority of patients.

The GlucoDay® has recently been employed by a Spanish medical group to determine the incidence of hypoglycaemia in patients after pancreas transplantation, concluding that unrecognized nocturnal hypoglycaemia is relatively frequent in these patients and 24 h of continuous glucose monitoring may be useful to detect these episodes [122].

In conclusion, the combination of the microdialysis sampling technique with continuous glucose monitoring achieved with a biosensor can provide very useful information relative to individual glucose metabolism patterns during the day. This can allow a significant improvement of the therapeutic management of the patient's glycaemic level, with a significant reduction in the risk of hypo- and hyperglycaemic episodes and of long-term complications from diabetes.

In the previously mentioned approach, a relatively high flow rate (20 μl/min) was used for the continuous monitoring; and this resulted in high time resolution as well as the previously mentioned extension in the linear range of the glucose biosensor. The absolute concentration of the interstitial fluid was not determined, but variations in glucose concentration in the subcutaneous tissue were well correlated with those of the blood values and in general there was no phase lag between the two series of measures.

An opposite approach was taken by the Dutch group of Korf [123,124]. In order to achieve a (near) quantitative in vivo recovery, they proposed an ultraslow microdialysis with flow rates of less than 300 nl/min. This was based on the principle that diffusion of substances from the interstitial fluids tends to reach equilibrium between dialysate and interstitial concentrations at extremely low perfusion flow rates [125,126]. In this way, long in vivo calibration procedures could be avoided.

The drawback of this approach turned out to be samples of submicrolitre level, which necessitated the development of a dedicated small measuring device with a very low dead volume. A portable lightweight apparatus, consisting in a miniaturized flow-through biosensor connected to a microdialysis probe at one side, and to a semi-vacuum pulse-free pump at the other was then built. A portable potentiostat equipped for data collection and storage was used to handle data.

Their biosensor consisted in a Pt wire on which GOD was immobilized by the electropolymerization of m-phenylenediamine. The advantage of this type of immobilization consists in creating an effective barrier against electrochemical interference due to the polymer formed onto the electrode. Moreover, an extended linearity for the glucose sensor was also obtained, and this was a requisite for the direct measurement of the subcutaneous glucose at diabetic levels, since in practice an extremely low dilution of the subcutaneous fluid was realized.

This device was then tested in vitro with good results in terms of sensitivity, reproducibility, and absence of interference. However, a lack of stability was observed. Optimization of electropolymerization procedures ultimately resulted in a biosensor suitable for measurements for up to 3 days in continuous mode.

In vivo measurements during oral glucose tolerance tests on healthy volunteers were performed and results showed a delay time amounting to several minutes. Moreover, despite the (near) complete recovery theoretically achievable with the ultraslow flow rate, and which should allow an in vitro calibration [65], a one-point in vivo calibration was still necessary to improve the correlation between the glucose determined in dialysate and blood glucose values.

Another interesting approach has been the coupling of microdialysis with a miniaturized glucose electrode embodied in a sensor chip as described by Steinkuhl et al. [127]. The working Pt electrode (located in a cavity filled with a PVA membrane containing the immobilized GOD) and the reference Ag/AgCl electrodes were contained in a specially developed flow-through cell fabricated of silicon with micro-machining techniques. This cell was designed as a full wafer process in order to be compatible with mass-production methods. The dimensions of the chip were 11 mm × 6 mm × 0.8 mm. An excellent functionality over a period of 5 days was demonstrated in vitro. In combination with a microdialysis probe (cuprophane, 4 mm length, 20,000 Da cut-off) and at a flow rate of 1.5 µl/min, the sensor showed a linear range up to 30 mM. In a further development of the system [128], the chip was coupled with a programmable medical pump set to work in a stop-flow procedure.

Several advantages arise from this analysis procedure: a near complete equilibration of the perfusate with the surrounding medium can be reached, as the solution remains in the probe for a reasonably long period (10 min stop) while the effective time for the analysis is not too long on account of the high flow rate transport after the sampling (21 µl/min). The pump inactivity during the stop period increases the lifetime of the batteries. All the components including pump, electronics, glucose biosensor, perfusate and waste containers were located in a wearable apparatus, which weighed about 140 g. Tested during glucose tolerance tests on human volunteers, the results obtained showed a good correlation with venous blood glucose levels, and this was confirmed by an "error-grid analysis".

The applications reported above are the ones that have reached the most advanced stages of development and even commercialization. However, numerous other examples of glucose monitoring by microdialysis can be found in the literature: some of them are briefly reported below.

Laurell [129] connected the outlet of a microdialysis probe to a GOD reactor that was coupled to an oxygen electrode in order to measure the subcutaneous glucose levels in a diabetic patient over a 4-h period. Since the oxygen sensor displays a rather strong temperature dependence, a temperature sensor was placed close to the biosensor during the measurements. The current output, after electronic filtration to suppress electronic disturbances and non-physiological artefacts, was well correlated with glucose determinations on fingertip blood samples.

Bioanalytical Systems (BAS), one of the leading companies in the field of the electrochemical and chromatographic analytical instrumentation, is also heavily involved in the development of microdialysis equipment. The apparatus they produced has demonstrated its suitability for real-time, on-line monitoring of glucose in subcutaneous tissue [130] and in brains of normal [131] and diabetic rats [132]. The dialysate was analysed both by an enzyme reactor (IMER) coupled to amperometric detection, and by a biosensor with GOD immobilized onto the surface of an Os(II)polyvinylpyridine wired horseradish peroxidase (HPR) glassy carbon electrode which was positioned in a radial flow cell. The use of the Os–gel–HPR polymer facilitates the reduction of H_2O_2 at an applied potential of 0 mV, minimizing the interference from electrooxidizable substances. However, the relative concentration of ascorbic acid in the brain is so high that additional barriers, such as a cellulose acetate layer or a pre-oxidation cell, were necessary to measure the analyte of interest [131].

12.6.2 In vivo glucose monitoring through microdialysis coupled to other biosensing detection techniques

All the above reported examples utilized electrochemical biosensors because these kinds of biosensors have numerous and still unsurpassed advantages in terms of versatility, reliability, low cost plus offering the possibility of fabrication in different materials and shapes and of miniaturization (also of the related instrumentation).

However, other interesting biosensing detection techniques have been explored in conjunction with microdialysis. Amine et al. [133] developed a miniaturized thermal flow injection analysis (FIA) glucose biosensing device, consisting in a small column containing the co-immobilized enzymes GOD and catalase. A 1 μl loop connected the microdialysis probe to the reactor. The glucose profile, obtained during a glucose load experiment on a volunteer, and using a one-point calibration procedure, was compared to blood values. Results showed that the glucose variations were closely followed but with an 18 min delay time.

The chemiluminescent detection method has also been employed for monitoring the glucose content of dialysate. This detection mode requires the addition of reagents such as luminol to obtain the analytical signal, and this requires a more complex system with additional pump lines, mixing coils, etc. Flow injection is considered the ideal sample introduction mode for chemiluminescence [134], but this makes the measuring system rather cumbersome and not easy to miniaturize.

Fang et al. [135] reported a very complex flow system for the in vivo on-line monitoring of glucose in intravenous and subcutaneous tissue. The system was also complicated by the presence of a second reagent, potassium hexacyanoferrate, in addition to luminol for the chemiluminescent detection of the H_2O_2 produced by the enzymatic reaction of glucose with GOD immobilized in a reactor. In addition, two sample loops were present, one for the dialysate and the other one for standard solutions, allowing an on-line control of the stability of the system. A syringe pump was used for the microdialysis perfusate, while an eight channel peristaltic pump was employed for buffer, standards and reagents. The intravenous glucose was measured by a sort of shunt probe, as the blood was withdrawn at low flow rate from the ear of a rabbit passing through a microtube where a linear microdialysis probe was located.

Despite the complexity of the system, this was the first apparatus that successfully used chemiluminescent detection for the on-line monitoring of glucose with in vivo microdialysis sampling.

Subsequently, a much simplified FIA system has been reported in which the co-immobilization of GOD and HPR in an enzymatic reactor permitted the use of luminol as a single reagent for monitoring blood glucose during a glucose load experiment in which the sample was obtained from a microdialysis probe in the jugular vein of a rabbit [136].

Recently, an alternative measuring principle has been proposed by Beyer et al. [137], who describes a viscosimetry affinity sensor, able to measure glucose because of its strong effect on the viscosity of a "sensing fluid" containing concanavalin A (Con-A) and dextran. The system consists of the microdialysis probe, two flow resisting capillaries, and two pressure transducers in a linear flow system. The Con-A is a carbohydrate binding protein and dextran serves as its polymer glycoligand; their binding can reversibly be weakened by glucose because Con-A binding sites have an affinity for both sugar and polysaccharide. The viscosity found in concentrated solutions of dextran and Con-A is glucose dependent, so its variation can be related to the glucose concentration within the physiological range in an approximately linear way. However, early attempts in using this system revealed a delay in the in vitro sensor response (5–10 min) and a lower read-out in plasma and subcutaneous tissue when the calibration was carried out with polymer free glucose solutions. Glucose standard solutions containing polyethyleneglycol (PEG) at nearly isotonic concentrations, used as calibrators, resulted in a good fit of the sensor response to the real glucose concentration in the physiological liquids [138].

Even though much work still remains to be done to achieve practical applications in glucose sensing, these examples demonstrate the interest and efforts of scientists in coupling the microdialysis sampling technique with the ever-increasing array of detection modes.

12.7 IN VIVO MONITORING OF OTHER METABOLITES

12.7.1 Glutamate detection

In addition to glucose, other metabolites have been continuously monitored in vivo. Glutamate is a prime example, being one of the most important excitatory neurotransmitters.

One of the first attempts to measure this metabolite in the brain was carried out by Zilka et al. [139] using microdialysis coupled to an electrochemical flow cell with the immobilized enzyme glutamate oxidase (GlOD). To avoid the interference of endogenous electroactive compounds, a film of 1,2 diaminobenzene was electropolymerized onto the Pt working electrode.

The problem of the interference is especially serious in the case of glutamate detection because this analyte is present in brain extracellular fluid in a much lower concentration compared to the other electroactive analytes, and often more than one strategy may be required for their exclusion from the measured signal.

So far, the Zilka group has successfully used this system to study changes in extracellular glutamate associated with several neurological diseases [140–144]. One drawback lies in the rapid deterioration of the electropolymerized membrane, which has to be used within a short time after the preparation to ensure the exclusion of electroactive interferents.

The use of electrochemical mediators is another strategy to solve the problem of interference because the electrochemical reaction of the mediators occurs at potentials lower than those useful for the direct oxidation of the H_2O_2 produced by the oxidase enzymes. The possibility for interfering compounds to react with the electrode is significantly reduced. Such an approach has been employed by Fillenz and colleagues [145–147] who followed changes in glutamate in brain by immobilizing the enzymes GlOD and HPR in a reactor, then using the mediator ferrocene to obtain the electrochemical signal. However, this approach was not adequate for the exclusion of all interference and thus the introduction of a tubular pre-oxidation cell before the enzymatic reactor was necessary. The sensitivity of the proposed method was quite good, and this group has reported several applications measuring glutamate in the striatum of freely moving rats. The same approach was used to measure changes in brain glucose [148] and lactate [149].

Recently, an on-line amplification system based on substrate recycling has been proposed for the measurement of glutamate [150]. By co-immobilizing GlOD together with glutamate dehydrogenase in a reactor, and then adding NADPH in the carrier solution, glutamate was assayed in vivo with about a 600-fold increase in sensitivity. In this case, selectivity was also ensured by covering Pt working electrode with a film of electropolymerized 1,2 diaminobenzene. In vivo assay of glutamate in the extracellular space of rat brain, following a continuous stimulation with KCl, demonstrated the reliability of this detection system coupled to a microdialysis probe.

12.7.2 Lactate detection

Lactate is another interesting metabolite whose continuous real-time monitoring could be useful for intensive care, cardiology and sports medicine. However, in contrast to glucose, only a few applications of lactate monitoring by microdialysis have been reported. Our group reported a first attempt in this direction by coupling microdialysis with a wall-jet flow cell in which an electrochemical lactate biosensor was located [151]. The system was fully characterized in vitro, and then some in vivo applications were attempted with the microdialysis probe placed subcutaneously in rabbits and also on a human volunteer who performed some physical exercise during the monitoring. Exercise caused a lactate increase, which depended on the intensity and length of the exercise. No decrease in the sensitivity of the system was found in the course of 4-h experiments.

Kissinger [130] carried out similar experiments following lactate changes observed in a rat subcutaneous microdialysate upon an intraperitoneal injection of lactate. The half-life of the sensor, assembled by coating the enzyme lactate oxidase (LOD) on a glassy carbon electrode with an Os-redox polymer and a Nafion overcoating, was about 24 h.

A different lactate biosensor was proposed by Pfeiffer et al. [152], who used an enzyme sandwich membrane that was commercially available for whole blood lactate analysers. The membrane was inserted into a flow cell connected to a microdialysis probe. This membrane showed a significant day-to-day variation in sensitivity ($\pm 50\%$), but no trend in sensitivity decrease. The problem of rejecting interference has not been completely solved by this system. However, the continuous monitoring of subcutaneous lactate was feasible at least in small rodents, and results were consistent with liquid chromatographic measurements performed on dialysate samples collected during the in vivo experiment.

12.7.3 Acetylcholine detection

Acetylcholine is another very important neurotransmitter in the central nervous system, but its detection is usually very difficult due to its low concentration, which falls in the nanomolar or even subnanomolar range. The enzymatic approach to its determination involves the use of two enzymes, acetylcholinesterase (AChE), which produces choline, and choline oxidase (ChOX) that yields H_2O_2, which is amperometrically detected. Problems arise from the concomitant presence of high amounts of endogenous choline with respect to acetylcholine, so a separation step or a dual enzymatic detection system, one for each metabolite will probably be necessary.

A completely satisfactory solution to this problem has not been achieved up to now, but attempts at resolution, also in combination with microdialysis techniques, have been reported.

A first report appeared in 1992 from Flentge [153] who developed an enzyme reactor for the amperometric detection of both metabolites, but only the detection system for choline was coupled with microdialysis. The enzymes (one or both) were physically entrapped between two cellulose nitrate filters, and the sandwich reactor was put on-line after an HPLC column for the separation step, or else placed directly after the microdialysis probe for the on-line measurement of choline. The sensitivity of this system was sufficient to measure acetylcholine and choline in rat striatum when used after the HPLC column, while the dilution due to the microdialysis sampling only permitted determination of choline, even if in a continuous mode.

A more complex biosensor for acetylcholine has been developed by Larsson et al. [154]. Three enzymes, AChE, ChOX, and HPR, have been co-immobilized in an Os-based redox polymer on solid graphite electrodes. After a careful optimization of the immobilization procedure, the biosensor, inserted into a flow cell of very small volume, was integrated into a flow injection system, and some samples of microdialysate, taken from rat brains before and after stimulation with KCl, were analysed. Even if a clear increase in signal could be noted, it was not possible to distinguish whether it was due to an increase in choline or in acetylcholine, since the biosensor responded to both metabolites.

A continuous monitoring of acetylcholine by microdialysis, avoiding a chromatographic separation step, could be achieved by the development of a dual biosensor having one part sensitive to choline, the other one to acetylcholine. In this case, the concentration of the latter could be obtained by subtraction, if the sensitivity, stability and rejection of interference by both biosensors turned out to be suitable for the concentration levels typical of the brain.

12.7.4 Uric acid and dopamine detection

Two different systems for the detection of dopamine and uric acid, respectively, have recently been reported in the literature. Both are based on the same detection technique, that of chemiluminescence (CL). In the first one, a chemiluminescence biosensor has been assembled using whole vegetable tissue (fresh mature potato root) as molecular recognition element [155].

Plant and animal tissues have long been successfully employed as biocatalytic components in the construction of biosensors, but it is quite unusual to find such a recognition system coupled to the microdialysis sampling techniques.

Tissue-based biosensors provide potential advantages of low cost, high stability, longer lifetime and a high level of activity. This is in part due also to the fact that the enzymes can find cofactors in the natural microenvironment.

The potato root tissue has been reported to be rich in both polyphenol oxidase (PPO) and peroxidase. Dopamine is oxidized by oxygen with the catalysis of PPO to produce hydrogen peroxide, which could then react immediately with luminol in the presence of peroxidase to generate CL signal. The PPO and peroxidase remain naturally immobilized in the vegetable tissue, and do not require any special immobilization procedures, also retaining high and relatively long-term activity. Combined with microdialysis sampling, this biosensor was applied to monitor the variation of dopamine levels in the blood of rabbits after the administration of dopamine. It demonstrated good resolution and reliability for in vivo on-line monitoring.

The second paper deals with the in vivo monitoring of uric acid (UA) using CL detection of enzymatic reaction product formation (H_2O_2), catalysed by uricase [156]. In this work, a new detection method for monitoring UA in vivo is developed, through microdialysis sampling coupled to bis(2,4,6-trichlorophenyl)oxalate (TCPO) initiated chemiluminescence detection. These workers also used an unusual enzyme immobilization site because the uricase enzyme was immobilized on the surface of a microdialysis probe through a streptavidin–biotin linker. Moreover, a polyurethane-trapped ferrocene film was used to exclude any potential interference from reductant molecules which are often present in the blood (e.g., ascorbic acid). Finally, through the use of an acute myocardial infarction (AMI) model this system was used for in vivo UA monitoring with good sensitivity and selectivity. Following AMI-induced oxidative stress, the UA level decreases continuously, thus suggesting that UA plays a protective role as a substitute antioxidant.

12.7.5 Simultaneous determination of different metabolites

The previously mentioned drawback of biosensing devices, i.e., the possibility to measure just one metabolite at a time, is nowadays partially overcome through the recent advances in miniaturization and the development of "arrays" of sensors. Each of the sensors constituting the array can consist in a different biosensor, allowing multi-analyte determinations. In some cases, if the array is located in a flow cell, it can be coupled to microdialysis sampling. Examples have been presented by several authors, even if the proposed systems up to now are limited to the measurement of two or three metabolites, principally glucose and lactate.

Even before the introduction of miniaturized biosensor arrays, however, some systems able to simultaneously measure glucose and lactate had been reported in the literature. One example is provided by Osborne et al. [157], who described plastic film carbon electrodes fabricated in a "split-disk" configuration and then modified to obtain a dual biosensor. They achieved a continuous monitoring of these metabolites by placing the dual electrode in a thin-layer radial flow cell coupled to a microdialysis probe. The stability of the sensors was sufficient for short-term in vivo experiments in which the cross-talk, i.e., the percentage of current measured by one biosensor but due to product generated by the partner biosensor, was acceptable for an in vivo application.

Recently, the same author [158] has again proposed the simultaneous determination of glucose and lactate. The method was applied to assess the optimal O_2 and CO_2 concentrations to be employed in the perfusate utilized in brain microdialysis to mimic the extracellular cerebral fluid (ECF). In this case, glucose and lactate were analysed using an immobilized enzyme reactor and FIA. Either GOD, LOD, or HRP enzymes were linked to silica beads via surface amino groups and the conjugated beads were used as contents of enzyme microbore reactors, serially connected between a Rheodyne valve and a radial flow cell containing a glassy carbon electrode. This study demonstrated that the concentration of these two gases, O_2 and CO_2, can influence the cerebral concentrations of both glucose and lactate, and that perfusion solutions containing them at their physiological levels, usually lower than in air-equilibrated solutions, are the most appropriate for microdialysis studies.

A very similar system has been previously presented by the group of Boutelle [159], who claim that for the first time, the levels of glucose and lactate have been measured simultaneously and on-line in patients during surgery and in the critical care unit. In this case too, a flow injection assay using enzyme-modified beds with mediated electrochemical detection is employed. This allowed simultaneous measurement of the levels of glucose and lactate present in the dialysate with high time resolution (30 s sampling intervals).

Other examples of flow injection systems are also found in the recent literature [160]. However, the problem of these apparatuses is that they are quite bulky, requiring pumps, valves and reactors, so that they are not easily miniaturized and are suitable only for experiments on animals or during surgery where the patient is immobilized.

An improvement with respect to the miniaturization of systems is represented by the chip sensors. In their work, Perdomo et al. [161] describe an implementation of the sensor chip summarized above for glucose measurements [127], adding a lactate sensor onto the same chip. The latter was integrated in a more complex flow-through system linked to a microdialysis probe, a micropump, a programmable temperature controller for the fluidic cell and a potentiostat connected to a computer. This system, which up to now has been tested only in vitro using human blood serum, showed a promising performance for in vivo monitoring, even if problems such as the response time in real samples and rejection of interferents still have to be solved.

Petrou et al. [162,163] have also reported the simultaneous determination of glucose and lactate. However, they used a microdevice consisting of a dialysis sampling system linked to a flow-through cell with a microfabricated biosensor array. The biosensor chip was fabricated by thin-film technology on a glass carrier. Three sensors, one for glucose, one for lactate, and one blank (no enzyme), were created on each chip. After the deposition of the specific enzyme membrane, the fabrication of the sensors was accomplished by subsequent deposition of a diffusion limiting and a catalase top membrane. The flow cell and the fluidics were created by appropriate photo-patterning of dry resist layers on a PC-Board that also contained conductive pads and the gold counter electrode.

The system showed short equilibration and response times, and was free of cross-talk, one of the major problems encountered when multiple amperometric biosensors are placed in an array within the same flow cell. Until now, the device has been shown to perform excellently in vitro, but several issues (such as the biocompatibility and the possibility of sterilization of the materials used) must be addressed before in vivo application.

All the current efforts directed at miniaturization of the analytical techniques are addressing the problem of interfacing the miniaturized probes used in microdialysis with suitable instrumentation, so that smaller sample volumes can be subjected to appropriate analyses, thus increasing the time resolution of the overall systems.

12.8 DIALYSIS ELECTRODES

A particular application of microdialysis can be found in the so-called "dialysis electrode". This device was first reported by Albery et al. [164] who constructed an electrochemical cell inside a concentric microdialysis probe, inserting three wire electrodes into the probe (Fig. 12.9).

Coupling of microdialysis sampling with biosensing detection modes

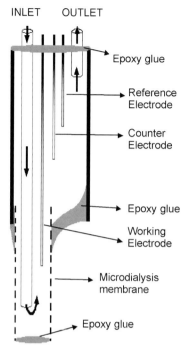

Fig. 12.9. Schematic representation of the dialysis electrode (from Ref. [164]).

The enzyme was introduced in solution by slow perfusion, thus permitting the measurement of the target analyte "in loco", as it enters the probe. By varying the enzyme, a number of different analytes could be studied, at least in theory. This arrangement combines the advantages of the microdialysis sampling with the time resolution of an implanted electrode. However, the presence of endogenous interferences is a major problem that has not yet been completely solved.

Many approaches have been explored to overcome this problem, e.g., further addition of enzymes such as ascorbate oxidase [165,166], or peroxidase and catalase [167], that would be able to destroy the interferents. Alternatively, the coverage of the Pt working electrode with electropolymerized membranes such as 1,2-diaminobenzene [168] has been tried. However, this last procedure can cause a decrease in sensitivity.

Dialysis electrodes have been applied to the in vivo measurements of ascorbate [169], glutamate [170], neurotransmitters such as catecholamines and indoleamines [165] and the capability of measuring glycerol in human serum samples in vitro [171] has been demonstrated.

A recent approach in solving the problem of interference has been presented by Kohno et al. [170], who added ferrocene mediator along with GlOX to the perfusion fluid for the measurement of glutamate. In order to prevent its outward diffusion through the probe, the mediator was covalently linked to a high molecular weight protein, BSA. With this set-up, the response was independent of the pO_2, thus allowing the investigation of the dynamics of glutamate during anoxia. More recently, Sugahara et al. [171] applied the same set-up to study glutamate changes during brain ischaemia.

A very interesting application of these dialysis electrodes has been reported by Cook [168], who transformed these electrodes into on-line immunosensors, immobilizing specific antibodies on the surface of a Pt electrode. The measurement procedure is the same as that carried out in typical immunochemical formats. It consists in the equilibration of the test solution at the antibody site, followed by a perfusion of the antigen conjugated to HRP (Ag–HRP), and finally by the amperometric detection of the product of Ag–HRP reaction after the addition of enzymatic substrates. It was also possible to regenerate the binding sites by adding a weak acid to the perfusion fluid. Each step was done in 30 or 60 s, and thus a complete cycle was made every 2 or 4 min. The outer surface of the dialysis membrane was also covered with phenylenediamine polymer to reduce the ascorbate interference. The probe gave a linear response in the range 0–280 μM, with a detection limit below 0.5 μM. It was tested in vivo allowing the on-line monitoring of γ-amino-4-butyric acid (GABA) [172], and, very recently, has been tested for the measurement of extracellular cortisol and corticotropin-releasing hormone (CRH) by the same author [173].

Complications related to these "dialysis probes" are the effect of the oxidative species generated within the probe which could theoretically damp down the variations in metabolite concentration if they back-diffuse across the membrane. Similarly, changes in the perfusing medium used to introduce different enzymes for measurement of other metabolites could perturb the extracellular fluid being sampled. This could result in local neuronal changes, thus limiting the physiological interpretation of the data. Moreover, the presence of high external concentrations of protein may reduce the sensitivity of the measurements due to the protein absorbing onto the probe.

These dialysis electrodes are produced by "Sycopel International Ltd" (only the e-mail address of this company is available: sycopelint@aol.com).

12.9 BIOTECHNOLOGICAL APPLICATIONS

In addition to the historical clinical applications of microdialysis, increasing interest in this technique is emerging in the field of biotechnology, especially in the work on bioprocess monitoring that is appearing in the literature. Monitoring through microdialysis probes can supply valuable information about the environment in which the bioprocess is occurring, also enabling optimal concentration of the key components to be maintained within a narrow range throughout the process. An accurate characterization of the physical, chemical and biological status of a bioprocess may require frequent analysis of biological variables such as extracellular concentration of substrates and products, activities of enzymes involved, etc. These analyses are usually carried out off-line. However, a delay in effecting required changes can occur if all the analyses are carried out at the end of the process. In addition, these procedures can be quite expensive, especially in terms of manpower.

Up to now, only a few physical parameters such as temperature, pressure, and agitation, or basic chemical parameters such as pH, dissolved oxygen and CO_2, are controlled on-line on a routine basis. Other important parameters, such as concentration of biomass, reaction products, substrates, and inhibitors, are more difficult to measure on-line, since suitable systems for their measurements have to fulfil strict requirements of selectivity, reliability, robustness, suitable detection range, and easy maintenance.

Biosensors may exhibit some of these characteristics. They could, in theory, be inserted in situ, in order to provide a near real-time continuous measurement of a given analyte and allow the selective determination of a single analyte in a complex solution, this detection being achieved by the use of specific enzymes or antibodies. However, from a practical point of view, biosensors suffer from poor stability, often associated with the harsh conditions of a bioreactor. In fact, they are also susceptible to negative effects from physicochemical features of the sample, such as the pH, temperature, and from components in the sample such as inhibitors or other active compounds. Recalibration is not likely to be feasible in in situ conditions. Moreover, the sterilization necessary for the in situ implant could destroy or merely reduce the activity of the enzyme.

A microdialysis probe could help to overcome some of these problems by acting as a protective interface between the bioreactor and the biosensing system, at the same time allowing on-line detection of the selected analyte. In fact, the microdialysis probe can be easily sterilized, thus avoiding problems of contamination. Membranes impermeable to bacteria would also act as a sterile barrier between the interior of the bioreactor and the analysis stream. The microdialysis probe provides a sample clean-up because the sample will be free of cells, but will also be free of enzymes, thus preventing any metabolic activity that could continue after sample removal and that could cause an erroneous interpretation of data. In general, the presence of a microdialysis probe does not disturb the culture; it can furnish a representative sample of the actual bioreactor content and sample dilutions can be achieved, broadening the detection range of the analytical system employed.

The microdialysis sampling technique has been applied to bioprocess monitoring, but also to the analysis of foods such as milk, yoghurt, biscuits, fruit juices, etc. The microdialysis probes used for such in vitro applications may require different features from those used for in vivo applications. The dimensions of the probe, for example, are not critical because there is no requirement for a low degree of invasiveness. On the contrary, larger probes are generally used in order to increase the recovery. As shown by the extensive research by the Swedish group of Gorton and Torto, a careful choice of the membrane is very critical [44] because the membranes have to meet requirements such as autoclavability, pH tolerance, and high temperature resistance. Moreover, recoveries can be influenced by the composition of the external medium (variable concentration of analytes, ionic strength and pH) but it has also been observed that sometimes membranes made of the same material and with the same MWCO can show different recoveries. Factors such as physical structure and charge of the polymeric material can affect the interaction between the membrane and the biomatrix, and most of the membranes show some degree of interaction with proteins of the biomatrix [174]. However, coating of membranes with suitable polymers such as PEG could help in reducing protein and cell adsorption [175–178].

In order to obtain reliable results, all parameters which can affect the microdialysis sampling should be carefully evaluated and the selection of the membrane should be preceded by an investigation of its performance when exposed to the conditions obtained in the bioprocess of interest [20,179].

In spite of all these difficulties, microdialysis has been successfully coupled to several analytical techniques for the bioprocess monitoring. HPLC is commonly used and such systems have been applied in the analysis of oligosaccharides [44,180], ethanol [181], enzymatic digestion of lactose in milk [182], studies on drug dissolution [183], saccharides in wastewater [184], starch hydrolysis products [185].

Recent trends in the application of microdialysis in bioprocess monitoring, such as innovative techniques for in vitro sampling, have been extensively reviewed by Torto et al. [20,186]. Thus, in the following section only a few examples of on-line coupling of microdialysis with biosensors will be focused on briefly, referring the readers to these reviews for a more extensive discussion of the overall applications to bioprocess monitoring.

Relatively few applications can be found in this field that involve the coupling of microdialysis to biosensors, especially electrochemical ones. The major problem is always the need to exclude electrochemically active substances, which can be present in the external medium, and, because of their low molecular weight, can easily cross the membrane of the microdialysis probe. Some attempts to overcome this problem have been reported by Csöregi [187], who coupled a microdialysis probe equipped with a polysulphone membrane with a 5000 Da cut-off to a carbon paste electrode with wired GOD, thus excluding many interfering substances. Similarly, Rohm [188] has assembled a screen-printed biosensor for lactate based on an ultraviolet-polymerizable enzyme paste and then applied it in an FIA system for the on-line monitoring of cell cultivation.

The continuous measurement of histamine released from rat basophilic leukaemia cells has been recently reported by Niwa [190], who coupled a microdialysis probe with a histamine oxidase-based carbon electrode obtained by coating the enzyme on the top of an Os–gel–HPR layer. For the exclusion of electrochemically active compounds, they suggested an overcoating of Nafion. In addition, a sensor using micro-machining techniques has been fabricated by the authors and applied to the monitoring of histamine release from a small mast cell colony.

Glucose and lactate were simultaneously monitored by Wei Min et al. [191] during the fermentation of *Lactococcus lactis*. They made use of a microdialysis probe connected to a dual flow-through cell in which two amperometric biosensors were located. The biosensors were based on the co-immobilization of the respective oxidase enzymes together with HRP in a carbon paste matrix. Both analytes were monitored on-line for about 14 h in a very complex aqueous two-phase fermentation process, and the results were in good agreement with independent off-line HPLC measurements.

In the field of food analysis, milk was the real matrix in which several metabolites have been analysed by the use of the microdialysis sampling technique. Acetylcholine, glucose and glutamate were analysed by Yao et al. [189], coupling a microdialysis probe with an on-line enzymatic reactor and a Pt electrode coated with a 1,2-diaminobenzene polymer.

Palmisano et al. [34] developed a lactate biosensor where the LOD enzyme was electrochemically immobilized in a bilayer membrane formed from 1,2-diaminobenzene and an overoxidized polypyrrole, thus achieving an effective exclusion of electrochemical interferents. This system allowed the measurement of lactate content in untreated, undiluted milk and yoghurt samples.

The microdialysis probe was used in a shunt configuration, with both flow rates, inside and outside the fibre, set at 50 µl/min. The connection of the probe to an FIA system contributed to the easy maintenance of the system because only plugs of milk and yoghurt were in contact with the fibre, which was continuously washed by the carrier solution.

These authors used the same set-up to follow, by the use of a glucose electrode, the growth of *Escherichia coli* in a glucose containing liquid culture medium. Results showed that a level of 10^6 colony forming units/ml could be detected after about 5 h [192].

Lactulose, a disaccharide that is produced by the heat treatment of milk to ensure its safety and is also considered as a quality index of milk, was determined by our group [193] using a loop microdialysis probe (formed by two parallel fibres, each 5 cm long) connected *in series* to a β-galactosidase reactor and a fructose biosensor. Two enzymes were necessary for the hydrolysis of lactulose into galactose and fructose, and the latter was determined by the fructose biosensor through the oxidation of ferrocyanide functioning as mediator. To allow the electrochemical reaction of the mediator, no permselective membranes were used. Thus, the measurement of lactulose was carried out by subtraction of signals with or without bypass of the β-galactosidase reactor. In this way, it was possible to circumvent the problem of the electrochemical interferent. Additionally, this study demonstrated that the lactulose recovery was greatly influenced by the ionic strength of the medium such that a 1:1 dilution of milk with buffer was necessary to render the ionic strength of milk more similar to that of the carrier buffer.

Finally, Mannino [194] reported the on-line monitoring of glucose in food samples by a microdialysis bioreactor realized by immobilizing the GOD onto the tungsten wire inserted into the lumen of a linear microdialysis probe.

12.10 CONCLUSION

This overview has presented a survey, certainly not exhaustive, that nevertheless shows some promising examples of the potential of microdialysis for future application in clinical, food and bioprocess analysis.

Recent clinical developments suggest that this technique will become useful for a broad range of diagnostic applications. In addition to its role as a research tool, microdialysis is becoming an integral part of neurointensive care monitoring and a standard for drug distribution studies in the course of drug development.

This sampling procedure, when coupled to biosensors, can help to improve their reliability, and the miniaturization and automation of the glucose monitoring systems, which have finally reached the stage of commercialization, provides a clear example of successful practical application of microdialysis.

This suitability of this technique has also been demonstrated for bioprocess monitoring, especially when it is coupled to biosensors through FIA systems, or when an on-line HPLC separation step is interposed.

In order to obtain reliable quantitative measurement and universally applicable systems based on microdialysis sampling, numerous problems still have to be solved. However, this is the most exciting part of the scientific research, which always goes on and provides a stimulus to scientists whose main duty is related to "problem solving". In fact, what would scientists do without problems to be solved?

Acknowledgements

The CNR Target Project on Biotechnology is acknowledged for financial support.

REFERENCES

1 J.M.R. Delgado, F.V. DeFeudis, R.H. Roth, D.K. Ryugo and B.M. Mitruka, *Arch. Int. Pharmacodyn. Ther.*, 198 (1972) 9.
2 J.H. Gaddum, *J. Physiol.*, 155 (1961) 1.
3 L. Bito, H. Davson, E.M. Levin, M. Murray and N. Sneider, *J. Neurochem.*, 13 (1966) 1057.
4 U. Ungerstedt and C. Pycock, *Bull. Schweiz. Akad. Med. Wiss.*, 30 (1974) 44.
5 U. Ungerstedt, M. Herrera-Marschitz, U. Jungnelius, L. Stanhle, U. Tossman and T. Zetterström, *Advances in Dopamine Research*. Pergamon Press, New York, 1982, pp. 219.
6 T. Zeterström, T. Sharp, C.A. Marsden and U. Ungerstedt, *J. Neurochem.*, 41 (1984) 1769.
7 U. Ungerstedt, *Measurements of Neurotransmitters Release In Vivo*. Wiley, Chichester, 1984, pp. 81.
8 U. Ungerstedt and U. Tossman, *Acta Physiol. Scand.*, 128 (1986) 9.
9 R.D. Johnson and J.B. Justice, *Brain Res. Bull.*, 10 (1983) 567.
10 M. Sandberg and S. Lindström, *J. Neurosci. Methods*, 9 (1983) 65.
11 J.M.R. Delgado, J. Lerma, R. Martin del Rio and J.M. Solis, *J. Neurochem.*, 42 (1984) 1218.
12 J. Korf and K. Venema, *J. Neurochem.*, 45 (1985) 1341.
13 L. Hernandez, B. Stanley and B.G. Hoebel, *Life Sci.*, 39 (1986) 2629.

14 M. Sandberg, B. Nyström and A. Hagberg, *J. Neurosci. Res.*, 13 (1985) 489.
15 H. Benveniste, *J. Neurochem.*, 52 (1989) 1667.
16 U. Ungerstedt, *J. Intern. Med.*, 230 (1991) 365.
17 M. Telting-Diaz, D.O. Scott and C.E. Lunte, *Anal. Chem.*, 64 (1992) 806.
18 K.M. Kendrick, *Methods Enzymol.*, 168 (1989) 182.
19 C.E. Lunte, D.O. Scott and P.T. Kissinger, *Trends Anal. Chem.*, 5 (1991) 171.
20 N. Torto, T. Laurell, L. Gorton and G. Marko-Varga, *Anal. Chim. Acta*, 379 (1999) 281.
21 R.K. Palsmeier and C.E. Lunte, *Life Sci.*, 55 (1994) 815.
22 H. Zuo, M. Ye and M.I. Davies, *Curr. Sep.*, 14 (1995) 54.
23 J.M. Ault, C.E. Lunte, N.M. Meltzer and C.M. Riley, *Pharm. Res.*, 9 (1992) 1256.
24 J.M. Ault, C.M. Riley, N.M. Meltzer and C.E. Lunte, *Pharm. Res.*, 11 (1994) 1631.
25 H. Zuo, M. Ye and M.I. Davies, *Curr. Sep.*, 15 (1996) 63.
26 M.I. Davies and C.E. Lunte, *Life Sci.*, 59 (1996) 1001.
27 H. Zuo, M. Ye and M.I. Davies, *Curr. Sep.*, 14 (1995) 54.
28 R.K. Palsmeier and C.E. Lunte, *Cancer Bull.*, 46 (1994) 58.
29 D. Devineni, A. Klein Szanto and J.M. Gallo, *Cancer Chemother. Pharmacol.*, 38 (1996) 499.
30 D. Devineni, A. Klein Szanto and J.M. Gallo, *Cancer Res.*, 56 (1996) 1983.
31 D.O. Scott and C.E. Lunte, *Pharm. Res.*, 10 (1993) 335.
32 K.E. Heppert, W.H. Flora and M.I. Davies, *Curr. Sep.*, 17 (1998) 61.
33 K.E. Heppert and M.I. Davies, *Curr. Sep.*, 18 (1999) 55.
34 F. Palmisano, D. Centonze, M. Quinto and P.G. Zambonin, *Biosens. Bioelectron.*, 11 (1996) 419.
35 M. Ulbritch, M. Riede and U. Marx, *J. Membr. Sci.*, 120 (1996) 239.
36 N. Wisniewski, F. Moussy and W.M. Reichert, *Fresenius J. Anal. Chem.*, 366 (2000) 611.
37 N. Wisniewski and W.M. Reichert, *Colloids Surf. B*, 18 (2000) 197.
38 N. Wisniewski, B. Klitzman, B. Miller and W.M. Reichert, *J. Biomed. Mater. Res.*, 57(4) (2001) 513.
39 Y. Maa and C. Hsu, *J. Pharm. Sci.*, 87 (1998) 808.
40 L. Millesime, J. Dulieu and B. Chaufer, *Bioseparations*, 6 (1996) 135.
41 V. Kuberkar, P. Czekaj and R. Davis, *Biotechnol. Bioeng.*, 60 (1998) 77.
42 H. Ridgway, A. Kelly, C. Justice and B. Olson, *Appl. Environ. Microbiol.*, 45 (1983) 1066.
43 N. Torto, T. Laurell, L. Gorton and G.S. Marko Varga, *J. Membr. Sci.*, 130 (1997) 239.
44 N. Torto, J. Bång, S. Richardson, G.S. Nilsson, L. Gorton, T. Laurell and G. Marko-Varga, *J. Chromatogr. A*, 806 (1998) 265.
45 R. Boyd and A. Zydney, *Biotechnol. Bioeng.*, 59 (1998) 451.
46 I. Jacobson, M. Sandberg and A. Hamberger, *J. Neurosci. Methods*, 15 (1985) 263.
47 P.M. Bungay, P.F. Morrison and R.L. Dedrick, *Life Sci.*, 46 (1990) 105.

48 J.K. Hsiao, B.A. Ball, P.F. Morrison, N.I. Mefford and P.M. Bungay, *J. Neurochem.*, 54 (1990) 1449.
49 S.I. Jeon and J.D. Andrade, *J. Colloid Interface Sci.*, 142 (1991) 159.
50 J.D. Andrade (Ed.), *Protein Adsorption*. Plenum Press, New York, 1985.
51 J.D. Andrade and V. Hlady, *J. Biomater. Sci. Polym. Educ.*, 2 (1991) 161.
52 J.D. Andrade, V. Hlady, A.-P. Wei, C.-H. Ho, A.S. Lea, S.I. Jeon, Y.S. Lin and E. Stroup, *Clin. Mater.*, 11 (1992) 67.
53 N. Torto, E. Mikeladze, L. Gorton, E. Csöregi and T. Laurell, *Anal. Commun.*, 36 (1999) 171.
54 N. Wisniewski and N. Torto, *Analyst*, 127 (2002) 1129.
55 P. Lönnroth, P.A. Jansson and U. Smith, *Am. J. Physiol. Endocrinol. Metab.*, 253(16) (1987) E228.
56 C.I. Larsson, *Life Sci.*, 49 (1991) 173.
57 R.A. Yokel, D.D. Allen, D.E. Burgio and P.J. McNamara, *J. Pharm. Methods*, 27 (1992) 135.
58 D. Scheller and J. Kolb, *J. Neurosci. Methods*, 40 (1991) 31.
59 A. LeQuellec, S. Dupin, A.E. Tufenkji, P. Genissel and G. Houin, *Pharm. Res.*, 11 (1994) 835.
60 J.A. Stenken, *Anal. Chim. Acta*, 379 (1999) 337.
61 C.M. de Lange, A.G. de Boer and D.D. Breimer, *Adv. Drug Delivery Rev.*, 45 (2000) 125.
62 L. Stahle, *Adv. Drug Delivery Rev.*, 45 (2000) 149.
63 H. Yang, J.L. Petres, C. Allen, S.S. Chern, R.D. Coalson and A.C. Michael, *Anal. Chem.*, 72 (2000) 2049.
64 V. Poitout, D. Moatti-Sirat, G. Reach, Y. Zhang, G.S. Wilson, F. Lemonnies and J.C. Klein, *Diabetologia*, 36 (1993) 658.
65 G. Reach and G.S. Wilson, *Anal. Chem.*, 64 (1992) 381A.
66 T.E. Robinson and J.B. Justice Jr., In: J.P. Huston (Ed.), *Microdialysis in the Neurosciences*. Elsevier, Amsterdam, 1991.
67 N. Torto, J. Mwatseteza and G. Sawula, *Anal. Chim. Acta*, 456 (2002) 253.
68 R.M. Rhemrev-Boom, R.G. Tiessen, A.A. Jonker, K. Venema, P. Vadgama and J. Korf, *Anal. Chim. Acta*, 316 (2002) 1.
69 L. Olsson, U. Schulze and J. Nielsen, *Trends Anal. Chem.*, 17 (1998) 88.
70 S.A. Wages, W.H. Church and J.B. Justice Jr., *Anal. Chem.*, 58 (1986) 1649.
71 S. Sarre, D. Deleu, K. Van Belle, G. Ebinger and Y. Michotte, *Trends Anal. Chem.*, 12 (1993) 67.
72 M.I. Davies and C.E. Lunte, *Chem. Soc. Rev.*, 26 (1997) 215.
73 K.M. Steele and C.E. Lunte, *J. Pharm. Biomed. Anal.*, 13 (1995) 149.
74 K.E. Heppert and M.I. Davies, *Curr. Sep.*, 18 (1999) 3.
75 M.I. Davies, J.D. Cooper, S.S. Desmond, C.E. Lunte and S.M. Lunte, *Adv. Drug Delivery Rev.*, 45 (2000) 169.
76 D.R. Thévenot, K. Toth, R.A. Durst and G.S. Wilson, *Biosens. Bioelectron.*, 16 (2001) 121.

77 A.P.F. Turner, I. Karube and G.S. Wilson (Eds.), *Biosensors—Fundamentals and Applications*. Oxford University Press, Oxford, 1987.
78 E.A.H. Hall, *Biosensors*. Open University Press, Maidenhead, UK, 1992.
79 A.E.G. Cass (Ed.), *Biosensors—A Practical Approach*. Oxford University Press, Oxford, 1990.
80 F. Scheller and F. Schubert, *Biosensors*. Elsevier, Amsterdam, 1992.
81 D.G. Buerk (Ed.), *Biosensors—Theory and Applications*. Technomic Publishing, Lancaster, PA, 1993.
82 D.L. Wise and L.B. Wingard Jr. (Eds.), *Biosensors with Fiberoptics*. Humana Press, Totowa, NJ, 1991.
83 W. Kerner, M. Kiwit, B. Linke, F.S. Kek, H. Zier and E.F. Pfeiffer, *Biosens. Bioelectron.*, 8 (1993) 473.
84 C.F. Mandenius, B. Danielsson and B. Mattiasson, *Anal. Chim. Acta*, 163 (1984) 135.
85 C.F. Mandenius, *Anal. Lett.*, 21 (1988) 1817.
86 D.T. Lemaitre, K.A. Martinelli and T.J. Lee, Diabetes Monitoring and Therapy: Opportunities for Investors on Pins and Needles. http://www.devicelink.com/mx/archive/02/01/flora.html.
87 H.M. Heise, *Horm. Metab. Res.*, 28 (1996) 527.
88 D.C. Klonoff, *Diabetes Care*, 20 (1997) 433.
89 J.N. Roe and B.R. Smoller, *Crit. Rev. Ther. Drug Carrier Syst.*, 15 (1998) 199.
90 M. Shichiri, R. Kawamori, N. Hakui, N. Askawa, Y. Yamasaki and H. Habe, *Biomed. Biochim. Acta*, 43 (1984) 4561.
91 M. Shichiri, N. Asakawa, Y. Yamasaki, R. Kawamori and H. Abe, *Diabetes Care*, 9 (1986) 298.
92 K.W. Johnson, J.J. Mastrototaro, D.C. Howey, R.L. Brunelle, P.L. Burden-Brady, N.A. Bryan, C.C. Andrew, H.M. Rowe, D.J. Allen, B.W. Noffke, W.C. McMahan, R.J. Morff, D. Lipson and R.S. Nevin, *Biosens. Bioelectron.*, 7 (1992) 709.
93 J.C. Pickup, D.J. Claremont and G.W. Shaw, *Acta Diabetol.*, 30 (1993) 143.
94 M. Ishikawa, D.W. Schmidtke, P. Raskin and C.A. Quinn, *J. Diabetes Complications*, 12 (1998) 295.
95 B.W. Bode, T.M. Gross, K.R. Thornton and J.J. Mastrototaro, *Diabetes Res. Clin. Pract.*, 46 (1999) 183.
96 G.S. Wilson and Y. Hu, *Chem. Rev.*, 100 (2000) 2693.
97 G. Velho, P. Proguel, R. Stenberg, D.R. Thévenot and G. Reach, *Diabetes*, 38 (1989) 164.
98 P. Lönnroth, P.A. Jansson and U. Smith, *Am. J. Physiol.*, 253 (1987) E228.
99 P.A. Jansson, J. Fowelin, U. Smith and P. Lönnroth, *Am. J. Physiol.*, 255 (1988) E218.
100 J. Bolinder, E. Hangström, U. Ungerstedt and P. Arner, *Scand. J. Clin. Lab. Invest.*, 49 (1989) 465.
101 P. Arner and J. Bolinder, *J. Intern. Med.*, 230 (1991) 381.
102 J. Bolinder, U. Ungerstedt and P. Arner, *Diabetologia*, 35 (1992) 1177.

103   L.J. Petersen, J.K. Kristensen and J. Bulow, *J. Invest. Dermatol.*, 99 (1992) 357.
104   J. Bolinder, U. Ungerstedt and P. Arner, *Lancet*, 342 (1993) 1080.
105   F.S. Keck, W. Kerner, C. Meyeroff and E.F. Pfeiffer, *Horm. Metab. Res.*, 23 (1991) 617.
106   F.S. Keck, C. Meyeroff, W. Kerner, T. Siegmund, H. Zier and E.F. Pfeiffer, *Horm. Metab. Res.*, 24 (1992) 492.
107   C. Meyerhoff, F. Bishop, F. Stenberg, H. Zier and E.F. Pfeiffer, *Diabetologia*, 35 (1992) 1087.
108   Y. Hashiguchi, M. Sakakida, K. Nishida, T. Uemuka, K. Kajiwara and M. Shichiri, *Diabetes Care*, 17 (1994) 387.
109   J. Bruckel, H. Zier, W. Kerner and E.F. Pfeiffer, *Horm. Metab. Res.*, 22 (1990) 382.
110   E.F. Pfeiffer, C. Meyerhoff, F. Bishof, F.S. Keck and W. Kerner, *Horm. Metab. Res.*, 25 (1993) 121.
111   C. Meyerhoff, F. Bishof, F.J. Mennel, F. Stenberg and E.F. Pfeiffer, *Int. J. Artif. Organs*, 16 (1993) 268.
112   E.F. Pfeiffer, *Horm. Metab. Res.*, 26 (1994) 510.
113   A.J. Schoonen, F.J. Schmidt, H. Hasper, D.A. Verbrugge, R.G. Tiessen and C.F. Lerk, *Biosens. Bioelectron.*, 5 (1990) 37.
114   A.L. Aalders, F.J. Schmidt, A.J.M. Shoonen, I.R. Broek and A.G.F.M. Maessen, *Int. J. Artif. Organs*, 14 (1991) 102.
115   K.K. Jungheim, K. Wientjes, L. Heinemann, V. Lodwig, T. Koschinsky and A.J. Schoonen, *Diabetes Care*, 24 (2001) 1696.
116   D. Moscone, M. Pasini and M. Mascini, *Talanta*, 8 (1992) 1039.
117   D. Moscone and M. Mascini, *Ann. Biol. Clin.*, 50 (1992) 323.
118   D. Moscone and M. Mascini, In: G.G. Guilbault and M. Mascini (Eds.), *Uses of Immobilized Biological Compounds*. Kluwer, Dordrecht, 1993, pp. 115–122.
119   A. Poscia, M. Mascini, D. Moscone, M. Luzzana, G. Caramenti, P. Cremonesi, F. Valgimigli, C. Bongiovanni and M. Varalli, *Biosens. Bioelectron.*, 18 (2003) 891.
120   M. Varalli, G. Marelli, A. Maran, S. Bistoni, M. Luzzana, P. Cremonesi, G. Caramenti, F. Valgimigli and A. Poscia, *Biosens. Bioelectron.*, 18 (2003) 899.
121   A. Maran, C. Crepaldi, A. Tiengo, G. Grassi, E. Vitali, G. Pagano, S. Bistoni, G. Calabrese, F. Santeusanio, F. Leonetti, M. Ribando, U. Di Mario, G. Annuzzi, S. Genovese, G. Riccardi, M. Previti, D. Cucinotta, F. Giorgino, A. Bellomo, R. Giorgino, A. Poscia and M. Varalli, *Diabetes Care*, 25 (2002) 347.
122   E. Esmatjes, L. Flores, M. Vidal, L. Rodriguez, A. Cortes, L. Almirall, M.J. Ricart and R. Gomis, *Clin. Transplant.*, 17 (2003) 534.
123   R.M. Rhemrev-Boom, J. Korf, K. Venema, G. Urban and P. Vadgama, *Biosens. Bioelectron.*, 16 (2001) 839.
124   R.M. Rhemrev-Boom, M.A. Jonker, K. Venema, G. Jobst, R. Tiessen and J. Korf, *Analyst*, 126 (2001) 1073.

125 M. Menacherry, W. Hubert and J.B. Justice, *Anal. Chem.*, 64 (1992) 577.
126 S.A. Wages, W.H. Church and J.B. Justice, *Anal. Chem.*, 58 (1986) 1649.
127 R. Steinkuhl, C. Sundermeier, H. Hinkers, C. Dumschat, K. Camman and M. Knoll, *Sens. Actuators B*, 33 (1996) 19.
128 T. Vering, S. Adams, H. Drewer, C. Dumschat, R. Steinkuhl, A. Schultze, G. Siegel and M. Knoll, *Analyst*, 123 (1998) 1605.
129 T. Laurell, *Sens. Actuators B*, 13–14 (1993) 323.
130 L. Yang, P.T. Kissinger, T. Ohara and E. Heller, *Curr. Sep.*, 14 (1995) 31.
131 P.G. Osborne, O. Niwa, T. Kato and K. Yamamoto, *Curr. Sep.*, 15 (1996) 19.
132 E.M. Janle, P.T. Kissinger and J.F. Pesek, *Curr. Sep.*, 14 (1995) 58.
133 A. Amine, K. Digua, B. Xie and B. Danielsson, *Anal. Lett.*, 28 (1995) 2275.
134 B.A. Peterson, E.H. Hansen and J. Ruzicka, *Anal. Lett.*, 19 (1986) 649.
135 Q. Fang, X. Shi, Y. Sun and Z. Fang, *Anal. Chem.*, 69 (1997) 3570.
136 B. Li, Z. Zhang and Y. Jin, *Anal. Chim. Acta*, 432 (2001) 95.
137 U. Beyer, D. Schäfer, A. Thomas, H. Aulich, U. Haueter, B. Reihl and R. Ehwald, *Diabetologia*, 44 (2001) 416.
138 U. Beyer, A. Fleischer, A. Kage, U. Haueter and R. Ehwald, *Biosens. Bioelectron.*, 18 (2003) 1391.
139 E. Zilka, A. Koshy, T.P. Obrenovich, H.P. Bennetto and L. Simon, *Anal. Lett.*, 27 (1994) 453.
140 T.P. Obrenovitch and E. Zilkha, In: J. Olesen and M.A. Moskowitz (Eds.), *Experimental Headache Models*. Raven Press, New York, 1995, pp. 113–117.
141 T.P. Obrenovitch, E. Zilkha and J. Urenjak, *J. Neurochem.*, 64 (1995) 1884.
142 T.P. Obrenovitch, J. Urenjak and E. Zilkha, *J. Neurochem.*, 66 (1996) 2446.
143 T.M. Jay, E. Zilkha and T.P. Obrenovitch, *J. Neurophysiol.*, 81 (1999) 1741.
144 N.G. Harris, J. Houseman, E. Zilkha, T.P. Obrenovitch and S.R. Williams, *Proc. ISMRM*, 1 (1998) 339.
145 M.G. Boutelle, L.K. Fellows and C. Cook, *Anal. Chem.*, 64 (1992) 1790.
146 M. Miele, M.G. Boutelle and M. Fillenz, *J. Physiol. (London)*, 497 (1996) 745.
147 M. Miele, M. Berners, M.G. Boutelle, H. Kusakabe and M. Fillenz, *Brain Res.*, 707 (1996) 131.
148 A.E. Fray, M. Boutelle and M. Fillenz, *J. Physiol. (London)*, 504 (1997) 721.
149 M. Demestre, M. Boutelle and M. Fillenz, *J. Physiol. (London)*, 499 (1997) 825.
150 T. Yao, Y. Nanjyo and H. Nishino, *Anal. Sci.*, 17 (2001) 703.
151 G. Volpe, D. Moscone, D. Compagnone and G. Palleschi, *Sens. Actuators B*, 24–25 (1995) 138.
152 D. Pfeiffer, L. Yang, F.W. Scheller and P.T. Kissinger, *Anal. Chim. Acta*, 351 (1997) 127.
153 F. Flentge, K. Venema, T. Koch and J. Korf, *Anal. Biochem.*, 204 (1992) 305.
154 N. Larsson, T. Ruzgas, L. Gorton, M. Kokaia, P. Kissinger and E. Csöregi, *Electrochim. Acta*, 43 (1998) 3541.
155 B. Li, Z. Zhang and Y. Jin, *Biosens. Bioelectron.*, 17 (2003) 585.

156 D. Yao, A.G. Vlessidis and N.P. Evmiridis, *Anal. Chim. Acta*, 478 (2003) 23.
157 P.G. Osborne, O. Niwa and K. Yamamoto, *Anal. Chem.*, 70 (1998) 1701.
158 L. Wang, Y. Li, H. Han, G. Liu and P.G. Osborne, *Neurosci. Lett.*, 344 (2003) 91.
159 D.A. Jones, M.C. Parkin, H. Langemann, H. Landolt, S.E. Hopwood, A.J. Strong and M.G. Boutelle, *J. Electroanal. Chem.*, 538/539 (2002) 243.
160 T. Yao, T. Yano, Y. Nanjyo and H. Nishino, *Anal. Sci.*, 19 (2003) 61.
161 J. Perdomo, H. Hinkers, C. Sundermeyer, W. Seifert, O. Martinez Morell and M. Knoll, *Biosens. Bioelectron.*, 15 (2000) 515.
162 P.S. Petrou, I. Moser and G. Jobst, *Biosens. Bioelectron.*, 18 (2003) 613.
163 P.S. Petrou, I. Moser and G. Jobst, *Biosens. Bioelectron.*, 17 (2002) 859.
164 W.J. Albery, M.G. Boutelle and P.T. Galley, *J. Chem. Soc., Chem. Commun.*, 12 (1992) 900.
165 M. Miele, M.G. Boutelle and M. Fillenz, *Neuroscience*, 82 (1994) 87.
166 M.C. Walker, P.T. Galley, M.L. Errington, S.D. Shorvon and J.G.R. Jefferys, *J. Neurochem.*, 65 (1995) 725.
167 L.J. Murphy and P.T. Galley, *Anal. Chem.*, 66 (1994) 4345.
168 C.J. Cook, *J. Neurosci. Methods*, 72 (1997) 161.
169 W.J. Albery, P.T. Galley and L.J. Murphy, *J. Electroanal. Chem.*, 344 (1993) 161.
170 T. Kohno, S. Asai, Y. Iribe, K. Shibata and K. Ishikawa, *J. Neurosci. Methods*, 81 (1998) 199.
171 M. Sugahara, S. Asai, H. Zaho, T. Nagata, T. Kunimatsu, Y. Ishii, T. Kohno and K. Ishikawa, *Neurochem. Int.*, 39 (2001) 65.
172 C.J. Cook, *J. Neurosci. Methods*, 72 (1997) 161.
173 C.J. Cook, *J. Neurosci. Methods*, 110 (2001) 95.
174 Z. Fang, Z. Liu and Q. Shen, *Anal. Chim. Acta*, 346 (1997) 135.
175 T. Butler, L. Gorton, H. Jarksog, G. Marko-Varga, B. Hahn-Hägerdal, N. Meinander and L. Olsson, *Biotechnol. Bioeng.*, 44 (1994) 322.
176 J. Ruzicka and E.H. Hansen, *Flow Injection Analysis*, 2nd ed., Wiley, New York, 1988.
177 J. Ruzicka and G.D. Marshall, *Anal. Chim. Acta*, 237 (1990) 329.
178 M. Carlsen, C. Johansen, R.W. Min and J. Nielsen, *Biotechnol. Tech.*, 8 (1994) 479.
179 T. Butler, H. Lidén, J.Å. Jönsson, L. Gorton, G. Marko-Varga and H. Jeppsson, *Anal. Chim. Acta*, 324 (1996) 103.
180 N. Torto, G. Marko-Varga, L. Gorton and T. Laurell, *J. Chromatogr. A*, 725 (1996) 165.
181 H. Lidén, T. Buttler, H. Jeppsson, G. Marko-Varga, J. Volc and L. Gorton, *Chromatographia*, 47 (1998) 501.
182 C.M. Zook and W.R. LaCourse, *Curr. Sep.*, 17 (1998) 41.
183 Q. Fang, S. Liu, J. Wu, Y. Sun and Z. Fang, *Talanta*, 49 (1999) 403.
184 N. Torto, B. Lobelo and L. Gorton, *Analyst*, 125 (2000) 1379.
185 G.S. Nilsson, S. Richardson, A. Huber, N. Torto, T. Laurell and L. Gorton, *Carbohydr. Polym.*, 46 (2001) 59.

186 N. Torto, L. Gorton, T. Laurell and G. Marko-Varga, *Trends Anal. Chem.*, 18 (1999) 252.
187 E. Csöregi, T. Laurell, I. Katakis, A. Heller and L. Gorton, *Mikrochim. Acta*, 121 (1995) 31.
188 I. Rohm, M. Genrich, W. Collier and U. Bilitewski, *Analyst*, 121 (1996) 877.
189 T. Yao, S. Suzuki, H. Nishino and T. Nakahara, *Electroanalysis*, 7 (1995) 395.
190 O. Niwa, R. Kurita, K. Hayashi, T. Horiuchi, K. Torimitsu, K. Maeyama and K. Tänizawa, *Sens. Actuators B*, 67 (2000) 43.
191 R. Wei Min, V. Rajendran, N. Larsson, L. Gorton, J. Planas and B. Hahn-Hägerdal, *Anal. Chim. Acta*, 366 (1998) 127.
192 F. Palmisano, A. De Santis, G. Tantillo, T. Volpicella and P.G. Zambonin, *Analyst*, 122 (1997) 1125.
193 D. Moscone, R.A. Bernardo, E. Marconi, A. Amine and G. Palleschi, *Analyst*, 124 (1999) 325.
194 S. Mannino, O. Brenna, S. Buratti and M.S. Cosio, *Electroanalysis*, 9 (1997) 1337.

Index

1,2-diaminobenzene, 159, 613, 617
1,3-diaminobenzene, 159
1,4 phenylene-diisothiocyanate, 290
1-methoxy-*N*-methylphenazonium, 463
2,4-dichlorophenoxyacetic acid (2,4-D), 297, 298, 348, 359, 396
2,6-anthraquinonedisulfonic acid, 22
2-hydroxypentamethylene sulfide, 8
3,3′-dithiodipropionic acid, 31
3,4-dichlorophenol, 479
4,4′-bipyridyl, 68
4,4′-dipyridyldisulfide, 50
4,4′-dithiodipyridine, 79
4-vinylpyridine, 297, 300
5-(octyldithio)-2-nitrobenzoic acid, 31
5-bromo-4-chloro-3-indolyl phosphate, 117, 188, 361
5-methylphenazonium methyl sulfate, 151
5′-nucleotidase, 559
5′-amino-terminated ssDNA, 18

acetaminophene, 134, 142, 150, 155
acetyl cholinesterase (AChE), 165, 166, 169, 170, 290, 310, 470, 486–488, 491, 608, 609
acetylcholine, 46, 131, 132, 144, 146, 151, 162, 165, 166, 169, 170, 272, 290, 298, 310, 483, 491, 505, 509, 608, 609, 617
Acinetobacter, 34
affinity binding, 12, 13, 15, 19–22, 24, 32, 34, 36, 39, 40, 42, 44–47, 49, 50, 237, 416
affinity biosensor, 1, 12–14, 24, 219, 308
affinity complex, 1, 6, 12–15, 18, 19, 21, 24, 32–37, 40, 42, 44–46, 49, 50, 237, 416, 463
alanine aminotransferase, 355
alanine dehydrogenase, 558
alcohol, 77, 84, 92, 94, 96, 97, 115, 145, 151, 152, 160, 170, 256, 300, 304, 314, 471, 481, 543–547, 549

alcohol dehydrogenase (ADH), 5, 75–77, 94, 96, 97, 145, 160, 164, 165, 167, 170, 187, 188, 234, 258, 304, 584, 596
alcohol oxidase (AlcOx), 152, 169, 481, 544, 546, 549
aldose, 131, 148
aldose dehydrogenase, 148
alkaline phosphatase, 117, 188, 197, 312, 350, 351, 356, 359, 360, 365, 467, 490, 492, 494
alkanethiol/alkylthiol SAM, 4–10, 13–19, 22–25, 31, 32, 34, 36, 39, 40, 42, 44, 45–47, 49–51
alkylthiols, 3–6
allylamine, 164, 166
amicyanin, 70
amine dehydrogenase, 74
amine oxidase (AO), 74, 97, 144, 152, 163, 169, 170
γ-amino-4-butyric acid, 614
amino acid oxidase, 32, 135, 508, 546
amino-containing ligands, 15
amino-terminated alkylthiol SAM, 7, 9
aminophenol, 23, 132, 347, 356, 362, 367, 449, 488, 551
aminopropyltri-ethoxysilane, 563
aminosilanes, 308
amphetamine, 366, 367
aniosotropic immobilization, 69
anisotropic protein films, 14
antibody, 1, 13, 21, 49, 117, 118, 210, 218, 219, 242, 266, 268, 271, 285, 299, 302, 307, 329–335, 339–343, 345, 346, 348–367, 376–378, 385–388, 390–394, 397, 399, 400–407, 409–414, 416, 419, 502, 503, 540, 550, 552–554, 589, 614
antiresonant reflecting optical waveguide (ARROW), 229
arabinose, 132
Arthromyces ramosus peroxidase, 92
Arxula adeninivorans, 447

627

Index

aryltriphosphate dialkylphosphohydrolase, 470
ascorbate/ascorbic acid, 46, 96, 98–100, 111, 131, 132, 134, 142, 145, 150, 152, 155, 159, 169, 304, 361, 494, 605, 610, 613, 614
ascorbate oxidase, 46, 72, 96, 98, 100, 494, 613
asparagine, 342
aspartate, 132, 136
Aspergillus nidulans, 539
Aspergillus niger, 463, 550
atomic force microscopy (AFM), 88, 183, 187, 188, 192–194, 252, 264, 269, 341
atrazine, 297, 361, 398, 400, 401, 408, 409, 431, 433, 494, 514
Au(111), 4, 6
avidin, 11, 15, 16, 19, 21–23, 46, 47, 143, 181, 183, 188, 189, 237, 240, 289–292, 296, 358, 410
avidin–biotin, 181, 183, 188, 290, 291, 296, 358, 410
azide, 10, 68, 100, 132, 144, 165, 170, 412, 415, 494
azurin, 70, 71, 98, 303, 463

Bacillus, 365, 366, 501, 539
Bacillus cereus, 365
Bacillus licheniformis, 551
Bacillus subtilis, 366, 539, 568, 569
benzenesulfonamide, 6, 7, 15
benzoquinone, 154, 168, 358, 444
Biacore, 211, 225, 240, 242, 270
bile acid, 132, 460
bilirubin, 132, 460
biotin, 6, 10, 11, 14–16, 18–23, 46, 47, 50, 51, 143, 170, 181, 184–186, 188, 189, 195, 237, 238, 240, 291, 292, 296, 298, 302, 304, 310, 341, 358, 564–566
biotin–streptavidin, 14, 184, 195
biotinyl, 10, 14–16, 18, 19, 21, 23, 46, 47, 50, 181, 184–186, 188, 237, 291, 292, 296, 298, 304, 310, 358, 564, 565
biotinylated dsDNA hybrids, 18
biotinylated glucose oxidase, 46
bis(*N*-hydroxysuccinimidyl) 3,3′-dithiopropionate, 32, 33
"blue" copper proteins, 99
"blue" oxidases, 98
boron-doped diamond, 159, 314, 457
bromoethylamine, 140
butyryl-cholinesterase (BuChE), 486–488, 506

Candida boidinii, 479
capillary electrophoresis, 201, 269, 554
carbodiimide, 7, 8, 99, 150, 240, 290, 296, 303, 312, 342, 357, 508, 564
carbon fibers, 265
carbon nanotubes (CNT), 69, 72, 76, 300–304, 317
  multi-walled nanotubes (MWNTs), 301, 303
  single-walled nanotubes (SWNTs), 300–304
carbon paste, 198, 309, 458
carbon paste electrode (CPE), 94, 96, 111, 137–139, 144–150, 152, 154, 155, 157, 160–162, 197, 198, 308, 309, 315, 349, 367, 476, 544, 545, 617
carbon-felt electrodes, 351
carbonic anhydrase, 6, 15, 494, 495
carboxylic-terminated alkylthiols SAMs, 8, 9
catalase, 7, 65, 69, 76–78, 90, 91, 111, 117, 303, 362, 440, 590, 594, 595, 605, 612, 613
catalytic amperometric biosensors, 24
catalytic antibodies, 502, 503, 516
catalytic biosensors, 1, 24
catalytic RNAs, 496
catechol, 50, 98, 198, 132, 304, 310, 354, 455, 459, 514, 613
catecholamine, 304, 613
cell growth, 5, 198, 549, 554
cellobiose, 25, 71, 74, 77, 95, 96, 132, 460
cellobiose dehydrogenase, 25, 70, 74, 76, 77, 95, 460
ceruloplasmin, 98
cetyltrimethylammonium bromide, 83
CHEMFET, 313
chemically-active surface groups, 13
chemically modified electrodes, 314
chemisorption of alkylthiols, 3, 5
chitosan, 154
chloroperoxidase, 73, 76, 92
chlorophenols, 459, 460, 494
cholesterol, 72, 100, 131, 132, 146, 162, 169
cholesterol esters, 132
choline, 46, 131, 132, 144, 151, 155, 169, 438, 470, 481–483, 486, 487, 489–491, 494, 506, 608, 609
choline oxidase (ClOx), 46, 155, 165, 166, 169, 170, 489, 608, 609
cholinesterase, 438, 470, 481–483, 486, 487, 489–491, 493, 494, 506
chronocoulometry, 21
Cibacron Blue, 36, 40

Index

Clark-type electrode, 67, 467
cocaine, 366
colloidal gold particles, 197, 303, 307
colloidal particles, 301, 305, 307
competitive assay, 118, 218, 224, 333, 334, 358, 387, 390, 394, 409, 411
competitive immunoassay, 334, 348, 353, 359, 386–388, 390, 391, 396, 397, 399, 401, 406, 413
complexes, 22, 45, 135, 136, 179, 180, 182, 196, 198, 298, 299, 302, 330, 349, 378, 386, 401, 411, 509
concanavalin A, 6, 46, 338, 606
copper proteins, 69, 98, 99
corticosteroids, 353
creatinine, 318, 342, 355–357
cresol, 71, 75, 77, 97, 132
cresolmethylhydrolase, 75, 77, 97
cyclic voltammetry, 19, 26, 28, 35, 88, 96, 293, 294, 311, 339
cystamine, 11, 21, 31, 33–35, 38, 39, 41, 45, 47, 48, 96, 310
cysteine tryptophylquinone, 97
cysteine, 48, 69, 79, 81, 86, 93, 97, 100, 107, 357, 405, 491, 507, 508
cystine, 159
cytochrome b, 31, 49, 77, 95
cytochrome b_2, 31, 49, 70
cytochrome $b5$, 70, 77
cytochrome $b562$, 70, 71
cytochrome c, 13, 25–31, 45, 49, 50, 68, 70, 71, 73, 74, 77–82, 84, 86, 87, 92, 95, 97, 98, 103–108, 111, 113, 114, 118, 132, 200, 303, 304, 462, 473, 474, 477, 478, 505, 511
cytochrome c', 71, 81, 82, 108
cytochrome $c3$, 70, 71, 97, 473
cytochrome $c550$, 70
cytochrome c oxidase, 25, 26, 28, 29, 31, 49, 50, 70, 74, 76, 77, 98, 307
cytochrome c peroxidase (Baker's yeast), 70, 76, 92
cytochrome P450, 76–78, 84, 86, 87, 87, 474, 477, 478, 505, 511, 513
cytochrome P450$_{cam}$, 70, 71, 73, 86–90, 477, 478, 513
cytokines, 398

dehydrogenase, 25, 31, 34–37, 45, 49, 67, 70, 77, 94–97, 144, 145, 148, 154, 159–161, 170, 302, 304, 348, 352, 366, 438, 446, 453, 459, 460, 465, 466, 479–481, 490, 505, 508, 557–559, 607
Desulfovibrio norvegicum, 474
Desulfovibrio vulgaris, 473
di-n-alkyl disulfides, 3, 4
diaphorase, 72, 76, 558
dichlorophenol, 479, 550
didodecyldimethylammoniumbromide, 83
digoxin, 355, 401
diphenols, 310, 455, 459
dithiothreitol, 9
DMSO-reductase, 76
DNA, 16, 18, 19, 21, 23, 24, 179–203, 210, 233, 239, 240, 243, 244, 251, 252, 266, 268–273, 275, 290–292, 299, 302, 304, 310, 311, 447, 449, 478, 497–499, 511, 541, 543, 550, 554, 555, 560–562, 564, 566–570
   DNA chips, 560, 562, 567, 570
   DNA hybridization assay, 24, 197, 271
   DNA-acoustic wave biosensors, 186
   double-stranded DNA (dsDNA), 16, 50, 51, 180, 291, 310, 311
   single-stranded DNA (ssDNA), 16, 180, 183
dopamine, 132, 303, 310, 609, 610

Eastman AQ, 81, 139
electrochemical enzyme immunoassay, 118, 350
electrochemical immunoassays, 100, 329, 337, 338, 389, 342, 348, 350
electron microscopy, 180, 294, 363
ellipsometry, 180, 184, 212, 288
ENFET, 458, 465, 480, 481, 490
enzyme, 1, 21, 23, 24, 26, 28, 30, 31, 34–37, 39, 40, 45, 46, 49, 65–71, 77, 81, 84, 86–88, 90–103, 111, 115, 117, 118, 131, 133–150, 152, 154–162, 171, 184, 186, 197, 198, 210, 245, 266, 268, 272, 285, 288, 290, 292, 295, 298, 300, 303, 308, 309, 310, 314, 315, 318, 333, 335, 336, 342, 343, 345, 347–352, 355–359, 361–364, 388, 400, 406, 409, 414–416, 429, 435, 437–443, 446–449, 452–457, 459, 461–467, 470, 471, 473, 474, 477–484, 486–491, 494–496, 498, 499, 501–512, 516, 517, 540, 541, 543–548, 550–552, 555–560, 581, 584, 587, 589–591, 594, 595, 604, 605, 607–618

enzyme-linked immunosorbent assay (ELISA), 288, 335, 343, 347, 356, 361, 362, 497, 550
Escherichia coli, 270, 362–364, 413, 444, 445, 447, 449, 462, 471, 477, 499, 501, 539, 551–553, 555, 560, 567, 568, 618
estradiol, 132, 403, 404, 511
ethanol, 75, 76, 96, 97, 103, 132, 145, 151, 152, 160, 161, 164, 165, 169–171, 304, 313, 357, 489, 543, 544, 569, 616
evanescent-wave spectroscopy (IR-EWS), 297

fatty acid hydroperoxides, 100
Fe_2O_3, 305
ferredoxin, 69, 86, 89, 90
ferritin, 303
ferrocene, 33, 46, 67, 135–139, 145–152, 155, 156, 158, 167, 169, 170, 197, 340, 347, 348, 567, 607, 610, 614
ferrocene derivatives, 67, 135, 136, 348
ferrocenemethanol, 46
ferrocenyl, 21, 23, 47, 48, 135, 148, 149, 165–167, 170
fiber optic (FO) biosensor, 216–219, 226
flavin, 67, 77, 94, 95, 97, 134, 453, 460, 474, 503, 508
flavin adenine dinucleotide (FAD), 32, 33, 49, 72, 74–76, 95–97, 133–138, 143, 146, 147, 150, 154, 155, 158, 340, 453, 460, 466, 468, 469, 477, 508
flavin mononucleotide (FMN), 72, 76, 443, 453, 477, 497
flavin monooxygenase, 474
flavocytochrome $b2$, 72, 75, 77
flavocytochrome $c3$, 76, 77, 97
flavocytochrome c_{522}, 76
flavodoxin, 70
flavoheme, 25, 94
flavohemoproteins, 77, 94
flavoproteins, 69, 86
fluorescence microscopy, 180
formaldehyde, 91, 131, 132, 144, 160, 164, 170, 448, 479–481
formaldehyde dehydrogenase, 160, 479, 480
fructose, 31, 70, 71, 75–77, 94, 96, 103, 132, 618
fructose dehydrogenase, 31, 70, 75, 77, 94, 96, 164, 170, 479, 480
fumarate, 31, 71, 76, 97, 132
fumarate reductase, 31, 76, 97

β-galactosidase, 46, 347, 363, 364, 409, 449, 551, 618
galactose, 132, 494, 543, 546, 618
genetically modified organisms (GMOs), 181
gluconate, 71, 77, 94, 103, 559
gluconate dehydrogenase, 31, 34, 35, 67, 76, 77, 94, 348, 459, 460
glucose, 21, 22, 32–35, 46–49, 65, 67, 70–72, 103, 112, 115, 117, 118, 131–144, 146–152, 154–159, 161, 162, 170, 171, 272–274, 295, 296, 300, 304, 314, 315, 340, 342, 346–348, 352, 356, 358, 361, 366, 441, 452, 459, 460, 466, 467, 490, 492, 493, 540, 542, 543, 546, 549, 550, 559, 569, 587, 590–600, 602–608, 611, 612, 617, 618
glucose dehydrogenase (GDH), 34, 35, 67, 161, 348, 459, 460
glucose oxidase (GOx), 21, 22, 32, 33, 46–49, 67, 72, 76, 117, 118, 133–140, 143, 144, 146–150, 152, 154, 155, 157–159, 163–170, 177, 295, 296, 340, 346–348, 352, 356, 358, 360, 361, 460, 467, 490, 492, 493, 550, 590–592, 594, 595, 599, 603–606, 611, 617, 618
glucose sensors, 139, 141, 143, 144, 146, 148, 149, 151, 152, 154–156, 158, 162, 274, 587
glucose-6-phosphate dehydrogenase, 366, 466, 559
glutamate, 115, 132, 136, 146, 152, 170, 465, 557, 558, 606, 607, 613, 614, 617
glutamate dehydrogenase, 465, 557, 607
glutamate oxidase, 166, 170, 557, 607, 614
glutaminase, 557
glutaraldehyde (GA), 4, 86, 99, 118, 143, 151, 185, 186, 188, 221, 256, 260, 275, 290, 294, 295, 298, 301, 303, 306, 312–314, 330, 338, 343, 347, 354, 361, 404, 410, 411, 435, 449, 452, 464, 465, 471, 475, 476, 478–480, 488, 494, 543, 545–547, 555, 559, 579, 611, 614, 618
glutathione, 32, 33, 100–102, 474, 479, 494, 505
glutathione peroxidase, 72, 76, 100
glutathione reductase, 32, 33, 505
glutathione transferases, 474
glycerol, 142, 151, 161, 164, 170, 171, 543, 544, 546, 549, 568, 613
glycerol-3-phosphate, 142, 151, 544, 546
glycerol-3-phosphate oxidase, 544, 546

Index

glycerophosphate, 142, 170
glycogen phosphorylase, 467
glycolate, 132, 146, 165, 170
glycolate oxidase, 165, 170
gold colloidal particles, 307
gold nanoparticle, 23, 497, 498, 566
graphite paste, 353

haemoglobin, 71, 77, 83–85, 108–111, 307, 356
hanging mercury drop electrode (HMDE), 198
Hansenula polymorpha, 481, 544
heme *c*, 96, 97
heme proteins, 69, 70, 77, 87
hemin, 36, 111
heterogeneous assays, 334, 410
heterogeneous-phase immunosensors, 349
hexokinase, 559
high affinity antibodies, 331
highly oriented pyrolytic graphite (HOPG), 73, 193–195
histamine, 71, 74, 97, 131, 132, 144, 163, 170, 617
histamine oxidase, 617
histidine-tagged proteins, 6, 15
horseradish peroxidase (HRP), 36, 48, 73, 76, 92–94, 111, 114–118, 142–144, 150–152, 170, 184, 197, 203, 298, 341, 342, 352–355, 358, 361, 362, 366, 414, 544, 566, 604, 606, 607, 609, 611, 614, 617
human chorionic gonadotropin, 352, 412
human heart fatty-acid binding, 356
human luteinizing hormone, 357
hydrogel, 140–144, 148–152, 157, 160–162, 170, 237, 239, 274, 299, 315, 459
hydrogen, 3, 25, 45, 47, 65, 81, 83, 86, 92, 100–102, 110, 111, 134, 142, 144, 148, 150, 151, 154, 155, 161, 162, 170, 182, 190, 302, 304, 305, 312, 313, 338, 341, 342, 348, 354, 356, 358, 377, 415, 459, 466, 467, 473, 489, 544, 557, 561, 610
hydrogen peroxide (H_2O_2), 65, 71–74, 76, 81, 83, 91, 92, 100–102, 106, 107, 110–112, 114, 115, 117, 133, 134, 142–144, 148, 150, 151, 154, 156, 161, 162, 165, 166, 168, 169, 302, 304, 341, 342, 346, 348, 354, 356, 358, 415, 459, 466–469, 489, 544, 557, 604, 605, 607, 608, 610
hydrogenase, 25, 47, 75, 76, 155, 170, 473
hydrophobic interactions, 5, 197, 288, 377, 378, 410, 413, 545

hydroquinone, 7, 8, 93, 99, 153, 154, 158, 167, 170, 358, 455, 566
hydroxy-terminated alkylthiol SAM, 14, 17, 21, 25
hypoxanthine, 103, 132, 558–560

immunoaffinity complex, 21
immunoassay, 71, 100, 117, 118, 218, 242, 244, 271, 273, 275, 304, 329, 330, 333–335, 337–343, 345, 348–351, 353–359, 361, 364, 366, 367, 375–388, 390, 391, 393–403, 405–413, 415–419
immunoglobuline G (IgG), 111, 292, 294, 295, 297, 302, 307, 332, 341, 342, 347, 376, 377, 399
immunosensing, 21, 342
immunosensors, 49, 99, 117, 242, 289, 307, 308, 329, 335, 337–339, 343, 349, 351, 352, 356, 360, 362, 364, 490, 495, 553, 614
impedance, 188, 198, 308, 337, 342, 346, 357, 364, 365
indium–tin oxide (ITO), 90, 158, 318
integrated optical (IO) sensors, 226
interdigitated electrodes (IDA), 79, 308, 347, 364
interdigitated ultramicroelectrode arrays (IDUAs), 263, 264
invertase, 493
ion-channel sensors, 19, 21, 49
IrO_2, 305
ISFET, 313, 478

Klebsiella pneumoniae, 363

L-lactate dehydrogenase (flavocytochrome b_2), 75
laccase, 68, 71, 72, 76, 98–100, 118, 144, 163, 170, 298, 348, 455, 456, 458–460, 515
lactate, 36, 37, 39, 41–46, 48, 49, 65, 71, 75, 76, 115, 131, 132, 141–144, 146, 150–152, 155, 160, 162–165, 170, 290, 493, 540, 569, 607, 608, 611, 612, 617
lactate dehydrogenase (LDH), 31, 36, 37, 39–45, 49, 75, 76, 160, 352, 493
lactate oxidase, 46, 48, 142, 143, 152, 155, 163–165, 170, 290, 493, 608, 611, 617
lactoperoxidase, 406
lactose, 74, 95, 132, 364, 501, 616
lactulose, 618
Langmuir–Blodgett (LB), 3, 187
layer-by-layer assembly, 10

631

ligand-terminated thiol, 15
lignin peroxidase, 76, 92
Listeria monocytogenes, 364
luciferase, 443, 449, 559

m-phenylenediamine, 603
Mach-Zehnder interferometer (MZI), 227–231
malate, 132, 500
maltose, 132, 463, 467, 468, 470
maltose phosphorylase, 467, 470
mannose, 46, 132
manganese peroxidase, 74
Marcus Theory, 68
meldola blue, 160, 170, 465
membrane receptor, 1, 7, 333
mercaptopropanol, 81
mercaptopropionic acid, 26–30, 38–42, 79, 81, 99, 100, 107, 365
mercaptopurine, 79
mercaptosuccinic acid, 82, 108
mercaptoundecanamine, 98
mercaptoundecanoic acid, 11, 40, 41, 48, 79, 365
mercaptoundecanol, 98
metal oxide semiconductor field effect transistors (MOSFETs), 312
methane monooxygenase, 73
methanol, 132, 448, 481, 540, 543, 549
methyl viologen, 462–464
methylamine dehydrogenase, 72, 97
methylene blue, 22, 197
microdialysis, 272, 353, 460, 548, 570, 579, 581–590, 592–600, 602–613, 615–619
microperoxidase, 45, 74, 76, 111, 462
miniaturized total analysis systems (μ-TAS), 182, 251, 252, 269, 272, 273
mixed alkylthiol SAM, 5, 6, 14, 15, 18, 19, 22, 31, 36
mixed monolayer, 15, 21, 106, 107, 340
molecular beacons, 24, 183
molecular lego, 504, 509
molecularly imprinted polymers (MIP), 273, 274, 297, 509, 516
molybdopterin, 25, 98
monoclonal antibody, 242, 302, 330, 332, 352, 355, 356, 358–361
monomolecular film, 2, 3
montmorillonite, 81, 83–85, 87–91, 114
MOSFET, 66, 312, 313
multilayered assembly, 11, 15, 46–48
mutarotase, 467

myoglobin, 13, 45, 69–71, 77, 82, 83, 108, 110, 114, 357, 478

N-(2-methylferrocene)caproic acid, 33
N-hydroxysuccinimide, 290, 359
N-succinimidyl *S*-acetylthiopropionate, 9
NAD-dependent dehydrogenases, 36
nicotinamide adenine dinucleotide
  oxidised form, NAD$^+$, 34, 36, 37, 39, 42–45, 67, 76, 86, 89, 94, 151, 159–161, 302, 304, 348, 352, 366, 446, 453, 460–462, 465–467, 474, 477–480, 490, 505, 508, 557–559, 607
  NAD-, 36, 37, 42, 45
  reduced form NADH, 34, 76, 89, 151, 160, 161, 170, 302, 304, 348, 352, 366, 446, 460–462, 465, 467, 469, 477, 480, 490, 505, 557–559
NADH-dependent, 477
NADH$_2$ dehydrogenase (ubiquinone), 446
nicotinamide adenine dinucleotide phosphate
  NAD(P)H, 86, 94, 159, 461, 462, 466
  oxidised form, NADP$^+$, 474
  reduced form, NADPH, 461, 474, 477, 478, 505, 607
Nafion, 140, 142, 349, 464, 547, 608, 617
NaN$_3$, 151
nanocomposite, 143, 161, 299, 305
nanomaterials, 300, 301, 305, 317
nanoparticles, 23, 84, 159, 197, 271, 294, 302, 305, 306, 480, 497, 498, 566
nanosize-scale array, 10
nanostructured gold electrode, 340
neurotransmitters, 304, 587, 606, 613
neutravidin, 237
nitrate, 45, 84, 108, 110, 131, 132, 155, 162, 170, 418, 434, 435, 460–465, 609
nitrate reductase, 45, 76, 155, 170, 461–465
nitric oxide, 1–6, 10, 12–19, 21–26, 31–33, 35–37, 39, 40, 45, 53, 65, 67, 69–71, 77, 79, 82–84, 86–89, 93–99, 101–103, 106–111, 117, 118, 133, 134, 137–140, 142–145, 147, 148, 150, 154, 155, 158, 159, 161, 162, 171, 180, 181, 183, 186, 187, 189, 193, 195–199, 201, 202, 209–213, 215, 218, 223–228, 231, 235, 237, 241, 242, 244, 251–254, 258–260, 262–270, 272–275, 286–288, 291, 294, 299, 301, 302, 305, 306, 308–315, 319, 330, 331, 334, 337–341, 343, 345, 346, 348–351, 353–357, 360, 361, 366, 367, 375, 377–379, 381, 383–386, 388, 390,

391, 393–395, 397–405, 409–413, 415–418, 429, 430, 435–437, 439–444, 446–448, 451–454, 456, 457, 459–467, 470, 471, 473, 474, 477–485, 487–491, 494–499, 501–512, 516, 517, 539–541, 543–550, 553–555, 559–563, 565–570, 579, 581, 582, 584, 587–598, 601, 602, 604, 605, 607–613, 615, 616, 618
nitrite, 83, 98, 110, 361, 460–465
nitrite reductase, 70, 76, 98, 460–463
nitromerocyanine, 49
nitrospiropyrane, 49
non-specific covalent binding, 13
N'-hydroxysuccinimide (NHS), 7, 8
(N-methyl-N'-(carboxyalkyl)-4,4′-bipyridinium salts), 33
nucleoside phosphorylase, 467, 470, 559

o-phenylenediamine, 93, 151
octacyanotungstens, 135
octadecyl mercaptane, 31
oligo(ethylene oxide), 5, 15
oligo(propylene sulfoxide), 5
oligosaccharide, 95, 144, 154, 162, 241, 348, 459, 616
oligosaccharide dehydrogenase (ODH), 144, 154, 164, 167, 170, 348, 459
optical evanescent field techniques, 1
optical waveguide lightmode spectroscopy (OWLS), 235
organosilanes, 3
organosiloxane, 299
organosulfur, 3
orientation of antibody, 13
orthoquinone, 50, 160
Os, 22, 48, 96, 135, 136, 140–142, 144, 145, 151, 152, 160, 161, 163–165, 169–171, 184, 185, 187, 196, 219, 262, 296–298, 341, 466, 467, 581, 604, 608, 609, 611, 617
  Os-based redox polymer, 298, 609
  Os(bipyridine)$_2$Cl, 140, 141
  Os(bpy), 96, 140, 152, 170, 341, 466
  Os(bpy)$_2$, 96, 140, 152, 170, 341, 466
  Os(bpy)$_2$Cl, 96, 140
  Os(bpy)$_2$Cl$_2$ redox polymer, 96
  Os(bpy)$_2$pyCl, 466
  Os(1,10-phenanthroline-5,6-dione)$_2$Cl$_2$, 467
  Os(II)polyvinylpyridine, 604
  Os complex, 48, 135, 145, 151, 161
osmium–bipyridyl complexes, 22

Os–gel–HPR, 604, 617
oxalate, 115, 132

p-aminophenyl phosphate, 356, 460
p-dihydroxybenzaldehyde, 97
p-diphenol, 98, 455, 456, 459
p-phenylenediamine, 151, 160
p-quinone, 347
Pichia stipitis, 544, 546
paracetamol, 151, 159, 274, 342
paraquinone, 50, 51
peanut peroxidase, 74, 92
penicillin G, 342
Penicillium notatum, 539
pentachlorophenol, 432, 445, 460
peptide nucleic acid (PNA), 21, 181
peroxidase, 23, 36, 48, 67–74, 77, 78, 87, 92–94, 100, 101, 111–113, 115–118, 142, 143, 150, 152, 163, 164, 184, 185, 197, 290, 298, 310, 335, 341, 352, 358–362, 366, 414, 415, 453, 458–460, 466, 468–470, 481, 489, 490, 544, 546, 566, 604, 610, 613
peroxidase chemiluminescence, 468, 469
persistent organic pollutant (POP), 430, 434
phenazine methosulfate, 39, 42–44
phenol, 94, 111, 131, 132, 144, 158, 168, 243, 292, 309, 310, 314, 315, 406, 455–457, 459, 460, 479, 517
phenolic compound, 94, 315, 455–457, 459, 460, 517
phenolics, 111, 131, 144, 165
phenosafranin, 463
phenothiazine, 158, 160, 201, 463
phenoxazine, 160
phenylalanine, 132
phenylenediamine, 614
phosphorylase, 466
photonic band gap (PBG) 313
phthalocyanine, 302, 488
Pichia pastoris, 91, 539, 543–545, 549, 553
plasmid DNA 543, 555
plastocyanin, 70, 98
poly(acrylic acid), 11
poly(ethylene glycol) diglycidyl ether, 142, 151, 161, 163, 164, 168, 169
polyacrylamide, 83, 289
polyallylamine, 101, 115, 116
polyaniline, 157, 295, 359
polychlorinated biphenyls (PCBs) 360
polyclonal, 331, 352, 353, 357, 391–393, 395, 397, 398

633

polyethyleneimine, 111, 144, 157, 165, 188, 358, 476, 545, 546
polylysine, 90, 76, 545, 563
polymerase chain reaction (PCR), 184, 185, 197, 201, 202, 268, 273, 505, 510, 555, 556, 561–564
polyphenol, 14, 165, 309, 455, 456, 610
polyphenol oxidase, 455, 610
polypyrrole, 96, 112, 145, 157–159, 168, 188, 289, 294, 295, 305, 355, 463, 617
polypyrrole-viologen, 463
polythiophene, 145, 157, 295, 462
polyvinyl alcohol, 84, 300
potassium hexacyanoferrate (III), 135, 444, 447
potentiometric stripping, 24, 197
propazine, 408, 409
protein adsorption, 5, 12, 14, 15, 40, 42, 84, 302
protein film, 13, 14, 26, 28, 32
protein-resistant, 5, 12, 14, 15, 51, 52
protein-resistant surface, 5, 12, 15, 52
Pseudomonas diminuta, 470
Pseudomonas fluorescens, 94, 451
Pseudomonas putida, 97, 144, 444, 445, 451, 479
Pseudomonas stutzeri, 463
putidaredoxin, 70, 71, 87–90, 477
putrescine, 132, 152, 169
pyranose oxidase, 544, 546
pyridine-nucleotide dependent oxidoreductase, 36
pyrrole, 145, 155, 157, 158, 165, 168, 188, 293, 294, 296, 298, 466
pyrroloquinoline quinone (PQQ), 34, 35, 39, 45, 67, 75, 77, 96, 113, 143, 148, 160, 161, 164, 167, 178, 348
 PQQ-dependent alcohol dehydrogenase (PQQ-ADH), 164
 PQQ-dependent glucose dehydrogenase (PQQ-GDH), 161, 164
 PQQ-dependent glycerol dehydrogenase (PQQ-GlyDH), 164
pyruvate, 39, 131, 132, 142, 143, 145, 150, 162, 165, 355, 466, 468–470, 558, 559
pyruvate kinase, 559
pyruvate oxidase, 145, 355, 466, 468–470
pyruvic acid, 160

quartz crystal microbalance (QCM) 12, 16, 18, 187, 188, 270, 288, 289, 297, 341
quinoheme, 25, 94, 97
quinohemoprotein, 77, 94–96, 145

quinohemoprotein alcohol dehydrogenase, 96, 145
quinone, 8, 34, 94, 132, 135, 143, 152, 154–156, 160, 161, 303, 348, 455, 456, 460
quinoprotein, 34, 94, 97, 459, 460

redox polymer, 23, 25, 67, 96, 136, 137, 139, 140, 142–147, 149–152, 155, 157, 164, 166, 168, 170, 298, 341, 455, 463, 466, 608, 609
redox protein, 25, 32, 67–70, 77, 118, 340, 453, 474
resonant mirror device (IAsys) 232
resorcinol, 151, 159
RNA, 182, 183, 185, 187, 269, 275, 299, 449, 496–498, 541, 560, 563, 566, 567
RuO_2, 305
ruthenium complexes, 135, 136
riboflavin, 90, 443

Saccharomyces cerevisiae, 478, 515, 539
salicylate hydrolase, 480
Salmonella, 364
sarcosine, 132
scanning near field optical, 183
scanning tunnel microscope, 4, 6
scanning tunnelling microscopy (STM), 180, 192
screen-printed carbon electrodes, 359
self-assembled gold electrode, 364
self-assembled monolayer (SAM), 68, 237, 258, 288, 316, 355, 359, 365, 455
self-organization, 2
SH-terminated alkylthiol SAM, 19
silicium oxide, 341
simazine, 118, 360–362, 398, 408, 409, 432
single-stranded DNA (ssDNA) 16, 180
single stranded nucleic acid, 1
SiO_2, 143, 185, 227, 232, 234, 239, 308, 312, 314, 340, 346
site specific affinity, 40
sodium azide, 132, 412
sodiumdodecyl sulphate, 83
sol–gel, 144, 145, 157, 165, 214, 234, 274, 287, 288, 299, 300, 317, 318, 353, 354, 471, 486
soybean peroxidase, 23, 74, 92, 142, 143, 164, 298
spectroelectrochemistry, 193
Staphylococcus aureus, 362
steroids, 132, 403, 405

634

Index

streptavidin, 6, 10, 14, 15, 18, 20, 50, 181, 184–186, 195, 197, 237, 240, 266, 271, 302, 304, 358, 564–566, 610
streptavidin–horseradish peroxidase, 185, 566
Streptomyces, 539, 568, 569
sucrose, 132
sugars, 71, 93, 144, 164, 167, 181, 190, 191, 197, 543, 545, 591, 594, 606
sulfhydryl-terminated alkylthiol SAMs, 7, 9
sulfite, 25, 98, 131, 132, 154, 168
sulfur-containing helical peptides, 10
sulphite dehydrogenase, 76
sulphite oxidase, 25, 76, 98, 154, 168
superoxide, 67, 69–72, 81, 100, 103, 104, 106–108
superoxide anion, 81, 100, 103, 106
superoxide dismutase, 69, 72, 100, 103, 107
surface acoustic wave (SAW), 288, 297, 313, 493
surface-bound affinity complex, 1, 6, 12, 14, 18, 19, 33, 36, 37, 40, 42, 45
surface-bound ferrocenyl molecules, 21
surface-bound ligand, 12, 14, 24, 40
surface-bound protein, 13, 31, 37, 40, 46
surface-enhanced Raman spectroscopy (SERS), 193
surface plasmon resonance (SPR), 12, 16, 18, 20, 26, 27, 42–44, 182, 185, 186, 211, 212, 215, 216, 219–226, 232, 240, 242, 243, 252, 265, 270, 271, 275, 288–290, 313, 314, 495, 496, 553, 555, 561, 563, 565, 566, 589
surface-bound carboxylic groups, 11, 17, 40
sweet potato peroxidase (SWPP), 74, 94, 115, 152, 178

tartarate, 132
testosterone, 132, 403, 404
testosterone/A, 404
tetracyanoquinodimethanide (TCNQ), 135, 468, 469
tetrathiafulvalene (TTF), 135, 155–157, 168
theophylline oxidase, 70
thickness-shear-mode (TSM) acoustic wave sensor, 187
thiolated ssDNA, 17, 20
thionine, 48, 158, 168, 354, 405
thiophene, 145, 165, 466

thyroid hormones thyroxine, 400
thyroxin/A, 404
tin-doped indium oxide, 68
TiO_2, 234, 305, 307, 308
TIR, 212, 213, 227–229, 232
TM, 213, 219, 234, 235
tobacco peroxidase, 74
toluidine blue O, 160, 170
topa quinone, 74
total internal reflection fluorescence (TIRF), 225, 235
triazine, 36–38, 40, 360, 361, 408, 514
tryptophan tryptophylquinone, 97
triple-helical DNA, 183
tween, 80, 88, 89, 91, 411, 412
two- or three-dimensional supramolecular structure, 2
tyrosinase, 98, 158, 165, 168, 198, 290, 300, 309, 315, 348, 452, 455–460, 481, 490, 492, 494, 517

urate, 134, 140, 142, 150, 151, 155, 159
urease, 66, 438, 487, 491–493, 495, 496
uricase, 469, 610
uric acid, 103, 106, 132, 468, 559, 609, 610

Vibrio fisheri, 445
viologen, 32, 47, 154–156, 168, 462–464
vitamin K, 132, 160
vitamin K_3, 170

whole cell, 1, 2, 10, 15, 118, 171, 216, 235, 237, 251, 356, 357, 375, 382, 392, 393, 397, 399, 400, 402, 409, 412, 418, 443, 446, 452, 464, 489, 499, 500, 510, 512, 550, 552, 555, 560, 565, 567, 597, 608, 609

xanthine, 103, 106, 132, 470, 558–560
xanthine oxidase, 559
xylose, 132, 167, 543, 546

zeolites, 314, 315
ZnO, 305

635